Vemuri Balakotaiah, Ram R. Ratnakar

Applied Linear Analysis for Chemical Engineers

Also of Interest

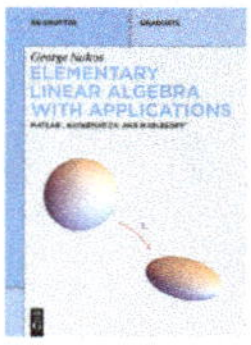

Elementary Linear Algebra with Applications
MATLAB®, Mathematica® and Maplesoft™
George Nakos, 2024
ISBN 978-3-11-133179-9, e-ISBN (PDF) 978-3-11-133185-0

Differential Equations
Solving Ordinary and Partial Differential Equations with Mathematica®
Marian Mureşan, 2024
ISBN 978-3-11-141109-5, e-ISBN (PDF) 978-3-11-141139-2

Process Engineering
Addressing the Gap between Study and Chemical Industry
Michael Kleiber, Gökce Adali, Michael Benje, Verena Haas, 2023
ISBN 978-3-11-102811-8, e-ISBN (PDF) 978-3-11-102814-9

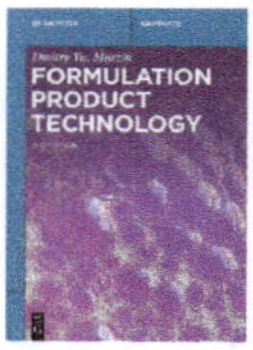

Formulation Product Technology
Dmitry Yu. Murzin, 2023
ISBN 978-3-11-078844-0, e-ISBN (PDF) 978-3-11-079796-1

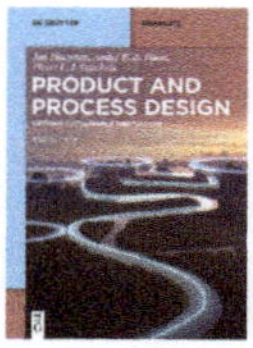

Product and Process Design
Driving Sustainable Innovation
Jan Harmsen, André B. de Haan, Pieter L. J. Swinkels, 2024
ISBN 978-3-11-078206-6, e-ISBN (PDF) 978-3-11-078212-7

Homogenization Methods
Effective Properties of Composites
Rainer Glüge, 2023
ISBN 978-3-11-079351-2, e-ISBN (PDF) 978-3-11-079352-9

Vemuri Balakotaiah, Ram R. Ratnakar

Applied Linear Analysis for Chemical Engineers

A Multi-scale Approach with Mathematica®

2nd, Revised Edition

DE GRUYTER

Authors
Prof. Vemuri Balakotaiah
University of Houston
Dept of Chemical and Biomolecular Engineering
4800 Calhoun Road
Houston TX 77204-4004
USA
bala@uh.edu

Dr. Ram R. Ratnakar
Shell International Exploration & Production
Houston TX 77082
USA
Ram.Ratnakar@shell.com

ISBN 978-3-11-159738-6
e-ISBN (PDF) 978-3-11-159805-5
e-ISBN (EPUB) 978-3-11-159847-5

Library of Congress Control Number: 2025932735

Bibliographic information published by the Deutsche Nationalbibliothek
The Deutsche Nationalbibliothek lists this publication in the Deutsche Nationalbibliografie; detailed bibliographic data are available on the Internet at http://dnb.dnb.de.

Cover image: selimcan / iStock / Getty Images Plus
Typesetting: VTeX UAB, Lithuania

www.degruyter.com
Questions about General Product Safety Regulation:
productsafety@degruyterbrill.com

Preface

This book is based on a course that the first author taught at the University of Houston for about 30 years. This course was a requirement for all first-year graduate students and was a prerequisite for two other optional courses, taken mostly by graduate students whose research involved modeling, computational and nonlinear analysis. As we state in the Introduction, while this book deals only with the solution of linear equations, linear analysis is the foundation of all numerical and nonlinear techniques.

Since there are many books already available on applied mathematics for chemical engineers, it is fair to ask the question why another book? For this, our response is that every author has a unique perspective that may be appealing to others. Further, the authors are not aware of any book that deals exclusively with the solution of various linear equations that arise in engineering in a unified manner, and with examples.

The senior author had the pleasure of taking the applied mathematics course from Professor Neal R. Amundson and later teaching the same course when Professor Amundson retired. Both authors have used the material extensively in their own research and would like to point out the following highlights of the material presented: (i) use of symbolic software (Mathematica®) for illustrating and enhancing the impact of physical parameter changes on solutions, (ii) multiscale analysis of chemical engineering problems with physical interpretation of time and length scales in terms of eigenvalues and eigenvectors/eigenfunctions, (iii) detailed discussion of compartment models for various finite- dimensional problems and their solution in phase spaces, (iv) evaluation and illustration of functions of matrices (and use of symbolic manipulation) to solve multicomponent diffusion-convection-reaction problems, (v) illustration of the techniques and interpretation of solutions to several classical chemical engineering and related problems, (vi) emphasis on the connection between discrete (matrix algebra) and continuum models (initial, boundary and initial-boundary value problems), (vii) physical interpretation of adjoint operator and adjoint systems and their application in solving inverse problems and (viii) use of complex analysis and algebra in the solution of practical engineering problems.

The senior author has taught most of the contents of Parts I, III, IV and V in a single semester (14 weeks or 28 lectures of 90 minutes duration) on most course offerings while parts II and VI were covered occasionally. However, the entire contents of the book can be taught in a two-semester course. For a single semester course, we recommend covering Chapters 1 to 5, 14, 17, selected sections of Chapters 18 to 21 and 23 to 25.

We wish to acknowledge many colleagues, former students and our mentors who over the years contributed to our understanding and organization of the subject.

https://doi.org/10.1515/9783111598055-202

We also want to thank Karin Sora, Nadja Schedensack, Ute Skambraks and Vilma Vaičeliūnienė of De Gruyter for their help during production.

The second author wishes to acknowledge the constant encouragement and support of his familty, especially his eldest brother Siddhesh Satyakar.

Finally, the first author wishes to acknowledge the patience and understanding of his wife, Nalini Vemuri, and dedicate it to her with affection and gratitude.

Introduction

This book deals with the solution of linear equations. We discuss the solutions of linear algebraic equations, linear initial value problems, linear boundary value problems, linear integral equations and linear partial differential equations along with their application to various chemical engineering problems.

It should be pointed out that most practical problems encountered by engineers are nonlinear and are often solved on a computer using numerical techniques. In most cases, the nonlinear problem is linearized around a known or approximate solution and a more accurate solution is obtained by solving a sequence of linear problems. The nonlinear methods of analysis as well as the numerical techniques used by engineers draw heavily from the linear analysis. In other words, linear analysis is the foundation of all nonlinear and numerical techniques.

Generally speaking, most linear problems that arise in applications may be classified into two groups: (i) problems describing the steady state or equilibrium state of a physical system and (ii) problems describing the dynamic or transient behavior of a physical system. The first type of problems are described by linear equations of the form

$$\mathbf{L}\mathbf{u} = \mathbf{f} \tag{1}$$

where $\mathbf{L}$ is a linear operator, $\mathbf{u}$ is a state vector and $\mathbf{f}$ is a source function. For example, in finite dimensions, equation (1) may be a set of n linear algebraic equations in n unknowns,

$$\mathbf{A}\mathbf{u} = \mathbf{b} \tag{2}$$

where $\mathbf{A}$ is a $n \times n$ matrix, $\mathbf{u}$ and $\mathbf{b}$ are $n \times 1$ vectors. When the state vector $\mathbf{u}$ belongs to an infinite-dimensional space, equation (1) may be a two-point boundary value problem such as

$$-\frac{d}{dx}\left(p(x)\frac{du}{dx}\right) + q(x)u = f(x), \quad a < x < b \tag{3}$$

$$u(a) = u(b) = 0 \tag{4}$$

or an integral equation such as the Fredholm integral equation of the first kind given by

$$\int_a^b K(x,s)u(s)\,ds = f(x) \tag{5}$$

or a partial differential equation such as the Poisson's equation

$$-\left(\frac{\partial^2 u}{\partial x^2} + \frac{\partial^2 u}{\partial y^2}\right) = f(x,y) \quad \text{in } \Omega \tag{6}$$

https://doi.org/10.1515/9783111598055-203

$$u = 0 \quad \text{on } \partial\Omega \tag{7}$$

where Ω is some domain in the x-y plane and $\partial\Omega$ is its boundary.

The second class of problems are of the form

$$\frac{d\mathbf{u}}{dt} = \mathbf{L}\mathbf{u}, \quad t > 0 \tag{8}$$

$$\mathbf{u} = \mathbf{u}^0 \quad \text{at } t = 0 \tag{9}$$

where t is the time and the evolution equation (8) describes the system behavior for $t > 0$, while equation (9) gives the initial condition. In the simpler case of the finite-dimensional problems, equations (8)–(9) may be of the form

$$\frac{d\mathbf{u}}{dt} = \mathbf{A}\mathbf{u} \tag{10}$$

$$\mathbf{u} = \mathbf{u}^0 \quad \text{at } t = 0 \tag{11}$$

where $\mathbf{A}$ is a constant coefficient $n \times n$ matrix, $\mathbf{u}$ is a $n \times 1$ vector of state variables and $\mathbf{u}^0$ is a $n \times 1$ vector of initial conditions. An example of an initial value problem in infinite dimensions is the heat equation in one spatial coordinate and time:

$$\frac{\partial u}{\partial t} = \frac{\partial^2 u}{\partial x^2}; \quad 0 < x < 1, \ t > 0 \tag{12}$$

$$u(0,t) = u(1,t) = 0 \quad \text{(Boundary conditions)} \tag{13}$$

$$u(x,0) = f(x) \quad \text{(Initial condition)} \tag{14}$$

We shall see that many of the concepts involved in the solution of linear ordinary and partial differential equations are generalizations of the ideas involved in the solution of the finite-dimensional problems represented by equations (2) and (10). Therefore, we shall focus first on the finite-dimensional case.

Properties of solutions to linear equations

When the matrix $\mathbf{A}$ is invertible, the solution of equation (2) may be expressed as

$$\mathbf{u} = \sum_{j=1}^{n} \frac{1}{\lambda_j} c_j \mathbf{x}_j \tag{15}$$

where the scalars λ_j (eigenvalues) and the (eigen)vectors $\mathbf{x}_j$ depend only on the matrix $\mathbf{A}$, while the constants c_j are given by

$$c_j = \frac{\langle \mathbf{b}, \mathbf{y}_j \rangle}{\langle \mathbf{x}_j, \mathbf{y}_j \rangle}, \tag{16}$$

where $\mathbf{y}_j$ are known as the left (adjoint) eigenvectors of $\mathbf{A}$. Here, $\langle \mathbf{x}, \mathbf{y} \rangle$ denotes the dot or inner product of vectors. When the matrix $\mathbf{A}$ is symmetric (self-adjoint), $\mathbf{x}_j = \mathbf{y}_j$ and the eigenvectors are normalized to have unit length ($\langle \mathbf{x}_j, \mathbf{x}_j \rangle = 1$), the expression for the constants c_j simplifies to

$$c_j = \langle \mathbf{b}, \mathbf{x}_j \rangle. \tag{17}$$

The above form of the solution has advantages over the direct solution (e. g., by Gaussian elimination) when n is large. For example, when $\mathbf{A}$ is symmetric and the eigenvalues are well separated ($0 < |\lambda_1| \ll |\lambda_2| \ll |\lambda_3| \ll \cdots \ll |\lambda_n|$), the first few terms may be sufficient to compute the solution if the desired accuracy is not high. A second advantage is that the solution has the same form for all linear equations of the form given by (1). For example, when the linear (differential/integral) operator $\mathbf{L}$ is symmetric, the same solution is applicable with a slight modification:

$$u = \sum_{j=1}^{\infty} \frac{1}{\lambda_j} c_j \phi_j; \quad c_j = \langle f, \phi_j \rangle, \tag{18}$$

where λ_j are the eigenvalues and ϕ_j are the normalized eigenfunctions of the operator $\mathbf{L}$.

The solution of the initial value problem, equations (10)–(11) may be expressed as

$$\mathbf{u}(t) = \sum_{j=1}^{n} c_j e^{\lambda_j t} \mathbf{x}_j; \quad c_j = \frac{\langle \mathbf{u}^0, \mathbf{y}_j \rangle}{\langle \mathbf{x}_j, \mathbf{y}_j \rangle} \tag{19}$$

which for the case of symmetric matrix simplifies to $c_j = \langle \mathbf{u}^0, \mathbf{x}_j \rangle$. The generalization of this result for the case of a symmetric differential operator is

$$u(t) = \sum_{j=1}^{\infty} c_j e^{\lambda_j t} \phi_j; \quad c_j = \langle f, \phi_j \rangle. \tag{20}$$

An important observation regarding the various solutions to the linear equations is that they are all expressed in terms of the eigenvalues and eigenfunctions of the operator appearing in the equation. These eigenvalues and eigenfunctions correspond to various time and length scales that are of interest in the physical system. An important task of linear analysis is the identification of these length and time scales and relating them to the parameters or dimensionless groups characterizing the physical system. We hope to illustrate this for various chemical engineering problems.

Contents

Part V: Fourier transforms and solution of boundary and initial-boundary value problems

Part I: **Applied matrix algebra**

1 Matrices and linear algebraic equations

1.1 Simultaneous linear equations

We consider m simultaneous linear equations in n unknowns:

$$\begin{aligned}
a_{11}u_1 + a_{12}u_2 + \cdots + a_{1n}u_n &= b_1 \\
a_{21}u_1 + a_{22}u_2 + \cdots + a_{2n}u_n &= b_2 \\
&\cdot \\
&\cdot \\
a_{m1}u_1 + a_{m2}u_2 + \cdots + a_{mn}u_n &= b_m
\end{aligned} \tag{1.1}$$

or in matrix notation,

$$\begin{bmatrix} a_{11} & a_{12} & a_{13} & \cdot & \cdot & a_{1n} \\ a_{21} & a_{22} & a_{23} & \cdot & \cdot & a_{2n} \\ \cdot & \cdot & \cdot & \cdot & \cdot & \cdot \\ \cdot & \cdot & \cdot & \cdot & \cdot & \cdot \\ \cdot & \cdot & \cdot & \cdot & \cdot & \cdot \\ a_{m1} & a_{m2} & a_{m3} & \cdot & \cdot & a_{mn} \end{bmatrix} \begin{bmatrix} u_1 \\ u_2 \\ \cdot \\ \cdot \\ \cdot \\ u_n \end{bmatrix} = \begin{bmatrix} b_1 \\ b_2 \\ \cdot \\ \cdot \\ \cdot \\ b_m \end{bmatrix}$$

$$\mathbf{A}\mathbf{u} = \mathbf{b} \tag{1.2}$$

where $\mathbf{A}$ is the coefficient matrix with m rows and n columns ($m \times n$ matrix), $\mathbf{u}$ is the unknown vector ($n \times 1$ matrix) and $\mathbf{b}$ is a $m \times 1$ vector of constants. The elements a_{ij} of the matrix $\mathbf{A}$ and b_i of the vector $\mathbf{b}$ may be real or complex numbers. The matrix

$$\begin{bmatrix} a_{11} & a_{12} & a_{13} & \cdot & \cdot & a_{1n} & b_1 \\ a_{21} & a_{22} & a_{23} & \cdot & \cdot & a_{2n} & b_2 \\ \cdot & \cdot & \cdot & \cdot & \cdot & \cdot & \cdot \\ \cdot & \cdot & \cdot & \cdot & \cdot & \cdot & \cdot \\ \cdot & \cdot & \cdot & \cdot & \cdot & \cdot & \cdot \\ a_{m1} & a_{m2} & a_{m3} & \cdot & \cdot & a_{mn} & b_m \end{bmatrix}$$

with m rows and $(n + 1)$ columns is called the *augmented matrix* and is denoted by

$$\operatorname{aug}\mathbf{A} = [\mathbf{A}\ \mathbf{b}]$$

and the matrix $\mathbf{A}$ will often be written as

$$\mathbf{A} = [a_{ij}]; \quad i = 1, 2, \ldots, m; \quad j = 1, 2, \ldots, n$$

where a_{ij} is the element of $\mathbf{A}$ in the i-th row and j-th column. When $\mathbf{b} = \mathbf{0}$, we obtain the *homogeneous system* of equations

https://doi.org/10.1515/9783111598055-002

$$\mathbf{Au} = \mathbf{0} \tag{1.3}$$

or

$$\begin{aligned}
a_{11}u_1 + a_{12}u_2 + \cdots + a_{1n}u_n &= 0 \\
a_{21}u_1 + a_{22}u_2 + \cdots + a_{2n}u_n &= 0 \\
&. \\
&. \\
a_{m1}u_1 + a_{m2}u_2 + \cdots + a_{mn}u_n &= 0
\end{aligned} \tag{1.4}$$

As stated in the Introduction, many of the ideas involved in the solution of linear differential equations are generalizations of those involved in the solution of the homogeneous algebraic equation (1.3) and the inhomogeneous algebraic equation (1.2). Generally speaking, linear equations have either 0, 1 or ∞ number of solutions. In what follows, we shall discuss the conditions under which equations (1.1) have no solution (inconsistent), a unique solution and an infinite number of solutions.

1.2 Review of basic matrix operations

We review here briefly some terminology, basic matrix operations and some special matrices. We shall refer to the $m \times n$ matrix

$$\mathbf{A} = \begin{bmatrix} a_{11} & a_{12} & a_{13} & . & . & a_{1n} \\ a_{21} & a_{22} & a_{23} & . & . & a_{2n} \\ . & . & . & . & . & . \\ . & . & . & . & . & . \\ . & . & . & . & . & . \\ a_{m1} & a_{m2} & a_{m3} & . & . & a_{mn} \end{bmatrix}$$

as real-valued if all its elements are real numbers or real-valued functions. It will be called complex-valued if one or more of the elements is a complex number or complex-valued function.

By convention, the elements of a matrix are double subscripted to denote location. For example, a_{ij} refers to the element appearing in the i-th row of the j-th column. If the number of rows equals to the number of columns ($m = n$), the matrix is referred to as a square matrix of order n. (Square matrices appear in most of our applications). In a square matrix, the elements a_{ii} ($i = 1, 2, 3, \ldots, n$) are called diagonal elements. For the special case $n = 1$, the matrix is called a column vector (with m elements) while for $m = 1$, we have a row vector. The transpose of an $m \times n$ matrix $\mathbf{A}$ is the $n \times m$ matrix obtained by interchanging the rows and columns of $\mathbf{A}$ and is denoted by $\mathbf{A}^T$.

1.2.1 Matrix addition and subtraction

Let $\mathbf{A} = [a_{ij}]$ be an $m_1 \times n_1$ matrix and $\mathbf{B} = [b_{ij}]$ be an $m_2 \times n_2$ matrix. Then $\mathbf{A}$ and $\mathbf{B}$ can be added only if $m_1 = m_2$ and $n_1 = n_2$, i. e., the number of rows and columns in $\mathbf{A}$ and $\mathbf{B}$ are equal. The sum is given by

$$\mathbf{C} = \mathbf{A} + \mathbf{B},$$

where

$$c_{ij} = a_{ij} + b_{ij}$$

i. e., the sum is obtained by adding the corresponding elements. Similarly, we define for any scalar k,

$$\mathbf{A} \pm k\mathbf{B} = [a_{ij} \pm kb_{ij}]$$

1.2.2 Matrix multiplication

Let $\mathbf{A} = [a_{ij}]$ be an $m \times n$ matrix and $\mathbf{B} = [b_{ij}]$ be another $p \times r$ matrix. If the number of columns of $\mathbf{A}$ equals to the number of rows of $\mathbf{B}$ (i. e., $n = p$) we say that $\mathbf{A}$ and $\mathbf{B}$ are conformable to multiplication or the product of $\mathbf{AB}$ is defined. We define

$$\mathbf{AB} = \mathbf{C}$$

where the elements of the $m \times r$ matrix $\mathbf{C}$ are given by

$$c_{ij} = \sum_{k=1}^{n} a_{ik} b_{kj}; \quad i = 1, 2, \ldots, m, \; j = 1, 2, \ldots, r$$

From this definition, it can be shown that matrix multiplication is associative and distributes over addition. However, it is not commutative. Thus,

$$\mathbf{A}(\mathbf{BC}) = (\mathbf{AB})\mathbf{C}$$
$$\mathbf{A}(\mathbf{B} + \mathbf{C}) = \mathbf{AB} + \mathbf{AC},$$

whenever the products are defined. In general,

$$\mathbf{AB} \neq \mathbf{BA}$$

even in the cases in which both the products are defined. Two square matrices $\mathbf{A}$ and $\mathbf{B}$ for which $\mathbf{AB} = \mathbf{BA}$ are said to *commute* with each other. Also, it may be shown that

$$(\mathbf{AB})^T = \mathbf{B}^T\mathbf{A}^T$$

where the superscript T on the matrix denotes the transpose.

1.2.3 Special matrices

We now review some special types of square matrices that play an important role in our applications.

A *diagonal matrix* is a square matrix of all zero elements except possibly those on the main diagonal ($a_{ij} = 0$ if $i \neq j$). The *zero matrix* is a matrix having all its elements equal to zero.

An *identity matrix* of order n is a diagonal matrix of order n having all its diagonal elements equal to one. It is usually denoted by $\mathbf{I}_n$, or simply by $\mathbf{I}$ when the order is not specified. Thus,

$$\mathbf{I}_4 = \begin{bmatrix} 1 & 0 & 0 & 0 \\ 0 & 1 & 0 & 0 \\ 0 & 0 & 1 & 0 \\ 0 & 0 & 0 & 1 \end{bmatrix}$$

A matrix with real elements is called *symmetric* if it is equal to its transpose, i. e.,

$$\mathbf{A} = \mathbf{A}^T$$

or

$$a_{ij} = a_{ji}$$

for a real symmetric matrix. A square matrix with complex elements is called *Hermitian* if it equals its conjugate transpose, i. e.,

$$\mathbf{A} = (\overline{\mathbf{A}})^T = \mathbf{A}^*$$

or

$$a_{ij} = \bar{a}_{ji},$$

where '$*$' stands for the transpose plus complex conjugation while the overbar stands for only complex conjugation. Thus,

$$\mathbf{A} = \begin{bmatrix} 1 & 2 \\ 2 & -4 \end{bmatrix}$$

is a (real) symmetric matrix while

$$\mathbf{B} = \begin{bmatrix} 1 & i & 2+3i \\ -i & -4 & 3 \\ 2-3i & 3 & 6 \end{bmatrix}$$

is a Hermitian matrix.

A square matrix is said to be *normal* if

$$\mathbf{A}\mathbf{A}^* = \mathbf{A}^*\mathbf{A},$$

i. e., if it commutes with its conjugate transpose. If **A** has real elements then $\mathbf{A}^T$ has real elements and **A** is normal if it commutes with its transpose.

A square matrix **A** is called *lower triangular* if $a_{ij} = 0$ for $j > i$, i. e., all the elements above the diagonal are zero. Similarly, **A** is called *upper triangular* if $a_{ij} = 0$ for $i > j$, or equivalently all the elements below the diagonal are zero. For example,

$$A = \begin{bmatrix} 1 & 3 & 5 \\ 0 & 7 & 9 \\ 0 & 0 & 8 \end{bmatrix}$$

is an upper triangular matrix of order 3.

A square matrix **A** is called *tridiagonal* if $a_{ij} = 0$ for $|i-j| > 1$, i. e., all elements except those on the diagonal and two main off-diagonals are zero. For example,

$$\mathbf{A} = \begin{bmatrix} -2 & 1 & 0 & 0 & 0 \\ 1 & -3 & 1 & 0 & 0 \\ 0 & 1 & -4 & 1 & 0 \\ 0 & 0 & 1 & -5 & 1 \\ 0 & 0 & 0 & 1 & -6 \end{bmatrix}$$

is a tridiagonal matrix of order 5.

There are many other special matrices that will appear in our applications. We shall discuss them as they arise.

1.3 Elementary row operations and row echelon form of a matrix

Consider the m simultaneous linear algebraic equations in n-unknowns

$$\mathbf{A}\mathbf{u} = \mathbf{b}$$

and recall the following operations that are used to simplify the system and obtain a solution:

(a) Rearrangement (or reordering) of the equations.
(b) Multiplication of any equation by a nonzero constant
(c) Multiplication of any equation by a constant and adding to another equation.

We note that these operations do not change the solution. Thus, we define *elementary row operations* (ERO) of three basic types on the rows of a matrix:
(E1): interchange of any two rows of a matrix
(E2): multiplication of any row by a nonzero scalar
(E3): multiplication of a row by a constant and add to another row, element by element

(Similarly, we can define elementary column operations on the columns of a matrix but that is not of interest here.) Given any matrix **A**, we can use elementary row operations on its rows to reduce it to *row echelon form*. A matrix is said to be in row echelon form if
(i) Any nonzero row is above that of any zero row
(ii) The first nonzero element in any nonzero row is unity
(iii) If the first nonzero element in a row appears in column r, then all elements in column r in succeeding rows are zero.
(iv) The first nonzero element in row j occurs to the right of the first nonzero element in row i if $j > i$.

Examples 1.1.

$$\mathbf{A} = \begin{pmatrix} 1 & 3 & 1 & 5 & 4 \\ 0 & 1 & 2 & 3 & 1 \\ 0 & 0 & 1 & 4 & 6 \\ 0 & 0 & 0 & 1 & 0 \\ 0 & 0 & 0 & 0 & 0 \end{pmatrix}$$

is in row echelon form.

$$\mathbf{B} = \begin{pmatrix} 1 & 0 & 5 \\ 2 & 0 & 0 \\ 0 & 0 & 1 \end{pmatrix}; \quad \mathbf{C} = \begin{pmatrix} 1 & 2 & 3 & 4 \\ 0 & 1 & 1 & 5 \\ 0 & 3 & 6 & 1 \end{pmatrix}$$

B and **C** are not in row echelon form.

1.3.1 Representation of elementary row operations

Elementary row operations on an $m \times n$ matrix may be represented by using matrix multiplication. To illustrate, we consider the 3×4 matrix

$$\mathbf{A} = \begin{pmatrix} a_{11} & a_{12} & a_{13} & a_{14} \\ a_{21} & a_{22} & a_{23} & a_{24} \\ a_{31} & a_{32} & a_{33} & a_{34} \end{pmatrix}$$

and show that all EROs on $\mathbf{A}$ can be performed by doing the same operations on the $m \times m$ identity matrix and premultiplying $\mathbf{A}$ by the resulting matrix. For this example, we take

$$\mathbf{I}_3 = \begin{pmatrix} 1 & 0 & 0 \\ 0 & 1 & 0 \\ 0 & 0 & 1 \end{pmatrix}$$

Suppose that we interchange rows 2 and 3 ($R_2 \longleftrightarrow R_3$). This transforms $\mathbf{I}_3$ to

$$\mathbf{E}_1 = \begin{pmatrix} 1 & 0 & 0 \\ 0 & 0 & 1 \\ 0 & 1 & 0 \end{pmatrix}$$

We note that

$$\mathbf{E}_1\mathbf{A} = \begin{pmatrix} a_{11} & a_{12} & a_{13} & a_{14} \\ a_{31} & a_{32} & a_{33} & a_{34} \\ a_{21} & a_{22} & a_{23} & a_{24} \end{pmatrix}$$

Similarly, let

$$\mathbf{E}_2 = \begin{pmatrix} 1 & 0 & 0 \\ 0 & k & 0 \\ 0 & 0 & 1 \end{pmatrix}, \quad k \neq 0$$

$$\mathbf{E}_3 = \begin{pmatrix} 1 & 0 & 0 \\ 0 & 1 & k \\ 0 & 0 & 1 \end{pmatrix}$$

Then

$$\mathbf{E}_2\mathbf{A} = \begin{pmatrix} a_{11} & a_{12} & a_{13} & a_{14} \\ ka_{21} & ka_{22} & ka_{23} & ka_{24} \\ a_{31} & a_{32} & a_{33} & a_{34} \end{pmatrix}$$

$$\mathbf{E}_3\mathbf{A} = \begin{pmatrix} a_{11} & a_{12} & a_{13} & a_{14} \\ a_{21} + ka_{31} & a_{22} + ka_{32} & a_{23} + ka_{33} & a_{24} + ka_{34} \\ a_{31} & a_{32} & a_{33} & a_{34} \end{pmatrix}$$

Thus, every elementary operation on $\mathbf{A}$ can be represented as a premultiplication of $\mathbf{A}$ by $\mathbf{E}_i$ ($i = 1, 2, 3, \ldots$). This property implies that given any $\mathbf{A}$, we can find a $m \times m$ matrix

P such that **PA** is in row echelon form. [The matrix **P** is the product of the elementary matrices $\mathbf{E}_i$].

1.4 Rank of a matrix and condition for existence of solutions

We define the rank (or more precisely the row rank) of a matrix **A** as the number of nonzero rows in its row echelon form (Note: There are many other equivalent definitions of rank and it can be shown that the row rank and column rank are identical).

Definition. If **A** is a square matrix of order n, then it is called nonsingular (or invertible) if rank $\mathbf{A} = n$. If rank of $\mathbf{A} < n$, then **A** is called singular.

Example 1.2. We consider the matrices

$$\mathbf{A} = \begin{pmatrix} 2 & 1 & 0 \\ 3 & 6 & 1 \\ 5 & 7 & 1 \end{pmatrix}, \quad \mathbf{B} = \begin{pmatrix} 1 & 2 & -1 \\ 3 & 8 & 9 \\ 2 & -1 & 2 \end{pmatrix}$$

and note that their row-echelon forms are given by

$$\mathbf{A}_R = \begin{pmatrix} 1 & \frac{1}{2} & 0 \\ 0 & 1 & \frac{2}{9} \\ 0 & 0 & 0 \end{pmatrix}, \quad \mathbf{B}_R = \begin{pmatrix} 1 & 2 & -1 \\ 0 & 1 & 6 \\ 0 & 0 & 1 \end{pmatrix}$$

Thus, rank $\mathbf{A} = 2$ while rank $\mathbf{B} = 3$. Thus, **A** is a singular matrix while **B** is nonsingular.

We now consider the linear equations $\mathbf{Au} = \mathbf{0}$ and $\mathbf{Au} = \mathbf{b}$ and state the conditions under which they have solutions.

1.4.1 The homogeneous system Au = 0

The following theorem may be stated for the homogeneous system:

Theorem. *Consider the m simultaneous linear homogeneous algebraic equations in n unknowns:*

$$\mathbf{Au} = \mathbf{0}, \quad \mathbf{u} \in \mathbb{R}^n/\mathbb{C}^n \tag{1.5}$$

A necessary and sufficient condition for (1.5) *to have a nontrivial (nonzero) solution is*

$$\text{rank}(\mathbf{A}) < n.$$

Proof. The necessity is clear, for suppose that rank $\mathbf{A} = n$. Then reducing **A** to echelon form gives the following equivalent set of equations:

$$
\begin{aligned}
u_1 + \gamma_{12}u_2 + \gamma_{13}u_3 + \cdots + \gamma_{1n}u_n &= 0\\
u_2 + \gamma_{23}u_3 + \cdots + \gamma_{2n}u_n &= 0\\
u_3 + \cdots + \gamma_{3n}u_n &= 0\\
&\cdot\\
&\cdot\\
u_{n-1} + \gamma_{n-1,n}u_n &= 0\\
u_n &= 0
\end{aligned}
\tag{1.6}
$$

Here, γ_{ij} are the elements in the echelon form of **A**. We note that the only solution to equations (1.6) is the trivial one.

To prove sufficiency (i. e., there is a nonzero solution when $\operatorname{rank}\mathbf{A} < n$), we let $\operatorname{rank}\mathbf{A} = r$. Then, based on the row echelon form of **A**, the reduced equivalent system may be written as

$$
\begin{aligned}
u_1 + \gamma_{12}u_2 + \gamma_{13}u_3 + \cdots + \gamma_{1n}u_n &= 0\\
u_2 + \gamma_{23}u_3 + \cdots + \gamma_{2n}u_n &= 0\\
&\cdot\\
&\cdot\\
u_r + \gamma_{r\,r+1}u_{r+1} + \cdots + \gamma_{r\,n}u_n &= 0
\end{aligned}
\tag{1.7}
$$

Now, we can choose nonzero values for $(u_{r+1}, \ldots, u_n)$ and evaluate $(u_1, u_2, \ldots, u_r)$ uniquely from equations (1.7). Hence, we get a nontrivial solution when $r < n$. □

Example 1.3. Consider the homogeneous system in three variables

$$
\begin{aligned}
u_1 - u_2 &= 0\\
u_2 - u_3 &= 0\\
u_1 + u_3 &= 0
\end{aligned}
$$

for which $\operatorname{rank}\mathbf{A} = 3$. Thus, the only solution is the trivial one.

Example 1.4. Consider the homogeneous system in four variables (with complex coefficients)

$$
\begin{aligned}
u_1 - iu_2 &= 0\\
u_2 + u_3 &= 0\\
u_1 + u_2 - u_4 &= 0\\
u_2 + iu_3 + iu_4 &= 0
\end{aligned}
$$

or $\mathbf{Au} = \mathbf{0}$ with

$$\mathbf{A} = \begin{pmatrix} 1 & -i & 0 & 0 \\ 0 & 1 & 1 & 0 \\ 1 & 1 & 0 & -1 \\ 0 & 1 & i & i \end{pmatrix}; \quad i = \sqrt{-1}.$$

It may be verified that $\operatorname{rank}\mathbf{A} = 3$ and

$$\mathbf{u} = \alpha \begin{pmatrix} i \\ 1 \\ -1 \\ 1+i \end{pmatrix}$$

is a solution for any α (real or complex constant).

Example 1.5. Consider the homogeneous system in four variables

$$\begin{aligned} u_1 - 2u_2 - u_4 &= 0 \\ -2u_1 + 3u_2 + 3u_3 &= 0 \\ -u_2 + 3u_3 - 2u_4 &= 0 \\ 3u_1 - 7u_2 + 3u_3 - 5u_4 &= 0 \end{aligned}$$

for which $\operatorname{rank}\mathbf{A} = 2$. It may be verified that

$$\mathbf{u} = c_1 \begin{pmatrix} 3 \\ 0 \\ 2 \\ 3 \end{pmatrix} + c_2 \begin{pmatrix} 0 \\ 1 \\ -1 \\ -2 \end{pmatrix}$$

is a solution for any constants c_1 and c_2. [We discuss in Chapter 3 how to obtain this solution.]

1.4.2 The inhomogeneous system Au = b

We now consider the inhomogeneous (or non-homogeneous) system and examine when it has 0, 1 or an infinite number of solutions. We use the elementary row operations to reduce the augmented matrix to row echelon form. Without loss of generality, we can assume the echelon form of the augmented matrix is given by

$$\begin{bmatrix} 1 & \gamma_{12} & \gamma_{13} & . & . & \gamma_{1r} & \gamma_{1r+1} & . & . & \gamma_{1n} & \alpha_1 \\ 0 & 1 & \gamma_{22} & . & . & \gamma_{2r} & \gamma_{2r+1} & . & . & \gamma_{2n} & \alpha_2 \\ . & . & . & . & . & . & . & . & . & . & . \\ . & . & . & . & . & . & . & . & . & . & . \\ 0 & 0 & 0 & . & . & 1 & \gamma_{r\,r+1} & . & . & \gamma_{r\,n} & \alpha_r \\ 0 & 0 & 0 & . & . & 0 & 0 & . & . & 0 & \alpha_{r+1} \\ . & . & . & . & . & . & . & . & . & . & . \\ 0 & 0 & 0 & . & . & . & . & . & . & 0 & \alpha_m \end{bmatrix} \tag{1.8}$$

We now consider various cases

Case 1: $r \leq m \leq n$ (more unknowns than equations). The equations are consistent only if

$$\begin{aligned} \alpha_{r+1} &= 0 \\ &. \\ \alpha_m &= 0 \end{aligned}$$

If $\alpha_i \neq 0$ for any $r + 1 \leq i \leq m$, the rank of **A** and aug **A** are different and the equations are inconsistent. Hence, no solution exists in this case. Thus, a necessary condition for solutions to exist is

$$\operatorname{rank} \mathbf{A} = \operatorname{rank}(\operatorname{aug} \mathbf{A}) \tag{1.9}$$

We now show that the above condition is also sufficient. Suppose that (1.9) is satisfied and let

$$\operatorname{rank} \mathbf{A} = r$$

We can rearrange (1.8) as follows:

$$\begin{aligned} u_r &= -\gamma_{r\,r+1}u_{r+1} - \gamma_{r\,r+2}u_{r+2} - \cdots - \gamma_{r\,n}u_n + \alpha_r \\ u_{r-1} &= -\gamma_{r-1\,r}u_r - \gamma_{r-1\,r+1}u_{r+1} - \cdots - \gamma_{r-1\,n}u_n + \alpha_{r-1} \\ &. \\ &. \\ u_1 &= -\gamma_{12}u_2 - \gamma_{13}u_3 - \cdots - \gamma_{1r+1}u_{r+1} - \cdots - \gamma_{1n}u_n + \alpha_1 \end{aligned} \tag{1.10}$$

Thus, we can choose $(n - r)$ of the variables $(u_{r+1}, \ldots, u_n)$ as we please and obtain the values of remaining variables using (1.10) above to get a solution. [In Chapter 3, we shall show that the solution space has dimension $(n - r)$].

Case 2: $r \leq n \leq m$ (more equations than unknowns) In this case again, for consistency, we require

$$a_{r+1} = 0$$
$$\vdots$$
$$a_m = 0$$

and the last $m - r$ equations are redundant. If $r = n$, then there is a unique solution. If $r < n$, we can assign the values of $(n - r)$ variables at pleasure and determine the values of the other variables using (1.10). Thus, we have the following theorem.

Theorem.
(a) *A necessary and sufficient condition for the system* $\mathbf{Au} = \mathbf{b}$ *to have solutions is* $\operatorname{rank}\mathbf{A} = \operatorname{rank}(\operatorname{aug}\mathbf{A})$
(b) *If* $\operatorname{rank}\mathbf{A} = \operatorname{rank}(\operatorname{aug}\mathbf{A}) = r$ *and n is the number of unknowns (with* $r \leq n$*), we can assign values of* $(n - r)$ *of the unknowns and determine the remaining r unknowns uniquely provided the matrix of coefficients of these unknowns has* rank r.

An important (and very useful) corollary that follows from this theorem is given below.

Corollary. *Suppose that is* $\mathbf{A}$ *a square matrix and consider the inhomogeneous system* $\mathbf{Au} = \mathbf{b}$. *This system has a unique solution for any* $\mathbf{b}$ *iff (if and only if) the only solution to the corresponding homogeneous system* $\mathbf{Au} = \mathbf{0}$ *is the trivial one.*

The generalization of the above theorem to the case in which $\mathbf{A}$ is replaced by a linear operator is called *Fredholm alternative (theorem)* and will be discussed later.

Example 1.6. Consider the inhomogeneous system in three variables

$$\begin{aligned} u_1 - u_2 &= b_1 \\ u_2 - u_3 &= b_2 \\ u_1 + u_3 &= b_3 \end{aligned}$$

for which $\operatorname{rank}(\mathbf{A}) = 3$ (see Example 1.3). Thus, there is a unique solution to the above equations for any choice of b_1, b_2 and b_3.

Example 1.7. Consider the inhomogeneous system in four variables

$$\begin{aligned} u_1 - 2u_2 - u_4 &= b_1 \\ -2u_1 + 3u_2 + 3u_3 &= b_2 \\ -u_2 + 3u_3 - 2u_4 &= b_3 \\ 3u_1 - 7u_2 + 3u_3 - 5u_4 &= b_4 \end{aligned}$$

for which $\operatorname{rank}(\mathbf{A}) = 2$. It may be verified that the system is consistent iff $b_3 = 2b_1 + b_2$ and $b_4 = 5b_1 + b_2$. We shall return to this example in Chapter 3.

1.5 Gaussian elimination and LU decomposition

We now discuss the Gaussian elimination algorithm for determining (or numerically computing) the solution(s) of the linear system $\mathbf{Ax} = \mathbf{b}$. [For notational convenience, we shall use $\mathbf{x}$ in place of $\mathbf{u}$ in this section]. Before we illustrate the Gaussian algorithm, we consider two special cases of the linear system

$$\mathbf{Ax} = \mathbf{b} \tag{1.11}$$

in which $\mathbf{A}$ is an $n \times n$ nonsingular upper or lower triangular matrix.

1.5.1 Lower and upper triangular systems

We first consider the case in which $\mathbf{A}$ is lower triangular and write equation (1.11) as

$$\mathbf{Lx} = \mathbf{b} \tag{1.12}$$

or in expanded form

$$\begin{aligned} l_{11}x_1 &= b_1 \\ l_{21}x_1 + l_{22}x_2 &= b_2 \\ l_{31}x_1 + l_{32}x_2 + l_{33}x_3 &= b_3 \\ &\cdot \\ &\cdot \\ l_{n1}x_1 + l_{n2}x_2 + \cdots + l_{nn}x_n &= b_n \end{aligned} \tag{1.13}$$

Since we assumed $\mathbf{L}$ is nonsingular, $l_{ii} \neq 0$ for any i and we can solve equations (1.13) by forward substitution:

$$\begin{aligned} x_1 &= b_1/l_{11} \\ x_2 &= (b_2 - l_{21}x_1)/l_{22} \\ &\cdot \\ &\cdot \\ x_k &= \frac{(b_k - \sum_{j=1}^{k-1} l_{kj}x_j)}{l_{kk}}; \quad k = 1, 2, \ldots, n \end{aligned} \tag{1.14}$$

It is of practical interest to count the number of arithmetic operations (additions or subtractions denoted by AS and multiplications or divisions denoted by MD) needed to obtain the solution. It follows from (1.14) that the operation count in solving the lower triangle system is given by

$$\begin{aligned}\#\text{AS} &= 0 + 1 + 2 + \cdots + (n-1) \\ &= \frac{n(n-1)}{2} \approx \frac{n^2}{2} \quad \text{for a large } n \end{aligned} \tag{1.15}$$

$$\begin{aligned}\#\text{MD} &= 1 + 2 + \cdots + n \\ &= \frac{n(n+1)}{2} \approx \frac{n^2}{2} \quad \text{for large } n \end{aligned} \tag{1.16}$$

Usually, when n is large, AS is equal to MD and hereafter, we shall only count MD. (Another reason for this is that multiplication or division on the computer takes much longer than addition or subtraction). Thus, the operation count (OC) for solving a lower triangular system, given by (1.12), by forward substitution is $0.5n^2$ ($n \gg 1$).

Next, we consider the upper triangular system

$$\mathbf{Ux} = \mathbf{c} \tag{1.17}$$

or in expanded form

$$\begin{aligned} u_{11}x_1 + u_{12}x_2 + \cdots + u_{1n}x_n &= c_1 \\ u_{22}x_2 + \cdots + u_{2n}x_n &= c_2 \\ &\cdot \\ u_{nn}x_n &= c_n \end{aligned} \tag{1.18}$$

Again, we assume that $\mathbf{U}$ is not singular, i. e., $u_{ii} \neq 0$ for any i. The solution of (1.18) can be obtained by back substitution as

$$\begin{aligned} x_n &= c_n/u_{nn} \\ x_{n-1} &= (c_{n-1} - u_{n-1\,n}x_n)/u_{n-1,n-1} \\ &\cdot \\ &\cdot \\ x_k &= \Big(c_k - \sum_{j=k+1}^{n} u_{k,j}x_j\Big)/u_{k,k}, \quad k = n, n-1, \ldots, 1 \end{aligned} \tag{1.19}$$

We note that the operation count for solving the upper triangular system is also $0.5n^2$ (for $n \gg 1$).

1.5.2 Gaussian elimination

We now describe the Gaussian elimination algorithm for solving the general linear system given by equation (1.11). In this method, we first reduce $\mathbf{Ax} = \mathbf{b}$ to an equivalent upper triangular system $\mathbf{Ux} = \mathbf{c}$ by using elementary row operations of type 3. The upper triangular system is then solved by back substitution.

Denote the augmented matrix of the original system by

$$\operatorname{aug}\mathbf{A}^{(1)} = [\mathbf{A}^{(1)}\ \mathbf{b}^{(1)}] = \begin{bmatrix} a_{11}^{(1)} & a_{12}^{(1)} & \cdot & \cdot & \cdot & a_{1n}^{(1)} & b_1^{(1)} \\ a_{21}^{(1)} & a_{22}^{(1)} & \cdot & \cdot & \cdot & a_{2n}^{(1)} & b_2^{(1)} \\ \cdot & \cdot & \cdot & \cdot & \cdot & \cdot & \cdot \\ \cdot & \cdot & \cdot & \cdot & \cdot & \cdot & \cdot \\ \cdot & \cdot & \cdot & \cdot & \cdot & \cdot & \cdot \\ a_{n1}^{(1)} & a_{n2}^{(1)} & \cdot & \cdot & \cdot & a_{nn}^{(1)} & b_n^{(1)} \end{bmatrix}$$

In the first step, we assume $a_{11}^{(1)} \neq 0$ and define row multipliers

$$m_{i1} = a_{i1}^{(1)}/a_{11}^{(1)}; \quad i = 2, 3, \ldots, n$$

Multiply row 1 by m_{i1} and subtract from row i ($i = 2, \ldots, n$). At the end of the step, the form of the augmented system is given by

$$\operatorname{aug}\mathbf{A}^{(2)} = \begin{bmatrix} a_{11}^{(1)} & a_{12}^{(1)} & \cdot & \cdot & \cdot & a_{1n}^{(1)} & b_1^{(1)} \\ 0 & a_{22}^{(2)} & \cdot & \cdot & \cdot & a_{2n}^{(2)} & b_2^{(2)} \\ \cdot & \cdot & \cdot & \cdot & \cdot & \cdot & \cdot \\ \cdot & \cdot & \cdot & \cdot & \cdot & \cdot & \cdot \\ \cdot & \cdot & \cdot & \cdot & \cdot & \cdot & \cdot \\ 0 & a_{n2}^{(2)} & \cdot & \cdot & \cdot & a_{nn}^{(2)} & b_n^{(2)} \end{bmatrix}$$

where $a_{ij}^{(2)} = a_{ij}^{(1)} - m_{i1}a_{1j}^{(1)}$; $i, j = 2, \ldots, n$. In the second step, we assume $a_{22}^{(2)} \neq 0$ and continue to eliminate the unknowns leaving the first row undisturbed. After $(n-1)$ steps, we obtain the upper triangular system

$$\operatorname{aug}\mathbf{A}^{(n)} = \begin{bmatrix} a_{11}^{(1)} & a_{12}^{(1)} & \cdot & \cdot & \cdot & a_{1n}^{(1)} & b_1^{(1)} \\ 0 & a_{22}^{(2)} & \cdot & \cdot & \cdot & a_{2n}^{(2)} & b_2^{(2)} \\ 0 & 0 & a_{33}^{(3)} & \cdot & \cdot & a_{3n}^{(3)} & b_3^{(3)} \\ \cdot & \cdot & \cdot & \cdot & \cdot & \cdot & \cdot \\ \cdot & \cdot & \cdot & \cdot & \cdot & \cdot & \cdot \\ \cdot & \cdot & \cdot & \cdot & \cdot & \cdot & \cdot \\ 0 & 0 & \cdot & \cdot & \cdot & a_{nn}^{(n)} & b_n^{(n)} \end{bmatrix}$$

or equivalently,

$$\mathbf{Ux} = \mathbf{c} \tag{1.20}$$

This completes the elimination procedure. The upper triangular system given by (1.20) can be solved by back substitution as shown earlier. [Remark: The element $a_{ii}^{(i)}$ which is at the upper left corner after $i-1$ steps is called the *pivot*. When the Gaussian elimination algorithm is implemented in practice, the rows are interchanged so that the pivot element has the maximum absolute value. This *partial pivoting* procedure minimizes round off errors when solving large systems of linear equations. However, this procedure does not preserve the initial matrix.]

1.5.3 LU decomposition/factorization

Let

$$\mathbf{U} = \begin{bmatrix} a_{11}^{(1)} & a_{12}^{(1)} & . & . & . & a_{1n}^{(1)} \\ 0 & a_{22}^{(2)} & . & . & . & a_{2n}^{(2)} \\ 0 & 0 & a_{33}^{(3)} & . & . & a_{3n}^{(3)} \\ . & . & . & . & . & . \\ . & . & . & . & . & . \\ 0 & 0 & . & . & . & a_{nn}^{(n)} \end{bmatrix}, \quad \mathbf{L} = \begin{bmatrix} 1 & 0 & 0 & 0 & . & . & 0 \\ m_{21} & 1 & 0 & 0 & . & . & 0 \\ m_{31} & m_{32} & 1 & . & . & . & 0 \\ . & . & . & . & . & . & . \\ . & . & . & . & . & . & . \\ m_{n1} & m_{n2} & m_{n3} & . & . & . & 1 \end{bmatrix}$$

where m_{ij} are the row multipliers determined in the elimination process (Remark: These row multipliers can be stored in place of zeros during the elimination process). A straightforward but tedious calculation shows that

$$\mathbf{A} = \mathbf{LU} \tag{1.21}$$

We also note that the number of operations (AS or MD) needed to factorize $\mathbf{A}$ as in (1.21) is given by

$$\begin{aligned} \text{OC} &= (n-1)^2 + (n-2)^2 + \cdots + 1^2 \\ &= \frac{(n-1)n(2n-1)}{6} \approx \frac{1}{3}n^3 \quad (\text{for } n \gg 1) \end{aligned}$$

Thus, the total operation count for solving $\mathbf{Ax} = \mathbf{b}$ is $\frac{1}{3}n^3 + n^2$ (for $n \gg 1$). Hence, for large n, the major part of the work is the $\mathbf{LU}$ decomposition. For comparison, we note that a matrix and vector multiplication involves n^2 operations while multiplication of two $n \times n$ matrices requires n^3 operations.

Example 1.8.

$$\begin{aligned} x_1 + 2x_2 + x_3 &= 3 \\ 2x_1 + 3x_2 - x_3 &= -6 \\ 3x_1 - 2x_2 - 4x_3 &= -2 \end{aligned}$$

$$\text{aug}\,\mathbf{A}^{(1)} = \begin{bmatrix} 1 & 2 & 1 & 3 \\ 2 & 3 & -1 & -6 \\ 3 & -2 & -4 & -2 \end{bmatrix}$$

$$\text{aug}\,\mathbf{A}^{(2)} = \begin{bmatrix} 1 & 2 & 1 & 3 \\ 0 & -1 & -3 & -12 \\ 0 & -8 & -7 & -11 \end{bmatrix}$$

$$\text{aug}\,\mathbf{A}^{(3)} = \begin{bmatrix} 1 & 2 & 1 & 3 \\ 0 & -1 & -3 & -12 \\ 0 & 0 & 17 & 85 \end{bmatrix}$$

$$17x_3 = 85, \quad x_3 = 5$$

$$-x_2 - 3x_3 = -12 \Rightarrow x_2 = 12 - 3x_3 = -3$$

$$x_1 + 2x_2 + x_3 = 3 \Longrightarrow x_1 = 3 + 6 - 5 = 4$$

We also note that

$$\mathbf{U} = \begin{bmatrix} 1 & 2 & 1 \\ 0 & -1 & -3 \\ 0 & 0 & 17 \end{bmatrix}, \quad \mathbf{L} = \begin{bmatrix} 1 & 0 & 0 \\ 2 & 1 & 0 \\ 3 & 8 & 1 \end{bmatrix}$$

$$\mathbf{LU} = \begin{bmatrix} 1 & 0 & 0 \\ 2 & 1 & 0 \\ 3 & 8 & 1 \end{bmatrix} \begin{bmatrix} 1 & 2 & 1 \\ 0 & -1 & -3 \\ 0 & 0 & 17 \end{bmatrix} = \begin{bmatrix} 1 & 2 & 1 \\ 2 & 3 & -1 \\ 3 & -2 & -4 \end{bmatrix} = \mathbf{A}$$

1.6 Inverse of a square matrix

If $\mathbf{A}$ is a square matrix of order n, the inverse of $\mathbf{A}$ is another square matrix $\mathbf{B}$ such that

$$\mathbf{AB} = \mathbf{BA} = \mathbf{I}_n \tag{1.22}$$

The inverse of $\mathbf{A}$ is often denoted by $\mathbf{A}^{-1}$. When $\mathbf{A}$ has an inverse, it is said to be nonsingular or invertible. If $\mathbf{A}$ does not have an inverse, it is said to be singular.

1.6.1 Properties of inverse

The following properties may be verified from the definition of the inverse:

1. $\mathbf{A}$ has an inverse if and only if it has rank n.
2. When it exists, the inverse of $\mathbf{A}$ is unique.
3. When $\mathbf{A}$ is nonsingular,

$$(\mathbf{A}^{-1})^{-1} = \mathbf{A} \tag{1.23}$$

4. If **A** and **B** are square matrices of same order and both have inverses, then

$$(\mathbf{AB})^{-1} = \mathbf{B}^{-1}\mathbf{A}^{-1} \tag{1.24}$$

5. If **A** is invertible, so is $\mathbf{A}^T$ and

$$(\mathbf{A}^T)^{-1} = (\mathbf{A}^{-1})^T \tag{1.25}$$

1.6.2 Calculation of inverse

Suppose that **A** and **B** are square matrices of order n and

$$\mathbf{AB} = \mathbf{I}. \tag{1.26}$$

Then it may be shown that **A** has rank n, **A** and **B** commute and

$$\mathbf{BA} = \mathbf{I}. \tag{1.27}$$

Thus, to calculate $\mathbf{A}^{-1}$, it is sufficient to satisfy the relation given by (1.26). Suppose that the columns of **B** are denoted by $\mathbf{b}_1, \mathbf{b}_2, \mathbf{b}_3, \ldots, \mathbf{b}_n$ and let $\mathbf{e}_j$ ($j = 1, 2, \ldots, n$) be the n-dimensional column vector having unity element in row j and zeros everywhere else. Then (1.26) is equivalent to

$$\mathbf{Ab}_j = \mathbf{e}_j; \quad j = 1, 2, 3, \ldots, n \tag{1.28}$$

and the j-th column of $\mathbf{A}^{-1}$ can be found by solving the linear equations given by equation (1.28). Thus, we have the following two methods for finding the inverse of a nonsingular matrix **A**.

Method 1: Use **LU** decomposition to factor $\mathbf{A} = \mathbf{LU}$. Then solve $\mathbf{LUb}_j = \mathbf{e}_j$; $j = 1, 2, \ldots, n$. We note that this procedure gives $\mathbf{A}^{-1}$ with a total of $\frac{1}{3}n^3$ operations (for **LU** decomposition) plus $n \times n^2$ operations (for solving equation (1.28)). Thus, total operation count is $\frac{4}{3}n^3$ for $n \gg 1$.

Method 2: We form the $n \times 2n$ augmented matrix $[\mathbf{A}\ \mathbf{I}]$ and use elementary row operations to transform it to the form $[\mathbf{I}\ \mathbf{B}]$, where $\mathbf{B} = \mathbf{A}^{-1}$. It can be shown that the operation count for this procedure is the same as that for method 1.

Example 1.9.

$$\mathbf{A} = \begin{bmatrix} 5 & 8 & 1 \\ 0 & 2 & 1 \\ 4 & 3 & -1 \end{bmatrix}$$

We use method 2 and form the augmented matrix

$$\begin{bmatrix} 5 & 8 & 1 & 1 & 0 & 0 \\ 0 & 2 & 1 & 0 & 1 & 0 \\ 4 & 3 & -1 & 0 & 0 & 1 \end{bmatrix}$$

$(-\frac{4}{5})R_1 + R_3$ gives

$$\begin{bmatrix} 5 & 8 & 1 & 1 & 0 & 0 \\ 0 & 2 & 1 & 0 & 1 & 0 \\ 0 & -\frac{17}{5} & -\frac{9}{5} & -\frac{4}{5} & 0 & 1 \end{bmatrix}$$

$\frac{R_1}{5}$, $(\frac{17}{10})R_2 + R_3$ and $\frac{R_2}{2}$ gives

$$\begin{bmatrix} 1 & \frac{8}{5} & \frac{1}{5} & \frac{1}{5} & 0 & 0 \\ 0 & 1 & \frac{1}{2} & 0 & \frac{1}{2} & 0 \\ 0 & 0 & -\frac{1}{10} & -\frac{4}{5} & \frac{17}{10} & 1 \end{bmatrix}$$

$R_3 \times 5 + R_2$ and $R_3 \times (-10)$ gives

$$\begin{bmatrix} 1 & \frac{8}{5} & \frac{1}{5} & \frac{1}{5} & 0 & 0 \\ 0 & 1 & 0 & -4 & 9 & 5 \\ 0 & 0 & 1 & 8 & -17 & -10 \end{bmatrix}$$

$R_3(-\frac{1}{5}) + R_1$ and $R_2(-\frac{8}{5}) + R_1$ gives

$$\begin{bmatrix} 1 & 0 & 0 & 5 & -11 & -6 \\ 0 & 1 & 0 & -4 & 9 & 5 \\ 0 & 0 & 1 & 8 & -17 & -10 \end{bmatrix}$$

Thus,

$$\mathbf{A}^{-1} = \begin{bmatrix} 5 & -11 & -6 \\ -4 & 9 & 5 \\ 8 & -17 & -10 \end{bmatrix}$$

1.7 Vector-matrix formulation of some chemical engineering problems

In this section, we consider some chemical engineering applications of elementary matrix concepts. First, we present the formulation of some flow and reaction problems in the vector-matrix notation. Next, we illustrate the application of the elementary matrix concepts discussed above. Further analysis of these and other similar models will be considered in later chapters.

1.7.1 Batch reactor: evolution equations with multiple reactions

Consider a batch reactor of constant volume in which the reactions

$$\sum_{j=1}^{S} \nu_{ij} A_j = 0; \quad i = 1, 2, \ldots, R \tag{1.29}$$

occur. There are R reactions among S species. Let V_R be the volume of reactor contents (assumed to be constant) and C_j be the molar concentration of species A_j. Further assumptions are: (i) the reactor contents are well mixed so that there are no spatial gradients and C_j is uniform throughout the tank, (ii) the density of the fluid is constant, (iii) isothermal system and (iv) the volume of fluid in the tank remains constant. Let $r_i(C_1, \ldots, C_S)$ be the rate of reaction i and ν_{ij} be the stoichiometric coefficient of species A_j in reaction i. The mole balance for species A_j is

$$\{\text{Rate of accumulation of moles of } A_j\} = \left\{ \begin{array}{c} \text{Rate of production of moles } A_j \\ \text{due to various chemical reactions} \end{array} \right\}$$

In the notation introduced above, this leads to

$$\frac{d}{dt}\{V_R C_j\} = \left(\sum_{i=1}^{R} \nu_{ij} r_i \right) V_R; \quad j = 1, 2, \ldots, S \tag{1.30}$$

Since V_R is assumed to be constant, the above balance may be simplified and written in the following vector form:

$$\frac{d\mathbf{c}}{dt} = \boldsymbol{\nu}^T \mathbf{r}(\mathbf{c}) \tag{1.31}$$

or in expanded form

$$\frac{d}{dt} \begin{bmatrix} C_1 \\ C_2 \\ \cdot \\ \cdot \\ \cdot \\ C_S \end{bmatrix} = \begin{bmatrix} \nu_{11} & \nu_{21} & \cdot & \cdot & \cdot & \nu_{R1} \\ \nu_{12} & \nu_{22} & \cdot & \cdot & \cdot & \nu_{R2} \\ \cdot & \cdot & \cdot & \cdot & \cdot & \cdot \\ \cdot & \cdot & \cdot & \cdot & \cdot & \cdot \\ \cdot & \cdot & \cdot & \cdot & \cdot & \cdot \\ \nu_{1S} & \nu_{2S} & \cdot & \cdot & \cdot & \nu_{RS} \end{bmatrix} \begin{bmatrix} r_1 \\ r_2 \\ \cdot \\ \cdot \\ \cdot \\ r_R \end{bmatrix} \tag{1.32}$$

Here, $\mathbf{c}$ is the $S \times 1$ vector of concentrations, $\mathbf{r}(\mathbf{c})$ is the $R \times 1$ vector of reaction rates (as a function of various concentrations) and $\boldsymbol{\nu}$ is the $R \times S$ matrix of stoichiometric coefficients. To complete the model, we also specify the initial condition corresponding to the species concentrations at time zero, i. e.,

$$\mathbf{c}(t = 0) = \mathbf{c}_0 \tag{1.33}$$

[Note that the same model is obtained for the case of an ideal isothermal tubular plug flow reactor (PFR) with time replaced by space time. In this case, the initial condition is the vector of inlet concentrations]. For the special case of linear kinetics, we have

$$\mathbf{r}(\mathbf{c}) = \widehat{\mathbf{K}}.\mathbf{c}, \tag{1.34}$$

where $\widehat{\mathbf{K}}$ is the $R \times S$ matrix of first-order rate constants. Defining the $S \times S$ matrix of rate constants $\mathbf{K}$ by

$$\mathbf{K} = \boldsymbol{\nu}^T.\widehat{\mathbf{K}} \tag{1.35}$$

we obtain the batch reactor evolution equation (initial value problem):

$$\frac{d\mathbf{c}}{dt} = \mathbf{K}.\mathbf{c}, \quad t > 0; \quad \mathbf{c}(t=0) = \mathbf{c}_0 \tag{1.36}$$

As an example, we consider the monomolecular reaction scheme (shown in Figure 1.1) where k_{ji} is the first-order rate constant for the formation of species A_j from A_i. Here, $S = 3$, $R = 6$ and ordering the six reactions as $(A_1 \to A_2, A_2 \to A_1, A_1 \to A_3, A_3 \to A_1, A_2 \to A_3, A_3 \to A_2)$ the various matrices may be expressed as

$$\boldsymbol{\nu}^T = \begin{pmatrix} -1 & 1 & -1 & 1 & 0 & 0 \\ 1 & -1 & 0 & 0 & -1 & 1 \\ 0 & 0 & 1 & -1 & 1 & -1 \end{pmatrix} \tag{1.37}$$

$$\widehat{\mathbf{K}} = \begin{pmatrix} k_{21} & 0 & 0 \\ 0 & k_{12} & 0 \\ k_{31} & 0 & 0 \\ 0 & 0 & k_{13} \\ 0 & k_{32} & 0 \\ 0 & 0 & k_{23} \end{pmatrix}; \quad \mathbf{K} = \begin{pmatrix} -(k_{21}+k_{31}) & k_{12} & k_{13} \\ k_{21} & -(k_{12}+k_{32}) & k_{23} \\ k_{31} & k_{32} & -(k_{13}+k_{23}) \end{pmatrix} \tag{1.38}$$

[Remark: The ordering of the reactions changes the matrices $\boldsymbol{\nu}$ and $\widehat{\mathbf{K}}$ but $\mathbf{K}$ depends only on the ordering of the species]. Since the total concentration is fixed for this specific reaction system, i. e., $C_1 + C_2 + C_3 = C_{10} + C_{20} + C_{30} = C_0$, we can define the mole fraction of species A_j as $x_j = C_j/C_0$ and write the evolution equation (1.36) as

$$\frac{d\mathbf{x}}{dt} = \mathbf{K}.\mathbf{x}, \quad t > 0; \quad \mathbf{x}(t=0) = \mathbf{x}_0. \tag{1.39}$$

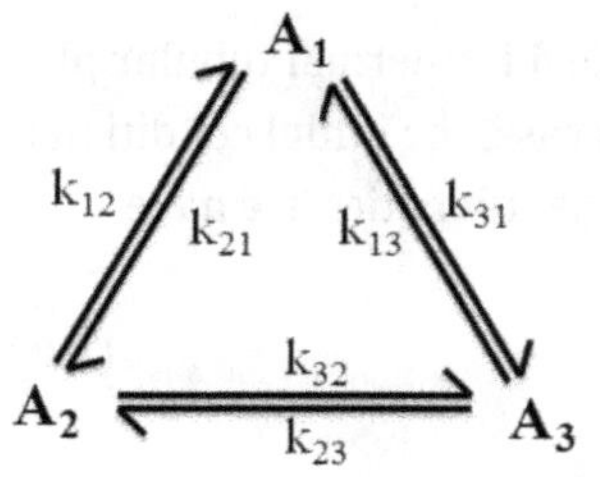

Figure 1.1: Schematic diagram of monomolecular first-order reaction scheme between 3 species.

1.7.2 Continuous-flow stirred tank reactor (CSTR): transient and steady-state models with multiple reactions

We now extend the above batch reactor model to include the flow terms. With the same assumptions as those in the batch reactor case, the species balance equations for the isothermal case may be expressed as

$$\left\{\begin{array}{c}\text{Rate of accumulation}\\ \text{of moles of } A_j\end{array}\right\} = \left\{\begin{array}{c}\text{Inlet molar}\\ \text{flow rate of } A_j\end{array}\right\} - \left\{\begin{array}{c}\text{Outlet molar}\\ \text{flow rate of } A_j\end{array}\right\} + \left\{\begin{array}{c}\text{Rate of production of moles } A_j\\ \text{due to various chemical reactions}\end{array}\right\},$$

which in mathematical form may be expressed as

$$\frac{d}{dt}[V_R C_j] = q_{\text{in}} C_{j,\text{in}}(t) - q_{\text{out}} C_j + \left(\sum_{i=1}^{R} \nu_{ij} r_i\right) V_R \tag{1.40}$$

For the special case of constant and equal in and out flow rates $q_{\text{in}} = q_{\text{out}} = q$ and constant V_R, the above equation simplifies to

$$\frac{d}{dt}[C_j] = \frac{C_{j,\text{in}}(t) - C_j}{\tau} + \left(\sum_{i=1}^{R} \nu_{ij} r_i\right), \tag{1.41}$$

where τ is the residence or space time, defined as the volume of reactor contents over the volumetric flow rate ($\tau = V_R/q$). In vector-matrix form, the above equations may be expressed as

$$\frac{d\mathbf{c}}{dt} = \frac{1}{\tau}(\mathbf{c}_{\text{in}}(t) - \mathbf{c}) + \boldsymbol{\nu}^T \mathbf{r}(\mathbf{c}), \quad t > 0 \tag{1.42}$$

along with the initial condition given by equation (1.33). Here, the $S \times 1$ vector $\mathbf{c}$ represents the species molar concentrations and $\boldsymbol{\nu}$ is the $R \times S$ stoichiometric coefficient matrix. If the inlet concentrations are independent of time, the steady-state reactor effluent concentrations are described by the following set of nonlinear algebraic equations:

$$\frac{1}{\tau}(\mathbf{c}_{\text{in}} - \mathbf{c}_s) + \mathbf{v}^T \mathbf{r}(\mathbf{c}_s) = \mathbf{0}. \tag{1.43}$$

For the special case of linear kinetics and constant feed/inlet concentrations, we obtain the linear system of equations

$$(\mathbf{I} + \mathbf{K}^* \tau)\mathbf{c}_s = \mathbf{c}_{\text{in}}; \quad \mathbf{K}^* = -\mathbf{v}^T.\widehat{\mathbf{K}} \tag{1.44}$$

where $\mathbf{c}_s$ is the vector of steady-state effluent concentrations.

1.7.3 Two interacting tank system: transient model for mixing with in- and outflows

Consider the interacting two tank system shown in Figure 1.2. To develop a mathematical model for describing the transient behavior of the system, we make the following assumptions: (i) each tank is an ideal mixer so that the concentration is uniform within each tank so that the concentration of species A (salt, tracer or a chemical) in the stream leaving each tank is equal to that in the tank, (ii) the flow rate entering each tank (q) is constant (independent of time) but the inlet concentration to tank 1, $C_{\text{in}}(t)$, may change with time, (iii) the exchange or circulation flow rate (q_e) between the tanks is constant, (iv) the density of the fluid is constant and the volume of fluid that each tank holds is constant at V_{R1} and V_{R2} [This assumption implies that the total volumetric flow rate of the streams entering must be equal to that of the streams leaving the tank], (v) no chemical reaction takes place in either tank. The notation for various quantities (volumes of tanks, flow rates and concentrations of species A in each tank) is as shown in the figure.

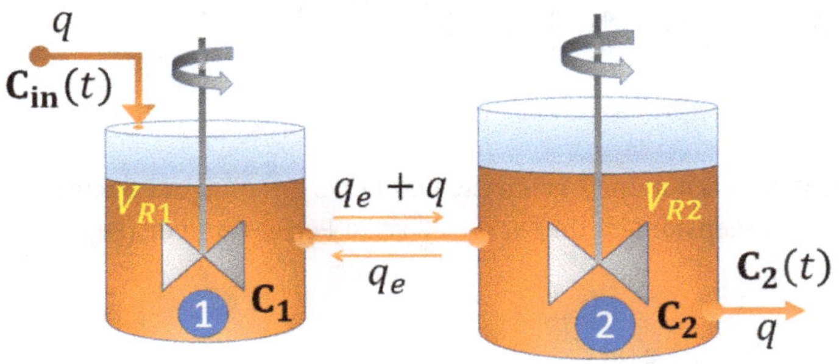

Figure 1.2: Schematic diagram of two interacting tanks with in and outflows.

Mass or mole balance of species A in tank 1 gives

$$V_{R1}\frac{dC_1}{dt} = qC_{\text{in}}(t) + q_eC_2 - (q + q_e)C_1 \tag{1.45}$$

Similarly, mass/mole balance of species A in tank 2 gives

$$V_{R2}\frac{dC_2}{dt} = (q + q_e)C_1 - (q + q_e)C_2 \tag{1.46}$$

[These are mass balances on species if concentration is measured in kg/m^3, and mole balances on species A if the concentration is measured in molar units, moles/m^3 or moles/liter].

To complete the model, we have to supplement it with initial conditions which specify the concentration of the species at time zero, i. e.,

$$C_1\ (\text{at } t = 0) = C_{10} \tag{1.47}$$

$$C_2\ (\text{at } t = 0) = C_{20} \tag{1.48}$$

In vector-matrix form, the above model may be written as

$$\frac{d\mathbf{c}}{dt} = \mathbf{A}\mathbf{c} + \mathbf{b}(t); \quad \mathbf{c}\ (\text{at } t = 0) = \mathbf{c}_0 = \begin{pmatrix} C_{10} \\ C_{20} \end{pmatrix} \tag{1.49}$$

$$\mathbf{c} = \begin{pmatrix} C_1 \\ C_2 \end{pmatrix}; \quad \mathbf{A} = \begin{pmatrix} -\frac{(q+q_e)}{V_{R1}} & \frac{q_e}{V_{R1}} \\ \frac{(q+q_e)}{V_{R2}} & -\frac{(q+q_e)}{V_{R2}} \end{pmatrix}; \quad \mathbf{b}(t) = \begin{pmatrix} \frac{q}{V_{R1}}C_{\text{in}}(t) \\ 0 \end{pmatrix} \tag{1.50}$$

Note that the interaction matrix can be written as the sum of diffusive (or exchange) matrix and a convective (flow) matrix:

$$\begin{aligned} \mathbf{A} &= \begin{pmatrix} -\frac{q_e}{V_{R1}} & \frac{q_e}{V_{R1}} \\ \frac{q_e}{V_{R2}} & -\frac{q_e}{V_{R2}} \end{pmatrix} + \begin{pmatrix} -\frac{q}{V_{R1}} & 0 \\ \frac{q}{V_{R2}} & -\frac{q}{V_{R2}} \end{pmatrix} \\ &= \mathbf{A}_d + \mathbf{A}_c \end{aligned}$$

One special case of this model is obtained when the two tanks are of equal volume. In this case, we can define a dimensionless time ($t' = \frac{q_e t}{V_R}$) and Peclet number ($\text{Pe}_D = \frac{q}{q_e}$; $q_e \neq 0$) and write it as

$$\frac{d\mathbf{c}}{dt'} = \widehat{\mathbf{A}}\mathbf{c} + \text{Pe}_D\,\widehat{\mathbf{b}}(t'); \quad \mathbf{c}\ (\text{at } t' = 0) = \mathbf{c}_0 \tag{1.51}$$

$$\widehat{\mathbf{A}} = \begin{pmatrix} -1 & 1 \\ 1 & -1 \end{pmatrix} + \text{Pe}_D \begin{pmatrix} -1 & 0 \\ 1 & -1 \end{pmatrix}; \quad \widehat{\mathbf{b}}(t') = \begin{pmatrix} C_{\text{in}}(t') \\ 0 \end{pmatrix}. \tag{1.52}$$

The model defined by equations (1.51)–(1.52) is the simplest example of a transient discrete diffusion-convection system. When there is no inflow or outflow from the system,

i. e., $\mathrm{Pe}_D = 0$, we obtain a homogeneous initial value problem describing transient mixing in the system. A second limiting case is that of equal volume tanks with no exchange (or diffusive) flow between them, i. e., $q_e = 0$. In this case, we define the total residence time (or space time as there is no change in moles or volumetric flow rate),

$$\tau = \frac{2V_R}{q}$$

and write the model in the form

$$\frac{\tau}{2}\frac{d\mathbf{c}}{dt} = \begin{pmatrix} -1 & 0 \\ 1 & -1 \end{pmatrix}\mathbf{c} + \begin{pmatrix} C_{\mathrm{in}}(t) \\ 0 \end{pmatrix}; \quad \mathbf{c}\,(\mathrm{at}\ t' = 0) = \mathbf{c}_0. \tag{1.53}$$

1.7.4 Models for transient diffusion, convection and diffusion-convection (compartment models)

The example above of two interacting tanks (or cells) can be generalized to any number of cells which interact through exchange (diffusion), imposed flow (convection) and with or without reaction. We consider here these models without reaction so that the structure of the models can be seen more clearly. These models are referred to as *cell or compartment models* and are discrete (or finite-dimensional) analogs of the continuous diffusion–convection–reaction models. With the same assumptions as above, the formulation of the transient models for these cases is similar to the two tank (cell) system. We provide here only the final model equations for different cases as their derivation is straightforward. [Remark: The compartment models formulated here for discrete interacting systems also appear when partial differential equations of diffusion–convection–reaction type are discretized using finite difference or finite volume methods.]

Discrete transient diffusion model

For the case of N identical (equal volume) interacting tanks arranged in a linear array with equal forward and backward exchange flows (and no imposed external flow), the evolution is described by

$$\frac{d\mathbf{c}}{dt'} = \mathbf{A}_d\mathbf{c}; \quad \mathbf{c}\,(\mathrm{at}\ t' = 0) = \mathbf{c}_0 \tag{1.54}$$

where $\mathbf{c}$ is the $N \times 1$ vector of species concentrations and the $N \times N$ dimensionless diffusion (exchange) matrix $\mathbf{A}_d$ is given by

$$\mathbf{A}_d = \begin{pmatrix} -1 & 1 & 0 & 0 & . & 0 & 0 & 0 \\ 1 & -2 & 1 & 0 & . & 0 & 0 & 0 \\ 0 & 1 & -2 & 1 & . & 0 & 0 & 0 \\ . & . & . & . & . & . & . & . \\ 0 & 0 & 0 & 0 & . & -2 & 1 & 0 \\ 0 & 0 & 0 & 0 & . & 1 & -2 & 1 \\ 0 & 0 & 0 & 0 & . & 0 & 1 & -1 \end{pmatrix} \tag{1.55}$$

Note that the matrix is symmetric and sum of each row and column is zero. The same matrix is obtained when the one-dimensional transient diffusion equation (with zero flux boundary conditions at the ends) is discretized using the second order finite difference or finite volume method. If the cells are arranged in a circular array (so that cell N is connected to cell $N-1$ as well as cell 1), the exchange matrix is modified to

$$\mathbf{A}_d = \begin{pmatrix} -2 & 1 & 0 & 0 & . & 0 & 0 & 1 \\ 1 & -2 & 1 & 0 & . & 0 & 0 & 0 \\ 0 & 1 & -2 & 1 & . & 0 & 0 & 0 \\ . & . & . & . & . & . & . & . \\ 0 & 0 & 0 & 0 & . & -2 & 1 & 0 \\ 0 & 0 & 0 & 0 & . & 1 & -2 & 1 \\ 1 & 0 & 0 & 0 & . & 0 & 1 & -2 \end{pmatrix}. \tag{1.56}$$

Again, this discrete model (with the symmetric circulant matrix) is obtained when the one-dimensional transient diffusion equation on a circle (and periodic boundary conditions) is discretized using second-order finite difference or finite volume methods.

Discrete transient diffusion-convection model

When convective flow is superimposed on the exchange flow, we have a generalization of the two-cell model to the N-cell system. The model equations for a linear array of N identical cells are given by equation (1.51) with

$$\widehat{\mathbf{A}} = \mathbf{A}_d + \mathrm{Pe}_D\,\mathbf{A}_c \tag{1.57}$$

where $\mathbf{A}_d$ is as defined by equation (1.55) and the $N \times N$ convective flow matrix $\mathbf{A}_c$ and the $N \times 1$ forcing vector $\widehat{\mathbf{b}}(t')$ (assuming that there is only a single inlet stream entering tank 1 and leaving tank N) are given by

$$\mathbf{A}_c = \begin{pmatrix} -1 & 0 & 0 & 0 & . & 0 & 0 & 0 \\ 1 & -1 & 0 & 0 & . & 0 & 0 & 0 \\ 0 & 1 & -1 & 0 & . & 0 & 0 & 0 \\ . & . & . & . & . & . & . & . \\ 0 & 0 & 0 & 0 & . & -1 & 0 & 0 \\ 0 & 0 & 0 & 0 & . & 1 & -1 & 0 \\ 0 & 0 & 0 & 0 & . & 0 & 1 & -1 \end{pmatrix} \tag{1.58}$$

$$\widehat{\mathbf{b}}(t') = C_{\mathrm{in}}(t')\mathbf{e}_1. \tag{1.59}$$

Here, $\mathbf{e}_1$ is the $N \times 1$ unit vector (corresponding to unity in the first element and zeros in all other elements). Once again, this discrete model may be obtained from the one dimensional continuum transient diffusion-convection model by using differencing methods.

Model for discrete transient convective loop

Consider a set of 3 identical cells arranged in a convective loop as shown in Figure 1.3. For simplicity, assume that all the cells have the same volume and the flow rate through the loop is constant (at q). The transient model of the system is given by

$$\frac{d\mathbf{c}}{dt'} = \mathbf{A}_L\mathbf{c}; \quad \mathbf{c}\,(\text{at } t' = 0) = \mathbf{c}_0 \tag{1.60}$$

where the 3×3 dimensionless loop connectivity matrix $\mathbf{A}_L$ is given by

$$\mathbf{A}_L = \begin{pmatrix} -1 & 0 & 1 \\ 1 & -1 & 0 \\ 0 & 1 & -1 \end{pmatrix} \tag{1.61}$$

Generalizing to the case of a loop consisting of N identical cells, we obtain equation (1.60) with

$$\mathbf{A}_L = \begin{pmatrix} -1 & 0 & 0 & 0 & . & 0 & 0 & 1 \\ 1 & -1 & 0 & 0 & . & 0 & 0 & 0 \\ 0 & 1 & -1 & 0 & . & 0 & 0 & 0 \\ . & . & . & . & . & . & . & . \\ 0 & 0 & 0 & 0 & . & -1 & 0 & 0 \\ 0 & 0 & 0 & 0 & . & 1 & -1 & 0 \\ 0 & 0 & 0 & 0 & . & 0 & 1 & -1 \end{pmatrix}. \tag{1.62}$$

We note that $\mathbf{A}_L$ is not a symmetric matrix but is a special case of a circulant matrix. A more general circulant matrix denotes the matrix

$$\begin{bmatrix} a_1 & a_2 & \cdots & \cdots & \cdots & a_N \\ a_N & a_1 & \cdots & \cdots & \cdots & a_{N-1} \\ \vdots & & \ddots & & & \vdots \\ \vdots & & & \ddots & & \vdots \\ \vdots & & & & \ddots & \vdots \\ a_2 & \cdots & \cdots & \cdots & \cdots & a_1 \end{bmatrix} \tag{1.63}$$

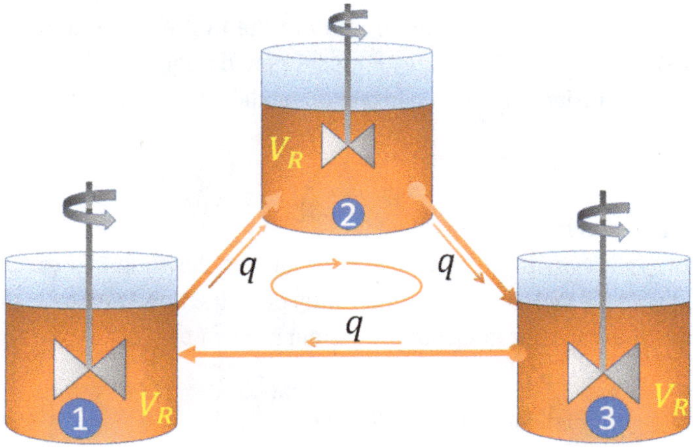

Figure 1.3: Schematic diagram of a discrete convective loop with 3 cells.

in which every row starting with the second can be obtained from the previous row by moving each of its elements one column to the right with the last element circling to become the first. Each diagonal of a circulant matrix consists of identical elements. It is a special case of a *Toeplitz matrix* for which elements on all the diagonals are constants. The matrices appearing in the above examples are also known as banded Toeplitz matrices.

1.8 Application of elementary matrix concepts

Example 1.10 (Mass transfer disguised reaction rate constant matrix). Consider a fluid-solid system in which reactions occur only on the solid (catalyst) surface (Figure 1.4).

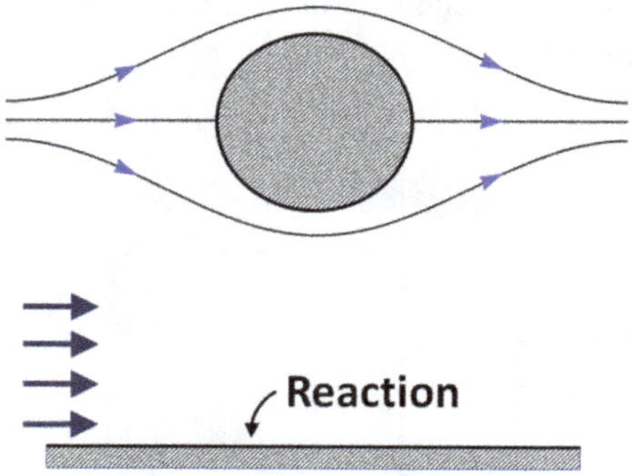

Figure 1.4: Schematic diagram of flow past a particle on the surface of which chemical reactions occur.

A simplified model of such a system at steady state is obtained by using the concept of the mass transfer coefficient between the fluid and solid. For the case of a single reaction of the form $A \rightarrow B$, the steady-state model is obtained by equating the external flux

$$j_e = k_c(C_{Ab} - C_{As}) \tag{1.64}$$

to the reaction rate (the rate of disappearance of A), which for the case of linear kinetics (first-order reaction) may be expressed as

$$r = k_s a_v C_{As} = k_v C_{As}. \tag{1.65}$$

Here, k_c is the external mass transfer coefficient, a_v is the solid-fluid exchange area per unit volume, k_s is the first-order surface reaction rate constant ($k_v = k_s a_v$ is the first-order rate constant) and C_{Ab} and C_{As} are the bulk (cup-mixing) and surface concentrations of the reactant species A, respectively. For the case of several reactions among N species with linear kinetics, the external flux vector may be expressed as

$$\mathbf{j}_e = \mathbf{K}_c(\mathbf{C}_b - \mathbf{C}_s) \tag{1.66}$$

while the reaction rate vector is of the form

$$\mathbf{r} = \mathbf{K}_{Rs}\mathbf{C}_s \tag{1.67}$$

where $\mathbf{K}_c$ is an $N \times N$ matrix of mass transfer coefficients, $\mathbf{K}_{Rs}$ is an $N \times N$ reaction rate constant matrix and $\mathbf{C}_b$ ($\mathbf{C}_s$) is the $N \times 1$ bulk (surface) concentration vector. The i-th component of the vector on the LHS of equation (1.66) is the rate of transport of the species i from the bulk to the solid (catalytic) surface while the i-th component of the vector ($\mathbf{K}_{Rs}\mathbf{C}_s$) is the net rate of consumption of species i in various chemical reactions on the surface. Eliminating the unknown surface concentration vector $\mathbf{C}_s$ gives the rate in terms of the bulk concentration as

$$\mathbf{r} = \mathbf{K}^*\mathbf{C}_b \tag{1.68}$$

where the apparent or mass transfer disguised rate constant matrix $\mathbf{K}^*$ is defined by

$$\mathbf{K}^* = \mathbf{K}_{Rs}(\mathbf{K}_{Rs} + \mathbf{K}_c)^{-1}\mathbf{K}_c. \tag{1.69}$$

For numerical calculations and physical interpretation of this result, it is convenient to define

$$\mathbf{K}_{Rs} = k_s\mathbf{A}, \quad \mathbf{K}_c = k_c\mathbf{M} \tag{1.70}$$

where **A** is the matrix of relative rate constants and **M** is the matrix of relative mass transfer coefficients. This allows us to write (1.69) as

$$\mathbf{K}^* = \mathbf{K}_{Rs}\mathbf{H} \tag{1.71}$$

where the external effectiveness factor matrix is defined by

$$\begin{aligned}\mathbf{H} &= (\mathrm{Da}_{pm}\,\mathbf{A} + \mathbf{M})^{-1}\mathbf{M} \\ &= (\mathbf{I} + \mathrm{Da}_{pm}\,\mathbf{M}^{-1}\mathbf{A})^{-1}\end{aligned} \tag{1.72}$$

and Da_{pm} is the particle (or local) Damköhler number defined by

$$\mathrm{Da}_{pm} = \frac{k_s}{k_c} = \frac{k_v}{k_c a_v}$$

[Remarks: The matrices **A**, **M** and **H** are dimensionless or have elements with no units. In simplifying the expression for **H**, we have assumed that **M** is invertible and used the property given by equation (1.24).] To illustrate how the external mass transfer can disguise the true reaction network (leading to the so called falsified kinetics), we consider the following numerical values:

$$\mathrm{Da}_{pm} = 1, \quad \mathbf{M} = \begin{pmatrix} 1 & 0 & 0 \\ 0 & 1 & 0 \\ 0 & 0 & 1 \end{pmatrix}, \quad \mathbf{A} = \begin{pmatrix} 1 & -\frac{1}{2} & 0 \\ -1 & 1 & -\frac{1}{4} \\ 0 & -\frac{1}{2} & \frac{1}{4} \end{pmatrix},$$

which gives

$$\mathbf{H} = \frac{1}{33}\begin{pmatrix} 19 & 5 & 1 \\ 10 & 20 & 4 \\ 4 & 8 & 28 \end{pmatrix}$$

$$\mathbf{K}^* = \frac{k_s}{33}\begin{pmatrix} 14 & -5 & -1 \\ -10 & 13 & -4 \\ -4 & -8 & 5 \end{pmatrix}$$

The true and disguised rate constants (and reaction networks) are shown in Figure 1.5. We note that two extra reactions ($P \to R$ and $R \to P$) (that are not present in the true reaction network) appear in the mass transfer disguised reaction network.

In the general case of several reactions among many species, the true reaction network may be represented by a sparse matrix of about N nonzero rate constants while the mass transfer disguised rate constant matrix may have N^2 nonzero rate constants. Thus, the number of spurious reactions that appear in the mass transfer disguised rate constant matrix is $N^2 - N$, which is quite large for large N. See below for further illustration

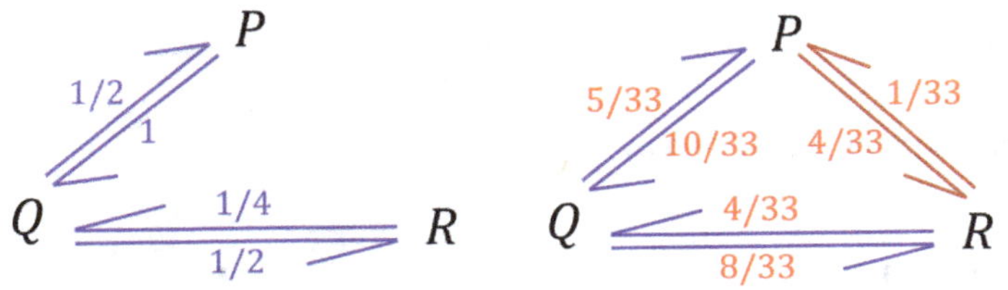

Figure 1.5: Schematic representation of the true (left) and mass transfer disguised reaction networks.

for the case of large N, where, symbolic manipulation and computer algebra software (such as Mathematica®, Matlab®, Fortran/Python or others) can be used, as solving by hand may take much longer time.

1.9 Application of computer algebra and symbolic manipulation

Consider a true reaction network among S species A_j ($j = 1, 2, \ldots, S$) given by

$$A_1 \underset{k_{-1}}{\overset{k_1}{\rightleftharpoons}} A_2 \underset{k_{-2}}{\overset{k_2}{\rightleftharpoons}} A_3 \ldots \underset{k_{-(i-1)}}{\overset{k_{i-1}}{\rightleftharpoons}} A_i \underset{k_{-i}}{\overset{k_i}{\rightleftharpoons}} A_{i+1} \ldots \underset{k_{-(s-1)}}{\overset{k_{s-1}}{\rightleftharpoons}} A_s \tag{1.73}$$

where k_i and k_{-i} are forward and backward rate constants between species A_i to A_{i+1} with linear kinetics. Thus, the net rate of formation of these species is given by

$$\frac{d[A_1]}{dt} = -k_1[A_1] + k_{-1}[A_2] \tag{1.74}$$

$$\frac{d[A_2]}{dt} = k_1[A_1] - (k_{-1} + k_2)[A_2] + k_{-2}[A_3] \tag{1.75}$$

$$\vdots$$

$$\frac{d[A_i]}{dt} = k_{i-1}[A_{i-1}] - (k_{-(i-1)} + k_i)[A_i] + k_{-i}[A_{i+1}] \tag{1.76}$$

$$\vdots$$

$$\frac{d[A_{s-1}]}{dt} = k_{s-2}[A_{s-2}] - (k_{-(s-2)} + k_{s-1})[A_{s-1}] + k_{-(s-1)}[A_s] \tag{1.77}$$

$$\frac{d[A_s]}{dt} = k_{s-1}[A_{s-1}] - k_{-(s-1)}[A_s] \tag{1.78}$$

Thus assuming x_i being the mole fraction:

$$x_i = \frac{[A_i]}{\sum_{j=1}^{S}[A_j]}, \tag{1.79}$$

we can write

$$\frac{d\mathbf{x}}{dt} = -\mathbf{K}_{Rs}\mathbf{x}, \quad t > 0 \quad \text{and} \quad \mathbf{x} = \mathbf{x}^0 \ @\, t = 0 \tag{1.80}$$

where the true rate constant matrix is given by

$$\mathbf{K}_{Rs} = \begin{pmatrix} k_1 & -k_{-1} & 0 & 0 & 0 & 0 & 0 \\ -k_1 & k_{-1} + k_2 & -k_{-2} & 0 & 0 & 0 & 0 \\ 0 & \ddots & \ddots & \ddots & 0 & 0 & 0 \\ 0 & 0 & -k_{(i-1)} & k_{-(i-1)} + k_i & -k_{-i} & 0 & 0 \\ 0 & 0 & 0 & \ddots & \ddots & \ddots & 0 \\ 0 & 0 & 0 & 0 & -k_{(s-2)} & k_{-(s-2)} + k_{s-1} & -k_{-(s-1)} \\ 0 & 0 & 0 & 0 & 0 & -k_{(s-1)} & k_{-(s-1)} \end{pmatrix} \tag{1.81}$$

and the mole fraction vector $\mathbf{x} = \begin{pmatrix} x_1 & x_2 & \cdots & x_s \end{pmatrix}^T$. (1.82)

Thus depending on the mass transfer coefficient matrix $\mathbf{K}_c$, the apparent (mass-transfer disguised) rate constant matrix $\mathbf{K}^*$ given by equation (1.69) can have all elements nonzero, i. e.,

$$\mathbf{K}^* = \{k_{ij}^*\} \quad \text{where } k_{ij}^* \neq 0 \tag{1.83}$$

and new reaction appears between species A_i and A_j (where $|i - j| > 1$) as

$$A_i \underset{k_{ij}^*}{\overset{k_{ji}^*}{\rightleftharpoons}} A_j, \quad |i - j| > 1. \tag{1.84}$$

The number of new reactions appearing for the reaction network (equation (1.73)) are shown in Table 1.1. It can be seen that for a large number of species in sequential reactions, the mass trasnfer disguised reaction network is complex as a significant number of new reactions appear.

Table 1.1: Number of new mass-transfer disguised reactions.

Number of species S	Number of reactions in true network (equation (1.73))	Number of reactions in mass-transfer disguised network	Spurious reactions (difference)
2	2	2	0
3	4	6	2
4	6	12	6
5	8	20	12
10	18	90	72
50	98	2450	2352
100	198	9900	9702
S	$2(S-1)$	$S(S-1)$	$(S-1)(S-2)$

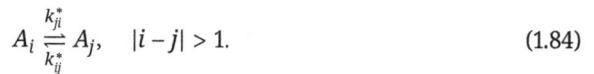

Note that when number of species (S) is large, the evaluation of mass-transfer disguised rate constants can be time consuming, and hence any symbolic/numerical programming software such as Mathematica®, Matlab® and Maple can be utilized for these purposes. Here, we use Mathematica® to demonstrate some examples.

1.9.1 Example 1: mass transfer disguised matrix for a five species system

Assuming $S = 5$ and the true rate constants (see equation (1.73)) to be

$$\begin{aligned} &k_1 = k_s; \quad k_{-1} = \frac{1}{2}k_s; \\ &k_2 = \frac{1}{2}k_s; \quad k_{-2} = \frac{1}{4}k_s; \\ &k_3 = \frac{1}{4}k_s; \quad k_{-3} = \frac{1}{8}k_s; \\ &k_4 = \frac{1}{10}k_s; \quad k_{-4} = \frac{1}{20}k_s. \end{aligned}$$

$\Rightarrow$

$$\mathbf{K}_{Rs} = k_s\mathbf{A} \quad \text{where } \mathbf{A} = \begin{pmatrix} 1 & -1/2 & 0 & 0 & 0 \\ -1 & 1 & -1/4 & 0 & 0 \\ 0 & -1/2 & 1/2 & -1/8 & 0 \\ 0 & 0 & -1/4 & 9/40 & -1/20 \\ 0 & 0 & 0 & -1/10 & 1/20 \end{pmatrix}$$

In addition, assuming that the mass transfer coefficient matrix $\mathbf{K}_c$ is diagonal with all diagonal elements same, i. e.,

$$\mathbf{K}_c = k_s\mathbf{M}, \quad \text{where } \mathbf{M} = \mathbf{I}_5 = \begin{pmatrix} 1 & 0 & 0 & 0 & 0 \\ 0 & 1 & 0 & 0 & 0 \\ 0 & 0 & 1 & 0 & 0 \\ 0 & 0 & 0 & 1 & 0 \\ 0 & 0 & 0 & 0 & 1 \end{pmatrix},$$

such that $\mathrm{Da}_{pm} = \frac{k_s}{k_c} = 1$. Then, the mass-transfer disguised rate constant matrix can be computed using equations (1.71) and (1.72) as

$$\mathbf{K}^* = \mathbf{K}_{Rs}\mathbf{H} = k_s\mathbf{A}(\mathbf{I} + \mathrm{Da}_{pm}\,\mathbf{M}^{-1}\mathbf{A})^{-1} = k_s\mathbf{A}(\mathbf{I}_5 + \mathbf{A})^{-1}$$

$$= \frac{k_s}{8053}\begin{pmatrix} 3422 & -1209 & -205 & -21 & -1 \\ -2418 & 3217 & -820 & -84 & -4 \\ -820 & -1640 & 2313 & -588 & -28 \\ -168 & -336 & -1176 & 1333 & -320 \\ -16 & -32 & -112 & -640 & 353 \end{pmatrix}$$

which has all elements nonzero. In other words, there are $S(S-1) = 20$ reactions can be observed due to mass transfer. The true and mass transfer disguised reaction networks are shown in Figure 1.6, where the additional reactions are depicted.

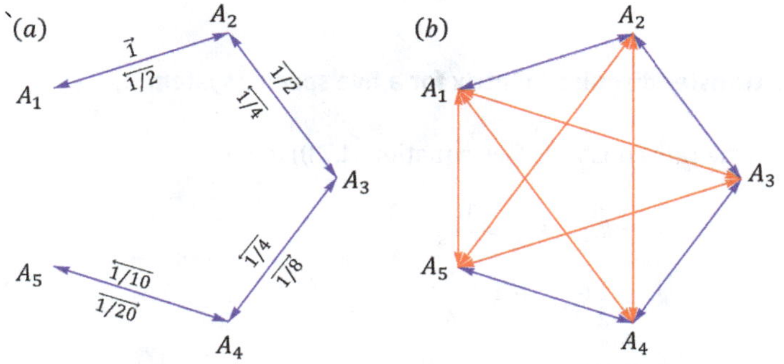

Figure 1.6: Effect of mass-transfer on observed kinetics: (left) true reaction network and (right) mass-transfer disguised reaction network for Example 1 (5 species leading to 12 spurious reactions).

1.9.2 Example 2: mass transfer disguised matrix for a ten species system

Assuming $S = 10$ and the true rate constants (see equation (1.73)) are

$$k_1 = k_s; \quad k_{-1} = 0.5k_s;$$
$$k_2 = 0.5k_s; \quad k_{-2} = 0.25k_s;$$
$$k_3 = 0.25k_s; \quad k_{-3} = 0.125k_s;$$
$$k_4 = 0.1k_s; \quad k_{-4} = 0.05k_s;$$
$$k_5 = 0.25k_s; \quad k_{-5} = 0.1k_s$$
$$k_6 = 0.1k_s; \quad k_{-6} = 0.15k_s$$
$$k_7 = 0.2k_s; \quad k_{-7} = 0.3k_s$$
$$k_8 = 0.1k_s; \quad k_{-8} = 0.05k_s$$
$$k_9 = 0.2k_s; \quad k_{-9} = 0.1k_s$$

$\Rightarrow \mathbf{K}_{Rs} = k_s\mathbf{A}$ where

$$\mathbf{A} = \begin{pmatrix} 1 & -0.5 & 0 & 0 & 0 & 0 & 0 & 0 & 0 & 0 \\ -1 & 1 & -0.25 & 0 & 0 & 0 & 0 & 0 & 0 & 0 \\ 0 & -0.5 & 0.5 & -0.125 & 0 & 0 & 0 & 0 & 0 & 0 \\ 0 & 0 & -0.25 & 0.225 & -0.05 & 0 & 0 & 0 & 0 & 0 \\ 0 & 0 & 0 & -0.1 & 0.3 & -0.1 & 0 & 0 & 0 & 0 \\ 0 & 0 & 0 & 0 & -0.25 & 0.2 & -0.15 & 0 & 0 & 0 \\ 0 & 0 & 0 & 0 & 0 & -0.1 & 0.35 & -0.3 & 0 & 0 \\ 0 & 0 & 0 & 0 & 0 & 0 & -0.2 & 0.4 & -0.05 & 0 \\ 0 & 0 & 0 & 0 & 0 & 0 & 0 & -0.1 & 0.25 & -0.1 \\ 0 & 0 & 0 & 0 & 0 & 0 & 0 & 0 & -0.2 & 0.1 \end{pmatrix}$$

In addition, assuming that the transfer coefficient matrix $\mathbf{K}_c$ is diagonal as given by $\mathbf{K}_c = k_s\mathbf{M}$, where

$$\mathbf{M} = \begin{pmatrix} 1 & 0 & 0 & 0 & 0 & 0 & 0 & 0 & 0 & 0 \\ 0 & 2 & 0 & 0 & 0 & 0 & 0 & 0 & 0 & 0 \\ 0 & 0 & 3 & 0 & 0 & 0 & 0 & 0 & 0 & 0 \\ 0 & 0 & 0 & 1 & 0 & 0 & 0 & 0 & 0 & 0 \\ 0 & 0 & 0 & 0 & 4 & 0 & 0 & 0 & 0 & 0 \\ 0 & 0 & 0 & 0 & 0 & 2 & 0 & 0 & 0 & 0 \\ 0 & 0 & 0 & 0 & 0 & 0 & 1 & 0 & 0 & 0 \\ 0 & 0 & 0 & 0 & 0 & 0 & 0 & 3 & 0 & 0 \\ 0 & 0 & 0 & 0 & 0 & 0 & 0 & 0 & 5 & 0 \\ 0 & 0 & 0 & 0 & 0 & 0 & 0 & 0 & 0 & 1 \end{pmatrix},$$

such that $\mathrm{Da}_{pm} = \frac{k_s}{k_c} = 2$. Then, the mass transfer disguised rate constant can be computed using equations (1.71) and (1.72) as

$$\mathbf{K}^* = \mathbf{K}_{Rs}\mathbf{H} = k_s\mathbf{A}(\mathbf{I} + Da_{pm}\mathbf{M}^{-1}\mathbf{A})^{-1} = k_s\mathbf{A}(\mathbf{I}_{10} + 2\mathbf{M}^{-1}\mathbf{A})^{-1}$$

$\Rightarrow$

$$\mathbf{K}^* = k_s \begin{pmatrix}
2.99\times10^{-1} & -1.04\times10^{-1} & -1.99\times10^{-2} & -1.15\times10^{-3} & -1.01\times10^{-4} & -4.27\times10^{-6} & -3.91\times10^{-7} & -1.85\times10^{-7} & -5.69\times10^{-9} & -1.9\times10^{-10} \\
-2.08\times10^{-1} & 3.76\times10^{-1} & -1.2\times10^{-1} & -6.89\times10^{-3} & -6.05\times10^{-4} & -2.56\times10^{-5} & -2.35\times10^{-6} & -1.11\times10^{-6} & -3.41\times10^{-8} & -1.14\times10^{-9} \\
-7.97\times10^{-2} & -2.39\times10^{-1} & 3.04\times10^{-1} & -6.89\times10^{-2} & -6.05\times10^{-3} & -2.56\times10^{-4} & -2.35\times10^{-5} & -1.11\times10^{-5} & -3.41\times10^{-7} & -1.14\times10^{-8} \\
-9.19\times10^{-3} & -2.76\times10^{-2} & -1.38\times10^{-1} & 1.46\times10^{-1} & -3.11\times10^{-2} & -1.31\times10^{-3} & -1.2\times10^{-4} & -5.71\times10^{-5} & -1.75\times10^{-6} & -5.84\times10^{-8} \\
-1.61\times10^{-3} & -4.84\times10^{-3} & -2.42\times10^{-2} & -6.21\times10^{-2} & 2.39\times10^{-1} & -7.45\times10^{-2} & -6.83\times10^{-3} & -3.24\times10^{-3} & -9.93\times10^{-5} & -3.31\times10^{-6} \\
-1.71\times10^{-4} & -5.12\times10^{-4} & -2.56\times10^{-3} & -6.57\times10^{-3} & -1.86\times10^{-1} & 1.46\times10^{-1} & -7.83\times10^{-2} & -3.71\times10^{-2} & -1.14\times10^{-3} & -3.79\times10^{-5} \\
-1.04\times10^{-5} & -3.13\times10^{-5} & -1.56\times10^{-4} & -4.01\times10^{-4} & -1.14\times10^{-2} & -5.22\times10^{-2} & 1.9\times10^{-1} & -1.47\times10^{-1} & -4.51\times10^{-3} & -1.5\times10^{-4} \\
-3.3\times10^{-6} & -9.89\times10^{-6} & -4.94\times10^{-5} & -1.27\times10^{-4} & -3.6\times10^{-3} & -1.65\times10^{-2} & -9.81\times10^{-2} & 2.68\times10^{-1} & -3.78\times10^{-2} & -1.26\times10^{-3} \\
-2.02\times10^{-7} & -6.07\times10^{-7} & -3.03\times10^{-6} & -7.78\times10^{-6} & -2.21\times10^{-4} & -1.01\times10^{-3} & -6.02\times10^{-3} & -7.56\times10^{-2} & 1.97\times10^{-1} & -7.68\times10^{-2} \\
-1.35\times10^{-8} & -4.04\times10^{-8} & -2.02\times10^{-7} & -5.19\times10^{-7} & -1.47\times10^{-5} & -6.75\times10^{-5} & -4.01\times10^{-4} & -5.04\times10^{-3} & -1.54\times10^{-1} & 7.82\times10^{-2}
\end{pmatrix}$$

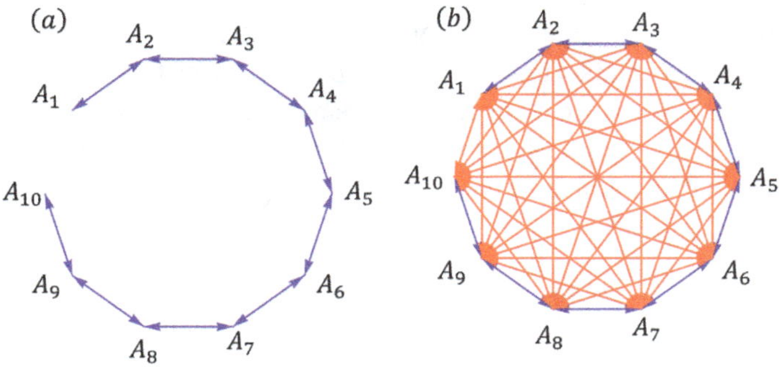

Figure 1.7: Effect of mass transfer on kinetics: (a) true reaction network and (b) mass-transfer disguised reaction network for example 2 (10 species leading to 72 spurious reactions).

which has all elements nonzero. Note that all diagonal elements are positive while off-diagonal elements are negative as expected. In this case, total number of observed reactions are $S(S-1) = 90$, i. e., 72 new reactions appear as shown in Figure 1.7, where the true and the mass-transfer disguised reactions networks are depicted.

Problems

1. *(Formulation of linear models)*: Consider the following simplified economic model of a small town represented by three major industries: coal mining, transportation and electricity. It is estimated that the production of a one dollar value of coal requires the purchase of 10 cents of electricity and 20 cents of transportation. Similarly, the production of one dollar output of transportation requires the purchase of 2 cents of coal and 35 cents of electricity, while production of one dollar value of electricity requires purchase of 10 cents of electricity, 50 cents of coal and 30 cents of transportation. The town has external contracts for $1,000,000 of coal, $1,600,000 of transportation and $500,000 of electricity. Show that the problem of determining how much coal, electricity and transportation is required to supply external demand without a surplus is equivalent to solving a system of (three) linear equations of the form

$$\mathbf{x} = \mathbf{A}\mathbf{x} + \mathbf{d} \quad \text{or} \quad (\mathbf{I} - \mathbf{A})\mathbf{x} = \mathbf{d}$$

   where $\mathbf{x}$ is called the production vector, $\mathbf{d}$ is the demand vector and $\mathbf{A}$ is the consumption matrix. Determine the production vector by solving the linear equations. [Remark: Models of this type were first developed by W. W. Leontief, who won the 1973 economics Nobel prize.]
2. *(Formulation of linear models)*: Paul, Jim and Mike decide to help each other build houses. Paul will spend half of his time on his own house and a quarter of his time on each of the houses of Jim and Mike. Jim will spend one-third of his time on each of the three houses under construction. Mike will spend one-sixth of his time on Paul's

house, one-third on Jim's house and one-half on his own house. For tax purposes, each must place a price on his labor, but they want to do so in a way that each will break even. Formulate the relevant equations, solve them and suggest a price on the labor of each person such that the hourly wages for each exceed the minimum wage.

3. *(Gaussian elimination and LU decomposition)*:
   (a) Consider the linear system

$$x_1 + 2x_2 - x_3 = 0$$
$$2x_1 + x_2 + 2x_3 = 8$$
$$6x_1 + 2x_2 + 2x_3 = 14$$

   i. Solve the above system by Gaussian elimination.
   ii. Determine the matrices **L** and **U** in the decomposition **A** = **LU**.
   iii. Use the result in (ii) to find $\mathbf{A}^{-1}$.

   (b)
   i. Develop an algorithm to decompose **A** = **LU** where **A** is a tridiagonal matrix.
   ii. Use the algorithm to solve **Ax** = **b** where

$$\mathbf{A} = \begin{pmatrix} 4 & -1 & 0 & 0 & 0 \\ -1 & 4 & -1 & 0 & 0 \\ 0 & -1 & 4 & -1 & 0 \\ 0 & 0 & -1 & 4 & -1 \\ 0 & 0 & 0 & -1 & 4 \end{pmatrix}, \quad \mathbf{b} = \begin{pmatrix} 1 \\ 0 \\ 0 \\ 0 \\ 0 \end{pmatrix}$$

4. *(Operation count for LU decomposition and matrix inverse calculations)*:
   (a) Show that the number of operations required to do the decomposition of **A** = **LU** of a square matrix is given by

$$\mathrm{MD} = \frac{(n-1)n(n+1)}{3} \approx \frac{1}{3}n^3 (n \gg 1)$$
$$\mathrm{AS} = \frac{n(n-1)(2n-1)}{6} \approx \frac{1}{3}n^3 (n \gg 1)$$

   where AS = additions/subtractions and MD = multiplications/divisions. How many additional operations are needed to solve **Ax** = **b**?
   (b) Show that (for $n \gg 1$) the calculation of $\mathbf{A}^{-1}$ is only four times the expense of solving **Ax** = **b**.
   (c) Obtain the results analogous to those in (a) for a tridiagonal system of equations.

5. *(Solution of linear homogeneous equations)*:
   (a) Solve the following system and determine the number of linearly independent solutions:

$$u_2 - 2u_3 - 4u_4 = 0$$

$$\begin{aligned} u_3 - 3u_4 &= 0 \\ 2u_1 + u_2 + 3u_3 + 7u_4 &= 0 \\ 6u_1 + 2u_2 + 10u_3 + 28u_4 &= 0 \end{aligned}$$

(b) Determine the relationship between the parameters k and Ra for which the following set of equations has a nontrivial solution:

$$\begin{aligned} c_1 - (\pi^2 + k^2)c_2 &= 0 \\ -(\pi^2 + k^2)c_1 + \mathrm{Ra}\, k^2 c_2 &= 0. \end{aligned}$$

[Remark: This relation is called the neutral stability curve and defines the onset of convection in a fluid layer heated from below. Here, k is the wave number and Ra is the Rayleigh number]. Determine the minimum value of Ra for which a nontrivial solution exists.

(c) Determine all of the nontrivial solutions for the following system of homogeneous equations:

$$\begin{aligned} u_1 - iu_2 &= 0 \\ u_2 + u_3 &= 0 \\ u_1 + u_2 - u_4 &= 0 \\ u_2 + iu_3 + iu_4 &= 0, \quad \text{where } i = \sqrt{-1} \end{aligned}$$

6. *(Solution of linear inhomogeneous equations):*

(a) Determine the values of λ for which the following system has (i) a unique solution, (ii) no solution and (iii) more than one solution:

$$\begin{aligned} u_1 - 3u_3 &= -3 \\ 2u_1 + \lambda u_2 - u_3 &= -2 \\ u_1 + 2u_2 + \lambda u_3 &= 1. \end{aligned}$$

(b) Verify that the following linear system is inconsistent, and hence has no solution:

$$\begin{aligned} u_1 + 4u_2 - u_3 &= -5 \\ 5u_1 + 2u_2 - 3u_3 &= -1 \\ -2u_1 + u_2 + u_3 &= -2 \\ -u_1 + 5u_2 &= -2 \end{aligned}$$

(c) Obtain a general solution to the following system of equations:

$$\begin{aligned} u_1 - u_3 + 2u_4 + u_5 + 6u_6 &= -3 \\ u_2 + u_3 + 3u_4 + 2u_5 + 4u_6 &= 1 \end{aligned}$$

$$u_1 - 4u_2 + 3u_3 + u_4 + 2u_6 = 0$$
$$2u_1 - 4u_2 + 2u_3 + 3u_4 + u_5 + 8u_6 = -3.$$

7. *(Simultaneous linear equations: Countercurrent extraction process):* The models used to describe the steady-state (and also transient) behavior of plate, gas absorbers, extraction units and rectifying and stripping sections of a distillation column are essentially the same. As an illustration, consider a 3-stage absorber (Figure 1.8) in which a heavy (liquid) phase and a light (gas) phase pass countercurrent to each other. The contacting may be assumed to be uniform so that equilibrium is attained in each stage and the equilibrium relationship is linear ($y = Kx$). (a) Show that the steady-state model of the system is of the form

$$0 = \alpha x_{j-1} - (\alpha + \beta)x_j + \beta x_{j+1} \quad j = 1, 2, 3$$

where $\alpha = \frac{L}{h}$, $\beta = \frac{GK}{h}$, L is the liquid (heavy-phase) flow rate, h is the holdup (which is assumed to be constant and same for all stages), G is the gas (light-phase) flow rate and x_j is the composition of the transferable component in the liquid stream leaving stage j. State any other assumptions involved. (b) Generalize the model in (a) to the case of N stages and put it in vector-matrix form. Assuming that the compositions x_0 and $x_{N+1} = \frac{y_{N+1}}{K}$, of the entering streams are known, show that the model may be written in the form

$$\mathbf{Ax} = \mathbf{b}$$

Identify the vectors $\mathbf{x}, \mathbf{b}$ and the matrix $\mathbf{A}$ (c) Compute the steady-state values for y_1 and x_3 when $N = 3$ (three-stage process) and other parameters are given as follows:

$$L = 5, \quad G = 3, \quad h = 1, \quad K = 1, \quad x_0 = 0, \quad y_4 = 0.5$$

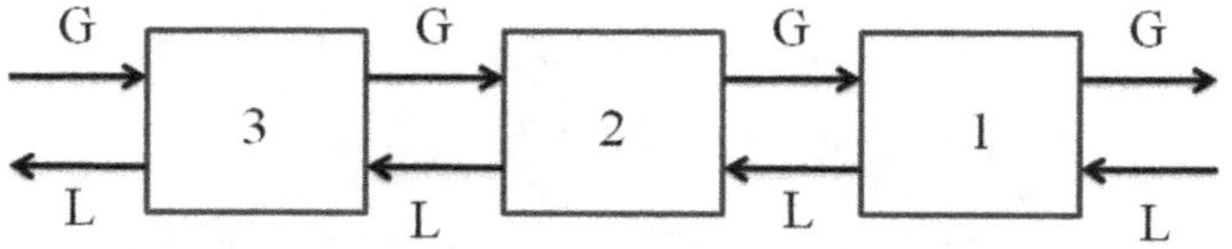

Figure 1.8: Schematic diagram of three stage extraction process.

8. *(Formulation of interacting cell/discrete diffusion models)*: Consider the flow system shown in Figure 1.9. Assume that each tank is well mixed and a unit mass (e. g., 1 kg) of species A is suddenly dumped into the larger tank at time zero and that all the tanks are free of species A at $t < 0$. Further assume that $V_R = 1\,\text{m}^3$ and $q_e = 1\,\text{m}^3/\text{min}$:
   (a) Formulate the differential equations describing the transient behavior of the system and put them in vector/matrix form.
   (b) Determine the steady-state concentrations in each tank.
   (c) Generalize the model for a system of N interacting tanks/cells of equal volume arranged in a circular array with all equal forward and reverse exchange flow

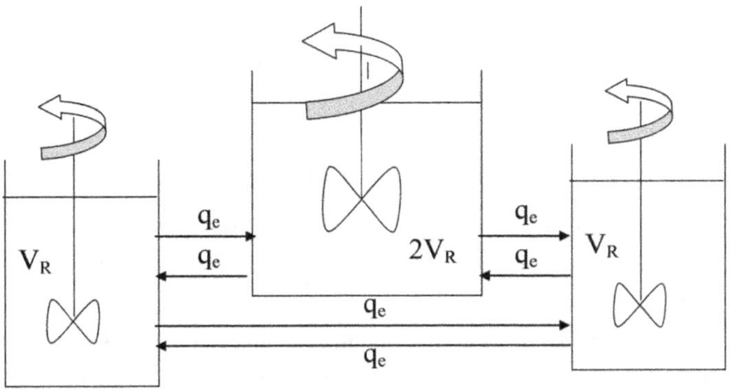

Figure 1.9: Schematic diagram of three interacting tanks.

rates. Put the equations in dimensionless form and identify the structure of the matrix that appears.

9. (*Mass transfer disguised rate constant matrix*): Determine the mass-transfer disguised rate constant matrix for the following reaction network (Figure 1.10):

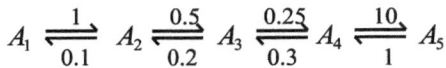

$$A_1 \underset{0.1}{\overset{1}{\rightleftharpoons}} A_2 \underset{0.2}{\overset{0.5}{\rightleftharpoons}} A_3 \underset{0.3}{\overset{0.25}{\rightleftharpoons}} A_4 \underset{1}{\overset{10}{\rightleftharpoons}} A_5$$

Figure 1.10: Sequential reversible reaction network with five species.

Assume that the mass transfer coefficients for all the species are equal and $\mathrm{Da}_{pm} = 1$. How many new reactions appear? Generalize the result for the case of $2N$ consecutive (reversible) reactions among $(N + 1)$ species. How does the result change for N consecutive irreversible reactions among (N+1) species? [Use of Mathematica® is recommended for this exercise.]

10. (*Discrete or compartmental loop diffusion–convection–reaction model*): Consider a discrete convective loop consisting of N identical cells of equal volume and assume that on the main convective flow (Q_L), we superimpose weak flows that correspond to the entry or exit of one or more streams containing reactants and/or products in any particular cell. Assume a constant density system with a single reaction occurring in each cell. Using the same notation as in Section 1.7, show that the reactant species balance in vector-matrix form is given by

$$\mathbf{Qc} = V_R\left[\frac{d\mathbf{c}}{dt} + \mathbf{r}(\mathbf{c})\right] - \mathbf{q}_{\text{in}}\mathbf{c}_{\text{in}}(t) + \mathbf{q}_e\mathbf{c} \tag{1.85}$$

with an appropriate initial condition. Here, $\mathbf{c}$ ($\mathbf{c}_{\text{in}}(t)$) is the vector representing the limiting reactant (inlet) concentrations in various cells, $\mathbf{q}_{\text{in}}$ is a diagonal matrix representing the inlet volumetric flow rates to the cells, $\mathbf{q}_e$ is a matrix representing the auxiliary flow rates leaving various cells (excluding the main convective flow), $\mathbf{Q}$ is the $N \times N$ (loop or cell connectivity) matrix defined by

$$\mathbf{Q} = Q_L \begin{pmatrix} -1 & 0 & \dots & 0 & 1 \\ 1 & -1 & \dots & 0 & 0 \\ 0 & 1 & \dots & 0 & 0 \\ \vdots & \vdots & \ddots & \ddots & \vdots \\ 0 & 0 & \dots & 1 & -1 \end{pmatrix}, \tag{1.86}$$

and $\mathbf{r}(\mathbf{c})$ is the vector of reaction rates. Cast the model in dimensionless form and identify the various matrices for the special case of one entering stream in cell 1 and one exit stream in cell j $(1 \le j \le N)$.

11. *Compartment models for 2D and 3D transient diffusion*: Consider the arrangement of cells shown in Figure 1.11. Assuming that all cells are of equal volume and exchange flow rates are identical in magnitude, determine the coupling matrix for each case.
12. *Discrete interacting convective loops*: Determine the coupling matrix for the discrete interacting convective loops shown in Figure 1.12. Assume all cells to be of identical volume and magnitude of all exchange/convective flows to be equal.

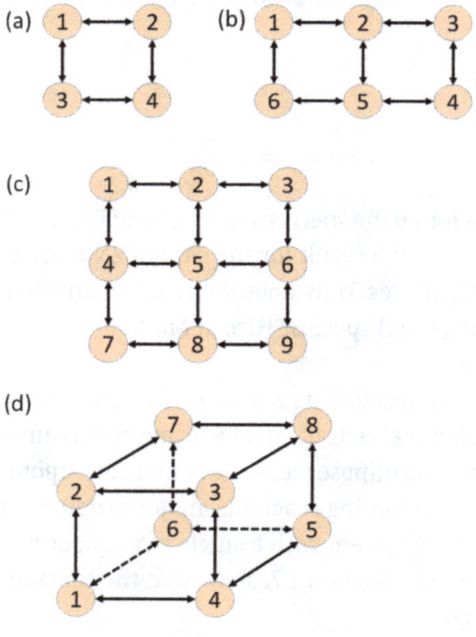

Figure 1.11: Schematic of interacting tanks/cells in two and three dimensions.

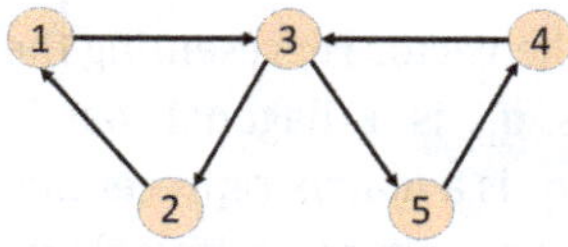

Figure 1.12: Schematic of interacting convective loops with a common cell.

2 Determinants

The theory of determinants plays a very important role in the solution of linear algebraic and differential equations. Specifically, the conditions for the existence and uniqueness of solutions to linear equations are often expressed in terms of determinants. The conditions for the existence of new (bifurcating) solutions to nonlinear equations may also be expressed in terms of the determinant of the linearized (Jacobian) matrix. In this chapter, we review the properties of determinants and illustrate their use with some examples.

2.1 Definition of determinant

Let $\mathbf{A}$ be an $n \times n$ square matrix with real or complex entries

$$\mathbf{A} = \begin{bmatrix} a_{11} & a_{12} & a_{13}\cdot & . & . & . & a_{1n} \\ a_{21} & a_{22} & a_{23} & . & . & . & a_{2n} \\ . & . & . & . & . & . & . \\ . & . & . & . & . & . & . \\ . & . & . & . & . & . & . \\ a_{n1} & a_{n2} & a_{n3} & . & . & . & a_{nn} \end{bmatrix}$$

The determinant of $\mathbf{A}$, denoted by $|\mathbf{A}|$, $\det \mathbf{A}$ or

$$\begin{vmatrix} a_{11} & a_{12} & a_{13}\cdot & . & . & . & a_{1n} \\ a_{21} & a_{22} & a_{23} & . & . & . & a_{2n} \\ . & . & . & . & . & . & . \\ . & . & . & . & . & . & . \\ . & . & . & . & . & . & . \\ a_{n1} & a_{n2} & a_{n3} & . & . & . & a_{nn} \end{vmatrix}$$

is defined by

$$\det \mathbf{A} = \sum (-1)^h a_{1k_1} a_{2k_2} \dots a_{nk_n} \tag{2.1}$$

where the summation is taken over all possible products $a_{1k_1} a_{2k_2} \cdots a_{nk_n}$ in which each product has n elements with exactly one element arising from each row and each column of $\mathbf{A}$. The value of h is the number of transpositions required to put the sequence $(k_1, k_2, \dots, k_n)$ in its natural order. Though h is not unique, it may be shown that it is always even or odd for a given sequence. Note that there are $n!$ terms in the summation in equation (2.1) and $(k_1, k_2, \dots, k_n)$ is a permutation of the sequence $(1, 2, 3, \dots, n)$.

https://doi.org/10.1515/9783111598055-003

Example 2.1. Consider the 2×2 matrix

$$\mathbf{A} = \begin{pmatrix} a_{11} & a_{12} \\ a_{21} & a_{22} \end{pmatrix}$$

$$\det \mathbf{A} = \sum (-1)^h a_{1k_1} a_{2k_2}$$

$$(k_1, k_2) = (1, 2) \Longrightarrow h = 0$$

$$(k_1, k_2) = (2, 1) \Longrightarrow h = 1$$

Therefore, we get

$$|\mathbf{A}| = a_{11}a_{22} - a_{12}a_{21}$$

Example 2.2. Consider the 3×3 matrix

$$\mathbf{A} = \begin{pmatrix} a_{11} & a_{12} & a_{13} \\ a_{21} & a_{22} & a_{23} \\ a_{31} & a_{32} & a_{33} \end{pmatrix}$$

$$|\mathbf{A}| = \sum (-1)^h a_{1k_1} a_{2k_2} a_{3k_3}$$

$$(k_1, k_2, k_3) = (1, 2, 3) \Rightarrow h = 0$$

$$(k_1, k_2, k_3) = (2, 3, 1) \Rightarrow h = 2$$

$$(k_1, k_2, k_3) = (3, 1, 2) \Rightarrow h = 2$$

$$(k_1, k_2, k_3) = (3, 2, 1) \Rightarrow h = 1$$

$$(k_1, k_2, k_3) = (1, 3, 2) \Rightarrow h = 1$$

$$(k_1, k_2, k_3) = (2, 1, 3) \Rightarrow h = 1$$

Thus, the determinant of a 3×3 matrix is given by

$$\begin{aligned} |\mathbf{A}| &= a_{11}a_{22}a_{33} + a_{12}a_{23}a_{31} + a_{13}a_{21}a_{32} - a_{13}a_{22}a_{31} - a_{11}a_{23}a_{32} - a_{12}a_{21}a_{33} \\ &= a_{11}(a_{22}a_{33} - a_{23}a_{32}) - a_{12}(a_{21}a_{33} - a_{23}a_{31}) + a_{13}(a_{21}a_{32} - a_{22}a_{31}) \end{aligned}$$

Example 2.3. What is the sign of the term $a_{13}a_{24}a_{31}a_{42}$ in the expansion of a 4×4 determinant?

Since the first subscripts are ordered, we need to look at the sequence formed by the second subscripts $(3, 4, 1, 2)$. Permuting 1 and 3 gives $(1, 4, 3, 2)$. Permuting 2 and 4 gives $(1, 2, 3, 4)$. Thus, $h = 2$ and the term $a_{13}a_{24}a_{31}a_{42}$ appears with a positive sign.

2.2 Properties of the determinant

The following properties of a determinant may be established from the definition.

1. (a) The determinant of a matrix **A** and its transpose are identical, i. e., $\det \mathbf{A} = \det \mathbf{A}^T$. From the definition, we have

$$\det \mathbf{A} = \sum (-1)^h a_{1k_1} a_{2k_2} a_{3k_3} \dots a_{nk_n} \quad \text{(rows ordered, columns permuted)}$$
$$\det \mathbf{A}^T = \sum (-1)^{h'} a_{k_1 1} a_{k_2 2} a_{k_3 3} \dots a_{k_n n} \quad \text{(columns ordered, rows permuted)}$$

The same $n!$ terms appear with the same sign.
(b) $\det(\overline{\mathbf{A}}) = \overline{(\det \mathbf{A})}$. This property follows from the following property of the complex numbers:

$$\overline{\alpha_1}.\overline{\alpha_2} = \overline{\alpha_1 \alpha_2}$$

For a square matrix with complex entries, we have

$$\det \mathbf{A}^* = \det(\overline{\mathbf{A}}^T) = \det(\overline{\mathbf{A}}) = \overline{\det \mathbf{A}}.$$

2. If **A** has a row (or column) of zeros, then $|\mathbf{A}| = 0$
3. When two rows of **A** are interchanged, $|\mathbf{A}|$ changes sign. Suppose rows i and j of **A** are interchanged, to obtain $\widetilde{\mathbf{A}}$.

$$\det \mathbf{A} = \sum (-1)^h a_{1k_1} \dots a_{ik_i} \dots a_{jk_j} \dots a_{nk_n} \tag{2.2}$$
$$\det \widetilde{\mathbf{A}} = \sum (-1)^{h'} a_{1k_1} \dots a_{jk_i} \dots a_{ik_j} \dots a_{nk_n}$$
$$= \sum (-1)^{h' \pm 1} a_{1k_1} \dots a_{ik_i} \dots a_{jk_j} \dots a_{nk_n} \tag{2.3}$$

Equations (2.2) and (2.3) differ by one transposition. Thus, $h = h' \pm 1$ and

$$|\widetilde{\mathbf{A}}| = -|\mathbf{A}|$$

4. If two rows (or columns) are identical, then

$$|\mathbf{A}| = 0$$

Using property (3), we get

$$\det \mathbf{A} = -\det \mathbf{A} \text{ (interchanging rows)}$$
$$\Rightarrow \det \mathbf{A} = 0$$

5. Multiplication of a row or column by a nonzero scalar:

$$\begin{aligned}\alpha|\mathbf{A}| &= \alpha\sum(-1)^h a_{1k_1}a_{2k_2}\dots a_{nk_n}\\ &= \sum(-1)^h a_{1k_1}a_{2k_2}\dots(\alpha a_{ik_i})\dots a_{nk_n}\end{aligned}$$

This implies that if all the elements in a row (or column) are multiplied by α then the determinant is multiplied by α.

6. If row j is multiplied by $\alpha(\neq 0)$ and added to row i, the determinant is unchanged

$$\begin{aligned}\det\mathbf{A} &= \sum(-1)^h a_{1k_1}\dots a_{ik_i}\dots a_{jk_j}\dots a_{nk_n}\\ \det\widetilde{\mathbf{A}} &= \sum(-1)^{h'} a_{1k_1}\dots(a_{ik_i}+\alpha a_{jk_j})\dots a_{jk_j}\dots a_{nk_n}\\ &= \sum(-1)^{h'} a_{1k_1}\dots a_{nk_n}+\alpha\sum(-1)^{h'} a_{1k_1}\dots a_{jk_j}\dots a_{jk_j}\dots a_{nk_n}\\ &= \det\mathbf{A}+0\end{aligned}$$

7. If $\mathbf{A}$ is upper or lower triangular matrix, then $\det\mathbf{A}$ is the product of all diagonal elements. From the definition, we have

$$\det\mathbf{A} = \sum(-1)^h a_{1k_1}\dots a_{ik_i}\dots a_{jk_j}\dots a_{nk_n}$$

Now, to get a nonzero value we must have $k_1 = 1, k_2 = 2, \dots, k_n = n$,

8. Let $\mathbf{I}_n = n\times n$ identity matrix. Then

$$\det\mathbf{I}_n = 1$$

(a) Let $\mathbf{E}_{1n}$ = elementary matrix obtained by performing row operation of type 1 (interchange of rows) on $\mathbf{I}_n$. Then

$$|\mathbf{E}_n| = -1$$

(b) Let $\mathbf{E}_{2n}$ = elementary matrix obtained by performing row operation of type 2 on $\mathbf{I}_n$. Then

$$|\mathbf{E}_{2n}| = k$$

(c) Let $\mathbf{E}_{3n}$ = matrix obtained by performing elementary row operation of type 3 on $\mathbf{I}_n$. Then

$$|\mathbf{E}_{3n}| = 1$$

Let $\mathbf{B} = \mathbf{E}_{in}\mathbf{A}$ ($i = 1, 2$ or 3). Then $|\mathbf{B}| = |\mathbf{E}_{in}||\mathbf{A}|$. Now suppose that

$$\mathbf{B} = \mathbf{E}_m\mathbf{E}_{m-1}\dots\mathbf{E}_1\mathbf{A}$$

Then

$$|\mathbf{B}| = |\mathbf{E}_m||\mathbf{E}_{m-1}|\dots|\mathbf{E}_1||\mathbf{A}|$$

It follows from this property that if a square matrix is not singular, then its echelon form is an upper triangular matrix with nonzero elements along the diagonal.

2.3 Computation of determinant by pivotal condensation

The above properties of the determinant may be used to calculate any n-th order determinant numerically. Let

$$\det \mathbf{A} = \begin{vmatrix} a_{11} & a_{12} & . & . & . & . & a_{1n} \\ a_{21} & a_{22} & . & . & . & . & a_{2n} \\ . & . & . & . & . & . & . \\ . & . & . & . & . & . & . \\ . & . & . & . & . & . & . \\ a_{n1} & a_{n2} & . & . & . & . & a_{nn} \end{vmatrix}$$

Then the pivotal condensation algorithm may be stated as follows:

1. Set $\det \mathbf{A} = 1$
2. Perform elementary row operations on $\mathbf{A}$ to reduce it to a triangular matrix. Reset $\det \mathbf{A}$ after each operation as follows:
   type 1 $\rightarrow$ multiply $\det \mathbf{A}$ by (-1)
   type 2 $\rightarrow$ multiply $\det \mathbf{A}$ by $\frac{1}{k}$ $(k \neq 0)$
   type 3 $\rightarrow$ multiply $\det \mathbf{A}$ by 1
3. Compute the determinant as the product of diagonal elements and all the factors in step (2).

Example 2.4.

$$\mathbf{A} = \begin{pmatrix} 1 & 3 & 5 \\ 2 & 0 & -1 \\ 1 & 4 & 3 \end{pmatrix}$$

Set $D_0 = \det \mathbf{A} = 1$. Multiply row 1 by 2 and subtract from row 2. Subtract row 1 from row 3. $D_1 = 1$.

$$\mathbf{A} \rightarrow \begin{pmatrix} 1 & 3 & 5 \\ 0 & -6 & -11 \\ 0 & 1 & -2 \end{pmatrix}$$

Multiply row 2 by $\frac{1}{6}$. $D_2 = 6$

$$\mathbf{A} \to \begin{pmatrix} 1 & 3 & 5 \\ 0 & -1 & -\frac{11}{6} \\ 0 & 1 & -2 \end{pmatrix}$$

Add row 2 to row 3. $D_3 = 6$

$$\mathbf{A} \to \begin{pmatrix} 1 & 3 & 5 \\ 0 & -1 & -\frac{11}{6} \\ 0 & 0 & -\frac{23}{6} \end{pmatrix}$$

Thus,

$$\begin{aligned} \det \mathbf{A} &= 6 \times 1 \times (-1) \times \left(\frac{-23}{6} \right) \\ &= 23 \end{aligned}$$

2.4 Minors, cofactors and Laplace's expansion

We consider the n-th order determinant

$$\begin{aligned} \det \mathbf{A} &= \begin{vmatrix} a_{11} & a_{12} & \cdot & a_{1j} & \cdot & a_{1n} \\ a_{21} & a_{22} & \cdot & a_{2j} & \cdot & a_{2n} \\ \cdot & \cdot & \cdot & \cdot & \cdot & \cdot \\ a_{i1} & a_{i2} & \cdot & a_{ij} & \cdot & a_{in} \\ \cdot & \cdot & \cdot & \cdot & \cdot & \cdot \\ a_{n1} & a_{n2} & \cdot & a_{nj} & \cdot & a_{nn} \end{vmatrix} \\ &= \sum (-1)^h a_{1k_1} a_{2k_2} \ldots a_{nk_n} \end{aligned}$$

and the $(n-1)\times(n-1)$ matrix obtained from **A** by deleting the i-th row and j-th column of **A**. Denote this matrix by $\mathbf{M}_{ij}$. This is called a *minor*. The *cofactor* of element a_{ij} is defined by

$$A_{ij} = (-1)^{i+j} |\mathbf{M}_{ij}|$$

Note that the cofactor is a number whereas minor is a matrix of order $(n-1) \times (n-1)$. Laplace's expansion of an n-th order determinant in terms of determinants of order $(n-1)$ may be stated as follows:

$$\det \mathbf{A} = \sum_{j=1}^{n} a_{ij} A_{ij} \quad \text{for any } i = 1, 2, 3, \ldots, n$$

That is, take any row and determine the cofactors of the elements of this row. Multiply the elements by the corresponding cofactors and sum to get the determinant. Similarly, the expansion in terms of columns is given by

$$\det \mathbf{A} = \sum_{i=1}^{n} a_{ij}A_{ij} \quad \text{for any } j = 1, 2, 3, \ldots, n$$

Proof of Laplace's expansion:

$$\det \mathbf{A} = \sum (-1)^h a_{1k_1} a_{2k_2} \ldots a_{ik_i} \ldots a_{nk_n} \tag{2.4}$$

$$= \sum_{j=1}^{n} a_{ij}(-1)^{i+j}|\mathbf{M}_{ij}| \tag{2.5}$$

To prove Laplace's expansion (equations (2.4) and (2.5) are identical), first we note that (2.5) has $n!$ terms and each term contains exactly one element from each row and column of $\mathbf{A}$. This follows from the fact that each $|\mathbf{M}_{ij}|$ has $(n-1)!$ terms, which do not include any elements from the i-th row or j-th column of $\mathbf{A}$. Thus, the sum in (2.5) has $(n)(n-1)! = n!$ terms. To account for the signs of the terms, we consider the matrix $\hat{\mathbf{A}}$ obtained by moving the i-th row of $\mathbf{A}$ to the last row and the j-th column of $\mathbf{A}$ to the last column. Then, when we expand the determinant of $\hat{\mathbf{A}}$, each term in it differs from that of $\mathbf{A}$ by exactly $(n-i)$ row transpositions and $(n-j)$ column transpositions. Equivalently, the sign of the term differs by a factor $(-1)^{(n-i)+(n-j)} = (-1)^{2n-(i+j)} = (-1)^{i+j}$. Thus, the expansion given by (2.5) gives the determinant of $\mathbf{A}$. □

Classical adjoint of a square matrix A: (also called the adjugate of A)

The matrix obtained by replacing each element a_{ij} of $\mathbf{A}$ by its corresponding cofactor and transposing the rows and columns is called the *classical adjoint* or adjugate of $\mathbf{A}$ and is denoted by $\operatorname{adj} \mathbf{A}$, i. e.,

$$\operatorname{adj} \mathbf{A} = \{A_{ji}\} \tag{2.6}$$

Alien cofactor expansion

Suppose we multiply the elements of row i by the cofactors of the k-th row ($k \neq i$) and sum the result. We claim

$$\sum_{j=1}^{n} a_{ij}A_{kj} = 0 \quad (k \neq i) \tag{2.7}$$

A_{kj} is called the *alien cofactor* of a_{ij} and equation (2.7) is called the alien cofactor expansion. To establish this expansion, we consider the identity

$$\begin{vmatrix} a_{11} & a_{12} & a_{13} & . & . & . & a_{1n} \\ a_{21} & a_{22} & a_{23} & . & . & . & a_{2n} \\ . & . & . & . & . & . & . \\ a_{i1} & a_{i2} & a_{i3} & . & . & . & a_{in} \\ . & . & . & . & . & . & . \\ a_{k1} & a_{k2} & a_{k3} & . & . & . & a_{kn} \\ . & . & . & . & . & . & . \\ a_{n1} & a_{n2} & a_{n3} & . & . & . & a_{nn} \end{vmatrix} = a_{k1}A_{k1} + a_{k2}A_{k2} + \cdots + a_{kn}A_{kn}$$

and replace $(a_{k1}, a_{k2}, \ldots, a_{kn})$ by $(a_{i1}, a_{i2}, \ldots, a_{in})$, i. e., replace the elements of the k-th row by the elements of the i-th row. Then, on the left-hand side (LHS) we have two identical rows, and hence LHS = 0. This replacement of the k-th row by elements of the i-th row does not change the cofactors A_{ki} $(i = 1, \ldots, n)$. Now, RHS = $\sum_{j=1}^{n} a_{ij}A_{kj}$ (we are multiplying the cofactors of the k-th row by elements of the i-th row). Therefore, we have $\sum_{j=1}^{n} a_{ij}A_{kj} = 0, i \neq k$.

2.4.1 Classical adjoint and inverse matrices

The Laplace and alien cofactor expansions may be used to prove the following theorem.

Theorem. *Let* adj **A** = *adjugate of the n × n matrix* **A** *(Classical adjoint). Then*

$$\mathbf{A}(\operatorname{adj}\mathbf{A}) = (\operatorname{adj}\mathbf{A})\mathbf{A} = (\det\mathbf{A})\mathbf{I}$$

Proof.

$$\operatorname{adj}\mathbf{A} = \begin{bmatrix} A_{11} & A_{21} & . & . & . & A_{n1} \\ A_{12} & A_{22} & . & . & . & A_{n2} \\ . & . & . & . & . & . \\ . & . & . & . & . & . \\ . & . & . & . & . & . \\ A_{1n} & A_{2n} & . & . & . & A_{nn} \end{bmatrix}$$

$$\begin{aligned} \mathbf{A}\operatorname{adj}\mathbf{A} &= \begin{bmatrix} a_{11} & a_{12} & . & a_{1n} \\ a_{21} & a_{22} & . & a_{2n} \\ . & . & . & . \\ a_{n1} & a_{n2} & . & a_{nn} \end{bmatrix} \begin{bmatrix} A_{11} & A_{21} & . & A_{n1} \\ A_{12} & A_{22} & . & A_{n2} \\ . & . & . & . \\ A_{1n} & A_{2n} & . & A_{nn} \end{bmatrix} \\ &= \begin{bmatrix} \det\mathbf{A} & 0 & . & 0 \\ 0 & \det\mathbf{A} & . & 0 \\ . & . & . & . \\ 0 & 0 & . & \det\mathbf{A} \end{bmatrix} = (\det\mathbf{A})\mathbf{I}_n \end{aligned}$$

Similarly,

$$(\operatorname{adj}\mathbf{A})\mathbf{A} = (\det\mathbf{A})\mathbf{I}_n \qquad \square$$

Theorem. *If* $\det\mathbf{A} \neq 0$, *then the inverse matrix* $\mathbf{A}^{-1}$ *exists and is given by*

$$\mathbf{A}^{-1} = \frac{1}{|\mathbf{A}|}(\operatorname{adj}\mathbf{A})$$

Proof. It follows from the definition and the previous theorem. □

2.5 Determinant of the product of two matrices

Suppose that $\mathbf{A}$ and $\mathbf{B}$ are two square matrices of order n and $\mathbf{C} = \mathbf{AB}$. Then $|\mathbf{C}| = |\mathbf{A}||\mathbf{B}|$, i. e., the determinant of a product of two matrices is equal to the product of determinants. This property can be established in several ways. We show it here more directly and discuss further implications.

Since the proof (for the general case) is notationally complicated (but otherwise straightforward), we sketch the procedure for the case $n = 2$. We let

$$\mathbf{P} = \begin{bmatrix} a_{11} & a_{12} & 0 & 0 \\ a_{21} & a_{22} & 0 & 0 \\ -1 & 0 & b_{11} & b_{12} \\ 0 & -1 & b_{21} & b_{22} \end{bmatrix}$$

and show that $|\mathbf{P}| = |\mathbf{A}||\mathbf{B}| = |\mathbf{AB}|$. We expand $\mathbf{P}$ by row 1 to get

$$|\mathbf{P}| = a_{11}\begin{vmatrix} a_{22} & 0 & 0 \\ 0 & b_{11} & b_{12} \\ -1 & b_{21} & b_{23} \end{vmatrix} - a_{12}\begin{vmatrix} a_{21} & 0 & 0 \\ -1 & b_{11} & b_{12} \\ 0 & b_{21} & b_{22} \end{vmatrix}$$

Expand again the 3×3 determinant by row 1 to get

$$\begin{aligned} |\mathbf{P}| &= (a_{11}a_{22} - a_{12}a_{21})\begin{vmatrix} b_{11} & b_{21} \\ b_{21} & b_{22} \end{vmatrix} \\ &= |\mathbf{A}||\mathbf{B}| \end{aligned}$$

To show $|\mathbf{P}| = |\mathbf{AB}|$, we use elementary row operation of type 3 to transform $\mathbf{P}$ to $\hat{\mathbf{P}}$. Since ERO of type 3 does not change the value of a determinant, $|\hat{\mathbf{P}}| = |\mathbf{P}|$. To get $\hat{\mathbf{P}}$, we multiply row 3 by a_{11} and add to row 1, row 4 by a_{12} and add to row 1, row 3 by a_{21} and add to row 2, row 4 by a_{22} and add to row 2. This gives

$$\hat{\mathbf{P}} = \begin{bmatrix} 0 & 0 & c_{11} & c_{12} \\ 0 & 0 & c_{21} & c_{22} \\ -1 & 0 & b_{11} & b_{12} \\ 0 & -1 & b_{21} & b_{22} \end{bmatrix}$$

where

$$c_{ij} = \sum_{k=1}^{2} a_{ik} b_{kj}; \quad i,j = 1, 2$$

or $\mathbf{C} = \mathbf{AB}$. Expanding $\hat{\mathbf{P}}$ by column 1, we get

$$|\hat{\mathbf{P}}| = (-1) \begin{vmatrix} 0 & c_{11} & c_{12} \\ 0 & c_{21} & c_{22} \\ -1 & b_{21} & b_{22} \end{vmatrix}$$

Expanding again by column 1 gives the result.

It follows from the above result that if $\mathbf{A}$ is nonsingular,

$$\det(\mathbf{A}^{-1}) = \frac{1}{(\det \mathbf{A})} \tag{2.8}$$

This relation is useful in many applications.

2.6 Rank of a matrix defined in terms of determinants

Recall our earlier definition of the rank of a matrix as the number of nonzero rows in the row echelon from of $\mathbf{A}$. Now, let $\mathbf{E}_i$ be the elementary matrix obtained by performing ERO of type i on the identity matrix $\mathbf{I}$. Then we have seen that

$$\begin{aligned} |\mathbf{E}_1| &= -1 \\ |\mathbf{E}_2| &= k (k \neq 0) \\ |\mathbf{E}_3| &= 1 \end{aligned}$$

Now, we let $\mathbf{A}$ be any $m \times n$ matrix and $\mathbf{A}_e$ be the row echelon form of $\mathbf{A}$. Then

$$\mathbf{A}_e = \mathbf{PA} \tag{2.9}$$

where $\mathbf{P}$ is the product of elementary matrices $\mathbf{E}_i$ of order m. Thus, $|\mathbf{P}| \neq 0$. Suppose that rank of $\mathbf{A} = r$. Then, without loss of generality, we may assume that $\mathbf{A}_e$ is of the form

$$\mathbf{A}_e = \begin{bmatrix} 1 & \hat{a}_{12} & \hat{a}_{13} & . & . & \hat{a}_{1r} & . & \hat{a}_{1n} \\ 0 & 1 & \hat{a}_{23} & . & . & \hat{a}_{2r} & . & \hat{a}_{2n} \\ 0 & 0 & . & . & . & \hat{a}_{3r} & . & \hat{a}_{3n} \\ . & . & . & . & . & . & . & . \\ 0 & 0 & 0 & . & . & 1 & . & \hat{a}_{rn} \\ 0 & 0 & 0 & . & . & 0 & . & 0 \\ 0 & 0 & 0 & . & . & 0 & . & 0 \end{bmatrix}$$

Here, $\mathbf{A}_e$ has $m - r$ zero rows and r nonzero rows, and the first nonzero element in row i ($\leq r$) appears in the i-th column (this can always be arranged by renumbering the columns of $\mathbf{A}$). Thus, $\mathbf{A}_e$ has a $r \times r$ minor whose determinant is not zero. From equation (2.9), it follows that $\mathbf{A}$ also has a $r \times r$ minor with a nonzero determinant. Thus, if $\mathbf{A}$ has rank r, there is at least one $r \times r$ minor of $\mathbf{A}$ whose determinant is not zero. Conversely, if $\mathbf{A}$ has rank r then all $k \times k$ minors ($k > r$) of $\mathbf{A}$ have a zero determinant.

2.7 Solution of Au = 0 and Au = b by Cramer's rule

If $\mathbf{A}$ is an $n \times n$ nonsingular matrix, the system of equations

$$\mathbf{Au} = \mathbf{b} \tag{2.10}$$

or the homogeneous system

$$\mathbf{Au} = \mathbf{0} \tag{2.11}$$

has a unique solution. This solution may be expressed in terms of determinants using Cramer's rule. Let $D = |\mathbf{A}| \neq 0$. Then we have

$$u_j D = \begin{vmatrix} a_{11} & a_{12} & . & . & a_{1j}u_j & . & . & a_{1n} \\ a_{21} & a_{22} & . & . & a_{2j}u_j & . & . & a_{2n} \\ . & . & . & . & . & . & . & . \\ . & . & . & . & . & . & . & . \\ a_{n1} & a_{n2} & . & . & a_{nj}u_j & . & . & a_{nn} \end{vmatrix} \tag{2.12}$$

Next, for each $k \neq j$, add u_k times column k to column j of the matrix in equation (2.12). This does not change the value of the determinant. Thus,

$$u_j D = \begin{vmatrix} a_{11} & a_{12} & . & . & a_{11}u_1 + a_{12}u_2 + \cdots + a_{1n}u_n & . & . & a_{1n} \\ a_{21} & a_{22} & . & . & a_{21}u_1 + a_{22}u_2 + \cdots + a_{2n}u_n & . & . & a_{2n} \\ . & . & . & . & . & . & . & . \\ . & . & . & . & . & . & . & . \\ a_{n1} & a_{n2} & . & . & a_{n1}u_1 + a_{n2}u_2 + \cdots + a_{nn}u_n & . & . & a_{nn} \end{vmatrix}$$

But

$$\sum_{k=1}^{n} a_{ik}u_k = b_i; \quad i = 1, 2, \ldots, n$$

Thus,

$$u_j D = \begin{vmatrix} a_{11} & a_{12} & . & . & b_1 & . & . & a_{1n} \\ a_{21} & a_{22} & . & . & b_2 & . & . & a_{2n} \\ . & . & . & . & . & . & . & . \\ . & . & . & . & . & . & . & . \\ a_{n1} & a_{n2} & . & . & b_n & . & . & a_{nn} \end{vmatrix}$$

We define D_j to be the determinant obtained by replacing the j-th column of $\mathbf{A}$ by the vector $\mathbf{b}$. Thus, if $D \neq 0$, we have

$$u_j = \frac{D_j}{D}, \quad j = 1, 2, \ldots, n \tag{2.13}$$

This is the explicit solution of the linear system given by equation (2.10). The result given by equation (2.13) is referred to as Cramer's rule.

Example 2.5. For a 2×2 system,

$$\begin{aligned} a_{11}u_1 + a_{12}u_2 &= b_1 \\ a_{21}u_1 + a_{22}u_2 &= b_2 \end{aligned}$$

we have

$$u_1 = \frac{\begin{vmatrix} b_1 & a_{12} \\ b_2 & a_{22} \end{vmatrix}}{\begin{vmatrix} a_{11} & a_{12} \\ a_{21} & a_{22} \end{vmatrix}}, \quad u_2 = \frac{\begin{vmatrix} a_{11} & b_1 \\ a_{21} & b_2 \end{vmatrix}}{\begin{vmatrix} a_{11} & a_{12} \\ a_{21} & a_{22} \end{vmatrix}} \tag{2.14}$$

For the 3×3 system,

$$\begin{aligned} a_{11}u_1 + a_{12}u_2 + a_{13}u_3 &= b_1 \\ a_{21}u_1 + a_{22}u_2 + a_{23}u_3 &= b_2 \\ a_{31}u_1 + a_{32}u_2 + a_{33}u_3 &= b_3 \end{aligned}$$

we have

$$u_1 = \frac{\begin{vmatrix} b_1 & a_{12} & a_{13} \\ b_2 & a_{22} & a_{23} \\ b_3 & a_{32} & a_{33} \end{vmatrix}}{\begin{vmatrix} a_{11} & a_{12} & a_{13} \\ a_{21} & a_{22} & a_{23} \\ a_{31} & a_{32} & a_{33} \end{vmatrix}}, \quad u_2 = \frac{\begin{vmatrix} a_{11} & b_1 & a_{13} \\ a_{21} & b_2 & a_{23} \\ a_{31} & b_3 & a_{33} \end{vmatrix}}{\begin{vmatrix} a_{11} & a_{12} & a_{13} \\ a_{21} & a_{22} & a_{23} \\ a_{31} & a_{32} & a_{33} \end{vmatrix}}, \quad u_3 = \frac{\begin{vmatrix} a_{11} & a_{12} & b_1 \\ a_{21} & a_{22} & b_2 \\ a_{31} & a_{32} & b_3 \end{vmatrix}}{\begin{vmatrix} a_{11} & a_{12} & a_{13} \\ a_{21} & a_{22} & a_{23} \\ a_{31} & a_{32} & a_{33} \end{vmatrix}}$$

It follows from Cramer's rule that for the special case of the homogeneous system ($\mathbf{b} = \mathbf{0}$), we obtain $\mathbf{u} = \mathbf{0}$. Thus, as already seen before, when $\det \mathbf{A} \neq 0$, the only solution to the homogeneous system is the trivial one.

It should also be noted that when $\det \mathbf{A} \neq 0$, the solution given by Cramer's rule is unique. To show this, suppose that there are two solutions and call them $\mathbf{u}$ and $\mathbf{y}$. We have

$$\mathbf{Au} = \mathbf{b}$$
$$\mathbf{Ay} = \mathbf{b}$$

Subtracting, we get

$$\mathbf{Az} = \mathbf{0} \tag{2.15}$$

where

$$\mathbf{z} = \mathbf{u} - \mathbf{y}$$

Since $D \neq 0$, the only solution to the homogeneous system (2.15) is the trivial one. Thus, $\mathbf{z} = \mathbf{0}$ and $\mathbf{u} = \mathbf{y}$.

Cramer's rule is not used in practice for higher order systems (e. g., $n > 10$) as it requires more computational time than the Gaussian elimination procedure.

2.8 Differentiation of a determinant

In our later applications, we will be dealing with square matrices whose elements depend continuously on a parameter t. We consider now determinants of such matrices and the derivatives of the determinant. First, we illustrate the problem with a 2×2 matrix. Let

$$\mathbf{A}(t) = \begin{bmatrix} a_{11}(t) & a_{12}(t) \\ a_{21}(t) & a_{22}(t) \end{bmatrix} \tag{2.16}$$

Then

$$D(t) \equiv \det \mathbf{A} = a_{11}(t)a_{22}(t) - a_{12}(t)a_{21}(t) \tag{2.17}$$

If we assume that $a_{ij}(t)$ is differentiable for all i and j, then

$$\begin{aligned}\frac{dD}{dt} &= a_{11}(t)\frac{da_{22}}{dt} + \frac{da_{11}}{dt}a_{22} - a_{12}\frac{da_{21}}{dt} - \frac{da_{12}}{dt}a_{21}\\ &= \begin{vmatrix} \frac{da_{11}}{dt} & \frac{da_{12}}{dt} \\ a_{21} & a_{22} \end{vmatrix} + \begin{vmatrix} a_{11} & a_{12} \\ \frac{da_{21}}{dt} & \frac{da_{22}}{dt} \end{vmatrix}\\ &= D_1 + D_2\end{aligned}$$

Thus, the derivative of a determinant is the sum of two determinants in which a single row is differentiated. This result is easily generalized to the n-th order determinant:

$$\begin{aligned}D &= \sum (-1)^h a_{1k_1}(t)a_{2k_2}(t)\dots a_{nk_n}(t)\\ \frac{dD}{dt} &= \sum_{j=1}^{n}\left(\sum (-1)^h a_{1k_1}(t)a_{2k_2}(t)\dots \frac{da_{jk_j}(t)}{dt}\dots a_{nk_n}(t)\right)\\ &= \sum_{j=1}^{n} D_j = D_1 + D_2 + \dots + D_n\end{aligned}$$

where

$$D_j = \begin{vmatrix} a_{11}(t) & a_{12}(t) & . & . & a_{1n}(t) \\ a_{21}(t) & a_{22}(t) & . & . & a_{2n}(t) \\ . & . & . & . & . \\ . & . & . & . & . \\ \frac{da_{j1}(t)}{dt} & \frac{da_{j2}(t)}{dt} & . & . & \frac{da_{jn}(t)}{dt} \\ . & . & . & . & . \\ a_{n1}(t) & a_{n2}(t) & . & . & a_{nn}(t) \end{vmatrix}$$

is the determinant obtained by differentiating only the j-th row.

2.9 Applications of determinants

In this section, we consider some elementary applications of determinants. As outlined above, the most important use of a determinant is in determining the conditions under which a system of homogeneous equations has a nontrivial solution. This is illustrated in the following examples as well as in later chapters.

Example 2.6 (Linear dependence/independence of functions). A set of functions $\{w_i(x); i = 1, 2, \dots, n\}$ is called linearly independent if the only solution to the homogeneous equation

$$c_1 w_1(x) + \dots + c_n w_n(x) = 0 \tag{2.18}$$

is $c_1 = c_2 = \cdots = c_n = 0$. Suppose that the set $\{w_i(x); i = 1, 2, \ldots, n\}$ is linearly independent. Differentiating equation (2.18) w. r. t. x once, twice, and $(n-1)$ times gives

$$c_1 w_1'(x) + \cdots + c_n w_n'(x) = 0 \tag{2.19}$$

$$c_1 w_1''(x) + \cdots + c_n w_n''(x) = 0 \tag{2.20}$$

$$\cdots$$

$$\cdots$$

$$\cdots$$

$$c_1 w_1^{[n-1]}(x) + \cdots + c_n w_n^{[n-1]}(x) = 0 \tag{2.21}$$

Now, equations (2.18) to (2.21) are a set of n homogeneous linear equations for the coefficients $\{c_1, c_2, \ldots, c_n\}$ whose only solution is the trivial one. Thus, the determinant of the coefficient matrix (also called the Wronskian determinant), defined by

$$W(x) = \begin{vmatrix} w_1(x) & w_2(x) & . & . & w_n(x) \\ w_1'(x) & w_2'(x) & . & . & w_n'(x) \\ . & . & . & . & . \\ . & . & . & . & . \\ . & . & . & . & . \\ w_1^{[n-1]} & w_2^{[n-1]} & . & . & w_n^{[n-1]} \end{vmatrix} \neq 0$$

Conversely, if the Wronskian determinant is zero, then there exists a nontrivial solution $\{c_1, c_2, \ldots, c_n\}$ and the set $\{w_i(x); i = 1, 2, \ldots, n\}$ is said to be linearly dependent. As an illustration, we note that the set $\{1, x, x^2\}$ is linearly independent since

$$W(x) = \begin{vmatrix} 1 & x & x^2 \\ 0 & 1 & 2x \\ 0 & 0 & 2 \end{vmatrix} = 2$$

while the set $\{2x, 1 + x^2, (1 - x)^2\}$ is linearly dependent since

$$W(x) = \begin{vmatrix} 2x & 1 + x^2 & (1 - x)^2 \\ 2 & 2x & -2(1 - x) \\ 0 & 2 & 2 \end{vmatrix} = 0.$$

Example 2.7 (Bifurcation of solutions to nonlinear equations). Another important application of the theory of determinants is in the solution of nonlinear equations. In contrast to the linear system $\mathbf{Au} = \mathbf{b}$, which can have either zero (inconsistent), one (when rank $\mathbf{A} = \text{rank}[\mathbf{A}\,\mathbf{b}] = n$) or an infinite number (when rank $\mathbf{A} = \text{rank}[\mathbf{A}\,\mathbf{b}] = r < n$) of solutions, the nonlinear parametrized system of equations

$$f_i(u_1, u_2, \ldots, u_n, \boldsymbol{\alpha}) = 0; \quad i = 1, 2, \ldots, n, \tag{2.22}$$

where $\boldsymbol{\alpha}$ is a vector of m parameters ($m \geq 1$) can have any number of solutions $(0, 1, 2, \ldots, \infty)$. However, if the functions f_i are continuous and have continuous derivatives, the implicit function theorem of multivariable calculus states that if the determinant of the (linearized) Jacobian matrix of equations (2.22) does not vanish, then the solution is a continuous function of the parameters $\boldsymbol{\alpha}$. Equivalently, the number of solutions to equations (2.22) can change only when the determinant of the Jacobian matrix

$$\mathbf{J} = \left[\frac{\partial f_i}{\partial u_j}(\mathbf{u}, \boldsymbol{\alpha})\right]; \quad i = 1, 2, \ldots, n;\ j = 1, 2, \ldots, n$$

vanishes, i. e.,

$$\det \mathbf{J} = 0 \tag{2.23}$$

The elimination of the state variables $\mathbf{u}$ from equations (2.22) and (2.23) gives a locus in the $\boldsymbol{\alpha}$ parameter space. This locus is called the *bifurcation set*, as new solutions can emerge (or bifurcate) only when $\boldsymbol{\alpha}$ values cross this set. [Remark: It can be shown that if the zero eigenvalue of $\mathbf{J}$ is simple, the number of solutions of equations (2.22) changes if and only if the parameters $\boldsymbol{\alpha}$ cross the bifurcation set. However, when the eigenvalue is not simple, the vanishing of the Jacobian determinant is necessary but not a sufficient condition for bifurcation.] Thus, the solution of the original set of equations (2.22) along with the vanishing of the Jacobian determinant can be used to determine the bifurcation set for nonlinear problems

As an example, we consider the steady-state equation describing the temperature (u) in an adiabatic CSTR with parameters $B > 0$ and $\mathrm{Da} > 0$:

$$f(u, B, \mathrm{Da}) = u - \frac{B\,\mathrm{Da}\,e^u}{1 + \mathrm{Da}\,e^u} = 0 \tag{2.24}$$

Differentiating equation (2.24) w. r. t. u and setting the derivatives to zero gives

$$\frac{df}{du} = 1 - \frac{B\,\mathrm{Da}\,e^u}{(1 + \mathrm{Da}\,e^u)^2} = 0. \tag{2.25}$$

Writing

$$t = \mathrm{Da}\ e^u, \tag{2.26}$$

the two equations may be solved for B and Da and the bifurcation set may be expressed in a parametric form as

$$B = \frac{(1+t)^2}{t} \tag{2.27}$$

$$\mathrm{Da} = t \exp\{-1 - t\}, \quad t > 0 \tag{2.28}$$

This is plotted in Figure 2.1. The bifurcation set consists of two branches, the upper ignition branch and the lower extinction branch. It divides the (B, Da) space into two

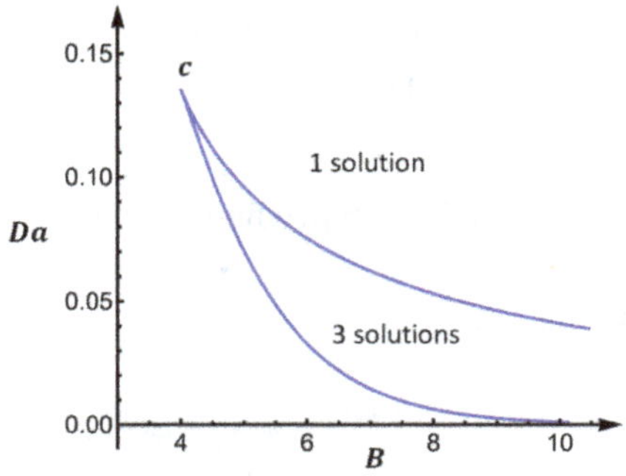

Figure 2.1: A schematic diagram of the bifurcation set of equation (2.24).

regions, corresponding to unique and three solutions of equation (2.24). [Remark: The cusp point c in Figure 2.1 where the ignition and extinction branches meet is at $B = 4$ and $\mathrm{Da} = e^{-2}$. We note that for $B < 4$, there is only a unique solution for any value of the dimensionless residence time, Da]. For higher-dimensional problems, symbolic manipulation or computer programming can be used to determine the bifurcation set.

Example 2.8 (Lorenz equations). Consider the following discrete model of convection (where the fluid density is assumed to vary only with temperature):

$$\frac{dx}{dt} = \mathrm{Pr}(y - x) = f_1 \tag{2.29}$$

$$\frac{dy}{dt} = -xz + Rx - y = f_2 \tag{2.30}$$

$$\frac{dz}{dt} = xy - \frac{8}{3}z = f_3 \tag{2.31}$$

where R is the (scaled) Rayleigh number and Pr is the Prandtl number ($\mathrm{Pr} > 0$).

Writing the vector of variables as $\psi = (x, y, z)^T$, then equations (2.29)–(2.31) can be written as

$$\frac{d\psi}{dt} = \mathbf{f}(\psi) = (f_1, f_2, f_3)^T. \tag{2.32}$$

It can easily be verified that equations (2.29)–(2.31) or (2.32) has a trivial steady-state solution, i. e., $\psi_s = (x, y, z)^T = \mathbf{0}$ is a steady-state solution for any R and Pr. [Remark: The steady-state solution does not depend on Pr.] The Jacobian of the function $\mathbf{f}$ can be calculated easily and is given by

$$\mathbf{J} = \left\{\frac{\partial f_i}{\partial \psi_j}\right\} = \begin{pmatrix} -\mathrm{Pr} & \mathrm{Pr} & 0 \\ R - z & -1 & -x \\ y & x & \frac{-8}{3} \end{pmatrix} \tag{2.33}$$

Evaluation of the Jacobian matrix at the trivial solution $\psi_s = \mathbf{0}$ gives

$$\Rightarrow \mathbf{J}_s = \mathbf{J}|_{\psi_s=\mathbf{0}} = \begin{pmatrix} -\mathrm{Pr} & \mathrm{Pr} & 0 \\ R & -1 & 0 \\ 0 & 0 & \frac{-8}{3} \end{pmatrix}. \tag{2.34}$$

Thus, the Jacobian matrix at the trivial steady-state is of rank 3 except when $R = 1$ (where it has rank 2, i. e., $\det \mathbf{J}_s = 0$ when $R = 1$).

Note that the linearized system of equations at $R = 1$,

$$\mathbf{J}\psi = \mathbf{0}$$

has a nontrivial solution $\psi = \begin{pmatrix} \alpha \\ \alpha \\ 0 \end{pmatrix}$ for any arbitrary α. Thus, the linearized matrix $\mathbf{J}_s$ has a simple zero eigenvalue and $R = 1$ is a bifurcation point, i. e., new solutions appear or disappear when R-value crosses unity.

In this specific case, we can determine all solutions since at steady-state (equations (2.29) and (2.31)) give

$$x = y \tag{2.35}$$

$$z = \frac{3}{8}xy = \frac{3}{8}x^2 \tag{2.36}$$

and equation (2.30) gives

$$-\frac{3}{8}x^3 - x + Rx = 0$$

or

$$x\left(R - 1 - \frac{3}{8}x^2\right) = 0$$

$\Rightarrow$

$$x = 0 \quad \text{or} \quad x = \pm\sqrt{\frac{8}{3}(R-1)} \tag{2.37}$$

In other words, for $R > 1$, the system has three steady-state solutions (including a trivial steady state).

The solution diagram (referred to as *pitchfork bifurcation* in the literature) is shown in Figure 2.2. In this figure, $x = 0$ is the trivial conduction solution while $x \neq 0$ correspond to convective solutions. As stated in the Introduction to this chapter, the condition for the existence of a nontrivial solution to a system of linearized equations (expressed in terms of a determinant) may be used to determine the possible bifurcation points of many nonlinear systems.

For further discussion on the theory of determinants, we refer to the books by Amundson [3] and Lipschutz and Lipson [22].

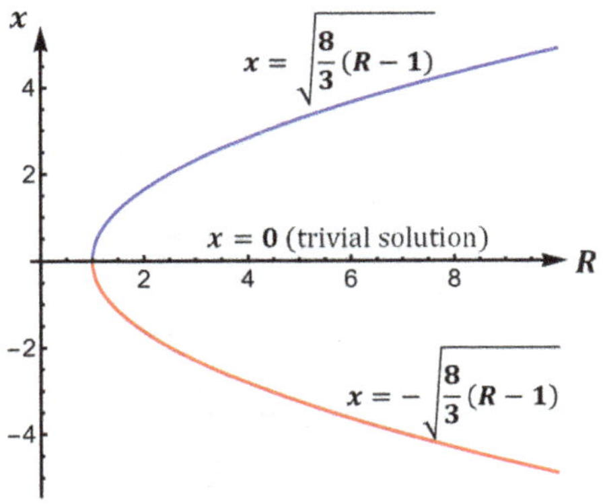

Figure 2.2: Steady-state solution diagram of Lorenz equations illustrating the pitchfork bifurcation.

Problems

1. *(Simplification of determinants):*

   (a) Show that

$$D_n = \begin{vmatrix} 1+c_1 & 1 & 1 & . & . & 1 \\ 1 & 1+c_2 & 1 & . & . & 1 \\ 1 & 1 & 1+c_3 & . & . & 1 \\ . & . & . & . & . & . \\ 1 & 1 & 1 & . & . & 1+c_n \end{vmatrix} = c_1c_2\ldots c_n\left(1+\frac{1}{c_1}+\frac{1}{c_2}+\cdots+\frac{1}{c_n}\right)$$

   Hint: First, show that D_n satisfies the recursion formula $D_n = c_nD_{n-1} + c_1c_2 \ldots c_{n-1}$.

   (b) Show without expanding that

$$D_n = \begin{vmatrix} x & 1 & 1 & . & . & . & 1 \\ 1 & x & 1 & . & . & . & 1 \\ 1 & 1 & x & . & . & . & 1 \\ . & . & . & . & . & . & . \\ 1 & 1 & 1 & . & . & . & x \end{vmatrix} = (x-1)^{n-1}(x+n-1)$$

   for this n-th order determinant.

   (c) Show without expanding the determinant that the Vandermonde determinant

$$\begin{vmatrix} x_1^{n-1} & x_1^{n-2} & . & . & . & x_1 & 1 \\ x_2^{n-1} & x_2^{n-2} & . & . & . & x_2 & 1 \\ . & . & . & . & . & . & . \\ x_n^{n-1} & x_n^{n-2} & . & . & . & x_n & 1 \end{vmatrix}$$

   has the value

$$D_n = (x_1 - x_2)(x_1 - x_3)\dots(x_1 - x_n).(x_2 - x_3)(x_2 - x_4)\dots(x_2 - x_n)$$
$$\dots(x_3 - x_4)\dots(x_3 - x_n)\dots(x_{n-1} - x_n)$$
$$= \prod_{\substack{i=1 \\ j>i}}^{n-1} (x_i - x_j)$$

2. *(Linear homogeneous equations and determinants)*: Let **A** be a square matrix of order n, **x** be the vector of n unknowns and consider the homogeneous equations $\mathbf{Ax} = \mathbf{0}$. Use the properties of determinant to prove that a necessary and sufficient condition for $\mathbf{Ax} = \mathbf{0}$ to possess a nontrivial solution is $\det(\mathbf{A}) = 0$. (Hint: Verify the result for $n = 1$ and use induction along with the properties of the determinant).
3. *(Determinant and neutral stability curve)*: Determine the relationship between the parameters k, Le, Ra_t and Ra_c for which the following set of equations has nontrivial solution:

$$c_1 - (\pi^2 + k^2)c_2 = 0$$
$$\mathrm{Le}\, c_1 - (\pi^2 + k^2)c_3 = 0$$
$$-(\pi^2 + k^2)c_1 + \mathrm{Ra}_t\, k^2 c_2 + \mathrm{Ra}_c\, k^2 c_3 = 0.$$

[Remark: This relation is called the neutral stability curve and defines the onset of convection in a fluid layer heated from below. Here, k is the wave number, Le is the Lewis number and $\mathrm{Ra}_t, \mathrm{Ra}_c$ are the thermal and concentration Rayleigh numbers, respectively. The unknowns c_1, c_2, c_3 are the amplitudes of the velocity, temperature and concentration modes, respectively.]

4. *(Equation for a plane and a circle in terms of determinants)*
   (a) Show that the equation of a plane passing through the points (x_i, y_i), $i = 1, 2, 3$ may be expressed as

$$\begin{vmatrix} x & y & z & 1 \\ x_1 & y_1 & z_1 & 1 \\ x_2 & y_2 & z_2 & 1 \\ x_3 & y_3 & z_3 & 1 \end{vmatrix} = 0$$

   (b) *(Equation for a circle in terms of determinants)* Show that the equation of a circle passing through the points (x_i, y_i), $i = 1, 2, 3$ may be expressed as

$$\begin{vmatrix} x^2 + y^2 & x & y & 1 \\ x_1^2 + y_1^2 & x_1 & y_1 & 1 \\ x_2^2 + y_2^2 & x_2 & y_2 & 1 \\ x_3^2 + y_3^2 & x_3 & y_3 & 1 \end{vmatrix} = 0$$

5. (*Common root condition for polynomial equations*) Show that the necessary and sufficient conditions for the equations

$$x^3 + ax^2 + bx + c = 0$$
$$x^2 + \alpha x + \beta = 0$$

to have a common root may be expressed as

$$\begin{vmatrix} 1 & a & b & c & 0 \\ 0 & 1 & a & b & c \\ 1 & \alpha & \beta & 0 & 0 \\ 0 & 1 & \alpha & \beta & 0 \\ 0 & 0 & 1 & \alpha & \beta \end{vmatrix} = 0$$

3 Vectors and vector expansions

In this chapter, we review some elementary concepts about vectors and vector expansions. A more general discussion will be given in Part II when we deal with abstract vector space concepts.

For the purpose of this chapter, we define a vector to be an n-tuple of real or complex numbers arranged in a single row or column:

$$\mathbf{u} = \begin{pmatrix} u_1 \\ u_2 \\ \cdot \\ \cdot \\ u_n \end{pmatrix} \quad \text{(column vector)}$$

$$\mathbf{u}^T = (u_1 \ \ u_2 \ \ u_3 \ \ \ldots \ \ u_n) \quad \text{(row vector)}$$

For simplicity, we shall deal with only column vectors in the discussion below. However, all the concepts and properties of column vectors are also applicable to row vectors. Also, when the elements of the column vector $\mathbf{u}$ are complex numbers, we define the corresponding row vector by

$$\mathbf{u}^* = (\overline{u_1} \ \ \overline{u_2} \ \ \overline{u_3} \ \ \ldots \ \ \overline{u_n}) \quad \text{(row vector)}$$

where $\overline{u_i}$ is the complex conjugate of u_i and $\mathbf{u}^*$ denotes complex conjugate transpose of the vector $\mathbf{u}$.

Let $\mathbf{V}$ be the collection of all such vectors with two operations specifying the *vector addition* and *multiplication of a vector by a scalar* be defined. It is assumed that these two operations are defined such that the usual rules (associative, commutative, and distributive) are satisfied. This set $\mathbf{V}$ is denoted by $\mathbb{R}^n$ (the space of n-tuples of real numbers) or $\mathbb{C}^n$ (the space of n-tuples of complex numbers). The sum of two vectors $\mathbf{u}$ and $\mathbf{v}$ is defined by

$$\mathbf{u} + \mathbf{v} = \begin{pmatrix} u_1 + v_1 \\ u_2 + v_2 \\ \cdot \\ u_n + v_n \end{pmatrix},$$

and the product (scalar multiplication) of a vector $\mathbf{u}$ by a real (or complex) number α by

$$\alpha\mathbf{u} = \begin{pmatrix} \alpha u_1 \\ \alpha u_2 \\ \cdot \\ \cdot \\ \alpha u_n \end{pmatrix}$$

https://doi.org/10.1515/9783111598055-004

The set V with the above operations is called a *vector space*. We now deal with the algebraic and geometric properties of this set.

3.1 Linear dependence, basis and dimension

Suppose that $\mathbf{V}$ is the collection of vectors, all having n elements. We define the *zero vector* in $\mathbf{V}$ as the n-tuple whose elements are all zero. It will be denoted by the symbol $\mathbf{0}_n$ or simply by $\mathbf{0}$, when the number of elements is clear. By definition, a nonzero vector contains at least one element which is not zero. Now, suppose that $\mathbf{S}$ is a subset of vectors $\{\mathbf{u}_1, \mathbf{u}_2, \ldots, \mathbf{u}_r\}$ in $\mathbf{V}$. This subset is called *linearly independent* if the relation

$$c_1\mathbf{u}_1 + c_2\mathbf{u}_2 + \cdots + c_r\mathbf{u}_r = \mathbf{0} \tag{3.1}$$

implies

$$c_1 = c_2 = \cdots = c_r = 0$$

Otherwise, the set is called *linearly dependent*. We note that equation (3.1) defines a system of n homogeneous equations in r unknowns.

Example 3.1. Consider the set $\mathbf{u}_1 = \binom{1}{2}$, $\mathbf{u}_2 = \binom{3}{5}$ in $\mathbb{R}^2$. It is linearly independent since $c_1\mathbf{u}_1 + c_2\mathbf{u}_2 = \mathbf{0}$ implies $c_1 + 3c_2 = 0$, $2c_1 + 5c_2 = 0$, whose only solution is $c_1 = c_2 = 0$.

Example 3.2. Consider the set $\mathbf{u}_1 = \begin{pmatrix} 2 \\ 6 \\ -2 \end{pmatrix}$, $\mathbf{u}_2 = \begin{pmatrix} 3 \\ 1 \\ 2 \end{pmatrix}$, $\mathbf{u}_3 = \begin{pmatrix} 8 \\ 16 \\ -3 \end{pmatrix}$ in $\mathbb{R}^3$. To check if this set is linearly independent, we form the homogeneous equations

$$c_1\mathbf{u}_1 + c_2\mathbf{u}_2 + c_3\mathbf{u}_3 = \mathbf{0},$$

which is equivalent to the system

$$\begin{aligned} 2c_1 + 3c_2 + 8c_3 &= 0 \\ 6c_1 + c_2 + 16c_3 &= 0 \\ -2c_1 + 2c_2 - 3c_3 &= 0 \end{aligned}$$

Using the elementary row operations, we reduce this system to the following echelon form:

$$\begin{pmatrix} 1 & \frac{3}{2} & 4 \\ 0 & 1 & 1 \\ 0 & 0 & 0 \end{pmatrix} \mathbf{c} = \mathbf{0}.$$

Thus, we have $c_2 = -c_3$, $c_1 = -\frac{5}{2}c_3$, and can get a nontrivial solution (e. g., by taking $c_3 = 2$, $c_2 = -2$ and $c_1 = -5$), and hence, the vectors are linearly dependent.

The following facts may be established from the above definition of linear dependence:

1. The zero vector is linearly dependent (since $1.\mathbf{0} = \mathbf{0}$).
2. Any single nonzero vector is linearly independent.
3. If a set of vectors is linearly dependent, then any larger set containing this set is also linearly dependent.
4. Any subset of a linearly independent set is also linearly independent.
5. Any set of vectors containing the zero vector is linearly dependent.
6. If $r > n$, the set $\{\mathbf{u}_1, \mathbf{u}_2, \ldots, \mathbf{u}_r\}$ is linearly dependent, i. e., there can be at most n linearly independent vectors in a set where all the vectors have n elements. As already noted, equation (3.1) defines a set of n linear homogeneous equations in r unknowns. When $r > n$, there are more unknowns than equations and we can always find a nontrivial solution.

The collection of all vectors, which are linear combinations of elements of the set $\mathbf{S} = \{\mathbf{u}_1, \mathbf{u}_2, \ldots, \mathbf{u}_r\}$ is called the *subspace spanned* by $\mathbf{S}$. A set $\mathbf{S} = \{\mathbf{u}_1, \mathbf{u}_2, \ldots, \mathbf{u}_r\}$ is called a *basis* for a vector space $\mathbf{V}$ if it is linearly independent and spans $\mathbf{V}$. The number of elements in a basis is called the *dimension* of the vector space $\mathbf{V}$. The following theorem may be established easily from the above properties of n-tuples.

Theorem. *The vector space* $\mathbf{V}$ *of all n-tuples of real numbers* $\mathbb{R}^n$ *(or complex numbers,* $\mathbb{C}^n$*) has dimension n.*

Example 3.3. Consider the set

$$e_1 = \begin{pmatrix} 1 \\ 0 \end{pmatrix}, \quad e_2 = \begin{pmatrix} 0 \\ 1 \end{pmatrix}$$

in $\mathbb{R}^2$. This is called the standard basis. The set in example (3.1) is another basis for $\mathbb{R}^2$.

3.2 Dot or scalar product of vectors

The vector space of n-tuples as defined above (i. e., with vector addition and multiplication of a vector by a scalar) has only algebraic structure. The concept of scalar or dot product of vectors allows us to introduce geometrical properties and extend the familiar geometric concepts such as distances, lengths, angles and orthogonality from two or three dimensions ($\mathbb{R}^2$ or $\mathbb{R}^3$) to other finite- (and also to infinite-) dimensional vector spaces.

Let $\mathbf{V}$ be a vector space consisting of n-tuples of real or complex numbers. Suppose that to each pair of vectors $\mathbf{u}, \mathbf{v} \in \mathbf{V}$ we assign a scalar denoted by $\mathbf{u}.\mathbf{v}$, or more generally (anticipating our later notation) $\langle \mathbf{u}, \mathbf{v} \rangle$, which is a real or complex number. This function is called scalar or dot product (or more generally, an inner product) if it satisfies the following three rules:

(i) $\langle \alpha\mathbf{u} + \beta\mathbf{v}, \mathbf{w}\rangle = \alpha\langle \mathbf{u}, \mathbf{w}\rangle + \beta\langle \mathbf{v}, \mathbf{w}\rangle$; for $\mathbf{u}, \mathbf{v}, \mathbf{w} \in \mathbf{V}$ and α, β are scalars

(ii) $\langle \mathbf{u}, \mathbf{v}\rangle = \overline{\langle \mathbf{v}, \mathbf{u}\rangle}$

(iii) $\langle \mathbf{u}, \mathbf{u}\rangle \geq 0$ and $\langle \mathbf{u}, \mathbf{u}\rangle = 0$ iff $\mathbf{u} = \mathbf{0}$

It is important to note that the scalar product maps pairs of vectors in $\mathbf{V}$ to the set of real or complex numbers. The first property requires linearity in the first variable. The second property is called Hermitian symmetry. For the case in which $\mathbf{u}$ and $\mathbf{v}$ contain real elements, this simply requires the scalar product to be symmetric. The third property known as the positive definiteness requires that the scalar product of a vector with itself to be positive for all vectors in $\mathbf{V}$ except the zero vector. A vector space in which a scalar product is defined has a geometric structure (and we can change this geometric structure by properly choosing the scalar product for a particular application. This will be demonstrated in Part II). We define the length of a vector by

$$\|\mathbf{u}\| = \sqrt{\langle \mathbf{u}, \mathbf{u}\rangle} \tag{3.2}$$

and the distance between two vectors $\mathbf{u}$ and $\mathbf{v}$ by

$$d(\mathbf{u}, \mathbf{v}) = \|\mathbf{u} - \mathbf{v}\| = \sqrt{\langle \mathbf{u} - \mathbf{v}, \mathbf{u} - \mathbf{v}\rangle} \tag{3.3}$$

Using the Schwarz's inequality,

$$\left|\langle \mathbf{u}, \mathbf{v}\rangle\right|^2 \leq \langle \mathbf{u}, \mathbf{u}\rangle\langle \mathbf{v}, \mathbf{v}\rangle \tag{3.4}$$

we can also define the angle between two vectors. [A proof of Schwarz's inequality is given in Part II]. When $\mathbf{V}$ is the set of n-tuples of real numbers, we define the angle between two vectors $\mathbf{u}, \mathbf{v} \in \mathbf{V}$ as

$$\cos\theta = \frac{\langle \mathbf{u}, \mathbf{v}\rangle}{\|\mathbf{u}\|\|\mathbf{v}\|}. \tag{3.5}$$

When $\mathbf{V}$ is the set of n-tuples of complex numbers, we define the angle between two vectors $\mathbf{u}, \mathbf{v} \in \mathbf{V}$ as

$$\cos\theta = \frac{|\langle \mathbf{u}, \mathbf{v}\rangle|}{\|\mathbf{u}\|\|\mathbf{v}\|}. \tag{3.6}$$

[Remark: It can be shown that the angle defined by equation (3.5) satisfies $0 \leq \theta \leq \pi$ while that defined by equation (3.6) satisfies $0 \leq \theta \leq \frac{\pi}{2}$.] The vectors $\mathbf{u}, \mathbf{v} \in V$ are said to be *orthogonal* if $\langle \mathbf{u}, \mathbf{v}\rangle = 0$. A vector $\mathbf{u}$ is said to be *normalized* (or is a *unit vector*) if $\|\mathbf{u}\| = 1$. If the set of vectors $\{\mathbf{u}_1, \mathbf{u}_2, \ldots, \mathbf{u}_n\}$ is linearly independent and forms a basis for V, then this basis is called an *orthonormal basis* if each vector in the set is orthogonal to other vectors and is normalized to have unit length. In terms of scalar product, an orthonormal basis satisfies the condition

$$\langle \mathbf{u}_i, \mathbf{u}_j \rangle = \delta_{ij} \tag{3.7}$$

where δ_{ij} is the Kronecker delta function ($\delta_{ij} = 1$ for $i = j$ and zero otherwise).

Example 3.4. Let $\mathbf{V} = \mathbb{R}^n$ and for $\mathbf{u}, \mathbf{v} \in \mathbf{V}$ define

$$\langle \mathbf{u}, \mathbf{v} \rangle = \sum_{i=1}^{n} u_i v_i$$

This is the usual inner (dot) product and it may be verified that it satisfies all the three axioms. The length of a vector with respect to this inner product is given by

$$\|\mathbf{u}\| = \sqrt{u_1^2 + u_2^2 + \cdots + u_n^2}$$

and the distance between the vectors $\mathbf{u}$ and $\mathbf{v}$ is given by

$$d(\mathbf{u}, \mathbf{v}) = \|\mathbf{u} - \mathbf{v}\| = \sqrt{(u_1 - v_1)^2 + (u_2 - v_2)^2 + \cdots + (u_n - v_n)^2}$$

The set consisting of the unit vectors $\mathbf{e}_1 = (1, 0, \ldots, 0)$, $\mathbf{e}_2 = (0, 1, \ldots, 0), \ldots, \mathbf{e}_n = (0, 0, \ldots, 1)$ is one possible orthonormal basis for this space. This vector space is often referred to as the n-dimensional Euclidean space.

Example 3.5. Let $\mathbf{V} = \mathbb{C}^n$ and for $\mathbf{u}, \mathbf{v} \in \mathbf{V}$, define

$$\langle \mathbf{u}, \mathbf{v} \rangle = \sum_{i=1}^{n} u_i \overline{v_i},$$

where the bar denotes complex conjugate. Again, it may be verified that all the three axioms are satisfied. The length of a vector with respect to this inner product is given by

$$\begin{aligned} \|\mathbf{u}\| &= \sqrt{u_1 \overline{u_1} + u_2 \overline{u}_2 + \cdots + u_n \overline{u_n}} \\ &= \sqrt{|u_1|^2 + |u_2|^2 + \cdots + |u_n|^2}, \end{aligned}$$

and the distance between the vectors $\mathbf{u}$ and $\mathbf{v}$ is given by

$$d(\mathbf{u}, \mathbf{v}) = \|\mathbf{u} - \mathbf{v}\| = \sqrt{|u_1 - v_1|^2 + |u_2 - v_2|^2 + \cdots + |u_n - v_n|^2},$$

where $|u_i|$ denotes the absolute value (modulus) of the complex number u_i. The vector space here is the space of n-tuples of complex numbers and has a geometric structure similar to that of the n-dimensional Euclidean space. It is an example of a finite-dimensional Hilbert space.

It may be shown that every finite-dimensional inner product space has an orthonormal basis. If $\{\mathbf{u}_1, \mathbf{u}_2, \dots, \mathbf{u}_n\}$ is a basis for $\mathbf{V}$ but is not orthogonal, the following *Gram–Schmidt procedure* may be used to transform it to an orthogonal basis. Define

$$\mathbf{v}_1 = \mathbf{u}_1$$

and for $k > 1$,

$$\mathbf{v}_k = \mathbf{u}_k - \sum_{i=1}^{k-1} \frac{\langle \mathbf{u}_k, \mathbf{v}_i \rangle}{\|\mathbf{v}_i\|^2} \mathbf{v}_i$$

Then it is easily verified that $\mathbf{v}_k$ is orthogonal to $\{\mathbf{v}_1, \mathbf{v}_2, \dots, \mathbf{v}_{k-1}\}$ and $\mathbf{v}_k \neq \mathbf{0}$ since this would mean that the vectors $\{\mathbf{u}_1, \mathbf{u}_2, \mathbf{u}_3, \dots, \mathbf{u}_k\}$ are linearly dependent. Thus, $\{\mathbf{v}_1, \mathbf{v}_2, \dots, \mathbf{v}_n\}$ is an orthogonal basis and by dividing each vector by its length we get an orthonormal basis.

3.3 Linear algebraic equations

We return again to the set of m linear equations in n unknowns:

$$\mathbf{Au} = \mathbf{0} \tag{3.8}$$

Suppose that the rank of $\mathbf{A}$ is $r (r \leq m, r \leq n)$. Then there is at least one $r \times r$ minor of $\mathbf{A}$ whose determinant is not zero. Without loss of generality, we can assume that the nonzero $r \times r$ minor is at the upper left corner. We can solve for the first r variables in terms of the remaining $(n - r)$ variables to obtain

$$\begin{aligned}
u_1 &= \gamma_{12} u_2 + \gamma_{13} u_3 + \cdots + \gamma_{1,r+1} u_{r+1} + \gamma_{1,r+2} u_{r+2} + \cdots + \gamma_{1,n} u_n \\
u_2 &= \gamma_{23} u_3 + \cdots + \gamma_{2,r+1} u_{r+1} + \gamma_{2,r+2} u_{r+2} + \cdots + \gamma_{2,n} u_n \\
&\cdot \\
&\cdot \\
u_r &= \gamma_{r,r+1} u_{r+1} + \gamma_{r,r+2} u_{r+2} + \cdots + \gamma_{r,n} u_n
\end{aligned} \tag{3.9}$$

Suppose that we choose values for the variables $\{u_{r+1}, u_{r+2}, \dots, u_n\}$ and calculate $\{u_1, u_2, \dots, u_r\}$ from equation (3.9). Suppose that we make $(n-r+1)$ choices for $\{u_{r+1}, u_{r+2}, \dots, u_n\}$ and arrange the solution in rows. Then, in this solution matrix, the first r columns are obtained as linear combination of the last $(n-r)$ columns. Hence, the rank of this matrix is at most $(n-r)$. If the choices of $\{u_{r+1}, u_{r+2}, \dots, u_n\}$ are such that the rank of the solution matrix is equal to $(n-r)$, then the last row is a linear combination of the first $(n-r)$ rows. Thus, there can be at most $(n - r)$ linearly independent solutions. The $(n - r)$ linearly independent solutions are called a *fundamental set of solutions*. The following theorem may be stated for the solutions of equation (3.8).

Theorem. *Every solution of the homogeneous system* (3.8) *is of the form*

$$\mathbf{u}_h = c_1\mathbf{u}_1 + c_2\mathbf{u}_2 + \cdots + c_{n-r}\mathbf{u}_{n-r} \tag{3.10}$$

where r is the rank of $\mathbf{A}$ *and* $\{\mathbf{u}_1, \mathbf{u}_2, \ldots, \mathbf{u}_{n-r}\}$ *is a set of fundamental (linearly independent) solutions and c_i are arbitrary constants.*

We have already seen that the inhomogeneous system $\mathbf{Au} = \mathbf{b}$ is consistent (has solutions) iff $\text{rank}(\mathbf{A}) = \text{rank}(\text{aug}\,\mathbf{A})$. When this condition is satisfied, the following theorem may be stated for the solutions of the inhomogeneous system.

Theorem. *Suppose that* $\text{rank}(\mathbf{A}) = \text{rank}(\text{aug}\,\mathbf{A}) = r$ *and* $1 \le r \le n$. *Then the general solution to the inhomogeneous system* $\mathbf{Au} = \mathbf{b}$ *may be written in the form*

$$\mathbf{u} = \mathbf{u}_h + \mathbf{u}_p \tag{3.11}$$

where $\mathbf{u}_h$ *is the general solution of the homogeneous system given by* (3.10) *and* $\mathbf{u}_p$ *is any (particular) solution of the inhomogeneous equations.*

[Remark: The theorem is also valid for the case $r = 0$ but this is omitted as it corresponds to zero equations in n unknowns.]

Example 3.6. Consider the homogeneous system in four variables:

$$u_1 - 2u_2 - u_4 = 0; \quad -2u_1 + 3u_2 + 3u_3 = 0$$
$$-u_2 + 3u_3 - 2u_4 = 0; \quad 3u_1 - 7u_2 + 3u_3 - 5u_4 = 0$$

for which $\text{rank}\,\mathbf{A} = 2$. We have already seen that [see Example 1.7] every solution to the homogeneous system is of the form

$$\mathbf{u}_h = c_1\begin{pmatrix} 3 \\ 0 \\ 2 \\ 3 \end{pmatrix} + c_2\begin{pmatrix} 0 \\ 1 \\ -1 \\ -2 \end{pmatrix}$$

where c_1 and c_2 are constants. The inhomogeneous system

$$\begin{aligned} u_1 - 2u_2 - u_4 &= b_1 \\ -2u_1 + 3u_2 + 3u_3 &= b_2 \\ -u_2 + 3u_3 - 2u_4 &= b_3 \\ 3u_1 - 7u_2 + 3u_3 - 5u_4 &= b_4 \end{aligned}$$

is consistent only if $b_3 = 2b_1 + b_2$ and $b_4 = 5b_1 + b_2$, or equivalently,

$$\mathbf{b} = b_1 \begin{pmatrix} 1 \\ 0 \\ 2 \\ 5 \end{pmatrix} + b_2 \begin{pmatrix} 0 \\ 1 \\ 1 \\ 1 \end{pmatrix}$$

Taking $\mathbf{b}^T = (0, 1, 1, 1)$, the general solution of the inhomogeneous system may be written as

$$\mathbf{u} = c_1 \begin{pmatrix} 3 \\ 0 \\ 2 \\ 3 \end{pmatrix} + c_2 \begin{pmatrix} 0 \\ 1 \\ -1 \\ -2 \end{pmatrix} + \begin{pmatrix} -2 \\ -1 \\ 0 \\ 0 \end{pmatrix}.$$

3.4 Applications of vectors and vector expansions

The vector space concepts discussed above find applications in several topics of interest to chemical engineers. We discuss here two of them: namely, stoichiometry and dimensional analysis.

3.4.1 Stoichiometry

A single chemical reaction occurring in a homogeneous system among S species denoted by $A_1, A_2, \ldots, A_S$ may be written as

$$\sum_{j=1}^{S} \nu_j A_j = 0 \tag{3.12}$$

where ν_j is the stoichiometric coefficient of species A_j in the reaction. By convention, ν_j is positive if A_j is a product and negative if A_j is a reactant. For example, consider the methanol synthesis reaction

$$CO + 2H_2 = CH_3OH, \tag{3.13}$$

and denote $A_1 = CH_3OH$, $A_2 = CO$ and $A_3 = H_2$. With this notation, we can write equation (3.13) in the form of equation (3.12) as

$$A_1 - A_2 - 2A_3 = 0 \tag{3.14}$$

with $\nu_1 = 1$, $\nu_2 = -1$ and $\nu_3 = -2$.

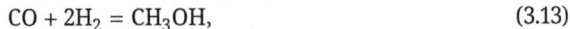

A set of R reactions occurring among S species may be written as

$$\sum_{j=1}^{S} \nu_{ij} A_j = 0; \quad (i = 1, 2, \ldots, R) \tag{3.15}$$

where ν_{ij} is the stoichiometric coefficient of species A_j in the i-th reaction. For obvious reasons, the $R \times S$ matrix $\{\nu_{ij}\}$ is called the *stoichiometric coefficient matrix*. For example, consider the above methanol synthesis reaction with the side reactions

$$CO_2 + H_2 = H_2O + CO \tag{3.16}$$

$$CO_2 + 3H_2 = H_2O + CH_3OH \tag{3.17}$$

Denoting $A_4 = CO_2$ and $A_5 = H_2O$, we can write these as

$$A_2 + A_5 - A_3 - A_4 = 0 \tag{3.18}$$

$$A_1 + A_5 - 3A_3 - A_4 = 0 \tag{3.19}$$

The three reactions may be written together as

$$\boldsymbol{\nu} \mathbf{a} = \mathbf{0}$$

where $\boldsymbol{\nu}$ is the 3×5 stoichiometric coefficient matrix defined by

$$\boldsymbol{\nu} = \begin{pmatrix} 1 & -1 & -2 & 0 & 0 \\ 0 & 1 & -1 & -1 & 1 \\ 1 & 0 & -3 & -1 & 1 \end{pmatrix}$$

and $\mathbf{a}$ is the species vector defined by $\mathbf{a}^T = (A_1 \ A_2 \ A_3 \ A_4 \ A_5)$. The two reactions (3.13) and (3.16) are independent of each other while the third reaction (3.17) is the sum of the other two reactions. It is important to know just how many independent reactions there are in a given system. This can be answered using the vector space concepts in two different ways:

1. When we already know the system of reactions, we can determine the number of independent reactions and pick one such set by looking at the stoichiometric coefficient matrix. In the above example, the rank of $\boldsymbol{\nu}$ is two and only two reactions are independent.
2. We can also determine the number of independent reactions between S species $(A_1, A_2, \ldots, A_S)$ by determining the rank of the atomic matrix. Suppose that each species A_j is made up of atoms a_i and let the number of atoms a_i in species A_j be denoted by λ_{ij}. A table may be made up listing the species A_j along the top row and the building blocks of the species (i. e., the atoms) a_i vertically at the left so that the element at the intersection of the i-th row and j-th column is λ_{ij}. The $n \times S$ matrix $\{\lambda_{ij}\}$ is called the *atomic matrix*:

	A_1	A_2	A_3	.	.	A_S
a_1	λ_{11}	λ_{12}	λ_{13}	.	.	λ_{1S}
a_2	λ_{21}	λ_{22}	λ_{23}	.	.	λ_{2S}
.	.	.	.	.	.	.
.	.	.	.	.	.	.
a_n	λ_{n1}	λ_{n2}	λ_{n3}	.	.	λ_{nS}

In this notation, each species A_j is represented as a vector in the n-dimensional atom space. The elements of the j-th column of the atomic matrix represent the various atoms in A_j in the vector representation of this species.

Suppose that the rank of the atomic matrix is r. Then the number of independent vectors is r and the remaining $(S - r)$ vectors (species) may be represented as a linear combination of r basis vectors (species). These $(S - r)$ relations are nothing but the independent reactions between the species.

Example 3.7. Consider a reaction mixture consisting of CH_3OH, CO, H_2, CO_2 and H_2O. There are five species and the three distinct atoms. We form the atomic matrix and see that it has rank 3. Thus, there are two independent reactions between these five species.

3.4.2 Dimensional analysis

Dimensional analysis is useful to analyze and correlate the behavior of a physical system when it is not possible to write down the governing equations explicitly or when they are too complicated to solve. In such cases, the Buckingham method may be used to determine the dimensionless groups that characterize the behavior of the system. In this method, one lists all the variables that are significant in a given problem and determines the number of independent dimensionless groups formed by these variables by using the Buckingham pi-theorem. This theorem states that the number of dimensionless groups used to describe a system involving n variables is equal to $n - r$, where r is the rank of the dimensional matrix of the variables. Thus,

$$i = n - r,$$

where i is number of independent dimensionless groups, n is the number of variables and r is the rank of the dimensional matrix of these variables. The dimensionless matrix is simply the matrix formed by tabulating the exponents of the fundamental dimensions (such as mass, M; length L; time, t; temperature, T; electric current, A and so on) of each variable.

To see how the pi-theorem arises, we assume that the n physical variables may be expressed in terms of m fundamental dimensions (usually with integer exponents). The exponents of the fundamental dimensions may be used to represent each variable as a vector in the m-dimensional space. Suppose that the rank of this matrix is $r(\leq m)$. Then only r of these vectors are linearly independent, and hence, $(n-r)$ of these vectors may be expressed as a linear combination of the r independent vectors. These $(n-r)$ relations are the dimensionless groups formed by the variables.

Example 3.8. Consider the motion of a solid body through a fluid. The drag force exerted by the fluid (F_D) depends on the velocity V_0 of this solid body, the size of the solid body (such as diameter, D), the fluid density (ρ) and the fluid viscosity (μ). Determine the relevant dimensionless groups.

Variable	Symbol	Dimensions	Vector representation
Drag Force	F_D	MLt^{-2}	$(1, 1, -2)$
Velocity	V_o	Lt^{-1}	$(0, 1, -1)$
Density	ρ	ML^{-3}	$(1, -3, 0)$
Viscosity	μ	$ML^{-1}t^{-1}$	$(1, -1, -1)$
Size of body	D	L	$(0, 1, 0)$

We note that there are five vectors (variables) in a three-dimensional space and only three of them can be linearly independent. We take the three linearly independent vectors to be V_0, ρ and D. The two dimensionless groups may be formed by expanding the remaining two vectors in terms of these three linearly independent ones. Equivalently, we can form the product of each of the other variables with these three and choose the exponents so that the resulting combination has no dimensions. (This is the pi-method.) To form the first group, we write

$$\begin{aligned}\pi_1 &= (F_D)^a (V_0)^b (\rho)^c (D)^d \\ &= (MLt^{-2})^a (Lt^{-1})^b (ML^{-3})^c (L)^d \\ &= M^{(a+c)} L^{(a+b-3c+d)} t^{(-2a-b)}\end{aligned}$$

To make π_1 dimensionless, each of the above exponents must be zero. Solving these three homogeneous equations, we get $a = -c$, $d = 2c$ and $b = 2c$. Thus,

$$\pi_1 = \left(\frac{\rho D^2 V_o^2}{F_D}\right)^c$$

The value for c is arbitrary and we take it to be -1 so that π_1 becomes the familiar Euler number:

$$\mathrm{Eu} = \frac{F_D}{\rho D^2 V_o^2}$$

To form the second group, we write

$$\pi_2 = (V_o)^a(\rho)^b(\mu)^c(D)^d$$
$$= M^{(b+c)}L^{(a-3b-c+d)}t^{(-a-c)}$$

To make π_2 dimensionless, we choose $c = -b$, $a = b$ and $d = b$. Thus,

$$\pi_2 = \left(\frac{DV_o\rho}{\mu}\right)^b$$

We choose $b = 1$ and identify the dimensionless group as the Reynolds number:

$$\mathrm{Re} = \frac{DV_o\rho}{\mu}$$

Thus, we can relate the five variables in terms of two dimensionless groups Eu and Re. A relationship of the form $\mathrm{Eu} = f(\mathrm{Re})$ may be determined experimentally. In the literature, the Euler number is often replaced by the drag coefficient, which is defined by

$$C_D = \frac{F_D}{\frac{1}{2}\rho V_0^2 A_c}$$

where A_c is the projected area of the body in the direction of flow (For a sphere, $A_c = \pi D^2/4$). Plots of experimentally determined drag coefficient curves (C_D versus Re) for various shapes (e. g., sphere) may be found in standard fluid mechanics textbooks.

3.5 Application of computer algebra and symbolic manipulation

3.5.1 Determination of independent reactions

It is often the case that many species are found in a chemical reactor and various reaction pathways are conjectured. For example, during oxidative dehydrogenation of methane in a catalytic reactor, following gas-phase species are found:

$$s^T = (CH_4, C_2H_6, H_2O, H_2, O_2, CO, CO_2, C_2H_4, C_2H_2) \tag{3.20}$$

and following reactions may occur:

$$2CH_4 + \frac{1}{2}O_2 \rightarrow C_2H_6 + H_2O \tag{3.21}$$
$$2CH_4 + O_2 \rightarrow C_2H_4 + 2H_2O \tag{3.22}$$
$$CH_4 + \frac{1}{2}O_2 \rightarrow CO + 2H_2 \tag{3.23}$$
$$CO + \frac{1}{2}O_2 \rightarrow CO_2 \tag{3.24}$$
$$2CH_4 \rightleftharpoons C_2H_6 + H_2 \tag{3.25}$$

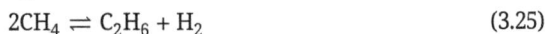

$$C_2H_6 \rightleftharpoons C_2H_4 + H_2 \quad (3.26)$$
$$C_2H_4 \rightleftharpoons C_2H_2 + H_2 \quad (3.27)$$
$$CH_4 + H_2O \rightleftharpoons CO + 3H_2 \quad (3.28)$$
$$CO + H_2O \rightleftharpoons CO_2 + H_2, \quad (3.29)$$

where not all of them are independent reactions. There are many ways to determine the independent reactions as described earlier, however, when number of species is large, the determination may be cumbersome and computer programming can be utilized for the task. As an example, we consider the above example of oxidative dehydrogenation of methane, where nine species (given in equation (3.20)) are made up of 3 atoms and leads to the following atomic matrix:

	CH_4	C_2H_6	H_2O	H_2	O_2	CO	CO_2	C_2H_4	C_2H_2
C	1	2	0	0	0	1	1	2	2
H	4	6	2	2	0	0	0	4	2
O	0	0	1	0	2	1	2	0	0

(3.30)

It may be verified that the rank of this matrix is 3 from its row echelon form:

$$\begin{pmatrix} 1 & 0 & 0 & 2 & -4 & -5 & -7 & -2 & -4 \\ 0 & 1 & 0 & -1 & 2 & 3 & 4 & 2 & 3 \\ 0 & 0 & 1 & 0 & 2 & 1 & 2 & 0 & 0 \end{pmatrix}. \quad (3.31)$$

Thus, the number of independent reactions is $9 - 3 = 6$. Similarly, the number of independent reactions can also be determined from stoichiometry. For example, the stoichiometric matrix $\boldsymbol{\nu}$ can be written from equations (3.21)–(3.29) as follows:

$$\boldsymbol{\nu} = \begin{pmatrix} -2 & 1 & 1 & 0 & -\frac{1}{2} & 0 & 0 & 0 & 0 \\ -2 & 0 & 2 & 0 & -1 & 0 & 0 & 1 & 0 \\ -1 & 0 & 0 & 2 & -\frac{1}{2} & 1 & 0 & 0 & 0 \\ 0 & 0 & 0 & 0 & -\frac{1}{2} & -1 & 1 & 0 & 0 \\ -2 & 1 & 0 & 1 & 0 & 0 & 0 & 0 & 0 \\ 0 & -1 & 0 & 1 & 0 & 0 & 0 & 1 & 0 \\ 0 & 0 & 0 & 1 & 0 & 0 & 0 & -1 & 1 \\ -1 & 0 & -1 & 3 & 0 & 1 & 0 & 0 & 0 \\ 0 & 0 & -1 & 1 & 0 & -1 & 1 & 0 & 0 \end{pmatrix} \quad (3.32)$$

where species numbers are assigned in the order they appear in equation (3.20). The row-echelon form of this matrix is given below:

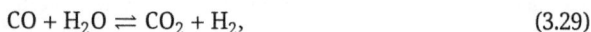

$$\hat{\nu} = \begin{pmatrix} 1 & 0 & 0 & 0 & 0 & 0 & 0 & \frac{-3}{2} & 1 \\ 0 & 1 & 0 & 0 & 0 & 0 & 0 & -2 & 1 \\ 0 & 0 & 1 & 0 & 0 & 0 & \frac{-1}{2} & \frac{-5}{4} & \frac{3}{2} \\ 0 & 0 & 0 & 1 & 0 & 0 & 0 & -1 & 1 \\ 0 & 0 & 0 & 0 & 1 & 0 & -1 & \frac{-1}{2} & 1 \\ 0 & 0 & 0 & 0 & 0 & 1 & \frac{-1}{2} & \frac{1}{4} & \frac{-1}{2} \\ 0 & 0 & 0 & 0 & 0 & 0 & 0 & 0 & 0 \\ 0 & 0 & 0 & 0 & 0 & 0 & 0 & 0 & 0 \\ 0 & 0 & 0 & 0 & 0 & 0 & 0 & 0 & 0 \end{pmatrix} \tag{3.33}$$

which shows that the rank of stoichiometric matrix is 6, i. e., number of independent reactions is 6. These independent reactions can be obtained by multiplying $\hat{\nu}$ (given in equation (3.33)) with the species vector given in equation (3.20).

Note that this set of independent reactions is not unique and can be determined in many other ways, e. g., another method is shown below, which is based on eliminating each atom as follows:

Step 1: Write each species in terms of atoms

$$\begin{aligned} CH_4 &= C + 4H \\ C_2H_6 &= 2C + 6H \\ H_2O &= 2H + O \\ H_2 &= 2H \\ O_2 &= 2O \\ CO &= C + O \\ CO_2 &= C + 2O \\ C_2H_4 &= 2C + 4H \\ C_2H_2 &= 2C + 2H \end{aligned}$$

Step 2: Eliminate the atoms

Eliminating H from the above nine equations, we get

$$\begin{aligned} CH_4 &= C + 2H_2 \\ C_2H_6 &= 2C + 3H_2 \\ H_2O &= H_2 + O \\ O_2 &= 2O \\ CO &= C + O \\ CO_2 &= C + 2O \\ C_2H_4 &= 2C + 2H_2 \\ C_2H_2 &= 2C + H_2 \end{aligned}$$

Eliminating O from the above eight equations, we get

$$\begin{aligned}
CH_4 &= C + 2H_2 \\
C_2H_6 &= 2C + 3H_2 \\
H_2O &= H_2 + \frac{1}{2}O_2 \\
CO &= C + \frac{1}{2}O_2 \\
CO_2 &= C + \frac{1}{2}O_2 \\
C_2H_4 &= 2C + 2H_2 \\
C_2H_2 &= 2C + H_2
\end{aligned}$$

Finally, eliminating C from the above seven equations, we get the six linearly independent reactions:

$$\begin{aligned}
2CH_4 &= C_2H_6 + H_2 : \text{dimerization/pyrolysis} \\
H_2 + \frac{1}{2}O_2 &= H_2O : \text{hydrogen oxidation} \\
CH_4 + \frac{1}{2}O_2 &= CO + 2H_2 : \text{partial oxidation to syngas} \\
CO + \frac{1}{2}O_2 &= CO_2 : \text{CO oxidation} \\
C_2H_6 &= C_2H_4 + H_2 : \text{dehydrogenation of ethane} \\
C_2H_4 &= C_2H_2 + H_2 : \text{dehydrogenation of ethylene}
\end{aligned}$$

Problems

1. *(Linear dependence and independence of vectors)*: Which of the following sets of vectors are linearly independent? Find the corresponding linear relations:
   (i) $(5,4,3)$, $(3,3,2)$, $(8,1,3)$
   (ii) $(4,-5,2,6)$, $(2,-2,1,3)$, $(6,-3,3,9)$, $(4,-1,5,6)$
   (iii) Suppose we have a set of vectors

   $$a_i^T = (a_{i1}, a_{i2}, \ldots, a_{in}) \quad i = 1,2,\ldots,n$$

   such that

   $$|a_{jj}| > \sum_{i=1;i\neq j}^{n} |a_{ij}| \quad j = 1,2,\ldots,n$$

   Show that this set of vectors is linearly independent.
2. *(Application of vector expansions to stoichiometry)*:

(a) Determine the number of independent reactions in the following set by examining the rank of the stoichiometric coefficient matrix:

$$\begin{aligned} 4NH_3 + 5O_2 &= 4NO + 6H_2O \\ 4NH_3 + 3O_2 &= 2N_2 + 6H_2O \\ O_2 + 2NO &= 2NO_2 \\ N_2 + O_2 &= 2NO \end{aligned}$$

(b) Determine the number of independent reactions in the above system by examining the rank of the atomic matrix.

(c) The following species are found to be present in the pyrolysis of a low molecular hydrocarbon:

$$C_2H_6, H, C_2H_5, CH_3, CH_4, H_2, C_2H_4, C_3H_8, C_4H_{10}.$$

Determine the number of independent reactions and write down a set of independent reactions.

3. *(Application of vector expansions to dimensional analysis)*:
   (a) When a gas and liquid flow simultaneously in a horizontal pipe, several different flow patterns are obtained (e. g., stratified flow, bubble flow, slug flow, annular flow, etc.). The type of flow pattern obtained in a particular system depends on the pipe diameter (D), the liquid and gas superficial velocities (U_{LS}, U_{GS}), the density and viscosities of the phases (ρ_L, ρ_G, μ_L, μ_G), the interfacial (surface) tension (σ) and the gravitational acceleration (g). Determine the relevant dimensionless groups and give a physical interpretation.
   (b) Small droplets of liquid are formed when a liquid jet breaks up in spray and fuel injection processes. Assume that the droplet diameter (d) depends on the liquid density, viscosity and surface tension, as well as the jet speed (V) and diameter (D). Determine the relationship between these quantities by dimensional analysis. Give a physical interpretation of the dimensionless groups.
4. *(Application of vector expansions to dimensional analysis)*: It was shown by G. I. Taylor that the energy (E) released in a nuclear explosion may be estimated from the relation

$$R = \left(\frac{E}{\rho_0}\right)^{1/5} ct^{2/5},$$

where R is the radius of the spherical shock wave generated by the explosion, ρ_0 is the ambient density, t is the time and c is a constant. Taylor suggested to determine the constant c (which turns out to be close to unity) by using experimentation with lighter explosives (such as TNT) and E by using photographic data of R as a function of time.

(a) Assuming that R depends on E, ρ_0, t and the ambient pressure p_0, derive the relevant dimensionless groups.
(b) Discuss the additional assumptions or approximations involved in obtaining Taylor's formula from the result in (a).

5. (*Gas phase microkinetics*): Consider a gas phase system consisting of molecules H_2, Br_2, HBr and free radicals H and Br.
   (a) Determine the number of independent reactions and write down one such set.
   (b) Determine the number of reactions if the system has no free radicals.
6. (*Catalytic microkinetics*) In the oxidation of CO on a catalytic site (s), the following gas phase and surface species are present:

$$\mathrm{CO}, \mathrm{CO_2}, \mathrm{O_2}, s, \mathrm{CO}.s, \mathrm{O_2}.s, \mathrm{O}.s, \mathrm{CO_2}.s$$

   Determine the number of independent reactions and write down one such set.

4 Solution of linear equations by eigenvector expansions

The main goal of this chapter is to solve the linear algebraic equations

$$\mathbf{Au} = \mathbf{b}, \tag{4.1}$$

the linear initial value problem

$$\frac{d\mathbf{u}}{dt} = \mathbf{Au}, \quad t > 0; \quad \mathbf{u}(t = 0) = \mathbf{u}_0, \tag{4.2}$$

and related equations containing a square matrix **A** by eigenvector expansions. As stated in the Introduction, the solution of these equations reveals the structure of the solutions of many other linear equations containing differential and integral operators.

4.1 The matrix eigenvalue problem

Let **A** be an $n \times n$ square matrix with real or complex entries. Consider the system of homogeneous equations

$$\mathbf{Ax} = \lambda \mathbf{x} \tag{4.3}$$

where λ is a scalar.

Definition. A real or complex number λ is called an *eigenvalue* of **A** if the system of homogeneous equations (4.3) has a nontrivial solution. The nontrivial solution is called the *eigenvector*, or more precisely, the *right eigenvector* of **A** corresponding to eigenvalue λ.

Eigenvalues are of fundamental importance in most physical systems as they represent the time or length scales (temporal or spatial frequencies) associated with the system. The eigenvectors corresponding to the eigenvalues describe the different modes (or independent states of the system). We give here a geometrical interpretation and defer their physical interpretation until we consider specific physical examples.

In order to interpret equation (4.3) and the concept of right eigenvectors geometrically, we consider the case of two dimensions. Let

$$\mathbf{x} = \begin{pmatrix} x_1 \\ x_2 \end{pmatrix}, \quad \mathbf{A} = \begin{pmatrix} a_{11} & a_{12} \\ a_{21} & a_{22} \end{pmatrix}$$

and

https://doi.org/10.1515/9783111598055-005

$$\mathbf{y} = \mathbf{A}\mathbf{x}$$
$$= \begin{pmatrix} a_{11} & a_{12} \\ a_{21} & a_{22} \end{pmatrix} \begin{pmatrix} x_1 \\ x_2 \end{pmatrix}$$
$$= \begin{pmatrix} a_{11}x_1 + a_{12}x_2 \\ a_{21}x_1 + a_{22}x_2 \end{pmatrix} \equiv \begin{pmatrix} y_1 \\ y_2 \end{pmatrix}$$

The matrix **A** operating on the vector **x** gives the vector **y**. In general, the length of **y** is different from that of **x** as the operator **A** stretches (or contracts) and rotates **x** to obtain **y**. However, when $\mathbf{y} = \lambda\mathbf{x}$ (with λ real) we see that when **A** operates on **x** we get only a stretching (or contraction) of **x** but there is no rotation (Figure 4.1 shows this for λ real and positive with magnitude greater than unity).

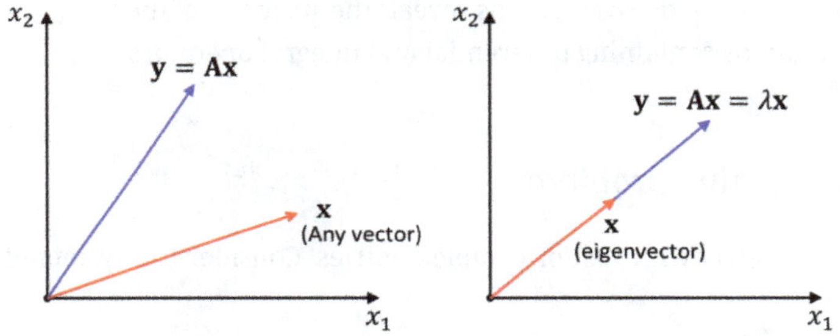

Figure 4.1: Schematic diagrams giving geometrical interpretation of **y** = **Ax**.

We note that equation (4.3) is a system of n homogeneous equations in n unknowns and may be written as

$$\begin{aligned} (a_{11} - \lambda)x_1 + a_{12}x_2 + \cdots + a_{1n}x_n &= 0 \\ a_{21}x_1 + (a_{22} - \lambda)x_2 + \cdots + a_{2n}x_n &= 0 \\ &. \\ &. \\ a_{n1}x_1 + a_{n2}x_2 + \cdots + (a_{nn} - \lambda)x_n &= 0 \end{aligned} \tag{4.4}$$

We have seen that this homogeneous system has a nontrivial solution iff $\det(\mathbf{A} - \lambda\mathbf{I}) = 0$.

$$\Rightarrow P_n(\lambda) \equiv \begin{vmatrix} a_{11} - \lambda & a_{12} & . & . & a_{1n} \\ a_{21} & a_{22} - \lambda & . & . & a_{2n} \\ . & . & . & . & . \\ . & . & . & . & . \\ a_{n1} & a_{n2} & . & . & a_{nn} - \lambda \end{vmatrix} = 0 \tag{4.5}$$

Equation (4.5) is called the *characteristic equation* of the square matrix **A**. The LHS of (4.5) is a polynomial of degree n in λ and may be written as

$$\begin{aligned} P_n(\lambda) &= (-\lambda)^n + a_1(-\lambda)^{n-1} + \cdots + a_{n-1}(-\lambda) + a_n = 0 \\ &= (\lambda_1 - \lambda)(\lambda_2 - \lambda)\dots(\lambda_n - \lambda) \end{aligned} \tag{4.6}$$

In order to determine the number of solutions of the characteristic equation, we invoke the fundamental theorem of algebra.

Theorem. *Every polynomial of degree n has exactly n roots (real or complex with counting of repetition or multiplicity).*
It follows from this theorem that a square matrix **A** of order n has n eigenvalues.

Definition. An eigenvalue λ_i of **A** is called simple if

$$P_n(\lambda_i) = 0 \quad \text{and} \quad P_n'(\lambda_i) = \left.\frac{dP_n(\lambda)}{d\lambda}\right|_{\lambda=\lambda_i} \neq 0$$

Theorem. *If λ_i is a simple eigenvalue of* **A**, *then the corresponding eigenvector obtained by solving*

$$(\mathbf{A} - \lambda_i \mathbf{I})\mathbf{x}_i = \mathbf{0}$$

is determined uniquely except for a nonzero multiplicative constant.

Proof.

$$P_n(\lambda) = \begin{vmatrix} a_{11}-\lambda & a_{12} & \cdot & \cdot & a_{1n} \\ a_{21} & a_{22}-\lambda & \cdot & \cdot & a_{2n} \\ \cdot & \cdot & \cdot & \cdot & \cdot \\ \cdot & \cdot & \cdot & \cdot & \cdot \\ a_{n1} & a_{n2} & \cdot & \cdot & a_{nn}-\lambda \end{vmatrix}$$

Using the formula for differentiation of a determinant, we have

$$P_n'(\lambda) = -\begin{vmatrix} a_{22}-\lambda & a_{23} & \cdot & \cdot & a_{2n} \\ \cdot & \cdot & \cdot & \cdot & a_{2n} \\ \cdot & \cdot & \cdot & \cdot & \cdot \\ \cdot & \cdot & \cdot & \cdot & \cdot \\ a_{n2} & a_{n2} & \cdot & \cdot & a_{nn}-\lambda \end{vmatrix} - \begin{vmatrix} a_{11}-\lambda & a_{13} & \cdot & \cdot & a_{1n} \\ a_{31} & a_{33}-\lambda & \cdot & \cdot & a_{3n} \\ \cdot & \cdot & \cdot & \cdot & \cdot \\ \cdot & \cdot & \cdot & \cdot & \cdot \\ a_{n1} & a_{n3} & \cdot & \cdot & a_{nn}-\lambda \end{vmatrix} - \cdots$$
$$- \begin{vmatrix} a_{11}-\lambda & a_{12} & \cdot & \cdot & a_{1\,n-1} \\ a_{21} & \cdot & \cdot & \cdot & a_{2\,n-1} \\ \cdot & \cdot & \cdot & \cdot & \cdot \\ \cdot & \cdot & \cdot & \cdot & \cdot \\ a_{n-11} & a_{n-12} & \cdot & \cdot & a_{n-1\,n-1}-\lambda \end{vmatrix}$$

If $P_n'(\lambda_i) \neq 0$, then at least one of the above $(n-1) \times (n-1)$ determinants is not zero.

$\Rightarrow \text{rank}(\mathbf{A} - \lambda_i \mathbf{I}) = n - 1$.
$\therefore$ There is only one linearly independent solution to the homogeneous system $(\mathbf{A} - \lambda_i \mathbf{I})\mathbf{x}_i = \mathbf{0}$.
$\therefore$ The result. □

Remark. The eigenvalues are also called characteristic values, characteristic roots or latent roots.

Example 4.1 (Characteristic equation for 2 × 2 and 3 × 3 matrices). We have

$$\begin{aligned} P_2(\lambda) &= \begin{vmatrix} a_{11} - \lambda & a_{12} \\ a_{21} & a_{22} - \lambda \end{vmatrix} \\ &= \lambda^2 - (a_{11} + a_{22})\lambda + (a_{11}a_{22} - a_{12}a_{21}) \\ &= \lambda^2 - (\operatorname{tr} \mathbf{A})\lambda + \det \mathbf{A} = (\lambda_1 - \lambda)(\lambda_2 - \lambda) \end{aligned}$$

where $\operatorname{tr} \mathbf{A}$ is the trace of $\mathbf{A}$ (sum of diagonal elements) and $\det \mathbf{A}$ is the determinant.

For the 3 × 3 case, we have

$$\begin{aligned} P_3(\lambda) &= \begin{vmatrix} a_{11} - \lambda & a_{12} & a_{13} \\ a_{21} & a_{22} - \lambda & a_{23} \\ a_{31} & a_{32} & a_{33} - \lambda \end{vmatrix} \\ &= -\lambda^3 + (\operatorname{tr} \mathbf{A})\lambda^2 - (a_{11}a_{22} + a_{11}a_{33} + a_{22}a_{33} - a_{12}a_{21} - a_{13}a_{31} - a_{23}a_{32})\lambda + \det \mathbf{A} \\ &= (\lambda_1 - \lambda)(\lambda_2 - \lambda)(\lambda_3 - \lambda) \end{aligned}$$

For the general case, the characteristic equation may be written as

$$\begin{aligned} P_n(\lambda) &= (\lambda_1 - \lambda)(\lambda_2 - \lambda) \dots (\lambda_n - \lambda) \\ &= \prod_{j=1}^{n} (\lambda_j - \lambda) \end{aligned}$$

and we have the following relations:

$$\prod_{j=1}^{n} \lambda_j = \det \mathbf{A}; \quad \sum_{j=1}^{n} \lambda_j = \operatorname{tr} \mathbf{A}.$$

4.2 Left eigenvectors and the adjoint eigenvalue problem (eigenrows)

Let $\mathbf{A}$ be an $n \times n$ square matrix with real or complex entries. Let $\mathbf{y}^*$ be a row vector:

$$\begin{aligned} \mathbf{y}^* &= (\overline{\mathbf{y}})^T \\ &= \begin{pmatrix} \overline{y}_1 & \overline{y}_2 & \overline{y}_3 & . & \overline{y}_n \end{pmatrix} \end{aligned}$$

where $*$ denotes complex conjugate transpose and

$$\mathbf{y} = \begin{pmatrix} y_1 \\ y_2 \\ . \\ . \\ y_n \end{pmatrix}$$

We consider the eigenvalue problem

$$\mathbf{y}^*\mathbf{A} = \mu\mathbf{y}^* \tag{4.7}$$

To illustrate, consider the case of $n = 2$. Then equation (4.7) gives

$$\begin{pmatrix} \overline{y_1} & \overline{y_2} \end{pmatrix}\begin{pmatrix} a_{11} & a_{12} \\ a_{21} & a_{22} \end{pmatrix} = \mu \begin{pmatrix} \overline{y_1} & \overline{y_2} \end{pmatrix}$$

Multiplying $\mathbf{A}$ on the left by a row vector gives another row vector. Thus, we get

$$\begin{aligned} \mathbf{y}^*\mathbf{A} &= \begin{pmatrix} a_{11}\overline{y_1} + a_{21}\overline{y_2} & a_{12}\overline{y_1} + a_{22}\overline{y_2} \end{pmatrix} \\ &= \begin{pmatrix} \mu\overline{y_1} & \mu\overline{y_2} \end{pmatrix} \end{aligned}$$

This gives the homogeneous equations:

$$\begin{aligned} a_{11}\overline{y_1} + a_{21}\overline{y_2} - \mu\overline{y_1} &= 0 \\ a_{12}\overline{y_1} + a_{22}\overline{y_2} - \mu\overline{y_2} &= 0 \end{aligned}$$

Definition. A real or complex number μ for which (4.7) has nontrivial solutions is called an eigenvalue of $\mathbf{A}$ and the nontrivial solution $\mathbf{y}^*$ is called *eigenrow* or more precisely *left eigenvector* of $\mathbf{A}$ corresponding to eigenvalue μ.

Taking the complex conjugate transpose operation, equation (4.7) may also be written as

$$\begin{aligned} &(\mathbf{y}^*\mathbf{A})^* = (\mu\mathbf{y}^*)^* \\ &\Rightarrow \mathbf{A}^*\mathbf{y} = \bar{\mu}\mathbf{y} \end{aligned} \tag{4.8}$$

Thus, the left eigenvalue problem for matrix $\mathbf{A}$ (also called the *adjoint eigenvalue problem*) is an eigenvalue problem for $\mathbf{A}^*$ [Considered as an operator, the matrix $\mathbf{A}^*$ is called the adjoint of $\mathbf{A}$]. We shall refer to the column vector $\mathbf{y}$ as the adjoint eigenvector. (The complex conjugate of the transpose of $\mathbf{y}$, namely $\mathbf{y}^*$ will be referred to as the eigenrow or left eigenvector of $\mathbf{A}$).

When $\mathbf{A}$ has real elements, equation (4.8) reduces to

$$\mathbf{A}^T\mathbf{y} = \bar{\mu}\mathbf{y} \tag{4.9}$$

Theorem. *The set of eigenvalues defined by equation* (4.8) *is identical to that defined by equation* (4.3).

Proof. The eigenvalues defined by equation (4.8) are the roots of the polynomial

$$\begin{aligned}
Q_n(\mu) &= |\mathbf{A}^* - \bar{\mu}\mathbf{I}| \\
&= |(\mathbf{A}^* - \bar{\mu}\mathbf{I})^T| \\
&= |\bar{\mathbf{A}} - \bar{\mu}\mathbf{I}| \\
&= \overline{|\mathbf{A} - \mu\mathbf{I}|} \\
&= \overline{\mathbf{P}_n(\mu)} \\
\overline{\mathbf{P}_n(\mu)} &= 0 \quad \text{iff} \quad \mathbf{P}_n(\mu) = 0
\end{aligned}$$

Thus, the adjoint problem has the same set of eigenvalues. This can be seen more directly from equation (4.7). This equation may be written as the homogeneous system

$$\mathbf{y}^*(\mathbf{A} - \mu\mathbf{I}) = \mathbf{0} \tag{4.10}$$

The condition for a nontrivial solution is

$$|\mathbf{A} - \mu\mathbf{I}| = \mathbf{P}_n(\mu) = 0. \tag{4.11}$$

Thus, the set of eigenvalues defined by (4.3) and (4.7) are the same. However, note that if (4.7) is written in the form given by equation (4.8), then the adjoint eigenvalue problem has eigenvalues

$$\mu = \bar{\lambda}. \tag{4.12}$$

Thus, if λ is an eigenvalue of $\mathbf{A}$, $\bar{\lambda}$ is an eigenvalue of $\mathbf{A}^*$. □

4.3 Properties of eigenvectors/eigenrows

We now consider some properties of eigenvectors and eigenrows that follow from the definition:

1.(a) If $\mathbf{x}_j$ is an eigenvector of $\mathbf{A}$ corresponding to eigenvalue λ_j, then so is $\alpha\mathbf{x}_j$ where α is any nonzero constant:

$$\begin{aligned}
\mathbf{A}\mathbf{x}_j &= \lambda_j\mathbf{x}_j \\
&\Rightarrow \mathbf{A}(\alpha\mathbf{x}_j) = \lambda_j(\alpha_j\mathbf{x}_j) \\
&\Rightarrow \alpha_j\mathbf{x}_j (\neq \mathbf{0}) \text{ is also an eigenvector.}
\end{aligned}$$

1.(b) If λ_i is a simple eigenvalue, the eigenvector is uniquely determined except for an arbitrary nonzero factor.

2.(a) If $\mathbf{x}_i$ and $\mathbf{x}_j$ are eigenvectors of $\mathbf{A}$ corresponding to eigenvalues λ_i and $\lambda_j(\lambda_j \neq \lambda_i)$, then $\mathbf{x}_i$ and $\mathbf{x}_j$ are linearly independent.

Proof. Suppose that $\mathbf{x}_i$ and $\mathbf{x}_j$ are linearly dependent. Then there exist constants c_i and c_j such that

$$c_i\mathbf{x}_i + c_j\mathbf{x}_j = \mathbf{0}$$

and at least one of these constants is not zero. Assume that $c_i \neq 0 \Rightarrow$

$$\begin{aligned}
\mathbf{x}_i &= -\frac{c_j}{c_i}\mathbf{x}_j \\
\Rightarrow \mathbf{A}\mathbf{x}_i &= -\frac{c_j}{c_i}\mathbf{A}\mathbf{x}_j \\
\Rightarrow \lambda_i\mathbf{x}_i &= -\frac{c_j}{c_i}\lambda_j\mathbf{x}_j \\
\Rightarrow \lambda_i\mathbf{x}_i &= \lambda_j\left(\frac{-c_j}{c_i}\mathbf{x}_j\right) = \lambda_j\mathbf{x}_i \\
\Rightarrow (\lambda_i - \lambda_j)\mathbf{x}_i &= \mathbf{0}
\end{aligned}$$

or

$$\mathbf{x}_i = \mathbf{0},$$

since we assumed that the eigenvalues are distinct, i. e., $\lambda_i \neq \lambda_j$. However, $\mathbf{x}_i$ cannot be the zero vector, and we arrive at a contradiction. $\Rightarrow$ $\mathbf{x}_i$ and $\mathbf{x}_j$ are linearly independent. □

2.(b) Suppose that $\mathbf{A}$ has simple and distinct eigenvalues $\{\lambda_1, \lambda_2, \ldots, \lambda_n\}$ with eigenvectors $\{\mathbf{x}_1, \mathbf{x}_2, \ldots, \mathbf{x}_n\}$. Then the set of eigenvectors $\{\mathbf{x}_1, \mathbf{x}_2, \mathbf{x}_3, \ldots, \mathbf{x}_n\}$ is linearly independent.

Proof. Suppose that $\{\mathbf{x}_1, \mathbf{x}_2, \mathbf{x}_3, \ldots, \mathbf{x}_n\}$ is linearly dependent. Then $\exists$ constants c_i not all zero such that

$$c_1\mathbf{x}_1 + c_2\mathbf{x}_2 + \cdots + c_n\mathbf{x}_n = \mathbf{0} \tag{4.13}$$

premultiply (4.13) by $(\mathbf{A} - \lambda_1\mathbf{I}) \Rightarrow$

$$c_1(\mathbf{A} - \lambda_1\mathbf{I})\mathbf{x}_1 + c_2(\mathbf{A} - \lambda_1\mathbf{I})\mathbf{x}_2 + \cdots + c_n(\mathbf{A} - \lambda_1\mathbf{I})\mathbf{x}_n = \mathbf{0}$$

Using the fact that

$$\mathbf{A}\mathbf{x}_i = \lambda_i\mathbf{x}_i \Rightarrow$$

$$\mathbf{0} + c_2(\lambda_2 - \lambda_1)\mathbf{x}_2 + \cdots + c_n(\lambda_n - \lambda_1)\mathbf{x}_n = \mathbf{0} \tag{4.14}$$

Premultiply (4.14) by $(\mathbf{A} - \lambda_2\mathbf{I}) \Rightarrow$

$$c_3(\lambda_3 - \lambda_2)(\lambda_3 - \lambda_1)\mathbf{x}_3 + \cdots + c_n(\lambda_n - \lambda_2)(\lambda_n - \lambda_1)\mathbf{x}_n = \mathbf{0}$$
$$\Rightarrow \sum_{j=3}^{n} c_j \left[\prod_{i=1}^{2}(\lambda_j - \lambda_i)\right]\mathbf{x}_j = \mathbf{0}$$

Continuing this procedure, we get after $(n-1)$ steps

$$c_n \left[\prod_{i=1}^{n-1}(\lambda_n - \lambda_i)\right]\mathbf{x}_n = \mathbf{0} \tag{4.15}$$

Since the eigenvalues are all distinct and $\mathbf{x}_n \neq \mathbf{0}$, (4.15) $\Rightarrow c_n = 0$. Repeating the same process with the remaining part of equation (4.13), we can show that

$$c_{n-1} = c_{n-2} = \cdots = c_1 = 0$$

Thus, all constants are zero and we have a contradiction. This implies that the eigenvectors are linearly independent. □

3. Properties (i) and (ii) are also valid for the eigenrows.

4.(a) If $\mathbf{x}_i$ is an eigenvector of $\mathbf{A}$ corresponding to eigenvalue λ_i and $\mathbf{y}_j^*$ is the eigenrow of $\mathbf{A}$ corresponding to eigenvalue $\lambda_j (\neq \lambda_i)$, then we have

$$\mathbf{y}_j^*\mathbf{x}_i = 0 \tag{4.16}$$

This important property is referred to as the *biorthogonality property*, i. e., eigenrows and eigenvectors corresponding to different eigenvalues are orthogonal.

Proof. We have from the definition,

$$\mathbf{A}\mathbf{x}_i = \lambda_i\mathbf{x}_i \tag{4.17}$$
$$\mathbf{y}_j^*\mathbf{A} = \lambda_j\mathbf{y}_j^* \tag{4.18}$$

Multiply equation (4.18) on the right by $\mathbf{x}_i \Rightarrow$

$$\mathbf{y}_j^*\mathbf{A}\mathbf{x}_i = \lambda_j\mathbf{y}_j^*\mathbf{x}_i$$

Using (4.17) $\Rightarrow$

$$\mathbf{y}_j^*\lambda_i\mathbf{x}_i = \lambda_j\mathbf{y}_j^*\mathbf{x}_i$$
$$\lambda_i\mathbf{y}_j^*\mathbf{x}_i = \lambda_j\mathbf{y}_j^*\mathbf{x}_i \quad \text{(since } \lambda_i \text{ is a scalar)}$$

$\Rightarrow$

$$(\lambda_i - \lambda_j)\mathbf{y}_j^*\mathbf{x}_i = 0$$

Since $\lambda_i \neq \lambda_j \Rightarrow \mathbf{y}_j^* \mathbf{x}_i = 0$. In terms of the dot product, this result may be written as $\langle \mathbf{x}_i, \mathbf{y}_j \rangle = 0; i \neq j$. □

4.(b) Suppose that $\mathbf{A}$ has simple and distinct eigenvalues $\lambda_1, \lambda_2, \ldots, \lambda_n$ with eigenvectors $\{\mathbf{x}_1, \mathbf{x}_2, \ldots, \mathbf{x}_n\}$ and eigenrows $\{\mathbf{y}_1^*, \mathbf{y}_2^*, \ldots, \mathbf{y}_n^*\}$. Then

$$\mathbf{y}_j^* \mathbf{x}_i = \langle \mathbf{x}_i, \mathbf{y}_j \rangle = 0 \quad i \neq j \tag{4.19}$$

$$\mathbf{y}_j^* \mathbf{x}_j \neq 0 \tag{4.20}$$

We have already proved equation (4.19). To prove (4.20), we use the property that the eigenvectors and eigenrows are linearly independent. Now, if $\mathbf{y}_j^* \mathbf{x}_j = 0$, $\mathbf{x}_j$ is orthogonal to n linearly independent vectors $\mathbf{y}_i$, $i = 1, \ldots, n$. However, the only such vector is the zero vector. But $\mathbf{x}_j \neq \mathbf{0} \Rightarrow \mathbf{y}_j^* \mathbf{x}_j \neq 0$. We can normalize the eigenrows (or eigenvectors) such that

$$\langle \mathbf{x}_i, \mathbf{y}_j \rangle = \delta_{ij} \tag{4.21}$$

[Note: The symbol δ_{ij} is the Kronecker delta, which takes a values of unity when the indices are equal and zero otherwise]. □

We now present some examples illustrating the calculation of the eigenvalues, eigenvectors and eigenrows.

Example 4.2.

$$\mathbf{A} = \begin{pmatrix} -3 & 2 \\ 4 & -5 \end{pmatrix}$$

This is a real matrix and is not symmetric.
Eigenvalues:

$$\begin{aligned} P(\lambda) &= \begin{vmatrix} -3-\lambda & 2 \\ 4 & -5-\lambda \end{vmatrix} \\ &= \lambda^2 + 8\lambda + 7 \\ &= (\lambda+1)(\lambda+7) \\ P(\lambda) = 0 &\Rightarrow \lambda_1 = -1, \quad \lambda_2 = -7 \end{aligned}$$

eigenvectors:

$$(\mathbf{A} - \lambda_1 \mathbf{I})\mathbf{x}_1 = \mathbf{0} \Rightarrow$$

$$\begin{pmatrix} -2 & 2 \\ 4 & -4 \end{pmatrix} \mathbf{x}_1 = \mathbf{0}$$

$$\Rightarrow \mathbf{x}_1 = \begin{pmatrix} 1 \\ 1 \end{pmatrix}$$

$$(\mathbf{A} - \lambda_2 \mathbf{I})\mathbf{x}_2 = \mathbf{0} \Rightarrow$$

$$\begin{pmatrix} 4 & 2 \\ 4 & 2 \end{pmatrix} \mathbf{x}_2 = \mathbf{0}$$

$$\Rightarrow \mathbf{x}_2 = \begin{pmatrix} 1 \\ -2 \end{pmatrix}$$

eigenrows:

$$\mathbf{y}_1^T(\mathbf{A} - \lambda_1 \mathbf{I}) = \mathbf{0}$$

$$\Rightarrow \mathbf{y}_1^T \begin{pmatrix} -2 & 2 \\ 4 & -4 \end{pmatrix} = \mathbf{0}$$

$$\mathbf{y}_1^T = (2, 1)$$

$$\mathbf{y}_2^T(\mathbf{A} - \lambda_2 \mathbf{I}) = \mathbf{0}$$

$$\Rightarrow \mathbf{y}_2^T \begin{pmatrix} 4 & 2 \\ 4 & 2 \end{pmatrix} = \mathbf{0}$$

$$\mathbf{y}_2^T = (1, -1)$$

Thus, we have
eigenvalues: $\lambda_1 = -1, \lambda_2 = -7$
eigenvectors: $\mathbf{x}_1 = \binom{1}{1}, \mathbf{x}_2 = \binom{1}{-2}$
eigenrows: $\mathbf{y}_1^T = (2 \;\; 1), \mathbf{y}_2^T = (1 \;\; -1)$
Biorthogonality relations:

$$\mathbf{y}_1^T \mathbf{x}_1 = 3 \neq 0 \quad \mathbf{y}_2^T \mathbf{x}_1 = 0$$
$$\mathbf{y}_1^T \mathbf{x}_2 = 0 \quad \mathbf{y}_2^T \mathbf{x}_2 = 3 \neq 0$$

Figure 4.2 shows a plot of the eigenvectors and eigenrows (dashed lines). The biorthogonality relationship can be seen clearly.

Normalizing the eigenrows such that

$$\mathbf{y}_i^T \mathbf{x}_j = \delta_{ij}$$

gives the normalized eigenrows as $\mathbf{y}_1^T = (\frac{2}{3} \;\; \frac{1}{3}), \mathbf{y}_2^T = (\frac{1}{3} \;\; -\frac{1}{3})$.

Example 4.3.

$$\mathbf{A} = \begin{pmatrix} -1 & 1 \\ -1 & -1 \end{pmatrix}$$

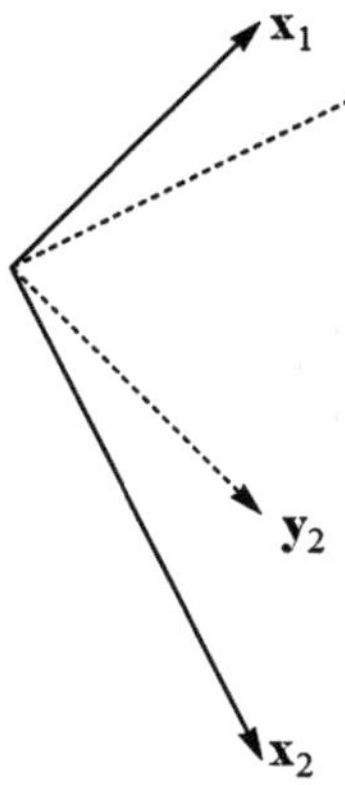

Figure 4.2: Schematic plot of the eigenvectors and eigenrows of the matrix in Example (4.2).

Here, $\mathbf{A}$ is a real matrix but not symmetric

$$\mathbf{A}^* = \mathbf{A}^T = \begin{pmatrix} -1 & -1 \\ 1 & -1 \end{pmatrix}$$

$$|\mathbf{A} - \lambda\mathbf{I}| = \begin{vmatrix} -1-\lambda & 1 \\ -1 & -1-\lambda \end{vmatrix} = \lambda^2 + 2\lambda + 2$$

$$P(\lambda) = 0 \Rightarrow \lambda_1 = -1 + i, \quad \lambda_2 = -1 - i$$

Eigenvectors:

$$(\mathbf{A} - \lambda_1\mathbf{I})\mathbf{x}_1 = \mathbf{0} \Rightarrow \begin{pmatrix} -i & 1 \\ -1 & -i \end{pmatrix}\mathbf{x}_1 = \begin{pmatrix} 0 \\ 0 \end{pmatrix} \Rightarrow \mathbf{x}_1 = \begin{pmatrix} 1 \\ i \end{pmatrix}$$

$$(\mathbf{A} - \lambda_2\mathbf{I})\mathbf{x}_2 = \mathbf{0} \Rightarrow \begin{pmatrix} i & 1 \\ -1 & i \end{pmatrix}\mathbf{x}_2 = \begin{pmatrix} 0 \\ 0 \end{pmatrix} \Rightarrow \mathbf{x}_2 = \begin{pmatrix} 1 \\ -i \end{pmatrix}$$

eigenrows:

$$\mathbf{y}_1^*(\mathbf{A} - \lambda_1\mathbf{I}) = \mathbf{0} \Rightarrow \begin{pmatrix} \bar{y}_{11} & \bar{y}_{12} \end{pmatrix}\begin{pmatrix} -i & 1 \\ -1 & -i \end{pmatrix} = \begin{pmatrix} 0 & 0 \end{pmatrix}$$

$$\Rightarrow \mathbf{y}_1^* = \begin{pmatrix} i & 1 \end{pmatrix}$$

$$\mathbf{y}_1 = \begin{pmatrix} -i \\ 1 \end{pmatrix}$$

$$\mathbf{y}_2^*(\mathbf{A} - \lambda_2\mathbf{I}) = \mathbf{0} \Rightarrow \begin{pmatrix} \bar{y}_{21} & \bar{y}_{22} \end{pmatrix}\begin{pmatrix} i & 1 \\ -1 & i \end{pmatrix} = \begin{pmatrix} 0 & 0 \end{pmatrix}$$

$$\Rightarrow \mathbf{y}_2^* = \begin{pmatrix} -i & 1 \end{pmatrix} \quad \text{or} \quad \mathbf{y}_2 = \begin{pmatrix} i \\ 1 \end{pmatrix}$$

To summarize, we have
eigenvalues

$$\lambda_1 = -1 + i, \quad \lambda_2 = -1 - i$$

eigenvectors

$$\mathbf{x}_1 = \begin{pmatrix} 1 \\ i \end{pmatrix}, \quad \mathbf{x}_2 = \begin{pmatrix} 1 \\ -i \end{pmatrix}$$

eigenrows

$$\mathbf{y}_1^* = \begin{pmatrix} i & 1 \end{pmatrix}, \quad \mathbf{y}_2^* = \begin{pmatrix} -i & 1 \end{pmatrix}$$

or adjoint eigenvectors

$$\mathbf{y}_1 = \begin{pmatrix} -i \\ 1 \end{pmatrix}, \quad \mathbf{y}_2 = \begin{pmatrix} i \\ 1 \end{pmatrix}$$

Biorthogonality

$$\begin{aligned} \mathbf{y}_1^*\mathbf{x}_1 &= \langle \mathbf{x}_1, \mathbf{y}_1 \rangle = i + i = 2i \neq 0 \\ \mathbf{y}_2^*\mathbf{x}_1 &= \langle \mathbf{x}_1, \mathbf{y}_2 \rangle = -i + i = 0 \\ \mathbf{y}_1^*\mathbf{x}_2 &= \langle \mathbf{x}_2, \mathbf{y}_1 \rangle = i - i = 0 \\ \mathbf{y}_2^*\mathbf{x}_2 &= \langle \mathbf{x}_2, \mathbf{y}_2 \rangle = -i - i = -2i \neq 0 \end{aligned}$$

Example 4.4.

$$\mathbf{A} = \begin{pmatrix} -1 & 1 \\ 1 & -1 \end{pmatrix}$$

Here, $\mathbf{A}$ is a real symmetric matrix.

$$P(\lambda) = \begin{vmatrix} -1-\lambda & 1 \\ 1 & -1-\lambda \end{vmatrix} = \lambda^2 + 2\lambda = 0 \Rightarrow \lambda_1 = 0, \quad \lambda_2 = -2 \quad .$$

$$(\mathbf{A} - \lambda_1\mathbf{I})\mathbf{x}_1 = \mathbf{0} \Rightarrow \begin{pmatrix} -1 & 1 \\ 1 & -1 \end{pmatrix}\mathbf{x}_1 = \mathbf{0} \Rightarrow \mathbf{x}_1 = \begin{pmatrix} 1 \\ 1 \end{pmatrix}$$

$$(\mathbf{A} - \lambda_2\mathbf{I})\mathbf{x}_2 = \mathbf{0} \Rightarrow \begin{pmatrix} 1 & 1 \\ 1 & 1 \end{pmatrix}\mathbf{x}_2 = \mathbf{0} \Rightarrow \mathbf{x}_2 = \begin{pmatrix} 1 \\ -1 \end{pmatrix}$$

since $\mathbf{A}^T = \mathbf{A}$, adjoint eigenvectors are the same as the eigenvectors, i. e., $\mathbf{y}_1 = \mathbf{x}_1$ and $\mathbf{y}_2 = \mathbf{x}_2$. Note also that $\mathbf{x}_1$ and $\mathbf{x}_2$ are orthogonal to each other. Normalizing these vectors such that $\|\mathbf{x}_i\| = 1$, we obtain the orthonormal set of eigenvectors (and adjoint eigenvectors):

$$\mathbf{x}_1 = \begin{pmatrix} \frac{1}{\sqrt{2}} \\ \frac{1}{\sqrt{2}} \end{pmatrix}, \quad \mathbf{x}_2 = \begin{pmatrix} \frac{1}{\sqrt{2}} \\ \frac{-1}{\sqrt{2}} \end{pmatrix}.$$

Example 4.5. We consider the Hermitian matrix

$$\mathbf{A} = \begin{pmatrix} 1 & 1-i \\ 1+i & 2 \end{pmatrix}$$

Recall that a Hermitian matrix is characterized by $\mathbf{A}^* = \mathbf{A}$,

$$\overline{\mathbf{A}} = \begin{pmatrix} 1 & 1+i \\ 1-i & 2 \end{pmatrix}$$

$$(\overline{\mathbf{A}})^T = \mathbf{A}^* = \begin{pmatrix} 1 & 1-i \\ 1+i & 2 \end{pmatrix} = \mathbf{A}$$

$$P(\lambda) = \begin{vmatrix} 1-\lambda & 1-i \\ 1+i & 2-\lambda \end{vmatrix} = \lambda^2 - 3\lambda$$

$$P(\lambda) = 0 \Rightarrow \lambda_1 = 0, \quad \lambda_2 = 3$$

Eigenvectors

$$(\mathbf{A} - \lambda_1 \mathbf{I})\mathbf{x}_1 = \mathbf{0} \Rightarrow \begin{pmatrix} 1 & 1-i \\ 1+i & 2 \end{pmatrix} \mathbf{x}_1 = \mathbf{0} \Rightarrow \mathbf{x}_1 = \begin{pmatrix} -1+i \\ 1 \end{pmatrix}$$

$$\mathbf{y}_1^* (\mathbf{A} - \lambda \mathbf{I}) = \mathbf{0}$$

$$(\bar{y}_{11} \ \bar{y}_{21}) \begin{pmatrix} 1 & 1-i \\ 1+i & 2 \end{pmatrix} = \begin{pmatrix} 0 & 0 \end{pmatrix}$$

$$\Rightarrow \mathbf{y}_1^* = \begin{pmatrix} -1-i & 1 \end{pmatrix} \text{ or } \mathbf{y}_1 = \begin{pmatrix} -1+i \\ 1 \end{pmatrix} = \mathbf{x}_1$$

$$(\mathbf{A} - \lambda_2 \mathbf{I})\mathbf{x}_2 = \mathbf{0}$$

$$\begin{pmatrix} -2 & 1-i \\ 1+i & -1 \end{pmatrix} \mathbf{x}_2 = \mathbf{0}$$

$$\Rightarrow \mathbf{x}_2 = \begin{pmatrix} 1-i \\ 2 \end{pmatrix} = \mathbf{y}_2$$

$$\mathbf{y}_2^* \mathbf{x}_1 = \begin{pmatrix} 1+i & 2 \end{pmatrix} \begin{pmatrix} -1+i \\ 1 \end{pmatrix} = 0, \quad \mathbf{y}_1^* \mathbf{x}_2 = 0.$$

□

We now prove an important theorem about the eigenvalues of a real symmetric matrix ($\mathbf{A}^T = \mathbf{A}$) or a complex Hermitian matrix ($\mathbf{A}^* = \mathbf{A}$).

Theorem. *Suppose that the square matrix* $\mathbf{A}$ *is such that* $\mathbf{A}^* = \mathbf{A}$. *Then the eigenvalues of* $\mathbf{A}$ *are real and the left and right eigenvectors of* $\mathbf{A}$ *are related by* $\mathbf{y}_i^* = \mathbf{x}_i^*$, *i. e., the* *eigenrows are the conjugate transposes of the eigenvectors (equivalently, the eigenvectors and adjoint eigenvectors are the same).*

Proof. (a) Let λ be an eigenvalue of $\mathbf{A}$ and $\mathbf{x}$ be the corresponding eigenvector. From the definition, we have

$$\mathbf{A}\mathbf{x} = \lambda\mathbf{x} \tag{4.22}$$

Premultiplying (4.22) by $\mathbf{x}^*$ (or taking the dot or inner product with $\mathbf{x}$) we get

$$\mathbf{x}^*\mathbf{A}\mathbf{x} = \lambda\mathbf{x}^*\mathbf{x} \tag{4.23}$$

Now, $\mathbf{x}^*\mathbf{A}\mathbf{x}$ is a scalar and equation (4.23) is a scalar identity. We take the $*$ operation (complex conjugation and transpose) on both sides of (4.23) $\Rightarrow$

$$\begin{aligned}
(\mathbf{x}^*\mathbf{A}\mathbf{x})^* &= (\lambda\mathbf{x}^*\mathbf{x})^* \\
\Rightarrow \mathbf{x}^*\mathbf{A}^*\mathbf{x} &= \bar{\lambda}\mathbf{x}^*\mathbf{x} \\
\Rightarrow \mathbf{x}^*\mathbf{A}\mathbf{x} &= \bar{\lambda}\mathbf{x}^*\mathbf{x} \quad (\text{since } \mathbf{A}^* = \mathbf{A}) \\
\Rightarrow \mathbf{x}^*\lambda\mathbf{x} &= \bar{\lambda}\mathbf{x}^*\mathbf{x} \\
\Rightarrow \lambda\mathbf{x}^*\mathbf{x} &= \bar{\lambda}\mathbf{x}^*\mathbf{x} \quad (\text{since } \lambda \text{ is a scalar}) \\
\Rightarrow (\lambda - \bar{\lambda})\mathbf{x}^*\mathbf{x} &= 0
\end{aligned} \tag{4.24}$$

But $\mathbf{x}^*\mathbf{x} = \|\mathbf{x}\|^2 \neq 0$ since $\mathbf{x}$ is an eigenvector

$$\Longrightarrow \lambda = \bar{\lambda}$$
$$\Longrightarrow \lambda \text{ is real}$$

(b) The eigenrows are defined by

$$\mathbf{y}_i^*\mathbf{A} = \lambda_i\mathbf{y}_i^* \tag{4.25}$$

(4.22)$\Longrightarrow$

$$\mathbf{x}_i^*\mathbf{A} = \bar{\lambda}_i\mathbf{x}_i^* \Longrightarrow \mathbf{x}_i^*\mathbf{A} = \lambda_i\mathbf{x}_i^* \quad (\text{since } \lambda \text{ is real}) \tag{4.26}$$

Comparing (4.25) and (4.26), we see that $\mathbf{y}_i^*$ may be chosen to be a scalar multiple of $\mathbf{x}_i^*$. Thus, we choose $\mathbf{y}_i^* = \mathbf{x}_i^*$. Because of this very important property (real eigenvalues and orthogonal set of eigenvectors) of real symmetric (or complex Hermitian) matrices, many problems involving such matrices can be solved using only orthogonal expansions. □

4.4 Orthogonal and biorthogonal expansions

4.4.1 Vector expansions

Let $\{\mathbf{x}_1, \mathbf{x}_2, \dots, \mathbf{x}_n\}$ be a set of n linearly independent vectors each containing n elements. Let $\mathbf{z}$ be any other vector with n elements. Then $\mathbf{z}$ can be expanded as

$$\mathbf{z} = \sum_{i=1}^{n} \alpha_i \mathbf{x}_i \tag{4.27}$$

This expansion is unique, i. e., the coefficients α_i are uniquely determined for each vector $\mathbf{z}$. These coefficients are called the coordinates of $\mathbf{z}$ w. r. t the basis $\{\mathbf{x}_1, \mathbf{x}_2, \dots, \mathbf{x}_n\}$. In general, we have to solve a set of n linear equations in n unknowns to determine $\{\alpha_i\}$.

Example 4.6. Consider $\mathbb{R}^2$ and take $\mathbf{x}_1 = \binom{1}{2}$, $\mathbf{x}_2 = \binom{2}{5}$ as a basis. Find the coordinates of $\mathbf{z} = \binom{1}{-1}$ with respect to this basis. Writing

$$\begin{pmatrix} 1 \\ -1 \end{pmatrix} = \alpha_1 \begin{pmatrix} 1 \\ 2 \end{pmatrix} + \alpha_2 \begin{pmatrix} 2 \\ 5 \end{pmatrix}$$

gives the linear equations

$$\begin{aligned} \alpha_1 + 2\alpha_2 &= 1 \\ 2\alpha_1 + 5\alpha_2 &= -1. \end{aligned}$$

Solving these equations, we obtain

$$\begin{aligned} \alpha_1 &= 7 \\ \alpha_2 &= -3. \end{aligned}$$

4.4.2 Orthogonal expansions

Now, suppose that $\{\mathbf{x}_1, \mathbf{x}_2, \dots, \mathbf{x}_n\}$ is an orthogonal set, i. e.,

$$\mathbf{x}_i^* \mathbf{x}_j = 0 \quad \text{if } i \neq j \tag{4.28}$$

Then the determination of the coefficients in equation (4.27) is simplified greatly as shown below. Multiply (4.27) by $\mathbf{x}_j^*$ on the left (or take scalar product of equation (4.27) with $\mathbf{x}_j$), $\Longrightarrow$

$$\mathbf{x}_j^* \mathbf{z} = \sum_{i=1}^{n} \alpha_i \mathbf{x}_j^* \mathbf{x}_i \tag{4.29}$$

Since $\mathbf{x}_j^*\mathbf{x}_j = \|\mathbf{x}_j\|^2 \neq 0$, there is only one nonzero term on the RHS of equation (4.29). Now, the linear equations for α_j are decoupled and we can solve for α_j as

$$\alpha_j = \frac{\mathbf{x}_j^*\mathbf{z}}{\mathbf{x}_j^*\mathbf{x}_j} \tag{4.30}$$

Thus, we obtain the expansion

$$\mathbf{z} = \sum_{j=1}^{n} \frac{\langle \mathbf{z}, \mathbf{x}_j \rangle}{\|\mathbf{x}_j\|^2}\mathbf{x}_j \tag{4.31}$$

Note that if each $\mathbf{x}_j$ is normalized so that $\|\mathbf{x}_j\| = 1$. Then we have

$$\alpha_j = \mathbf{x}_j^*\mathbf{z} = \langle \mathbf{z}, \mathbf{x}_j \rangle.$$

Thus, the coordinates of any vector $\mathbf{z}$ w. r. t. an orthonormal basis can be obtained by simply taking the dot product of $\mathbf{z}$ with each basis vector.

Example 4.7.

(a) Consider $\mathbb{R}^2$ and take $\mathbf{e}_1 = \binom{1}{0}$, $\mathbf{e}_2 = \binom{0}{1}$ as the orthonormal set. Taking

$$\mathbf{z} = \begin{pmatrix} 1 \\ -1 \end{pmatrix}$$

we have $\mathbf{z} = \alpha_1\mathbf{e}_1 + \alpha_2\mathbf{e}_2$,

$$\alpha_1 = \mathbf{e}_1^T\mathbf{z} = \begin{pmatrix} 1 & 0 \end{pmatrix}\begin{pmatrix} 1 \\ -1 \end{pmatrix} = 1 \quad \text{(first element of } \mathbf{z}\text{)}$$

$$\alpha_2 = \mathbf{e}_2^T\mathbf{z} = \begin{pmatrix} 0 & 1 \end{pmatrix}\begin{pmatrix} 1 \\ -1 \end{pmatrix} = -1 \quad \text{(second element of } \mathbf{z}\text{)}$$

(b) Consider $\mathbb{R}^2$ and take $\mathbf{x}_1 = \binom{1/\sqrt{2}}{1/\sqrt{2}}$, $\mathbf{x}_2 = \binom{1/\sqrt{2}}{-1/\sqrt{2}}$

$$\mathbf{z} = \alpha_1\mathbf{x}_1 + \alpha_2\mathbf{x}_2$$

$$\alpha_1 = \mathbf{x}_1^T\mathbf{z} = \begin{pmatrix} \frac{1}{\sqrt{2}} & \frac{1}{\sqrt{2}} \end{pmatrix}\begin{pmatrix} 1 \\ -1 \end{pmatrix} = 0$$

$$\alpha_2 = \mathbf{x}_2^T\mathbf{z} = \begin{pmatrix} \frac{1}{\sqrt{2}} & \frac{-1}{\sqrt{2}} \end{pmatrix}\begin{pmatrix} 1 \\ -1 \end{pmatrix} = \sqrt{2}$$

4.4.3 Biorthogonal expansions

Let $\{\mathbf{x}_1, \mathbf{x}_2, \ldots, \mathbf{x}_n\}$ be a set of n linearly independent column vectors and $\{\mathbf{y}_1^*, \mathbf{y}_2^*, \ldots, \mathbf{y}_n^*\}$ be another set of n linearly independent row vectors. Suppose that

$$\mathbf{y}_j^* \mathbf{x}_i = 0 \quad \text{if } i \neq j \tag{4.32}$$

These vectors satisfy the biorthogonality relation stated above. Now, let $\mathbf{z}$ be any vector and consider the expansion of $\mathbf{z}$ in terms of the set $\{\mathbf{x}_1, \mathbf{x}_2, \mathbf{x}_3, \ldots, \mathbf{x}_n\}$:

$$\mathbf{z} = \sum_{i=1}^{n} \alpha_i \mathbf{x}_i \tag{4.33}$$

To determine α_i, we multiply equation (4.33) on the left by $\mathbf{y}_j^* \implies$

$$\mathbf{y}_j^* \mathbf{z} = \sum_{i=1}^{n} \alpha_i \mathbf{y}_j^* \mathbf{x}_i$$

Again, due to the biorthogonality property, there is only one nonzero term (corresponding to $i = j$) in the sum and we get

$$\alpha_j = \frac{\mathbf{y}_j^* \mathbf{z}}{\mathbf{y}_j^* \mathbf{x}_j} \tag{4.34}$$

Substituting (4.34) in (4.33), we get the identity

$$\mathbf{z} = \sum_{j=1}^{n} \left(\frac{\mathbf{y}_j^* \mathbf{z}}{\mathbf{y}_j^* \mathbf{x}_j} \right) \mathbf{x}_j \tag{4.35}$$

Given the two sets of vectors $\{\mathbf{x}_i\}$ and $\{\mathbf{y}_j^*\}$, we can define a transformation that converts the vector $\mathbf{z}$ into the vector $\boldsymbol{\alpha}$ through equation (4.34). Given the vector $\boldsymbol{\alpha}$, we can recover $\mathbf{z}$ through equation (4.33). Thus, we have a transform:

$$\mathbf{z} \to \boldsymbol{\alpha}, \quad \alpha_i = \frac{\mathbf{y}_i^* \mathbf{z}}{\mathbf{y}_i^* \mathbf{x}_i}, \tag{4.36}$$

and an inverse transform:

$$\boldsymbol{\alpha} \to \mathbf{z}, \quad \mathbf{z} = \sum_{i=1}^{n} \alpha_i \mathbf{x}_i \tag{4.37}$$

This procedure may be used to decouple and solve many linear equations containing a square matrix $\mathbf{A}$. This is illustrated in the next section.

4.5 Solution of linear equations using eigenvector expansions

In this section, we show how the eigenvector expansions may be used to solve many types of linear equations containing a square matrix **A**. Here, we shall assume that the eigenvalues of **A** are simple, and hence there are n eigenvectors (and n eigenrows when **A** is not symmetric). The case of repeated eigenvalues will be considered in Chapter 6.

4.5.1 Solution of linear algebraic equations Au = b

Consider the solution of the linear system

$$\mathbf{A}\mathbf{u} = \mathbf{b} \tag{4.38}$$

where **A** is a square matrix of order n and **u**, **b** are $n \times 1$ vectors. First, we consider the case in which rank $\mathbf{A} = n$. Premultiply equation (4.38) by $\mathbf{y}_j^*$, $\Longrightarrow$

$$\mathbf{y}_j^* \mathbf{A}\mathbf{u} = \mathbf{y}_j^* \mathbf{b}$$

$\Longrightarrow$

$$\lambda_j \mathbf{y}_j^* \mathbf{u} = \mathbf{y}_j^* \mathbf{b} \tag{4.39}$$

[Remark: Taking the dot product of equation (4.38) with $\mathbf{y}_j$ or multiplying on the left by $\mathbf{y}_j^*$ and using the fact that $\mathbf{y}_j^*$ is a left eigenvector of **A**, decouples the equations.] Now, equation (4.39) $\Longrightarrow$

$$\mathbf{y}_j^* \mathbf{u} = \mathbf{y}_j^* \mathbf{b}/\lambda_j \quad (\text{assuming } \lambda_j \neq 0)$$

$\Longrightarrow$

$$\frac{\mathbf{y}_j^* \mathbf{u}}{\mathbf{y}_j^* \mathbf{x}_j} = \frac{\mathbf{y}_j^* \mathbf{b}}{\mathbf{y}_j^* \mathbf{x}_j} \frac{1}{\lambda_j}$$

$\Longrightarrow$

$$\mathbf{u} = \sum_{j=1}^{n} \left(\frac{\mathbf{y}_j^* \mathbf{b}}{\mathbf{y}_j^* \mathbf{x}_j} \right) \frac{1}{\lambda_j} \mathbf{x}_j \tag{4.40}$$

is the solution to the linear equations (4.38) in terms of the eigenvalues, eigenvectors and eigenrows of the matrix **A**. We shall see later that this formula is also applicable for many other types of linear equations. A special case of the above result for the case of a symmetric matrix **A** with normalized eigenvectors (orthonormal set) is

$$\mathbf{u} = \sum_{j=1}^{n} \frac{\langle \mathbf{b}, \mathbf{x}_j \rangle}{\lambda_j} \mathbf{x}_j \tag{4.41}$$

where $\langle \mathbf{b}, \mathbf{x}_j \rangle = \mathbf{x}_j^T \mathbf{b}$ is the standard dot (inner) product. The above solution in terms of the eigenvector expansion should be compared with that of direct solution methods (e. g., by Gaussian elimination). When n is large and the eigenvalues are well separated (e. g., $|\lambda_1| \ll |\lambda_2| \ll |\lambda_3| \cdots \ll |\lambda_n|$), only a few terms of the expansion may be sufficient to compute the solution to the desired accuracy (the extreme case being one eigenvalue being very small in magnitude compared to all others, requiring only one term in the expansion). In such cases, the eigenvector expansion is more useful than the direct solution method, especially for large values of n, where the number of operations required to solve the system varies as $\frac{1}{3}n^3$. A second application of the solution in terms of the eigenfunctions is in the development of the so-called multigrid methods for the solution of large sparse systems (obtained by discretization of Laplace–Poisson-type equations).

Example 4.8. Consider the symmetric 3×3 matrix

$$\begin{pmatrix} 4 & -1 & 0 \\ -1 & 4 & -1 \\ 0 & -1 & 4 \end{pmatrix}$$

that arises in the solution of discretized Laplace/Poisson equation. The eigenvalues are given by $\lambda_1 = 4 - \sqrt{2}, \lambda_2 = 4, \lambda_3 = 4 + \sqrt{2}$ while the corresponding normalized eigenvectors are

$$\mathbf{x}_1 = \begin{pmatrix} 1/2 \\ 1/\sqrt{2} \\ 1/2 \end{pmatrix}, \quad \mathbf{x}_2 = \begin{pmatrix} -1/\sqrt{2} \\ 0 \\ 1/\sqrt{2} \end{pmatrix}, \quad \mathbf{x}_3 = \begin{pmatrix} 1/2 \\ -1/\sqrt{2} \\ 1/2 \end{pmatrix}$$

Taking $\mathbf{b}^T = (1, 1, 1)$ and calculating the three terms in equation (4.41), we get

$$\mathbf{u} = \begin{pmatrix} 0.3301 \\ 0.4668 \\ 0.3301 \end{pmatrix} + \begin{pmatrix} 0.0 \\ 0.0 \\ 0.0 \end{pmatrix} + \begin{pmatrix} 0.0270 \\ -0.0382 \\ 0.0270 \end{pmatrix} = \begin{pmatrix} 0.3571 \\ 0.4286 \\ 0.3571 \end{pmatrix} = \begin{pmatrix} 5/14 \\ 3/7 \\ 5/14 \end{pmatrix}.$$

In this specific case, though the separation between the eigenvalues is not extreme ($\lambda_3/\lambda_1 = 2.09$), the first term still gives the solution to within about 10 % error.

4.5.2 (*Fredholm alternative*): solution of linear algebraic equations $\mathbf{Au} = \mathbf{b}$ when $\mathbf{A}$ is singular

If rank $\mathbf{A} < n$, then the solution given by equation (4.40) needs to be modified. Suppose that rank $\mathbf{A} = r$ ($r \geq 1$). Then we have seen that the homogeneous system

$$\mathbf{Au} = \mathbf{0}$$

has $(n-r)$ linearly independent solutions. This implies that $\mathbf{A}$ has $(n-r)$ eigenvectors corresponding to the zero eigenvalue. This implies that $\mathbf{A}$ has $(n-r)$ zero eigenvalues. Assume that $\lambda_{r+1} = \lambda_{r+2} = \cdots = \lambda_n = 0$. Then it follows from equation (4.39) that

$$\mathbf{y}_j^* \mathbf{b} = 0, \quad j = r+1, \ldots, n, \tag{4.42}$$

where $\mathbf{y}_j$ are the linearly independent solutions of the adjoint homogeneous system

$$\mathbf{A}^* \mathbf{y}_j = \mathbf{0}, \quad j = r+1, \ldots, n. \tag{4.43}$$

Equation (4.42) is another way of expressing the consistency of the linear system, i. e., if $\mathbf{A}$ has rank r ($< n$), then $\mathbf{Au} = \mathbf{b}$ is consistent iff $\mathbf{b}$ is orthogonal to the adjoint eigenvectors corresponding to the zero eigenvalue. In this case, the general solution of

$$\mathbf{Au} = \mathbf{b}$$

is of the form

$$\begin{aligned} \mathbf{u} &= \sum_{j=1}^{r} \left(\frac{\mathbf{y}_j^* \mathbf{b}}{\mathbf{y}_j^* \mathbf{x}_j} \right) \frac{1}{\lambda_j} \mathbf{x}_j + \sum_{j=r+1}^{n} c_j \mathbf{x}_j \\ &= \mathbf{u}_p + \mathbf{u}_h \end{aligned} \tag{4.44}$$

Here, $\mathbf{u}_h$ is the solution of the homogeneous equations $\mathbf{Au} = \mathbf{0}$ containing arbitrary constants c_j ($j = r+1, \ldots, n$) and $\mathbf{u}_p$ is a particular solution defined by the first summation in equation (4.44). [A common case that occurs in solving nonlinear equations near bifurcation points is $r = n-1$, corresponding to a simple zero eigenvalue of the linearized matrix. In this case, $\mathbf{u}_h$ is a scalar multiple of the eigenvector corresponding to the zero eigenvalue].

Example 4.9. Consider the linear system (from Example 3.6)

$$\begin{aligned} u_1 - 2u_2 - u_4 &= b_1 \\ -2u_1 + 3u_2 + 3u_3 &= b_2 \\ -u_2 + 3u_3 - 2u_4 &= b_3 \\ 3u_1 - 7u_2 + 3u_3 - 5u_4 &= b_4. \end{aligned}$$

We have seen that the rank of the 4×4 coefficient matrix is two, implying that it has two zero eigenvalues. We note that the adjoint homogeneous system

$$\mathbf{A}^T \mathbf{y} = \begin{pmatrix} 1 & -2 & 0 & 3 \\ -2 & 3 & -1 & -7 \\ 0 & 3 & 3 & 3 \\ -1 & 0 & -2 & -5 \end{pmatrix} \mathbf{y} = \mathbf{0}$$

has two linearly independent solutions

$$\mathbf{y}_1 = \begin{pmatrix} -5 \\ -1 \\ 0 \\ 1 \end{pmatrix}, \quad \mathbf{y}_2 = \begin{pmatrix} -2 \\ -1 \\ 1 \\ 0 \end{pmatrix}$$

Thus, the solvability conditions, equations (4.42) lead to the relations $b_4 = 5b_1 + b_2$ and $b_3 = 2b_1 + b_2$. Equivalently, the system is consistent and has solutions if and only if $\mathbf{b}$ is of the form

$$\mathbf{b} = b_1 \begin{pmatrix} 1 \\ 0 \\ 2 \\ 5 \end{pmatrix} + b_2 \begin{pmatrix} 0 \\ 1 \\ 1 \\ 1 \end{pmatrix}.$$

Taking $\mathbf{b}^T = (0, 1, 1, 1)$, the general solution of the inhomogeneous system may be written as

$$\mathbf{u} = c_1 \begin{pmatrix} 3 \\ 0 \\ 2 \\ 3 \end{pmatrix} + c_2 \begin{pmatrix} 0 \\ 1 \\ -1 \\ -2 \end{pmatrix} + \begin{pmatrix} -2 \\ -1 \\ 0 \\ 0 \end{pmatrix}.$$

4.5.3 Linear coupled first-order differential equations with constant coefficients

Consider the initial value problem

$$\frac{d\mathbf{u}}{dt} = \mathbf{A}\mathbf{u} \tag{4.45}$$

$$\mathbf{u} = \mathbf{u}_0 \ @\ t = 0 \tag{4.46}$$

Equation (4.45) defines a set of n coupled linear equations while (4.46) gives the initial conditions. For the case of $n = 2$, equation (4.45) in component form is

$$\frac{du_1}{dt} = a_{11}u_1 + a_{12}u_2$$

$$\frac{du_2}{dt} = a_{21}u_1 + a_{22}u_2$$

while the initial condition (4.46) takes the form

$$u_1 = u_{10}, \quad u_2 = u_{20} \ @\ t = 0.$$

To solve equations (4.45) and (4.46), we use the biorthogonal expansion. Multiply (4.45) on the left by $\mathbf{y}_j^* \Longrightarrow$

$$\mathbf{y}_j^* \frac{d\mathbf{u}}{dt} = \mathbf{y}_j^* \mathbf{A}\mathbf{u}$$

since $\mathbf{y}_j^*$ is a constant vector (independent of t)$\Longrightarrow$

$$\frac{d}{dt}(\mathbf{y}_j^* \mathbf{u}) = \lambda_j \mathbf{y}_j^* \mathbf{u}$$

This is a scalar differential equation for $\mathbf{y}_j^* \mathbf{u}$. Solving we get

$$\mathbf{y}_j^* \mathbf{u} = c_j e^{\lambda_j t}$$

At $t = 0$, $\mathbf{u} = \mathbf{u}_0 \Longrightarrow c_j = \mathbf{y}_j^* \mathbf{u}_0$

$\therefore$

$$\mathbf{y}_j^* \mathbf{u} = (\mathbf{y}_j^* \mathbf{u}_0) e^{\lambda_j t}$$

$\Longrightarrow$

$$\frac{\mathbf{y}_j^* \mathbf{u}}{\mathbf{y}_j^* \mathbf{x}_j} = \frac{(\mathbf{y}_j^* \mathbf{u}_0)}{\mathbf{y}_j^* \mathbf{x}_j} e^{\lambda_j t}$$

$\Longrightarrow$

$$\mathbf{u} = \sum_{j=1}^{n} \left(\frac{\mathbf{y}_j^* \mathbf{u}_0}{\mathbf{y}_j^* \mathbf{x}_j} \right) e^{\lambda_j t} \mathbf{x}_j \tag{4.47}$$

This is the formal solution. Let

$$\hat{c}_j = \frac{\mathbf{y}_j^* \mathbf{u}_0}{\mathbf{y}_j^* \mathbf{x}_j}$$

Then (4.47) gives

$$\mathbf{u} = \sum_{j=1}^{n} \hat{c}_j \mathbf{x}_j e^{\lambda_j t} \tag{4.48}$$

Thus, the solution is a linear combination of terms of the form $\mathbf{x}_j e^{\lambda_j t}$. Equivalently, the state of the system at any time is a linear combination of the eigenvectors. For this reason, the eigenvectors are also called the *fundamental modes* or basic (linearly independent) states of the system. Note that if $\mathbf{u}_0 = \alpha \mathbf{x}_i$ then equation (4.47) simplifies to

$$\mathbf{u} = \mathbf{u}_0 e^{\lambda_i t}$$

Thus, if the initial state corresponds to one of the eigenvectors then the system will be in that state at all later times. For this reason, the eigenspaces are called invariants for the flow (or trajectory) defined by equation (4.45). Note also that the reciprocal of the eigenvalue λ_j determines the time constant for the system evolution for this special initial condition.

4.5.4 Linear coupled inhomogeneous equations

Consider the initial value problem

$$\frac{d\mathbf{u}}{dt} = \mathbf{A}\mathbf{u} + \mathbf{b}, \quad t > 0$$
$$\mathbf{u} = \mathbf{u}_0 \,@\, t = 0$$

We first determine the steady-state solution by setting the time derivative to zero. Assuming that $\mathbf{A}$ is invertible,

$$\mathbf{A}\mathbf{u}_s + \mathbf{b} = 0 \Rightarrow \mathbf{u}_s = -\mathbf{A}^{-1}\mathbf{b}$$

Define

$$\begin{aligned} \mathbf{z} &= \mathbf{u} - \mathbf{u}_s \\ \frac{d\mathbf{z}}{dt} &= \mathbf{A}(\mathbf{u}_s + \mathbf{z}) + \mathbf{b} \\ &= \mathbf{A}\mathbf{z} \\ \mathbf{z} &= \mathbf{z}_0 \,@\, t = 0; \quad \mathbf{z}_0 = \mathbf{u}_0 - \mathbf{u}_s \end{aligned}$$

Using the result of equation (4.47), we have

$$\mathbf{z} = \sum_{j=1}^{n} \left(\frac{\mathbf{y}_j^* \mathbf{z}_0}{\mathbf{y}_j^* \mathbf{x}_j} \right) e^{\lambda_j t} \mathbf{x}_j$$

$$\Longrightarrow \mathbf{u} = \mathbf{u}_s + \sum_{j=1}^{n} \left(\frac{\mathbf{y}_j^* \mathbf{z}_0}{\mathbf{y}_j^* \mathbf{x}_j} \right) \mathbf{x}_j e^{\lambda_j t} \tag{4.49}$$

Physical interpretation of the solution (4.49):

If $\operatorname{Re}\lambda_j < 0$ for all j, then $\mathbf{u} \to \mathbf{u}_s$ for $t \to \infty$, i. e., the system approaches the steady-state. Suppose that the eigenvalues are all simple and real and arranged in the order

$$\lambda_1 < \lambda_2 < \lambda_3 < \cdots < \lambda_n$$

If $\lambda_n > 0$, the term containing $e^{\lambda_n t}$ will be the dominant term in the solution as it increases without bound for $t \to \infty$. If $\lambda_n < 0$ (and hence is smaller in magnitude than all other λ_i), then the term containing $e^{\lambda_n t}$ will determine the time taken by the system to approach the steady state. All other terms decay to zero more rapidly than this term. Thus, the time constant of the system (time required for the system to approach steady state with say < 5 % deviation $\approx \frac{3}{|\lambda_n|}$) is determined by the eigenvalue having the smallest magnitude.

Now consider the case of complex eigenvalues. If $\lambda_j = a_j + ib_j$,

$$e^{\lambda_j t} = e^{a_j t}(\cos b_j t + i \sin b_j t), \quad i = \sqrt{-1}$$

If $a_j < 0$, then the approach to steady state is oscillatory. Once again, the eigenvalue with the smallest real part (in absolute value) determines the time constant of the system.

4.5.5 A second-order vector initial value problem

The vibrations of many systems such as coupled point masses and springs, molecules and structures are described by the equations of the form

$$\mathbf{M}\frac{d^2\mathbf{u}}{dt^2} = -\mathbf{K}\mathbf{u} \tag{4.50}$$

$$\mathbf{u} = \mathbf{u}_0 \;@\; t = 0 \quad \text{(initial displacement)} \tag{4.51}$$

$$\frac{d\mathbf{u}}{dt} = \mathbf{v}_0 \;@\; t = 0 \quad \text{(initial velocity)} \tag{4.52}$$

where $\mathbf{M}$ is an $n \times n$ matrix (called the inertia matrix) and $\mathbf{K}$ is $n \times n$ matrix (called the stiffness matrix or matrix of spring constants) and $\mathbf{u}$ is the displacement vector. We assume that $\mathbf{M}$ is invertible and let $\mathbf{M}^{-1}\mathbf{K} = \mathbf{A}$,

$$\Longrightarrow \frac{d^2\mathbf{u}}{dt^2} = -\mathbf{A}\mathbf{u} \tag{4.53}$$

Multiplying (4.53) on the left by $\mathbf{y}_j^* \Longrightarrow$

$$\mathbf{y}_j^* \frac{d^2\mathbf{u}}{dt^2} = -\mathbf{y}_j^* \mathbf{A}\mathbf{u}$$

$$\frac{d^2}{dt^2}(\mathbf{y}_j^*\mathbf{u}) = -\lambda_j \mathbf{y}_j^* \mathbf{u}$$

$\Longrightarrow$

$$\mathbf{y}_j^*\mathbf{u} = c_{1j} \sin \sqrt{\lambda_j} t + c_{2j} \cos \sqrt{\lambda_j} t \tag{4.54}$$

$$\mathbf{u} = \mathbf{u}_0 \;@\; t = 0 \Longrightarrow c_{2j} = \mathbf{y}_j^* \mathbf{u}_0$$

$$\frac{d\mathbf{u}}{dt} = \mathbf{v}_0 \;@\; t = 0 \Longrightarrow c_{1j}\sqrt{\lambda_j} = \mathbf{y}_j^* \mathbf{v}_0$$

∴

$$\mathbf{y}_j^*\mathbf{u} = (\mathbf{y}_j^*\mathbf{v}_0)\frac{\sin\sqrt{\lambda_j}t}{\sqrt{\lambda_j}} + (\mathbf{y}_j^*\mathbf{u}_0)\cos\sqrt{\lambda_j}t$$

$$\frac{\mathbf{y}_j^*\mathbf{u}}{\mathbf{y}_j^*\mathbf{x}_j} = \left(\frac{\mathbf{y}_j^*\mathbf{v}_0}{\mathbf{y}_j^*\mathbf{x}_j}\right)\frac{\sin\sqrt{\lambda_j}t}{\sqrt{\lambda_j}} + \left(\frac{\mathbf{y}_j^*\mathbf{u}_0}{\mathbf{y}_j^*\mathbf{x}_j}\right)\cos\sqrt{\lambda_j}t$$

$$\mathbf{u} = \sum_{j=1}^{n}\left[\left(\frac{\mathbf{y}_j^*\mathbf{v}_0}{\mathbf{y}_j^*\mathbf{x}_j}\right)\frac{\sin\sqrt{\lambda_j}t}{\sqrt{\lambda_j}} + \left(\frac{\mathbf{y}_j^*\mathbf{u}_0}{\mathbf{y}_j^*\mathbf{x}_j}\right)\cos\sqrt{\lambda_j}t\right]\mathbf{x}_j \tag{4.55}$$

This is the formal solution to the initial value problem defined by equations (4.50) to (4.52).

Remarks and physical interpretation

1. Suppose that the initial conditions are given by

$$\mathbf{u}^0 = \alpha\mathbf{x}_i, \quad \mathbf{v}^0 = \mathbf{0}$$

   Then (4.55) gives

$$\mathbf{u} = (\cos\sqrt{\lambda_i}t)\mathbf{u}^0 \tag{4.56}$$

   Thus, if the system is initially in a state corresponding to eigenvector $\mathbf{x}_i$ then it will be in that state for all $t > 0$. Also, the solution given by equation (4.56) is periodic with a period $T = \frac{2\pi}{\sqrt{\lambda_i}}$. Equivalently, the frequency of vibration is given by $f_i = \frac{\sqrt{\lambda_i}}{2\pi}$. Therefore, the eigenvalues of **A** give the frequencies of oscillation (vibration) while the eigenvectors correspond to pure modes of vibration.

2. The solution of the inhomogeneous system

$$\frac{d^2\mathbf{u}}{dt^2} = -\mathbf{A}\mathbf{u} + \mathbf{b}$$

$$@\, t = 0, \quad \begin{cases} \mathbf{u} = \mathbf{u}^0 \\ \frac{d\mathbf{u}}{dt} = \mathbf{v}^0 \end{cases}$$

   is given by

$$\mathbf{u} = \mathbf{u}_s + \sum_{j=1}^{n}\left[\left(\frac{\mathbf{y}_j^*\mathbf{v}^0}{\mathbf{y}_j^*\mathbf{x}_j}\right)\frac{\sin\sqrt{\lambda_j}t}{\sqrt{\lambda_j}} + \left(\frac{\mathbf{y}_j^*\mathbf{z}^0}{\mathbf{y}_j^*\mathbf{x}_j}\right)\cos\sqrt{\lambda_j}t\right]\mathbf{x}_j \tag{4.57}$$

where

$$\mathbf{u}_s = \mathbf{A}^{-1}\mathbf{b} \quad \text{and} \quad \mathbf{z}^0 = \mathbf{u}^0 - \mathbf{u}_s$$

is the equilibrium solution.

4.5.6 Multicomponent diffusion and reaction in a catalyst pore

The problem of multicomponent diffusion and reaction in a catalyst pore (with no radial and only longitudinal gradients) or a slab (or plate) of catalyst is described by the vector boundary value problem

$$\mathbf{D}\frac{d^2\mathbf{c}}{ds^2} = \mathbf{K}\mathbf{c}, \quad 0 < s < L$$

with boundary conditions

$$\begin{aligned} \mathbf{c} &= \mathbf{c}_0 \ @\ s = 0 \quad \text{(bulk or surface condition)} \\ \frac{d\mathbf{c}}{ds} &= \mathbf{0} \ @\ s = L \quad \text{(no flux through pore end or mid-plane symmetry)} \end{aligned}$$

Here, $\mathbf{K}$ is the matrix of first-order rate constants, $\mathbf{c}$ is the vector of species concentrations and $\mathbf{D}$ is a (positive definite matrix) of diffusivities. Define

$$\xi = \frac{s}{L}, \quad \mathbf{A} = \mathbf{D}^{-1}\mathbf{K}L^2 (= \mathbf{\Phi}^2; \mathbf{\Phi} = \text{Thiele matrix})$$

$$\Longrightarrow \frac{d^2\mathbf{c}}{d\xi^2} = \mathbf{A}\mathbf{c} \tag{4.58}$$

$$\mathbf{c} = \mathbf{c}_0 \ @\ \xi = 0 \tag{4.59}$$

$$\frac{d\mathbf{c}}{d\xi} = \mathbf{0} \ @\ \xi = 1 \tag{4.60}$$

Let $\{\lambda_1, \lambda_2, \ldots, \lambda_n\}$ be the eigenvalues, $\{\mathbf{x}_1, \mathbf{x}_2, \ldots, \mathbf{x}_n\}$ be the eigenvectors and $\{\mathbf{y}_1^*, \mathbf{y}_2^*, \ldots, \mathbf{y}_n^*\}$ be the eigenrows of $\mathbf{A}$, respectively. Then, from equation (4.58) we obtain after pre-multiplying by $\mathbf{y}_j^*$,

$$\frac{d^2}{d\xi^2}(\mathbf{y}_j^*\mathbf{c}) = \mathbf{y}_j^*\mathbf{A}\mathbf{c} = \lambda_j(\mathbf{y}_j^*\mathbf{c})$$

$$\mathbf{y}_j^*\mathbf{c} = \alpha_{1j}\cosh\sqrt{\lambda_j}\xi + \alpha_{2j}\cosh\sqrt{\lambda_j}(1-\xi)$$

Equation (4.60) gives

$$\alpha_{1j} = 0$$

Boundary condition (4.59) $\Longrightarrow \alpha_{2j} = (\mathbf{y}_j^* \mathbf{c}_0)/\cosh\sqrt{\lambda_j}$

$$\Longrightarrow \mathbf{c}(\xi) = \sum_{j=1}^{n} \left(\frac{\mathbf{y}_j^* \mathbf{c}_0}{\mathbf{y}_j^* \mathbf{x}_j} \right) \frac{\cosh\sqrt{\lambda_j}(1-\xi)}{\cosh\sqrt{\lambda_j}} \mathbf{x}_j \tag{4.61}$$

This is the formal solution to the concentration vector in the multicomponent diffusion–reaction problem defined by equations (4.58) to (4.60). This solution may be used to determine the average reaction rate in the pore in terms of the bulk concentrations. The average (or observed) reaction rate vector is defined by

$$\begin{aligned}
\mathbf{r_{obs}} &= \frac{1}{L} \int_0^L \mathbf{K}\mathbf{c}(s)\, ds \\
&= \frac{\mathbf{D}}{L^2} \left(- \frac{d\mathbf{c}}{d\xi}\bigg|_{\xi=0} \right) \\
&= \frac{\mathbf{D}}{L^2} \sum_{j=1}^{n} \frac{\mathbf{y}_j^* \mathbf{c}_0}{\mathbf{y}_j^* \mathbf{x}_j} (\sqrt{\lambda_j} \tanh\sqrt{\lambda_j}) \mathbf{x}_j
\end{aligned} \tag{4.62}$$

The *diffusion disguised rate constant matrix* $\mathbf{K}^*$ is defined by

$$\mathbf{r}_{\text{obs}} = \mathbf{K}^* \mathbf{c}_0 \tag{4.63}$$

Comparing equations (4.62) and (4.63), we get

$$\begin{aligned}
\mathbf{K}^* &= \frac{\mathbf{D}}{L^2} \sum_{j=1}^{n} \frac{\mathbf{x}_j \mathbf{y}_j^*}{\mathbf{y}_j^* \mathbf{x}_j} (\sqrt{\lambda_j} \tanh\sqrt{\lambda_j}) \\
&= \frac{\mathbf{D}}{L^2} \sqrt{\mathbf{D}^{-1}\mathbf{K}L^2} \tanh(\sqrt{\mathbf{D}^{-1}\mathbf{K}L^2})
\end{aligned} \tag{4.64}$$

The second equality follows from the spectral theorem to be discussed in the next chapter. In terms of the Thiele matrix, equation (4.64) may be expressed as

$$\mathbf{K}^* = \frac{\mathbf{D}}{L^2} \mathbf{\Phi} \tanh(\mathbf{\Phi}) \tag{4.65}$$

It follows from equation (4.64) that when the pore diffusional effects are negligible ($L \to 0$) the diffusion disguised rate constant matrix is equal to the true rate constant matrix ($\mathbf{K}^* = \mathbf{K}$) while for the case of strong pore diffusional limitations ($L \to \infty$) or more precisely $|\lambda_j| \gg 1$ for all j, we have $\mathbf{K}^* = \frac{1}{L}\mathbf{D}\sqrt{\mathbf{D}^{-1}\mathbf{K}}$. (Remark: The square root of a matrix is uniquely defined only for positive definite matrices. See Chapter 7 for further detail. Also note that $\mathbf{D}\sqrt{\mathbf{D}^{-1}\mathbf{K}} \neq \sqrt{\mathbf{D}\mathbf{K}}$ unless $\mathbf{D}$ is a scalar multiple times the identity matrix).

4.6 Diagonalization of matrices and similarity transforms

Definition. Let $\mathbf{A}$ and $\mathbf{B}$ be two square matrices of order n. They are called similar if ∃ an invertible matrix $\mathbf{T}$ such that

$$\mathbf{B} = \mathbf{T}\mathbf{A}\mathbf{T}^{-1} \tag{4.66}$$

From this definition, the following properties may be established:

1. If $\mathbf{B}$ is similar to $\mathbf{A}$, then $\mathbf{A}$ is similar to $\mathbf{B}$,

$$\mathbf{A} = \mathbf{T}^{-1}\mathbf{B}\mathbf{T} \tag{4.67}$$

2. Similar matrices have the same eigenvalues. To prove this, let λ be an eigenvalue of $\mathbf{A}$ and $\mathbf{x}$ be the corresponding eigenvector. Then

$$\mathbf{A}\mathbf{x} = \lambda\mathbf{x} \tag{4.68}$$

   $\mathbf{A}$ and $\mathbf{B}$ are similar $\Longrightarrow$

$$\mathbf{T}^{-1}\mathbf{B}\mathbf{T}\mathbf{x} = \lambda\mathbf{x}$$

   Since $\mathbf{T}$ is nonsingular, we can rewrite this expression as

$$\mathbf{B}(\mathbf{T}\mathbf{x}) = \lambda(\mathbf{T}\mathbf{x})$$

   or

$$\mathbf{B}\mathbf{y} = \lambda\mathbf{y}, \quad \mathbf{y} = \mathbf{T}\mathbf{x} \tag{4.69}$$

   Thus, if λ is an eigenvalue of $\mathbf{A}$ with eigenvector $\mathbf{x}$, then λ is also an eigenvalue of $\mathbf{B}$ with eigenvector $\mathbf{T}\mathbf{x}$.

3. If $\mathbf{A}$ has n distinct eigenvalues, then it is similar to a diagonal matrix. Let

$$\mathbf{T} = (\mathbf{x}_1, \mathbf{x}_2, \ldots, \mathbf{x}_n) = \text{modal matrix},$$

   whose columns are eigenvectors of $\mathbf{A}$. It may be shown (using biorthogonality property) that

$$\mathbf{T}^{-1}\mathbf{A}\mathbf{T} = \mathbf{\Lambda} \quad \text{(diagonal matrix)}.$$

   In applications, similar matrices appear when the same physical phenomenon (described by linear equations) is modeled using different coordinate systems (not necessarily orthogonal). The eigenvalues (which represent the system time constants or frequencies) do not change but the eigenvectors change with the coordinate system used. The matrix $\mathbf{T}$ usually defines the relationship between the two coordinate systems.

4.6.1 Examples of similarity transforms

Consider the 3-tank interacting system shown in Figure 4.3. The model describing the transient behavior of this system may be expressed as

$$\mathbf{V}_T \frac{d\mathbf{c}}{dt'} = \mathbf{Q}\mathbf{c}, \tag{4.70}$$

where $\mathbf{V}_T$ is the diagonal capacitance matrix (of tank volumes) and $\mathbf{Q}$ is the (symmetric) matrix of exchange flows. With proper normalization of time and dividing by the respective capacitances, the dimensionless form of the model may be written as

$$\frac{d\mathbf{c}}{dt} = \mathbf{A}\mathbf{c}, \quad t > 0 \quad \text{and} \quad \mathbf{c} = \mathbf{c}^0 \ @\, t = 0, \tag{4.71}$$

$$\mathbf{A} = \begin{pmatrix} -2 & 1 & 1 \\ 1/2 & -1 & 1/2 \\ 1 & 1 & -2 \end{pmatrix}. \tag{4.72}$$

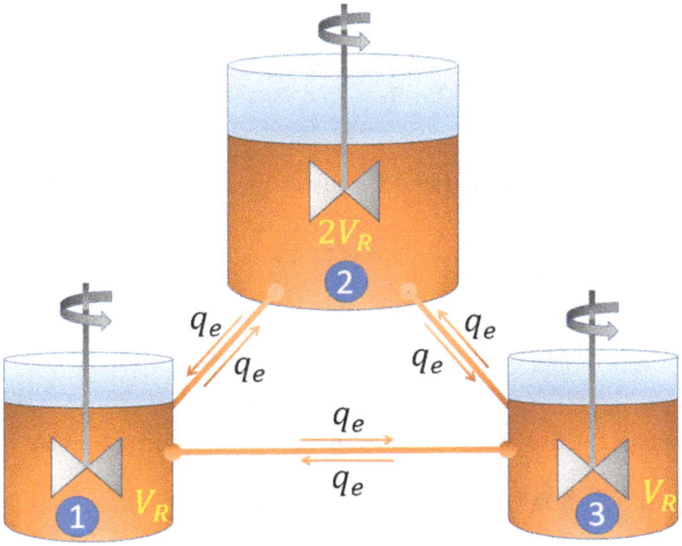

Figure 4.3: Three interacting tanks: Configuration 1.

[Note that even though $\mathbf{Q}$ is symmetric, the matrix $\mathbf{A}$ is not symmetric.] If we label the tanks differently, i. e., denoting the large tank by subscript 1 as shown in Figure 4.4, the system is now described by

$$\frac{d\mathbf{c}}{dt} = \mathbf{B}\mathbf{c}, \quad t > 0 \quad \text{and} \quad \mathbf{c} = \mathbf{c}^0 \ @\, t = 0, \tag{4.73}$$

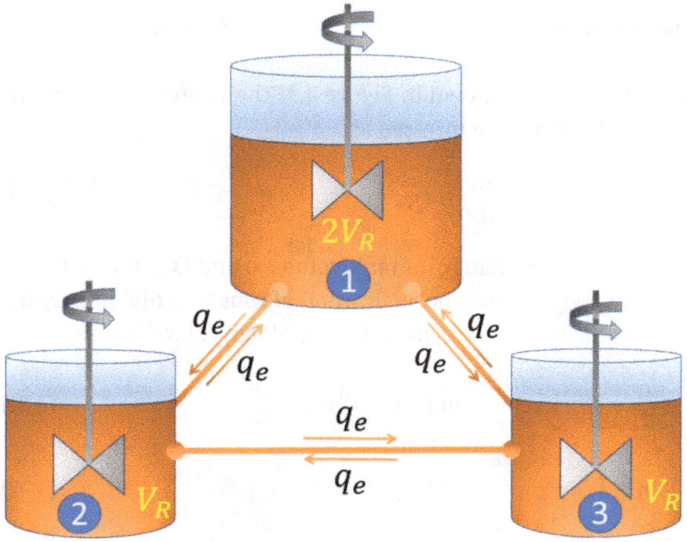

Figure 4.4: Three interacting tanks: Configuration 2.

$$\mathbf{B} = \begin{pmatrix} -1 & 1/2 & 1/2 \\ 1 & -2 & 1 \\ 1 & 1 & -2 \end{pmatrix}. \tag{4.74}$$

It can be seen that **A** and **B** are similar matrices. They have the same eigenvalues but different eigenvectors.

The difference between the two cases is flipping of tanks 1 and 2. Thus, if we take

$$\mathbf{T} = \begin{pmatrix} 0 & 1 & 0 \\ 1 & 0 & 0 \\ 0 & 0 & 1 \end{pmatrix} = \mathbf{T}^{-1}, \tag{4.75}$$

which is obtained by flipping the first and second row of the identity matrix $\mathbf{I}_3$, then

$$\begin{aligned} \mathbf{T}^{-1}\mathbf{A}\mathbf{T} &= \begin{pmatrix} 0 & 1 & 0 \\ 1 & 0 & 0 \\ 0 & 0 & 1 \end{pmatrix} \begin{pmatrix} -2 & 1 & 1 \\ 1/2 & -1 & 1/2 \\ 1 & 1 & -2 \end{pmatrix} \begin{pmatrix} 0 & 1 & 0 \\ 1 & 0 & 0 \\ 0 & 0 & 1 \end{pmatrix} \\ &= \begin{pmatrix} -1 & 1/2 & 1/2 \\ 1 & -2 & 1 \\ 1 & 1 & -2 \end{pmatrix} = \mathbf{B}. \end{aligned}$$

In addition, we can relate the eigenvectors as follows:

$$\mathbf{A}\mathbf{x} = \lambda\mathbf{x} \Longrightarrow \mathbf{T}\mathbf{A}\mathbf{x} = \lambda\mathbf{T}\mathbf{x} \tag{4.76}$$

If we assume $\mathbf{y} = \mathbf{T}\mathbf{x} \Longleftrightarrow \mathbf{x} = \mathbf{T}^{-1}\mathbf{y}$, and we can write

$$\mathbf{T}\mathbf{A}\mathbf{T}^{-1}\mathbf{y} = \lambda\mathbf{y} \Longrightarrow \mathbf{B}\mathbf{y} = \lambda\mathbf{y} \tag{4.77}$$

In other words, eigenvalues of $\mathbf{A}$ and $\mathbf{B}$ are the same but eigenvectors are related by flipping the first and second element of $\mathbf{x}$. The eigenvalues of $\mathbf{A}$ are given by $\lambda_1 = 0$, $\lambda_2 = -2$ and $\lambda_3 = -3$ with corresponding eigenvectors $\mathbf{x}_1 = \begin{pmatrix} 1 \\ 1 \\ 1 \end{pmatrix}$, $\mathbf{x}_2 = \begin{pmatrix} 1 \\ -1 \\ 1 \end{pmatrix}$ and $\mathbf{x}_3 = \begin{pmatrix} 1 \\ 0 \\ -1 \end{pmatrix}$. Similarly, the eigenvalues of $\mathbf{B}$ are the same as $\mathbf{A}$ but corresponding eigenvectors have flipped first and second element as $\mathbf{y}_1 = \begin{pmatrix} 1 \\ 1 \\ 1 \end{pmatrix}$, $\mathbf{y}_2 = \begin{pmatrix} -1 \\ 1 \\ 1 \end{pmatrix}$ and $\mathbf{y}_3 = \begin{pmatrix} 0 \\ 1 \\ -1 \end{pmatrix}$.

Remark. Though we use the same symbol, the matrix $\mathbf{T}$ in the above example is not the modal matrix. Here, it relates the two coordinate systems.

4.6.2 Canonical form

Now, consider again the initial value problem

$$\frac{d\mathbf{c}}{dt} = \mathbf{A}\mathbf{c}, \quad t > 0 \quad \text{and} \quad \mathbf{c} = \mathbf{c}^0 \ @\ t = 0.$$

If we change the coordinates by using the transform

$$\mathbf{c} = \mathbf{T}\hat{\mathbf{c}}; \quad \hat{\mathbf{c}} = \mathbf{T}^{-1}\mathbf{c} \tag{4.78}$$

where $\mathbf{T} = (\mathbf{x}_1, \mathbf{x}_2, \mathbf{x}_3) = \begin{pmatrix} 1 & 1 & 1 \\ 1 & -1 & 0 \\ 1 & 1 & -1 \end{pmatrix}$ is the *modal* matrix, i. e., matrix whose columns are eigenvectors $\mathbf{x}_j$ of $\mathbf{A}$. Then

$$\begin{aligned} \frac{d\mathbf{c}}{dt} &= \mathbf{A}\mathbf{c} \Longrightarrow \mathbf{T}\frac{d\hat{\mathbf{c}}}{dt} = \mathbf{A}\mathbf{T}\hat{\mathbf{c}} \\ \Longrightarrow \frac{d\hat{\mathbf{c}}}{dt} &= \mathbf{T}^{-1}\mathbf{A}\mathbf{T}\hat{\mathbf{c}} = \Lambda\hat{\mathbf{c}}, \end{aligned} \tag{4.79}$$

which is in canonical (decoupled) form:

$$\begin{aligned} \frac{d\hat{c}_1}{dt} &= 0, \\ \frac{d\hat{c}_2}{dt} &= -2\hat{c}_2 \\ \frac{d\hat{c}_3}{dt} &= -3\hat{c}_3 \end{aligned} \tag{4.80}$$

Here, $\hat{c}_i$ are the canonical variables in which the original system becomes diagonal. Thus, the solution can be expressed in these canonical variables as

$$\hat{c}_1 = \hat{c}_{10}$$

$$
\begin{aligned}
\widehat{c}_2 &= \widehat{c}_{20}e^{-2t} \\
\widehat{c}_3 &= \widehat{c}_{30}e^{-3t}.
\end{aligned}
\tag{4.81}
$$

Further, since the rows of $\mathbf{T}^{-1}$ define the left eigenvectors, each left eigenvector defines a canonical variable:

$$
\mathbf{T}^{-1} = \begin{pmatrix} \mathbf{y}_1^T \\ \mathbf{y}_2^T \\ \mathbf{y}_3^T \end{pmatrix} = \frac{1}{4}\begin{pmatrix} 1 & 2 & 1 \\ 1 & -2 & 1 \\ 2 & 0 & -2 \end{pmatrix}
$$

where $\mathbf{y}_i^T$ are (normalized) eigenrows of matrix $\mathbf{A}$. Thus,

$$
\widehat{\mathbf{c}} = \mathbf{T}^{-1}\mathbf{c} = \frac{1}{4}\begin{pmatrix} 1 & 2 & 1 \\ 1 & -2 & 1 \\ 2 & 0 & -2 \end{pmatrix}\begin{pmatrix} c_1 \\ c_2 \\ c_3 \end{pmatrix},
$$

which leads to

$$
\begin{aligned}
\widehat{c}_1 &= \frac{1}{4}(c_1 + 2c_2 + c_3) \\
\widehat{c}_2 &= \frac{1}{4}(c_1 - 2c_2 + c_3) \\
\widehat{c}_3 &= \frac{1}{2}(c_1 - c_3).
\end{aligned}
\tag{4.82}
$$

Similarly,

$$
\begin{aligned}
\widehat{c}_{10} &= \frac{1}{4}(c_{10} + 2c_{20} + c_{30}) \\
\widehat{c}_{20} &= \frac{1}{4}(c_{10} - 2c_{20} + c_{30}) \\
\widehat{c}_{30} &= \frac{1}{2}(c_{10} - c_{30})
\end{aligned}
$$

Thus, we can also write

$$
\begin{aligned}
c_1 + 2c_2 + c_3 &= 4\widehat{c}_1 = 4\widehat{c}_{10} = c_{10} + 2c_{20} + c_{30} \\
(c_1 - 2c_2 + c_3) &= 4\widehat{c}_2 = 4\widehat{c}_{20}e^{-2t} = (c_{10} - 2c_{20} + c_{30})e^{-2t} \\
(c_1 - c_3) &= 2\widehat{c}_1 = 2\widehat{c}_{30}e^{-3t} = (c_{10} - c_{30})e^{-3t}.
\end{aligned}
$$

These relations also define the initial conditions in the phase space for which the transient behavior is confined to a subspace. For example, if $c_{10} = c_{30}$, the concentration vector at any time is in the subspace spanned by the first two eigenvectors. Since the system is decoupled in the variables $\widehat{c}_j$, we refer to them as the "canonical variables."

4.6.3 Similarity transform when $\mathbf{A}^T = \mathbf{A}$

When $\mathbf{A}$ is real and symmetric, the eigenvalues are real and eigenvectors can be normalized to have unit length. They form an orthonormal set, i. e.,

$$\mathbf{x}_i^T \mathbf{x}_j = \delta_{ij}$$

In this case, the modal matrix is an orthogonal matrix, i. e.,

$$\mathbf{T}\mathbf{T}^{-1} = \mathbf{T}^{-1}\mathbf{T} = \mathbf{I} \quad \text{with} \quad \mathbf{T}^{-1} = \mathbf{T}^T,$$

and represents either a rotation of the axes, i. e., the two coordinate systems are either related by a pure rotation of the axes (about the origin) or a combination of rotations and reflections.

Example 4.10. Consider a 2×2 real symmetric matrix $\mathbf{A} = \begin{pmatrix} -1 & 1 \\ 1 & -1 \end{pmatrix}$. The eigenvalues of this matrix are real: $\lambda_1 = 0$ and $\lambda_2 = -2$ corresponding to normalized eigenvectors $\mathbf{x}_1 = \frac{1}{\sqrt{2}}\begin{pmatrix} 1 \\ 1 \end{pmatrix}$ and $\mathbf{x}_2 = \frac{1}{\sqrt{2}}\begin{pmatrix} 1 \\ -1 \end{pmatrix}$. Thus, the modal matrix is given by

$$\mathbf{T} = (\mathbf{x}_1 \ \mathbf{x}_2) = \frac{1}{\sqrt{2}}\begin{pmatrix} 1 & 1 \\ 1 & -1 \end{pmatrix} \tag{4.83}$$

$\Longrightarrow$

$$\mathbf{T}^{-1} = \frac{1}{\sqrt{2}}\begin{pmatrix} 1 & 1 \\ 1 & -1 \end{pmatrix} = \mathbf{T}^T. \tag{4.84}$$

Now, the initial value problem

$$\frac{d\mathbf{c}}{dt} = \mathbf{A}\mathbf{c} \tag{4.85}$$

can be converted into canonical form as discussed earlier by using the transform

$$\mathbf{c} = \mathbf{T}\widehat{\mathbf{c}} \Longrightarrow \widehat{\mathbf{c}} = \mathbf{T}^{-1}\mathbf{c}$$

$\Longrightarrow$

$$\widehat{c}_1 = \frac{c_1 + c_2}{\sqrt{2}}$$
$$\widehat{c}_2 = \frac{c_1 - c_2}{\sqrt{2}}.$$

We note that this transformation represents a 45° counterclockwise rotation of axes as shown in Figure 4.5.

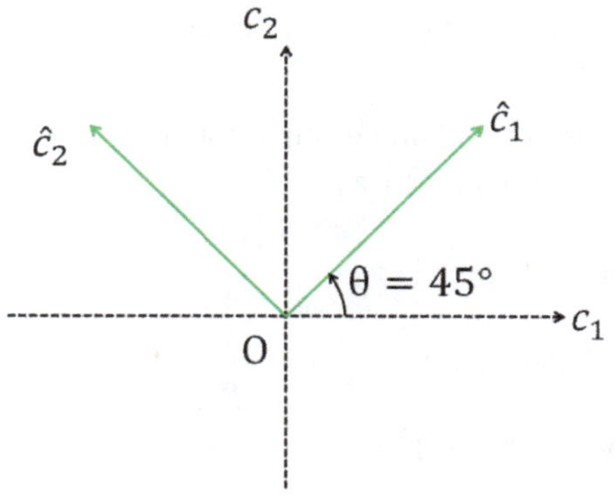

Figure 4.5: Canonical transform and rotation of axes for initial value problems.

For example, in $\mathbb{R}^2$, the matrix

$$\mathrm{Rot}(\theta) = \begin{pmatrix} \cos\theta & -\sin\theta \\ \sin\theta & \cos\theta \end{pmatrix} \tag{4.86}$$

represents a (counterclockwise) rotation about the origin by an angle θ, while the matrix

$$\mathrm{Ref}(\phi) = \begin{pmatrix} \cos 2\phi & \sin 2\phi \\ \sin 2\phi & -\cos 2\phi \end{pmatrix} \tag{4.87}$$

represents a reflection about a line through the origin that makes an angle ϕ with the x-axis. The set of all 2×2 orthogonal matrices describing rotations and reflections form an orthogonal group, denoted by $O(2)$. As discussed in Part II, orthogonal or unitary matrices represent rotations and reflections.

Problems

1. (a) Given the matrix

$$A = \begin{pmatrix} -3 & 0 & 2 \\ 1 & -2 & 1 \\ 1 & 1 & -4 \end{pmatrix}$$

   i. Determine the eigenvalues.
   ii. Determine the eigenvectors and a set of eigenrows (adjoint eigenvectors) and verify by direct computation that they form two biorthogonal sets.
   iii. Show by direct computation that the sets of eigenvectors and eigenrows are linearly independent.

(b) Given the matrix

$$A = \begin{pmatrix} -1 & 2 & 2 \\ -5 & -3 & -1 \\ -3 & -2 & -2 \end{pmatrix}$$

(c) Repeat parts (i), (ii) and (iii).

2. Given the matrix

$$A = \begin{pmatrix} -3 & 0 & 2 \\ 1 & -2 & 1 \\ 1 & 1 & -4 \end{pmatrix}$$

(a) Solve the initial value problem

$$\frac{d\mathbf{u}}{dt} = \mathbf{Au}; \quad \mathbf{u}(t=0) = \begin{pmatrix} 1 \\ 0 \\ -1 \end{pmatrix}$$

(b) Solve the initial value problem

$$\frac{d^2\mathbf{y}}{dt^2} = \mathbf{Ay}; \quad \mathbf{y}(\mathbf{0}) = \begin{pmatrix} 2 \\ 2 \\ 0 \end{pmatrix} \quad \frac{d\mathbf{y}}{dt}(0) = \begin{pmatrix} 0 \\ 0 \\ 0 \end{pmatrix}$$

3. Consider the system of first-order reactions (shown in Figure 4.6) occurring in a batch reactor of constant volume:

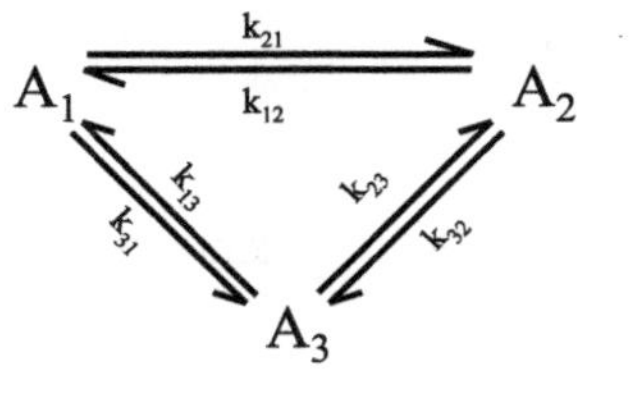

Figure 4.6: Monomolecular reaction network.

$$k_{12} = 0.5s^{-1} \quad k_{21} = 0.25s^{-1} \quad k_{13} = 0.2s^{-1}$$
$$k_{31} = 0.05s^{-1} \quad k_{23} = 0.3s^{-1} \quad k_{32} = 0.15s^{-1}$$

(a) Formulate the differential equations that describe the evolution of the concentration vector.

(b) Determine the eigenvalues and eigenvectors of the matrix in (a).

(c) Give a physical interpretation to the eigenvalues and eigenvectors (Hint: Consider what happens when the initial concentration vector is equal to the equilibrium vector plus a constant multiple of one of the other two eigenvectors).

4. Consider the vibration of the spring-mass system shown in Figure 4.7. Suppose that the masses are equal and the springs are identical.

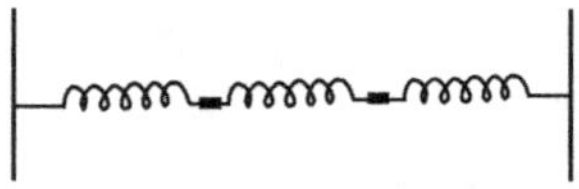

Figure 4.7: Schematic of spring-mass system.

(a) Formulate the Newton's equations of motion and cast them in dimensionless form.
(b) Determine the eigenvalues and eigenvectors of the matrix and give a physical interpretation. Sketch the different modes of vibration.
(c) Generalize the above results to N equal masses with identical springs.

5. The thermal conductivity tensor of an anisotropic solid is given by

$$\mathbf{K} = \begin{pmatrix} 6 & 2 & -2 \\ 2 & 6 & -2 \\ -2 & -2 & 10 \end{pmatrix}$$

(a) Determine the principal conductivities (eigenvalues) and the principal axes of conductivity (eigenvectors).
(b) Write the expanded form of the heat conduction equation

$$\rho c \frac{\partial T}{\partial t} = \nabla.(\mathbf{K}. \nabla T)$$

in the two coordinate systems.
(c) Show that with proper scaling and transformations, the heat conduction equation can be reduced to the form

$$\rho c \frac{\partial T}{\partial t} = k \nabla^2 T$$

where k is a scalar.

6. (a) A real square matrix $\mathbf{A}$ is called *idempotent* if $\mathbf{A}^2 = \mathbf{A}$. Show that the eigenvalues of an idempotent matrix are equal to either 0 or 1.
(b) A real square matrix $\mathbf{A}$ is called *orthogonal* if $\mathbf{A}^{-1} = \mathbf{A}^t$ (transpose of $\mathbf{A}$). Show that the *real eigenvalues* of an orthogonal matrix are equal ± 1. Show also that the complex eigenvalues (if any) of an orthogonal matrix have an absolute value of unity.

(c) A complex square matrix $\mathbf{A}$ is called *unitary* if $\mathbf{A}^{-1} = \mathbf{A}^*$ (conjugate transpose of $\mathbf{A}$). Prove that the eigenvalues of a unitary matrix are on the unit circle in the complex plane.

(d) A complex square matrix $\mathbf{A}$ is called *skew-Hermitian* if $\mathbf{A}^* = -\mathbf{A}$. Prove that the eigenvalues of a skew-Hermitian matrix are purely imaginary.

(e) A complex square matrix $\mathbf{A}$ is called *positive definite* if $\mathbf{A} = \mathbf{S}^*\mathbf{S}$ where $\mathbf{S}$ is nonsingular ($\det \mathbf{S} \neq 0$). Prove that any eigenvalue of $\mathbf{A}$ must be real and positive.

Remark. The above special matrices play a very important role in group theory, tensor analysis, numerical computation of eigenvalues, etc.

7. Let $\mathbf{A}$ and $\mathbf{B}$ be real $k \times k$ matrices and consider the matrices

$$\mathbf{L}_1 = \begin{pmatrix} \mathbf{A} & \mathbf{B} \\ \mathbf{B} & \mathbf{A} \end{pmatrix}$$

$$\mathbf{L}_2 = \begin{pmatrix} \mathbf{A} & \mathbf{B} & \mathbf{B} \\ \mathbf{B} & \mathbf{A} & \mathbf{B} \\ \mathbf{B} & \mathbf{B} & \mathbf{A} \end{pmatrix}$$

(a) Show that the eigenvalues of the $2k \times 2k$ matrix $\mathbf{L}_1$ are the same as those of $k \times k$ matrices $(\mathbf{A} + \mathbf{B})$ and $(\mathbf{A} - \mathbf{B})$.

(b) Show that the eigenvalues of the $3k \times 3k$ matrix $\mathbf{L}_2$ are the same as those of $(\mathbf{A} + 2\mathbf{B})$ and $(\mathbf{A} - \mathbf{B})$ (repeated twice).

(c) Discuss how you would find the eigenvectors of $\mathbf{L}_1$ and $\mathbf{L}_2$ from the lower-dimensional matrices.

Remark. Matrices of the above type appear in the transient analysis of coupled systems such as cells (catalyst particles), reactors, distillation columns, etc.

8. Consider the differential algebraic equation (DAE) system

$$\mathbf{C}\frac{d\mathbf{u}}{dt} = \mathbf{A}\mathbf{u} \quad \mathbf{u}(t = 0) = \mathbf{u}^0$$

where $\mathbf{A}$ and $\mathbf{C}$ are real $n \times n$ matrices and $\mathbf{C}$ is not necessarily invertible.

(a) Show that the substitution $\mathbf{u} = \mathbf{y}e^{\lambda t}$ leads to the eigenvalue problem $\mathbf{A}\mathbf{y} = \lambda\mathbf{C}\mathbf{y}$ If the rank of $\mathbf{C}$ is $r (0 \leq r \leq n)$, how many eigenvalues are there?

(b) What is the adjoint eigenvalue problem? Determine the form of the biorthogonality relations.

(c) Obtain a formal solution of the DAE in terms of the eigenvalues, eigenvectors, etc.

(d) Can $\mathbf{u}_0$ be arbitrary? What conditions must be satisfied by $\mathbf{u}_0$ so that the DAE is consistent and has a unique solution?

(e) Solve the DAE

$$\begin{aligned} 0 &= u_1 + 2u_2 + u_3 \\ \frac{du_2}{dt} &= u_2 + 6u_3 \\ \frac{du_3}{dt} &= -u_1 + u_2 + 3u_3. \end{aligned}$$

9. Circulant matrices play an important role in digital signal processing and in the computation of discrete and fast Fourier transforms. A circulant is a constant diagonal matrix with the special form

$$\mathbf{A} = \begin{pmatrix} f_0 & f_1 & f_2 & \cdot & \cdot & f_{n-1} \\ f_{n-1} & f_0 & f_1 & \cdot & \cdot & f_{n-2} \\ f_{n-2} & f_{n-1} & f_0 & \cdot & \cdot & f_{n-3} \\ \cdot & \cdot & \cdot & \cdot & \cdot & \cdot \\ f_1 & f_2 & f_3 & \cdot & \cdot & f_0 \end{pmatrix}$$

(a) Show that the eigenvalues of **A** are given by

$$\lambda = f_0 + f_1\omega + f_2\omega^2 + \cdots + f_{n-1}\omega^{n-1},$$

where ω is the k-th root of unity, i. e.,

$$\omega^k = 1; \quad k = 1, 2, \ldots, n$$

i. Determine the eigenvectors. Let **F** (for Fourier) be the matrix of eigenvectors of **A**. Determine **F** and show that it is independent of the numbers $(f_0, f_1, f_2, \ldots, f_{n-1})$, i. e., all circulant matrices of a given order have the same eigenvectors.

ii. Show that $\mathbf{F}\overline{\mathbf{F}} = n\mathbf{I}$. Here, $\overline{\mathbf{F}}$ is the complex conjugate of **F**.

(b) the discrete Fourier transform **c** of a sequence $\mathbf{f} = (f_0, f_1, f_2, \ldots, f_{n-1})$ is a set of n-numbers defined by

$$\mathbf{Fc} = \mathbf{f}.$$

Show that the discrete transform (and its inverse) can be computed by a simple matrix multiplication. Write the explicit form of these formulas (for the transform and its inverse) in terms of the components of the Fourier matrix.

Note: The discrete transform is extremely useful in applications. Its computation involves n^2 multiplications. However, by taking advantage of the special structure of the Fourier matrix and choosing n as a power of 2, we can reduce the number of multiplications to $n \log_2 n$. This is the basis for the fast Fourier transform.

10. Consider the initial value problem:

$$\frac{d\mathbf{u}}{dt} = \mathbf{A}\mathbf{u}, \quad \mathbf{u}(t = 0) = \mathbf{u}^0$$

where $\mathbf{A}$ is a $n \times n$ matrix with real elements and $\mathbf{u}$ is a $n \times 1$ vector. Suppose that all the eigenvalues of $\mathbf{A}$ are simple. (a) Write the general form of the solution to the initial value problem in terms of eigenvalues, eigenvectors and eigenrows of $\mathbf{A}$. (b) Discuss the asymptotic form ($t \to \infty$) of the solution if (i) all eigenvalues have negative real part, (ii) one zero eigenvalue while all others have negative real part and (iii) if $\mathbf{A}$ has a pair of purely imaginary eigenvalues while all others have negative real part.
11. Consider the linear system $\mathbf{A}\mathbf{x} = \mathbf{b}$, where $\mathbf{A}$ is an $m \times n$ matrix and $\mathbf{x}, \mathbf{b}$ are $n \times 1$ and $m \times 1$ vectors, respectively. Reason that a necessary and sufficient condition for the system $\mathbf{A}\mathbf{x} = \mathbf{b}$ to have a solution is that every solution of the system $\mathbf{y}^*\mathbf{A} = \mathbf{0}$ (where $\mathbf{y}^*$ is a row vector) should also satisfy $\mathbf{y}^*\mathbf{b} = \mathbf{0}$.
12. A square matrix $\mathbf{A}$ is called normal if $\mathbf{A}\mathbf{A}^* = \mathbf{A}^*\mathbf{A}$. Show that a normal matrix has a complete set of eigenvectors.
13. Suppose that $\mathbf{A} = \mathbf{X}\Lambda\mathbf{X}^{-1}$, where Λ is a diagonal matrix. Find a similarity transformation $\mathbf{Y}$ that diagonalizes the matrix

$$\mathbf{B} = \begin{pmatrix} \mathbf{0} & \mathbf{A} \\ \mathbf{A} & \mathbf{0} \end{pmatrix}.$$

What is the diagonal form of $\mathbf{B}$? Here, all matrices except $\mathbf{B}$ are $n \times n$ while $\mathbf{B}$ is $2n \times 2n$.
14. The same similarity transformation diagonalizes both matrices $\mathbf{A}$ and $\mathbf{B}$. Show that $\mathbf{A}$ and $\mathbf{B}$ must commute. (a) Two Hermitian matrices $\mathbf{A}$ and $\mathbf{B}$ have the same eigenvalues. Show that $\mathbf{A}$ and $\mathbf{B}$ are related by unitary similarity transformation.
15. The transient response of a two-phase system (containing a solid and fluid) is described by the equation

$$\frac{d\mathbf{c}}{dt} = \mathbf{A}\mathbf{c} + \mathbf{b}(\mathbf{c}, t) \quad (1)$$

where $\mathbf{c} = \begin{pmatrix} c_f \\ c_s \end{pmatrix}$ (subscript f refers to the fluid and s to the solid) $\mathbf{A} = \begin{pmatrix} -\frac{1}{\epsilon_f} & \frac{1}{\epsilon_f} \\ \frac{1}{1-\epsilon_f} & -\frac{1}{1-\epsilon_f} \end{pmatrix}$, $\mathbf{b} = \begin{pmatrix} b_1(c_f, c_s, t) \\ b_2(c_f, c_s, t) \end{pmatrix}$, ϵ_f = volume fraction/capacitance of fluid phase and $\mathbf{b}$ is some nonlinear function of $\mathbf{c}$ as well as time. (a) Determine the eigenvalues and eigenvectors of $\mathbf{A}$ and give physical interpretation. (b) What is the transformation that will reduce the linear part of (1) to a diagonal form? Perform this transformation and write equation (1) in terms of the canonical variables. Give a physical interpretation of the canonical variables.
16. Consider the linear system $\mathbf{A}\mathbf{u} = \mathbf{b}$, where $\mathbf{A}$ is an $n \times n$ matrix and $\mathbf{u}$ and $\mathbf{b}$ are $n \times 1$ vectors. Suppose that rank of $\mathbf{A} = n - 2$ (a) Write the conditions the vector $\mathbf{b}$ has

to satisfy so that the system is consistent and has solutions. (b) Assuming that the conditions in (a) are satisfied, write down the general form of the solution in terms of the eigenvalues, eigenvectors and eigenrows of **A**. (c) What is the general form of the solution to the initial value problem $\frac{d\mathbf{u}}{dt} = \mathbf{A}\mathbf{u} - \mathbf{b}, \mathbf{u}(0) = \mathbf{0}$, where **A** and **b** are as above?

17. Consider the 3-tank interacting system with the tanks having different volumes but the exchange flows being all identical. (a) Formulate the transient model and cast it in dimensionless form (but without dividing by the capacitances). (b) Rewrite the model in (a) by dividing each equation by the respective capacitance term and reason that the resulting model may be interpreted as three equal sized tanks but with different exchange flow rates. (c) Give a physical interpretation of the transient system obtained when the matrix appearing in (b) is replaced by its transpose (or adjoint matrix).

5 Solution of linear equations containing a square matrix

While the biorthogonal expansions were useful to express the solution of linear equations in terms of eigenvectors and eigenrows, they may not be convenient for numerical calculations, especially when the order of the matrix is large. In this chapter, we discuss other important properties of square matrices and alternate methods for determining the solutions of linear equations containing a square matrix **A**. The methods discussed here also give a procedure to calculate functions defined on square matrices.

5.1 Cayley–Hamilton theorem

A very important property of a square matrix is given by the following Cayley–Hamilton theorem.

Theorem. *Every square matrix satisfies its own characteristic equation, i. e.,*

$$\mathbf{P}_n(\mathbf{A}) = \mathbf{0}$$

where

$$\begin{aligned} \mathbf{P}_n(\lambda) &= |\mathbf{A} - \lambda\mathbf{I}| \\ &= (-\lambda)^n + a_1(-\lambda)^{n-1} + a_2(-\lambda)^{n-2} + \cdots + a_{n-1}(-\lambda) + a_n \end{aligned}$$

is the characteristic polynomial.

Before we give a proof of this theorem, we illustrate it with a simple example.

Example 5.1. Let

$$\mathbf{A} = \begin{pmatrix} 1 & 2 \\ 3 & 4 \end{pmatrix}$$

$$\begin{aligned} P_2(\lambda) &= \begin{vmatrix} 1-\lambda & 2 \\ 3 & 4-\lambda \end{vmatrix} \\ &= \lambda^2 - 5\lambda - 2 \end{aligned}$$

Cayley–Hamilton theorem states that

$$\mathbf{A}^2 - 5\mathbf{A} - 2\mathbf{I} = \mathbf{0}.$$

To verify this, we compute

https://doi.org/10.1515/9783111598055-006

$$\mathbf{A}^2 = \begin{pmatrix} 1 & 2 \\ 3 & 4 \end{pmatrix}\begin{pmatrix} 1 & 2 \\ 3 & 4 \end{pmatrix} = \begin{pmatrix} 7 & 10 \\ 15 & 22 \end{pmatrix}$$

$$5\mathbf{A} + 2\mathbf{I} = \begin{pmatrix} 5 & 10 \\ 15 & 20 \end{pmatrix} + \begin{pmatrix} 2 & 0 \\ 0 & 2 \end{pmatrix} = \begin{pmatrix} 7 & 10 \\ 15 & 22 \end{pmatrix}$$

$\therefore$

$$\mathbf{A}^2 - 5\mathbf{A} - 2\mathbf{I} = \begin{pmatrix} 7 & 10 \\ 15 & 22 \end{pmatrix} - \begin{pmatrix} 7 & 10 \\ 15 & 22 \end{pmatrix} = \begin{pmatrix} 0 & 0 \\ 0 & 0 \end{pmatrix} = \mathbf{0}$$

Thus, we have verified the Cayley–Hamilton theorem.

Proof. Let

$$\mathbf{P}_n(\lambda) = (-\lambda)^n + a_1(-\lambda)^{n-1} + a_2(-\lambda)^{n-2} + \cdots + a_{n-1}(-\lambda) + a_n \tag{5.1}$$

Consider the identity

$$\begin{aligned}(\mathbf{A} - \lambda\mathbf{I})\,\mathrm{adj}(\mathbf{A} - \lambda\mathbf{I}) &= \det(\mathbf{A} - \lambda\mathbf{I}) \cdot \mathbf{I} \\ &= \mathbf{P}_n(\lambda) \cdot \mathbf{I}\end{aligned} \tag{5.2}$$

where, $\mathrm{adj}(\mathbf{A} - \lambda\mathbf{I})$ = classical adjoint of $(\mathbf{A} - \lambda\mathbf{I})$. As seen in Chapter 2, this is the matrix formed from $(n-1) \times (n-1)$ determinants of the minors of $(\mathbf{A} - \lambda\mathbf{I})$. Thus, each element of the matrix $\mathrm{adj}(\mathbf{A} - \lambda\mathbf{I})$ is at most a polynomial of degree $(n-1)$ and by collecting the coefficients of various powers of λ, we can write

$$\mathrm{adj}(\mathbf{A} - \lambda\mathbf{I}) = \mathbf{B}_0\lambda^{n-1} + \mathbf{B}_1\lambda^{n-2} + \cdots + \mathbf{B}_{n-2}\lambda + \mathbf{B}_{n-1} \tag{5.3}$$

where $\mathbf{B}_0, \mathbf{B}_1, \ldots, \mathbf{B}_{n-1}$ are $n \times n$ constant matrices. Substituting (5.1) and (5.3) in (5.2) $\Longrightarrow$

$$(\mathbf{A}-\lambda\mathbf{I})(\mathbf{B}_0\lambda^{n-1}+\mathbf{B}_1\lambda^{n-2}+\cdots+\mathbf{B}_{n-2}\lambda+\mathbf{B}_{n-1}) = [(-\lambda)^n+a_1(-\lambda)^{n-1}+a_2(-\lambda)^{n-2}+\cdots+a_n]\mathbf{I} \tag{5.4}$$

Comparing the coefficients of various powers of λ, we get

$$\lambda^n: \quad -\mathbf{B}_0 = (-1)^n\mathbf{I} \tag{5.5}$$

$$\lambda^{n-1}: \quad \mathbf{A}\mathbf{B}_0 - \mathbf{B}_1 = (-1)^{n-1}a_1\mathbf{I} \tag{5.6}$$

$$\lambda^{n-2}: \quad \mathbf{A}\mathbf{B}_1 - \mathbf{B}_2 = (-1)^{n-2}a_2\mathbf{I} \tag{5.7}$$

.

.

$$\lambda: \quad \mathbf{A}\mathbf{B}_{n-2} - \mathbf{B}_{n-1} = (-1)a_{n-1}\mathbf{I} \tag{5.8}$$

$$\lambda^0: \quad \mathbf{A}\mathbf{B}_{n-1} = a_n\mathbf{I} \tag{5.9}$$

Multiply (5.5) by $\mathbf{A}^n$, (5.6) by $\mathbf{A}^{n-1}, \ldots$, and (5.9) by $\mathbf{I}$ to obtain

$$
\begin{aligned}
-\mathbf{A}^n\mathbf{B}_0 &= (-1)^n\mathbf{A}^n \\
\mathbf{A}^n\mathbf{B}_0 - \mathbf{A}^{n-1}\mathbf{B}_1 &= (-1)^{n-1}a_1\mathbf{A}^{n-1} \\
\mathbf{A}^{n-1}\mathbf{B}_1 - \mathbf{A}^{n-2}\mathbf{B}_2 &= (-1)^{n-2}a_2\mathbf{A}^{n-2} \\
&\cdot \\
&\cdot \\
\mathbf{A}^2\mathbf{B}_{n-2} - \mathbf{A}\mathbf{B}_{n-1} &= (-1)a_{n-1}\mathbf{A} \\
\mathbf{A}\mathbf{B}_{n-1} &= a_n\mathbf{I}
\end{aligned}
$$

Adding all these equations, we have

$$\mathbf{0} = (-\mathbf{A})^n + a_1(-\mathbf{A})^{n-1} + a_2(-\mathbf{A})^{n-2} + \cdots + a_{n-1}(-\mathbf{A}) + a_n\mathbf{I}$$

Thus, $\mathbf{P}_n(\mathbf{A}) = \mathbf{0}$ and the Cayley–Hamilton theorem is proved. □

Consequences of Cayley–Hamilton theorem

Let

$$\mathbf{P}_n(\lambda) = (-\lambda)^n + a_1(-\lambda)^{n-1} + a_2(-\lambda)^{n-2} + \cdots + a_{n-1}(-\lambda) + a_n$$

be the characteristic equation of $\mathbf{A}$. Then Cayley–Hamilton theorem gives

$$\mathbf{P}_n(\mathbf{A}) = \mathbf{0} \tag{5.10}$$

Rewrite equation (5.10) as

$$\mathbf{A}^n = \alpha_1\mathbf{A}^{n-1} + \alpha_2\mathbf{A}^{n-2} + \cdots + \alpha_{n-1}\mathbf{A} + \alpha_n\mathbf{I}, \tag{5.11}$$

where $\alpha_i = (-1)^{1-i}a_i$. Thus, $\mathbf{A}^n$ can be expressed as a polynomial of degree $(n-1)$ in $\mathbf{A}$. Equation (5.11) gives

$$
\begin{aligned}
\mathbf{A}^{n+1} &= \alpha_1\mathbf{A}^n + \alpha_2\mathbf{A}^{n-1} + \cdots + \alpha_n\mathbf{A} \\
&= \alpha_1(\alpha_1\mathbf{A}^{n-1} + \cdots + \alpha_n\mathbf{I}) + \alpha_2\mathbf{A}^{n-1} + \cdots + \alpha_n\mathbf{A} \\
&= \beta_1\mathbf{A}^{n-1} + \beta_2\mathbf{A}^{n-2} + \cdots + \beta_n\mathbf{I}
\end{aligned} \tag{5.12}
$$

where $\beta_1 = \alpha_1^2 + \alpha_2, \ldots, \beta_n = \alpha_1\alpha_n$ are some constants. It follows from equation (5.12) that $\mathbf{A}^{n+1}$ can be expressed as a polynomial of degree $(n-1)$ in $\mathbf{A}$. Continuing this procedure, we see that if

$$Q(\lambda) = \sum_{i=0}^{\infty} c_i\lambda^i \tag{5.13}$$

is any function that is bounded and has a Maclaurin's series expansion, then $Q(\mathbf{A})$ can be expressed as a polynomial of degree $(n-1)$ in $\mathbf{A}$.

Now, suppose that $\mathbf{A}$ is invertible, i. e., $\mathbf{A}^{-1}$ exists. Then, multiplying both sides of (5.11) by $\mathbf{A}^{-1}$, we get

$$\mathbf{A}^{n-1} = \alpha_1\mathbf{A}^{n-2} + \alpha_2\mathbf{A}^{n-3} + \cdots + \alpha_{n-1}\mathbf{I} + \alpha_n\mathbf{A}^{-1}$$

Assuming that $\alpha_n \neq 0$, (this is the case if $\mathbf{A}$ is invertible), we can rewrite the above equation as

$$\mathbf{A}^{-1} = \gamma_1\mathbf{A}^{n-1} + \gamma_2\mathbf{A}^{n-2} + \cdots + \gamma_{n-1}\mathbf{A} + \gamma_n\mathbf{I} \tag{5.14}$$

Thus, $\mathbf{A}^{-1}$ can be expressed as a polynomial of degree $(n-1)$ in $\mathbf{A}$. Continuing this procedure, we see that $\mathbf{A}^{-k}(k \geq 1)$ can be expressed as a polynomial of degree $(n-1)$ in $\mathbf{A}$.

$\Longrightarrow$ If $Q(\lambda) = \sum_{i=-\infty}^{\infty} c_i\lambda^i$ is any bounded/convergent function, then $Q(\mathbf{A})$ can be expressed as a polynomial of degree $(n-1)$ in $\mathbf{A}$.

This is a very profound result as it implies that any function of $\mathbf{A}$ can be expressed as a polynomial of degree $(n-1)$ in $\mathbf{A}$. It is used in the following section for defining and computing functions of a square matrix.

5.2 Functions of matrices

Suppose that $\mathbf{A}$ is a square matrix with eigenvalues $\{\lambda_1, \lambda_2, \ldots, \lambda_n\}$. The set of eigenvalues of $\mathbf{A}$ is called the *spectrum* of $\mathbf{A}$. Let $f(\lambda)$ be any function for which $f(\lambda_j), j = 1, 2, \ldots, n$ is defined. Then we say that the function f is defined on the spectrum of $\mathbf{A}$. It follows from Cayley–Hamilton theorem that $f(\mathbf{A})$ is at most a polynomial of degree $(n-1)$ in $\mathbf{A}$, i. e.,

$$f(\mathbf{A}) = c_1\mathbf{A}^{n-1} + c_2\mathbf{A}^{n-2} + \cdots + c_{n-1}\mathbf{A} + c_n\mathbf{I} \tag{5.15}$$

for some constants $\{c_1, c_2, \ldots, c_n\}$. This result can be used to extend the definition of familiar scalar functions to functions of matrices as well as to compute them. For example, we define the exponential matrix function by the series

$$e^{\mathbf{A}} = \sum_{i=0}^{\infty} \frac{\mathbf{A}^i}{i!} = \mathbf{I} + \mathbf{A} + \frac{\mathbf{A}^2}{2!} + \frac{\mathbf{A}^3}{3!} + \cdots \tag{5.16}$$

or the trigonometric function by

$$\sin\mathbf{A} = \sum_{i=0}^{\infty} \frac{(-1)^i\mathbf{A}^{2i+1}}{(2i+1)!} = \mathbf{A} - \frac{\mathbf{A}^3}{3!} + \frac{\mathbf{A}^5}{5!} + \cdots \tag{5.17}$$

or the Bessel function of order zero by

$$\mathbf{J}_0(\mathbf{A}) = \sum_{i=0}^{\infty} \frac{(-1)^i(\frac{\mathbf{A}}{2})^{2i}}{(i!)^2} = \mathbf{I} - \frac{(\mathbf{A}/2)^2}{(1!)^2} + \frac{(\mathbf{A}/2)^4}{(2!)^2} - \cdots \tag{5.18}$$

It follows from Cayley–Hamilton theorem that the above matrix functions can be expressed as polynomials of degree $(n-1)$ in $\mathbf{A}$. These functions may also be computed if we can evaluate the coefficients in equation (5.15).

Procedure for the calculation of $f(\mathbf{A})$

Let

$$\mathbf{B} = f(\mathbf{A}) = c_1\mathbf{A}^{n-1} + c_2\mathbf{A}^{n-2} + \cdots + c_{n-1}\mathbf{A} + c_n\mathbf{I} \tag{5.19}$$

We can calculate $\mathbf{B}$ if we can determine the n-unknown constants in equation (5.19). We now develop a procedure for calculating these constants. The following lemmas are useful in this procedure.

Lemma 5.1. *Let λ_j be an eigenvalue of $\mathbf{A}$ with eigenvector $\mathbf{x}_j$. Then λ_j^k is an eigenvalue of $\mathbf{A}^k$ with the same eigenvector. Here, k is any positive or negative integer (including zero).*

Proof. For $k = 0, \mathbf{A}^0\mathbf{x}_j = \mathbf{I}\mathbf{x}_j = \mathbf{x}_j = \lambda_j^0\mathbf{x}_j$.

For $k > 1$, we have

$$\begin{aligned}\mathbf{A}^2\mathbf{x}_j &= \mathbf{A}(\mathbf{A}\mathbf{x}_j) = \mathbf{A}(\lambda_j\mathbf{x}_j) = \lambda_j(\mathbf{A}\mathbf{x}_j) = \lambda_j^2\mathbf{x}_j\\ \mathbf{A}^3\mathbf{x}_j &= \mathbf{A}(\mathbf{A}^2\mathbf{x}_j) = \lambda_j^3\mathbf{x}_j\end{aligned}$$

and in general,

$$\mathbf{A}^k\mathbf{x}_j = \lambda_j^k\mathbf{x}_j \quad \text{for any } k \geq 0.$$

Now, consider the equation

$$\mathbf{A}\mathbf{x}_j = \lambda_j\mathbf{x}_j$$

Assuming that $\mathbf{A}$ is invertible, multiply both sides by $\mathbf{A}^{-1}$.

$\Longrightarrow$

$$\begin{aligned}\mathbf{A}^{-1}\mathbf{A}\mathbf{x}_j &= \lambda_j\mathbf{A}^{-1}\mathbf{x}_j\\ \mathbf{x}_j &= \lambda_j\mathbf{A}^{-1}\mathbf{x}_j\\ (\lambda_j)^{-1}\mathbf{x}_j &= \mathbf{A}^{-1}\mathbf{x}_j\end{aligned}$$

Similarly,

$$(\lambda_j)^{-k}\mathbf{x}_j = \mathbf{A}^{-k}\mathbf{x}_j, \quad k \geq 1$$

Thus, $\mathbf{A}^k\mathbf{x}_j = \lambda_j^k\mathbf{x}_j$ for all integer values of k. □

Lemma 5.2. *If $f(x)$ is any function of the form,*

$$f(x) = \sum_{k=-\infty}^{\infty} c_k x^k$$

that is defined on the spectrum of **A**, *then*

$$f(\mathbf{A})\mathbf{x}_j = f(\lambda_j)\mathbf{x}_j.$$

This lemma follows directly from the previous one. Now, let

$$\mathbf{B} = f(\mathbf{A}) \Longrightarrow \mathbf{B}\mathbf{x}_j = f(\mathbf{A})\mathbf{x}_j = f(\lambda_j)\mathbf{x}_j \tag{5.20}$$

But C-H theorem gives equation (5.19). From this equation, we get

$$\begin{aligned} f(\mathbf{A})\mathbf{x}_j &= (c_1\mathbf{A}^{n-1} + c_2\mathbf{A}^{n-2} + \cdots + c_n\mathbf{I})\mathbf{x}_j \\ &= (c_1\lambda_j^{n-1} + c_2\lambda_j^{n-2} + \cdots + c_n)\mathbf{x}_j \end{aligned} \tag{5.21}$$

Comparing (5.20) and (5.21) gives

$$f(\lambda_j)\mathbf{x}_j = (c_1\lambda_j^{n-1} + c_2\lambda_j^{n-2} + \cdots + c_n)\mathbf{x}_j$$

This can be true only if

$$\{f(\lambda_j) - [c_1\lambda_j^{n-1} + c_2\lambda_j^{n-2} + \cdots + c_n]\}\mathbf{x}_j = \mathbf{0}$$

Since $\mathbf{x}_j \neq \mathbf{0} \Longrightarrow$

$$f(\lambda_j) = (c_1\lambda_j^{n-1} + c_2\lambda_j^{n-2} + \cdots + c_n); j = 1, \ldots, n \tag{5.22}$$

Thus, the n constants $(c_1, \ldots, c_n)$ can be determined by solving the linear equations defined by equation (5.22). If an eigenvalue λ_i is repeated r times ($r \geq 1$), then we consider the relation

$$f(\lambda) = (c_1\lambda^{n-1} + c_2\lambda^{n-2} + \cdots + c_n) \tag{5.23}$$

and differentiate it k times and set $\lambda = \lambda_i$ for $k = 0, 1, \ldots, r-1$. This gives r linear equations for the constants $(c_1, \ldots, c_n)$.

Example 5.2. Develop a formula for $f(\mathbf{A})$ when **A** is a 2×2 matrix with distinct eigenvalues.

From the Cayley–Hamilton theorem, we can write

$$f(\mathbf{A}) = c_1\mathbf{A} + \mathbf{c}_2\mathbf{I},$$

where c_1 and c_2 are determined by the equations

$$f(\lambda_1) = c_1\lambda_1 + c_2$$
$$f(\lambda_2) = c_1\lambda_2 + c_2$$

Solving these two linear equations gives

$$c_1 = \frac{f(\lambda_1) - f(\lambda_2)}{(\lambda_1 - \lambda_2)}, \quad c_2 = \frac{\lambda_2 f(\lambda_1) - \lambda_1 f(\lambda_2)}{(\lambda_2 - \lambda_1)}$$

Thus, for any 2×2 matrix with distinct eigenvalues we obtain the formula

$$f(\mathbf{A}) = \frac{f(\lambda_1) - f(\lambda_2)}{(\lambda_1 - \lambda_2)}\mathbf{A} + \frac{\lambda_2 f(\lambda_1) - \lambda_1 f(\lambda_2)}{(\lambda_2 - \lambda_1)}\mathbf{I} \tag{5.24}$$

Example 5.3. Develop a formula for $f(\mathbf{A})$ when $\mathbf{A}$ is a 2×2 matrix with repeated eigenvalues.

Now, the constants c_1 and c_2 are given by

$$f(\lambda_1) = c_1\lambda_1 + c_2$$
$$f'(\lambda_1) = c_1 \Longrightarrow c_2 = f(\lambda_1) - \lambda_1 f'(\lambda_1)$$

Therefore, for a 2×2 matrix with repeated eigenvalues we have the formula

$$f(\mathbf{A}) = [f'(\lambda_1)]\mathbf{A} + [f(\lambda_1) - \lambda_1 f'(\lambda_1)]\mathbf{I}$$

Remark. This formula may also be obtained from equation (5.24) by taking the limit $\lambda_2 \to \lambda_1$.

Example 5.4. Develop a formula for $f(\mathbf{A}t)$ when A is a 2×2 matrix with distinct eigenvalues.

First, we note that if λ_1 and λ_2 are eigenvalues of $\mathbf{A}$, then $\lambda_1 t$ and $\lambda_2 t$ are eigenvalues of $\mathbf{A}t$. This follows from the fact that

$$|\mathbf{A}t - \lambda t\mathbf{I}| = |t(\mathbf{A} - \lambda\mathbf{I})| = t^2|(\mathbf{A} - \lambda\mathbf{I})| = 0$$

where the last equality assumes that λ_i is an eigenvalue of $\mathbf{A}$, and hence satisfies the characteristic equation. Now replace λ_1 by $\lambda_1 t$ and λ_2 by $\lambda_2 t$ and $\mathbf{A}$ by $\mathbf{A}t$ in the formula of Example (5.2). This gives

$$f(\mathbf{A}t) = \frac{f(\lambda_1 t) - f(\lambda_2 t)}{(\lambda_1 - \lambda_2)}\mathbf{A} + \frac{\lambda_2 f(\lambda_1 t) - \lambda_1 f(\lambda_2 t)}{(\lambda_2 - \lambda_1)}\mathbf{I}$$

For $\lambda_1 = \lambda_2$, this formula may be simplified to

$$f(\mathbf{A}t) = f'(\lambda t)\mathbf{A}t + [f(\lambda t) - \lambda t f'(\lambda t)]\mathbf{I}$$

As a specific example, when $\lambda_1 \neq \lambda_2$, we have

$$\begin{aligned} e^{\mathbf{A}t} &= \frac{e^{\lambda_1 t} - e^{\lambda_2 t}}{(\lambda_1 - \lambda_2)}\mathbf{A} + \frac{\lambda_1 e^{\lambda_2 t} - \lambda_2 e^{\lambda_1 t}}{(\lambda_1 - \lambda_2)}\mathbf{I} \\ &= e^{\lambda_1 t}\left[\frac{1}{(\lambda_1 - \lambda_2)}\mathbf{A} - \frac{\lambda_2}{(\lambda_1 - \lambda_2)}\mathbf{I}\right] + e^{\lambda_2 t}\left[-\frac{1}{(\lambda_1 - \lambda_2)}\mathbf{A} + \frac{\lambda_1}{(\lambda_1 - \lambda_2)}\mathbf{I}\right] \end{aligned}$$

while for $\lambda_1 = \lambda_2 = \lambda$, we get

$$\begin{aligned} e^{\mathbf{A}t} &= e^{\lambda t}\mathbf{A}t + [e^{\lambda t} - \lambda t e^{\lambda t}]\mathbf{I} \\ &= e^{\lambda t}[\mathbf{A}t + \mathbf{I} - \lambda t\mathbf{I}] \\ &= e^{\lambda t}\mathbf{I} + t e^{\lambda t}[\mathbf{A} - \lambda\mathbf{I}]. \end{aligned}$$

5.3 Formal solutions of linear differential equations containing a square matrix

It follows from Cayley–Hamilton theorem that $f(\mathbf{A})$ commutes with $\mathbf{A}$, i. e.,

$$\mathbf{A}f(\mathbf{A}) = f(\mathbf{A})\mathbf{A} \tag{5.25}$$

This property may be used to write the solution of many vector differential equations containing a square matrix $\mathbf{A}$ with constant coefficients in terms of functions of $\mathbf{A}$ using the solution for the scalar case. This is illustrated in this section with several examples.

Example 5.5. Consider the initial value problem

$$\frac{d\mathbf{u}}{dt} = \mathbf{A}\mathbf{u}, \quad \mathbf{u} = \mathbf{u}_0 \ @\ t = 0 \tag{5.26}$$

We claim that

$$\mathbf{u} = e^{\mathbf{A}t}\mathbf{u}_0 \tag{5.27}$$

is a solution to the above initial value problem. To prove this, first we verify that (5.27) satisfies the initial condition:

$$\mathbf{u}(t = 0) = e^{\mathbf{A}.\mathbf{0}}\mathbf{u}_0 = e^{\mathbf{0}}\mathbf{u}_0 = \mathbf{I}\mathbf{u}_0 = \mathbf{u}_0$$

To show that (5.27) satisfies the differential equation, we differentiate it w. r. t. time to obtain

$$\frac{d\mathbf{u}}{dt} = e^{\mathbf{A}t}\mathbf{A}\mathbf{u}_0 = \mathbf{A}e^{\mathbf{A}t}\mathbf{u}_0 \quad \text{(since } \mathbf{A} \text{ commutes with } e^{\mathbf{A}t}\text{)}$$
$$= \mathbf{A}\mathbf{u}$$

Thus, the expression given by equation (5.27) is the solution to the initial value problem defined by (5.26).

Example 5.6. We consider the second-order vector initial value problem

$$\frac{d^2\mathbf{u}}{dt^2} = -\mathbf{A}\mathbf{u}, \tag{5.28}$$

$$\mathbf{u}(0) = \mathbf{u}_0, \quad \frac{d\mathbf{u}}{dt}(0) = \mathbf{v}_0 \tag{5.29}$$

Using the Cayley–Hamilton theorem, the general solution of equation (5.28) may be written as

$$\mathbf{u} = [\cos\sqrt{\mathbf{A}}t]\mathbf{c}_1 + [\sin\sqrt{\mathbf{A}}t]\mathbf{c}_2 \tag{5.30}$$

The constant vectors $\mathbf{c}_1$ and $\mathbf{c}_2$ may be determined from the initial conditions:

$$\mathbf{u}(0) = \mathbf{u}_0 \Longrightarrow \mathbf{c}_1 = \mathbf{u}_0$$
$$\frac{d\mathbf{u}}{dt} = -\sqrt{\mathbf{A}}\sin(\sqrt{\mathbf{A}}t)\mathbf{c}_1 + \sqrt{\mathbf{A}}\cos(\sqrt{\mathbf{A}}t)\mathbf{c}_2$$
$$\frac{d\mathbf{u}}{dt}(0) = \mathbf{v}_0 \Longrightarrow \sqrt{\mathbf{A}}\mathbf{c}_2 = \mathbf{v}_0 \Longrightarrow \mathbf{c}_2 = (\sqrt{\mathbf{A}})^{-1}\mathbf{v}_0$$

Thus,

$$\mathbf{u} = [\cos\sqrt{\mathbf{A}}t]\mathbf{u}_0 + [\sin\sqrt{\mathbf{A}}t](\sqrt{\mathbf{A}})^{-1}\mathbf{v}_0$$

is a formal solution. Note that $\cos\sqrt{\mathbf{A}}t$ and $[\sin\sqrt{\mathbf{A}}t](\sqrt{\mathbf{A}})^{-1}$ contain only integral powers of $\mathbf{A}$, and hence are polynomials of degree $(n-1)$ in $\mathbf{A}$.

Example 5.7. We consider the vector two-point boundary value problem

$$\frac{d^2\mathbf{c}}{d\xi^2} = \mathbf{\Phi}^2\mathbf{c}, \quad 0 < \xi < 1$$
$$\mathbf{c} = \mathbf{c}_0 \ @\ \xi = 0$$
$$\frac{d\mathbf{c}}{d\xi} = \mathbf{0} \ @\ \xi = 1$$

representing diffusion and reaction in a flat plate geometry (see Section 4.5.6). Here, $\mathbf{c}$ is the vector of species concentrations, $\mathbf{c}_0$ is the concentration vector at the interface, and

$\mathbf{\Phi}^2$ is the square of the Thiele matrix. The solution may be written as

$$\Longrightarrow \mathbf{c} = [\cosh \mathbf{\Phi}\xi]\boldsymbol{\alpha}_1 + [\cosh \mathbf{\Phi}(1-\xi)]\boldsymbol{\alpha}_2$$
$$\frac{d\mathbf{c}}{d\xi} = [\mathbf{\Phi} \sinh \mathbf{\Phi}\xi]\boldsymbol{\alpha}_1 + [\sinh \mathbf{\Phi}(1-\xi)](-\mathbf{\Phi})\boldsymbol{\alpha}_2$$

The two boundary conditions can be used to obtain the vector constants $\boldsymbol{\alpha}_1$ and $\boldsymbol{\alpha}_2$ as follows:

$$\frac{d\mathbf{c}}{d\xi} = \mathbf{0} \ @\ \xi = 1 \Longrightarrow \boldsymbol{\alpha}_1 = \mathbf{0}$$
$$\mathbf{c} = \mathbf{c}_0 \ @\ \xi = 0 \Longrightarrow \mathbf{c}_0 = [\cosh \mathbf{\Phi}]\boldsymbol{\alpha}_2$$
$$\Longrightarrow \mathbf{c}(\xi) = [\cosh \mathbf{\Phi}(1-\xi)](\cosh \mathbf{\Phi})^{-1}\mathbf{c}_0$$

is the solution. The quantity of practical interest is the average (or observed) reaction rate vector in the pore defined by (see Section 4.5.6)

$$\begin{aligned}\mathbf{r}_{\text{obs}} &= \int_0^1 \mathbf{K}.\mathbf{c}(\xi)\, d\xi \\ &= \mathbf{K}\mathbf{\Phi}^{-1}\tanh(\mathbf{\Phi})\mathbf{c}_0 \\ &= \mathbf{K}^*\mathbf{c}_0\end{aligned}$$

where $\mathbf{\Phi}^2 = L^2\mathbf{D}^{-1}\mathbf{K}$ and $\mathbf{K}^*$ is the diffusion disguised rate constant matrix given by

$$\mathbf{K}^* = \mathbf{K}\mathbf{\Phi}^{-1}\tanh(\mathbf{\Phi}) = \mathbf{K}\mathbf{H} \tag{5.31}$$

[Remark: $\mathbf{H} = \mathbf{\Phi}^{-1}\tanh(\mathbf{\Phi})$ is the so-called effectiveness factor matrix]. Writing $\mathbf{D} = d_m\mathbf{M}$ ($\mathbf{M}$ = matrix of relative species diffusivities), $\mathbf{K} = k\mathbf{A}$ ($\mathbf{A}$ = matrix of relative rate constants), $\mathbf{K}^* = k\mathbf{A}^*$ ($\mathbf{A}^*$ = matrix of diffusion-disguised relative rate constants) and $\phi^2 = \frac{L^2k}{d_m}$, $\mathbf{\Psi}^2 = \mathbf{M}^{-1}\mathbf{A}$, we have

$$\mathbf{A}^* = \mathbf{A}(\phi\mathbf{\Psi})^{-1}\tanh(\phi\mathbf{\Psi}) \tag{5.32}$$

The two limiting cases of equation (5.32) can be seen more easily now. For the case of negligible pore diffusional limitations ($\phi \to 0$), we have $\mathbf{A}^* = \mathbf{A}$ while for the case of strong pore diffusional limitations ($\phi \to \infty$), we have $\mathbf{A}^* = \frac{1}{\phi}\mathbf{A}\sqrt{\mathbf{A}^{-1}\mathbf{M}}$.

5.4 Sylvester's theorem

Sylvester's theorem and its generalization, known as the spectral theorem are important in understanding and computing the solutions of linear equations in which the square matrix $\mathbf{A}$ appears. Consider a scalar polynomial of degree $(n-1)$ in Lagrangian form

$$Q_{n-1}(x) = c_1(x - x_2)(x - x_3)\dots(x - x_n) + c_2(x - x_1)(x - x_3)\dots(x - x_n) + \cdots \\ + c_n(x - x_1)(x - x_2)\dots(x - x_{n-1}) \tag{5.33}$$

If we require that the polynomial has to pass through the point $(x_j, Q_{n-1}(x_j))$, we get

$$c_j = \frac{Q_{n-1}(x_j)}{\prod_{i=1,j}^{n}(x_j - x_i)} \tag{5.34}$$

Here, the notation $\prod_{i=1,j}^{n}$ means that the product excludes the term corresponding to $i = j$. Thus,

$$Q_{n-1}(x) = \sum_{j=1}^{n} Q_{n-1}(x_j)\frac{\prod_{i=1,j}^{n}(x - x_i)}{\prod_{i=1,j}^{n}(x_j - x_i)} \tag{5.35}$$

We now convert equation (5.35) into a matrix identity by the following assumptions: (i) Assume that the square matrix **A** has n distinct eigenvalues $\{\lambda_1, \lambda_2, \dots, \lambda_n\}$ and replace x_j by λ_j and (ii) Replace x by **A** on both sides of equation (5.35). This gives

$$Q_{n-1}(\mathbf{A}) = \sum_{j=1}^{n} Q_{n-1}(\lambda_j)\frac{\prod_{i=1,j}^{n}(\mathbf{A} - \lambda_i\mathbf{I})}{\prod_{i=1,j}^{n}(\lambda_j - \lambda_i)} \tag{5.36}$$

Note that the order of the products on the RHS. of this equation is immaterial since $(\mathbf{A} - \lambda_i\mathbf{I})$ and $(\mathbf{A} - \lambda_j\mathbf{I})$ commute. As it stands, equation (5.36) is valid for any polynomial of degree $(n - 1)$ in **A**. However, we note that any arbitrary polynomial in **A** and $\mathbf{A}^{-1}$ may be expressed as a polynomial of degree $(n - 1)$ in **A** (Cayley–Hamilton theorem). Now, if $f(\lambda)$ is any bounded function of the form

$$f(\lambda) = \sum_{k=-\infty}^{\infty} c_k\lambda^k \tag{5.37}$$

then $f(\lambda_j)$ can be expressed as a polynomial of degree $(n - 1)$ in λ_j using the relation $P_n(\lambda_j) = 0$. Thus, we can replace Q_{n-1} in equation (5.36) by any arbitrary function of the form given by equation (5.37). Thus, we have

$$f(\mathbf{A}) = \sum_{j=1}^{n} f(\lambda_j)\frac{\prod_{i=1,j}^{n}(\mathbf{A} - \lambda_i\mathbf{I})}{\prod_{i=1,j}^{n}(\lambda_j - \lambda_i)} \tag{5.38}$$

This is Sylvester's formula. We can simplify it further by using the identity (see proof below).

$$\prod_{i=1,j}^{n}(\mathbf{A} - \lambda_i\mathbf{I}) = (-1)^{n-1}\,\mathrm{adj}(\mathbf{A} - \lambda_j\mathbf{I}) \tag{5.39}$$

We note that

$$\prod_{i=1,j}^{n}(\lambda_j - \lambda_i) = (-1)^{n-1}\prod_{i=1,j}^{n}(\lambda_i - \lambda_j) \tag{5.40}$$

Substituting equations (5.39) and (5.40) into (5.38) gives

$$\begin{aligned} f(\mathbf{A}) &= \sum_{j=1}^{n} f(\lambda_j)\frac{(-1)^{n-1}\operatorname{adj}(\mathbf{A}-\lambda_j\mathbf{I})}{(-1)^{n-1}\prod_{i=1,j}^{n}(\lambda_i-\lambda_j)} \\ &= \sum_{j=1}^{n} f(\lambda_j)\frac{\operatorname{adj}(\mathbf{A}-\lambda_j\mathbf{I})}{\prod_{i=1,j}^{n}(\lambda_i-\lambda_j)} \end{aligned} \tag{5.41}$$

Defining

$$\mathbf{E}_j = \frac{\operatorname{adj}(\mathbf{A}-\lambda_j\mathbf{I})}{\prod_{i=1,j}^{n}(\lambda_i-\lambda_j)} \tag{5.42}$$

We get the final form of Sylvester's formula

$$f(\mathbf{A}) = \sum_{j=1}^{n} f(\lambda_j)\mathbf{E}_j \tag{5.43}$$

It can be shown that the matrices $\mathbf{E}_j$ ($j = 1, 2, 3, \ldots, n$) have rank one and satisfy the relations

$$\mathbf{E}_i\mathbf{E}_j = \mathbf{0}, \quad i \neq j \quad \text{and} \quad \sum_{i=1}^{n}\mathbf{E}_i = \mathbf{I} \tag{5.44}$$

The matrix $\mathbf{E}_i$ is called a projection. We shall prove the above relations as well as establish other properties of projections in the next section when we deal with the spectral theorem.

Proof of the identity given by equation (5.39).

$$\prod_{i=1,j}^{n}(\mathbf{A}-\lambda_i\mathbf{I}) = (-1)^{n-1}\operatorname{adj}(\mathbf{A}-\lambda_j\mathbf{I})$$

Write the characteristic polynomial as $P_n(\lambda) = (-1)^n[\lambda^n + \alpha_1\lambda^{n-1} + \cdots + \alpha_{n-1}\lambda + \alpha_n]$

$$\begin{aligned} \Longrightarrow P_n(y) - P_n(x) &= (-1)^n[y^n - x^n + \alpha_1(y^{n-1} - x^{n-1}) + \cdots + \alpha_{n-1}(y-x)] \\ &= (-1)^n(y-x)[(y^{n-1} + y^{n-2}x + \cdots + yx^{n-2} + x^{n-1}) \\ &\quad + \alpha_1(y^{n-2} + y^{n-3}x + \cdots + yx^{n-3} + x^{n-2}) + \cdots + \alpha_{n-1}] \\ &\equiv (-1)^n(y-x)\Phi(y,x) \end{aligned} \tag{5.45}$$

where $\Phi(y,x) = \Phi(x,y)$ is of degree $(n-1)$. Convert this scalar polynomial identity to a matrix identity by letting $x = \mathbf{I}\lambda, y = \mathbf{A} \Longrightarrow$

$$\begin{aligned} P_n(\mathbf{A}) - P_n(\mathbf{I}\lambda) &= \mathbf{0} - P_n(\mathbf{I}\lambda) = -\mathbf{I}P_n(\lambda) \\ &= (-1)^n(\mathbf{A} - \lambda\mathbf{I})\Phi(\mathbf{A}, \mathbf{I}\lambda) \end{aligned} \tag{5.46}$$

But

$$(\mathbf{A} - \lambda\mathbf{I})\,\mathrm{adj}(\mathbf{A} - \lambda\mathbf{I}) = P_n(\lambda)\mathbf{I} \tag{5.47}$$

Comparing (5.46) and (5.47) $\Longrightarrow$

$$(-1)^{n-1}(\mathbf{A} - \lambda\mathbf{I})\Phi(\mathbf{A}, \lambda\mathbf{I}) = (\mathbf{A} - \lambda\mathbf{I})\,\mathrm{adj}(\mathbf{A} - \lambda\mathbf{I})$$

Since this is an identity for all values of λ, it follows that

$$\mathrm{adj}(\mathbf{A} - \lambda\mathbf{I}) = (-1)^{n-1}\Phi(\mathbf{A}, \lambda\mathbf{I}) \tag{5.48}$$

Now, let $x = \lambda_j$ and $y = \lambda$ in (5.45)$\Longrightarrow$

$$\begin{aligned} P_n(\lambda) - P_n(\lambda_j) &= P_n(\lambda) = (-1)^n(\lambda - \lambda_j)\Phi(\lambda, \lambda_j) \\ &= (-1)^n(\lambda - \lambda_1)(\lambda - \lambda_2)\dots(\lambda - \lambda_n) \\ \Longrightarrow \Phi(\lambda, \lambda_j) &= \prod_{i=1,j}^{n}(\lambda - \lambda_i) \end{aligned} \tag{5.49}$$

Convert this to a matrix identity by letting $\lambda = \mathbf{A}$, $\lambda_i = \mathbf{I}\lambda$:

$$\Longrightarrow \Phi(\mathbf{A}, \lambda_j\mathbf{I}) = \prod_{i=1,j}^{n}(\mathbf{A} - \lambda_i\mathbf{I}) \tag{5.50}$$

From (5.48) and (5.50), it follows that

$$\mathrm{adj}(\mathbf{A} - \lambda\mathbf{I}) = (-1)^{n-1}\prod_{i=1,j}^{n}(\mathbf{A} - \lambda_i\mathbf{I})$$ □

Example 5.8 (Illustration of Sylvester's formula). Consider the 2×2 matrix

$$\mathbf{A} = \begin{pmatrix} 1 & 1 \\ 4 & 1 \end{pmatrix}$$

whose eigenvalues are

$$\lambda_1 = -1, \quad \lambda_2 = 3.$$

The eigenvectors and normalized eigenrows are given by

$$\mathbf{x}_1 = \begin{pmatrix} 1 \\ -2 \end{pmatrix}; \quad \mathbf{x}_2 = \begin{pmatrix} 1 \\ 2 \end{pmatrix}$$

$$\mathbf{y}_1^* = \begin{pmatrix} \frac{1}{2} & -\frac{1}{4} \end{pmatrix}; \quad \mathbf{y}_2^* = \begin{pmatrix} \frac{1}{2} & \frac{1}{4} \end{pmatrix}$$

$$(\mathbf{A} - \lambda_1 \mathbf{I}) = \begin{pmatrix} 2 & 1 \\ 4 & 2 \end{pmatrix} \Rightarrow \text{adj}(\mathbf{A} - \lambda_1 \mathbf{I}) = \begin{pmatrix} 2 & -1 \\ -4 & 2 \end{pmatrix}$$

$$(\mathbf{A} - \lambda_2 \mathbf{I}) = \begin{pmatrix} -2 & 1 \\ 4 & -2 \end{pmatrix} \Longrightarrow \text{adj}(\mathbf{A} - \lambda_2 \mathbf{I}) = \begin{pmatrix} -2 & -1 \\ -4 & -2 \end{pmatrix}$$

$$\mathbf{E}_1 = \frac{\text{adj}(\mathbf{A} - \lambda_1 \mathbf{I})}{(\lambda_2 - \lambda_1)} = \begin{pmatrix} \frac{1}{2} & \frac{-1}{4} \\ -1 & \frac{1}{2} \end{pmatrix} = \mathbf{x}_1 \mathbf{y}_1^*$$

$$\mathbf{E}_2 = \frac{\text{adj}(\mathbf{A} - \lambda_2 \mathbf{I})}{(\lambda_1 - \lambda_2)} = \begin{pmatrix} \frac{1}{2} & \frac{1}{4} \\ 1 & \frac{1}{2} \end{pmatrix} = \mathbf{x}_2 \mathbf{y}_2^*$$

$$\mathbf{E}_1^2 = \begin{pmatrix} \frac{1}{2} & \frac{-1}{4} \\ -1 & \frac{1}{2} \end{pmatrix} \begin{pmatrix} \frac{1}{2} & \frac{-1}{4} \\ -1 & \frac{1}{2} \end{pmatrix} = \begin{pmatrix} \frac{1}{4} + \frac{1}{4} & \frac{-1}{8} - \frac{1}{8} \\ -\frac{1}{2} - \frac{1}{2} & \frac{1}{4} + \frac{1}{4} \end{pmatrix} = \begin{pmatrix} \frac{1}{2} & \frac{-1}{4} \\ -1 & \frac{1}{2} \end{pmatrix} = \mathbf{E}_1$$

$$\mathbf{E}_2^2 = \mathbf{E}_2$$

$$\mathbf{E}_1 + \mathbf{E}_2 = \mathbf{I}$$

$$\mathbf{A} = \lambda_1 \mathbf{E}_1 + \lambda_2 \mathbf{E}_2$$

$$f(\mathbf{A}) = f(\lambda_1)\mathbf{E}_1 + f(\lambda_2)\mathbf{E}_2$$

For example,

$$\mathbf{A}^{100} = (-1)^{100} \begin{pmatrix} \frac{1}{2} & \frac{-1}{4} \\ -1 & \frac{1}{2} \end{pmatrix} + (3)^{100} \begin{pmatrix} \frac{1}{2} & \frac{1}{4} \\ 1 & \frac{1}{2} \end{pmatrix}$$

5.5 Spectral theorem

Before we state the spectral theorem, we review the concept of a projection.

Definition. A square matrix $\mathbf{P}$ is called a projection if $\mathbf{P}^2 = \mathbf{P}$.

Two projections $\mathbf{P}_1$ and $\mathbf{P}_2$ are called orthogonal if $\mathbf{P}_1\mathbf{P}_2 = \mathbf{P}_2\mathbf{P}_1 = \mathbf{0}$.

Let $\mathbf{A}$ be a square matrix with simple eigenvalue λ_j. Let $\mathbf{x}_j$ and $\mathbf{y}_j^*$ be the corresponding eigenvector and eigenrow. Consider the dyadic product of $\mathbf{x}_j$ and $\mathbf{y}_j^*$ defined by

$$\mathbf{E}_j = \frac{\mathbf{x}_j \mathbf{y}_j^*}{\mathbf{y}_j^* \mathbf{x}_j} \tag{5.51}$$

[Remark: Note that $\mathbf{y}_j^* \mathbf{x}_j = \langle \mathbf{x}_j, \mathbf{y}_j \rangle$ is a scalar and is a normalization constant while $\mathbf{x}_j \mathbf{y}_j^*$ is an $n \times n$ matrix of rank one.] We now show that $\mathbf{E}_j$ is a projection using the fact that matrix (or vector) multiplication is associative:

$$\mathbf{E}_j^2 = \frac{\mathbf{x}_j\mathbf{y}_j^*}{\mathbf{y}_j^*\mathbf{x}_j}\frac{\mathbf{x}_j\mathbf{y}_j^*}{\mathbf{y}_j^*\mathbf{x}_j}$$

$$= \frac{\mathbf{x}_j(\mathbf{y}_j^*\mathbf{x}_j)\mathbf{y}_j^*}{(\mathbf{y}_j^*\mathbf{x}_j)(\mathbf{y}_j^*\mathbf{x}_j)}$$

$$= \frac{\mathbf{x}_j\mathbf{y}_j^*}{\mathbf{y}_j^*\mathbf{x}_j} = \mathbf{E}_j \tag{5.52}$$

Let $\mathbf{u}$ be any vector in $\mathbb{R}^n/\mathbb{C}^n$ and consider the expansion

$$\mathbf{u} = \sum_{j=1}^{n} \alpha_j \mathbf{x}_j \tag{5.53}$$

where $\{\mathbf{x}_1, \mathbf{x}_2, \ldots, \mathbf{x}_n\}$ are the eigenvectors of $\mathbf{A}$. From the biorthogonal expansion, we have

$$\alpha_j = \frac{\mathbf{y}_j^*\mathbf{u}}{\mathbf{y}_j^*\mathbf{x}_j} \tag{5.54}$$

Now,

$$\mathbf{E}_i\mathbf{u} = \sum_{j=1}^{n} \alpha_j \left(\frac{\mathbf{x}_i\mathbf{y}_i^*}{\mathbf{y}_i^*\mathbf{x}_i}\right)\mathbf{x}_j$$

$$= \alpha_i \mathbf{x}_i \tag{5.55}$$

It follows from equations (5.54) and (5.55) that $\mathbf{E}_i\mathbf{u}$ is the component of $\mathbf{u}$ in the space spanned by the eigenvector $\mathbf{x}_i$, i. e., $\mathbf{E}_i$ is the projection operator onto the eigenspace spanned by $\mathbf{x}_i$. If $i \neq j$, we have

$$\mathbf{E}_i\mathbf{E}_j = \frac{\mathbf{x}_i\mathbf{y}_i^*}{\mathbf{y}_i^*\mathbf{x}_i}\frac{\mathbf{x}_j\mathbf{y}_j^*}{\mathbf{y}_j^*\mathbf{x}_j} = \mathbf{0}$$

Similarly,

$$\mathbf{E}_j\mathbf{E}_i = \mathbf{0}.$$

We can now state the spectral theorem.

Theorem (spectral). *Let $\mathbf{A}$ be a square matrix with distinct eigenvalues $\lambda_1, \lambda_2, \ldots, \lambda_n$. Then there exist projections $\mathbf{E}_j$ such that:*

1. $\sum_{i=1}^{n} \mathbf{E}_i = \mathbf{I}$ *(resolution of the identity)*
2. (a) $\mathbf{E}_i^2 = \mathbf{E}_i$
   (b) $\mathbf{E}_i\mathbf{E}_j = \mathbf{E}_j\mathbf{E}_i = \mathbf{0}$ $(i \neq j)$
3. $\mathbf{A} = \sum_{i=1}^{n} \lambda_i \mathbf{E}_i$ *(spectral resolution of $\mathbf{A}$)*
4. $f(\mathbf{A}) = \sum_{i=1}^{n} f(\lambda_i)\mathbf{E}_i$, *where f is any function defined on the spectrum of $\mathbf{A}$.*

Proof. Let $\mathbf{x}_1, \mathbf{x}_2, \mathbf{x}_3, \dots, \mathbf{x}_n$ be the eigenvectors and $\mathbf{y}_1^*, \mathbf{y}_2^*, \dots, \mathbf{y}_n^*$ be the eigenrows of $\mathbf{A}$. Defining

$$\mathbf{E}_i = \frac{\mathbf{x}_i \mathbf{y}_i^*}{\mathbf{y}_i^* \mathbf{x}_i}$$

we have already proved the relations given by (2-a) and (2-b).
Proof of (3). Consider the matrix

$$\mathbf{T} = (\mathbf{x}_1\, \mathbf{x}_2 \dots \mathbf{x}_n)$$

whose columns are the eigenvectors of $\mathbf{A}$. Then we have

$$\begin{aligned}
\mathbf{AT} &= \mathbf{A}(\mathbf{x}_1\, \mathbf{x}_2 \dots \mathbf{x}_n) \\
&= (\mathbf{Ax}_1\, \mathbf{Ax}_2 \dots \mathbf{Ax}_n) \\
&= (\lambda_1 \mathbf{x}_1\, \lambda_2 \mathbf{x}_2 \dots \lambda_n \mathbf{x}_n) \\
&= (\mathbf{x}_1\, \mathbf{x}_2 \dots \mathbf{x}_n) \begin{pmatrix} \lambda_1 & 0 & 0 & . & . & 0 \\ 0 & \lambda_2 & 0 & . & . & 0 \\ 0 & 0 & \lambda_3 & . & . & 0 \\ . & . & . & . & . & . \\ 0 & 0 & 0 & . & . & \lambda_n \end{pmatrix} \\
&= \mathbf{T\Lambda},
\end{aligned} \tag{5.56}$$

where $\mathbf{\Lambda}$ is a diagonal matrix of eigenvalues (called the spectral matrix). Since $\mathbf{T}$ is invertible, equation (5.56) may be written as

$$\mathbf{A} = \mathbf{T\Lambda T}^{-1} \tag{5.57}$$

We now show that the rows of the matrix $\mathbf{T}^{-1}$ are the normalized eigenrows of $\mathbf{A}$. Let

$$\mathbf{S} = \begin{bmatrix} \mathbf{y}_1^*/(\mathbf{y}_1^* \mathbf{x}_1) \\ \mathbf{y}_2^*/(\mathbf{y}_2^* \mathbf{x}_2) \\ . \\ . \\ \mathbf{y}_n^*/(\mathbf{y}_n^* \mathbf{x}_n) \end{bmatrix}$$

Then

$$\mathbf{ST} = \begin{pmatrix} \mathbf{y}_1^*/(\mathbf{y}_1^* \mathbf{x}_1) \\ \mathbf{y}_2^*/(\mathbf{y}_2^* \mathbf{x}_2) \\ . \\ . \\ \mathbf{y}_n^*/(\mathbf{y}_n^* \mathbf{x}_n) \end{pmatrix} (\mathbf{x}_1\, \mathbf{x}_2 \dots \mathbf{x}_n)$$

$$= \begin{pmatrix} 1 & 0 & . & . & 0 \\ 0 & 1 & . & . & 0 \\ . & . & . & . & . \\ 0 & 0 & . & . & 1 \end{pmatrix} = \mathbf{I}$$

Since the inverse of a matrix is unique, we have

$$\mathbf{S} = \mathbf{T}^{-1} \tag{5.58}$$

Thus,

$$\begin{aligned} \mathbf{A} &= (\lambda_1\mathbf{x}_1\ \lambda_2\mathbf{x}_2 \ldots \lambda_n\mathbf{x}_n) \begin{pmatrix} \mathbf{y}_1^*/(\mathbf{y}_1^*\mathbf{x}_1) \\ \mathbf{y}_2^*/(\mathbf{y}_2^*\mathbf{x}_2) \\ . \\ . \\ \mathbf{y}_n^*/(\mathbf{y}_n^*\mathbf{x}_n) \end{pmatrix} \\ &= \sum_{i=1}^{n} \lambda_i \frac{\mathbf{x}_i\mathbf{y}_i^*}{(\mathbf{y}_i^*\mathbf{x}_i)} \\ &= \sum_{i=1}^{n} \lambda_i \mathbf{E}_i \end{aligned}$$

□

Alternate proof of (3). Define

$$\begin{aligned} \mathbf{A}_1 &= \mathbf{A} - \lambda_1\mathbf{E}_1 \qquad (5.59) \\ \mathbf{A}_1\mathbf{x}_1 &= (\mathbf{A} - \lambda_1\mathbf{E}_1)\mathbf{x}_1 \\ &= \mathbf{0} \\ \mathbf{A}_1\mathbf{x}_j &= \lambda_j\mathbf{x}_j - \lambda_1\mathbf{E}_1\mathbf{x}_j \\ &= \lambda_j\mathbf{x}_j; \quad j = 2, \ldots, n \end{aligned}$$

Thus, the matrix $\mathbf{A}_1$ has eigenvalues $0, \lambda_2, \lambda_3, \ldots, \lambda_n$ with eigenvectors $\mathbf{x}_1, \mathbf{x}_2, \ldots, \mathbf{x}_n$. Now, let

$$\mathbf{A}_2 = \mathbf{A} - \lambda_1\mathbf{E}_1 - \lambda_2\mathbf{E}_2 \tag{5.60}$$

⇒

$$\begin{aligned} \mathbf{A}_2\mathbf{x}_1 &= \mathbf{0}, \quad \mathbf{A}_2\mathbf{x}_2 = \mathbf{0} \\ \mathbf{A}_2\mathbf{x}_j &= \lambda_j\mathbf{x}_j; \quad j = 3, \ldots, n \end{aligned}$$

The matrix $\mathbf{A}_2$ has eigenvalues $0, 0, \lambda_3, \ldots, \lambda_n$ and eigenvectors $\mathbf{x}_1, \mathbf{x}_2, \ldots, \mathbf{x}_n$. Similarly, we see that

$$\mathbf{A}_n = \mathbf{A} - \sum_{i=1}^{n} \lambda_i \mathbf{E}_i \tag{5.61}$$

has eigenvalues $0, 0, \ldots, 0$ and eigenvectors $\mathbf{x}_1, \mathbf{x}_2, \ldots, \mathbf{x}_n$, i. e.,

$$\mathbf{A}_n \mathbf{x}_j = \mathbf{0}; \quad j = 1, 2, 3, \ldots, n \tag{5.62}$$

We now show that equation (5.62) implies that $\mathbf{A}_n = \mathbf{0}$ ($n \times n$ zero matrix), and hence

$$\mathbf{A} = \sum_{i=1}^{n} \lambda_i \mathbf{E}_i$$

Now, (5.62) gives

$$\begin{bmatrix} \boldsymbol{a}_1^T \\ \boldsymbol{a}_2^T \\ \cdot \\ \cdot \\ \boldsymbol{a}_n^T \end{bmatrix} \mathbf{x}_j = 0$$

where $\boldsymbol{a}_k^T = k$-th row of $\mathbf{A}_n$. Since the eigenvectors are linearly independent, the only solution to the system of equations

$$\boldsymbol{a}_k^T \mathbf{x}_j = 0, \quad k = 1, 2, 3, \ldots, n$$

is the trivial one, i. e.,

$$\boldsymbol{a}_k = \mathbf{0} \quad \text{for each } k = 1, 2, 3, \ldots, n$$

$\Longrightarrow$

$$\mathbf{A}_n = \mathbf{0}$$ □

Proof of (1). We have shown

$$\mathbf{u} = \sum_{i=1}^{n} \mathbf{E}_i \mathbf{u}$$

for any vector **u**. To show that (1) implies

$$\sum_{i=1}^{n} \mathbf{E}_i = \mathbf{I},$$

we let

$$\mathbf{S} = \sum_{i=1}^{n} \mathbf{E}_i$$

and rewrite (1) as

$$(\mathbf{S} - \mathbf{I})\mathbf{u} = \mathbf{0} \tag{5.63}$$

Equation (5.63) implies that each row of the matrix $\mathbf{S} - \mathbf{I}$ is orthogonal to n linearly independent vectors $(\mathbf{u}_1, \mathbf{u}_2, \dots, \mathbf{u}_n)$.
$\Longrightarrow$ Each row of $(\mathbf{S} - \mathbf{I})$ is a zero row.
$\therefore$ The result. □

Proof of (4). To prove this, we use (2)

$$\begin{aligned} \mathbf{A} &= \sum_{j=1}^{n} \lambda_j \mathbf{E}_j \\ \Longrightarrow \mathbf{A}^2 &= \left(\sum_{j=1}^{n} \lambda_j \mathbf{E}_j\right)\left(\sum_{i=1}^{n} \lambda_i \mathbf{E}_i\right) \\ &= \sum_{j=1}^{n} \lambda_j^2 \mathbf{E}_j \end{aligned}$$

Similarly,

$$\mathbf{A}^k = \sum_{j=1}^{n} \lambda_j^k \mathbf{E}_j, \quad \text{for } k = 0, 1, 2, \dots$$

Thus, if $f(\lambda)$ is any function that may be expressed in the form

$$f(\lambda) = \sum_{i=0}^{\infty} c_i \lambda^i,$$

we have

$$f(\mathbf{A}) = \sum_{j=1}^{n} f(\lambda_j) \mathbf{E}_j$$

If $\mathbf{A}$ is invertible, it follows from Cayley–Hamilton theorem that $\mathbf{A}^{-1}$ is a polynomial of degree $(n-1)$ in $\mathbf{A}$. Thus, for any arbitrary function $f(\lambda)$, the function $f(\mathbf{A})$ is at most a polynomial of degree $(n-1)$ in $\mathbf{A}$, i. e.,

$$\begin{aligned} f(\mathbf{A}) &= \gamma_0 \mathbf{I} + \gamma_1 \mathbf{A} + \gamma_2 \mathbf{A}^2 + \dots + \gamma_{n-1} \mathbf{A}^{n-1} \\ &= \gamma_0 \left(\sum_{j=1}^{n} \mathbf{E}_j\right) + \gamma_1 \left(\sum_{j=1}^{n} \lambda_j \mathbf{E}_j\right) + \dots + \gamma_{n-1} \left(\sum_{j=1}^{n} \lambda_j^{n-1} \mathbf{E}_j\right) \\ &= \left(\sum_{j=1}^{n} f(\lambda_j) \mathbf{E}_j\right) \end{aligned}$$

Other and more general forms of the spectral theorem may be found in the books on linear algebra (Halmos [20]; Naylor and Sell [24]; Lipschutz and Lipson [22]). □

Example 5.9 (Spectral decomposition). Consider the 2×2 matrix

$$\mathbf{A} = \begin{pmatrix} -3 & 2 \\ 4 & -5 \end{pmatrix}$$

whose eigenvalues are $\lambda_1 = -1$, $\lambda_2 = -7$. We have eigenvectors and normalized eigenrows:

$$\begin{aligned}
\mathbf{x}_1 &= \begin{pmatrix} 1 \\ 1 \end{pmatrix}; \quad \mathbf{x}_2 = \begin{pmatrix} 1 \\ -2 \end{pmatrix} \\
\mathbf{y}_1^* &= \begin{pmatrix} \frac{2}{3} & \frac{1}{3} \end{pmatrix}; \quad \mathbf{y}_2^* = \begin{pmatrix} \frac{1}{3} & -\frac{1}{3} \end{pmatrix} \\
\mathbf{E}_1 &= \frac{\mathbf{x}_1\mathbf{y}_1^*}{\mathbf{y}_1^*\mathbf{x}_1} = \begin{pmatrix} \frac{2}{3} & \frac{1}{3} \\ \frac{2}{3} & \frac{1}{3} \end{pmatrix} \\
\mathbf{E}_2 &= \frac{\mathbf{x}_2\mathbf{y}_2^*}{\mathbf{y}_2^*\mathbf{x}_2} = \begin{pmatrix} \frac{1}{3} & -\frac{1}{3} \\ -\frac{2}{3} & \frac{2}{3} \end{pmatrix} \\
\mathbf{E}_1^2 &= \begin{pmatrix} \frac{2}{3} & \frac{1}{3} \\ \frac{2}{3} & \frac{1}{3} \end{pmatrix}\begin{pmatrix} \frac{2}{3} & \frac{1}{3} \\ \frac{2}{3} & \frac{1}{3} \end{pmatrix} = \begin{pmatrix} \frac{2}{3} & \frac{1}{3} \\ \frac{2}{3} & \frac{1}{3} \end{pmatrix} = \mathbf{E}_1 \\
\mathbf{E}_2^2 &= \begin{pmatrix} \frac{1}{3} & -\frac{1}{3} \\ -\frac{2}{3} & \frac{2}{3} \end{pmatrix}\begin{pmatrix} \frac{1}{3} & -\frac{1}{3} \\ -\frac{2}{3} & \frac{2}{3} \end{pmatrix} = \begin{pmatrix} \frac{1}{3} & -\frac{1}{3} \\ -\frac{2}{3} & \frac{2}{3} \end{pmatrix} = \mathbf{E}_2 \\
\mathbf{E}_1\mathbf{E}_2 &= \begin{pmatrix} \frac{2}{3} & \frac{1}{3} \\ \frac{2}{3} & \frac{1}{3} \end{pmatrix}\begin{pmatrix} \frac{1}{3} & -\frac{1}{3} \\ -\frac{2}{3} & \frac{2}{3} \end{pmatrix} = \begin{pmatrix} 0 & 0 \\ 0 & 0 \end{pmatrix} \\
\mathbf{E}_1 + \mathbf{E}_2 &= \begin{pmatrix} \frac{2}{3} & \frac{1}{3} \\ \frac{2}{3} & \frac{1}{3} \end{pmatrix} + \begin{pmatrix} \frac{1}{3} & -\frac{1}{3} \\ -\frac{2}{3} & \frac{2}{3} \end{pmatrix} = \begin{pmatrix} 1 & 0 \\ 0 & 1 \end{pmatrix} \\
\lambda_1\mathbf{E}_1 + \lambda_2\mathbf{E}_2 &= -1\begin{pmatrix} \frac{2}{3} & \frac{1}{3} \\ \frac{2}{3} & \frac{1}{3} \end{pmatrix} - 7\begin{pmatrix} \frac{1}{3} & -\frac{1}{3} \\ -\frac{2}{3} & \frac{2}{3} \end{pmatrix} \\
&= \begin{pmatrix} -\frac{2}{3} - \frac{7}{3} & -\frac{1}{3} + \frac{7}{3} \\ -\frac{2}{3} + \frac{14}{3} & -\frac{1}{3} - \frac{14}{3} \end{pmatrix} \\
&= \begin{pmatrix} -3 & 2 \\ 4 & -5 \end{pmatrix} = \mathbf{A} \\
f(\mathbf{A}) &= f(\lambda_1)\mathbf{E}_1 + f(\lambda_2)\mathbf{E}_2.
\end{aligned}$$

To show that the same expression is obtained when we use the Cayley–Hamilton theorem, we note that

$$f(\mathbf{A}) = \alpha_0\mathbf{I} + \alpha_1\mathbf{A},$$

where α_0 and α_1 are given by the two linear equations

$$f(\lambda_1) = \alpha_0 - \alpha_1 = f(-1)$$
$$f(\lambda_2) = \alpha_0 - 7\alpha_1 = f(-7).$$

Solving, we get $\alpha_1 = \frac{f(-1)-f(-7)}{6}$; $\alpha_0 = \frac{7f(-1)-f(-7)}{6}$. Thus, we have

$$\begin{aligned}
f(\mathbf{A}) &= \frac{7f(-1)-f(-7)}{6}\begin{pmatrix} 1 & 0 \\ 0 & 1 \end{pmatrix} + \frac{f(-1)-f(-7)}{6}\begin{pmatrix} -3 & 2 \\ 4 & -5 \end{pmatrix} \\
&= \begin{pmatrix} \frac{7f(-1)-f(-7)}{6} & 0 \\ 0 & \frac{7f(-1)-f(-7)}{6} \end{pmatrix} + \begin{pmatrix} \frac{3f(-7)-3f(-1)}{6} & \frac{2f(-1)-2f(-7)}{6} \\ \frac{4f(-1)-4f(-7)}{6} & \frac{5f(-7)-5f(-1)}{6} \end{pmatrix} \\
&= \begin{pmatrix} \frac{4f(-1)+2f(-7)}{6} & \frac{2f(-1)-2f(-7)}{6} \\ \frac{4f(-1)-4f(-7)}{6} & \frac{2f(-1)+4f(-7)}{6} \end{pmatrix} \\
&= f(-1)\begin{pmatrix} \frac{2}{3} & \frac{1}{3} \\ \frac{2}{3} & \frac{1}{3} \end{pmatrix} + f(-7)\begin{pmatrix} \frac{1}{3} & \frac{-1}{3} \\ -\frac{2}{3} & \frac{2}{3} \end{pmatrix} \\
&= f(\lambda_1)\mathbf{E}_1 + f(\lambda_2)\mathbf{E}_2.
\end{aligned}$$

5.6 Projection operators and vector projections

We have already seen that the projection operators E_j appear in the calculation of function of matrices and spectral theorem. These also have geometrical interpretation that is useful in applications. We illustrate the geometrical features briefly in $\mathbb{R}^2$.

5.6.1 Standard basis and projection in $\mathbb{R}^2$

Consider the two elementary orthonormal basis vectors $\mathbf{e}_1 = \binom{1}{0}$ and $\mathbf{e}_2 = \binom{0}{1}$ as shown in Figure 5.1, and projection matrices

$$\mathbf{E}_1 = \mathbf{e}_1\mathbf{e}_1^T = \begin{pmatrix} 1 & 0 \\ 0 & 0 \end{pmatrix}$$

and

$$\mathbf{E}_2 = \mathbf{e}_2\mathbf{e}_2^T = \begin{pmatrix} 0 & 0 \\ 0 & 1 \end{pmatrix}.$$

If $\mathbf{z} = \binom{z_1}{z_2}$ be any vector, then we can write

$$\begin{aligned}
\mathbf{z} &= z_1\mathbf{e}_1 + z_2\mathbf{e}_2 \\
&= \mathbf{E}_1\mathbf{z} + \mathbf{E}_2\mathbf{z},
\end{aligned}$$

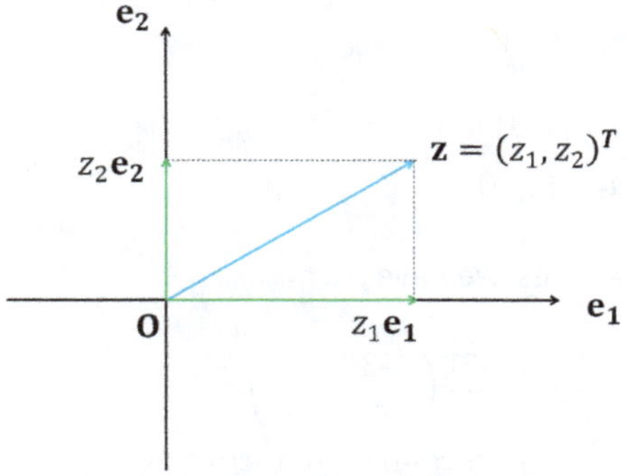

Figure 5.1: Projection in standard basis in $\mathbb{R}^2$.

where

$$\mathbf{E}_1\mathbf{z} = \begin{pmatrix} 1 & 0 \\ 0 & 0 \end{pmatrix}\begin{pmatrix} z_1 \\ z_2 \end{pmatrix} = \begin{pmatrix} z_1 \\ 0 \end{pmatrix} = z_1\mathbf{e}_1$$

is the projection of $\mathbf{z}$ on basis vector $\mathbf{e}_1$ and

$$\mathbf{E}_2\mathbf{z} = \begin{pmatrix} 0 & 0 \\ 0 & 1 \end{pmatrix}\begin{pmatrix} z_1 \\ z_2 \end{pmatrix} = \begin{pmatrix} 0 \\ z_2 \end{pmatrix} = z_2\mathbf{e}_2$$

is the projection of $\mathbf{z}$ on basis vector $\mathbf{e}_2$ (see Figure 5.1).

It can easily be verified that the matrices $\mathbf{E}_1$ and $\mathbf{E}_2$ are orthogonal projection matrices, i. e.,

$$\mathbf{E}_j^2 = \mathbf{E}_j, \quad j = 1, 2$$
$$\mathbf{E}_1\mathbf{E}_2 = \mathbf{E}_2\mathbf{E}_1 = \mathbf{0},$$

and satisfy the resolution of identity, i. e.,

$$\sum_{j=1}^{2} \mathbf{E}_j = \mathbf{I} = \begin{pmatrix} 1 & 0 \\ 0 & 1 \end{pmatrix}.$$

5.6.2 Nonorthogonal projections

Consider a two-dimensional vector $\mathbf{z} = \begin{pmatrix} 2 \\ -1 \end{pmatrix}$ split it into two nonorthogonal independent vectors $\mathbf{x}_1 = \begin{pmatrix} 1 \\ 1 \end{pmatrix}$ and $\mathbf{x}_2 = \begin{pmatrix} 1 \\ -2 \end{pmatrix}$ as shown in Figure 5.2, i. e., $\mathbf{z} = \mathbf{x}_1 + \mathbf{x}_2$. Note that $\mathbf{x}_2^T\mathbf{x}_1 = -1 \neq 0$ (i. e., not orthogonal). These two vectors also form the basis for $\mathbb{R}^2$ with $\mathbf{y}_1^T = \begin{pmatrix} \frac{2}{3} & \frac{1}{3} \end{pmatrix}$ and $\mathbf{y}_2^T = \begin{pmatrix} \frac{1}{3} & \frac{-1}{3} \end{pmatrix}$ forming normalized biorthonormal basis. Note that

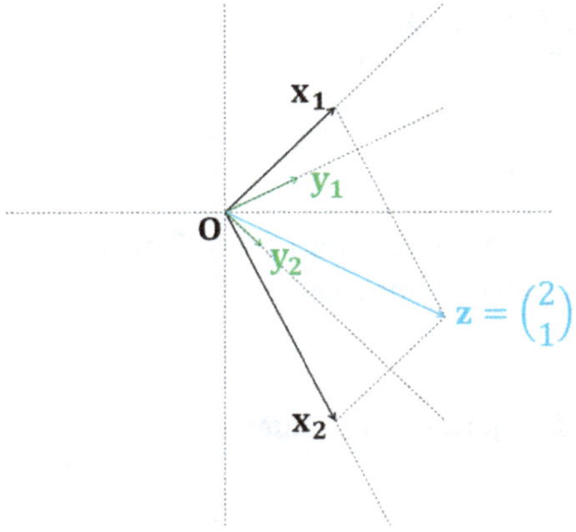

Figure 5.2: Nonorthogonal projections in two-dimensional space.

$\mathbf{y}_2^T\mathbf{x}_1 = \mathbf{y}_1^T\mathbf{x}_2 = 0$ and $\mathbf{y}_1^T\mathbf{x}_1 = \mathbf{y}_2^T\mathbf{x}_2 = 1$. Here, the nonorthogonal projection matrices are given by

$$\mathbf{E}_1 = \mathbf{x}_1\mathbf{y}_1^T = \frac{1}{3}\begin{pmatrix} 2 & 1 \\ 2 & 1 \end{pmatrix}$$

$$\mathbf{E}_2 = \mathbf{x}_2\mathbf{y}_2^T = \frac{1}{3}\begin{pmatrix} 1 & -1 \\ -2 & 2 \end{pmatrix}$$

It is easily verified that

$$\mathbf{E}_1\mathbf{z} = \frac{1}{3}\begin{pmatrix} 2 & 1 \\ 2 & 1 \end{pmatrix}\begin{pmatrix} 2 \\ -1 \end{pmatrix} = \begin{pmatrix} 1 \\ 1 \end{pmatrix} = \mathbf{x}_1,$$

i. e., projection of $\mathbf{z}$ onto $\mathbf{x}_1$ and

$$\mathbf{E}_2\mathbf{z} = \frac{1}{3}\begin{pmatrix} 1 & -1 \\ -2 & 2 \end{pmatrix}\begin{pmatrix} 2 \\ -1 \end{pmatrix} = \begin{pmatrix} 1 \\ -2 \end{pmatrix} = \mathbf{x}_2,$$

i. e., projection of $\mathbf{z}$ onto $\mathbf{x}_2$ (see Figure 5.2).

Here, the projection matrices $\mathbf{E}_1$ and $\mathbf{E}_2$ satisfy:

$$\mathbf{E}_1^2 = \begin{pmatrix} 2/3 & 1/3 \\ 2/3 & 1/3 \end{pmatrix}\begin{pmatrix} 2/3 & 1/3 \\ 2/3 & 1/3 \end{pmatrix} = \begin{pmatrix} 2/3 & 1/3 \\ 2/3 & 1/3 \end{pmatrix} = \mathbf{E}_1$$

$$\mathbf{E}_2^2 = \begin{pmatrix} 1/3 & -1/3 \\ -2/3 & 2/3 \end{pmatrix}\begin{pmatrix} 1/3 & -1/3 \\ -2/3 & 2/3 \end{pmatrix} = \begin{pmatrix} 1/3 & -1/3 \\ -2/3 & 2/3 \end{pmatrix} = \mathbf{E}_2$$

$$\mathbf{E}_1 + \mathbf{E}_2 = \begin{pmatrix} 2/3 & 1/3 \\ 2/3 & 1/3 \end{pmatrix} + \begin{pmatrix} 1/3 & -1/3 \\ -2/3 & 2/3 \end{pmatrix} = \begin{pmatrix} 1 & 0 \\ 0 & 1 \end{pmatrix} = \mathbf{I}$$

$$\mathbf{E}_1\mathbf{E}_2 = \begin{pmatrix} 2/3 & 1/3 \\ 2/3 & 1/3 \end{pmatrix} \begin{pmatrix} 1/3 & -1/3 \\ -2/3 & 2/3 \end{pmatrix} = \begin{pmatrix} 0 & 0 \\ 0 & 0 \end{pmatrix} = \mathbf{0}$$

$$\mathbf{E}_2\mathbf{E}_1 = \begin{pmatrix} 1/3 & -1/3 \\ -2/3 & 2/3 \end{pmatrix} \begin{pmatrix} 2/3 & 1/3 \\ 2/3 & 1/3 \end{pmatrix} = \begin{pmatrix} 0 & 0 \\ 0 & 0 \end{pmatrix} = \mathbf{0}$$

Hence, for the case of real eigenvalues, the eigenvectors and normalized left eigenvectors define the projections onto the one-dimensional eigenspaces.

5.6.3 Geometric interpretation with real and negative eigenvalues

Consider the initial value problem:

$$\frac{d\mathbf{u}}{dt} = \mathbf{A}\mathbf{u}, \quad t > 0; \quad \mathbf{u} = \mathbf{u}^0 \ @ \ t = 0$$

with $\mathbf{A} = \left(\begin{smallmatrix} -3 & 2 \\ 4 & -5 \end{smallmatrix}\right)$. Eigenvalues of $\mathbf{A}$ are $\lambda_1 = -1$ and $\lambda_2 = -7$ with corresponding eigenvectors

$$\mathbf{x}_1 = \begin{pmatrix} 1 \\ 1 \end{pmatrix} \quad \text{and} \quad \mathbf{x}_2 = \begin{pmatrix} 1 \\ -2 \end{pmatrix},$$

respectively and normalized eigenrows as

$$\mathbf{y}_1^T = \begin{pmatrix} 2/3 & 1/3 \end{pmatrix} \quad \text{and} \quad \mathbf{y}_2^T = \begin{pmatrix} 1/3 & -1/3 \end{pmatrix},$$

respectively. As shown in the previous section,

$$\mathbf{E}_1 = \mathbf{x}_1\mathbf{y}_1^T = \begin{pmatrix} 2/3 & 1/3 \\ 2/3 & 1/3 \end{pmatrix}$$

and

$$\mathbf{E}_2 = \mathbf{x}_2\mathbf{y}_2^T = \begin{pmatrix} 1/3 & -1/3 \\ -2/3 & 2/3 \end{pmatrix}$$

are the projection matrices. Thus,

$$\begin{aligned} \lambda_1\mathbf{E}_1 + \lambda_2\mathbf{E}_2 &= (-1)\begin{pmatrix} 2/3 & 1/3 \\ 2/3 & 1/3 \end{pmatrix} + (-7)\begin{pmatrix} 1/3 & -1/3 \\ -2/3 & 2/3 \end{pmatrix} \\ &= \begin{pmatrix} -3 & 2 \\ 4 & -5 \end{pmatrix} = \mathbf{A}. \end{aligned}$$

The solution of above differential equation can be expressed as

$$\mathbf{u}(t) = e^{\mathbf{A}t}\mathbf{u}^0 = \alpha_1\mathbf{x}_1 e^{\lambda_1 t} + \alpha_2\mathbf{x}_2 e^{\lambda_2 t}$$

where

$$\mathbf{u}^0 = \alpha_1\mathbf{x}_1 + \alpha_2\mathbf{x}_2 \Longrightarrow \alpha_1 = \mathbf{y}_1^T\mathbf{u}^0 \quad \text{and} \quad \alpha_2 = \mathbf{y}_2^T\mathbf{u}^0.$$

Assuming $\mathbf{u}^0 = \binom{u_{10}}{u_{20}}$,

$$\alpha_1 = \mathbf{y}_1^T\mathbf{u}^0 = \frac{2u_{10} + u_{20}}{3}$$

and

$$\alpha_2 = \mathbf{y}_2^T\mathbf{u}^0 = \frac{u_{10} - u_{20}}{3}.$$

Thus, the solution can be expressed as

$$\mathbf{u}(t) = \left(\frac{2u_{10} + u_{20}}{3}\right)\mathbf{x}_1 e^{-t} + \left(\frac{u_{10} - u_{20}}{3}\right)\mathbf{x}_2 e^{-7t}.$$

Physical interpretation of eigenvectors

Assuming $\alpha_2 = 0$, i. e.,

$$u_{10} = u_{20} = \beta \quad \text{or} \quad \mathbf{u}^0 = \beta\begin{pmatrix} 1 \\ 1 \end{pmatrix} = \beta\mathbf{x}_1$$

then

$$\mathbf{u}(t) = \beta\mathbf{x}_1 e^{-t}.$$

In other words, when initial condition is in the eigenstate $\mathbf{x}_1$, it always remains in the same state (see Figure 5.3). Similarly, when initial state is in $\mathbf{x}_2$, i. e.,

$$\alpha_1 = 0 \quad \text{or} \quad u_{10} = -\frac{u_{20}}{2} = \gamma,$$

then

$$\mathbf{u}^0 = \gamma\begin{pmatrix} 1 \\ -2 \end{pmatrix} = \gamma\mathbf{x}_2$$

and

$$\mathbf{u}(t) = \gamma\mathbf{x}_2 e^{-7t},$$

i. e., the solution vector always remains in eigenstate $\mathbf{x}_2$.

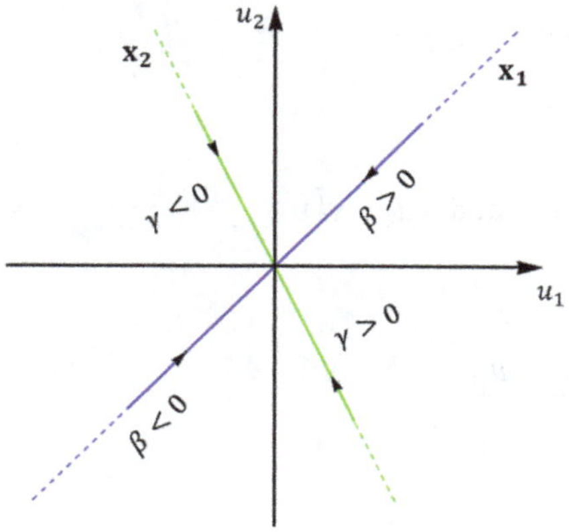

Figure 5.3: Eigenspaces of the initial value problem in the example.

Note that if $\mathbf{u}^0 = \binom{\beta}{0}$, i. e., on u_1-axis, the solution can be expressed as

$$\mathbf{u}(t) = \frac{2}{3}\beta\mathbf{x}_1 e^{-t} + \frac{1}{3}\beta\mathbf{x}_2 e^{-7t}.$$

Since, $e^{-7t} \longrightarrow 0$ much faster than e^{-t} (i. e., the component in eigenstate $\mathbf{x}_2$ vanishes faster), the solution for $t \gg 1$ simplifies to

$$\mathbf{u}(t) = \frac{2}{3}\beta\mathbf{x}_1 e^{-t}, \quad t \gg 1.$$

In other words, the solution approaches the steady state along the $\mathbf{x}_1$ direction. This is also true for any initial condition except when the initial condition is along $\mathbf{x}_2$ (see Figures 5.3 and 5.4).

The trajectory of solution in the (u_1, u_2) plane can be determined for any general initial condition corresponding to the point (u_{10}, u_{20}) in phase plane. Note that solution in two dimensional phase plane can be represented as

$$\mathbf{u}(t) = \left(\frac{2u_{10} + u_{20}}{3}\right)\mathbf{x}_1 e^{-t} + \left(\frac{u_{10} - u_{20}}{3}\right)\mathbf{x}_2 e^{-7t}$$

or

$$\begin{aligned} u_1(t) &= \left(\frac{2u_{10} + u_{20}}{3}\right)e^{-t} + \left(\frac{u_{10} - u_{20}}{3}\right)e^{-7t} \\ u_2(t) &= \left(\frac{2u_{10} + u_{20}}{3}\right)e^{-t} - 2\left(\frac{u_{10} - u_{20}}{3}\right)e^{-7t} \end{aligned}$$

parametrically in t, which can be eliminated by solving for e^{-t} and e^{-7t} that leads to the equation for the trajectory in the form of

$$g(u_1, u_2, u_{10}, u_{20}) = 0$$

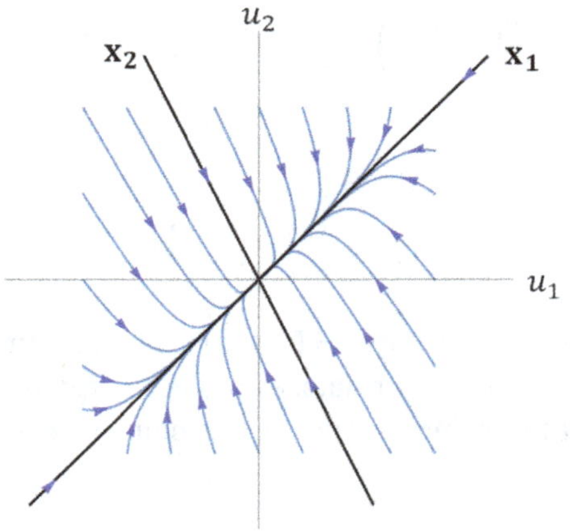

Figure 5.4: Trajectory of solution with stable node in the phase-plane for various initial points. The component of solution along $\mathbf{x}_2$ direction (with large eigenvalue) vanishes first, and solution approaches steady state in $\mathbf{x}_1$ direction (smaller eigenvalue).

in the phase-plane. Such trajectories for various initial points are demonstrated in Figure 5.4. The trivial steady-state here is stable because of negative real eigenvalues (and is referred to as node in the dynamical systems literature).

5.6.4 Geometrical interpretation with complex eigenvalues with negative real part

Consider a matrix $\mathbf{A} = \begin{pmatrix} -1 & 1 \\ -1 & -1 \end{pmatrix}$, whose eigenvalues are $\lambda_1 = -1 + i$ and $\lambda_2 = -1 - i$ with corresponding eigenvectors

$$\mathbf{x}_1 = \begin{pmatrix} 1 \\ i \end{pmatrix} \quad \text{and} \quad \mathbf{x}_2 = \begin{pmatrix} 1 \\ -i \end{pmatrix},$$

and normalized eigenrows

$$\mathbf{y}_1^* = (1/2, -i/2) \quad \text{and} \quad \mathbf{y}_2^* = (1/2, i/2),$$

respectively. Thus, the solution of initial value problem:

$$\frac{d\mathbf{u}}{dt} = \mathbf{A}\mathbf{u}, \quad t > 0 \quad \text{and} \quad \mathbf{u} = \mathbf{u}^0 = (u_{10}, u_{20})^T \ @\, t = 0$$

can be expressed as

$$\begin{aligned} \mathbf{u} &= c_1\mathbf{x}_1 e^{\lambda_1 t} + c_2\mathbf{x}_2 e^{\lambda_2 t} \\ &= \mathbf{y}_1^*\mathbf{u}^0 e^{\lambda_1 t}\mathbf{x}_1 + \mathbf{y}_2^*\mathbf{u}^0 e^{\lambda_2 t}\mathbf{x}_2 \end{aligned}$$

$$= \left(\frac{u_{10} - iu_{20}}{2}\right)\begin{pmatrix} 1 \\ i \end{pmatrix} e^{(-1+i)t} + \left(\frac{u_{10} + iu_{20}}{2}\right)\begin{pmatrix} 1 \\ -i \end{pmatrix} e^{(-1-i)t}$$

$$= e^{-t}\begin{pmatrix} u_{10}\cos t + u_{20}\sin t \\ u_{20}\cos t - u_{10}\sin t \end{pmatrix}$$

$$= e^{-t}\begin{pmatrix} \cos t & \sin t \\ -\sin t & \cos t \end{pmatrix}\begin{pmatrix} u_{10} \\ u_{20} \end{pmatrix}$$

The solution trajectory for this example is shown in Figure 5.5 for various initial conditions. As can be seen from this figure, the solution is oscillatory in time as expected because of complex eigenvalues but has stable focus since real part of the complex eigenvalue is negative.

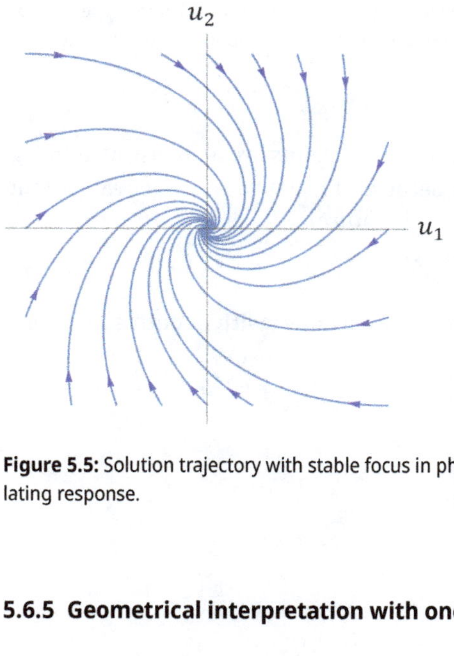

Figure 5.5: Solution trajectory with stable focus in phase-plane with complex eigenvalues leading to oscillating response.

5.6.5 Geometrical interpretation with one zero eigenvalue

Consider a matrix $\mathbf{A} = \begin{pmatrix} -1 & 1 \\ 1 & -1 \end{pmatrix}$, whose eigenvalues are $\lambda_1 = 0$ and $\lambda_2 = -2$ with corresponding eigenvectors

$$\mathbf{x}_1 = \begin{pmatrix} 1 \\ 1 \end{pmatrix} \quad \text{and} \quad \mathbf{x}_2 = \begin{pmatrix} 1 \\ -1 \end{pmatrix},$$

and normalized eigenrows $\mathbf{y}_1^T = (1/2, 1/2)$ and $\mathbf{y}_2^T = (1/2, -1/2)$, respectively. The initial value problem:

$$\frac{d\mathbf{u}}{dt} = \mathbf{Au}, \quad t > 0 \quad \text{and} \quad \mathbf{u} = \mathbf{u}^0 = (u_{10}, u_{20})^T \ @\, t = 0$$

represents the mass exchange between two identical tanks in absence of convection and reaction. The solution for this example can be expressed as

$$\begin{aligned}\mathbf{u} &= c_1\mathbf{x}_1 e^{\lambda_1 t} + c_2\mathbf{x}_2 e^{\lambda_2 t}\\ &= \mathbf{y}_1^*\mathbf{u}^0 e^{\lambda_1 t}\mathbf{x}_1 + \mathbf{y}_2^*\mathbf{u}^0 e^{\lambda_2 t}\mathbf{x}_2\\ &= \left(\frac{u_{10}+u_{20}}{2}\right)\begin{pmatrix}1\\1\end{pmatrix} + \left(\frac{u_{10}-u_{20}}{2}\right)\begin{pmatrix}1\\-1\end{pmatrix}e^{-2t}\end{aligned}$$

Note that at steady-state ($t \longrightarrow \infty$), the solution reduces to

$$\mathbf{u}(t \longrightarrow \infty) = \mathbf{u}_s = \left(\frac{u_{10}+u_{20}}{2}\right)\begin{pmatrix}1\\1\end{pmatrix} = \alpha\mathbf{x}_1, \quad \alpha = \frac{u_{10}+u_{20}}{2}$$

The term $\mathbf{y}_1^T\mathbf{u} = \frac{u_1+u_2}{2} = \frac{u_{10}+u_{20}}{2}$ is constant and denotes the mass conservation of species. The solution trajectories for this example are shown in Figure 5.6 for various initial conditions. The bullet points in this figure represents the equilibrium state ($\mathbf{x}_1$) corresponding to zero eigenvalue. Thus, for any initial condition, the component of the solution along $\mathbf{x}_2$ direction decreases with time (due to negative eigenvalue λ_2) and approaches the equilibrium composition at steady-state, which is practically for all $t \gg 2$.

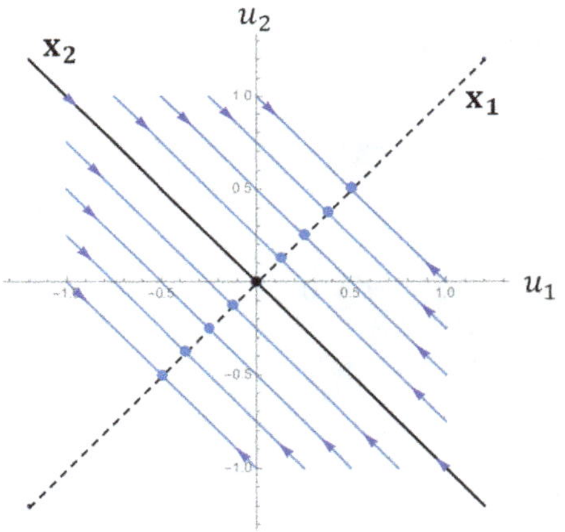

Figure 5.6: Solution trajectory in phase-plane for various initial conditions: the bullet points representing the equilibrium state (corresponding to zero eigenvalue).

5.6.6 Physical and geometrical interpretation of transient behavior of interacting tank systems for various initial conditions

Consider the flow system as shown in Figure 5.7 containing three interacting tanks with volumes $V_1 = \frac{1}{2}V_2 = V_3 = V_R$, and a fixed exchange rates q_e. Assuming no net convection

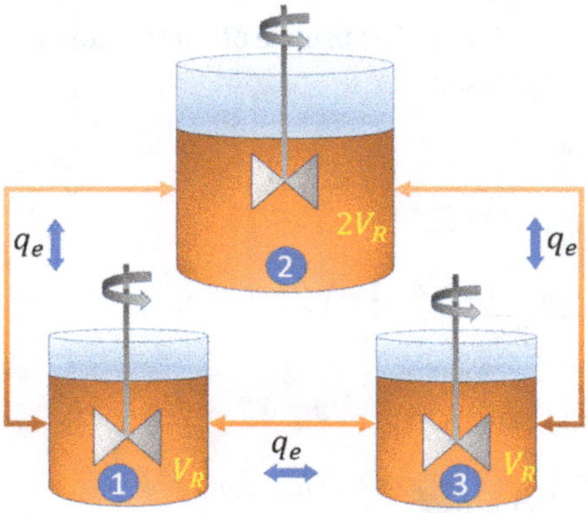

Figure 5.7: Flow system containing three interacting tanks with volumes $V_1 = \frac{1}{2}V_2 = V_3 = V_R$, with a fixed mass-exchange rates in absence of net convection and reaction.

and reaction, and tanks being well mixed, the model equations describing the species balances in these tanks can be expressed as

$$
\begin{aligned}
V_R \frac{dc_1}{dt'} &= -2q_e c_1 + q_e c_2 + q_e c_3 \\
2V_R \frac{dc_1}{dt'} &= q_e c_1 - 2q_e c_2 + q_e c_3 \\
V_R \frac{dc_1}{dt'} &= q_e c_1 + q_e c_2 - 2q_e c_3,
\end{aligned}
$$

with initial conditions

$$c_1 = c_{10}, \quad c_2 = c_{20} \quad \text{and} \quad c_3 = c_{30} \; @ \, t' = 0.$$

Nondimensionalizing real time t' as

$$t = \frac{q_e}{V_R} t'$$

the governing model can be expressed as follows:

$$
\begin{aligned}
&\frac{d\mathbf{c}}{dt} = \mathbf{A}\mathbf{c}, \quad t > 0 \quad \text{and} \quad \mathbf{c} = \mathbf{c}^0 = (c_{10}, c_{20}, c_{30})^T \; @ \, t = 0 \\
&\mathbf{A} = \begin{pmatrix} -2 & 1 & 1 \\ 1/2 & -1 & 1/2 \\ 1 & 1 & -2 \end{pmatrix}.
\end{aligned}
$$

Note that the matrix **A** is not symmetric, but has zero row sums, implying the existence of zero eigenvalue, which also denotes the mass conservation of species.

Eigensystem

The eigenvalues of matrix $\mathbf{A}$ are $\lambda_1 = 0$, $\lambda_2 = -2$ and $\lambda_3 = -3$ corresponding to the eigenvectors

$$\mathbf{x}_1 = \begin{pmatrix} 1 \\ 1 \\ 1 \end{pmatrix}, \quad \mathbf{x}_2 = \begin{pmatrix} 1 \\ -1 \\ 1 \end{pmatrix} \quad \text{and} \quad \mathbf{x}_3 = \begin{pmatrix} 1 \\ 0 \\ -1 \end{pmatrix}$$

and normalized eigenrows

$$\mathbf{y}_1^T = \frac{1}{4}(1,2,1), \quad \mathbf{y}_2^T = \frac{1}{4}(1,-2,1) \quad \text{and} \quad \mathbf{y}_3^T = \frac{1}{2}(1,0,-1).$$

Interpretation of eigenvectors

The eigenvector $\mathbf{x}_1 = \left(\begin{smallmatrix} 1 \\ 1 \\ 1 \end{smallmatrix}\right)$ corresponding to the eigenvalue $\lambda_1 = 0$ indicates the steady state or equilibrium composition $\mathbf{c}_s$:

$$@\, t \longrightarrow \infty, \quad \frac{d\mathbf{c}}{dt} = \mathbf{0} \Longrightarrow \mathbf{A}\mathbf{c}_s = \mathbf{0} \Longrightarrow \mathbf{c}_s = \alpha_1 \mathbf{x}_1 = \alpha_1 \begin{pmatrix} 1 \\ 1 \\ 1 \end{pmatrix}.$$

In order to obtain α_1, initial condition can be utilized. Since $\mathbf{y}_1^T$ is the eigenrow corresponding to zero eigenvalue, we have

$$\begin{aligned} \mathbf{y}_1^T \frac{d\mathbf{c}}{dt} &= \mathbf{y}_1^T \mathbf{A}\mathbf{c} = \mathbf{0}^T \mathbf{c} = 0 \Longrightarrow \mathbf{y}_1^T \mathbf{c} = \text{constant} \\ &\Longrightarrow \mathbf{y}_1^T \mathbf{c}_s = \mathbf{y}_1^T \mathbf{c}^0 \\ &\Longrightarrow \alpha_1 = \mathbf{y}_1^T \mathbf{c}^0 = \frac{c_{10} + 2c_{20} + c_{30}}{4} \end{aligned}$$

Thus, the steady-state concentration in each tank is given by

$$c_{1s} = c_{2s} = c_{3s} = \alpha_1 = \frac{c_{10} + 2c_{20} + c_{30}}{4}$$

$\Longrightarrow$

$$V_R c_{1s} + 2V_R c_{2s} + V_R c_{3s} = 4V_R \alpha_1 = V_R c_{10} + 2V_R c_{20} + V_R c_{30},$$

which represents the conservation of mass.

The other eigenvectors are transient modes, e. g., $\mathbf{x}_2 = \left(\begin{smallmatrix} 1 \\ -1 \\ 1 \end{smallmatrix}\right)$ is the slow transient (asymmetric) mode with eigenvalue $\lambda_2 = -2$ and $\mathbf{x}_3 = \left(\begin{smallmatrix} 1 \\ 0 \\ -1 \end{smallmatrix}\right)$ is the fast transient (skew-symmetric) mode with eigenvalue $\lambda_3 = -3$.

Interpretation of eigenrows

To see the meaning of eigenrows, we express the solution as

$$\mathbf{c}(t) = \sum_{i=1}^{3} \frac{\mathbf{y}_i^T \mathbf{c}^0}{\mathbf{y}_i^T \mathbf{x}_i} \mathbf{x}_i e^{\lambda_i t} = \sum_{i=1}^{3} (\mathbf{y}_i^T \mathbf{c}^0) \mathbf{x}_i e^{\lambda_i t} = \sum_{i=1}^{3} \alpha_i \mathbf{x}_i e^{\lambda_i t}$$
$$= \alpha_1 \mathbf{x}_1 + \alpha_2 \mathbf{x}_2 e^{-2t} + \alpha_3 \mathbf{x}_3 e^{-3t}, \quad \alpha_i = \mathbf{y}_i^T \mathbf{c}^0$$

We note that

$$\mathbf{y}_1^T \mathbf{c} = \alpha_1 = \mathbf{y}_1^T \mathbf{c}^0.$$

Thus, the first eigenrow corresponds to the overall mass-conservation and determines the equilibrium/steady-state composition as shown earlier, i. e., $\mathbf{c}_s = \alpha_1 \begin{pmatrix} 1 \\ 1 \\ 1 \end{pmatrix}$, where

$$\alpha_1 = \mathbf{y}_1^T \mathbf{c}^0 = \frac{c_{10} + 2c_{20} + c_{30}}{4}.$$

We also note that $\alpha_2 = \mathbf{y}_2^T \mathbf{c}^0$. Thus, if $\alpha_2 = 0$, then

$$\mathbf{c}(t) = \alpha_1 \mathbf{x}_1 + \alpha_3 \mathbf{x}_3 e^{-3t},$$

which implies that $\mathbf{y}_2^T \mathbf{c} = 0$. In other words, if the initial concentration vector $\mathbf{c}^0$ is such that $\alpha_2 = 0$, i. e.,

$$\mathbf{c}^0 = \alpha_1 \mathbf{x}_1 + \alpha_3 \mathbf{x}_3,$$

then concentration $\mathbf{c}(t)$ at all times remains a linear combination of $\mathbf{x}_1$ and $\mathbf{x}_3$ with $\mathbf{c}(t = 0) = \mathbf{c}^0$ and $\mathbf{c}(t \longrightarrow \infty) = \alpha_1 \mathbf{x}_1 = \mathbf{c}_s$. It can be seen by combining the condition of $\alpha_2 = 0$ with mass balance constraint as follows:

$$\alpha_2 = \mathbf{y}_2^T \mathbf{c}^0 = 0 \Longrightarrow c_{10} - 2c_{20} + c_{30} = 0$$
$$\alpha_1 = \mathbf{y}_1^T \mathbf{c}^0 \Longrightarrow \frac{c_{10} + 2c_{20} + c_{30}}{4} = \alpha_1.$$

The above two equations lead to

$$c_{20} = \alpha_1 \quad \text{and} \quad c_{10} + c_{30} = 2\alpha_1$$

Thus, the initial condition can be expressed in the form

$$\mathbf{c}^0 = \begin{pmatrix} \alpha_1 + \alpha_3 \\ \alpha_1 \\ \alpha_1 - \alpha_3 \end{pmatrix} = \alpha_1 \begin{pmatrix} 1 \\ 1 \\ 1 \end{pmatrix} + \alpha_3 \begin{pmatrix} 1 \\ 0 \\ -1 \end{pmatrix}$$
$$= \alpha_1 \mathbf{x}_1 + \alpha_3 \mathbf{x}_3 = \mathbf{c}_s + \alpha_3 \mathbf{x}_3$$

With such initial condition, the solution is given by

$$\mathbf{c}(t) = \alpha_1\mathbf{x}_1 + \alpha_3\mathbf{x}_3 e^{-3t} = \mathbf{c}_s + \alpha_3\mathbf{x}_3 e^{-3t},$$

which approaches to steady state along a straight line on the triangular diagram as shown in Figure 5.8 (see the line DEF). For example, for special initial condition $\mathbf{c}^0 = \begin{pmatrix} 0.5 \\ 0.25 \\ 0 \end{pmatrix}$ corresponding to the point D, we have

$$\alpha_1 = \mathbf{y}_1^T\mathbf{c}^0 = \frac{1}{4}, \quad \alpha_2 = \mathbf{y}_2^T\mathbf{c}^0 = 0 \quad \text{and} \quad \alpha_3 = \mathbf{y}_3^T\mathbf{c}^0 = \frac{1}{4},$$

and the solution is given by

$$\mathbf{c}(t) = \frac{1}{4}\mathbf{x}_1 + \frac{1}{4}\mathbf{x}_3 e^{-3t} = \frac{1}{4}\begin{pmatrix} 1 + e^{-3t} \\ 1 \\ 1 - e^{-3t} \end{pmatrix}.$$

In this case, the trajectory goes from point D (@ $t = 0$) to point E (@ $t \longrightarrow \infty$) as shown in Figure 5.8.

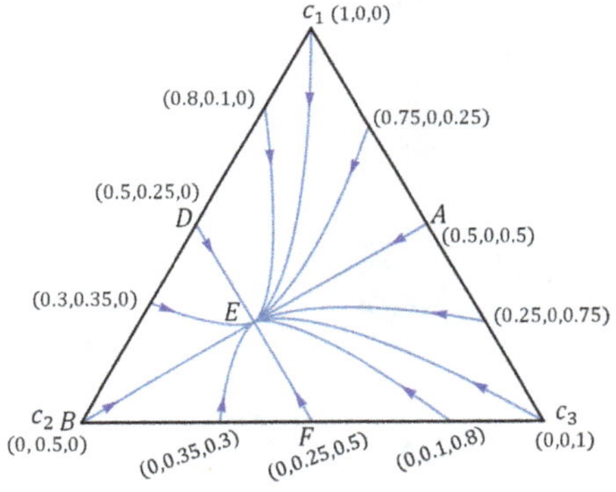

Figure 5.8: Solution trajectory in triangular phase-plane: E is the equilibrium state, AEB is the slow transient state ($\mathbf{x}_2$) and DEF is the fast transient state ($\mathbf{x}_3$).

Similarly, $\alpha_3 = 0$ along with the constraint $\mathbf{y}^1\mathbf{c}^0 = \alpha_1$ corresponds to all initial composition of the form

$$\mathbf{c}^0 = \begin{pmatrix} \alpha_1 + \alpha_2 \\ \alpha_1 - \alpha_2 \\ \alpha_1 + \alpha_2 \end{pmatrix} = \alpha_1 \begin{pmatrix} 1 \\ 1 \\ 1 \end{pmatrix} + \alpha_2 \begin{pmatrix} 1 \\ -1 \\ 1 \end{pmatrix}$$
$$= \alpha_1\mathbf{x}_1 + \alpha_2\mathbf{x}_2 = \mathbf{c}_s + \alpha_2\mathbf{x}_2,$$

which leads to the transient solution as

$$\mathbf{c}(t) = \alpha_1\mathbf{x}_1 + \alpha_2\mathbf{x}_2 e^{-2t} = \mathbf{c}_s + \alpha_2\mathbf{x}_2 e^{-2t}.$$

Thus, the solution remains a linear combination of $\mathbf{x}_1$ and $\mathbf{x}_2$ with

$$\mathbf{c}(t = 0) = \mathbf{c}^0 \quad \text{and} \quad \mathbf{c}(t \longrightarrow \infty) = \alpha_1\mathbf{x}_1 = \mathbf{c}_s,$$

and approaches to steady state along the straight line BEA as shown in triangular phase-trajectory in Figure 5.8. For example, for the special initial condition $\mathbf{c}^0 = \begin{pmatrix} 0 \\ 0.5 \\ 0 \end{pmatrix}$ corresponding to the point B, we have

$$\alpha_1 = \mathbf{y}_1^T\mathbf{c}^0 = \frac{1}{4}, \quad \alpha_2 = \mathbf{y}_2^T\mathbf{c}^0 = \frac{-1}{4} \quad \text{and} \quad \alpha_3 = \mathbf{y}_3^T\mathbf{c}^0 = 0.$$

Thus in this case, the solution simplifies to

$$\mathbf{c}(t) = \frac{1}{4}\mathbf{x}_1 - \frac{1}{4}\mathbf{x}_2 e^{-2t} = \frac{1}{4}\begin{pmatrix} 1 - e^{-2t} \\ 1 + e^{-2t} \\ 1 - e^{-2t} \end{pmatrix},$$

and the trajectory goes from point B (@ $t = 0$) to point E (@ $t \longrightarrow \infty$) as shown in Figure 5.8. Similarly, the trajectories can be determined for other initial conditions as shown in Figure 5.8.

Problems

1. Given the matrix

$$\mathbf{A} = \begin{pmatrix} 0.15 & -0.01 \\ -0.25 & 0.15 \end{pmatrix}$$

evaluate the following:
(i) $e^{\mathbf{A}}$, (ii) $\cos \mathbf{A}$, (iii) $\sinh \mathbf{A}$, (iv) $\ln \mathbf{A}$, (v) $\sum_{j=0}^{\infty} \mathbf{A}^j$, (vi) $J_0(\mathbf{A})$

2. (a) Show that the solution of the inhomogeneous system

$$\frac{d\mathbf{u}}{dt} = \mathbf{A}\mathbf{u} + \mathbf{f}(t), \quad \mathbf{u}(t = 0) = \mathbf{u}_0$$

is given by

$$\mathbf{u} = \mathbf{e}^{\mathbf{A}t}\mathbf{u}_0 + \int_0^t \mathbf{e}^{\mathbf{A}(t-\tau)}\mathbf{f}(\tau)d\tau$$

(b) Determine a similar formula for the solution of the inhomogeneous system

$$\frac{d^2\mathbf{u}}{dt^2} = -\mathbf{A}\mathbf{u} + \mathbf{f}(t), \quad \mathbf{u}(t = 0) = \mathbf{u}_0, \quad \frac{d\mathbf{u}}{dt}(t = 0) = \mathbf{v}_0$$

3. Given the matrix

$$\mathbf{A} = \begin{pmatrix} 3 & 1 & 1 \\ 2 & 4 & 2 \\ 1 & 1 & 3 \end{pmatrix}$$

(a) Compute the eigenvalues and eigenvectors.
(b) What matrix in a similarity transform will reduce **A** to diagonal form.
(c) Determine the spectral decomposition of **A**.
(d) If $f(\lambda)$ is any function defined on the spectrum of **A**, develop a formula for determining $f(\mathbf{A})$.

4. We had three representations for the solution of

$$\frac{d\mathbf{u}}{dt} = \mathbf{A}\mathbf{u}, \mathbf{u}(t = 0) = \mathbf{u}_0$$

(a)

$$\mathbf{u} = \sum_{j=1}^{n} \frac{\mathbf{y}_j^* \mathbf{u}_0}{\mathbf{y}_j^* \mathbf{x}_j} \exp(\lambda_j t)\mathbf{x}_j$$

(b)

$$\mathbf{u} = e^{\mathbf{A}t}\mathbf{u}_0$$

(c)

$$\mathbf{u} = \sum_{j=1}^{n} \exp(\lambda_j t)\mathbf{E}_j \mathbf{u}_0$$

Show that these representations are the same.

5. Consider a cascade of **N** ideal stirred tank reactors (CSTRs) in which the following first-order reactions occur:

$$\mathbf{A} \underset{k_1}{\overset{k_2}{\leftrightarrows}} \mathbf{B} \underset{k_3}{\overset{k_4}{\leftrightarrows}} \mathbf{C}$$

where

$$k_1 = k, \quad k_2 = \frac{k}{2}, \quad k_3 = \frac{k}{2}, \quad k_4 = \frac{k}{4}.$$

(a) Formulate the species balances for steady-state operation and show that the concentration vector in the stream leaving tank **N** is given by

$$\mathbf{c}_N = \left[\mathbf{I} + \frac{k\tau}{N}\mathbf{A}\right]^{-N} \mathbf{c}_0,$$

where $\mathbf{c}_0$ is the feed concentration vector, τ is the space time and **A** is the matrix of relative rate constants defined by

$$\mathbf{A} = \begin{pmatrix} 1 & -\frac{1}{2} & 0 \\ -1 & 1 & -\frac{1}{4} \\ 0 & -\frac{1}{2} & \frac{1}{4} \end{pmatrix}$$

(b) If the reactor is an ideal plug flow reactor, show that the exit concentration vector is given by

$$\mathbf{c}_N = \exp\{-k\tau\mathbf{A}\}\mathbf{c}_0.$$

(c) Compute the exit concentration vector for cases (a) and (b) above for the following parameter values:

$$N = 10, \quad k\tau = 2, \quad \mathbf{c}_0 = \begin{pmatrix} 1 \\ 0.2 \\ 0 \end{pmatrix}$$

6. Consider the problem of diffusion and reaction in an isothermal catalyst particle. When several first-order reactions occur, the concentration vector in a slab geometry satisfies the equations

$$\begin{aligned} \mathbf{D}\frac{d^2\mathbf{c}}{dx^2} &= \mathbf{K}\mathbf{c}, \quad 0 < x < L \\ \mathbf{D}\frac{d\mathbf{c}}{dx} &= \mathbf{k}_c(\mathbf{c} - \mathbf{c}_0) \;@\, x = 0 \\ \frac{d\mathbf{c}}{dx} &= \mathbf{0} \;@\, x = L \end{aligned}$$

where $\mathbf{D}$ is a positive definite matrix of diffusivities, $\mathbf{K}$ is a nonnegative definite matrix of rate constants, $\mathbf{c}_0$ is the vector of ambient concentrations and $\mathbf{k}_c$ is a diagonal matrix of mass transfer coefficients.

(a) Cast the equations in dimensionless form and obtain a formal solution.

(b) If the quantity of interest is the observed (or diffusion/mass transfer disguised) rate vector defined by

$$\mathbf{r}_{\text{obs}} = \frac{1}{L}\int_0^L \mathbf{K}\mathbf{c}(x)\,dx,$$

show that

$$\mathbf{r}_{\text{obs}} = \mathbf{K}^*\mathbf{c}_0,$$

where the diffusion/mass transfer disguised rate constant matrix $\mathbf{K}^*$ is given by

$$\mathbf{K}^* = \frac{\mathbf{D}}{L^2}\mathbf{\Phi}\tanh\mathbf{\Phi}(\mathbf{Bi} + \mathbf{\Phi}\tanh\mathbf{\Phi})^{-1}\mathbf{Bi},$$

where $\mathbf{\Phi}^2 = \mathbf{D}^{-1}\mathbf{K}L^2$; $\mathbf{Bi} = \mathbf{D}^{-1}\mathbf{k}_c L$

(c) If we write

$$\mathbf{K}^* = \mathbf{KH}$$

where $\mathbf{H}$ is called the effectiveness matrix, show that when $\mathbf{D}$ is diagonal and $\mathbf{K}$ is nonsingular, $\mathbf{H}$ is given by $\mathbf{H} = \mathbf{\Phi}^{-1}\tanh\mathbf{\Phi}(\mathbf{I} + \mathbf{Bi}^{-1}\mathbf{\Phi}\tanh\mathbf{\Phi})^{-1}$

(d) Use the result in (b) to calculate $\mathbf{H}$ for the reaction networks (i) $A \xrightarrow{k_1} B \xrightarrow{k_2} C$, (ii) $A \xrightarrow{k_1} B \xrightarrow{k_2} C, A \xrightarrow{k_3} D$ when the diffusivities and mass-transfer coefficients of all the species are identical. Give a physical interpretation of the diagonal elements of $\mathbf{H}$.

(e) Consider again the special case in which

$$\mathbf{D} = d\mathbf{I}; \quad \mathbf{Bi} = \mathrm{Bi}_m\mathbf{I}; \quad \mathrm{Bi}_m = \frac{Lk_c}{d}$$

and we write

$$\mathbf{K} = k\mathbf{A}; \quad \mathbf{K}^* = k\mathbf{A}^*; \quad \mathbf{\Phi}^2 = \phi^2\mathbf{A}; \quad \phi^2 = \frac{kL^2}{d}$$

where $\mathbf{A}$ ($\mathbf{A}^*$) is the relative rate constant (diffusion disguised relative rate constant) matrix. Show that the result in (b) may be expressed in dimensionless form as

$$\mathbf{A}^* = \frac{\sqrt{\mathbf{A}}\tanh\phi\sqrt{\mathbf{A}}}{\phi}\left[\mathbf{I} + \frac{\phi\sqrt{\mathbf{A}}\tanh\phi\sqrt{\mathbf{A}}}{\mathrm{Bi}_m}\right]^{-1}$$

Discuss the two limiting cases of negligible internal resistance ($\phi \to 0$) and negligible external resistance ($\mathrm{Bi}_m \to \infty$). Calculate the diffusion-disguised relative rate constant matrix when $\phi = 10$ and $\mathrm{Bi}_m = \infty$ for the following reaction network:

$$\mathbf{A} \underset{k_1}{\overset{k_2}{\leftrightarrows}} \mathbf{B} \underset{k_3}{\overset{k_4}{\leftrightarrows}} \mathbf{C}$$

where

$$k_1 = k, \quad k_2 = \frac{k}{2}, \quad k_3 = \frac{k}{2}, \quad k_4 = \frac{k}{4}.$$

Draw a schematic diagram of the diffusion-disguised reaction network and compare it with the above true reaction network.

7. The concentration vector in a tubular reactor with axial dispersion satisfies the equations

$$\frac{1}{\text{Pe}}\frac{d^2\mathbf{c}}{d\xi^2} - \frac{d\mathbf{c}}{d\xi} - \mathbf{Dac} = \mathbf{0} \quad 0 < \xi < 1$$

$$\frac{1}{\text{Pe}}\frac{d\mathbf{c}}{d\xi} = \mathbf{c} - \mathbf{c}_0 \ @\ \xi = 0$$

$$\frac{d\mathbf{c}}{d\xi} = \mathbf{0} \ @\ \xi = 1$$

where $\mathbf{c}_0$ is the feed concentration vector, Pe is the Peclet number and $\mathbf{Da} = \mathbf{K}\tau$ is the Damköhler matrix, $\mathbf{K}$ is the rate constant matrix and τ is the space time.

(a) Outline a procedure for solving the above equations using eigenvector expansions. Obtain the formal solution.

(b) Show that the exit concentration vector may be written as

$$\mathbf{c}(\xi = 1) = f(\text{Pe}, \mathbf{Da})\mathbf{c}_0$$

where the function f is defined by

$$f(\text{Pe}, \mathbf{Da}) = 4\exp\left(\frac{\text{Pe}}{2}\right)\mathbf{Y}\left[(\mathbf{I}+\mathbf{Y})^2\exp\left(\frac{\text{Pe}}{2}\mathbf{Y}\right) - (\mathbf{I}-\mathbf{Y})^2\exp\left(-\frac{\text{Pe}}{2}\mathbf{Y}\right)\right]^{-1}$$

and

$$\mathbf{Y} = \left(\mathbf{I} + \frac{4\mathbf{Da}}{\text{Pe}}\right)^{1/2}$$

(c) Use the result in (b) to calculate $\mathbf{c}(\xi = 1)$ for the reaction network in problem #5 and numerical values:

$$\text{Pe} = 2.0; \quad k\tau = 2; \quad \mathbf{c}_0 = \begin{pmatrix} 1 \\ 0.2 \\ 0 \end{pmatrix};$$

$$\mathbf{Da} = k\tau\mathbf{A}; \quad \mathbf{A} = \begin{pmatrix} 1 & -\frac{1}{2} & 0 \\ -1 & 1 & -\frac{1}{4} \\ 0 & -\frac{1}{2} & \frac{1}{4} \end{pmatrix}.$$

6 Generalized eigenvectors and canonical forms

When an $n \times n$ matrix $\mathbf{A}$ that is not symmetric has repeated eigenvalues and fewer than n eigenvectors, it is not possible to find a matrix $\mathbf{T}$ that reduces $\mathbf{A}$ to a diagonal form using a similarity transformation. However, Jordan's theorem states that any $n \times n$ matrix can be reduced to a canonical form called the Jordan canonical form. This chapter gives a brief introduction to generalized eigenvectors and theory of Jordan forms.

6.1 Repeated eigenvalues and generalized eigenvectors

Consider the eigenvalue problem

$$\mathbf{A}\mathbf{x} = \lambda \mathbf{x} \tag{6.1}$$

where $\mathbf{A}$ is a matrix of constants and the solution $\mathbf{x}$ is the eigenvector that depends on eigenvalue λ. Rewriting equation (6.1) as

$$\mathbf{A}\mathbf{x}(\lambda) = \lambda \mathbf{x}(\lambda) \tag{6.2}$$

and differentiating both sides of equation (6.2) w. r. t. λ, leads to

$$\mathbf{A}\mathbf{x}'(\lambda) = \lambda \mathbf{x}'(\lambda) + \mathbf{x}(\lambda) \tag{6.3}$$

Note that if $\lambda = \lambda_1$ is a repeated root of $P_n(\lambda) = 0$ and the rank of $(\mathbf{A} - \lambda_1 \mathbf{I})$ is $n-1$, then there is only one eigenvector that satisfies the equation (6.2) corresponding to repeated root λ_1. Thus, in this case, we can define

$$\mathbf{x}(\lambda) = \mathbf{x}_1 = \text{regular eigenvector corresponding to } \lambda_1; \quad (\mathbf{A} - \lambda_1 \mathbf{I})\mathbf{x}_1 = \mathbf{0}$$
$$\mathbf{x}'(\lambda) = \frac{d\mathbf{x}(\lambda)}{d\lambda} = \mathbf{x}_2 = \text{generalized eigenvector corresponding to } \lambda_1$$

where

$$\mathbf{A}\mathbf{x}_2 = \lambda_1 \mathbf{x}_2 + \mathbf{x}_1 \tag{6.4}$$

or

$$(\mathbf{A} - \lambda_1 \mathbf{I})\mathbf{x}_2 = \mathbf{x}_1 \quad \text{or} \quad (\mathbf{A} - \lambda_1 \mathbf{I})^2 \mathbf{x}_2 = \mathbf{0}. \tag{6.5}$$

It follows from equation (6.5) that $\mathbf{x}_2 \neq \mathbf{0}$ (since this would imply $\mathbf{x}_1 = \mathbf{0}$). Similarly, it can be shown that $\mathbf{x}_1$ and $\mathbf{x}_2$ are linearly independent. Now, we have two eigenvectors: one regular ($\mathbf{x}_1$) and one generalized eigenvector ($\mathbf{x}_2$) of rank 2. This procedure can be

https://doi.org/10.1515/9783111598055-007

generalized if the eigenvalues $\lambda = \lambda_1$ has multiplicity greater than 2 (say $r \geq 2$) and the rank of $(\mathbf{A} - \lambda_1\mathbf{I})$ is smaller than $(n - r)$.

Differentiating equation (6.3) again w. r. t. λ leads to the relation

$$\mathbf{A}\mathbf{x}''(\lambda) = \lambda\mathbf{x}''(\lambda) + 2\mathbf{x}'(\lambda). \tag{6.6}$$

Defining

$$\mathbf{x}_3 = \frac{1}{2!}\mathbf{x}''(\lambda) \tag{6.7}$$

Equation (6.6) at $\lambda = \lambda_1$ can be written as

$$(\mathbf{A} - \lambda_1\mathbf{I})\mathbf{x}_3 = \mathbf{x}_2 \quad \text{or} \quad (\mathbf{A} - \lambda_1\mathbf{I})^3\mathbf{x}_3 = \mathbf{0}, \tag{6.8}$$

and defines a generalized eigenvector of rank 3. Similarly, differentiating equation (6.6) again w. r. t. λ and defining

$$\mathbf{x}_4 = \frac{1}{3!}\mathbf{x}'''(\lambda) \tag{6.9}$$

gives

$$(\mathbf{A} - \lambda_1\mathbf{I})\mathbf{x}_4 = \mathbf{x}_3 \quad \text{or} \quad (\mathbf{A} - \lambda_1\mathbf{I})^4\mathbf{x}_4 = \mathbf{0}, \tag{6.10}$$

and so forth.

6.1.1 Linearly independent solutions of $\frac{d\mathbf{u}}{dt} = \mathbf{A}\mathbf{u}$ with repeated eigenvalues

Let λ be an eigenvalue of $\mathbf{A}$ with eigenvector $\mathbf{x}$. Then we have seen that $\mathbf{u}(t) = \mathbf{x}e^{\lambda t}$ is a solution of $\frac{d\mathbf{u}}{dt} = \mathbf{A}\mathbf{u}$.

Lemma. *Suppose that λ_1 is a repeated eigenvalue of $\mathbf{A}$ (with multiplicity $r = 2$) with eigenvector $\mathbf{x}_1$ and generalized eigenvector (GEV) $\mathbf{x}_2$ such that*

$$(\mathbf{A} - \lambda_1\mathbf{I})\mathbf{x}_1 = \mathbf{0}$$
$$(\mathbf{A} - \lambda_1\mathbf{I})\mathbf{x}_2 = \mathbf{x}_1$$

then

$$\mathbf{u}_1(t) = \mathbf{x}_1 e^{\lambda_1 t}$$
$$\mathbf{u}_2(t) = \mathbf{x}_2 e^{\lambda_1 t} + \mathbf{x}_1 t e^{\lambda_1 t}$$

are the two linearly independent solutions of the vector equation $\frac{d\mathbf{u}}{dt} = \mathbf{A}\mathbf{u}$.

Proof. It is easily verified that $\frac{d\mathbf{u}_1}{dt} = \mathbf{A}\mathbf{u}_1$ and $\frac{d\mathbf{u}_2}{dt} = \mathbf{A}\mathbf{u}_2$.

To show that $\mathbf{u}_1$ and $\mathbf{u}_2$ are linearly independent, let

$$\begin{aligned} c_1\mathbf{u}_1 + c_2\mathbf{u}_2 &= \mathbf{0} \\ \Longrightarrow c_1\mathbf{x}_1 e^{\lambda_1 t} + c_2(\mathbf{x}_2 e^{\lambda_1 t} + \mathbf{x}_1 t e^{\lambda_1 t}) &= \mathbf{0} \end{aligned}$$

Evaluating at $t = 0$,

$$\Longrightarrow c_1\mathbf{x}_1 + c_2\mathbf{x}_2 = \mathbf{0}$$

$\Longrightarrow c_1 = 0$ and $c_2 = 0$ since $\mathbf{x}_1$ and $\mathbf{x}_2$ are linearly independent.

[Remark: If we write $\mathbf{u}_1(t) = \mathbf{u}(\lambda, t) = \mathbf{x}(\lambda)e^{\lambda t}$, then $\mathbf{u}_2(t) = \frac{d}{d\lambda}\mathbf{u}(\lambda, t)$ when λ is a repeated eigenvalue]. □

Theorem. *If* $\mathbf{A}$ *is a symmetric matrix with real elements, then it cannot have any generalized eigenvectors of rank 2 or higher.*

Proof. If $\mathbf{A}^T = \mathbf{A} \Longrightarrow$ eigenvalues are real.

Suppose that λ_1 is repeated and $\mathbf{x}$ is a GEV of $\mathbf{A}$ of rank 2, i. e.,

$$(\mathbf{A} - \lambda_1\mathbf{I})^2\mathbf{x} = \mathbf{0} \tag{6.11}$$

$$(\mathbf{A} - \lambda_1\mathbf{I})\mathbf{x} \neq \mathbf{0} \tag{6.12}$$

Multiplying equation (6.11) on the left by $\mathbf{x}^T$ (or take dot product with $\mathbf{x}$) $\Longrightarrow$

$$\begin{aligned} \mathbf{x}^T(\mathbf{A} - \lambda_1\mathbf{I})^2\mathbf{x} &= \mathbf{0} \\ \Longrightarrow \mathbf{x}^T(\mathbf{A} - \lambda_1\mathbf{I})^T(\mathbf{A} - \lambda_1\mathbf{I})\mathbf{x} &= \mathbf{0} \end{aligned}$$

Since $\mathbf{A}^T = \mathbf{A}$ and λ_1 is real,

$$\begin{aligned} &\Longrightarrow \mathbf{y}^T\mathbf{y} = \mathbf{0}, \quad \text{where } \mathbf{y} = (\mathbf{A} - \lambda_1\mathbf{I})\mathbf{x} \\ &\Longrightarrow \mathbf{y} = \mathbf{0} \Longrightarrow (\mathbf{A} - \lambda_1\mathbf{I})\mathbf{x} = \mathbf{0}, \end{aligned}$$

which is a contradiction (see equation (6.12)). Thus, all eigenvectors of $\mathbf{A}$ are of rank 1. □

6.1.2 Examples of repeated EVs and GEVs

Consider the matrix $\mathbf{A} = \begin{pmatrix} -2 & 1 \\ -1 & -4 \end{pmatrix}$. Its eigenvalues can be obtained by solving

$$|\mathbf{A} - \lambda\mathbf{I}| = 0$$

$\Longrightarrow$

$$\lambda^2 + 6\lambda + 9 = 0 \Longrightarrow \lambda_{1,2} = -3, -3.$$

Note that

$$(\mathbf{A} - \lambda_1\mathbf{I}) = \begin{pmatrix} 1 & 1 \\ -1 & -1 \end{pmatrix} \Longrightarrow \text{rank}(\mathbf{A} - \lambda_1\mathbf{I}) = 1,$$

i. e., only one eigenvector and one generalized eigenvector exist. In addition,

$$(\mathbf{A} - \lambda_1\mathbf{I})^2 = \begin{pmatrix} 0 & 0 \\ 0 & 0 \end{pmatrix}.$$

Thus,

$$(\mathbf{A} - \lambda_1\mathbf{I})^2\mathbf{x}_2 = \mathbf{0} \Longrightarrow \mathbf{x}_2 = \begin{pmatrix} a \\ b \end{pmatrix}.$$

Since $\mathbf{x}_2 \neq \mathbf{0}$, both a and b cannot be simultaneously 0. Thus,

$$\begin{aligned}
\mathbf{x}_1 &= \text{regular eigenvector} \\
&= (\mathbf{A} - \lambda_1\mathbf{I})\mathbf{x}_2 = \begin{pmatrix} 1 & 1 \\ -1 & -1 \end{pmatrix}\begin{pmatrix} a \\ b \end{pmatrix} \\
&= (a+b)\begin{pmatrix} 1 \\ -1 \end{pmatrix}, \quad (a+b) \neq 0
\end{aligned}$$

If we take $a = 1$ and $b = 0$, then the GEV $\mathbf{x}_2 = \begin{pmatrix} 1 \\ 0 \end{pmatrix}$ and regular eigenvector $\mathbf{x}_1 = \begin{pmatrix} 1 \\ -1 \end{pmatrix}$. Using these two eigenvectors (one GEV and one regular eigenvector), the modal matrix can be constructed as

$$\mathbf{T} = (\mathbf{x}_1, \mathbf{x}_2) = \begin{pmatrix} 1 & 1 \\ -1 & 0 \end{pmatrix} \Longrightarrow \mathbf{T}^{-1} = \begin{pmatrix} 0 & -1 \\ 1 & 1 \end{pmatrix}$$

$\Longrightarrow$

$$\begin{aligned}
\mathbf{T}^{-1}\mathbf{A}\mathbf{T} &= \begin{pmatrix} 0 & -1 \\ 1 & 1 \end{pmatrix}\begin{pmatrix} -2 & 1 \\ -1 & -4 \end{pmatrix}\begin{pmatrix} 1 & 1 \\ -1 & 0 \end{pmatrix} \\
&= \begin{pmatrix} 1 & 4 \\ -3 & -3 \end{pmatrix}\begin{pmatrix} 1 & 1 \\ -1 & 0 \end{pmatrix} \\
&= \begin{pmatrix} -3 & 1 \\ 0 & -3 \end{pmatrix} = \mathbf{J} = \text{Jordan form of } \mathbf{A}
\end{aligned}$$

Also, note that the solution of $\frac{d\mathbf{u}}{dt} = \mathbf{A}\mathbf{u}$, $\mathbf{u} = \mathbf{u}_0$ @ $t = 0$ is given by $\mathbf{u} = e^{\mathbf{A}t}\mathbf{u}_0$. But

$$e^{\mathbf{A}t} = \mathbf{T}e^{\mathbf{J}t}\mathbf{T}^{-1},$$

where $\mathbf{J} = \begin{pmatrix} -3 & 1 \\ 0 & -3 \end{pmatrix}$ is the Jordan form of $\mathbf{A}$. Since $e^{\mathbf{J}t} = e^{-3t}\begin{pmatrix} 1 & t \\ 0 & 1 \end{pmatrix}$,

$$\begin{aligned} e^{\mathbf{A}t} &= \mathbf{T}e^{\mathbf{J}t}\mathbf{T}^{-1}, \\ &= \begin{pmatrix} 1 & 1 \\ -1 & 0 \end{pmatrix} e^{-3t} \begin{pmatrix} 1 & t \\ 0 & 1 \end{pmatrix} \begin{pmatrix} 0 & -1 \\ 1 & 1 \end{pmatrix} \\ &= e^{-3t} \begin{pmatrix} 1 & t+1 \\ -1 & -t \end{pmatrix} \begin{pmatrix} 0 & -1 \\ 1 & 1 \end{pmatrix} \\ &= e^{-3t} \begin{pmatrix} t+1 & t \\ -t & -t+1 \end{pmatrix} \end{aligned}$$

$\Longrightarrow$

$$\mathbf{u} = e^{\mathbf{A}t}\mathbf{u}_0 = e^{-3t} \begin{pmatrix} t+1 & t \\ -t & -t+1 \end{pmatrix} \begin{pmatrix} u_{10} \\ u_{20} \end{pmatrix}.$$

6.2 Jordan canonical forms

Definition. An upper (lower) Jordan block of order m is an $m \times m$ matrix with the eigenvalues along the diagonal and unity in the upper (lower) diagonal.

Examples of upper and lower Jordan blocks of order two and three are given below:

$$\mathbf{J}(\lambda) = \begin{pmatrix} \lambda & 1 \\ 0 & \lambda \end{pmatrix} \quad \text{(upper Jordan block of order 2)}$$

$$\mathbf{J}(\lambda) = \begin{pmatrix} \lambda & 1 & 0 \\ 0 & \lambda & 1 \\ 0 & 0 & \lambda \end{pmatrix} \quad \text{(upper Jordan block of order 3)}$$

$$\mathbf{J}(\lambda) = \begin{pmatrix} \lambda & 0 \\ 1 & \lambda \end{pmatrix} \quad \text{(lower Jordan block of order 2)}$$

$$\mathbf{J}(\lambda) = \begin{pmatrix} \lambda & 0 & 0 \\ 1 & \lambda & 0 \\ 0 & 1 & \lambda \end{pmatrix} \quad \text{(lower Jordan block of order 3)}$$

The theory for upper and lower Jordan blocks is identical, the only difference being the arrangement of the generalized eigenvectors. Here, we present the theory for upper Jordan blocks.

Jordan's theorem. *Given a square matrix* $\mathbf{A}$, *there exists a similarity transformation, i. e., a matrix* $\mathbf{T}$ *such that*

$$\mathbf{T}^{-1}\mathbf{AT} = \mathbf{B}, \quad \mathbf{B} = \begin{pmatrix} \mathbf{J}(\lambda_1) & 0 & . & 0 \\ 0 & \mathbf{J}(\lambda_2) & . & 0 \\ 0 & 0 & . & 0 \\ 0 & 0 & . & \mathbf{J}(\lambda_r) \end{pmatrix}$$

and the number of Jordan blocks is equal to the number of linearly independent eigenvectors of **A** *and there may be more than one block with the same eigenvalue along the diagonal.*

The proof of this theorem may be found in standard matrix algebra books (Bronson and Costa [9]; Gantmacher [18]).

Example 6.1. Consider a 5×5 matrix **A** with the following characteristic polynomial:

$$\mathbf{P}_5(\lambda) = (\lambda_1 - \lambda)^3(\lambda_2 - \lambda)^2.$$

Then the following possible Jordan forms may exist depending on the number of eigenvectors:

1. There are 5 eigenvectors (3 corresponding to λ_1 and 2 corresponding to λ_2). In this case, **A** can be diagonalized and

$$\mathbf{T}^{-1}\mathbf{AT} = \begin{pmatrix} \lambda_1 & 0 & 0 & 0 & 0 \\ 0 & \lambda_1 & 0 & 0 & 0 \\ 0 & 0 & \lambda_1 & 0 & 0 \\ 0 & 0 & 0 & \lambda_2 & 0 \\ 0 & 0 & 0 & 0 & \lambda_2 \end{pmatrix}$$

2. There are 4 eigenvectors, 2 corresponding to λ_1, 2 corresponding to λ_2. In this case, Jordan's theorem implies that there exists a **T**, such that

$$\mathbf{B} = \mathbf{T}^{-1}\mathbf{AT} = \begin{pmatrix} \lambda_1 & 1 & 0 & 0 & 0 \\ 0 & \lambda_1 & 0 & 0 & 0 \\ 0 & 0 & \lambda_1 & 0 & 0 \\ 0 & 0 & 0 & \lambda_2 & 0 \\ 0 & 0 & 0 & 0 & \lambda_2 \end{pmatrix} \quad \text{(4 Jordan blocks)}$$

3. There are 4 eigenvectors, 3 corresponding to λ_1, 1 corresponding to λ_2:

$$\mathbf{B} = \mathbf{T}^{-1}\mathbf{AT} = \begin{pmatrix} \lambda_1 & 0 & 0 & 0 & 0 \\ 0 & \lambda_1 & 0 & 0 & 0 \\ 0 & 0 & \lambda_1 & 0 & 0 \\ 0 & 0 & 0 & \lambda_2 & 1 \\ 0 & 0 & 0 & 0 & \lambda_2 \end{pmatrix} \quad \text{(4 Jordan blocks)}$$

4. There are 3 eigenvectors, 2 corresponding to λ_1, 1 corresponding to λ_2:

$$\mathbf{B} = \mathbf{T}^{-1}\mathbf{A}\mathbf{T} = \begin{pmatrix} \lambda_1 & 1 & 0 & 0 & 0 \\ 0 & \lambda_1 & 0 & 0 & 0 \\ 0 & 0 & \lambda_1 & 0 & 0 \\ 0 & 0 & 0 & \lambda_2 & 1 \\ 0 & 0 & 0 & 0 & \lambda_2 \end{pmatrix} \quad \text{(3 Jordan blocks)}$$

5. There are 3 eigenvectors, 1 corresponding to λ_1, 2 corresponding to λ_2:

$$\mathbf{B} = \mathbf{T}^{-1}\mathbf{A}\mathbf{T} = \begin{pmatrix} \lambda_1 & 1 & 0 & 0 & 0 \\ 0 & \lambda_1 & 1 & 0 & 0 \\ 0 & 0 & \lambda_1 & 0 & 0 \\ 0 & 0 & 0 & \lambda_2 & 0 \\ 0 & 0 & 0 & 0 & \lambda_2 \end{pmatrix} \quad \text{(3 Jordan blocks)}$$

6. There are 2 eigenvectors, each corresponding to λ_1 and λ_2. In this case, the canonical form of $\mathbf{A}$ consists of two Jordan blocks and is of the form,

$$\mathbf{B} = \mathbf{T}^{-1}\mathbf{A}\mathbf{T} = \begin{pmatrix} \lambda_1 & 1 & 0 & 0 & 0 \\ 0 & \lambda_1 & 1 & 0 & 0 \\ 0 & 0 & \lambda_1 & 0 & 0 \\ 0 & 0 & 0 & \lambda_2 & 1 \\ 0 & 0 & 0 & 0 & \lambda_2 \end{pmatrix} \quad \text{(2 Jordan blocks)}$$

Note that Jordan's theorem tells us that $\mathbf{A}$ can be reduced to a canonical form $\mathbf{B}$ but does not tell us how to find the matrix $\mathbf{T}$ in the similarity transformation. We now focus on this aspect.

6.3 Multiple eigenvalues and generalized eigenvectors

In the general case, let the characteristic polynomial of $\mathbf{A}$ be of the form

$$P_n(\lambda) = (\lambda_1 - \lambda)^{m_1}(\lambda_2 - \lambda)^{m_2} \dots (\lambda_r - \lambda)^{m_r} \tag{6.13}$$

where, $\lambda_1, \lambda_2, \dots, \lambda_r$ are distinct eigenvalues and

$$m_1 + m_2 + m_3 + \dots + m_r = n, \tag{6.14}$$

m_i is called the *algebraic multiplicity* of the eigenvalue λ_i. Let the number of linearly independent eigenvectors corresponding to λ_i be M_i. We note that if $\text{rank}(\mathbf{A} - \lambda_i\mathbf{I}) = n - M_i$, then there are M_i independent solutions of the homogenous equations $(\mathbf{A} - \lambda_i\mathbf{I})\mathbf{x} = \mathbf{0}$.

The positive integer M_i is called the *geometric multiplicity* of the eigenvalue λ_i. In general, $M_i \leq m_i$. If $M_i = m_i$ for all $i = 1, 2, \ldots, r$, then **A** is diagonalizable, i. e., **B** is a diagonal matrix as in case (i) of Example (6.1). If $M_i < m_i$ for some i, then **A** can only be reduced to a Jordan canonical form **B**.

To see how to find **T** that reduces **A** to a Jordan form, consider first the special case in which **A** is $n \times n$ and has a single eigenvalue λ_1 of multiplicity n. Let $\mathbf{x}_1, \mathbf{x}_2, \ldots, \mathbf{x}_n$ be the columns of **T**. Then, by Jordan's theorem,

$$\mathbf{T}^{-1}\mathbf{A}\mathbf{T} = \mathbf{J}_n(\lambda_1) \Longrightarrow \mathbf{A}\mathbf{T} = \mathbf{T}\mathbf{J}_n(\lambda_1) \tag{6.15}$$

$\Longrightarrow$

$$\mathbf{A}(\mathbf{x}_1\,\mathbf{x}_2\,\ldots\mathbf{x}_n) = (\mathbf{x}_1\,\mathbf{x}_2\,\ldots\mathbf{x}_n)\begin{pmatrix} \lambda_1 & 1 & 0 & . & . & 0 \\ 0 & \lambda_1 & 1 & . & . & 0 \\ 0 & 0 & \lambda_1 & . & . & 0 \\ . & . & . & . & . & 1 \\ 0 & 0 & 0 & . & . & \lambda_1 \end{pmatrix}$$

$$\Longrightarrow \mathbf{A}\mathbf{x}_1 = \lambda_1\mathbf{x}_1 \quad \text{or} \quad (\mathbf{A} - \lambda_1\mathbf{I})\mathbf{x}_1 = \mathbf{0} \tag{6.16}$$

$$\mathbf{A}\mathbf{x}_2 = \mathbf{x}_1 + \lambda_1\mathbf{x}_2 \quad \text{or} \quad (\mathbf{A} - \lambda_1\mathbf{I})\mathbf{x}_2 = \mathbf{x}_1 \tag{6.17}$$

$$\mathbf{A}\mathbf{x}_3 = \mathbf{x}_2 + \lambda_1\mathbf{x}_3 \quad \text{or} \quad (\mathbf{A} - \lambda_1\mathbf{I})\mathbf{x}_3 = \mathbf{x}_2 \tag{6.18}$$

$$.$$

$$.$$

$$\mathbf{A}\mathbf{x}_n = \mathbf{x}_{n-1} + \lambda_1\mathbf{x}_n \quad \text{or} \quad (\mathbf{A} - \lambda_1\mathbf{I})\mathbf{x}_n = \mathbf{x}_{n-1} \tag{6.19}$$

These equations define the columns of **T**. The vector $\mathbf{x}_1$ is the regular eigenvector. We call the vectors $\mathbf{x}_2, \mathbf{x}_3, \ldots, \mathbf{x}_n$ the *generalized eigenvectors*. To see the properties of these vectors, premultiplying equation (6.17) by $(\mathbf{A} - \lambda_1\mathbf{I})$, we get

$$(\mathbf{A} - \lambda_1\mathbf{I})^2\mathbf{x}_2 = \mathbf{0}$$

Similarly, it can be seen that

$$\begin{aligned} &(\mathbf{A} - \lambda_1\mathbf{I})^3\mathbf{x}_3 = \mathbf{0} \\ &\ldots \\ &\ldots \\ &(\mathbf{A} - \lambda_1\mathbf{I})^n\mathbf{x}_n = \mathbf{0} \end{aligned} \tag{6.20}$$

These are the equations for determining the generalized eigenvectors. The vector $\mathbf{x}_2$ is called a generalized eigenvector of rank 2, $\mathbf{x}_3$ is a generalized eigenvector of rank 3, etc. Note that once $\mathbf{x}_n$ is determined, all the others can be determined by simply using the above equations in reverse order, i. e.,

$$\mathbf{x}_{n-1} = (\mathbf{A} - \lambda_1 \mathbf{I})\mathbf{x}_n$$
$$\mathbf{x}_{n-2} = (\mathbf{A} - \lambda_1 \mathbf{I})\mathbf{x}_{n-1}$$
$$\vdots$$
$$\mathbf{x}_1 = (\mathbf{A} - \lambda_1 \mathbf{I})\mathbf{x}_2$$

Definition. A vector $\mathbf{x}_m$ is called a generalized eigenvector (GEV) of rank m for $\mathbf{A}$ corresponding to eigenvalue λ_1 if

$$(\mathbf{A} - \lambda_1 \mathbf{I})^m \mathbf{x}_m = \mathbf{0}$$

and

$$(\mathbf{A} - \lambda_1 \mathbf{I})^{m-1} \mathbf{x}_m \neq \mathbf{0}$$

Chains

A chain generated by a GEV $\mathbf{x}_m$ of rank m associated with eigenvalue λ_1 is a set of vectors $\{\mathbf{x}_m, \mathbf{x}_{m-1}, \ldots, \mathbf{x}_1\}$ defined recursively as

$$\mathbf{x}_j = (\mathbf{A} - \lambda_1 \mathbf{I})\mathbf{x}_{j+1}, \quad j = m-1, m-2, \ldots, 1$$

The number of vectors in the set is called the length of the chain.

The procedure for finding $\mathbf{T}$ may be summarized as follows: Let $P_n(\lambda)$ given by equation (6.13) be the characteristic polynomial. Determine the chain of GEVs corresponding to each eigenvalue λ_j. Arrange these as columns of $\mathbf{T}$, i. e.,

$$\mathbf{T} = (\mathbf{x}_{1,1}\mathbf{x}_{1,2} \ldots \mathbf{x}_{1,m_1}\mathbf{x}_{2,1} \ldots \mathbf{x}_{2,m_2} \ldots \mathbf{x}_{r,1} \ldots \mathbf{x}_{r,m_r})$$

We now focus on the determination of each chain of the GEV.

Procedure for determining the chain of the GEV corresponding to λ_1 (multiplicity m_1)

1. Determine first, rank $(\mathbf{A} - \lambda_1 \mathbf{I}) = r_1$. If, $r_1 = n - m_1$, then we have m_1 eigenvectors of rank 1 and we are finished. If, $r_1 > n - m_1$, then determine the smallest integer p_1 for which $\operatorname{rank}(\mathbf{A} - \lambda_1 \mathbf{I})^{p_1} = n - m_1$.
2. For each integer k between 1 and p_1, inclusive, compute the eigenvalue rank number N_k as

$$N_k = \operatorname{rank}(\mathbf{A} - \lambda_1 \mathbf{I})^{k-1} - \operatorname{rank}(\mathbf{A} - \lambda_1 \mathbf{I})^k$$

Each N_k is the number of generalized eigenvectors of rank k that will appear in $\mathbf{T}$.

3. Determine a GEV of rank p_1 by solving the equations

$$(\mathbf{A}-\lambda_1\mathbf{I})^{p_1}\mathbf{x}_{p_1}=\mathbf{0},\quad (\mathbf{A}-\lambda_1\mathbf{I})^{p_1-1}\mathbf{x}_{p_1}\neq\mathbf{0}$$

and construct the chain generated by this vector. Each of these vectors is part of the canonical basis.
4. Reduce each positive N_k by one. If all N_k are zero, we are finished. If not, find the highest value of k for which N_k is not zero and determine a GEV of that rank, which is linearly independent of all previously determined GEV associated with λ_1. Determine the chain generated by this vector and include this in the canonical basis.
5. Repeat step 4 until all GEVs are found.

Example 6.2.

$$\mathbf{A}=\begin{pmatrix}3&2&0&1\\0&3&0&0\\0&0&3&-1\\0&0&0&3\end{pmatrix}$$

Here, the eigenvalue $\lambda=3$ has multiplicity $m_1=4$. We have

$$(\mathbf{A}-3\mathbf{I})=\begin{pmatrix}0&2&0&1\\0&0&0&0\\0&0&0&-1\\0&0&0&0\end{pmatrix};\quad \operatorname{rank}(\mathbf{A}-3\mathbf{I})=2$$

Thus, there are two eigenvectors, and hence two generalized eigenvectors. This can also be confirmed from the following calculation:

$$(\mathbf{A}-3\mathbf{I})^2=\begin{pmatrix}0&0&0&0\\0&0&0&0\\0&0&0&0\\0&0&0&0\end{pmatrix};\quad \operatorname{rank}[(\mathbf{A}-3\mathbf{I})^2]=0\Rightarrow p_1=2$$

$N_2=\operatorname{rank}(\mathbf{A}-\lambda_1\mathbf{I})-\operatorname{rank}(\mathbf{A}-\lambda_1\mathbf{I})^2=2-0=2$ and $N_1=\operatorname{rank}(\mathbf{A}-\lambda_1\mathbf{I})^0-\operatorname{rank}(\mathbf{A}-\lambda_1\mathbf{I})=4-2=2$. Hence, there are two generalized eigenvectors of rank 2 and two of rank 1.

Now, to determine the generalized eigenvectors of rank 2, we solve

$$(\mathbf{A}-\lambda_1\mathbf{I})^2\mathbf{x}_2=\mathbf{0}\Rightarrow \text{any } \mathbf{x}_2 \text{ satisfies the equations.}$$

$$(\mathbf{A}-\lambda_1\mathbf{I})\mathbf{x}_2=\begin{pmatrix}2x_{22}+x_{42}\\0\\-x_{42}\\0\end{pmatrix}\neq\mathbf{0}\Rightarrow 2x_{22}+x_{42}\neq 0 \text{ or } x_{42}\neq 0$$

Hence, we take the two linearly independent generalized eigenvectors of rank 2 as

$$\mathbf{x}_2 = \begin{pmatrix} 0 \\ 0 \\ 0 \\ 1 \end{pmatrix}, \quad \mathbf{y}_2 = \begin{pmatrix} 0 \\ 1 \\ 0 \\ 0 \end{pmatrix}$$

and obtain the eigenvectors of the corresponding chains as

$$\mathbf{x}_1 = \begin{pmatrix} 1 \\ 0 \\ -1 \\ 0 \end{pmatrix}$$

and

$$\mathbf{y}_1 = \begin{pmatrix} 2 \\ 0 \\ 0 \\ 0 \end{pmatrix}$$

Hence,

$$\mathbf{T} = \begin{pmatrix} 2 & 0 & 1 & 0 \\ 0 & 1 & 0 & 0 \\ 0 & 0 & -1 & 0 \\ 0 & 0 & 0 & 1 \end{pmatrix}$$

$$\mathbf{T}^{-1}\mathbf{AT} = \begin{pmatrix} 3 & 1 & 0 & 0 \\ 0 & 3 & 0 & 0 \\ 0 & 0 & 3 & 1 \\ 0 & 0 & 0 & 3 \end{pmatrix}$$

Example 6.3.

$$\mathbf{A} = \begin{pmatrix} 4 & 2 & 1 & 0 & 0 & 0 \\ 0 & 4 & -1 & 0 & 0 & 0 \\ 0 & 0 & 4 & 0 & 0 & 0 \\ 0 & 0 & 0 & 4 & 2 & 0 \\ 0 & 0 & 0 & 0 & 4 & 0 \\ 0 & 0 & 0 & 0 & 0 & 7 \end{pmatrix} \Rightarrow P_6(\lambda) = (4-\lambda)^5(7-\lambda)$$

Thus, $\lambda_1 = 4$, $m_1 = 5$, $n - m_1 = 1$; $\lambda_2 = 7$, $m_2 = 1$. The eigenvector corresponding to λ_2 is given by

$$\begin{pmatrix} -3 & 2 & 1 & 0 & 0 & 0 \\ 0 & -3 & -1 & 0 & 0 & 0 \\ 0 & 0 & -3 & 0 & 0 & 0 \\ 0 & 0 & 0 & -3 & 2 & 0 \\ 0 & 0 & 0 & 0 & -3 & 0 \\ 0 & 0 & 0 & 0 & 0 & 0 \end{pmatrix} \begin{pmatrix} x_1 \\ x_2 \\ x_3 \\ x_4 \\ x_5 \\ x_6 \end{pmatrix} = \begin{pmatrix} 0 \\ 0 \\ 0 \\ 0 \\ 0 \\ 0 \end{pmatrix}$$

$$\Rightarrow \mathbf{z}_1 = \begin{pmatrix} 0 \\ 0 \\ 0 \\ 0 \\ 0 \\ 1 \end{pmatrix} \quad \text{is an eigenvector corresponding to } \lambda_2.$$

To determine the chain of GEV corresponding to λ_1, we note that

$$(\mathbf{A} - 4\mathbf{I}) = \begin{pmatrix} 0 & 2 & 1 & 0 & 0 & 0 \\ 0 & 0 & -1 & 0 & 0 & 0 \\ 0 & 0 & 0 & 0 & 0 & 0 \\ 0 & 0 & 0 & 0 & 2 & 0 \\ 0 & 0 & 0 & 0 & 0 & 0 \\ 0 & 0 & 0 & 0 & 0 & 3 \end{pmatrix} \Rightarrow \text{rank}(\mathbf{A} - 4\mathbf{I}) = 4$$

$$(\mathbf{A} - 4\mathbf{I})^2 = \begin{pmatrix} 0 & 0 & -2 & 0 & 0 & 1 \\ 0 & 0 & 0 & 0 & 0 & -1 \\ 0 & 0 & 0 & 0 & 0 & 0 \\ 0 & 0 & 0 & 0 & 0 & 0 \\ 0 & 0 & 0 & 0 & 0 & 0 \\ 0 & 0 & 0 & 0 & 0 & 9 \end{pmatrix} \Rightarrow \text{rank}(\mathbf{A} - 4\mathbf{I})^2 = 2$$

$$(\mathbf{A} - 4\mathbf{I})^3 = \begin{pmatrix} 0 & 0 & 0 & 0 & 0 & -2 \\ 0 & 0 & 0 & 0 & 0 & 0 \\ 0 & 0 & 0 & 0 & 0 & 0 \\ 0 & 0 & 0 & 0 & 0 & 0 \\ 0 & 0 & 0 & 0 & 0 & 0 \\ 0 & 0 & 0 & 0 & 0 & 27 \end{pmatrix} \Rightarrow \text{rank}(\mathbf{A} - 4\mathbf{I})^3 = 1$$

Thus, $p_1 = 3$ and $N_3 = \text{rank}(\mathbf{A} - 4\mathbf{I})^2 - \text{rank}(\mathbf{A} - 4\mathbf{I})^3 = 2 - 1 = 1$,

$$N_2 = \text{rank}(\mathbf{A} - 4\mathbf{I}) - \text{rank}(\mathbf{A} - 4\mathbf{I})^2 = 4 - 2 = 2$$

$$N_1 = \text{rank}(\mathbf{A} - 4\mathbf{I})^0 - \text{rank}(\mathbf{A} - 4\mathbf{I}) = 6 - 4 = 2$$

The chain corresponding to λ_1 will consist of one GEV of rank 3, two GEV of rank 2 and two GEVs of rank 1. We first determine the GEV of rank 3 by solving

$$(\mathbf{A}-4\mathbf{I})^3\mathbf{x}_3=\mathbf{0}; \quad (\mathbf{A}-4\mathbf{I})^2\mathbf{x}_3\neq\mathbf{0}$$

If $\mathbf{x}_3^T=(x_1\ x_2\ x_3\ x_4\ x_5\ x_6)$, then the above equations give $x_6=0$ while $x_1,\dots,x_5$ are arbitrary. However, for $\mathbf{x}_3$ to be GEV of rank 3, we should have $(\mathbf{A}-4\mathbf{I})^2\mathbf{x}_3\neq\mathbf{0}\Rightarrow x_3\neq 0$. Hence, we take

$$\mathbf{x}_3=\begin{pmatrix}0\\0\\1\\0\\0\\0\end{pmatrix}$$

The remaining vectors of the chain generated by $\mathbf{x}_3$ are

$$\mathbf{x}_2=(\mathbf{A}-4\mathbf{I})\mathbf{x}_3=\begin{pmatrix}-1\\-1\\0\\0\\0\\0\end{pmatrix}$$

$$\mathbf{x}_1=(\mathbf{A}-4\mathbf{I})\mathbf{x}_2=\begin{pmatrix}-2\\0\\0\\0\\0\\0\end{pmatrix}$$

We now reduce each N_k by one, obtaining $N_3=0, N_2=1, N_1=1$. This implies that we still need to find one GEV of rank 2 and one GEV of rank 1. If we denote the GEV of rank 2 by $\mathbf{y}_2^T=(y_{12}\ y_{22}\ y_{32}\ y_{42}\ y_{52}\ y_{62})$, it is given by solving

$$(\mathbf{A}-4\mathbf{I})^2\mathbf{y}_2=\mathbf{0}; \quad (\mathbf{A}-4\mathbf{I})\mathbf{y}_2\neq\mathbf{0}$$

$\Rightarrow y_{32}=y_{62}=0$ and $y_{12}, y_{22}, y_{42}, y_{52}$ are arbitrary but must be chosen such that $(\mathbf{A}-4\mathbf{I})\mathbf{y}_2\neq\mathbf{0}$ and $\mathbf{y}_2$ is independent of $\mathbf{x}_3, \mathbf{x}_2$ and $\mathbf{x}_1$ determined above:

$$(\mathbf{A}-4\mathbf{I})\begin{pmatrix}y_{12}\\y_{22}\\0\\y_{42}\\y_{52}\\0\end{pmatrix}\neq\mathbf{0}\Rightarrow y_{22}\neq 0 \quad \text{or} \quad y_{52}\neq 0$$

Hence, we take

$$\mathbf{y}_2 = \begin{pmatrix} 0 \\ 0 \\ 0 \\ 0 \\ 1 \\ 0 \end{pmatrix},$$

which is independent of $\mathbf{x}_3, \mathbf{x}_2$ and $\mathbf{x}_1$. The remaining vector of this chain is given by

$$\mathbf{y}_1 = (\mathbf{A} - 4\mathbf{I})\mathbf{y}_2 = \begin{pmatrix} 0 \\ 0 \\ 0 \\ 2 \\ 0 \\ 0 \end{pmatrix}$$

Now we have found all the GEV of $\mathbf{A}$. If we take

$$\mathbf{T} = (\mathbf{x}_1\,\mathbf{x}_2\,\mathbf{x}_3\,\mathbf{y}_1\,\mathbf{y}_2\,\mathbf{z}_1),$$

then

$$\mathbf{T}^{-1}\mathbf{A}\mathbf{T} = \begin{pmatrix} 4 & 1 & 0 & 0 & 0 & 0 \\ 0 & 4 & 1 & 0 & 0 & 0 \\ 0 & 0 & 4 & 0 & 0 & 0 \\ 0 & 0 & 0 & 4 & 1 & 0 \\ 0 & 0 & 0 & 0 & 4 & 0 \\ 0 & 0 & 0 & 0 & 0 & 7 \end{pmatrix}$$

6.4 Determination of $f(\mathbf{A})$ when A has repeated eigenvalues

If $\mathbf{A}$ is a real symmetric matrix, then it can be diagonalized and the procedure for finding $f(\mathbf{A})$ is same as before, i. e.,

$$f(\mathbf{A}) = \mathbf{T}f(\mathbf{\Lambda})\mathbf{T}^{-1}$$

where $\mathbf{\Lambda}$ is the diagonal matrix containing the eigenvalues. Now, let $\mathbf{T}$ be such that

$$\mathbf{T}^{-1}\mathbf{A}\mathbf{T} = \mathbf{B} = \begin{pmatrix} \mathbf{J}_1(\lambda) & . & 0 \\ . & . & . \\ 0 & 0 & \mathbf{J}_r(\lambda) \end{pmatrix} = \text{Jordan canonical form of } \mathbf{A}.$$

Then $\mathbf{A} = \mathbf{TBT}^{-1}$ and

$$f(\mathbf{A}) = \mathbf{T}f(\mathbf{\Lambda})\mathbf{T}^{-1} = \mathbf{T}\begin{pmatrix} f(\mathbf{J}_1) & . & 0 \\ . & . & . \\ 0 & 0 & f(\mathbf{J}_r) \end{pmatrix}\mathbf{T}^{-1}.$$

Thus, if we can evaluate $f(\mathbf{J})$ where **J** is a Jordan block, then we can evaluate $f(\mathbf{A})$.

Lemma. *Let $\mathbf{J}_m(\lambda)$ be a Jordan block of order m, i. e.,*

$$\mathbf{J}_m(\lambda) = \begin{pmatrix} \lambda & 1 & 0 & . & . & 0 \\ 0 & \lambda & 1 & . & . & 0 \\ 0 & 0 & \lambda & . & . & 0 \\ . & . & . & . & 1 & . \\ . & . & . & . & . & . \\ 0 & 0 & 0 & . & . & \lambda \end{pmatrix},$$

$$f(\mathbf{J}_m(\lambda)) = \begin{pmatrix} f(\lambda) & \frac{f'(\lambda)}{1!} & \frac{f''(\lambda)}{2!} & . & . & . & \frac{f^{[m-1]}(\lambda)}{(m-1)!} \\ 0 & f(\lambda) & \frac{f'(\lambda)}{1!} & . & . & . & \frac{f^{[m-2]}(\lambda)}{(m-2)!} \\ . & . & . & . & . & . & . \\ . & . & .. & . & . & . & . \\ 0 & 0 & 0 & . & . & . & f(\lambda) \end{pmatrix}$$

□

To illustrate the use of this lemma, we compute $\exp(\mathbf{J}_m(\lambda)t)$. Let,

$$\mathbf{U} = \mathbf{J}_m(\lambda).t = \begin{pmatrix} \lambda t & t & 0 & 0 & . & 0 \\ 0 & \lambda t & t & 0 & . & 0 \\ 0 & 0 & \lambda t & t & . & 0 \\ . & . & . & . & . & . \\ . & . & . & . & . & . \\ 0 & 0 & 0 & 0 & . & \lambda t \end{pmatrix}$$

Note that **U** is not a Jordan block because it does not have ones on the super diagonal. It is easily verified that **U** may be written as

$$\mathbf{U} = \mathbf{V}\mathbf{J}_m(\lambda t)\mathbf{V}^{-1}$$

where

$$\mathbf{V} = \begin{pmatrix} t^{m-1} & 0 & 0 & . & . & 0 \\ 0 & t^{m-2} & 0 & . & . & 0 \\ . & . & . & . & . & . \\ . & . & . & . & . & . \\ . & . & . & . & . & . \\ 0 & 0 & 0 & . & . & 1 \end{pmatrix}$$

$$\mathbf{J}_m(\lambda t) = \begin{pmatrix} \lambda t & 1 & 0 & . & . & 0 \\ 0 & \lambda t & 1 & . & . & 0 \\ 0 & 0 & \lambda t & 1 & . & 0 \\ . & . & . & . & . & . \\ . & . & . & . & \lambda t & 1 \\ 0 & 0 & 0 & . & . & \lambda t \end{pmatrix}$$

$$\mathbf{V}^{-1} = \begin{pmatrix} t^{1-m} & 0 & 0 & . & . & 0 \\ 0 & t^{2-m} & 0 & . & . & 0 \\ . & . & . & . & . & . \\ . & . & . & . & . & . \\ . & . & . & . & . & . \\ 0 & 0 & 0 & . & . & 1 \end{pmatrix}$$

Hence,

$$\exp(\mathbf{U}) = \exp(\mathbf{J}_m(\lambda)t) = \mathbf{V}\exp(\mathbf{J}_m(\lambda t))\mathbf{V}^{-1}.$$

Thus,

$$\exp(\mathbf{J}_m(\lambda)t) = \exp(\lambda t)\begin{pmatrix} 1 & \frac{t}{1!} & \frac{t^2}{2!} & . & . & \frac{t^{m-1}}{(m-1)!} \\ 0 & 1 & \frac{t}{1!} & . & . & \frac{t^{m-2}}{(m-2)!} \\ 0 & 0 & 1 & . & . & \frac{t^{m-3}}{(m-3)!} \\ . & . & . & . & . & . \\ . & . & . & . & . & . \\ 0 & 0 & 0 & . & . & 1 \end{pmatrix}$$

6.5 Application of Jordan canonical form to differential equations

Consider the solution of the first-order equations

$$\frac{d\mathbf{x}}{dt} = \mathbf{Ax}, \quad \mathbf{x(t} = 0) = \mathbf{x}_0 \tag{6.21}$$

Substituting $\mathbf{x} = \mathbf{Tz}$, (6.21) transforms into

$$\frac{d\mathbf{z}}{dt} = \mathbf{T}^{-1}\mathbf{ATz}, \quad \mathbf{z(t} = 0) = \mathbf{T}^{-1}\mathbf{x}_0 \equiv \mathbf{z}_0 \tag{6.22}$$

If $\mathbf{T}^{-1}\mathbf{AT}$ is a diagonal matrix, then (6.22) is a completely uncoupled system of n equations, i. e.,

$$\frac{dz_i}{dt} = \lambda_i z_i \Rightarrow z_i = z_{i0}\exp(\lambda_i t)$$

Since $\mathbf{x} = \mathbf{Tz}$, each component of the solution to (6.21) is of the form,

$$x_i = c_{1i}\exp(\lambda_1 t) + \cdots + c_{ni}\exp(\lambda_n t).$$

If $\mathbf{T}^{-1}\mathbf{AT}$ consists of Jordan blocks, then the equations (6.22) are uncoupled only partially (or equivalently in blocks). In this case, each component of the solution to (6.21) consists of terms like $\exp(\lambda_i t)$ as well as terms of the form $t\exp(\lambda_i t)$, $t^2\exp(\lambda_i t), \ldots$, etc. To illustrate this, we consider a simple case in which the Jordan canonical form is

$$\mathbf{T}^{-1}\mathbf{AT} = \begin{pmatrix} 4 & 1 & 0 & 0 & 0 & 0 \\ 0 & 4 & 1 & 0 & 0 & 0 \\ 0 & 0 & 4 & 0 & 0 & 0 \\ 0 & 0 & 0 & 4 & 1 & 0 \\ 0 & 0 & 0 & 0 & 4 & 0 \\ 0 & 0 & 0 & 0 & 0 & 7 \end{pmatrix} = \mathbf{B}$$

Then the equations for z_i are

$$\begin{aligned} \frac{dz_1}{dt} &= 4z_1 + z_2 \\ \frac{dz_2}{dt} &= 4z_2 + z_3 \\ \frac{dz_3}{dt} &= 4z_3 \\ \frac{dz_4}{dt} &= 4z_4 + z_5 \\ \frac{dz_5}{dt} &= 4z_5 \\ \frac{dz_6}{dt} &= 7z_6 \end{aligned}$$

or,

$$\frac{d\mathbf{z}}{dt} = \mathbf{Bz}; \quad \mathbf{z} = \mathbf{z}_0 \mathbin{@} t = 0 \tag{6.23}$$

The solution of (6.23) is given by

$$\mathbf{z} = \exp(\mathbf{Bt})\mathbf{z}_0,$$

where

$$\exp(\mathbf{Bt}) = \begin{pmatrix} e^{4t} & \frac{t}{1!}e^{4t} & \frac{t^2}{2!}e^{4t} & 0 & 0 & 0 \\ 0 & e^{4t} & \frac{t}{1!}e^{4t} & 0 & 0 & 0 \\ 0 & 0 & e^{4t} & 0 & 0 & 0 \\ 0 & 0 & 0 & e^{4t} & \frac{t}{1!}e^{4t} & 0 \\ 0 & 0 & 0 & 0 & e^{4t} & 0 \\ 0 & 0 & 0 & 0 & 0 & e^{7t} \end{pmatrix}.$$

We close this topic by stating a general theorem on the solution of the linear initial value problem,

$$\frac{d\mathbf{u}}{dt} = \mathbf{A}\mathbf{u}, \quad \mathbf{u}(@\, t = 0) = \mathbf{u}_0 \tag{6.24}$$

Theorem. *Let $\lambda_1, \lambda_2, \dots, \lambda_r$ be the distinct eigenvalues of $\mathbf{A}$ with multiplicities $m_1, m_2, \dots, m_r$ and*

$$\sum_{i=1}^{r} m_i = n \tag{6.25}$$

Let $\mathbf{W}_i(\mathbf{A})$ = generalized eigenspace of $\mathbf{A}$ corresponding to eigenvalue λ_i (This is the vector space spanned by all the eigenvectors corresponding to λ_i) and

$$\mathbb{R}^n = \mathbf{W}_1(\mathbf{A}) \oplus \mathbf{W}_2(\mathbf{A}) \oplus \cdots \oplus \mathbf{W}_r(\mathbf{A}) \tag{6.26}$$

Then the solution of the initial value problem, defined by equation (6.24) *is given by*

$$\mathbf{u}(t) = \sum_{j=1}^{r} \left[\sum_{k=0}^{m_j - 1} (\mathbf{A} - \lambda_j \mathbf{I})^k \frac{t^k}{k!} \exp(\lambda_j t) \right] \mathbf{u}_{0,j}$$

where

$$\mathbf{u}_0 = \sum_{j=1}^{r} \mathbf{u}_{0,j} \quad \textit{and} \quad \mathbf{u}_{0,j} \in \mathbf{W}_j(\mathbf{A}).$$

Problems

1. Given the matrix $\mathbf{A} = \begin{pmatrix} -2 & 1 \\ -9 & 4 \end{pmatrix}$
   (a) Determine the eigenvalues, generalized eigenvectors and generalized adjoint eigenvectors.
   (b) What is the Jordan canonical form of $\mathbf{A}$?
   (c) Determine the matrix $\exp[\mathbf{A}t]$.
2. Given the matrix

$$\mathbf{A} = \begin{pmatrix} 8 & -2 & -2 & 0 \\ 0 & 6 & 2 & -4 \\ -2 & 0 & 8 & -2 \\ 2 & -4 & 0 & 6 \end{pmatrix}$$

   (a) Determine the eigenvalues, generalized eigenvectors and generalized adjoint eigenvectors.
   (b) What is the Jordan canonical form of $\mathbf{A}$?

(c) Given the set of differential equations

$$\frac{d\mathbf{u}}{dt} = \mathbf{A}\mathbf{u}, \quad \mathbf{u}(t = 0) = \mathbf{u}_0,$$

with **A** as defined as above, determine the solution.

3. Consider the case of n-consecutive first-order reactions occurring in a tubular plug flow reactor and the special case in which the rate constants for all the reactions are equal.
   (a) Formulate the balance equations and put them in matrix/vector form.
   (b) If the feed to the reactor contains only the first species, determine the exit concentration of any intermediate species.
4. Determine the eigenvalues, left and right eigenvectors and generalized eigenvectors (if any) of the matrix

$$\mathbf{A} = \begin{pmatrix} 7 & -1 & -3 & 1 \\ -1 & 7 & 1 & -3 \\ -3 & 1 & 7 & -1 \\ 1 & -3 & -1 & 7 \end{pmatrix}.$$

7 Quadratic forms, positive definite matrices and other applications

Quadratic forms appear in many applications such as the determination of maxima or minima of functions of several variables, optimization theory, solution of linear and nonlinear equations by iterative methods, tensor analysis and coordinate transformations, stability and control theory, definitions of metric or inner products, classification of partial differential equations, etc. The first four sections of this chapter give a brief introduction to this topic, and the rest of the chapter deals with various other applications of matrix methods.

7.1 Quadratic forms

To illustrate one example where quadratic forms appear, we consider a scalar function of n-variables $f(x_1, x_2, \dots, x_n)$ and expand it in a Taylor (Maclaurin's) series around the origin $\mathbf{x} = \mathbf{0}$ and keep only the constant, linear and quadratic terms. This gives

$$f(\mathbf{x}) = c + \mathbf{b}^T\mathbf{x} + \mathbf{x}^T\mathbf{A}\mathbf{x}, \tag{7.1}$$

where

$$\begin{aligned} c &= f(\mathbf{x} = \mathbf{0}) \\ b_i &= \frac{\partial f}{\partial x_i}(\mathbf{x} = \mathbf{0}) \\ \{a_{ij}\} = \{a_{ji}\} &= \frac{1}{2!}\frac{\partial^2 f}{\partial x_i \partial x_j}(\mathbf{x} = \mathbf{0}). \end{aligned}$$

The real symmetric matrix $2\mathbf{A}$ (of second partial derivatives of $f(x_1, x_2, \dots, x_n)$) is also called the *Hessian matrix*. We can remove the constant and linear terms in equation (7.1) by a translation of the origin. Defining

$$\mathbf{y} = \mathbf{x} - \boldsymbol{\alpha} \tag{7.2}$$

we get

$$\begin{aligned} f(\mathbf{x}) &= c + \mathbf{b}^T(\mathbf{y} + \boldsymbol{\alpha}) + (\mathbf{y} + \boldsymbol{\alpha})^T\mathbf{A}(\mathbf{y} + \boldsymbol{\alpha}) \\ &= (c + \mathbf{b}^T\boldsymbol{\alpha} + \boldsymbol{\alpha}^T\mathbf{A}\boldsymbol{\alpha}) + (2\boldsymbol{\alpha}^T\mathbf{A} + \mathbf{b}^T)\mathbf{y} + \mathbf{y}^T\mathbf{A}\mathbf{y}. \end{aligned}$$

Suppose that we can choose $\boldsymbol{\alpha}$ such that

$$\mathbf{A}\boldsymbol{\alpha} = -\frac{1}{2}\mathbf{b} \tag{7.3}$$

https://doi.org/10.1515/9783111598055-008

Then the linear terms vanish. Defining $Q(\mathbf{x}) = f(\mathbf{x}) - (c + \mathbf{b}^T\boldsymbol{a} + \boldsymbol{a}^T\mathbf{A}\boldsymbol{a})$, the quadratic form simplifies to

$$Q(\mathbf{y}) = \mathbf{y}^T\mathbf{A}\mathbf{y} \tag{7.4}$$

Note that equation (7.3) can be solved for any $\mathbf{b}$ only if $\mathbf{A}$ is not singular (or if the vector $-\mathbf{b}/2$ is in the column space of $\mathbf{A}$). For the case of $n = 2$ with

$$\mathbf{A} = \begin{pmatrix} a & b \\ b & c \end{pmatrix}, \tag{7.5}$$

the linear terms can be removed only if $b^2 - ac \neq 0$, i. e., the quadratic form is not of parabolic type.

The quadratic form given by equation (7.4) can be put in canonical form by noting that for a real symmetric matrix $\mathbf{A}$, there exists an orthogonal matrix $\mathbf{U}$ such that $\mathbf{U}^T\mathbf{A}\mathbf{U} = \boldsymbol{\Lambda}$ (diagonal). Making the coordinate transformation (which is a rotation)

$$\mathbf{y} = \mathbf{U}\mathbf{z} \tag{7.6}$$

Equation (7.4) reduces to its canonical form

$$\begin{aligned} Q &= (\mathbf{U}\mathbf{z})^T\mathbf{A}\mathbf{U}\mathbf{z} \\ &= \mathbf{z}^T\mathbf{U}^T\mathbf{A}\mathbf{U}\mathbf{z} \\ &= \mathbf{z}^T\boldsymbol{\Lambda}\mathbf{z} \\ &= \lambda_1 z_1^2 + \lambda_2 z_2^2 + \cdots + \lambda_n z_n^2. \end{aligned}$$

Example 7.1. Examine the nature of the curve defined by $5x_1^2 - 8x_1x_2 + 5x_2^2 = 10$.

The quadratic form $Q = 5x_1^2 - 8x_1x_2 + 5x_2^2$ may be written as

$$Q = \begin{pmatrix} x_1 & x_2 \end{pmatrix} \begin{pmatrix} 5 & -4 \\ -4 & 5 \end{pmatrix} \begin{pmatrix} x_1 \\ x_2 \end{pmatrix} = 10$$

The eigenvalues and eigenvectors of $\mathbf{A} = \begin{pmatrix} 5 & -4 \\ -4 & 5 \end{pmatrix}$ are $\lambda_1 = 9$, $\lambda_2 = 1$, $\mathbf{x}_1 = \begin{pmatrix} 1/\sqrt{2} \\ -1/\sqrt{2} \end{pmatrix}$; $\mathbf{x}_2 = \begin{pmatrix} 1/\sqrt{2} \\ 1/\sqrt{2} \end{pmatrix}$. Thus, making the substitution (rotation) $\mathbf{x} = \mathbf{U}\mathbf{z}$, where

$$\mathbf{U} = \begin{pmatrix} \frac{1}{\sqrt{2}} & \frac{1}{\sqrt{2}} \\ -\frac{1}{\sqrt{2}} & \frac{1}{\sqrt{2}} \end{pmatrix} \tag{7.7}$$

the quadratic form takes the canonical form

$$9z_1^2 + z_2^2 = 10$$

or equivalently,

$$\frac{z_1^2}{(10/9)} + \frac{z_2^2}{10} = 1$$

Now, since

$$\mathbf{z} = \begin{pmatrix} z_1 \\ z_2 \end{pmatrix} = \mathbf{U}^T\mathbf{x} = \begin{pmatrix} \frac{1}{\sqrt{2}} & -\frac{1}{\sqrt{2}} \\ \frac{1}{\sqrt{2}} & \frac{1}{\sqrt{2}} \end{pmatrix}\begin{pmatrix} x_1 \\ x_2 \end{pmatrix} = \begin{pmatrix} (x_1 - x_2)/\sqrt{2} \\ (x_1 + x_2)/\sqrt{2} \end{pmatrix},$$

in the original x-coordinates, the quadratic form may be written as

$$9\left(\frac{x_1 - x_2}{\sqrt{2}}\right)^2 + \left(\frac{x_1 + x_2}{\sqrt{2}}\right)^2 = 10.$$

This represents an ellipse with semiaxis lengths of $\sqrt{10/9}$ and $\sqrt{10}$, respectively. Figure 7.1 shows the two coordinate systems as well as the curve (ellipse) represented by the quadratic form. In this case, equation (7.7) represents a $7\pi/4$ rotation of the axes in the positive (or counterclockwise) direction (or $\pi/4$ in the clockwise direction).

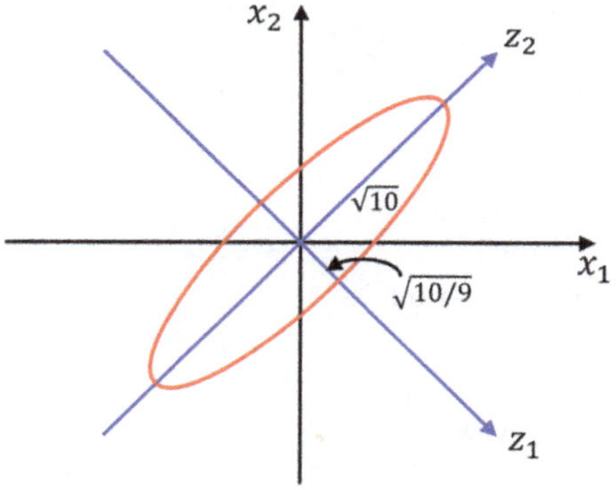

Figure 7.1: Ellipse represented by $5x_1^2 - 8x_1x_2 + 5x_2^2 = 10$ with standard and canonical coordinate systems.

Example 7.2. The quadratic form defined by $5x_1^2 + 8x_1x_2 + 5x_2^2 = 10$ is also an ellipse identical to the one in the above example but the major (longer) axis makes an angle $3\pi/4$ with the positive x_1-axis (see Figure 7.2).

Example 7.3. The quadratic form defined by $3x_1^2 - 8x_1x_2 - 3x_2^2 = 10$ is a hyperbola, as seen from the following analysis. The eigenvalues and eigenvectors of $\mathbf{A} = \begin{pmatrix} 3 & -4 \\ -4 & -3 \end{pmatrix}$ are

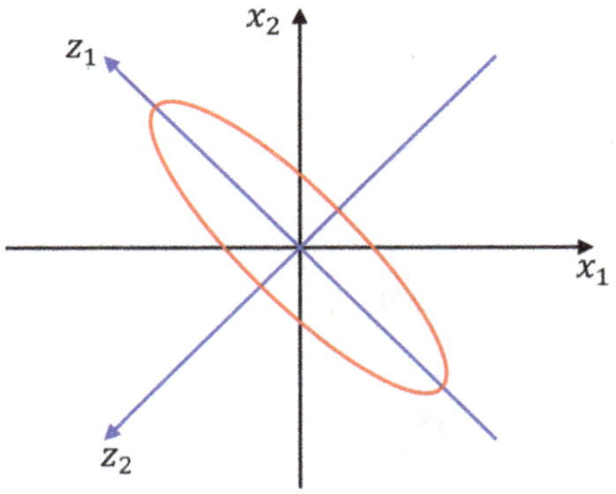

Figure 7.2: Ellipse represented by $5x_1^2 + 8x_1x_2 + 5x_2^2 = 10$ with standard and canonical coordinate systems.

$\lambda_1 = -5, \lambda_2 = 5, \mathbf{x}_1 = \binom{1/\sqrt{5}}{2/\sqrt{5}}; \mathbf{x}_2 = \binom{2/\sqrt{5}}{-1/\sqrt{5}}$. Making the substitution (rotation) $\mathbf{x} = \mathbf{U}\mathbf{z}$, where

$$\mathbf{U} = \begin{pmatrix} \frac{1}{\sqrt{5}} & \frac{2}{\sqrt{5}} \\ \frac{2}{\sqrt{5}} & -\frac{1}{\sqrt{5}} \end{pmatrix} \tag{7.8}$$

reduces it to the canonical form

$$-5z_1^2 + 5z_2^2 = 10,$$

or equivalently,

$$\frac{z_2^2}{2} - \frac{z_1^2}{2} = 1$$

Now, since

$$\mathbf{z} = \begin{pmatrix} z_1 \\ z_2 \end{pmatrix} = \mathbf{U}^T\mathbf{x} = \begin{pmatrix} \frac{1}{\sqrt{5}} & \frac{2}{\sqrt{5}} \\ \frac{2}{\sqrt{5}} & -\frac{1}{\sqrt{5}} \end{pmatrix}\begin{pmatrix} x_1 \\ x_2 \end{pmatrix} = \begin{pmatrix} (x_1 + 2x_2)/\sqrt{5} \\ (2x_1 - x_2)/\sqrt{5} \end{pmatrix},$$

in the original x-coordinates, the quadratic form may be written as

$$\left(\frac{2x_1 - x_2}{\sqrt{5}}\right)^2 - \left(\frac{x_1 + 2x_2}{\sqrt{5}}\right)^2 = 2.$$

Figure 7.3 shows the curve represented by the quadratic form along with the axes and asymptotes.

Another application of quadratic forms is in obtaining the nature of the extrema of multivariable functions (i. e., maxima, minima, saddle points, etc.), which we will discuss in Section 7.4.

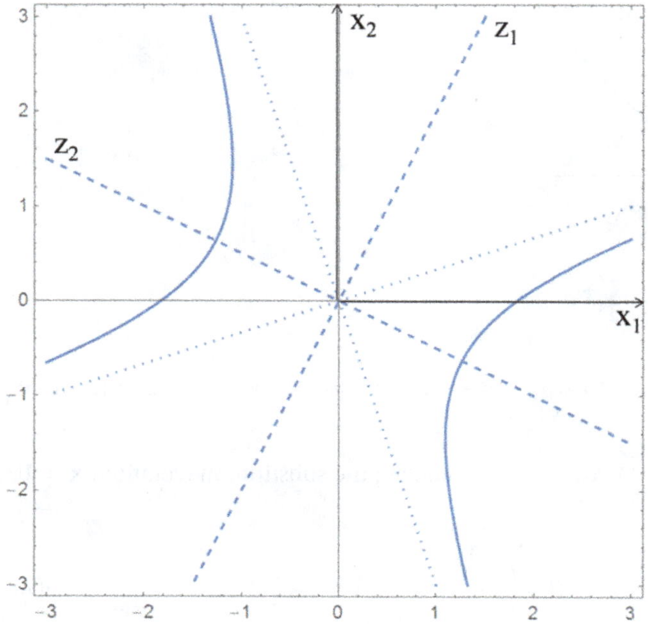

Figure 7.3: Hyperbola represented by $3x_1^2 - 8x_1x_2 - 3x_2^2 = 10$ with standard and canonical coordinate systems.

7.2 Positive definite matrices

Consider the quadratic form

$$Q(\mathbf{x}) = \mathbf{x}^T\mathbf{A}\mathbf{x} \tag{7.9}$$

where is **A** a real symmetric matrix.

Definition. $Q(\mathbf{x})$ is called *positive definite* if it takes only positive values for any choice of $\mathbf{x} \neq \mathbf{0}$ and is zero only for $\mathbf{x} = \mathbf{0}$, i. e., $\mathbf{x}^T\mathbf{A}\mathbf{x} > \mathbf{0}$ for all $\mathbf{x} \neq \mathbf{0}$. Similarly, $Q(\mathbf{x})$ is called negative definite if $\mathbf{x}^T\mathbf{A}\mathbf{x} < \mathbf{0}$ for all $\mathbf{x} \neq \mathbf{0}$.
For example, $Q(\mathbf{x}) = 5x_1^2 - 8x_1x_2 + 5x_2^2 = 9(\frac{x_1-x_2}{\sqrt{2}})^2 + (\frac{x_1+x_2}{\sqrt{2}})^2$ is positive definite.

The following tests may be used to check for the positive definiteness of a quadratic form or symmetric (or Hermitian) matrix:

(i) **A** is positive definite if and only if it can be reduced to an upper triangular form using only elementary row (or column) operations of type 3 and the diagonal elements of the resulting matrix (the pivots) are all positive.

(ii) **A** is positive definite if and only if all its principal minors are positive. A principal minor of **A** is the determinant of any submatrix obtained from **A** by deleting its last k rows and k columns ($k = 0, 1, \ldots, n-1$).

(iii) **A** is positive definite if and only if all its eigenvalues are positive.

The proof of the first two statements may be found in the book by Bronson [8]. Statement (iii) is proved in the next section.

Example 7.4. Test the matrix

$$\mathbf{A} = \begin{pmatrix} 6 & 2 & -2 \\ 2 & 6 & -2 \\ -2 & -2 & 10 \end{pmatrix}$$

for positive definiteness using criteria (i), (ii) and (iii) above.

(i) In step 1, we perform the row operations $R_2 \to R_2 - \frac{1}{3}R_1$ and $R_3 \to R_3 + \frac{1}{3}R_1$, which gives

$$\mathbf{A}_1 = \begin{pmatrix} 6 & 2 & -2 \\ 0 & \frac{16}{3} & -\frac{4}{3} \\ 0 & -\frac{4}{3} & \frac{28}{3} \end{pmatrix}.$$

In step 2, we perform the row operation $R_3 \to R_3 + \frac{1}{4}R_2 \implies$

$$\mathbf{A}_2 = \begin{pmatrix} 6 & 2 & -2 \\ 0 & \frac{16}{3} & -\frac{4}{3} \\ 0 & 0 & 9 \end{pmatrix}.$$

Since all the diagonal elements (pivots) of $\mathbf{A}_2$ are positive, $\mathbf{A}$ is positive definite.

(ii) The principal minors of $\mathbf{A}$ are

$$d_1 = \det[6] = 6,$$
$$d_2 = \det\begin{pmatrix} 6 & 2 \\ 2 & 6 \end{pmatrix} = 32,$$

and

$$d_3 = \det\begin{pmatrix} 6 & 2 & -2 \\ 2 & 6 & -2 \\ -2 & -2 & 10 \end{pmatrix} = 288.$$

Since all three principal minors are positive, the matrix is positive definite.

(iii) The eigenvalues of $\mathbf{A}$ are $\lambda_1 = 4, \lambda_2 = 6$ and $\lambda_3 = 12$. Since all eigenvalues are positive, the matrix is postive definite.

7.3 Rayleigh quotient

Suppose that $\mathbf{A}$ is a real symmetric (or a complex Hermitian) matrix and the eigenvalues of $\mathbf{A}$ (which are real) are arranged such that

$$\lambda_1 \le \lambda_2 \le \cdots \le \lambda_n.$$

We define the *Rayleigh quotient* (which defines a mapping from $\mathbb{R}^n/\mathbb{C}^n$ to the field of real numbers) as

$$R(\mathbf{x}) = \frac{\langle \mathbf{Ax}, \mathbf{x} \rangle}{\langle \mathbf{x}, \mathbf{x} \rangle}.$$

Then, the Rayleigh principle may be stated as

$$\lambda_1 \le R(\mathbf{x}) \le \lambda_n$$

i. e., the Rayleigh quotient attains its minimum value (equal to the smallest eigenvalue of $\mathbf{A}$) when $\mathbf{x}$ is the eigenvector corresponding to λ_1. Similarly, $R(\mathbf{x})$ attains its maximum value when $\mathbf{x}$ is the eigenvector corresponding to λ_n. This may be shown as follows:

Suppose that $\mathbf{U}$ is the orthogonal (unitary) matrix that diagonalizes $\mathbf{A}$. Then

$$\mathbf{U}^* \mathbf{A} \mathbf{U} = \mathbf{\Lambda} = \text{diag}\,.(\lambda_1, \ldots, \lambda_n)$$

We assume that the columns of $\mathbf{U}$ have been ordered so that $\lambda_1 \le \lambda_2 \le \cdots \le \lambda_n$. Setting $\mathbf{x} = \mathbf{Uy}$ in the Rayleigh quotient, we get

$$R(\mathbf{x}) = \frac{\langle \mathbf{AUy}, \mathbf{Uy} \rangle}{\langle \mathbf{Uy}, \mathbf{Uy} \rangle} = \frac{\langle \mathbf{U}^* \mathbf{AUy}, \mathbf{y} \rangle}{\langle \mathbf{y}, \mathbf{y} \rangle} = \frac{\langle \mathbf{\Lambda y}, \mathbf{y} \rangle}{\langle \mathbf{y}, \mathbf{y} \rangle} = \frac{\lambda_1 |y_1|^2 + \lambda_2 |y_2|^2 + \cdots + \lambda_n |y_n|^2}{|y_1|^2 + |y_2|^2 + \cdots + |y_n|^2}$$

The upper and lower bounds follow from this expression and the assumption that $\lambda_1 \le \lambda_2 \le \cdots \le \lambda_n$.

7.4 Maxima/minima for a function of several variables

Recall from calculus that for a scalar valued function $f(x) \in \mathbb{R}$, a necessary condition for a point x_0 to be an extremum is

$$\left.\frac{df}{dx}\right|_{x_0} = 0 \tag{7.10}$$

A sufficient condition is $\frac{d^2 f}{dx^2}|_{x_0} \neq 0$. Further, the extremum is a local minimum if

$$\left.\frac{d^2 f}{dx^2}\right|_{x_0} > 0, \tag{7.11}$$

while it is a local maximum if

$$\left.\frac{d^2 f}{dx^2}\right|_{x_0} < 0. \tag{7.12}$$

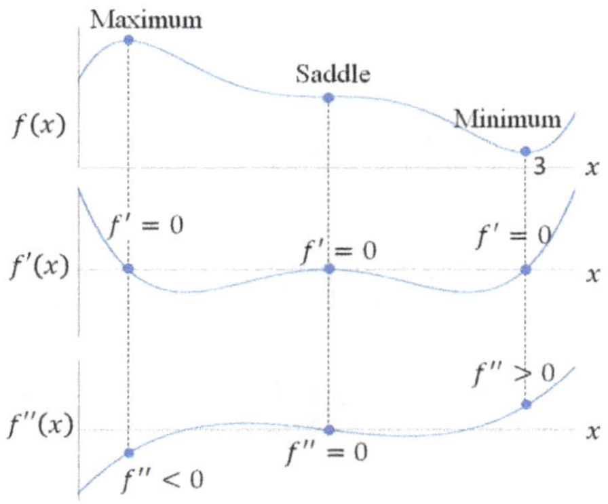

Figure 7.4: Schematic of extremum points: minimum, maximum and saddle points.

The case of $\frac{d^2f}{dx^2}|_{x_0} = 0$ corresponds to neither maximum nor minimum, but a saddle point (see Figure 7.4).

Now consider a scalar function of n variables $f(x_1, x_2, \ldots, x_n)$. Expanding this function in a Taylor series around a point $\mathbf{x}^0$ gives

$$f(\mathbf{x}) = f(\mathbf{x}^0) + \nabla f(\mathbf{x}^0)^T(\mathbf{x} - \mathbf{x}^0) + \frac{1}{2!}(\mathbf{x} - \mathbf{x}^0)^T\mathbf{A}(\mathbf{x} - \mathbf{x}^0) + \text{higher order terms} \quad (7.13)$$

where

$$\nabla f(\mathbf{x}^0) = \begin{pmatrix} \frac{\partial f}{\partial x_1} \\ \frac{\partial f}{\partial x_2} \\ \vdots \\ \frac{\partial f}{\partial x_n} \end{pmatrix}_{\mathbf{x}=\mathbf{x}^0} = \text{gradient vector of } f(\mathbf{x}) \text{ at } \mathbf{x} = \mathbf{x}^0 \quad (7.14)$$

and

$$\mathbf{A} = \{a_{ij}\}_{n\times n} = \left\{ \frac{\partial^2 f}{\partial x_i \partial x_j}\bigg|_{\mathbf{x}=\mathbf{x}^0} \right\}_{n\times n} = \text{Hessian matrix}$$
$$= \text{Symmetric matrix of second partial derivatives of } f(\mathbf{x}) \text{ evaluated at } \mathbf{x} = \mathbf{x}^0 \quad (7.15)$$

From equations (7.13)–(7.15), it can be seen that a necessary condition for $f(\mathbf{x})$ to have an extremum at $\mathbf{x} = \mathbf{x}^0$ is

$$\nabla f(\mathbf{x}^0) = \mathbf{0}, \quad (7.16)$$

while a sufficient condition for $\mathbf{x} = \mathbf{x}^0$ to be a minimum is that

$$(\mathbf{x} - \mathbf{x}^0)^T \mathbf{A}(\mathbf{x} - \mathbf{x}^0) > 0 \tag{7.17}$$

and for a maximum

$$(\mathbf{x} - \mathbf{x}^0)^T \mathbf{A}(\mathbf{x} - \mathbf{x}^0) < 0 \tag{7.18}$$

for all $\mathbf{x}$ in the neighborhood of $\mathbf{x}^0$. Otherwise, $\mathbf{x} = \mathbf{x}^0$ is neither a minimum nor a maximum.

For simplicity, we can assume $\mathbf{x}^0 = \mathbf{0}$. If this is not the case, we can shift the origin. Thus, the sufficient condition reduces to examining the quadratic form:

$$Q(\mathbf{x}) = \mathbf{x}^T \mathbf{A}\mathbf{x}$$

where $\mathbf{A}^T = \mathbf{A}$ is a real symmetric matrix. If $Q(\mathbf{x}) > 0$ for all $\mathbf{x}$ near the origin, then $f(\mathbf{x})$ has a local minimum, while if $Q(\mathbf{x}) < 0$ for all $\mathbf{x}$ near the origin ($\mathbf{0}$), then $f(\mathbf{x})$ has a local maximum. Otherwise, the extremum is a saddle point.

Let $\lambda_1, \lambda_2, \ldots, \lambda_n$ be the eigenvalues (real) and $\mathbf{T}$ be the orthogonal matrix that diagonalizes $\mathbf{A}$. Let

$$\mathbf{x} = \mathbf{T}\mathbf{z} \Longrightarrow \mathbf{z} = \mathbf{T}^{-1}\mathbf{x} = \mathbf{T}^T\mathbf{x} \tag{7.19}$$

$\Longrightarrow$

$$\begin{aligned} Q &= \mathbf{z}^T \mathbf{T}^T \mathbf{A}\mathbf{T}\mathbf{z} = \mathbf{z}^T \Lambda \mathbf{z} \\ &= \sum_{i=1}^{n} \lambda_i z_i^2 \end{aligned} \tag{7.20}$$

Thus, if all $\lambda_i > 0$ ($i = 1, 2, \ldots, n$), Q only takes positive values, and hence the extremum is a minimum. If $\lambda_i < 0$ for all i, Q is negative and the extremum is a maximum. If $\lambda_i = 0$ for some i or if eigenvalues of $\mathbf{A}$ are both positive and negative, the extremum is neither a maximum nor a minimum.

As stated earlier, a symmetric matrix $\mathbf{A}$ has positive eigenvalues (or is positive definite) if all the principal minors have positive determinants. A principle minor is obtained by striking out the last k rows and k columns ($k = 1, 2, \ldots, n-1$) of $\mathbf{A}$. For example, for 2×2 case,

$$\mathbf{A} = \begin{pmatrix} a_{11} & a_{12} \\ a_{21} & a_{22} \end{pmatrix}, \quad a_{12} = a_{21}$$

is positive definite if

$$\begin{aligned} a_{11} &> 0 \\ a_{11}a_{22} - a_{12}a_{21} &= a_{11}a_{22} - a_{12}^2 > 0. \end{aligned}$$

Similarly, for the 3×3 case,

$$\mathbf{A} = \begin{pmatrix} a_{11} & a_{12} & a_{13} \\ a_{21} & a_{22} & a_{23} \\ a_{31} & a_{32} & a_{33} \end{pmatrix}, \qquad \begin{aligned} a_{12} &= a_{21} \\ a_{13} &= a_{31} \\ a_{23} &= a_{32} \end{aligned}$$

is positive definite if

$$a_{11} > 0,$$
$$\begin{vmatrix} a_{11} & a_{12} \\ a_{21} & a_{22} \end{vmatrix} > 0,$$
$$\begin{vmatrix} a_{11} & a_{12} & a_{13} \\ a_{21} & a_{22} & a_{23} \\ a_{31} & a_{32} & a_{33} \end{vmatrix} > 0.$$

Remark. $\mathbf{A}$ is negative definite if $(-\mathbf{A})$ is positive definite. Then, for the 2×2 case, $\mathbf{A}$ is negative definite if

$$a_{11} < 0,$$
$$\begin{vmatrix} a_{11} & a_{12} \\ a_{21} & a_{22} \end{vmatrix} > 0.$$

Example 7.5. State the conditions for a function of two variables $f(x_1, x_2)$ to have a local maximum.

For $f(x_1, x_2)$ to have a local maximum at (x_{10}, x_{20}), a necessary condition is that the first partial derivatives of $f(x_1, x_2)$ w. r. t. x_1 and x_2 vanish, i. e.,

$$\frac{\partial f}{\partial x_1}(x_{10}, x_{20}) = 0; \quad \frac{\partial f}{\partial x_2}(x_{10}, x_{20}) = 0$$

The local extremum is a maximum if the quadratic form

$$\begin{aligned} Q = \frac{1}{2!}\Bigg[&\frac{\partial^2 f}{\partial x_1^2}(x_{10}, x_{20})(x_1 - x_{10})^2 + 2\frac{\partial^2 f}{\partial x_1 \partial x_2}(x_{10}, x_{20})(x_1 - x_{10})(x_2 - x_{20}) \\ &+ \frac{\partial^2 f}{\partial x_2^2}(x_{10}, x_{20})(x_2 - x_{20})^2 \Bigg] \end{aligned}$$

is negative definite. This is the case if the following two conditions are satisfied:

$$\frac{\partial^2 f}{\partial x_1^2}(x_{10}, x_{20}) < 0$$
$$\frac{\partial^2 f}{\partial x_1^2}(x_{10}, x_{20}) \frac{\partial^2 f}{\partial x_2^2}(x_{10}, x_{20}) - \left(\frac{\partial^2 f}{\partial x_1 \partial x_2}(x_{10}, x_{20}) \right)^2 > 0$$

Example 7.6. Consider the function $f(x,y) = 2x^2 - 2xy + 5y^2 - 18y + 23$, and examine it for extremum values.

The extremum points can be obtained by setting $\nabla f = \mathbf{0} \Longrightarrow$

$$\frac{\partial f}{\partial x} = 4x - 2y = 0$$
$$\frac{\partial f}{\partial y} = -2x + 10y - 18 = 0,$$

which after solving leads to $(x_0, y_0) = (1, 2)$ as a possible extremum point. We evaluate the Hessian matrix (matrix of second derivatives),

$$\mathbf{A} = \begin{pmatrix} \frac{\partial^2 f}{\partial x^2} & \frac{\partial^2 f}{\partial x \partial y} \\ \frac{\partial^2 f}{\partial y \partial x} & \frac{\partial^2 f}{\partial y^2} \end{pmatrix}_{(x_0,y_0)=(1,2)} = \begin{pmatrix} 4 & -2 \\ -2 & 10 \end{pmatrix}.$$

The eigenvalues of $\mathbf{A}$ are given by $|\mathbf{A} - \lambda \mathbf{I}| = \lambda^2 - 14\lambda + 36 = 0 \Longrightarrow \lambda_1 = 7 - \sqrt{13} > 0$ and $\lambda_2 = 7 + \sqrt{13} > 0$. Since both eigenvalues are positive, the extremum is a minimum. This can also be seen from the principal minors of $\mathbf{A}$:

$$d_1 = |4| = 4 > 0,$$
$$d_2 = \begin{vmatrix} 4 & -2 \\ -2 & 10 \end{vmatrix} = 36 > 0.$$

We also note that $f_{\min} = f(1, 2) = 5$.

Example 7.7. Consider the function $g(x,y) = 2x^2 - 8xy - y^2 + 18y - 9$. The extremum points can be obtained by setting $\nabla g = \mathbf{0} \Longrightarrow$

$$\frac{\partial g}{\partial x} = 4x - 8y = 0$$
$$\frac{\partial g}{\partial y} = -8x - 2y + 18 = 0.$$

Solving these linear equations gives a possible extremum point $(x_0, y_0) = (2, 1)$. We examine the Hessian matrix

$$\mathbf{A} = \begin{pmatrix} \frac{\partial^2 g}{\partial x^2} & \frac{\partial^2 g}{\partial x \partial y} \\ \frac{\partial^2 g}{\partial y \partial x} & \frac{\partial^2 g}{\partial y^2} \end{pmatrix}_{(x_0,y_0)=(2,1)} = \begin{pmatrix} 4 & -8 \\ -8 & -2 \end{pmatrix}.$$

The eigenvalues of $\mathbf{A}$ are $\lambda_1 = 1 + \sqrt{73} > 0$ and $\lambda_2 = 1 - \sqrt{73} < 0$. Thus, the point $x_0 = 2$, $y_0 = 1$ is neither a maximum nor a minimum. It is a saddle point.

7.5 Linear difference equations

In many applications, such as discrete dynamical systems, stage operations, Markov processes, etc., the governing equations may be expressed in vector-matrix form as

$$\mathbf{u}_{k+1} = \mathbf{A}\mathbf{u}_k, \quad k \geq 0; \quad \mathbf{u}_0 \text{ (i. e., initial state) given} \tag{7.21}$$

Here, $\mathbf{u}_k$ is the system state at time k, which is assumed to be discrete, i. e., taking values $k = 0, 1, \ldots \infty$ while $\mathbf{A}$ is the connectivity, coupling or transition probability matrix.

Scalar to vector representation

Scalar difference equations of any order can easily be represented in vector-matrix form given in equation (7.21). For example, let us consider a second-order difference equation:

$$u_{n+1} = au_n + bu_{n-1}, \quad n = 1, 2, \ldots \tag{7.22}$$

where u_0 and u_1 are given. Defining two-dimensional vectors

$$\mathbf{x}_n = \begin{pmatrix} u_{n-1} \\ u_n \end{pmatrix} \quad \text{and} \quad \mathbf{x}_{n+1} = \begin{pmatrix} u_n \\ u_{n+1} \end{pmatrix}, \tag{7.23}$$

the above second-order difference equation can be expressed as

$$\mathbf{x}_{n+1} = \begin{pmatrix} 0 & 1 \\ b & a \end{pmatrix} \mathbf{x}_n = \mathbf{A}\mathbf{x}_n \quad \text{with } \mathbf{x}_1 \text{ given.} \tag{7.24}$$

Similarly, a third-order difference equation:

$$u_{n+1} = au_n + bu_{n-1} + cu_{n-2}, \quad n = 2, 3, \ldots \tag{7.25}$$

with given u_0, u_1 and u_2, can be converted to vector-matrix form:

$$\mathbf{x}_{n+1} = \mathbf{A}\mathbf{x}_n \quad \text{where } \mathbf{x}_{n+1} = \begin{pmatrix} u_{n-1} \\ u_n \\ u_{n+1} \end{pmatrix} \quad \text{and} \quad \mathbf{A} = \begin{pmatrix} 0 & 1 & 0 \\ 0 & 0 & 1 \\ c & b & a \end{pmatrix} \tag{7.26}$$

with given $\mathbf{x}_1 = \begin{pmatrix} u_0 \\ u_1 \\ u_2 \end{pmatrix}$. Thus, higher-order scalar difference equations can be converted to vector equations.

Formal solution of difference equation in vector-matrix form

It follows from equation (7.21) that

$$\begin{aligned}\mathbf{u}_1 &= \mathbf{A}\mathbf{u}_0\\ \mathbf{u}_2 &= \mathbf{A}\mathbf{u}_1 = \mathbf{A}^2\mathbf{u}_0\\ &\vdots\\ \mathbf{u}_k &= \mathbf{A}^k\mathbf{u}_0, \quad k = 0, 1, 2, \ldots\end{aligned} \tag{7.27}$$

Thus, the computation of $\mathbf{u}_k$ requires the evaluation of $\mathbf{A}^k$, which is given by the spectral theorem as follows:

$$\mathbf{A}^k = \sum_{j=1}^{n} \lambda_j^k \mathbf{E}_j \tag{7.28}$$

where $\mathbf{E}_j$ are projection matrices given in terms of eigenvectors and eigenrows as

$$\mathbf{E}_j = \frac{\mathbf{x}_j \mathbf{y}_j^*}{\mathbf{y}_j^* \mathbf{x}_j} = \mathbf{x}_j \mathbf{y}_j^* \quad (\text{if } \mathbf{y}_j^* \text{ are normalized}). \tag{7.29}$$

Thus, the solution (equation (7.27)) becomes

$$\mathbf{u}_k = \sum_{j=1}^{n} \lambda_j^k \mathbf{E}_j \mathbf{u}_0, \quad k = 0, 1, 2, \ldots \tag{7.30}$$

We consider some special cases of this solution based on the magnitude of the eigenvalues.

Case 1: $|\lambda_j| < 1$ for all $j = 1, 2, \ldots, n$

In this case, $\lim_{k\to\infty} |\lambda_j|^k \to 0$, and thus, $\mathbf{u}_k \to \mathbf{0}$ for $k \to \infty$ and the system approaches the trivial state $\mathbf{u} = \mathbf{0}$ for $k \to \infty$.

Case 2: $\lambda_1 = 1$, $|\lambda_j| < 1$ for all $j = 2, 3, \ldots, n$

In this case, $\lim_{k\to\infty} \mathbf{u}_k \to \mathbf{E}_1\mathbf{u}_0 = (\mathbf{y}_1^*\mathbf{u}_0)\mathbf{x}_1$, and the system approaches a scalar multiple of the state given by the eigenvector $\mathbf{x}_1$ corresponding to $\lambda_1 = 1$ (eigenvalue of unity).

Case 3: $|\lambda_j| > 1$ for some j

In this case, $\lim_{k\to\infty} u_k \to \infty$ and the solution is not bounded.

Case 4: A pair of complex eigenvalues of unit modulus and all other eigenvalues inside the unit circle, i. e., $\lambda_{1,2} = \alpha \pm i\beta$ with $|\lambda_1| = |\lambda_2| = \alpha^2 + \beta^2 = 1$ while $|\lambda_j| < 1$ for $j = 3, 4, \ldots, n$. The solution is again bounded and lies on an invariant circle.

It should be pointed out that all these special cases may occur in the solution of nonlinear algebraic equations by local linearization, e. g., Newton–Raphson or other iterative methods.

Example 7.8 (Two-stage Markov process). Consider the case in which the state vector $\mathbf{u}$ is of the form

$$\mathbf{u}_k = \begin{pmatrix} a_k \\ b_k \end{pmatrix}$$

where a_k is the fraction of the population in state a_k and b_k is the fraction of the population in state B at time k. For example, a_k can be the fraction of students in a class having a mobile phone of type-I while b_k as the fraction having phone of type-II. Let

$$\mathbf{A} = \{p_{ij}\} = \mathbf{P}$$

where $\mathbf{P}$ being the transition/switching probability matrix, i. e., p_{ij} is the probability of switching from state j to i then

$$\sum_{i=1}^{n} p_{ij} = 1 \quad \text{for } j = 1, 2, \ldots, n$$

or the columns of $\mathbf{P}$ sum to unity for a Markov matrix. For example, for $n = 2$, $\mathbf{P} = \begin{pmatrix} \frac{2}{3} & \frac{1}{2} \\ \frac{1}{3} & \frac{1}{2} \end{pmatrix}$ is a Markov matrix where $p_{11} = \frac{2}{3}$ is the probability of a student in state A staying in state A, $p_{21} = \frac{1}{3}$ is the probability of student switching from state A to state B, $p_{12} = \frac{1}{2}$ is the probability of switching from state B to state A, and $p_{22} = \frac{1}{2}$ is the probability of students staying in state B (here $p_{11} + p_{21} = 1$, $p_{12} + p_{22} = 1$).

We note that if $\mathbf{P}$ is a Markov matrix then $\lambda_1 = 1$ is an eigenvalue of $\mathbf{P}$ with left eigenvector $\mathbf{y}_1^T = (1\ 1\ \ldots\ 1)$. This follows from the fact that $\sum_{i=1}^{n} p_{ij} = 1$ for each j. Thus, for a Markov process with initial state $\mathbf{u}_0$, we have

$$u_k = \sum_{j=1}^{n} \lambda_j^k \mathbf{E}_j \mathbf{u}_0$$

$\Rightarrow$

$$\begin{aligned} \lim_{k\to\infty} \mathbf{u}_k &= \mathbf{E}_1 \mathbf{u}_0 \quad \text{if } |\lambda_j| < 1 \text{ for } j = 2, 3, \ldots, n \\ &= (\mathbf{y}_1^T \mathbf{u}_0)\mathbf{x}_1 \\ &= \mathbf{x}_1 \quad \text{if } \mathbf{y}_1^* \mathbf{u}_0 = 1, \quad \text{i. e., sum of all initial probabilities is unity} \end{aligned}$$

In the example above, the eigenvalues are $\lambda_1 = 1$ and $\lambda_2 = \frac{1}{6}$ corresponding to the eigenvectors: $\mathbf{x}_1 = \begin{pmatrix} 3/5 \\ 2/5 \end{pmatrix}$ and $\mathbf{x}_2 = \begin{pmatrix} 1 \\ -1 \end{pmatrix}$ and eigenrows: $\mathbf{y}_1^T = (1\ 1)$ and $\mathbf{y}_2^T = (\frac{2}{5}\ \frac{-3}{5})$. Thus, assuming the components of $\mathbf{u}_0$ sum to unit (i. e., $\mathbf{y}_1^T \mathbf{u}_0 = 1$)

$$\begin{aligned} \mathbf{u}_k &= (\mathbf{y}_1^T \mathbf{u}_0)\mathbf{x}_1 + (\lambda_2)^k (\mathbf{y}_2^T \mathbf{u}_0)\mathbf{x}_2 \\ \Rightarrow \mathbf{u}_\infty &= \lim_{k\to\infty} \mathbf{u}_k = (\mathbf{y}_1^T \mathbf{u}_0)\mathbf{x}_1 = \mathbf{x}_1 = \begin{pmatrix} 3/5 \\ 2/5 \end{pmatrix}. \end{aligned}$$

The rate of convergence to the steady state or equilibrium solution depends on λ_2. In this example, since $\lambda_2 = \frac{1}{6}$, which is much smaller than unity, the rate of convergence is fast as shown by iterations below:

Taking $\mathbf{u}_0 = \begin{pmatrix} 0.5 \\ 0.5 \end{pmatrix} \Rightarrow \mathbf{u}_1 = \begin{pmatrix} \frac{2}{3} & \frac{1}{2} \\ \frac{1}{3} & \frac{1}{2} \end{pmatrix} \mathbf{u}_0 = \begin{pmatrix} 0.583 \\ 0.417 \end{pmatrix}$. Similarly, $\mathbf{u}_2 = \mathbf{P}\mathbf{u}_1 = \begin{pmatrix} 0.597 \\ 0.403 \end{pmatrix}$, $\mathbf{u}_3 = \begin{pmatrix} 0.5996 \\ 0.4004 \end{pmatrix}$, and so on. The convergence of Markov process is demonstrated in Figure 7.5 with $\mathbf{P} = \begin{pmatrix} \frac{2}{3} & \frac{1}{2} \\ \frac{1}{3} & \frac{1}{2} \end{pmatrix}$ and $\mathbf{u}_0 = \begin{pmatrix} 0.05 \\ 0.95 \end{pmatrix}$, along with eigenvectors and eigenrows of $\mathbf{P}$. It can be seen that the steady-state solution $\mathbf{u}_\infty = \mathbf{x}_1$ is achieved in few (less than 4) iterations to very good accuracy.

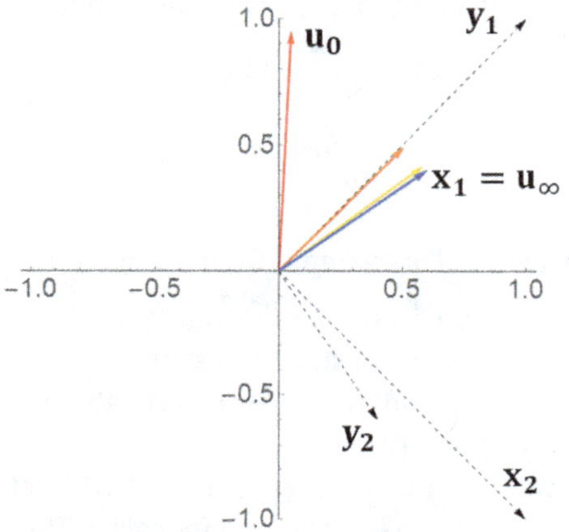

Figure 7.5: Convergence in Markov chains with $\mathbf{P} = \begin{pmatrix} \frac{2}{3} & \frac{1}{2} \\ \frac{1}{3} & \frac{1}{2} \end{pmatrix}$ and $\mathbf{u}_0 = \begin{pmatrix} 0.05 \\ 0.95 \end{pmatrix}$.

Remark 7.1. For the case of $n = 2$ or 2×2 Markov matrix, the $\lim_{k\to\infty} \mathbf{u}_k$ can also be found by solving a simple algebraic equation. Let $\mathbf{u}_0 = \begin{pmatrix} \beta \\ 1-\beta \end{pmatrix}$ and $\mathbf{u}_\infty = \begin{pmatrix} \alpha \\ 1-\alpha \end{pmatrix}$. Now since

$$\begin{aligned} \mathbf{u}_\infty = \mathbf{P}\mathbf{u}_\infty \Rightarrow \begin{pmatrix} \alpha \\ 1-\alpha \end{pmatrix} &= \begin{pmatrix} \frac{2}{3} & \frac{1}{2} \\ \frac{1}{3} & \frac{1}{2} \end{pmatrix} \begin{pmatrix} \alpha \\ 1-\alpha \end{pmatrix} \\ \Rightarrow \alpha = \frac{2}{3} + \frac{1}{2}(1-\alpha) &\Rightarrow \alpha = \frac{3}{5} \\ \Rightarrow \mathbf{u}_\infty &= \begin{pmatrix} \frac{3}{5} \\ \frac{2}{5} \end{pmatrix}, \end{aligned}$$

which is independent of β, i. e., the initial point (provided the sum of the two components of initial condition is unity).

Remark 7.2. For $n > 2$, $\mathbf{u}_\infty$ can be found by solving the equation

$$\mathbf{u}_\infty = \mathbf{P}\mathbf{u}_\infty$$

along with the constraints that the sum of all $\mathbf{u}_\infty$ components is unity. Note that this is proportional to the eigenvector $\mathbf{x}_1$ corresponding to the eigenvalue λ_1 where the proportionality constant can be obtained with the constraints stated earlier, i. e., $\mathbf{y}_1^T\mathbf{u}_\infty = 1$.

Example 7.9 (Fibonacci equation). Consider the Fibonacci equation

$$\begin{aligned} x_{n+1} &= x_n + x_{n-1}, \quad n = 1, 2, \ldots \\ x_0 &= x_1 = 1 \end{aligned}$$

which is a second-order linear difference equation. One way to solve this equation is by assuming the solution of the form

$$x_n = r^n,$$

which leads to the characteristic equation:

$$r^2 = r + 1 \Rightarrow r_{1,2} = \frac{1 \pm \sqrt{5}}{2} \quad (r_1 = 1.618, r_2 = -0.618).$$

Thus, the general solution can be expressed as

$$x_n = c_1 r_1^n + c_2 r_2^n$$

where c_1 and c_2 can be solved from initial points by setting for $n = 0$ and 1, respectively, i. e.,

$$\begin{aligned} n = 0 &\Rightarrow c_1 + c_2 = 1 \quad \text{and} \quad n = 1 \Rightarrow c_1 r_1 + c_2 r_2 = 1 \\ \Rightarrow c_{1,2} &= \frac{\sqrt{5} \pm 1}{2\sqrt{5}} \end{aligned}$$

Thus, the general solution of Fibonacci equation is given by

$$x_n = \frac{1}{\sqrt{5}}\left(\frac{(1+\sqrt{5})^{n+1} - (1-\sqrt{5})^{n+1}}{2^{n+1}}\right).$$

Vector-matrix form: The Fibonacci equation can also be represented in vector-matrix form by assuming

$$\mathbf{x}_n = \begin{pmatrix} x_{n-1} \\ x_n \end{pmatrix}$$

$$\Rightarrow \mathbf{x}_{n+1} = \begin{pmatrix} x_n \\ x_{n+1} \end{pmatrix} = \begin{pmatrix} 0 & 1 \\ 1 & 1 \end{pmatrix} \begin{pmatrix} x_{n-1} \\ x_n \end{pmatrix}$$

$$\Rightarrow \mathbf{x}_{n+1} = \mathbf{A}\mathbf{x}_n, \quad \mathbf{x}_1 = \begin{pmatrix} 1 \\ 1 \end{pmatrix}.$$

Thus,

$$\mathbf{x}_2 = \mathbf{A}\mathbf{x}_1, \quad \mathbf{x}_3 = \mathbf{A}^2\mathbf{x}_1, \quad \ldots, \quad \mathbf{x}_n = \mathbf{A}^{n-1}\mathbf{x}_1,$$

where $\mathbf{A} = \left(\begin{smallmatrix} 0 & 1 \\ 1 & 1 \end{smallmatrix}\right)$, which is a symmetric matrix with real eigenvalues given by

$$\lambda^2 - \lambda - 1 = 0,$$

$\Rightarrow \lambda_{1,2} = \frac{1 \pm \sqrt{5}}{2}$, and corresponding to eigenvectors:

$$\mathbf{x}_{1,2} = \begin{pmatrix} 1 \\ \lambda_{1,2} \end{pmatrix},$$

and normalized eigenrows:

$$\mathbf{y}_{1,2}^T = \frac{1}{\sqrt{1 + \lambda_{1,2}^2}} \begin{pmatrix} 1 & \lambda_{1,2} \end{pmatrix}.$$

Thus, the projection matrices can be obtained by

$$\mathbf{E}_{1,2} = \mathbf{x}_{1,2}\mathbf{y}_{1,2}^T = \frac{1}{1 + \lambda_{1,2}^2} \begin{pmatrix} 1 & \lambda_{1,2} \\ \lambda_{1,2} & \lambda_{1,2}^2 \end{pmatrix}$$

$\Rightarrow$

$$\begin{aligned} \mathbf{A}^{n-1} &= \lambda_1^{n-1}\mathbf{E}_1 + \lambda_2^{n-1}\mathbf{E}_2 \\ &= \frac{\lambda_1^{n-1}}{1 + \lambda_1^2} \begin{pmatrix} 1 & \lambda_1 \\ \lambda_1 & \lambda_1^2 \end{pmatrix} + \frac{\lambda_2^{n-1}}{1 + \lambda_2^2} \begin{pmatrix} 1 & \lambda_2 \\ \lambda_2 & \lambda_2^2 \end{pmatrix} \end{aligned}$$

$\Rightarrow$

$$\begin{aligned} \begin{pmatrix} x_{n-1} \\ x_n \end{pmatrix} &= \mathbf{x}_n = \mathbf{A}^{n-1}\mathbf{x}_1 = \mathbf{A}^{n-1} \begin{pmatrix} 1 \\ 1 \end{pmatrix} \\ &= \begin{pmatrix} \frac{(1+\lambda_1)\lambda_1^{n-1}}{1+\lambda_1^2} + \frac{(1+\lambda_2)\lambda_2^{n-1}}{1+\lambda_2^2} \\ \frac{(1+\lambda_1)\lambda_1^{n}}{1+\lambda_1^2} + \frac{(1+\lambda_2)\lambda_2^{n}}{1+\lambda_2^2} \end{pmatrix} \end{aligned}$$

$\Rightarrow$

$$x_n = \frac{(1 + \lambda_1)}{1 + \lambda_1^2}\lambda_1^n + \frac{(1 + \lambda_2)}{1 + \lambda_2^2}\lambda_2^n$$

But $\lambda_{1,2} = \frac{1\pm\sqrt{5}}{2}$ and $\lambda_{1,2}^2 = \lambda_{1,2} + 1 \Rightarrow$

$$\begin{aligned}\frac{(1+\lambda_{1,2})}{1+\lambda_{1,2}^2}\lambda_{1,2}^n &= \frac{\lambda_{1,2}^2}{1+\lambda_{1,2}^2}\lambda_{1,2}^n = \frac{\lambda_{1,2}^{n+2}}{1+\lambda_{1,2}^2}\\ &= \frac{(\frac{1\pm\sqrt{5}}{2})^{n+2}}{1+\frac{3\pm\sqrt{5}}{2}} = \frac{1}{2^{n+1}}\frac{(1\pm\sqrt{5})^{n+2}}{5\pm\sqrt{5}}\end{aligned}$$

$\Rightarrow$

$$\begin{aligned}x_n &= \frac{1}{2^{n+1}}\frac{(1+\sqrt{5})^{n+2}}{5+\sqrt{5}} + \frac{1}{2^{n+1}}\frac{(1-\sqrt{5})^{n+2}}{5-\sqrt{5}}\\ &= \frac{1}{\sqrt{5}}\left(\frac{(1+\sqrt{5})^{n+1}-(1-\sqrt{5})^{n+1}}{2^{n+1}}\right),\end{aligned}$$

which is the same result obtained earlier. Since $|\lambda_2| = 0.618 < 1$ and $\lambda_2^n \to 0$ for $n \to \infty$, x_n may be approximated for large n as

$$x_n \approx \frac{1+\lambda_1}{1+\lambda_1^2}\lambda_1^n = 0.7236(1.618)^n.$$

For $n = 5$, the approximation gives $x_5 = 8.02$ while the exact value is 8.

7.6 Generalized inverse and least square solutions

Consider the system of linear equations:

$$\mathbf{Au} = \mathbf{b} \tag{7.31}$$

where $\mathbf{A}$ is a real $m \times n$ matrix, $\mathbf{u}$ is $n \times 1$ vector and $\mathbf{b}$ is $m \times 1$ vector. When m (number of equations) $> n$ (number of unknowns), the augmented matrix $[\mathbf{A}\ \mathbf{b}]$ may have rank $> n$, in which case, the system is inconsistent and has no solution. Similarly, when $m < n$, even if $\text{rank}(\mathbf{A}) = \text{rank}[\mathbf{A}\ \mathbf{b}] = m$, there are fewer equations than unknowns and there is no unique solution. In many practical cases, a solution that satisfies equation (7.31) in the best possible way, often called the "least square solution," can be obtained by minimizing the scalar function:

$$\begin{aligned}f(\mathbf{u}) &= (\mathbf{Au}-\mathbf{b})^T(\mathbf{Au}-\mathbf{b})\\ &= \mathbf{uA}^T\mathbf{Au} - \mathbf{b}^T\mathbf{Au} - \mathbf{u}^T\mathbf{A}^T\mathbf{b} + \mathbf{b}^T\mathbf{b}.\end{aligned} \tag{7.32}$$

If we denote the j-th component of $(\mathbf{Au} - \mathbf{b})$ as e_j, which represents the error in j-th equation, then

$$f(\mathbf{u}) = \mathbf{e}^T\mathbf{e} = \sum_{j=1}^{m} e_j^2 = \text{sum of squares of residuals} \tag{7.33}$$

To minimize f, we set

$$\frac{\partial f}{\partial u_k} = 0, \quad k = 1, 2, \ldots, n$$
$$\Rightarrow \nabla f = \text{gradient of } f = \mathbf{0} \tag{7.34}$$

the gradient of f can be obtained from equation (7.32) as

$$\nabla f = 2\mathbf{A}^T\mathbf{A}\mathbf{u} - 2\mathbf{A}^T\mathbf{b} = \mathbf{0}$$
$$\Rightarrow \mathbf{A}^T\mathbf{A}\mathbf{u} = \mathbf{A}^T\mathbf{b}$$
$$\Rightarrow \mathbf{u} = (\mathbf{A}^T\mathbf{A})^{-1}\mathbf{A}^T\mathbf{b}. \tag{7.35}$$

In arriving at equation (7.35), it is assumed that $\mathbf{A}^T\mathbf{A}$ is invertible.

Definition. $\mathbf{A}^\dagger = (\mathbf{A}^T\mathbf{A})^{-1}\mathbf{A}^T$ is called the generalized inverse (or Moore–Penrose inverse) of $\mathbf{A}$.

Properties of $\mathbf{A}^\dagger$

(i) If $m = n$ and $\mathbf{A}$ is an invertible square matrix, then $\mathbf{A}^\dagger = \mathbf{A}^{-1}$.
(ii) $\mathbf{A}^\dagger\mathbf{A} = \mathbf{I}_n$ = Identity matrix.
(iii) $\mathbf{A}\mathbf{A}^\dagger\mathbf{A} = \mathbf{A}$.

The equations

$$\mathbf{A}^T\mathbf{A}\mathbf{u} = \mathbf{A}^T\mathbf{b} \tag{7.36}$$

are often referred to as the "*least squares equations*" and can be solved for $\mathbf{u}$ if $\mathbf{A}^T\mathbf{A}$ is invertible. If we let

$$\mathbf{B} = \mathbf{A}^T\mathbf{A}, \tag{7.37}$$

then

$$\mathbf{B}^T = \mathbf{A}^T\mathbf{A} = \mathbf{B} \tag{7.38}$$

$\Rightarrow$ $\mathbf{B}$ is a symmetric matrix, and hence the eigenvalues of $\mathbf{B}$ are real and nonnegative. The positive square roots of the eigenvalues of $\mathbf{B}$ are called "*singular values*" of $\mathbf{A}$.

It may be shown that

$$\mathbf{A} = \mathbf{U}\begin{pmatrix} \mathbf{D} & \mathbf{0} \\ \mathbf{0} & \mathbf{0} \end{pmatrix}\mathbf{V}^T \tag{7.39}$$

where **U** and **V** are orthogonal matrices and **D** is a diagonal matrix having all the positive singular values of **A** as its diagonal elements. Equation (7.39) is referred to as "*singular-value decomposition*" of **A**.

Example 7.10. Determine the least square solution to the equations

$$x_1 + 3x_2 = 5$$
$$x_1 - x_2 = 1$$
$$x_1 + x_2 = 0.$$

Here, we have

$$\mathbf{A} = \begin{pmatrix} 1 & 3 \\ 1 & -1 \\ 1 & 1 \end{pmatrix}; \quad b = \begin{pmatrix} 5 \\ 1 \\ 0 \end{pmatrix},$$

$\Longrightarrow$

$$\mathbf{A}^T\mathbf{A} = \begin{pmatrix} 3 & 3 \\ 3 & 11 \end{pmatrix}, \quad \mathbf{A}^T\mathbf{b} = \begin{pmatrix} 6 \\ 14 \end{pmatrix}.$$

Solving the normal equations (7.36)–(7.38) gives

$$\hat{x}_1 = 1 = \hat{x}_2.$$

Note that the least square solution is equivalent to fitting the line $y = \alpha + \beta x$ through the data points (3, 5), (−1, 1) and (1, 0). The least squares solution $\alpha = 1$ and $\beta = 1$ gives the line closest to the data points as shown in Figure 7.6.

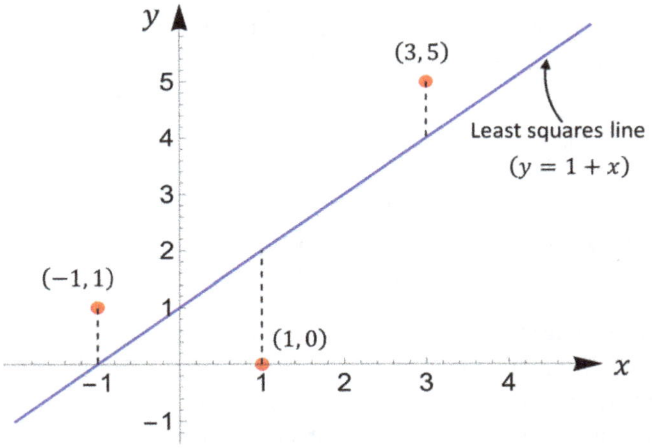

Figure 7.6: Geometric demonstration of least square solution.

Problems

1. Given the quadratic form $F = 6x_1^2 + 5x_2^2 + 7x_3^2 - 4x_1x_2 + 4x_1x_3$:
   (a) Reduce it to its canonical form.
   (b) If $F = 1$, what are the lengths of the semiaxes?
   (c) What are the directions of the principal axes with respect to the original axes?
2. Find the values of the parameter λ for which the following quadratic forms are positive definite:
   (a) $2x_1^2 + x_2^2 + 3x_3^2 + 2\lambda x_1x_2 + 2x_1x_3$
   (b) $x_1^2 + 4x_2^2 + x_3^2 + 2\lambda x_1x_2 + 10x_1x_3 + 6x_2x_3$
3. If $\mathbf{A}$ is an $n \times n$ real symmetric positive definite matrix and $\mathbf{x}$ is an n-vector, show that

$$\int_{-\infty}^{\infty}\int_{-\infty}^{\infty}\int_{-\infty}^{\infty}\cdots\int_{-\infty}^{\infty} \exp(-\mathbf{x}^t\mathbf{A}\mathbf{x})\, d\mathbf{x} = \frac{\pi^{n/2}}{(\det \mathbf{A})^{1/2}}.$$

4. Examine the following functions for relative extremum values:
   (a) $F = x^2y + 2x^2 - 2xy + 3y^2 - 4x + 7y$
   (b) $F = x^3 + 4x^2 + 3y^2 + 5x - 6y$
   (c) $F = x_1^2 + x_2^2$ subject to the constraint $5x_1^2 + 8x_1x_2 + 5x_2^2 = 10$. Give a physical interpretation of the Lagrange multipliers.
5. *(Quadratic forms):* Consider the following linear second-order partial differential equation (PDE) satisfied by a scalar function $u(x, y, z)$:

$$a\frac{\partial^2 u}{\partial x^2} + b\frac{\partial^2 u}{\partial y^2} + c\frac{\partial^2 u}{\partial z^2} + 2f\frac{\partial^2 u}{\partial x\partial y} + 2g\frac{\partial^2 u}{\partial y\partial z} + 2h\frac{\partial^2 u}{\partial z\partial x} + l\frac{\partial u}{\partial x} + m\frac{\partial u}{\partial y} + n\frac{\partial u}{\partial z} + qu = 0$$

where a, b, c, f, g, h, l, m, n and q are constants. The mathematical properties of the solutions of the PDE are largely determined by the algebraic properties of the quadratic form:

$$Q = ax^2 + by^2 + cz^2 + 2fxy + 2gyz + 2hzx = \begin{pmatrix} x & y & z \end{pmatrix}\begin{pmatrix} a & f & h \\ f & b & g \\ h & g & c \end{pmatrix}\begin{pmatrix} x \\ y \\ z \end{pmatrix} = \mathbf{x}^T\mathbf{A}\mathbf{x}$$

or equivalently, the eigenvalues of the symmetric matrix $\mathbf{A}$. If $\mathbf{A}$ has a zero eigenvalue, then the quadratic form (and the PDE) is called parabolic, if $\mathbf{A}$ has all three strictly positive or negative eigenvalues, then the quadratic form is termed elliptic while if $\mathbf{A}$ has three nonzero eigenvalues with sign of one eigenvalue distinct from the other two, then the quadratic form and the PDE is called hyperbolic.

Based on this definition, classify the following PDEs: (i) $u_{xx}+6u_{xy}+u_{yy}+2u_{yz}+u_{zz}=0$, (ii) $u_{xx}+u_{yy}+u_{zz}=0$, (iii) $u_z-u_{xx}-u_{yy}=0$, (iv) $u_{zz}-u_{xx}-u_{yy}=0$ [Note: Here, the subscripts stand for differentiation with respect to the subscripted variable].

6. (a) Consider the case of a linear PDE in two variables

$$a\frac{\partial^2 u}{\partial x^2}+2b\frac{\partial^2 u}{\partial x\partial y}+c\frac{\partial^2 u}{\partial y^2}+g\frac{\partial u}{\partial x}+h\frac{\partial u}{\partial y}+du=0$$

state explicitly (in terms of a, b and c) the conditions under which the PDE is parabolic, elliptic or hyperbolic.

(b) The dispersion of a tracer in a capillary is described by the equation

$$\frac{\partial C}{\partial t}+\frac{\partial C}{\partial z}+P\frac{\partial^2 C}{\partial t\partial z}-\frac{1}{\mathrm{Pe}}\frac{\partial^2 C}{\partial z^2}=0$$

where P and Pe are constants known as the transverse and axial Peclet numbers, respectively. Determine whether this equation is parabolic, elliptic or hyperbolic.

7. (*Scalar differential equation with constant coefficients*)

(a) Consider the scalar second-order equation

$$a\frac{d^2u}{dt^2}+b\frac{du}{dt}+cu=f(t),\quad t>0, a\neq 0$$
$$u(0)=\alpha,$$
$$u'(0)=\beta,$$

write the above equation in vector form:

$$\frac{d\mathbf{u}}{dt}=\mathbf{A}\mathbf{u}+\mathbf{b}(t),\quad t>0;\quad \text{and}\quad \mathbf{u}(t=0)=\mathbf{u}^0$$

where $\mathbf{u}=\binom{u_1}{u_2}$, and identify the coefficient matrix $\mathbf{A}$ and vector $\mathbf{b}$.

(b) Show that the nth order scalar initial value problem can also be expressed in the same vector form. Identify the coefficient matrix $\mathbf{A}$ and vector $\mathbf{b}$.

8. (*Solution of inhomogeneous vector IVP*)

Consider a general IVP in vector form

$$\frac{d\mathbf{u}}{dt}=\mathbf{A}\mathbf{u}+\mathbf{b}(t),\quad t>0;\quad \text{and}\quad \mathbf{u}(t=0)=\mathbf{u}^0$$

where $\mathbf{A}$ is a general constant coefficient matrix.

(a) Obtain a formal solution of above equation using the integrating factor $e^{-\mathbf{A}t}$

(b) Show that when $\mathbf{b}(t)=\mathbf{b}_0$ is a constant vector for $t>0$ and $\mathbf{A}$ is invertible, the solution simplifies to

$$\mathbf{u}=e^{\mathbf{A}t}\mathbf{u}^0+(e^{\mathbf{A}t}-\mathbf{I})\mathbf{A}^{-1}\mathbf{b}_0.$$

9. (*Real and complex canonical forms*)
Consider the initial value problem:

$$\frac{d\mathbf{u}}{dt} = \mathbf{A}\mathbf{u}, \quad t > 0; \quad \text{and} \quad \mathbf{u} = \mathbf{u}_0 @ t = 0$$

where $\mathbf{A}$ is a 3×3 matrix with one real eigenvalue λ_1 and a pair of complex eigenvalues $\lambda_2 = \alpha + i\beta$, $\lambda_3 = \alpha - i\beta$.

(a) Show that the eigenvector $\mathbf{x}_1$ has real elements, while $\mathbf{x}_2$ and $\mathbf{x}_3$ may be expressed as complex conjugates:

$$\mathbf{x}_2 = \mathbf{x}_{2R} + i\mathbf{x}_{2I} \quad \text{and} \quad \mathbf{x}_3 = \mathbf{x}_{2R} - i\mathbf{x}_{2I}$$

where $\mathbf{x}_{2R}$ and $\mathbf{x}_{2I}$ are real vectors obtained by real and imaginary part of $\mathbf{x}_2$.

(b) If we define

$$\mathbf{T} = \begin{pmatrix} \mathbf{x}_1 & \mathbf{x}_{2R} & \mathbf{x}_{2I} \end{pmatrix}$$

verify that

i.

$$\mathbf{T}^{-1}\mathbf{A}\mathbf{T} = \widehat{\mathbf{\Lambda}} = \begin{pmatrix} \lambda_1 & 0 & 0 \\ 0 & \alpha & \beta \\ 0 & -\beta & \alpha \end{pmatrix}$$

or

$$\mathbf{A} = \mathbf{T}\widehat{\mathbf{\Lambda}}\mathbf{T}^{-1},$$

ii.

$$f(\mathbf{A}) = \mathbf{T}f(\widehat{\mathbf{\Lambda}})\mathbf{T}^{-1},$$

for any analytical function f.

iii. Use the results of $(b - ii)$ to show that

$$e^{\mathbf{A}t} = \mathbf{T}e^{\widehat{\mathbf{\Lambda}}t}\mathbf{T}^{-1} = \mathbf{T}\begin{pmatrix} e^{\lambda_1 t} & 0 & 0 \\ 0 & e^{\alpha t}\cos(\beta t) & e^{\alpha t}\sin(\beta t) \\ 0 & -e^{\alpha t}\sin(\beta t) & e^{\alpha t}\cos(\beta t) \end{pmatrix}\mathbf{T}^{-1}.$$

[Remark: Here, $\widehat{\mathbf{\Lambda}}$ is referred to as the real canonical form of $\mathbf{A}$].

10. Find a least squares solution of $\mathbf{A}\mathbf{x} = \mathbf{b}$, when

$$\mathbf{A} = \begin{pmatrix} -1 & 2 \\ 2 & -3 \\ -1 & 3 \end{pmatrix}, \quad \mathbf{b} = \begin{pmatrix} 4 \\ 1 \\ 2 \end{pmatrix}.$$

11. Consider a set of data points $[(x_1,y_1),(x_2,y_2),\dots,(x_N,y_N)]$ and you want to fit a linear model $y = a_0 + a_1 x$.
    (a) Show that this is equivalent to solving the system of equations $\mathbf{A}\boldsymbol{a} = \mathbf{y}$, where

$$\mathbf{A} = \begin{pmatrix} 1 & x_1 \\ \cdot & \cdot \\ \cdot & \cdot \\ 1 & x_N \end{pmatrix}, \quad \boldsymbol{a} = \begin{pmatrix} a_0 \\ a_1 \end{pmatrix}, \quad \mathbf{y} = \begin{pmatrix} y_1 \\ \cdot \\ \cdot \\ y_N \end{pmatrix}.$$

    (b) Show that the least squares solution is given by solving the normal equations:

$$(\mathbf{A}^T\mathbf{A})\boldsymbol{a} = \mathbf{A}^T\mathbf{y}.$$

12. *(Application of matrix methods to linear difference equations):* Linear difference equations of the following type appear in many applications:
    (a) $u_{n+1} = au_n + bu_{n-1}$; $n \geq 1$; u_0 and u_1 given
    (b) $u_{n+2} = au_{n+1} + bu_n + cu_{n-1}$; $n \geq 1$; u_0, u_1 and u_2 given
    Formulate the above equations in vector-matrix form $\mathbf{x}_{k+1} = \mathbf{A}\mathbf{x}_k$ and identify the matrix that appears.
    (c) Solve the Fibonacci equation $u_{n+1} = u_n + u_{n-1}$; $n \geq 1$; $u_0 = u_1 = 1$ and determine a formula for u_n when n is large.
13. [*Discrete dynamical systems and Markov matrices*]: The manufacturer of a product (P) currently controls 20 % of the market in a particular country, while the rival brand (Q) controls the rest of the market. Data from previous year show that 75 % of P's customers remained loyal, while 25 % switched to the rival brand. In addition, 50 % of the competition's customers did not switch to P during the year while the other 50 % did. (a) Representing the market share of each brand as a vector (whose components are fractions), formulate the problem of determining the market share next year, given the market share this year, as a matrix-vector problem and (b) Use the spectral theorem (or any other method) to determine the market share of the two products in the long run.
14. Let $F(x,y,z)$ be a single valued and twice differential function of the variables x,y and z. State the conditions (in terms of partial derivatives) under which a point (x_0,y_0,z_0) can be a local maximum or minimum.
15. (a) The same similarity transformation diagonalizes both matrices **A** and **B**. Show that **A** and **B** must commute.
    (b) Two Hermitian matrices **A** and **B** have the same eigenvalues. Show that **A** and **B** are related by a unitary similarity transformation.

Part II: **Abstract vector space concepts**

Introduction

As stated in the introduction to the book, most of the theories of linear differential equations and other linear operators are generalizations to an infinite number of dimensions of the properties of matrices and vectors. For example, we have seen that the solution of the set of coupled linear first-order differential equations

$$\frac{d\mathbf{u}}{dt} = \mathbf{Au}, \tag{1}$$

with the initial condition

$$\mathbf{u}(t = 0) = \mathbf{u}^0, \tag{2}$$

is given by

$$\mathbf{u} = \sum_{j=1}^{n} \frac{\mathbf{y}_j^* \mathbf{u}^0}{\mathbf{y}_j^* \mathbf{x}_j} \mathbf{x}_j e^{\lambda_j t}, \tag{3}$$

where $\mathbf{A}$ is a constant coefficient $n \times n$ matrix, λ_j, $\mathbf{x}_j$ and $\mathbf{y}_j^*$ are the eigenvalues, eigenvectors and eigenrows of $\mathbf{A}$, respectively. If $\mathbf{A}$ is a real symmetric (self-adjoint) matrix, we have shown that $\mathbf{y}_j = \mathbf{x}_j$ and the eigenvectors may be normalized to form an orthonormal basis. In this case, the solution given by Eq. (3) simplifies to

$$\mathbf{u} = \sum_{j=1}^{n} \langle \mathbf{u}^0, \mathbf{x}_j \rangle \mathbf{x}_j e^{\lambda_j t}, \tag{4}$$

where $\langle \mathbf{u}^0, \mathbf{x}_j \rangle = \mathbf{x}_j^T \mathbf{u}^0$ is the scalar(dot) product. Now, consider the linear partial differential equation

$$\frac{\partial u}{\partial t} = \frac{\partial^2 u}{\partial \xi^2}; \quad 0 < \xi < 1,\ t > 0 \tag{5}$$

with boundary conditions

$$u(0, t) = u(1, t) = 0 \tag{6}$$

and initial condition

$$u(\xi, 0) = u^0(\xi). \tag{7}$$

The solution of Eqs. (5)–(7) may be written as

$$u(\xi, t) = \sum_{j=1}^{\infty} \langle u^0(\xi), x_j(\xi) \rangle x_j(\xi) e^{\lambda_j t} \tag{8}$$

https://doi.org/10.1515/9783111598055-009

where the dot (inner) product is defined by

$$\langle u^0(\xi), x_j(\xi)\rangle = \int_0^1 u^0(\xi)x_j(\xi)d\xi, \quad x_j(\xi) = \sqrt{2}\sin j\pi\xi. \tag{9}$$

Here, $\lambda_j = -j^2\pi^2$ ($j = 1, 2, \ldots$) are the eigenvalues and $x_j(\xi) = \sqrt{2}\sin j\pi\xi$ are the orthonormal set of eigenfunctions of the linear differential operator

$$Lv = \frac{\partial^2 v}{\partial \xi^2}; \quad v(0) = v(1) = 0. \tag{10}$$

The striking similarity between the two solutions is due to the fact that they are both linear equations and contain linear operators **A** and L which are symmetric or self-adjoint. Thus, it is useful to study the general properties of such linear operators.

In what follows, we consider some abstract vector space concepts that are useful in solving linear equations. The advantage of the abstract formalism is the unified treatment of various cases that arise in applications.

8 Vector space over a field

8.1 Definition of a field

A field F is a collection of objects called *scalars* (or numbers) such that two binary operations called *addition* and *multiplication* are defined with the following properties:

If $\alpha \in F, \beta \in F$, then

(i) $\alpha + \beta \in F$ (addition)

(ii) $\alpha.\beta \in F$ (multiplication)

Further, the following axiomatic laws of addition and multiplication must hold:

1. $\alpha + \beta = \beta + \alpha$ and $\alpha.\beta = \beta.\alpha$ (Commutativity)
2. $(\alpha + \beta) + \gamma = \alpha + (\beta + \gamma)$ and $(\alpha.\beta).\gamma = \alpha.(\beta.\gamma)$ (Associativity)
3. $\alpha.(\beta + \gamma) = \alpha.\beta + \alpha.\gamma$ (Distributivity)
4. There are two distinct elements, denoted 0 and 1, respectively, with the properties

   $$\alpha + 0 = \alpha, \quad \alpha.1 = \alpha \quad \forall \alpha \in F$$

   The element 0 is called the identity element for addition while 1 is called the identity element for multiplication.
5. For every $\alpha \in F, \exists$ an element x in $F \ni \alpha + x = 0$.
   x is called the additive inverse of α and is denoted by $(-\alpha)$.
6. For every $\alpha \in F(\alpha \neq 0), \exists$ an element x in $F \ni$

   $$\alpha.x = 1.$$

   x is called the multiplicative inverse of α and is denoted as α^{-1}.

The operations of subtraction and division are merely extensions of addition and multiplication, respectively. For example, $b + (-a)$ is denoted as $b - a$ and $b.a^{-1}$ is denoted as $b/a(a \neq 0)$.

Examples.

1. The set of rational numbers $(\frac{p}{q}, q \neq 0)$, (p and q integers) forms a field.
2. The set of real numbers (positive, negative and zero) forms a field (denoted by $\mathbb{R}$).
3. The set of complex numbers forms a field (denoted by $\mathbb{C}$).
4. The set of integers is not a field since no multiplicative inverse exists.
5. The set of quaternions is not a field. Quaternions are hypercomplex numbers of the form $a + ib + jc + kd$, where a, b, c and d are real and $ij = k, jk = i, ki = j, i^2 = j^2 = k^2 = -1$. This set has a structure similar to a field but multiplication is not commutative.

In almost all of our applications, the field is either $\mathbb{R}$ or $\mathbb{C}$.

https://doi.org/10.1515/9783111598055-010

8.2 Definition of an abstract vector or linear space:

An abstract vector or linear space consists of the following:

1. a field F;
2. a set $\mathbf{V}$ of objects called vectors;
3. a rule (or operation) called *vector addition*, which associates with each pair of vectors $\mathbf{u}, \mathbf{v}$ in $\mathbf{V}$ a vector $\mathbf{u} + \mathbf{v} \in \mathbf{V}$, ($\mathbf{u} + \mathbf{v}$ is called the sum of $\mathbf{u}$ and $\mathbf{v}$) in such a way that
   (a) $\mathbf{u} + \mathbf{v} = \mathbf{v} + \mathbf{u}$ (commutative)
   (b) $(\mathbf{u} + \mathbf{v}) + \mathbf{w} = \mathbf{u} + (\mathbf{v} + \mathbf{w})$, $\mathbf{u}, \mathbf{v}, \mathbf{w} \in \mathbf{V}$ (associative)
   (c) there is a unique vector $\mathbf{0}$ in $\mathbf{V}$ (called zero vector) such that

$$\mathbf{u} + \mathbf{0} = \mathbf{u} \quad \forall \mathbf{u} \in \mathbf{V}$$

   (d) For each $\mathbf{u}$ in $\mathbf{V}$, there is a unique vector $-\mathbf{u}$ in $\mathbf{V}$ such that $\mathbf{u} + (-\mathbf{u}) = \mathbf{0}$.
4. A rule (or operation) called *scalar multiplication*, which associates with each scalar $\alpha \in F$ and $\mathbf{u} \in \mathbf{V}$, a vector $\alpha\mathbf{u}$, called the product of α and $\mathbf{u}$, in such a way that
   (a) $1.\mathbf{u} = \mathbf{u}\ \forall \mathbf{u}$ in $\mathbf{V}$;
   (b) $0.\mathbf{u} = \mathbf{0}$.
   (1 is multiplicative identity in F and 0 is additive identity in F while $\mathbf{0}$ is additive identity in $\mathbf{V}$.)
   (c) $(\alpha\beta)\mathbf{u} = \alpha(\beta\mathbf{u})$;
   (d) $\alpha(\mathbf{u} + \mathbf{v}) = \alpha\mathbf{u} + \alpha\mathbf{v}$
   (e) $(\alpha + \beta)\mathbf{u} = \alpha\mathbf{u} + \beta\mathbf{u}$

Note that, as the definition states, a *vector space* is a *composite object* consisting of a field, a set of vectors and two operations with the above special properties.

Examples of vector spaces

1. The space of n-tuples
   Let F be $\mathbb{R}$ or $\mathbb{C}$ and $\mathbf{u} = (\alpha_1 \quad \alpha_2 \ldots \alpha_n)$, $\mathbf{v} = (\beta_1 \quad \beta_2 \ldots \beta_n)$. Define vector addition by

$$\mathbf{u} + \mathbf{v} = (\alpha_1 + \beta_1, \ldots, \alpha_n + \beta_n)$$

   and scalar multiplication ($\gamma \in F$) by

$$\gamma\mathbf{u} = (\gamma\alpha_1, \gamma\alpha_2, \ldots, \gamma\alpha_n)$$

   Then it can be shown that $\mathbf{V}$ has all the above properties. Hence, $\mathbf{V}$ is a vector space and is denoted by $\mathbb{R}^n/\mathbb{C}^n$.

2. The space of all $m \times n$ matrices over the field F
   Let $\mathbf{V} = \{\text{set of all } m \times n \text{ matrices over } F\}$. For $\mathbf{A} \in \mathbf{V}, \mathbf{B} \in \mathbf{V}$, define

$$(\mathbf{A} + \mathbf{B})_{ij} = \mathbf{A}_{ij} + \mathbf{B}_{ij} \quad \text{(vector addition)}$$

   $\gamma \in F$

$$(\gamma \mathbf{A})_{ij} = \gamma \mathbf{A}_{ij} \quad \text{(scalar multiplication)}$$

   This finite-dimensional vector space is denoted by $\mathbb{R}^{m \times n}$ or $\mathbb{C}^{m \times n}$.
3. Let $F = \mathbb{R}$ and $V = \{\text{continuous functions } f(x) \text{ defined in the interval } [a,b]\} = C[a,b]$. Define

$$(f+g)(x) = f(x) + g(x) \in V$$
$$(\gamma f)(x) = \gamma f(x) \in V$$

   This is a vector space (of infinite dimension).
4. Let F be a field ($\mathbb{R}$ or $\mathbb{C}$) and $\mathbf{V}$ the set of all polynomials $p(x)$ of the form

$$p(x) = a_0 + a_1 x + a_2 x^2 + \cdots + a_N x^N, \quad a_j \in F$$

   Then $\mathbf{V}$ is a vector space if we define

$$(p_1 + p_2)(x) = p_1(x) + p_2(x)$$
$$(\gamma p_1)(x) = \gamma p_1(x), \quad \gamma \in F$$

5. The set of all solutions to the system $\mathbf{Ax} = \mathbf{0}$, where $\mathbf{A}$ is an $m \times n$ matrix and $\mathbf{x}$ is $n \times 1$ vector is a vector space.
6. The set of all solutions to

$$Ly = 0$$

   where L is a linear n-th order differential operator with $Ly = a_0(x)\frac{d^n y}{dx^n} + a_1(x)\frac{d^{n-1}y}{dx^{n-1}} + \cdots + a_n(x)y = 0$ is a vector space.

8.2.1 Subspaces

Definition. Let $\mathbf{V}$ be a vector space over the field F. A *subspace* of $\mathbf{V}$ is a subset $\mathbf{W}$ of $\mathbf{V}$, which is itself a vector space over F with the operations of vector addition and scalar multiplication.

Examples. Let $\mathbf{V}$ be a vector space over a field F. Then:

1. The subspace containing the single element or the zero vector $\{\mathbf{0}\}$ is a subspace. This is called the zero subspace of $\mathbf{V}$
2. Let $\mathbf{V}$ be the set of n-tuples defined over F and

$$\mathbf{x} = (a_1\ a_2 \dots a_n) \in \mathbf{V}.$$

Then the set of n-tuples with $a_1 = 0$ is a vector space, which is a subspace of $\mathbf{V}$.
3. Let $\mathbf{V}$ be the set of $n\times n$ matrices over F and $\mathbf{W}$ be the set of symmetric $n\times n$ matrices over F. Then $\mathbf{W} \subset \mathbf{V}$
4. Let $\mathbf{V}$ be the set of all Euclidean vectors with n elements, i. e., $\mathbf{V} = \mathbb{R}^n$ and $\mathbf{W}$ be the set of all solutions to

$$\mathbf{A}\mathbf{x} = \mathbf{0},$$

where $\mathbf{A}$ is $m \times n$ and $\mathbf{x}$ is $n \times 1$. Then $\mathbf{W}$ is a subspace of $\mathbf{V}$.
5. $\mathbf{V}$ = set of all continuous functions defined in (a, b), $\mathbf{W}$ = space of all polynomial functions. Then $\mathbf{W} \subset \mathbf{V}$.

Definition. Let $\mathbf{V}$ be a vector space over F and $\mathbf{S}$ be a set of vectors in $\mathbf{V}$. Then the *subspace spanned* by $\mathbf{S}$ is defined to be the intersection $\mathbf{W}$ of all subspaces of $\mathbf{V}$, which contain $\mathbf{S}$.

8.2.2 Bases and dimension

Let $\mathbf{V}$ be a vector space over F. A subset $\mathbf{S} = \{\mathbf{x}_1, \mathbf{x}_2, \dots, \mathbf{x}_n\}$ of $\mathbf{V}$ is said to be *linearly dependent* if there exists a set of scalars $a_i (i = 1, 2, \dots, n) \in F$, not all zero, such that

$$a_1\mathbf{x}_1 + a_2\mathbf{x}_2 + \cdots + a_n\mathbf{x}_n = \mathbf{0}.$$

If no such set of scalars exists, i. e., if $a_i = 0$ for all i, then the set $\mathbf{S}$ is said to be *linearly independent*.

Definition. Let $\mathbf{V}$ be a vector space over F. A *basis* for $\mathbf{V}$ is a linearly independent set of vectors in $\mathbf{V}$, which spans $\mathbf{V}$.

Definition. The *dimension* of a vector space is the largest number of linearly independent vectors in that space. If there is no largest number, then we say that the vector space is of infinite-dimensional.

Example. The vector space of continuous functions over the unit interval $C[0, 1]$ is of infinite dimension.

8.2.3 Coordinates

Let **V** be a finite-dimensional vector space over the field F. Let $(\mathbf{x}_1, \mathbf{x}_2, \ldots, \mathbf{x}_n)$ be a basis for **V**. Let **z** be any vector in **V**. Then we have

$$\mathbf{z} = \alpha_1\mathbf{x}_1 + \alpha_2\mathbf{x}_2 + \cdots + \alpha_n\mathbf{x}_n \tag{8.1}$$

This representation is unique. We call $(\alpha_1, \alpha_2, \ldots, \alpha_n)$ the coordinates of **z** w. r. t. the basis $(\mathbf{x}_1, \mathbf{x}_2, \ldots, \mathbf{x}_n)$. Now, let $(\mathbf{y}_1, \mathbf{y}_2, \ldots, \mathbf{y}_n)$ be another basis for **V** and

$$\mathbf{z} = \beta_1\mathbf{y}_1 + \beta_2\mathbf{y}_2 + \cdots + \beta_n\mathbf{y}_n \tag{8.2}$$

$(\beta_1, \beta_2, \ldots, \beta_n)$ are coordinates of **z** w. r. t. the **y**-basis. To find the relationship between the coordinates, we expand the **y**-basis in terms of **x**-basis

$$\mathbf{y}_1 = p_{11}\mathbf{x}_1 + p_{21}\mathbf{x}_2 + \cdots + p_{n1}\mathbf{x}_n = \begin{bmatrix} \mathbf{x}_1 & \mathbf{x}_2 & \cdots & \mathbf{x}_n \end{bmatrix} \begin{bmatrix} p_{11} \\ p_{21} \\ \vdots \\ p_{n1} \end{bmatrix}$$

$$\vdots$$

$$\begin{aligned}
\mathbf{y}_n &= p_{1n}\mathbf{x}_1 + p_{2n}\mathbf{x}_2 + \cdots + p_{nn}\mathbf{x}_n \\
\mathbf{z} &= \beta_1(p_{11}\mathbf{x}_1 + p_{21}\mathbf{x}_2 + \cdots + p_{n1}\mathbf{x}_n) + \beta_2(p_{12}\mathbf{x}_1 + p_{22}\mathbf{x}_2 + \cdots + p_{n2}\mathbf{x}_n) \\
&\quad + \cdots + \beta_n(p_{1n}\mathbf{x}_1 + p_{2n}\mathbf{x}_2 + \cdots + p_{nn}\mathbf{x}_n) \\
&= \mathbf{x}_1(p_{11}\beta_1 + p_{12}\beta_2 + \cdots + p_{1n}\beta_n) + \mathbf{x}_2(p_{21}\beta_1 + p_{22}\beta_2 + \cdots + p_{2n}\beta_n) \\
&\quad + \cdots + \mathbf{x}_n(p_{n1}\beta_1 + p_{n2}\beta_2 + \cdots + p_{nn}\beta_n)
\end{aligned} \tag{8.3}$$

Comparing (8.1) and (8.3), we get

$$\begin{aligned}
\alpha_1 &= p_{11}\beta_1 + p_{12}\beta_2 + \cdots + p_{1n}\beta_n \\
&\;\;\vdots \\
\alpha_n &= p_{n1}\beta_1 + p_{n2}\beta_2 + \cdots + p_{nn}\beta_n
\end{aligned}$$

$\Rightarrow$

$$\boldsymbol{\alpha} = \mathbf{P}\boldsymbol{\beta} \Rightarrow \boldsymbol{\beta} = \mathbf{P}^{-1}\boldsymbol{\alpha} \ (\because \mathbf{P} \text{ is nonsingular})$$

The matrix **P** is called the change of basis matrix [from **x** to **y** basis].

Theorem. *Let* **V** *be a n-dimensional vector space over the field F, let* **B** *and* **B**$'$ *be two ordered bases of* **V**. *Then there is a unique, necessarily invertible,* $n \times n$ *matrix* **P** *with entries in F such that*

1.

$$[\mathbf{z}]_B = \mathbf{P}[\mathbf{z}]_{B'}$$

where $\mathbf{z}_B$ = *coordinate vector w. r. t. basis* **B** *and* $\mathbf{z}_{B'}$ = *coordinate vector w. r. t. basis* $\mathbf{B}'$

2. $[\mathbf{z}]_{B'} = \mathbf{P}^{-1}[\mathbf{z}]_B$ *for every vector* **z** *in* **V**.

Example. Consider $\mathbb{R}^2$ with standard basis $\mathbf{e}_1 = \binom{1}{0}$ and $\mathbf{e}_2 = \binom{0}{1}$. A second basis is $\mathbf{y}_1 = \binom{1}{2}$ and $\mathbf{y}_2 = \binom{2}{5}$. The coordinates of the vector $\mathbf{z} = \binom{1}{-1}$ in the standard basis and second **y** basis, $\boldsymbol{\beta} = \binom{7}{-3}$ are related by the matrices $\mathbf{P} = \begin{pmatrix} 1 & 2 \\ 2 & 5 \end{pmatrix}$ and $\mathbf{P}^{-1} = \begin{pmatrix} 5 & -2 \\ -2 & 1 \end{pmatrix}$.

An important point to note is that once a set of basis vectors is selected, coordinates can be defined and all algebraic operations in the abstract vector space (of finite dimension) can be reduced to operations on matrices and n-tuples.

Problems

1. Consider the vector space of polynomials of degree ≤ 3. Are the following four a linearly independent set?

$$1 - 2t + t^2 - 3t^3, \quad -2 + t - 4t^2 + 5t^3, \quad -1 - 4t - 5t^2 + t^3, \quad 3 + t - 4t^2 + 3t^3$$

2. Consider the vector space of 3×3 symmetric matrices. What is its dimension? Find a basis.
3. Consider the vector space of complex numbers over the real field. What is its dimension? Consider the vector space of complex numbers over the complex number field. What is its dimension?
4. Given the vector space $\mathbb{R}^4$, suppose $\mathbf{x}^T = (a, b, c, d)$ where $a, b, c, d \in \mathbb{R}$. Consider the subset with $a + c = 0, b = 3d$. Is this subset a vector space?
5. Consider a linear homogeneous n-th order differential equation with constant coefficients. Consider the space of polynomials of degree $\leq N$ where N is arbitrary but fixed. By operating in the coordinate space show that no finite polynomial can ever be a solution.
6. Let **V** be the vector space generated by the polynomials

$$p_1 = x^3 - 2x^2 + 4x + 1$$
$$p_2 = x^3 + 6x - 5$$
$$p_3 = 2x^3 - 3x^2 + 9x - 1$$
$$p_4 = 2x^3 - 5x^2 + 7x + 5$$

Find a basis and dimension of **V**.

7. Let $\mathbf{y}_1, \mathbf{y}_2, \ldots \mathbf{y}_n$ be independent vectors in a vector space $\mathbf{V}$. Let $\mathbf{w} \in \mathbf{V}$ be given by

$$\mathbf{w} = \alpha_1 \mathbf{y}_1 + \alpha_2 \mathbf{y}_2 + \cdots + \alpha_n \mathbf{y}_n$$

where α_i are constants (scalars in the field). Show that the representation of $\mathbf{w}$ is unique.

8. Let $\mathbf{U}$ be the vector space of all 2×3 matrices over field $\mathbb{R}$.
   (a) Determine a basis for $\mathbf{U}$.
   (b) Determine if the following matrices in $\mathbf{U}$ are dependent or independent:

$$\mathbf{A} = \begin{pmatrix} 1 & 2 & -3 \\ 4 & 0 & 1 \end{pmatrix}$$

$$\mathbf{B} = \begin{pmatrix} 1 & 3 & -4 \\ 6 & 5 & 4 \end{pmatrix}$$

$$\mathbf{C} = \begin{pmatrix} 3 & 8 & -11 \\ 16 & 10 & 9 \end{pmatrix}.$$

9 Linear transformations

9.1 Definition of a linear transformation

Recall from calculus the definition of a function. A function consists of the following:

1. a set **X** called the domain of the function
2. a set **Y** called the codomain of the function
3. a rule (or correspondence) f, which associates with each element x of **X** a single element $f(x)$ of **Y**.

We write $f : \mathbf{X} \to \mathbf{Y}$ to indicate the function. Figure 9.1 shows the key features of a function schematically.

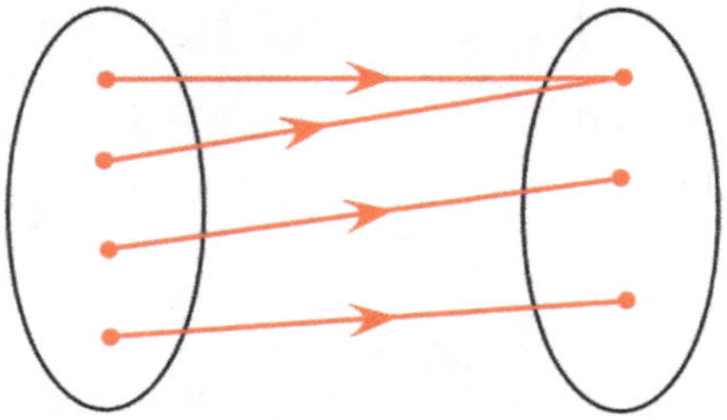

Figure 9.1: Schematic diagram illustrating the domain and codomain of a function or a linear transformation.

The function $f : \mathbf{X} \to \mathbf{Y}$ is said to be one-to-one or *injective* if different elements of **X** have distinct images, i. e.,

$$x \neq x' \Rightarrow f(x) \neq f(x') \quad \text{or equivalently} \quad f(x) = f(x') \Rightarrow x = x'$$

$f : \mathbf{X} \to \mathbf{Y}$ is *onto* or *surjective* if every $y \in \mathbf{Y}$ is the image of some $x \in \mathbf{X}$. The range of f or Image of $f = \{f(x)/x \in \mathbf{X}\} \Longrightarrow \text{Range } f \subset \mathbf{Y}$. The range of f consists of all elements y of **Y** such that $y = f(x)$ for some $x \in \mathbf{X}$. If the range of f is all of **Y** then we say f is a *function from* **X** *onto* **Y**, or simply f is *onto*. If each element of **X** is assigned to a distinct element of **Y**, then we say f is *one-to-one*. If f is *one-to-one* and *onto*, i. e., the inverse function $f^{-1} : \mathbf{Y} \to \mathbf{X}$ exists, then we say f is a *bijection*.

Definition (Linear transformation). Let **V** and **W** be vector spaces over a field F. A *linear transformation from* **V** *into* **W** is a function **T** from **V** into **W** such that

$$\mathbf{T}(\alpha\mathbf{u} + \mathbf{v}) = \alpha\mathbf{T}(\mathbf{u}) + \mathbf{T}(\mathbf{v}) = \alpha\mathbf{T}\mathbf{u} + \mathbf{T}\mathbf{v}$$

for all **u** and **v** in **V** and α in F.

https://doi.org/10.1515/9783111598055-011

Examples.

1. Let $\mathbf{V}$ be any vector space. Then the identity transformation $\mathbf{I}$ defined by

$$\mathbf{Iu} = \mathbf{u},$$

is a linear transformation from $\mathbf{V}$ into $\mathbf{V}$.

2. Let $\mathbf{V}$ be the vector space of $n \times 1$ (column) matrices defined over F and let $\mathbf{A}$ be a fixed $m \times n$ matrix over F. Then the function $\mathbf{T}$ on $\mathbf{V}$ defined by

$$\mathbf{T}(\mathbf{u}) = \mathbf{Au}$$

is a linear transformation from $\mathbf{V}$ into the space $\mathbf{W}$ of $m \times 1$ matrices over F, i. e., $\mathbf{T} : \mathbf{V} \to \mathbf{W}$ or $\mathbb{R}^n \to \mathbb{R}^m$ is a linear transformation.

3. Let F be a field and $\mathbf{V}$ be the space of polynomial functions, f, given by

$$f(x) = c_0 + c_1 x + \cdots + c_k x^k$$

Let $(Df)(x) = c_1 + 2c_2 x + \cdots + kc_k x^{k-1}$. Then D is a linear transformation from $\mathbf{V}$ into $\mathbf{V}$. It is the differentiation transformation.

4. Let $\mathbf{V}$ be the vector space of polynomials in t defined over the field $\mathbb{R}$ and $\mathbf{T}$ be integration operation from 0 to 1, i. e.,

$$\mathbf{T} : \mathbf{V} \to \mathbb{R}$$

For any $f \in \mathbf{V}$, define

$$\mathbf{T}f = \int_0^1 f(t)\, dt$$

This is a linear transformation.

5. Let $\mathbf{V}$ be the space of all $m \times n$ matrices defined over a field F, let $\mathbf{P}$ be a fixed $m \times m$ matrix over F and let $\mathbf{Q}$ be a fixed $n \times n$ matrix over F. Define $\mathbf{T} : \mathbf{V} \to \mathbf{V}$ by

$$\mathbf{T}(\mathbf{A}) = \mathbf{PAQ}.$$

Then $\mathbf{T}$ is a linear transformation from $\mathbf{V}$ into $\mathbf{V}$.

Lemma 9.1. *If* $\mathbf{T}$ *is a linear transformation, then*

$$\mathbf{T}(\mathbf{0}_v) = \mathbf{0}_w$$

Proof. Since $\mathbf{T}(\mathbf{u} + \mathbf{v}) = \mathbf{T}(\mathbf{u}) + \mathbf{T}(\mathbf{v})$, let $\mathbf{u} = \mathbf{0}_v$ and $\mathbf{v} = \mathbf{0}_v$

⇒

$$\begin{aligned}\mathbf{T}(\mathbf{0}_v + \mathbf{0}_v) &= \mathbf{T}(\mathbf{0}_v) + \mathbf{T}(\mathbf{0}_v)\\ \Rightarrow \mathbf{T}(\mathbf{0}_v) &= T(\mathbf{0}_v) + \mathbf{T}(\mathbf{0}_v)\\ \Rightarrow \mathbf{T}(\mathbf{0}_v) &= \mathbf{0}_w\end{aligned}$$

□

Lemma 9.2. *If* $\mathbf{T} : \mathbf{V} \to \mathbf{W}$ *is a linear transformation, then*

$$\mathbf{T}(a_1\mathbf{u}_1 + a_2\mathbf{u}_2 + \cdots + a_n\mathbf{u}_n) = a_1\mathbf{T}(\mathbf{u}_1) + a_2\mathbf{T}(\mathbf{u}_2) + \cdots + a_n\mathbf{T}(\mathbf{u}_n)$$

Definition. If $\mathbf{V}$ is a vector space over the field F, a *linear operator* on $\mathbf{V}$ is a linear transformation from $\mathbf{V}$ into $\mathbf{V}$.

Examples.
1. Let $\mathbf{V} = \mathbb{R}^2$ and define $\mathbf{T}(x_1, x_2) = (x_2, x_1)$, then $\mathbf{T}$ is a linear operator on $\mathbf{V}$
2. Let $\mathbf{V}$ = space of 2×2 matrices defined on $\mathbb{R}$. For $\mathbf{A} \in \mathbf{V}$, define $\mathbf{T}(\mathbf{A}) = \mathbf{AB} - \mathbf{BA}$ where $\mathbf{B}$ is a fixed 2×2 matrix, e. g.,

$$\mathbf{B} = \begin{pmatrix} 1 & 2 \\ 3 & 4 \end{pmatrix}$$

Then $\mathbf{T}$ is a linear operator.

9.2 Matrix representation of a linear transformation

Let $\mathbf{V}$ and $\mathbf{W}$ be vector spaces over F. If $\mathbf{V}$ and $\mathbf{W}$ are finite-dimensional, then any linear transformation from $\mathbf{V}$ into $\mathbf{W}$ has a matrix representation. Let $\mathbf{e}_1, \mathbf{e}_2, \dots, \mathbf{e}_n$ be a basis for $\mathbf{V}$ and $\mathbf{f}_1, \mathbf{f}_2, \dots, \mathbf{f}_m$ be a set of basis vectors for $\mathbf{W}$. Consider

$$\begin{aligned}\mathbf{Te}_1 &= a_{11}\mathbf{f}_1 + a_{21}\mathbf{f}_2 + \cdots + a_{m1}\mathbf{f}_m\\ \mathbf{Te}_2 &= a_{12}\mathbf{f}_1 + a_{22}\mathbf{f}_2 + \cdots + a_{m2}\mathbf{f}_m\\ &\ \ \vdots\\ \mathbf{Te}_n &= a_{1n}\mathbf{f}_1 + a_{2n}\mathbf{f}_2 + \cdots + a_{mn}\mathbf{f}_m\end{aligned}$$

$a_{ij} \in F$. These equations contain the essential information about $\mathbf{T}$, i. e., they tell us how $\mathbf{T}$ transforms each of the basis vectors $(\mathbf{e}_1, \mathbf{e}_2, \dots, \mathbf{e}_n)$. Now, let $\mathbf{u} \in \mathbf{V}$ be an arbitrary vector. Then

$$\mathbf{u} = a_1\mathbf{e}_1 + \cdots + a_n\mathbf{e}_n$$

where $a_i \in F$. We call the $n \times 1$ column vector

$$\boldsymbol{\alpha} = \begin{pmatrix} \alpha_1 \\ \cdot \\ \cdot \\ \alpha_n \end{pmatrix}$$

as the coordinate vector of **u** w. r. t. the e-basis. Now, let

$$\mathbf{Tu} = \mathbf{w} \in \mathbf{W}$$

⇒

$$\mathbf{w} = \beta_1\mathbf{f}_1 + \cdots + \beta_m\mathbf{f}_m$$

$$\boldsymbol{\beta} = \begin{pmatrix} \beta_1 \\ \cdot \\ \cdot \\ \beta_m \end{pmatrix}, \quad \beta_j \in F$$

is the coordinate vector of the image of **u** w. r. t. the basis $(\mathbf{f}_1, \mathbf{f}_2, \ldots, \mathbf{f}_m)$. Consider

$$\begin{aligned}
\mathbf{Tu} &= \alpha_1\mathbf{Te}_1 + \cdots + \alpha_n\mathbf{Te}_n \\
&= \alpha_1[a_{11}\mathbf{f}_1 + a_{21}\mathbf{f}_2 + \cdots + a_{m1}\mathbf{f}_m] + \cdots \\
&\quad + \alpha_n[a_{1n}\mathbf{f}_1 + a_{2n}\mathbf{f}_2 + \cdots + a_{mn}\mathbf{f}_m] \\
&= \mathbf{f}_1[a_{11}\alpha_1 + a_{12}\alpha_2 + \cdots + a_{1n}\alpha_n] + \cdots \\
&\quad + \mathbf{f}_m[a_{m1}\alpha_1 + a_{m2}\alpha_2 + \cdots + a_{mn}\alpha_n]
\end{aligned}$$

⇒

$$\begin{aligned}
\beta_1 &= a_{11}\alpha_1 + \cdots + a_{1n}\alpha_n \\
&\cdot \\
&\cdot \\
\beta_m &= a_{m1}\alpha_1 + \cdots + a_{mn}\alpha_n
\end{aligned}$$

or $\boldsymbol{\beta} = \mathbf{A}\boldsymbol{\alpha}$, where the $m \times n$ matrix $\mathbf{A} = \{a_{ij}\}$ is called the *matrix representation of the linear transformation* **T** with respect to the bases **e** (in **V**) and **f** (in **W**). Thus, if $\mathbf{u} \in \mathbf{V}$, $\mathbf{Tu} \in \mathbf{W}$ and once we choose a basis for **V** and **W** then the linear transformation $\mathbf{T} : \mathbf{V} \to \mathbf{W}$ (in abstract spaces) is equivalent to the transformation

$$\mathbf{T} : \mathbb{R}^n \to \mathbb{R}^m \Rightarrow \begin{bmatrix} \beta_1 \\ \cdot \\ \cdot \\ \beta_m \end{bmatrix} = \begin{bmatrix} a_{11}\cdot & \cdot & \cdot & a_{1n} \\ \cdot & \cdot & \cdot & \cdot \\ \cdot & \cdot & \cdot & \cdot \\ a_{m1} & \cdot & \cdot & a_{mn} \end{bmatrix} \begin{bmatrix} \alpha_1 \\ \cdot \\ \cdot \\ \alpha_n \end{bmatrix}$$

defined by $\mathbf{T}(\mathbf{u}) = \mathbf{A}\boldsymbol{\alpha} = \boldsymbol{\beta}$, where $\boldsymbol{\alpha}$ is the coordinate vector of $\mathbf{u}$ and $\boldsymbol{\beta}$ is the coordinate vector of $\mathbf{Tu}$.

Examples.

1. Let $\mathbf{V} = \mathbb{R}^2$ and consider the linear operator $\mathbf{T}(u_1, u_2) = (u_2, u_1)$. Choose $\mathbf{e}_1 = (1, 0)$, $\mathbf{e}_2 = (0, 1)$ as a basis for $\mathbf{V}$.

$$\mathbf{Te}_1 = (0, 1) = 0.(1, 0) + 1.(0, 1)$$
$$\mathbf{Te}_2 = (1, 0) = 1.(1, 0) + 0.(0, 1)$$

$\Rightarrow$

$$\mathbf{A} = \begin{pmatrix} 0 & 1 \\ 1 & 0 \end{pmatrix}$$

2. Let $\mathbf{V} = \mathbb{R}^2$, and $\mathbf{T}(u_1, u_2) = (u_1 - u_2, 2u_1 + 4u_2)$.
In this case, it is easily seen that

$$\mathbf{A} = \begin{pmatrix} 1 & -1 \\ 2 & 4 \end{pmatrix}$$

w. r. t. standard basis. Another basis

$$\hat{\mathbf{e}}_1 = \begin{pmatrix} 1 \\ -1 \end{pmatrix}, \quad \hat{\mathbf{e}}_2 = \begin{pmatrix} 1 \\ -2 \end{pmatrix}$$

gives

$$\mathbf{A} = \begin{pmatrix} 2 & 0 \\ 0 & 3 \end{pmatrix}.$$

3. Let $\mathbf{V}$ = space of all polynomials of degree ≤ 3. Consider the linear operator $\mathbf{T} : \mathbf{V} \to \mathbf{V}$, which is represented by the differentiation operator, $\frac{d}{dt}$. Take $(1, t, t^2, t^3)$ as a basis for $\mathbf{V}$. Now,

$$\frac{d}{dt}(1) = 0 = 0.1 + 0.t + 0.t^2 + 0.t^3$$
$$\frac{d}{dt}(t) = 1 = 1.1 + 0.t + 0.t^2 + 0.t^3$$
$$\frac{d}{dt}(t^2) = 2t = 0.1 + 2.t + 0.t^2 + 0.t^3$$
$$\frac{d}{dt}(t^3) = 3t^2 = 0.1 + 0.t + 3.t^2 + 0.t^3$$

$\Rightarrow$

$$\mathbf{A} = \begin{pmatrix} 0 & 1 & 0 & 0 \\ 0 & 0 & 2 & 0 \\ 0 & 0 & 0 & 3 \\ 0 & 0 & 0 & 0 \end{pmatrix}.$$

Let

$$f = \alpha_0 + \alpha_1 t + \alpha_2 t^2 + \alpha_3 t^3$$

i. e., coordinate of f is $\boldsymbol{\alpha} = [\alpha_0\, \alpha_1\, \alpha_2\, \alpha_3]^t$. Then the coordinate of $\frac{df}{dt}$ can be obtained by multiplying the matrix representation of T by the coordinates of f as

$$\mathbf{A}\boldsymbol{\alpha} = \begin{pmatrix} 0 & 1 & 0 & 0 \\ 0 & 0 & 2 & 0 \\ 0 & 0 & 0 & 3 \\ 0 & 0 & 0 & 0 \end{pmatrix} \begin{pmatrix} \alpha_0 \\ \alpha_1 \\ \alpha_2 \\ \alpha_3 \end{pmatrix} = \begin{pmatrix} \alpha_1 \\ 2\alpha_2 \\ 3\alpha_3 \\ 0 \end{pmatrix},$$

i. e.

$$\frac{df}{dt} = \alpha_1.1 + 2\alpha_2.t + 3\alpha_3.t^2 + 0.t^3$$

4. Consider same **V** as in example 3 above, **W** = ℝ, and **T** = Integration from 0 to 1;

$$\int_0^1 1.\,dt = 1, \quad \int_0^1 t.\,dt = \frac{t^2}{2}\bigg|_0^1 = \frac{1}{2}$$

$$\int_0^1 t^2.\,dt = \frac{1}{3}, \quad \int_0^1 t^3.\,dt = \frac{1}{4}$$

Thus, $\mathbf{T} : \mathbf{V} \to \mathbb{R}$ has matrix representation

$$\mathbf{A} = [\, 1 \quad \tfrac{1}{2} \quad \tfrac{1}{3} \quad \tfrac{1}{4} \,]$$

If

$$f = \alpha_0 + \alpha_1 t + \alpha_1 t^2 + \alpha_1 t^3 \in V$$

then $\mathbf{T}f$ is obtained by multiplying the matrix representation of **T** by the coordinates of f as

$$\mathbf{T}f = \mathbf{A}\boldsymbol{\alpha} = [\, 1 \quad \tfrac{1}{2} \quad \tfrac{1}{3} \quad \tfrac{1}{4} \,] \begin{bmatrix} \alpha_0 \\ \alpha_1 \\ \alpha_2 \\ \alpha_3 \end{bmatrix} = \alpha_0 + \frac{\alpha_1}{2} + \frac{\alpha_2}{3} + \frac{\alpha_3}{4}$$

5. Let **V** be the vector space of all 2 × 2 real matrices and **T** be a linear operator on **V** defined by

$$\mathbf{T}(\mathbf{A}) = \mathbf{A}\begin{pmatrix} 0 & 1 \\ 1 & 0 \end{pmatrix} - \begin{pmatrix} 0 & 1 \\ 1 & 0 \end{pmatrix}\mathbf{A}$$

Determine the matrix representation of **T**.
We choose as basis for $\mathbf{V} = \{\mathbf{e}_1 = \left(\begin{smallmatrix} 1 & 0 \\ 0 & 0 \end{smallmatrix}\right), \mathbf{e}_2 = \left(\begin{smallmatrix} 0 & 1 \\ 0 & 0 \end{smallmatrix}\right), \mathbf{e}_3 = \left(\begin{smallmatrix} 0 & 0 \\ 1 & 0 \end{smallmatrix}\right), \mathbf{e}_4 = \left(\begin{smallmatrix} 0 & 0 \\ 0 & 1 \end{smallmatrix}\right)\}$

⇒

$$\begin{aligned}
\mathbf{Te}_1 &= \begin{pmatrix} 1 & 0 \\ 0 & 0 \end{pmatrix}\begin{pmatrix} 0 & 1 \\ 1 & 0 \end{pmatrix} - \begin{pmatrix} 0 & 1 \\ 1 & 0 \end{pmatrix}\begin{pmatrix} 1 & 0 \\ 0 & 0 \end{pmatrix} \\
&= \begin{pmatrix} 0 & 1 \\ 0 & 0 \end{pmatrix} - \begin{pmatrix} 0 & 0 \\ 1 & 0 \end{pmatrix} \\
&= \begin{pmatrix} 0 & 1 \\ -1 & 0 \end{pmatrix} = 0.\mathbf{e}_1 + 1.\mathbf{e}_2 - 1.\mathbf{e}_3 + 0.\mathbf{e}_4 \\
\mathbf{Te}_2 &= \begin{pmatrix} 0 & 1 \\ 0 & 0 \end{pmatrix}\begin{pmatrix} 0 & 1 \\ 1 & 0 \end{pmatrix} - \begin{pmatrix} 0 & 1 \\ 1 & 0 \end{pmatrix}\begin{pmatrix} 0 & 1 \\ 0 & 0 \end{pmatrix} \\
&= \begin{pmatrix} 1 & 0 \\ 0 & 0 \end{pmatrix} - \begin{pmatrix} 0 & 0 \\ 0 & 1 \end{pmatrix} \\
&= \begin{pmatrix} 1 & 0 \\ 0 & -1 \end{pmatrix} = 1.\mathbf{e}_1 + 0.\mathbf{e}_2 + 0.\mathbf{e}_3 - 1.\mathbf{e}_4 \\
\mathbf{Te}_3 &= \begin{pmatrix} 0 & 0 \\ 1 & 0 \end{pmatrix}\begin{pmatrix} 0 & 1 \\ 1 & 0 \end{pmatrix} - \begin{pmatrix} 0 & 1 \\ 1 & 0 \end{pmatrix}\begin{pmatrix} 0 & 0 \\ 1 & 0 \end{pmatrix} \\
&= \begin{pmatrix} 0 & 0 \\ 0 & 1 \end{pmatrix} - \begin{pmatrix} 1 & 0 \\ 0 & 0 \end{pmatrix} \\
&= \begin{pmatrix} -1 & 0 \\ 0 & 1 \end{pmatrix} = -1.\mathbf{e}_1 + 0.\mathbf{e}_2 + 0.\mathbf{e}_3 + 1.\mathbf{e}_4 \\
\mathbf{Te}_4 &= \begin{pmatrix} 0 & 0 \\ 0 & 1 \end{pmatrix}\begin{pmatrix} 0 & 1 \\ 1 & 0 \end{pmatrix} - \begin{pmatrix} 0 & 1 \\ 1 & 0 \end{pmatrix}\begin{pmatrix} 0 & 0 \\ 0 & 1 \end{pmatrix} \\
&= \begin{pmatrix} 0 & 0 \\ 1 & 0 \end{pmatrix} - \begin{pmatrix} 0 & 1 \\ 0 & 0 \end{pmatrix} \\
&= \begin{pmatrix} 0 & -1 \\ 1 & 0 \end{pmatrix} = 0.\mathbf{e}_1 - 1.\mathbf{e}_2 + 1.\mathbf{e}_3 + 0.\mathbf{e}_4 \\
[\mathbf{T}]_{\{\mathbf{e}_i\}} &= \begin{pmatrix} 0 & 1 & -1 & 0 \\ 1 & 0 & 0 & -1 \\ -1 & 0 & 0 & 1 \\ 0 & -1 & 1 & 0 \end{pmatrix}
\end{aligned}$$

As shown in the next section, it may be noted that

$$\text{range of } \mathbf{T}: \{\mathbf{e}_2 - \mathbf{e}_3, \mathbf{e}_1 - \mathbf{e}_4\}, \quad \ker \mathbf{T} = \{\mathbf{e}_2 + \mathbf{e}_3, \mathbf{e}_1 + \mathbf{e}_4\}$$

9.2.1 Change of basis

Theorem. *Let* $\mathbf{V}$ *be a finite-dimensional vector space over the field* F*, and let* $\mathbf{B}_1 = (\mathbf{e}_1, \mathbf{e}_2, \ldots, \mathbf{e}_n)$*,* $\mathbf{B}_2 = (\mathbf{u}_1, \mathbf{u}_2, \ldots, \mathbf{u}_n)$ *be two sets of ordered bases for* $\mathbf{V}$*. Suppose that* $\mathbf{T}$ *is a linear operator on* $\mathbf{V}$*. If* $\mathbf{P}$ *is the* $n \times n$ *transition matrix, which expresses the coordinates of each vector in* $\mathbf{V}$ *relative to* $\mathbf{B}_1$ *in terms of its coordinates relative to* $\mathbf{B}_2$*, then*

$$\mathbf{A}_2 = \mathbf{P}^{-1}\mathbf{A}_1\mathbf{P}$$

where $\mathbf{A}_i$ *is the matrix representation of* $\mathbf{T}$ *w. r. t. the ordered bases* $\mathbf{B}_i$ $(i = 1, 2)$*.*

Sketch of the proof. $\mathbf{A}_1$ is the matrix representation of $\mathbf{T}$ w. r. t. $\mathbf{B}_1$. Express the new basis in terms of the old one

$$\mathbf{u}_1 = p_{11}\mathbf{e}_1 + p_{21}\mathbf{e}_2 + p_{31}\mathbf{e}_3 + \cdots + p_{n1}\mathbf{e}_n$$
$$\vdots$$
$$\mathbf{u}_n = p_{1n}\mathbf{e}_1 + p_{2n}\mathbf{e}_2 + p_{3n}\mathbf{e}_3 + \cdots + p_{nn}\mathbf{e}_n$$

If $\mathbf{x} \in \mathbf{V}$, then

$$\begin{aligned}\mathbf{x} &= \alpha_1\mathbf{e}_1 + \cdots + \alpha_n\mathbf{e}_n \\ &= \alpha_1'\mathbf{u}_1 + \cdots + \alpha_n'\mathbf{u}_n\end{aligned}$$

$\Rightarrow$

$$\begin{aligned}&\boldsymbol{\alpha} = \mathbf{P}\boldsymbol{\alpha}' \\ &[\mathbf{Tx}] = \mathbf{A}_1\boldsymbol{\alpha} = \mathbf{y} = \beta_1\mathbf{e}_1 + \cdots + \beta_n\mathbf{e}_n \\ &\Rightarrow \boldsymbol{\beta} = \mathbf{A}_1\boldsymbol{\alpha}\end{aligned}$$

Now,

$$\mathbf{y} = \beta_1'\mathbf{u}_1 + \cdots + \beta_n'\mathbf{u}_n \Rightarrow \boldsymbol{\beta} = \mathbf{P}\boldsymbol{\beta}'$$

$\Rightarrow$

$$\mathbf{P}\boldsymbol{\beta}' = \mathbf{A}_1\mathbf{P}\boldsymbol{\alpha}' \Rightarrow \boldsymbol{\beta}' = \mathbf{P}^{-1}\mathbf{A}_1\mathbf{P}\boldsymbol{\alpha}' \Rightarrow \boldsymbol{\beta}' = \mathbf{A}_2\boldsymbol{\alpha}'$$

$\therefore$ $\mathbf{A}_2 = \mathbf{P}^{-1}\mathbf{A}_1\mathbf{P}$ ($\mathbf{A}_2$ and $\mathbf{A}_1$ are similar matrices.) □

9.2.2 Kernel and range of a linear transformation

Definition. Let $\mathbf{V}$ and $\mathbf{W}$ be vector spaces over the field F and $\mathbf{T} : \mathbf{V} \to \mathbf{W}$ be a linear transformation. Then,

$$\text{range of } \mathbf{T} = \text{Image of } \mathbf{T} = \{\mathbf{w} \in \mathbf{W} : \mathbf{w} = \mathbf{T}(\mathbf{u}) \text{ for some } \mathbf{u} \in \mathbf{V}\}$$
$$\text{kernel of } \mathbf{T} = \text{null space of } \mathbf{T} = \{\mathbf{u} \in \mathbf{V} : \mathbf{Tu} = \mathbf{0}_w\}$$

Theorem. *Let* $\mathbf{T} : \mathbf{V} \to \mathbf{W}$ *be a linear mapping, then*
1. *range* $\mathbf{T}$ *is a subspace of* $\mathbf{W}$
2. *the kernel of* $\mathbf{T}$ *is a subspace of* $\mathbf{V}$

Proof.
1. Let $\mathbb{R}_T$ denote the range of $\mathbf{T}$. Let $\mathbf{w}_1 \in \mathbb{R}_T$ and $\mathbf{w}_2 \in \mathbb{R}_T$. Then $\exists \mathbf{u}_1, \mathbf{u}_2 \in \mathbf{V}$ such that

$$\mathbf{Tu}_1 = \mathbf{w}_1, \quad \mathbf{Tu}_2 = \mathbf{w}_2$$

Now consider

$$\begin{aligned}\mathbf{T}(\alpha\mathbf{u}_1 + \mathbf{u}_2) &= \alpha\mathbf{T}(\mathbf{u}_1) + \mathbf{T}(\mathbf{u}_2) \quad (\text{since } \mathbf{T} \text{ is linear})\\ &= \alpha\mathbf{Tu}_1 + \mathbf{Tu}_2\\ &= \alpha\mathbf{w}_1 + \mathbf{w}_2\end{aligned}$$

Thus, (1) $\Rightarrow \alpha\mathbf{w}_1 + \mathbf{w}_2 \in \mathbb{R}_T$
$\therefore$ $\mathbb{R}_T$ is a subspace of $\mathbf{W}$

2. Let $\mathbf{N}_T$ denote the null space of $\mathbf{T}$. If $\mathbf{u}_1, \mathbf{u}_2 \in \mathbf{N}_T$ and $\alpha \in F$
$\Rightarrow$

$$\mathbf{T}(\alpha\mathbf{u}_1 + \mathbf{u}_2) = \alpha\mathbf{Tu}_1 + \mathbf{Tu}_2 = \mathbf{0} \Rightarrow \alpha\mathbf{u}_1 + \mathbf{u}_2 \in \mathbf{N}_T$$

$\therefore$ $\mathbf{N}_T$ is a subspace.

□

Definition. Let $\mathbf{V}$ and $\mathbf{W}$ be finite-dimensional vector spaces defined over a field F. Then

$$\text{Rank of } \mathbf{T} = \dim(\text{range of } \mathbf{T})$$
$$\text{Nullity of } \mathbf{T} = \dim(\text{kernel of } \mathbf{T}) = \dim(\text{null space of } \mathbf{T})$$

Theorem. *Let* $\mathbf{T} : \mathbf{V} \to \mathbf{W}$ *be a linear transformation from* $\mathbf{V}$ *into* $\mathbf{W}$. *Suppose that* $\mathbf{V}$ *is finite-dimensional. Then*

$$\text{Rank}(\mathbf{T}) + \text{Nullity}(\mathbf{T}) = \dim(\mathbf{V})$$

Proof. Let $\mathbf{u}_1, \ldots, \mathbf{u}_k$ be a basis for $\mathbf{N}_T$ (null space of $\mathbf{T}$). Then there are vectors $(\mathbf{u}_{k+1}, \ldots, \mathbf{u}_n)$ in $\mathbf{V}$ such that $(\mathbf{u}_1, \ldots, \mathbf{u}_n)$ is a basis for $\mathbf{V}$. We shall prove that $\{\mathbf{T}\mathbf{u}_{k+1}, \ldots, \mathbf{T}\mathbf{u}_n\}$ is a basis for the range of $\mathbf{T}$. The vectors $\{\mathbf{T}\mathbf{u}_1, \ldots, \mathbf{T}\mathbf{u}_n\}$ certainly span the range of $\mathbf{T}$ and since $\mathbf{T}\mathbf{u}_j = \mathbf{0}, j = 1, \ldots, k$, we see that $\{\mathbf{T}\mathbf{u}_{k+1}, \ldots, \mathbf{T}\mathbf{u}_n\}$ span the range of $\mathbf{T}$. To see that these vectors are linearly independent, suppose we have scalars α_i such that

$$\sum_{i=k+1}^{n} \alpha_i(\mathbf{T}\mathbf{u}_i) = \mathbf{0} \Rightarrow \mathbf{T}\left(\sum_{i=k+1}^{n} \alpha_i \mathbf{u}_i\right) = \mathbf{0}$$

or the vector $\mathbf{y} = \sum_{i=k+1}^{n} \alpha_i \mathbf{u}_i$ is in the null space of $\mathbf{T}$. Since $\{\mathbf{u}_1, \mathbf{u}_2, \ldots, \mathbf{u}_k\}$ is a basis for $\mathbf{N}_T$, we can express

$$\mathbf{y} = \sum_{i=1}^{k} \beta_i \mathbf{u}_i$$

Thus,

$$\sum_{i=1}^{k} \beta_i \mathbf{u}_i - \sum_{i=k+1}^{n} \alpha_i \mathbf{u}_i = \mathbf{0}$$

Since, the vectors $(\mathbf{u}_1, \ldots, \mathbf{u}_n)$ are linearly independent $\Rightarrow$

$$\alpha_{k+1} = \cdots \alpha_n = 0, \quad \beta_1 = \cdots = \beta_k = 0$$

Thus, if k is the nullity of $\mathbf{T}$, the fact that the set $\{\mathbf{T}\mathbf{u}_{k+1}, \ldots, \mathbf{T}\mathbf{u}_n\}$ forms a basis for range of $\mathbf{T}$ implies that rank of

$$\mathbf{T} = n - k = \dim(\text{Range of } \mathbf{T})$$

∴ The result. □

9.2.3 Relation to linear equations

$$\mathbf{A}\mathbf{x} = \mathbf{b}, \quad \mathbf{x} \in \mathbb{R}^n,\ \mathbf{b} \in \mathbb{R}^m$$

The matrix $\mathbf{A}$ may be viewed as the linear transformation $\mathbf{A} : \mathbb{R}^n \to \mathbb{R}^m$ and the solution to $\mathbf{A}\mathbf{x} = \mathbf{b}$ is the preimage of $\mathbf{b} \in \mathbb{R}^m$. Furthermore, the solution of the associated homogeneous equation $\mathbf{A}\mathbf{x} = \mathbf{0}$ may be viewed as the kernel of the linear mapping $\mathbf{A} : \mathbb{R}^n \to \mathbb{R}^m$. Figure 9.2 illustrates the kernel and range of a linear transformation schematically.

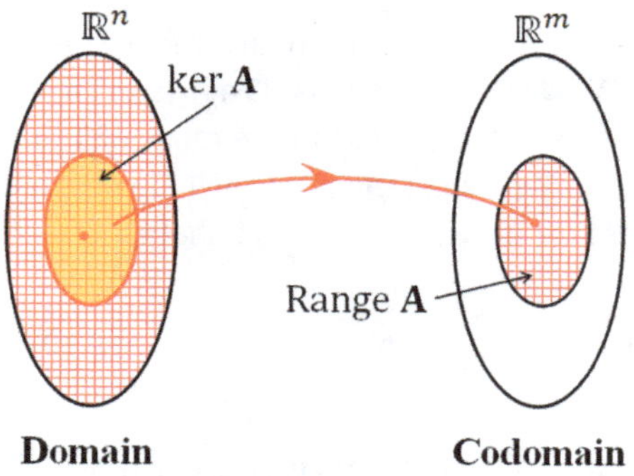

Figure 9.2: Schematic diagram illustrating the kernel and range of a linear transformation. The dot represents the zero vector.

From the above theorem, we have

$$\begin{aligned}\dim(\ker \mathbf{A}) &= \dim(\mathbb{R}^n) - \dim(\text{Image of } \mathbf{A}) \\ &= n - \operatorname{rank}(\mathbf{A})\end{aligned}$$

Thus, we have the following theorem.

Theorem. *The dimension of the solution space of the linear equations*

$$\mathbf{Ax} = \mathbf{0}$$

is $n - r$, where r is the rank of **A** *and n is the number of unknowns.*

9.2.4 Isomorphism

If **V** and **W** are vector spaces over the field F, any one-to-one linear transformation **T** of **V** onto **W** (i. e., any bijective linear transformation of **V** into **W**) is called an *isomorphism of* **V** *into* **W**. If there exists an isomorphism of **V** into **W**, we say that **V** is isomorphic to **W**.

Theorem. *Every n-dimensional space over F is isomorphic to the space* $\mathbf{F}^n$

Proof. Let **V** be the n-dimensional space over F and let $\mathbf{B} = \{\mathbf{e}_1, \mathbf{e}_2, \dots, \mathbf{e}_n\}$ be a basis for **V**. Define a function $\mathbf{T} : \mathbf{V} \to \mathbf{F}^n$ as follows:

If $\mathbf{x} \in \mathbf{V}$, let $\mathbf{Tx}$ be the n-tuple of coordinates $(a_1, \dots, a_n)$ of $\mathbf{x}$ relative to the basis **B**. Now it is easily verified that **T** is linear, one-to-one and maps **V** onto $\mathbf{F}^n$.
$\therefore$ The result.

For many purposes one often regards isomorphic vector spaces as being "the same," although the vectors and operations in the spaces may be quite different. □

9.2.5 Inverse of a linear transformation

Definition. Let $\mathbf{V}$ and $\mathbf{W}$ be vector spaces over a field F and $\mathbf{T} : \mathbf{V} \to \mathbf{W}$ be a linear transformation. $\mathbf{T}$ is said to be *singular* if $\exists$ a nonzero vector $\mathbf{x} \in \mathbf{V} \ni \mathbf{Tx} = \mathbf{0}_w$. If the kernel of $\mathbf{T}$ consists of only the zero vector in $\mathbf{V}$, then we say that $\mathbf{T}$ is nonsingular.

Theorem. $\mathbf{T}$ *is nonsingular* $\Longrightarrow$ $\mathbf{T}$ *is one-to-one and onto.*

Proof. Note that $\mathbf{T}$ is nonsingular implies that $\mathbf{T}$ is one-to-one and onto. To show this, let

$$\mathbf{Tx} = \mathbf{Ty}$$

$\Longrightarrow$

$$\mathbf{T}(\mathbf{x} - \mathbf{y}) = \mathbf{0}_w \quad (\text{since } \mathbf{T} \text{ is linear})$$

$\Longrightarrow$

$$\mathbf{x} - \mathbf{y} = \mathbf{0}_v \quad (\text{since } \mathbf{T} \text{ is nonsingular})$$

$\Longrightarrow \mathbf{x} = \mathbf{y} \therefore \mathbf{Tx} = \mathbf{Ty} \Longrightarrow \mathbf{x} = \mathbf{y} \therefore \mathbf{T}$ is one-to-one.

To show that $\mathbf{T}$ is onto, it is sufficient if we show that if $\{\mathbf{e}_1, \dots, \mathbf{e}_n\}$ is a basis for $\mathbf{V}$ then $\{\mathbf{Te}_1, \dots, \mathbf{Te}_n\}$ is a basis for $\mathbf{W}$. We claim that $\mathbf{Te}_1, \dots, \mathbf{Te}_n$ are an independent set of vectors; for suppose

$$\alpha_1 \mathbf{Te}_1 + \cdots + \alpha_n \mathbf{Te}_n = \mathbf{0}_w$$

$\Longrightarrow$

$$\mathbf{T}(\alpha_1 \mathbf{e}_1 + \cdots + \alpha_n \mathbf{e}_n) = \mathbf{0}_w \quad \text{since } \mathbf{T} \text{ is linear}$$

$\Rightarrow$

$$\alpha_1 \mathbf{e}_1 + \cdots + \alpha_n \mathbf{e}_n = \mathbf{0}_v \quad \text{since } \mathbf{T} \text{ is nonsingular}$$

$\Rightarrow \alpha_1 = \cdots = \alpha_n = 0$ since $\mathbf{e}_1, \dots, \mathbf{e}_n$ are independent. Thus, $\{\mathbf{Te}_1, \dots, \mathbf{Te}_n\}$ are independent, which implies that $\mathbf{T}$ is onto. □

Theorem. $\mathbf{T} : \mathbf{V} \to \mathbf{W}$ *is an isomorphism if and only if it is non-singular.*

Definition. Let $\mathbf{V}$ be a vector space defined over field F and $\mathbf{T} : \mathbf{V} \to \mathbf{W}$ is a linear transformation. We say that $\mathbf{T}$ is invertible if $\exists$ a transformation $\mathbf{T}^{-1}$ such that

$$\mathbf{TT}^{-1} = \mathbf{I}_w = \text{identity on } \mathbf{W}$$
$$\mathbf{T}^{-1}\mathbf{T} = \mathbf{I}_v = \text{identity on } \mathbf{V}$$

Theorem. *Let* $\mathbf{V}$ *and* $\mathbf{W}$ *be finite-dimensional vector spaces over the field* F *such that* dim $\mathbf{V}$ = dim $\mathbf{W}$. *If* $\mathbf{T}$ *is a linear transformation from* $\mathbf{V}$ *into* $\mathbf{W}$, *the following are equivalent:*
1. $\mathbf{T}$ *is invertible*
2. $\mathbf{T}$ *is nonsingular*
3. *The range of* $\mathbf{T}$ *is* $\mathbf{W}$
4. *If* $\{\mathbf{e}_1, \ldots, \mathbf{e}_n\}$ *is a basis for* $\mathbf{V}$, *then* $\{\mathbf{Te}_1, \ldots, \mathbf{Te}_n\}$ *is a basis for* $\mathbf{W}$.

Theorem. *If* $\mathbf{T}$ *is nonsingular, then the inverse function* $\mathbf{T}^{-1}$ *is a linear transformation from* $\mathbf{W}$ *onto* $\mathbf{V}$.

Proof. $\mathbf{T}$ is nonsingular $\Rightarrow$ $\mathbf{T}$ is one-to-one and onto. There is a uniquely determined inverse function $\mathbf{T}^{-1}$ from $\mathbf{W}$ to $\mathbf{V}$ such that $\mathbf{T}^{-1}\mathbf{T}$ is identity on $\mathbf{V}$ and $\mathbf{T}\mathbf{T}^{-1}$ is identity on $\mathbf{W}$. To prove that $\mathbf{T}^{-1}$ is linear, let $\mathbf{y}_1, \mathbf{y}_2 \in \mathbf{W}$, $c \in F$. Then $\exists$ unique $\mathbf{x}_i \in \mathbf{V}$ such that

$$\begin{aligned} \mathbf{Tx}_1 &= \mathbf{y}_1 \quad \text{or} \quad \mathbf{x}_1 = \mathbf{T}^{-1}\mathbf{y}_1 \\ \mathbf{Tx}_2 &= \mathbf{y}_2 \quad \text{or} \quad \mathbf{x}_2 = \mathbf{T}^{-1}\mathbf{y}_2 \end{aligned}$$

Now,

$$\begin{aligned} \mathbf{T}(c\mathbf{x}_1 + \mathbf{x}_2) &= c\mathbf{Tx}_1 + \mathbf{Tx}_2 \\ &= c\mathbf{y}_1 + \mathbf{y}_2 \end{aligned}$$

Thus,

$$\begin{aligned} \mathbf{T}^{-1}(c\mathbf{y}_1 + \mathbf{y}_2) &= c\mathbf{x}_1 + \mathbf{x}_2 \\ &= c\mathbf{T}^{-1}\mathbf{y}_1 + \mathbf{T}^{-1}\mathbf{y}_2, \end{aligned}$$

which implies that $\mathbf{T}^{-1}$ is linear. □

Application

Consider the solution of $\mathbf{Ax} = \mathbf{b}$ where $\mathbf{A}$ is $n \times n$, $\mathbf{x}, \mathbf{b} \in \mathbb{R}^n$. The above theorems can be used to prove the following theorem.

Theorem (Linear algebraic equations).
1. *If the homogeneous system* $\mathbf{Ax} = \mathbf{0}$ *(*$\mathbf{A}$ *is* $n \times n$ *and* $\mathbf{x}$ *is* $n \times 1$*) has only the trivial solution, then the inhomogeneous system has a unique solution for any* $\mathbf{b} \in \mathbb{R}^n$.
2. *If* $\mathbf{Ax} = \mathbf{0}$ *has nonzero solution, then* $\exists \mathbf{b} \in \mathbb{R}^n$ *for which* $\mathbf{Ax} = \mathbf{b}$ *has no solution. Furthermore, if a solution exists for some* $\mathbf{b}$, *it is not unique.*

Example 9.1. Consider the linear equations

$$\begin{aligned} u_1 - 2u_2 &= b_1 \\ 2u_1 - 4u_2 &= b_2 \end{aligned}$$

or $\mathbf{Au} = \mathbf{b}$ with $\mathbf{A} = \begin{pmatrix} 1 & -2 \\ 2 & -4 \end{pmatrix}$ and $\mathbf{b} = \begin{pmatrix} b_1 \\ b_2 \end{pmatrix}$.

In this case, the homogeneous equations $\mathbf{Au} = \mathbf{0}$ have the solution

$$\mathbf{u}_h = c\mathbf{x}_1; \quad \mathbf{x}_1 = \begin{pmatrix} 2 \\ 1 \end{pmatrix},$$

where c is any arbitrary constant. The inhomogeneous equations are inconsistent if $b_2 \neq 2b_1$. When $b_2 = 2b_1 = -6$, the general solution of $\mathbf{Au} = \mathbf{b}$ may be expressed as

$$\mathbf{u} = c_1\mathbf{x}_1 + \mathbf{x}_2; \quad \mathbf{x}_2 = \begin{pmatrix} 1 \\ 2 \end{pmatrix},$$

where c_1 is an arbitrary constant. Figure 9.3 shows the solution space of the homogeneous and (consistent) inhomogeneous system.

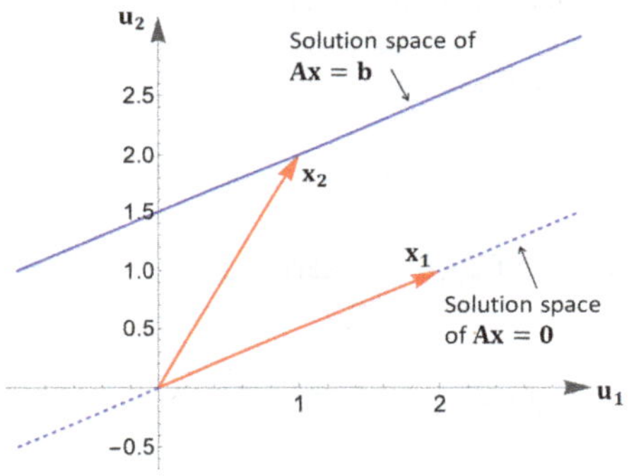

Figure 9.3: The eigenvectors of **A** and the solution spaces of the homogeneous equation **Au** = **0** and inhomogeneous equation **Au** = **b** (when they are consistent).

Remark. The vectors $\mathbf{x}_1$ and $\mathbf{x}_2$ are the eigenvectors of the matrix $\mathbf{A}$, with $\mathbf{x}_1$ corresponding to the zero eigenvalue. The solution space of $\mathbf{Au} = \mathbf{b}$ is a translation of that of $\mathbf{Au} = \mathbf{0}$ by $\mathbf{x}_2$. Such a space is called an *affine space*.

Further discussion and proofs of theorems on linear transformations may be found in the books by Halmos [20], Naylor and Sell [24] and Lipschutz and Lipson [22].

Problems

1. Determine which of the transformations are linear
   (a) $\mathbf{T} : \mathbb{R}^2 \to \mathbb{R}^2$ defined by $\mathbf{T}(x, y) = (x + y, x)$
   (b) $\mathbf{T} : \mathbb{R}^2 \to \mathbb{R}^3$ defined by $\mathbf{T}(x, y) = (x + 1, 2y, x + y)$

(c) $\mathbf{T} : C[0,1] \to C[0,1]$ defined by

$$g(t) = \int_0^1 K(t,s) f(s)\, ds,$$

where $g(t)$ and $f(t)$ are elements in $C[0,1]$ and $K(t,s)$ is continuous in $[0,1] \times [0,1]$.

2. Let $\mathbf{T}$ be a linear operator on $\mathbb{R}^3$ defined by

$$\mathbf{T}(x_1, x_2, x_3) = (3x_1 + x_2 + x_3, 2x_1 + 4x_2 + 2x_3, x_1 + x_2 + 3x_3)$$

(a) What is the matrix of $\mathbf{T}$ in the standard basis for $\mathbb{R}^3$?
(b) What is the matrix of $\mathbf{T}$ in the ordered basis $\mathbf{u}_1 = (1,-1,0)$, $\mathbf{u}_2 = (1,0,-1)$ and $\mathbf{u}_3 = (1,2,1)$?
(c) Show that $\mathbf{T}$ is invertible and give a rule for $\mathbf{T}^{-1}$ like the one which defines $\mathbf{T}$.
(d) Determine the spectral resolution of $\mathbf{T}$ and a formula for computing $f(\mathbf{T})$ where f is any function defined on the spectrum of $\mathbf{T}$.

3. Let $\mathbf{T} : \mathbb{R}^4 \to \mathbb{R}^3$ be the linear transformation defined by

$$\mathbf{T}(x,y,z,w) = (x - y + z + w, x + 2z - w, x + y + 3z - 3w).$$

Find a basis and the dimension of (a) the range of $\mathbf{T}$, (b) the kernel of $\mathbf{T}$.

4. Consider the linear operator $\mathbf{T}$ on $\mathbb{R}^3$ defined by

$$\mathbf{T}(x,y,z) = (2x, 4x - y, 2x + 3y - z).$$

(a) Show that $\mathbf{T}$ is invertible
(b) Determine formulas for $\mathbf{T}^{-1}$, $\mathbf{T}^2$ and $\mathbf{T}^{-2}$.

5. Let $\mathbf{E}$ be a linear operator on $\mathbf{V}$ for which $\mathbf{E}^2 = \mathbf{E}$ (such an operator is called a projection). Let $\mathbf{U}$ be the image of $\mathbf{E}$ and $\mathbf{W}$ be the kernel. Show that
(a) if $\mathbf{u} \in \mathbf{U}$, then $\mathbf{E}(\mathbf{u}) = \mathbf{u}$
(b) if $\mathbf{E} \neq \mathbf{I}$, then $\mathbf{E}$ is singular
(c) $\mathbf{V} = \mathbf{U} \oplus \mathbf{W}$.

10 Normed and inner product vector spaces

The problem so far is that the abstract vector space defined in Section 8.2 has only algebraic structure. In order for these concepts to be useful, we must add some geometrical structure by introducing the concepts of length of a vector, distance and angle between vectors and orthogonality. This is done by introducing the concepts of norm and scalar (dot) or inner product.

10.1 Definition of normed linear spaces

A *norm* on a vector space $\mathbf{V}$ is a rule which for every $\mathbf{x} \in \mathbf{V}$, specifies a real number $\|\mathbf{x}\|$, called the norm or length of $\mathbf{x}$ such that:

1. $\|\mathbf{x}\| \geq 0$, $\|\mathbf{x}\| = 0$ iff $\mathbf{x} = \mathbf{0}$ (positivity)
2. $\|\alpha\mathbf{x}\| = |\alpha|\|\mathbf{x}\|$, $\alpha \in \mathbb{R}/\mathbb{C}$ (homogeneity)
3. $\|\mathbf{x} + \mathbf{y}\| \leq \|\mathbf{x}\| + \|\mathbf{y}\|$ (triangle inequality)

In a normed linear space, in addition to length, we can also measure the distance between vectors by defining a distance function

$$d(\mathbf{x}, \mathbf{y}) = \|\mathbf{x} - \mathbf{y}\|.$$

Examples.

1. Let $\mathbf{V} = \mathbb{R}^n$ and for $\mathbf{x} \in \mathbf{V}$, define

$$\|\mathbf{x}\|_p = \left(\sum_{i=1}^{n} |x_i|^p\right)^{1/p}, \quad p \geq 1$$

   It can be shown that this definition satisfies all three rules of a norm:
   (a) For $p = 1$, $\|\mathbf{x}\|_1 = \sum_{i=1}^{n} |x_i|$.
   (b) For $p = 2$, $\|\mathbf{x}\|_2 = (\sum_{i=1}^{n} |x_i|^2)^{1/2}$, which is the standard Euclidean norm.
   (c) For $p = \infty$, $\|\mathbf{x}\|_\infty = \max_{1 \leq i \leq n} |x_i|$, which is also referred to as the *supremum norm*.
2. Let $\mathbf{V} = R[a, b]$, the space of piecewise continuous (and hence Riemann integrable) functions defined on $[a, b]$. For $f(x) \in \mathbf{V}$, define

$$\|f(x)\|_1 = \int_a^b |f(x)|\, dx$$

$$\|f(x)\|_2 = \left(\int_a^b |f(x)|^2\, dx\right)^{1/2}$$

$$\|f(x)\|_\infty = \sup_{a \leq x \leq b} |f(x)|$$

https://doi.org/10.1515/9783111598055-012

Again, it may be shown that these three definitions satisfy the three rules of a norm. [Note that the vector space $R[a,b]$ is infinite-dimensional.]
If $f(x), g(x) \in \mathbf{V}$, the distance functions corresponding to the above norms are given by

$$d_1(f,g) = \int_a^b |f(x) - g(x)|\, dx$$

$$d_2(f,g) = \left(\int_a^b |f(x) - g(x)|^2\, dx \right)^{1/2}$$

$$d_\infty(f,g) = \sup_{a \le x \le b} |f(x) - g(x)|,$$

which leads to the following observations:
(a) If $d(f,g) = 0$ for any of the above norms, f and g are equal at all $x \in [a,b]$, i. e., pointwise or uniform convergence.
(b) If we remove the restriction of piecewise continuity, the Riemann integral may not exist. Further, $d_2(f,g) = 0$ may not imply that f and g are equal at all $x \in [a,b]$. As discussed below, the vector space $R[a,b]$ is not complete under the usual inner product leading to this norm.

Limits of sequences and convergence in vector spaces
If we consider the sequence,

$$S_n = 1 + \frac{1}{1!} + \frac{1}{2!} + \frac{1}{3!} + \cdots + \frac{1}{n!} = \sum_{k=1}^{n} \frac{1}{k!};$$

then S_n is a rational number for any finite value of n. However,

$$\lim_{n \longrightarrow \infty} S_n = e \quad \text{(an irrational number)}$$

Similarly, the functions

$$f_n(x) = \exp(-nx), \quad 0 \le x \le 1$$
$$g_n(x) = \tanh(nx), \quad -1 \le x \le 1$$

are continuous, i. e., $f_n \in C[0,1]$ and $g_n \in C[-1,1]$ for any finite n, but for $n \longrightarrow \infty$,

$$f_\infty(x) = \begin{cases} 1, & x = 0 \\ 0, & x \neq 0 \end{cases} \notin C[0,1]$$

$$g_\infty(x) = \begin{cases} -1, & x < 0 \\ 0, & x = 0 \\ 1, & x > 0 \end{cases} \notin C[-1,1]$$

Thus, the limits of sequences of continuous functions may not be continuous and the space $C[a,b]$ (or the space $R[a,b]$) may not be complete depending on how $d(f,g)$ is defined. Similarly, in infinite-dimensional vector spaces, the functions (or vectors) may have uncountable number of discontinuities as illustrated by the example below.

Dirichlet function, Riemann and Lebesque integration

Consider the function defined on the closed unit interval $[0,1]$

$$f_D(x) = \begin{cases} 1, & x \text{ is rational} \\ 0, & x \text{ is irrational} \end{cases}$$

The Riemann integral $\int_0^1 f_D(x)\,dx$ does not exist. However, since the set of rational numbers have zero measure in $[0,1]$, the Lebesque integral exists. Further, there is no difference between $f_D(x)$ and function $\widehat{f}(x) = 0 \forall x \in [0,1]$ in the Lebesque theory of integration. The Lebesque integral of $f_D(x)$ is

$$\int_0^1 f_D(x)\,dx = 0 \quad \text{(Lebesque integral);} \quad \widehat{f}(x) \in C[0,1] \quad \text{and} \quad f_D(x) \in \mathcal{L}[0,1].$$

The Fourier coefficients of f_D and $\widehat{f}$ are identical and $d_2(\widehat{f}, f_D) = 0$. We return to this example in Chapter 21 when we discuss the theory of convergence in function spaces.

10.2 Inner product vector spaces

Definition. Let $\mathbf{V}$ be a vector space defined over a field $F(\mathbb{R}$ or $\mathbb{C})$. Suppose that to each pair of vectors, $\mathbf{u}, \mathbf{v} \in \mathbf{V}$, we assign a scalar $\langle \mathbf{u}, \mathbf{v} \rangle \in F$. This function is called an *inner product* if it satisfies the following axioms:

1. Linearity in the first component

$$\langle \alpha\mathbf{u} + \beta\mathbf{w}, \mathbf{v} \rangle = \alpha\langle \mathbf{u}, \mathbf{v} \rangle + \beta\langle \mathbf{w}, \mathbf{v} \rangle; \quad \text{for all } \mathbf{u}, \mathbf{v}, \mathbf{w} \in \mathbf{V} \text{ and } \alpha, \beta \in F$$

2. Hermitian symmetry

$$\langle \mathbf{u}, \mathbf{v} \rangle = \overline{\langle \mathbf{v}, \mathbf{u} \rangle}; \quad \text{for all } \mathbf{u}, \mathbf{v} \in \mathbf{V},$$

the bar denoting complex conjugation.

3. Positive definiteness

$$\langle \mathbf{u}, \mathbf{u} \rangle \geq 0 \quad \text{and} \quad \langle \mathbf{u}, \mathbf{u} \rangle = 0 \quad \text{iff} \quad \mathbf{u} = \mathbf{0}$$

An abstract vector space on which an inner product is defined is called an *inner product space*.

Remarks.
1. If $F = \mathbb{R}$, then the bar denoting complex conjugation is superflous.
2. Inner product is a generalization to an abstract vector space of the dot (scalar) product in two and three dimensions.

Examples.
1. Let $F = \mathbb{R}$ and $\mathbf{V} = \mathbb{R}^n$. For $\mathbf{u}, \mathbf{v} \in \mathbf{V}$, define

$$\langle \mathbf{u}, \mathbf{v} \rangle = u_1v_1 + u_2v_2 + \cdots + u_nv_n = \mathbf{u}.\mathbf{v} \quad \text{(standard dot product)}$$

This is an inner product as it satisfies all the axioms. Note that

$$\langle \mathbf{u}, \mathbf{u} \rangle = u_1^2 + u_2^2 + \cdots + u_n^2$$

The norm or length of the vector w. r. t. this inner product is

$$\|\mathbf{u}\| = +\sqrt{\langle \mathbf{u}, \mathbf{u} \rangle} = \sqrt{u_1^2 + u_2^2 + \cdots + u_n^2}.$$

The distance between two points (vectors) is defined as $d(\mathbf{u}, \mathbf{v}) = \|\mathbf{u} - \mathbf{v}\|$. These are the standard dot product and distance function in the Euclidean space $\mathbb{R}^n$.

2. Let $F = \mathbb{C}$ and $\mathbf{V} = \mathbb{C}^n$. For $\mathbf{u}, \mathbf{v} \in \mathbf{V}$ define

$$\langle \mathbf{u}, \mathbf{v} \rangle = u_1\overline{v_1} + u_2\overline{v_2} + \cdots + u_n\overline{v_n}$$

Then

(a)

$$\begin{aligned}\langle \alpha\mathbf{u} + \beta\mathbf{w}, \mathbf{v} \rangle &= (\alpha u_1 + \beta w_1)\overline{v_1} + \cdots + (\alpha u_n + \beta w_n)\overline{v_n} \\ &= \alpha(u_1\overline{v_1} + \cdots + u_n\overline{v_n}) + \beta(w_1\overline{v_1} + \cdots + w_n\overline{v_n}) \\ &= \alpha\langle \mathbf{u}, \mathbf{v} \rangle + \beta\langle \mathbf{w}, \mathbf{v} \rangle\end{aligned}$$

(b)

$$\langle \mathbf{v}, \mathbf{u} \rangle = v_1\overline{u_1} + \cdots + v_n\overline{u_n}$$

$\Rightarrow$

$$\begin{aligned}\overline{\langle \mathbf{v}, \mathbf{u} \rangle} &= \overline{v_1\overline{u_1} + \cdots + v_n\overline{u_n}} \\ &= \overline{v_1}u_1 + \cdots + \overline{v_n}u_n \\ &= \langle \mathbf{u}, \mathbf{v} \rangle\end{aligned}$$

(c)

$$\begin{aligned}\langle \mathbf{u}, \mathbf{u} \rangle &= u_1\overline{u_1} + \cdots + u_n\overline{u_n} \\ &= |u_1|^2 + \cdots + |u_n|^2 \ge 0.\end{aligned}$$

The space $\mathbb{C}^n$ with this dot/inner product is an example of a finite dimensional "Hilbert space."

3. Let $F = \mathbb{R}$ and $\mathbf{V} = \mathbb{R}^n$. Let $\mathbf{G}$ be any fixed $n \times n$ symmetric positive definite matrix. Define

$$\langle \mathbf{u}, \mathbf{v} \rangle = \mathbf{v}^T \mathbf{G} \mathbf{u}$$

Then

(a)

$$\begin{aligned} \langle \alpha \mathbf{u} + \beta \mathbf{w}, \mathbf{v} \rangle &= \mathbf{v}^T \mathbf{G}(\alpha \mathbf{u} + \beta \mathbf{w}) \\ &= \alpha \mathbf{v}^T \mathbf{G} \mathbf{u} + \beta \mathbf{v}^T \mathbf{G} \mathbf{w} \\ &= \alpha \langle \mathbf{u}, \mathbf{v} \rangle + \beta \langle \mathbf{w}, \mathbf{v} \rangle \end{aligned}$$

(b)

$$\begin{aligned} \langle \mathbf{u}, \mathbf{v} \rangle &= \mathbf{v}^T \mathbf{G} \mathbf{u} = (\mathbf{v}^T \mathbf{G} \mathbf{u})^T \quad \text{since it is a scalar} \\ &= \mathbf{u}^T \mathbf{G}^T \mathbf{v} \\ &= \mathbf{u}^T \mathbf{G} \mathbf{v} \quad (\text{since } \mathbf{G}^T = \mathbf{G}) \\ &= \langle \mathbf{v}, \mathbf{u} \rangle \end{aligned}$$

(c)

$$\langle \mathbf{u}, \mathbf{u} \rangle = \mathbf{u}^T \mathbf{G} \mathbf{u} \geq 0$$

since $\mathbf{G}$ is positive definite. $\mathbf{G}$ is called the matrix of the inner product (or *metric* of the inner product) space. For the standard (Euclidean) inner product in Example 1,

$$\mathbf{G} = \begin{pmatrix} 1 & 0 & . & 0 \\ . & . & . & . \\ 0 & 0 & . & 1 \end{pmatrix} = \mathbf{I}$$

4. Let $F = \mathbb{R}$ and V = space of all continuous real valued functions in the interval $a \leq t \leq b$. For $f(t), g(t) \in V$, define

$$\langle f, g \rangle = \int_a^b f(t) g(t)\, dt$$

This satisfies the axioms of an inner product. Note that the space V in this example is infinite-dimensional. This space is denoted by $C[a, b]$:

$$\|f\| = 0 \Rightarrow \int_a^b f(t)^2\, dt = 0$$

If $f(t)$ is continuous, the only way the integral can be zero is $f(t) \equiv 0, a \le t \le b$. This inner product is very useful in applications. Very often, we are interested in solving nonlinear equations of the form:

$$N(y) = 0$$

Since an exact solution is not possible, $y(t)$ is approximated by $f(t)$. The closeness of this approximation to the exact solution can be found only if an inner product (or a norm) is defined on the space.
There exist a variety of inner products on the space $C[a, b]$. For example,

$$\langle f, g \rangle = \int_a^b \rho(t) f(t) g(t)\, dt; \quad \rho(t) > 0 \text{ in } (a, b)$$

is also an inner product. One chooses an inner product that is convenient in a given application.

5. V = space of continuous complex valued functions in $[a, b]$. For $f(t), g(t) \in V$, define

$$\langle f, g \rangle = \int_a^b f(t)\overline{g(t)}\, dt$$

It can be shown that this satisfies all the axioms of an inner product.

Definition. Let $\mathbf{V}$ be an inner product space. The length of a vector $\mathbf{u} \in \mathbf{V}$ (also called the norm of $\mathbf{u}$ denoted by $\|\mathbf{u}\|$) is defined by $\|\mathbf{u}\| = \sqrt{\langle \mathbf{u}, \mathbf{u} \rangle}$.

Theorem (Schwarz inequality). *Let* $\mathbf{V}$ *be an inner product space and* $\mathbf{u}, \mathbf{v} \in \mathbf{V}$. *Then*

$$|\langle \mathbf{u}, \mathbf{v} \rangle|^2 \le \langle \mathbf{u}, \mathbf{u} \rangle . \langle \mathbf{v}, \mathbf{v} \rangle$$

Proof. If $\mathbf{v} = \mathbf{0}$, both sides of the inequality are zero and it is satisfied. Assume $\mathbf{v} \neq \mathbf{0}$. Then, by property (3) of inner product,

$$\langle \mathbf{w}, \mathbf{w} \rangle \ge 0, \quad \mathbf{w} \in \mathbf{V}$$

Let $\mathbf{w} = \mathbf{u} - \alpha \mathbf{v}$, where $\alpha \in F$; $\Rightarrow$

$$\begin{aligned}
&\langle \mathbf{u} - \alpha\mathbf{v}, \mathbf{u} - \alpha\mathbf{v} \rangle \ge 0 \\
&\quad \Rightarrow \langle \mathbf{u}, \mathbf{u} - \alpha\mathbf{v} \rangle - \alpha \langle \mathbf{v}, \mathbf{u} - \alpha\mathbf{v} \rangle \ge 0 \\
&\quad \Rightarrow \overline{\langle \mathbf{u} - \alpha\mathbf{v}, \mathbf{u} \rangle} - \alpha \overline{\langle \mathbf{u} - \alpha\mathbf{v}, \mathbf{v} \rangle} \ge 0 \\
&\quad \Rightarrow \overline{\langle \mathbf{u}, \mathbf{u} \rangle - \alpha \langle \mathbf{v}, \mathbf{u} \rangle} - \alpha \overline{[\langle \mathbf{u}, \mathbf{v} \rangle - \alpha \langle \mathbf{v}, \mathbf{v} \rangle]} \ge 0 \\
&\quad \Rightarrow \langle \mathbf{u}, \mathbf{u} \rangle - \overline{\alpha} \langle \mathbf{u}, \mathbf{v} \rangle - \alpha \langle \mathbf{v}, \mathbf{u} \rangle + \overline{\alpha}\alpha \langle \mathbf{v}, \mathbf{v} \rangle \ge 0
\end{aligned}$$

$$\Rightarrow \langle \mathbf{u}, \mathbf{u} \rangle + \bar{\alpha}\alpha \langle \mathbf{v}, \mathbf{v} \rangle \geq \alpha \langle \mathbf{v}, \mathbf{u} \rangle + \bar{\alpha} \langle \mathbf{u}, \mathbf{v} \rangle$$

Let $\alpha = \frac{\langle \mathbf{u},\mathbf{v} \rangle}{\langle \mathbf{v},\mathbf{v} \rangle}$, $\Longrightarrow$

$$\langle \mathbf{u}, \mathbf{u} \rangle + \frac{|\langle \mathbf{u}, \mathbf{v} \rangle|^2}{\langle \mathbf{v}, \mathbf{v} \rangle} \geq 2 \frac{|\langle \mathbf{u}, \mathbf{v} \rangle|^2}{\langle \mathbf{v}, \mathbf{v} \rangle}$$

$$\langle \mathbf{u}, \mathbf{u} \rangle \geq \frac{|\langle \mathbf{u}, \mathbf{v} \rangle|^2}{\langle \mathbf{v}, \mathbf{v} \rangle}$$

$\Rightarrow$

$$\langle \mathbf{u}, \mathbf{u} \rangle . \langle \mathbf{v}, \mathbf{v} \rangle \geq \left| \langle \mathbf{u}, \mathbf{v} \rangle \right|^2$$

$\therefore$ The result. □

Definition. The angle between two vectors $\mathbf{u}, \mathbf{v}$ in an inner product vector space $\mathbf{V}$ is defined by

$$\cos\theta = \frac{|\langle \mathbf{u}, \mathbf{v} \rangle|}{\|\mathbf{u}\|.\|\mathbf{v}\|}, \; F = \mathbb{C}; \quad \cos\theta = \frac{\langle \mathbf{u}, \mathbf{v} \rangle}{\|\mathbf{u}\|.\|\mathbf{v}\|}, \; F = \mathbb{R}.$$

Remark. According to the above definition, $0 \leq \theta \leq \frac{\pi}{2}$ if $F = \mathbb{C}$, while $0 \leq \theta \leq \pi$ if $F = \mathbb{R}$.

Definition. Let $\mathbf{V}$ be an inner product space over a field F. Two vectors $\mathbf{u}, \mathbf{v} \in \mathbf{V}$ are said to be *orthogonal* w. r. t. its inner product if $\langle \mathbf{u}, \mathbf{v} \rangle = 0$.

A vector is said to be normalized if it has unit length, i. e., $\|\mathbf{u}\| = 1$.

Remark. If $\mathbf{V}$ is an inner product space, we can define (1) distances between vectors (2) lengths of vectors (3) angles between vectors, i. e., an inner product space has a geometrical structure.

A vector space in which only distances are defined is called a *metric space.*

A vector space in which lengths are defined is called a *normed linear space.* The schematic diagram of Figure 10.1 shows the relationship between these spaces.

Theorem. *Let* $\mathbf{V}$ *be a finite dimensional inner product vector space and* $\{\mathbf{u}_1, \mathbf{u}_2, \ldots, \mathbf{u}_n\}$ *be a set of orthogonal vectors. Then, this set is linearly independent provided it does not include the zero vector.*

Proof. Let $\mathbf{v} \in \mathbf{V}$ be any vector that is in the subspace spanned by $\{\mathbf{u}_1, \mathbf{u}_2, \ldots, \mathbf{u}_n\}$

$$\mathbf{v} = \sum_{i=1}^{n} \alpha_i \mathbf{u}_i$$

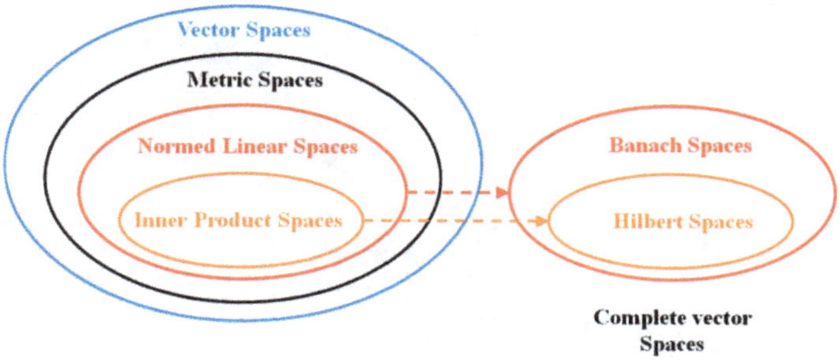

Figure 10.1: Schematic diagrams illustrating the structure of vector spaces.

⇒

$$\langle \mathbf{v}, \mathbf{u}_j \rangle = \left\langle \sum_{i=1}^{n} \alpha_i \mathbf{u}_i, \mathbf{u}_j \right\rangle = \alpha_j \langle \mathbf{u}_j, \mathbf{u}_j \rangle$$

If $\langle \mathbf{u}_j, \mathbf{u}_j \rangle \neq 0 \Rightarrow$

$$\alpha_j = \frac{\langle \mathbf{v}, \mathbf{u}_j \rangle}{\|\mathbf{u}_j\|^2}$$

∴

$$\mathbf{v} = \sum_{i=1}^{n} \frac{\langle \mathbf{v}, \mathbf{u}_i \rangle}{\|\mathbf{u}_i\|^2} \mathbf{u}_i$$

Now suppose that $\mathbf{v} = \mathbf{0} \Rightarrow \alpha_j = 0, j = 1, 2, \ldots, n \Rightarrow$ the set $\{\mathbf{u}_1, \mathbf{u}_2, \ldots, \mathbf{u}_n\}$ is linearly independent. □

Theorem. *Every finite-dimensional inner product space has an orthonormal basis.*

Proof. Let $\{\mathbf{u}_1, \mathbf{u}_2, \ldots, \mathbf{u}_n\}$ be a basis for $\mathbf{V}$. From this basis, we shall show how to obtain an orthogonal basis $\{\mathbf{v}_1, \mathbf{v}_2, \ldots, \mathbf{v}_n\}$. When each of the vectors in this orthogonal basis is normalized to have unit length, then we obtain an orthonormal basis:

$$\mathbf{e}_i = \frac{\mathbf{v}_i}{\|\mathbf{v}_i\|}, \quad i = 1, 2, \ldots, n$$

$\Rightarrow \{\mathbf{e}_1, \mathbf{e}_2, \ldots, \mathbf{e}_n\}$ is an orthonormal basis □

10.2.1 Gram–Schmidt orthogonalization procedure

Given a set of linearly independent vectors $(\mathbf{u}_1, \mathbf{u}_2, \ldots, \mathbf{u}_n)$, we can construct an orthogonal or an orthonormal basis from this set by using the following Gram–Schmidt procedure:

Let

$$\mathbf{v}_1 = \mathbf{u}_1$$
$$\mathbf{v}_2 = \mathbf{u}_2 - \frac{\langle \mathbf{u}_2, \mathbf{v}_1 \rangle}{\|\mathbf{v}_1\|^2}\mathbf{v}_1$$

Then

$$\langle \mathbf{v}_2, \mathbf{v}_1 \rangle = \langle \mathbf{u}_2, \mathbf{v}_1 \rangle - \frac{\langle \mathbf{u}_2, \mathbf{v}_1 \rangle}{\|\mathbf{v}_1\|^2}\langle \mathbf{v}_1, \mathbf{v}_1 \rangle = 0$$

$\therefore$ $\mathbf{v}_2$ is orthogonal to $\mathbf{v}_1$. Let

$$\mathbf{v}_3 = \mathbf{u}_3 - \sum_{i=1}^{2} \frac{\langle \mathbf{u}_3, \mathbf{v}_i \rangle}{\|\mathbf{v}_i\|^2}\mathbf{v}_i$$

Then

$$\langle \mathbf{v}_3, \mathbf{v}_1 \rangle = \langle \mathbf{u}_3, \mathbf{v}_1 \rangle - \langle \mathbf{u}_3, \mathbf{v}_1 \rangle = 0$$
$$\langle \mathbf{v}_3, \mathbf{v}_2 \rangle = \langle \mathbf{u}_3, \mathbf{v}_2 \rangle - \langle \mathbf{u}_3, \mathbf{v}_2 \rangle = 0$$

$\therefore$ $\mathbf{v}_3$ is orthogonal to both $\mathbf{v}_1$ and $\mathbf{v}_2$. At the k-th step, let

$$\mathbf{v}_k = \mathbf{u}_k - \sum_{i=1}^{k-1} \frac{\langle \mathbf{u}_k, \mathbf{v}_i \rangle}{\|\mathbf{v}_i\|^2}\mathbf{v}_i$$

Then $\mathbf{v}_k$ is orthogonal to $\mathbf{v}_1, \mathbf{v}_2, \ldots, \mathbf{v}_{k-1}$. If $\mathbf{v}_k = 0 \Rightarrow$

$$\mathbf{u}_k = \sum_{i=1}^{k-1} \frac{\langle \mathbf{u}_k, \mathbf{v}_i \rangle}{\|\mathbf{v}_i\|^2}\mathbf{v}_i$$

$\Rightarrow$ $\mathbf{u}_k$ is a linear combination of $\mathbf{v}_1, \mathbf{v}_2, \ldots, \mathbf{v}_{k-1}$ or equivalently, $\mathbf{u}_k$ is a linear combination of $(\mathbf{u}_1, \mathbf{u}_2, \ldots, \mathbf{u}_{k-1})$. But this cannot be true since $(\mathbf{u}_1, \mathbf{u}_2, \ldots, \mathbf{u}_n)$ is linearly independent. Therefore, the above procedure cannot fail and we obtain an orthogonal set.

If $\{\mathbf{e}_1, \mathbf{e}_2, \ldots, \mathbf{e}_n\}$ is an orthonormal basis and $\mathbf{x} \in \mathbf{V}$ is an arbitrary vector,

$$\mathbf{x} = \sum_{i=1}^{n} a_i \mathbf{e}_i$$

$\Rightarrow$

$$\langle \mathbf{x}, \mathbf{e}_j \rangle = \left\langle \sum_{i=1}^{n} a_i \mathbf{e}_i, \mathbf{e}_j \right\rangle = a_j$$

∴

$$\mathbf{x} = \sum_{i=1}^{n} \langle \mathbf{x}, \mathbf{e}_i \rangle \mathbf{e}_i$$

Remark. The above procedure also indicates to us how to define the inner product so that $\{\mathbf{u}_1, \mathbf{u}_2, \ldots, \mathbf{u}_n\}$ is an orthonormal basis. Thus, we may formulate a theorem as follows.

Theorem. *Let* $\mathbf{V}$ *be a finite dimensional vector space over a field* F. *Let* $\{\mathbf{u}_1, \mathbf{u}_2, \ldots, \mathbf{u}_n\}$ *be a basis for* $\mathbf{V}$. *Then,* $\exists$ *an inner product on* $\mathbf{V}$ *such that* $\{\mathbf{u}_1, \mathbf{u}_2, \ldots, \mathbf{u}_n\}$ *is an orthonormal basis.*

Proof. Let $\mathbf{x}, \mathbf{y} \in \mathbf{V}$ be any two linearly independent vectors. Expand $\mathbf{x}$ and $\mathbf{y}$ in terms of the basis $\{\mathbf{u}_i\}$

$$\mathbf{x} = \sum_{i=1}^{n} \alpha_i \mathbf{u}_i, \quad \mathbf{y} = \sum_{i=1}^{n} \beta_i \mathbf{u}_i$$

Define

$$\langle \mathbf{x}, \mathbf{y} \rangle = \sum_{i=1}^{n} \alpha_i \overline{\beta_i}$$

Now it is obvious that $\{\mathbf{u}_1, \mathbf{u}_2, \ldots, \mathbf{u}_n\}$ is an orthonormal basis w. r. t. this inner product. Suppose that $\{\mathbf{e}_1, \mathbf{e}_2, \ldots, \mathbf{e}_n\}$ is also a basis for $\mathbf{V}$. Let $\boldsymbol{\alpha}'$ be the coordinate vector of $\mathbf{x}$ w. r. t. the $\mathbf{e}$-basis. Then we have

$$\boldsymbol{\alpha} = \mathbf{P}\boldsymbol{\alpha}', \quad \boldsymbol{\beta} = \mathbf{P}\boldsymbol{\beta}'$$

where $\mathbf{P}$ = transition matrix. Now,

$$\langle \mathbf{x}, \mathbf{y} \rangle = \boldsymbol{\alpha}^T \overline{\boldsymbol{\beta}} = \boldsymbol{\alpha}'^T \mathbf{P}^T \overline{\mathbf{P}\boldsymbol{\beta}'}$$

This is the inner product in the $\mathbf{e}$-basis that makes $\{\mathbf{u}_i\}$ an orthonormal set. Note that $\mathbf{P}^T\mathbf{P}$ is a positive definite matrix. □

10.3 Linear functionals and adjoints

Definition. Let $\mathbf{V}$ be a vector space defined over a field F. A *linear functional* on $\mathbf{V}$ is a linear transformation from $\mathbf{V}$ into F (the field) $f : \mathbf{V} \to F$

Examples.

1. Let $\mathbf{V} = C[a,b]$ = space of continuous real valued functions over the field $\mathbb{R}$. For $g(t) \in V$, define

$$f(g(t)) = \int_a^b g(t)\,dt$$

This linear functional maps the space V into the real line.

2. Let $F = \mathbb{R}$ and $\mathbf{V} = \mathbb{R}^n$ and let $\{\mathbf{e}_1, \mathbf{e}_2, \dots, \mathbf{e}_n\}$ be a basis for $\mathbf{V}$. Let $f : \mathbb{R}^n \to \mathbb{R}$ be a linear functional on $\mathbf{V}$ and $f(\mathbf{e}_j) = a_j$. Then the matrix of f in the basis $\{\mathbf{e}_1, \mathbf{e}_2, \dots, \mathbf{e}_n\}$ is a row vector $[a_1, a_2, \dots, a_n]$. If $\mathbf{x} \in \mathbf{V}$ is any vector and

$$\mathbf{x} = \sum_{j=1}^{n} \beta_j \mathbf{e}_j$$

Then

$$\begin{aligned} f(\mathbf{x}) &= f\left(\sum_{j=1}^{n} \beta_j \mathbf{e}_j\right) \\ &= \sum_{j=1}^{n} \beta_j f(\mathbf{e}_j) \quad \text{since } f \text{ is linear} \\ &= \sum_{j=1}^{n} \beta_j a_j = \sum_{j=1}^{n} \beta_j \overline{\overline{a_j}} = \langle \boldsymbol{\beta}, \overline{\mathbf{a}} \rangle \end{aligned}$$

This appears like the standard inner product of (the coordinates of) $\mathbf{x}$ with a fixed vector in $\mathbf{V}$, i. e., $f(\mathbf{x}) = \langle \boldsymbol{\beta}, \overline{\mathbf{a}} \rangle$.

Any linear functional f on a finite-dimensional inner product space is "inner product with a fixed vector in that space," i. e., f has the form $f(\mathbf{x}) = \langle \mathbf{x}, \mathbf{y} \rangle$ for some fixed $\mathbf{y} \in \mathbf{V}$. This result may be used to prove the existence of the "adjoint" of a linear operator $\mathbf{T}$ on $\mathbf{V}$, this being a linear operator $\mathbf{T}^*$ such that

$$\langle \mathbf{T}\mathbf{x}, \mathbf{y} \rangle = \langle \mathbf{x}, \mathbf{T}^* \mathbf{y} \rangle \quad \text{for all } \mathbf{x}, \mathbf{y} \in \mathbf{V}$$

Through the use of an orthonormal basis, this adjoint operation on linear operators is identified with the operation of forming the conjugate transpose of a matrix. These ideas are illustrated below.

Theorem. *Given a linear functional f on a finite dimensional inner product space $\mathbf{V}$, $\exists$ in $\mathbf{V}$ a unique vector $\mathbf{y} \ni f(\mathbf{x}) = \langle \mathbf{x}, \mathbf{y} \rangle$ for all $\mathbf{x} \in \mathbf{V}$.*

Proof. Let $\{\mathbf{e}_1, \mathbf{e}_2, \dots, \mathbf{e}_n\}$ be an orthonormal basis for $\mathbf{V}$.

Let

$$\mathbf{y} = \sum_{j=1}^{n} \overline{f(\mathbf{e}_j)}\mathbf{e}_j$$

We shall show that this $\mathbf{y}$ is the $\mathbf{y}$ of the theorem. Let $\hat{f}$ be the linear functional on $\mathbf{V}$ defined by

$$\begin{aligned}\hat{f}(\mathbf{x}) &= \langle \mathbf{x}, \mathbf{y} \rangle \quad \text{for all } \mathbf{x} \in \mathbf{V} \\ &= \left\langle \mathbf{x}, \sum_{j=1}^{n} \overline{f(\mathbf{e}_j)}\mathbf{e}_j \right\rangle\end{aligned}$$

$\Rightarrow$

$$\begin{aligned}\hat{f}(\mathbf{e}_k) &= \langle \mathbf{e}_k, \sum_{j=1}^{n} \overline{f(\mathbf{e}_j)}\mathbf{e}_j \rangle \\ &= \overline{\left\langle \sum_{j=1}^{n} \overline{f(\mathbf{e}_j)}\mathbf{e}_j, \mathbf{e}_k \right\rangle} \\ &= \overline{\overline{f(\mathbf{e}_k)}} \\ &= f(\mathbf{e}_k)\end{aligned}$$

Thus,

$$\hat{f}(\mathbf{e}_k) = f(\mathbf{e}_k) \quad \text{for } k = 1, 2, \ldots, n$$

Since $\hat{f}$ and f agree on each basis vector, we have $f = \hat{f}$, and hence the theorem is proved. Now suppose that there are two such vectors (say $\mathbf{y}$ and $\mathbf{z}$). Then

$$f(\mathbf{x}) = \langle \mathbf{x}, \mathbf{y} \rangle = \langle \mathbf{x}, \mathbf{z} \rangle$$

$\Rightarrow$

$$\begin{aligned}\langle \mathbf{x}, \mathbf{y} \rangle - \langle \mathbf{x}, \mathbf{z} \rangle &= 0 \quad \text{for all } \mathbf{x} \\ \Rightarrow \overline{\langle \mathbf{y}, \mathbf{x} \rangle} - \overline{\langle \mathbf{z}, \mathbf{x} \rangle} &= 0 \\ \Rightarrow \overline{\langle \mathbf{y} - \mathbf{z}, \mathbf{x} \rangle} &= 0 \\ \Rightarrow \langle \mathbf{x}, \mathbf{y} - \mathbf{z} \rangle &= 0\end{aligned}$$

Take $\mathbf{x} = \mathbf{y} - \mathbf{z} \Rightarrow$

$$\langle \mathbf{y} - \mathbf{z}, \mathbf{y} - \mathbf{z} \rangle = 0 \Rightarrow \mathbf{y} - \mathbf{z} = \mathbf{0} \quad \text{or} \quad \mathbf{y} = \mathbf{z}$$

Thus, such a vector is unique. □

Theorem. *For any linear operator* $\mathbf{T}$ *on a finite-dimensional inner product space* $\mathbf{V}$, $\exists$ *a unique linear operator* $\mathbf{T}^*$ *in* $\mathbf{V}$ *such that*

$$\langle \mathbf{Tx}, \mathbf{y} \rangle = \langle \mathbf{x}, \mathbf{T}^* \mathbf{y} \rangle \quad \textit{for all } \mathbf{x}, \mathbf{y} \in \mathbf{V}$$

Proof.

1. $\mathbf{T}^*$ exists: Let $\mathbf{y} \in \mathbf{V}$ be a vector. We shall define $\mathbf{T}^*\mathbf{y}$ to prove its existence. Now,

$$f(\mathbf{x}) = \langle \mathbf{Tx}, \mathbf{y} \rangle \in F(\text{field}), \quad \mathbf{x}, \mathbf{y} \in \mathbf{V}$$

is a linear functional on $\mathbf{V}$. From the previous theorem, $\exists$ a unique $\hat{\mathbf{y}}$ in $\mathbf{V}$ such that $f(\mathbf{x}) = \langle \mathbf{x}, \hat{\mathbf{y}} \rangle$ for all $\mathbf{x}$ in $\mathbf{V} \Rightarrow \langle \mathbf{Tx}, \mathbf{y} \rangle = \langle \mathbf{x}, \hat{\mathbf{y}} \rangle$. $\hat{\mathbf{y}}$ is uniquely determined by $\mathbf{y}$ and we define $\mathbf{T}^*$ as the rule that associates $\hat{\mathbf{y}}$ for each $\mathbf{y}$, i. e.,

$$\hat{\mathbf{y}} = \mathbf{T}^* \mathbf{y}$$

[Note: $\mathbf{T}^*$ is called the adjoint operator, so that $\langle \mathbf{Tx}, \mathbf{y} \rangle = \langle \mathbf{x}, \mathbf{T}^* \mathbf{y} \rangle$ for $\mathbf{x}, \mathbf{y} \in \mathbf{V}$].

2. $\mathbf{T}^*$ is a linear operator: Consider

$$\begin{aligned}
\langle \mathbf{x}, \mathbf{T}^*(\alpha \mathbf{z} + \beta \mathbf{w}) \rangle &= \langle \mathbf{Tx}, \alpha \mathbf{z} + \beta \mathbf{w} \rangle \quad \text{(from the above definition of } \mathbf{T}^*) \\
&= \overline{\langle \alpha \mathbf{z} + \beta \mathbf{w}, \mathbf{Tx} \rangle} \\
&= \overline{\langle \alpha \mathbf{z}, \mathbf{Tx} \rangle + \langle \beta \mathbf{w}, \mathbf{Tx} \rangle} \\
&= \overline{\alpha \langle \mathbf{z}, \mathbf{Tx} \rangle + \beta \langle \mathbf{w}, \mathbf{Tx} \rangle} \\
&= \overline{\alpha \overline{\langle \mathbf{Tx}, \mathbf{z} \rangle} + \beta \overline{\langle \mathbf{Tx}, \mathbf{w} \rangle}} \\
&= \overline{\alpha} \langle \mathbf{Tx}, \mathbf{z} \rangle + \overline{\beta} \langle \mathbf{Tx}, \mathbf{w} \rangle \\
&= \overline{\alpha} \langle \mathbf{x}, \mathbf{T}^* \mathbf{z} \rangle + \overline{\beta} \langle \mathbf{x}, \mathbf{T}^* \mathbf{w} \rangle \\
&= \overline{\alpha} \overline{\langle \mathbf{T}^* \mathbf{z}, \mathbf{x} \rangle} + \overline{\beta} \overline{\langle \mathbf{T}^* \mathbf{w}, \mathbf{x} \rangle} \\
&= \overline{\alpha \langle \mathbf{T}^* \mathbf{z}, \mathbf{x} \rangle + \beta \langle \mathbf{T}^* \mathbf{w}, \mathbf{x} \rangle} \\
&= \overline{\langle \alpha \mathbf{T}^* \mathbf{z}, \mathbf{x} \rangle + \langle \beta \mathbf{T}^* \mathbf{w}, \mathbf{x} \rangle} \\
&= \langle \mathbf{x}, \alpha \mathbf{T}^* \mathbf{z} \rangle + \langle \mathbf{x}, \beta \mathbf{T}^* \mathbf{w} \rangle \\
&= \langle \mathbf{x}, \alpha \mathbf{T}^* \mathbf{z} + \beta \mathbf{T}^* \mathbf{w} \rangle
\end{aligned}$$

$\therefore$ $\mathbf{T}^*$ is a linear operator. □

Theorem. *Let* $\mathbf{V}$ *be a finite-dimensional inner product space with an orthonormal basis* $\{\mathbf{e}_1, \mathbf{e}_2, \ldots, \mathbf{e}_n\}$. *Let* $\mathbf{T}$ *be a linear operator on* $\mathbf{V}$. *Then*

(a) *the matrix of* $\mathbf{T}$ *with respect to the above basis is* $\mathbf{A} = \{a_{ij}\}$ *where*

$$a_{ij} = \langle \mathbf{T}\mathbf{e}_j, \mathbf{e}_i \rangle$$

(b) *the matrix of* $\mathbf{T}^*$ *with respect to the same basis is* $\mathbf{A}^*$ *(conjugate transpose of* $\mathbf{A}$*).*

Proof. Let

$$\mathbf{T}\mathbf{e}_j = \sum_{i=1}^{n} \alpha_{ij}\mathbf{e}_i \tag{10.1}$$

If we expand $\mathbf{T}\mathbf{e}_j$ in terms of the basis vectors, we get the j-th column of $\mathbf{A}$,
$\Rightarrow$

$$\langle \mathbf{T}\mathbf{e}_j, \mathbf{e}_i \rangle = \alpha_{ij}$$

If we expand $\mathbf{T}^*\mathbf{e}_j$ in terms of the basis, we get the j-th column of the matrix of $\mathbf{T}^*$,
$\Rightarrow$

$$\langle \mathbf{T}^*\mathbf{e}_j, \mathbf{e}_i \rangle = \left\langle \sum_{k=1}^{n} \beta_{kj}\mathbf{e}_k, \mathbf{e}_i \right\rangle = \beta_{ij}$$

$\Rightarrow$

$$\beta_{ij} = \langle \mathbf{T}^*\mathbf{e}_j, \mathbf{e}_i \rangle = \langle \mathbf{e}_j, \mathbf{T}\mathbf{e}_i \rangle = \overline{\langle \mathbf{T}\mathbf{e}_i, \mathbf{e}_j \rangle} = \overline{\alpha_{ji}}$$

$\therefore$ The result. □

Remark. If the basis of $\mathbf{V}$ is not orthonormal, the relationship between the matrix of $\mathbf{T}$ and $\mathbf{T}^*$ is more complicated than given in the theorem above.

Definition. $\mathbf{T}$ is called a self-adjoint operator if $\mathbf{T}^* = \mathbf{T}$. If the field is $\mathbb{R}$ then self-adjointness means the matrix of $\mathbf{T}$ in an orthonormal basis is symmetric. If $F = \mathbb{C}$, then self-adjointness means that the matrix of $\mathbf{T}$ is Hermitian, i. e. it is equal to its conjugate transpose.

Remark. Since $\mathbf{T}^*$ is defined by

$$\langle \mathbf{T}\mathbf{x}, \mathbf{y} \rangle = \langle \mathbf{x}, \mathbf{T}^*\mathbf{y} \rangle$$

if $\mathbf{T} = \mathbf{T}^*$ (self-adjoint) $\Rightarrow$

$$\langle \mathbf{T}\mathbf{x}, \mathbf{y} \rangle = \langle \mathbf{x}, \mathbf{T}\mathbf{y} \rangle$$

Thus, self-adjointness or the symmetry property of a linear operator or its matrix representation very much depends on the definition of inner product (or equivalently, adjointness depends on the definition of inner product).

Definition. A linear operator $\mathbf{T}$ is called *normal* if it commutes with its adjoint, i. e.,

$$\mathbf{T}\mathbf{T}^* = \mathbf{T}^*\mathbf{T}.$$

Normal operators are generalization of symmetric operators on real inner product spaces to complex inner product spaces.

Characteristic values

Let $\mathbf{T} : \mathbf{V} \to \mathbf{V}$ is a linear operator over a field F. A scalar $\lambda \in F$ is called a characteristic value or eigenvalue of $\mathbf{T}$ if $\mathbf{Tx} = \lambda\mathbf{x}$, $\mathbf{x} \in \mathbf{V}$, $\lambda \in F$. This definition is crucial for in some cases because of the limitation on the field there would be no characteristic values. If we choose an orthonormal basis for $\mathbf{V}$, the eigenvalues of $\mathbf{T}$ and $\mathbf{T}^*$(adjoint) as well as the corresponding eigenvectors can be found from the matrix representations. Suppose that $\{\mathbf{e}_1, \mathbf{e}_2, \ldots, \mathbf{e}_n\}$ is an orthonormal basis for $\mathbf{V}$, and $\mathbf{A} = [\mathbf{T}]_e = n \times n$ matrix with $a_{ij} \in F$. Then the eigenvalues of $\mathbf{T}$ are given by the algebraic equations

$$(\mathbf{A} - \lambda\mathbf{I})\mathbf{x} = \mathbf{0} \tag{10.2}$$

and of $\mathbf{T}^*$ by

$$(\mathbf{A}^* - \eta\mathbf{I})\mathbf{y} = \mathbf{0} \tag{10.3}$$

Since $\mathbf{A}^*$ is the conjugate transpose of $\mathbf{A}$, equations (10.2) and (10.3) $\Rightarrow$

$$\lambda \text{ are roots of } \det(\mathbf{A} - \lambda\mathbf{I}) = 0$$
$$\eta \text{ are roots of } \det(\overline{\mathbf{A}} - \eta\mathbf{I})^T = 0.$$

Now,

$$\det(\overline{\mathbf{A}} - \eta\mathbf{I})^T = 0 \Rightarrow \det(\overline{\mathbf{A} - \overline{\eta}\mathbf{I}})^T = 0 \Rightarrow \det(\overline{\mathbf{A} - \overline{\eta}\mathbf{I}}) = 0$$

Thus, if λ is a characteristic value of $\mathbf{T}$, $\eta = \overline{\lambda}$ is a characteristic value of $\mathbf{T}^*$.

Consider the eigenvalue problems:

$$\mathbf{Tx} = \lambda\mathbf{x} \quad \text{or} \quad \mathbf{Ax} = \lambda\mathbf{x}$$
$$\mathbf{T}^*\mathbf{y} = \eta\mathbf{y} \quad \text{or} \quad \mathbf{A}^*\mathbf{y} = \eta\mathbf{y} = \overline{\lambda}\mathbf{y}.$$

The following theorem on the nature of the eigenvalues may be stated.

Theorem. *Let* $\mathbf{T} : \mathbf{V} \to \mathbf{V}$ *be a linear operator and* λ *be an eigenvalue of* $\mathbf{T}$. *Then*

1. (a) *If* $\mathbf{T}$ *is self-adjoint, i. e.,* $\mathbf{T}^* = \mathbf{T}$, *then* λ *is real*
   (b) *the eigenvectors corresponding to different eigenvalues are orthogonal.*
2. *If* $\mathbf{T}^* = \mathbf{T}^{-1}$ *(i. e.* $\mathbf{T}$ *is unitary), then* $|\lambda| = 1$. *[The eigenvalues are located on the unit circle in the complex plane.]*
3. *If* $\mathbf{T}^* = -\mathbf{T}$, *then* λ *is purely imaginary* ($\mathbf{T}$ *is skew adjoint).*
4. *If* $\mathbf{T} = \mathbf{S}^*\mathbf{S}$ *with* $\mathbf{S}$ *nonsingular, then* λ *is real and positive.*

Proof.

1. (a) Let $\mathbf{x} \in \mathbf{V}$

$$\begin{aligned}\lambda\langle\mathbf{x},\mathbf{x}\rangle &= \langle\lambda\mathbf{x},\mathbf{x}\rangle\\ &= \langle\mathbf{T}\mathbf{x},\mathbf{x}\rangle\\ &= \langle\mathbf{x},\mathbf{T}^*\mathbf{x}\rangle\\ &= \langle\mathbf{x},\mathbf{T}\mathbf{x}\rangle\\ &= \langle\mathbf{x},\lambda\mathbf{x}\rangle\\ &= \overline{\langle\lambda\mathbf{x},\mathbf{x}\rangle}\\ &= \overline{\lambda\langle\mathbf{x},\mathbf{x}\rangle}\\ &= \bar{\lambda}\langle\mathbf{x},\mathbf{x}\rangle\end{aligned}$$

Since $\langle\mathbf{x},\mathbf{x}\rangle \neq 0 \Rightarrow \lambda = \bar{\lambda}$ or λ is real.

(b) Let λ_i, λ_j be two distinct values and $\mathbf{x}_i, \mathbf{x}_j$ be the corresponding eigenvectors:

$$\mathbf{T}\mathbf{x}_i = \lambda_i\mathbf{x}_i$$

$\Rightarrow$

$$\begin{aligned}\langle\mathbf{T}\mathbf{x}_i,\mathbf{x}_j\rangle &= \lambda_i\langle\mathbf{x}_i,\mathbf{x}_j\rangle\\ \langle\mathbf{T}\mathbf{x}_i,\mathbf{x}_j\rangle &= \langle\mathbf{x}_i,\mathbf{T}^*\mathbf{x}_j\rangle\\ &= \langle\mathbf{x}_i,\mathbf{T}\mathbf{x}_j\rangle\\ &= \langle\mathbf{x}_i,\lambda_j\mathbf{x}_j\rangle\\ &= \lambda_j\langle\mathbf{x}_i,\mathbf{x}_j\rangle \quad \text{since } \lambda_j \text{ is real}\end{aligned}$$

as $\lambda_i \neq \lambda_j \Rightarrow \langle\mathbf{x}_i,\mathbf{x}_j\rangle = 0$

2.

$$\begin{aligned}\lambda\bar{\lambda}\langle\mathbf{x},\mathbf{x}\rangle &= \langle\lambda\mathbf{x},\lambda\mathbf{x}\rangle\\ &= \langle\mathbf{T}\mathbf{x},\mathbf{T}\mathbf{x}\rangle\\ &= \langle\mathbf{x},\mathbf{T}^*\mathbf{T}\mathbf{x}\rangle\\ &= \langle\mathbf{x},\mathbf{x}\rangle\end{aligned}$$

Since $\langle\mathbf{x},\mathbf{x}\rangle \neq 0 \Rightarrow \lambda\bar{\lambda} = 1 \Rightarrow |\lambda| = 1$. Thus, eigenvalues of a unitary operator lie on the unit circle in the complex plane.

3.

$$\begin{aligned}\lambda\langle\mathbf{x},\mathbf{x}\rangle &= \langle\lambda\mathbf{x},\mathbf{x}\rangle\\ &= \langle\mathbf{T}\mathbf{x},\mathbf{x}\rangle\end{aligned}$$

$$
\begin{aligned}
&= \langle \mathbf{x}, \mathbf{T}^* \mathbf{x} \rangle \\
&= \langle \mathbf{x}, -\mathbf{T}\mathbf{x} \rangle \\
&= \langle \mathbf{x}, -\lambda \mathbf{x} \rangle \\
&= \overline{-\lambda \langle \mathbf{x}, \mathbf{x} \rangle} \\
&= -\bar{\lambda} \langle \mathbf{x}, \mathbf{x} \rangle
\end{aligned}
$$

$\Rightarrow \lambda = -\bar{\lambda} \Rightarrow \lambda$ is purely imaginary. Thus, the eigenvalues of a skew-adjoint operator are on the imaginary axis.

4.

$$
\begin{aligned}
\lambda \langle \mathbf{x}, \mathbf{x} \rangle &= \langle \lambda \mathbf{x}, \mathbf{x} \rangle \\
&= \langle \mathbf{T}\mathbf{x}, \mathbf{x} \rangle \\
&= \langle \mathbf{S}^* \mathbf{S}\mathbf{x}, \mathbf{x} \rangle \\
&= \langle \mathbf{S}\mathbf{x}, S\mathbf{x} \rangle
\end{aligned}
$$

But $\langle \mathbf{x}, \mathbf{x} \rangle$ and $\langle \mathbf{S}\mathbf{x}, S\mathbf{x} \rangle$ are positive.
$\Rightarrow \lambda$ is real and positive. □

Theorem. *Let* **V** *be a finite dimensional inner product space defined over F and let* **T** *be a self-adjoint linear operator. Then* **T** *has n eigenvectors.*

Proof. It is sufficient if we prove the theorem for a Hermitian matrix **A**. We need only to show that a Hermitian matrix does not have any generalized eigenvectors of rank 2. This implies there are no generalized eigenvectors of rank > 2. If there was a GEV of rank $k (k \geq 3)$, then it generates a chain of GEV of rank $k, k-1, \ldots, 2, 1$. Thus, it is sufficient to prove the theorem for $k = 2$. Assume **x** is a GEV of rank 2 with eigenvalue λ, $\Rightarrow$

$$
(\mathbf{A} - \lambda \mathbf{I})^2 \mathbf{x} = \mathbf{0}, \quad (\mathbf{A} - \lambda \mathbf{I})\mathbf{x} \neq \mathbf{0}
$$

We already proved that λ is real $\Rightarrow \mathbf{A} - \lambda \mathbf{I}$ is also Hermitian,
$\therefore$

$$
\begin{aligned}
0 &= \langle \mathbf{x}, \mathbf{0} \rangle \\
&= \langle \mathbf{x}, (\mathbf{A} - \lambda \mathbf{I})^2 \mathbf{x} \rangle \\
&= \overline{\langle (\mathbf{A} - \lambda \mathbf{I})^2 \mathbf{x}, \mathbf{x} \rangle} \\
&= \overline{\langle (\mathbf{A} - \lambda \mathbf{I})\mathbf{x}, (\mathbf{A} - \lambda \mathbf{I})^* \mathbf{x} \rangle} \\
&= \overline{\langle (\mathbf{A} - \lambda \mathbf{I})\mathbf{x}, (\mathbf{A} - \lambda \mathbf{I})\mathbf{x} \rangle} \\
&= \langle (\mathbf{A} - \lambda \mathbf{I})\mathbf{x}, (\mathbf{A} - \lambda \mathbf{I})\mathbf{x} \rangle
\end{aligned}
$$

$\Rightarrow (\mathbf{A} - \lambda \mathbf{I})\mathbf{x} = \mathbf{0}$ from property (iii) of inner product.
$\therefore$ Contradiction $\Rightarrow$ **x** cannot be a GEV of rank 2. □

Corollary. *If* $\mathbf{T}$ *is self-adjoint, then* $\exists$ *an orthonormal basis w. r. t., which the matrix of* $\mathbf{T}$ *is diagonal, i. e.,* $\mathbf{T}$ *can be diagonalized.*

Proof. It follows from previous two theorems. Let $\lambda_1, \lambda_2, \ldots, \lambda_r$ be the distinct eigenvalues. If $r = n$, then there are n orthogonal eigenvectors. If $r < n$, there are repeated eigenvalues. Suppose λ_i is an eigenvalue of multiplicity m_i. We showed that there cannot be any GEV of rank ≥ 2.
$\Rightarrow$ There are m_i eigenvectors corresponding to λ_i. Apply the Gram–Schmidt procedure to make these orthogonal. Then these are not only orthogonal to each other but also to other eigenvectors.
$\therefore$ The result.

If $\mathbf{x}$ is an arbitrary vector, then

$$\mathbf{x} = \sum_{i=1}^{n} \alpha_i \mathbf{x}_i$$

where $\{\mathbf{x}_1, \mathbf{x}_2, \ldots, \mathbf{x}_n\}$ is an orthonormal set, and $\alpha_i = \langle \mathbf{x}, \mathbf{x}_i \rangle$. Define $\mathbf{E}_j = \mathbf{x}_j \mathbf{x}_j^*$ (Note: $\mathbf{E}_j^* = \mathbf{E}_j$ or $\mathbf{E}_j$ is self-adjoint)
$\Rightarrow$

$$\mathbf{E}_j \mathbf{x} = \sum_{i=1}^{n} \alpha_i \mathbf{x}_j \mathbf{x}_j^* \mathbf{x}_i = \alpha_j \mathbf{x}_j \Rightarrow \mathbf{E}_i \mathbf{E}_j \mathbf{x} = \mathbf{0} \quad (\text{if } i \neq j)$$

$$\mathbf{E}_j^2 \mathbf{x} = \sum_{i=1}^{n} \alpha_i \mathbf{x}_j \mathbf{x}_j^* \mathbf{x}_j \mathbf{x}_j^* \mathbf{x}_i = \alpha_j \mathbf{x}_j \mathbf{x}_j^* \mathbf{x}_j = \alpha_j \mathbf{x}_j = \mathbf{E}_j \mathbf{x}$$

$\therefore$

$$\mathbf{x} = \sum_{i=1}^{n} \mathbf{E}_i \mathbf{x} \Rightarrow \mathbf{E}_1 + \mathbf{E}_2 + \cdots + \mathbf{E}_n = \mathbf{I} \tag{10.4}$$

$\therefore$

$$\begin{aligned} \mathbf{T}\mathbf{x} &= \sum_{i=1}^{n} \alpha_i \mathbf{T} \mathbf{x}_i \\ &= \sum \alpha_i \lambda_i \mathbf{x}_i \\ &= \sum_{i=1}^{n} \lambda_i \mathbf{E}_i \mathbf{x} \end{aligned}$$

$\Rightarrow$

$$\lambda_1 \mathbf{E}_1 + \lambda_2 \mathbf{E}_2 + \cdots + \lambda_n \mathbf{E}_n = \mathbf{T} \tag{10.5}$$

Group repeated eigenvalues. Then (10.4) and (10.5) become $\mathbf{E}_1 + \mathbf{E}_2 + \cdots + \mathbf{E}_r = \mathbf{I}$ and $\lambda_1 \mathbf{E}_1 + \cdots + \lambda_r \mathbf{E}_r = \mathbf{T}$. □

Theorem. *Let* $\mathbf{V}$ *be a finite-dimensional vector space and* $\{\mathbf{w}_1, \mathbf{w}_2, \ldots, \mathbf{w}_n\}$ *be a basis for* $\mathbf{V}$. *Let* $\mathbf{W}_i$ = *space spanned by* $\mathbf{w}_i = \{\mathbf{x}/\mathbf{x} = \alpha\mathbf{w}_i, \alpha \in F\}$. *Then*

$$\mathbf{V} = \mathbf{W}_1 \oplus \mathbf{W}_2 \oplus \cdots \oplus \mathbf{W}_n.$$

This is known as direct sum decomposition of $\mathbf{V}$. □

Invariant subspaces

Let $\mathbf{V}$ be a finite-dimensional vector space and $\mathbf{T}$ be a linear operator on $\mathbf{V}$. Let $\mathbf{W}$ be a subspace of $\mathbf{V}$. $\mathbf{W}$ is called an *invariant subspace* (w. r. t. $\mathbf{T}$) if $\mathbf{T}$ maps $\mathbf{W}$ into itself, i. e., $\mathbf{x} \in \mathbf{W} \Rightarrow \mathbf{Tx} \in \mathbf{W}$. A schematic diagram of such invariant subspaces is shown in Figure 10.2 for $\mathbf{V} = \mathbf{W}_1 \oplus \mathbf{W}_2$.

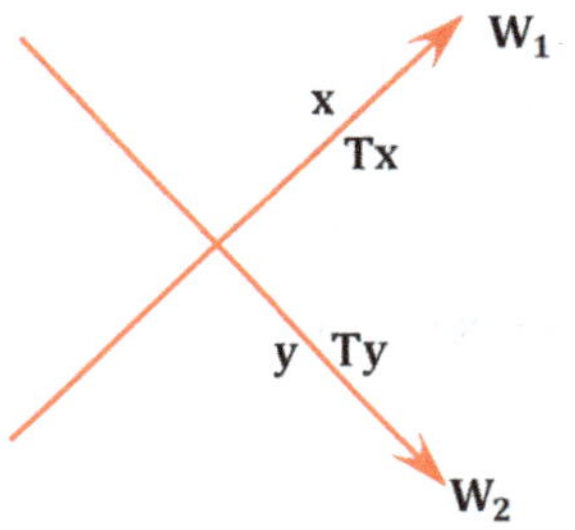

Figure 10.2: Schematic of invariant subspaces.

Example. If $\mathbf{T}$ is any linear operator on $\mathbf{V}$, then $\mathbf{V}$ is invariant under $\mathbf{T}$, as is the zero subspace. The range of $\mathbf{T}$ and the null space of $\mathbf{T}$ are also invariant under $\mathbf{T}$. If we can write

$$\mathbf{V} = \mathbf{W}_1 \oplus \mathbf{W}_2 \oplus \cdots \oplus \mathbf{W}_k,$$

where each $\mathbf{W}_i (i = 1, 2, \ldots, k)$ is invariant under $\mathbf{T}$, then the decomposition is called *invariant direct-sum decomposition*. In order to state the major result of finite-dimensional linear algebra (i. e., the *spectral theorem*), we need to introduce three additional concepts:

(a) direct sum decomposition
(b) invariant subspaces
(c) projections and orthogonal projections

Direct sum decomposition

Let $\mathbf{V}$ be a finite-dimensional vector space and $\mathbf{W}_1, \mathbf{W}_2, \ldots, \mathbf{W}_k$ be subspaces of $\mathbf{V}$. Let $\mathbf{x} \in \mathbf{V}$ be any vector. Suppose that we can expand $\mathbf{x}$ as

$$\mathbf{x} = \sum_{i=1}^{k} \alpha_i \mathbf{w}_i \quad \text{where } \mathbf{w}_i \in \mathbf{W}_i,$$

and $\{\alpha_i\}$ are uniquely determined. Then we say that $\mathbf{V}$ is a direct sum of $\mathbf{W}_1, \ldots, \mathbf{W}_k$ and write as

$$\mathbf{V} = \mathbf{W}_1 \oplus \mathbf{W}_2 \oplus \cdots \oplus \mathbf{W}_k$$

Theorem. $\dim \mathbf{V} = \dim \mathbf{W}_1 + \dim \mathbf{W}_2 + \cdots + \dim \mathbf{W}_k$.

Example.

1. Let $\mathbf{V} = \mathbb{R}^2$, $\mathbf{W}_1$ = space spanned by $\mathbf{e}_1 = (1, 0)$, $\mathbf{W}_2$ = space spanned by $\mathbf{e}_2 = (0, 1)$ If $\mathbf{x} \in \mathbf{V}$, then

$$\mathbf{x} = \alpha_1 \mathbf{e}_1 + \alpha_2 \mathbf{e}_2$$

   For any given $\mathbf{x}$, α_1 and α_2 are uniquely determined.
   $\therefore$

$$\mathbf{V} = \mathbf{W}_1 \oplus \mathbf{W}_2$$

2. Let $\mathbf{V} = \mathbb{R}^3$, $\mathbf{W}_1 = \{(1, 0, 0), (0, 1, 0)\}$, $\mathbf{W}_2 = \{(0, 0, 1)\}$. Then we can write

$$\mathbf{V} = \mathbf{W}_1 \oplus \mathbf{W}_2$$

A schematic diagram of the spaces $\mathbf{W}_1$ and $\mathbf{W}_2$ is shown in Figure 10.3. Here, $\mathbf{W}_1$ is the (x, y) plane of dimension 2 and $\mathbf{W}_2$ is the z-axis of dimension 1.

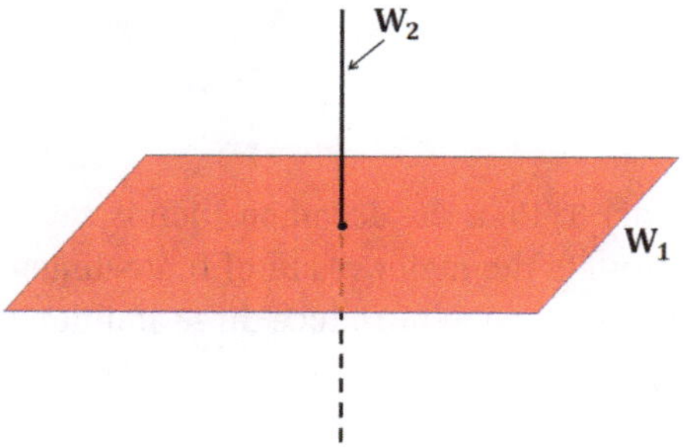

Figure 10.3: Schematic illustrating direct sum decomposition.

Projections

Let $\mathbf{E} : \mathbf{V} \to \mathbf{V}$ be a linear operator on a finite-dimensional vector space $\mathbf{V}$. Then $\mathbf{E}$ is called a projection if $\mathbf{E}^2\mathbf{x} = \mathbf{E}\mathbf{x}$ for all $\mathbf{x}$ in $\mathbf{V}$.

Theorem. *Let* $\mathbf{E}$ *be a projection. Let* $\mathbf{R}$ = *range of* $\mathbf{E} = \{\mathbf{y}/\mathbf{y} = \mathbf{E}\mathbf{x}\}$ *and* $\mathbf{N}$ = *null space of* $\mathbf{E} = \{\mathbf{x}/\mathbf{E}\mathbf{x} = \mathbf{0}\}$. *Then* $\mathbf{V} = \mathbf{R} \oplus \mathbf{N}$.

Proof. If $\mathbf{x} \in \mathbf{V}$, then

$$\mathbf{x} = \mathbf{x} + \mathbf{E}\mathbf{x} - \mathbf{E}\mathbf{x} = (\mathbf{x} - \mathbf{E}\mathbf{x}) + \mathbf{E}\mathbf{x}$$

$\Rightarrow$

$$\mathbf{E}\mathbf{x} = \mathbf{E}(\mathbf{x} - \mathbf{E}\mathbf{x}) + \mathbf{E}^2\mathbf{x} \Rightarrow \mathbf{E}(\mathbf{x} - \mathbf{E}\mathbf{x}) = \mathbf{0}$$

$\Rightarrow \mathbf{x} - \mathbf{E}\mathbf{x} \in \mathbf{N}$. Also $\mathbf{E}\mathbf{x} \in \mathbf{R}$. Thus, we can write

$$\mathbf{x} = \mathbf{y} + \mathbf{z}, \quad \mathbf{y} \in \mathbf{R},\ \mathbf{z} \in \mathbf{N}$$

We will now show that $\mathbf{R} \cap \mathbf{N} = \mathbf{0}$, i. e., $\mathbf{R}$ and $\mathbf{N}$ are disjoint. Suppose $\mathbf{x} \in \mathbf{R} \cap \mathbf{N}$,

$$\mathbf{x} = \mathbf{E}\mathbf{y} \quad \text{for some } \mathbf{y} \in \mathbf{V}$$

$\Rightarrow$

$$\mathbf{E}\mathbf{x} = \mathbf{E}(\mathbf{E}\mathbf{y}) = \mathbf{E}\mathbf{y} = \mathbf{x}$$

But $\mathbf{x}$ is also in $\mathbf{N} \Rightarrow \mathbf{E}\mathbf{x} = \mathbf{0}$
$\Rightarrow \mathbf{x} = \mathbf{0}$
$\therefore\ \mathbf{V} = \mathbf{R} \oplus \mathbf{N}$. □

Definition. $\mathbf{E}_1$ and $\mathbf{E}_2$ are called orthogonal projections on the vector space $\mathbf{V}$ if $\mathbf{E}_1\mathbf{E}_2\mathbf{x} = \mathbf{E}_2\mathbf{E}_1\mathbf{x} = \mathbf{0}$.

In general, let $\{\mathbf{E}_1, \mathbf{E}_2, \dots, \mathbf{E}_k\}$ be orthogonal projections on a finite-dimensional inner product space $\mathbf{V}$. Let

$$\mathbf{E}_j\mathbf{x} = \mathbf{x}_j, \quad j = 1, 2, \dots, k$$

Then

$$\mathbf{E}_j^2\mathbf{x} = \mathbf{x}_j, \dots, \mathbf{E}_j^n\mathbf{x} = \mathbf{x}_j, \quad n = 3, 4, \dots$$

Theorem (Spectral). *Let $\mathbf{T}$ be a normal (symmetric) operator on a complex (real) finite-dimensional inner product space $\mathbf{V}$. Then there exist orthogonal projections $\mathbf{E}_1, \mathbf{E}_2, \dots, \mathbf{E}_r$ on $\mathbf{V}$ and scalars $\lambda_1, \lambda_2, \dots, \lambda_r$ such that:*
(i) $\mathbf{E}_1 + \mathbf{E}_2 + \cdots + \mathbf{E}_r = \mathbf{I}$
(ii) $\lambda_1\mathbf{E}_1 + \lambda_2\mathbf{E}_2 + \cdots + \lambda_r\mathbf{E}_r = \mathbf{T}$
(iii) $\mathbf{E}_i\mathbf{E}_j = 0, i \neq j$
(iv) $Q(\mathbf{T}) = \sum_{i=1}^{r} Q(\lambda_i)\mathbf{E}_i$

where Q is any function defined on the spectrum of $\mathbf{T}$.

This form of the spectral theorem is a generalization of that stated in Part I. A proof of this may be found in the book by Halmos [20].

Finally, the following diagonalization theorems can be stated.

Theorem 1. *Let* $\mathbf{T}$ *be a self-adjoint operator on a real finite-dimensional inner product space* $\mathbf{V}$*. Then* $\exists$ *an orthonormal basis of* $\mathbf{V}$ *consisting of eigenvectors of* $\mathbf{T}$*, i. e.,* $\mathbf{T}$ *can be represented by a diagonal matrix relative to the orthonormal basis.*

Theorem 2. *Let* $\mathbf{T}$ *be an orthogonal operator on a real finite-dimensional inner product space* $\mathbf{V}$*. Then there is an orthonormal basis w. r. t., which* $\mathbf{T}$ *has the following form:*

$$\begin{bmatrix} 1 &&&&&&&&&&&& \\ & 1 &&&&&&&&&&& \\ && 1 &&&&&&&&&& \\ &&& \ddots &&&&&&&&& \\ &&&& 1 &&&&&&&& \\ &&&&& -1 &&&&&&& \\ &&&&&& -1 &&&&&& \\ &&&&&&& \ddots &&&&& \\ &&&&&&&& -1 &&&& \\ &&&&&&&&& \cos\theta_1 & -\sin\theta_1 && \\ &&&&&&&&& \sin\theta_1 & \cos\theta_1 && \\ &&&&&&&&&&& \ddots & \\ &&&&&&&&&&&& \begin{matrix} \cos\theta_n & -\sin\theta_n \\ \sin\theta_n & \cos\theta_n \end{matrix} \end{bmatrix}$$

Theorem 3. *Let* $\mathbf{T}$ *be a normal operator on a complex finite-dimensional inner product space* $\mathbf{V}$*. Then* $\exists$ *an orthonormal basis of* $\mathbf{V}$ *consisting of eigenvectors of* $\mathbf{T}$*, i. e.,* $\mathbf{T}$ *can be represented by a diagonal matrix w. r. t. an orthonormal basis.*

Theorem 4. *Let* $\mathbf{T}$ *be an arbitrary operator on a complex finite-dimensional inner product space* $\mathbf{V}$*. Then* $\mathbf{T}$ *can be represented by a triangular matrix w. r. t. an orthonormal basis of* $\mathbf{V}$*.*

The proof of these theorems may be found in the books by Halmos [20], Naylor and Sell [24] and Lipschutz and Lipson [22].

Problems

1. Given a vector space $\mathbf{V}$ of polynomials of degree at most N defined over the interval (a, b) with an inner product defined as

$$\langle f, g \rangle = \int_a^b \rho(t) f(t) g(t)\, dt$$

where $\rho(t) > 0$ for $a < t < b$. Indicate how one may determine an orthogonal basis set by applying the Gram–Schmidt process to the basis $\{1, t, t^2, \ldots, t^N\}$. Find the first three members for the following cases:
(a) $\rho(t) = 1, a = 0, b = 1$ (Legendre polynomials on the unit interval)
(b) $\rho(t) = 1, a = -1, b = 1$ (classical Legendre polynomials)
(c) $\rho(t) = \exp(-t), a = 0, b = \infty$ (Laguerre polynomials)
(d) $\rho(t) = \exp(-t^2), a = -\infty, b = \infty$ (Hermite polynomials)
(e) $\rho(t) = [t(1-t)]^{-\frac{1}{2}}, a = 0, b = 1$ (Chebyshev polynomials on the unit interval)

2. Consider the space $\mathbb{C}^n$ of n-tuples of complex numbers. Let $\mathbf{W}$ be a nonsingular $n \times n$ matrix. For $\mathbf{u}, \mathbf{v} \in \mathbb{C}^n$, define

$$(\mathbf{u}, \mathbf{v})_{\mathbf{W}} = \mathbf{v}^* \mathbf{W}^* \mathbf{W} \mathbf{u}$$

where the superscript * denotes complex conjugate transpose.
(a) Show that this satisfies the requirements of an inner product
(b) Prove the Cauchy–Schwarz inequality for this inner product
(c) Let $\mathbf{T}$ be a linear operator in $\mathbb{C}^n$ and $\mathbf{A}$ and $\mathbf{B}$ be the matrices of $\mathbf{T}$ and its adjoint $\mathbf{T}^*$ with respect to some orthonormal basis. Show that $\mathbf{W}^*\mathbf{W}\mathbf{B} = \mathbf{A}^*\mathbf{W}^*\mathbf{W}$

3. Let $\mathbf{V}$ be a finite-dimensional inner product vector space, and let $\mathbf{E}$ be an idempotent linear operator on $\mathbf{V}$, i. e., $\mathbf{E}^2 = \mathbf{E}$. Prove that $\mathbf{E}$ is self-adjoint if and only if $\mathbf{E}\mathbf{E}^* = \mathbf{E}^*\mathbf{E}$.

4. Let $\mathbf{V}$ be a finite-dimensional complex inner product space and $\mathbf{T}$ be a linear operator on $\mathbf{V}$. Prove that $\mathbf{T}$ is self-adjoint if and only if $\langle \mathbf{T}\mathbf{u}, \mathbf{u} \rangle$ is real for every $\mathbf{u}$ in $\mathbf{V}$.

5. Let $\mathbf{V}$ be a finite-dimensional inner product space and $\mathbf{T}$ be a linear operator on $\mathbf{V}$. Show that the range of $\mathbf{T}^*$ is the orthogonal complement of the null space of $\mathbf{T}$.

6. Let $\mathbf{V}$ be a finite-dimensional inner product space and $\mathbf{T}$ be a self-adjoint operator on $\mathbf{V}$. Show that if the eigenvalues of $\mathbf{T}$ are arranged so that $\lambda_1 \leq \lambda_2 \leq \cdots \leq \lambda_n$, then

$$\lambda_1 \leq \frac{(\mathbf{T}\mathbf{u}, \mathbf{u})}{(\mathbf{u}, \mathbf{u})} \leq \lambda_n$$

Here, $(\mathbf{u}, \mathbf{v})$ is the inner product on $\mathbf{V}$.

7. Let $\mathbf{V}$ be the vector space 2×2 matrices over $\mathbb{C}$ and let

$$\mathbf{M} = \begin{pmatrix} 1 & -1 \\ -2 & 2 \end{pmatrix}.$$

Define a linear operator $\mathbf{T}$ on $\mathbf{V}$ by

$$\mathbf{T}(\mathbf{A}) = \mathbf{M}\mathbf{A} - \mathbf{A}\mathbf{M} \quad \text{for } \mathbf{A} \in \mathbf{V}.$$

(a) Find a basis and the dimension of the kernel and image of $\mathbf{T}$.

(b) Show that $\langle \mathbf{A}, \mathbf{B} \rangle = \mathrm{tr}(\mathbf{B}^* \mathbf{A})$, where tr stands for the trace (sum of diagonal elements) satisfies the requirements of an inner product.

(c) Find the adjoint operator.

(d) Determine the eigenvalues and eigenvectors of $\mathbf{T}$.

8. In the numerical solution of transport and reaction problems in a tube in which the flow is laminar, we need a set of polynomial trial functions (to approximate the unknown solution) on the unit interval $0 < r < 1$ such that each function vanishes at $r = 1$ while its derivative vanishes at $r = 0$. The functions should also be orthogonal w. r. t. the weight function

$$\rho(r) = 4r(1 - r^2)$$

and normalized to have unit length (w. r. t. this weight function).

(a) Determine the first four of these functions and plot them.

(b) Determine the first four coefficients in the expansion of unity in terms of these normalized eigenfunctions.

(c) Verify that the first two coefficients contain 97 % of the energy while the first four contains 99.55 % (sum of squares).

9. Consider the space $\mathbb{R}^2$. Let $\mathbf{D}_p (p = 1, 2, \infty)$ be the set of all vectors (points in $\mathbb{R}^2$) having unit length w. r. t. the p-norm. Make a plot of $\mathbf{D}_1$, $\mathbf{D}_2$ and $\mathbf{D}_\infty$.

10. Suppose that $\mathbf{u}_1, \mathbf{u}_2, \ldots, \mathbf{u}_r$ form an orthogonal set of nonzero vectors in a vector space $\mathbf{V}$. Let $\mathbf{w}$ be any vector in $\mathbf{V}$ and define

$$\mathbf{w}' = \mathbf{w} - (c_1 \mathbf{u}_1 + c_2 \mathbf{u}_2 + \cdots + c_r \mathbf{u}_r)$$

where

$$c_j = \frac{\langle \mathbf{w}, \mathbf{u}_j \rangle}{\langle \mathbf{u}_j, \mathbf{u}_j \rangle}; \quad j = 1, 2, \ldots, r$$

Show that $\mathbf{w}'$ is orthogonal to the space spanned by $\{\mathbf{u}_j, j = 1, \cdots, r\}$. [Remark: The coefficient c_j is called Fourier coefficient and represents the component of $\mathbf{w}$ along $\mathbf{u}_j$.]

11 Applications of finite-dimensional linear algebra

This chapter is an introduction to some applications of abstract vector space concepts to problems of interest in transport phenomena, separations and kinetics.

11.1 Weighted dot/inner product in $\mathbb{R}^n$

Let $\mathbf{V}$ be a finite-dimensional inner product space over a field F (Hilbert space) and $\{\mathbf{e}_1, \mathbf{e}_2, \ldots, \mathbf{e}_n\}$ be an orthonormal basis for $\mathbf{V}$. Let $\mathbf{T} : \mathbf{V} \to \mathbf{V}$ be a linear operator and $[\mathbf{T}]_e = \mathbf{A}$ = matrix of $\mathbf{T}$ in the $\{\mathbf{e}_i\}$ basis. When $\mathbf{V} = \mathbb{R}^n$, the standard inner product is defined by

$$\langle \mathbf{x}, \mathbf{y} \rangle = \sum_{i=1}^{n} x_i y_i = \mathbf{y}^T \mathbf{x}; \quad \mathbf{x}, \mathbf{y} \in \mathbf{V} \tag{11.1}$$

Now suppose that $\mathbf{T}$ is not self-adjoint with respect to the standard inner product. Then the question is if we can make $\mathbf{T}$ self-adjoint w. r. t. the inner product

$$\langle \mathbf{x}, \mathbf{y} \rangle = \sum_{i=1}^{n} \sum_{j=1}^{n} g_{ij} x_j y_i = \mathbf{y}^T \mathbf{G} \mathbf{x} \tag{11.2}$$

where $\mathbf{G} = \{g_{ij}\}$ is a symmetric positive definite matrix. For $\mathbf{T}$ to be self-adjoint, we should have

$$\langle \mathbf{T}\mathbf{x}, \mathbf{y} \rangle = \langle \mathbf{x}, \mathbf{T}\mathbf{y} \rangle$$

or equivalently,

$$\langle \mathbf{A}\mathbf{x}, \mathbf{y} \rangle = \langle \mathbf{x}, \mathbf{A}\mathbf{y} \rangle \Rightarrow \mathbf{y}^T \mathbf{G}\mathbf{A}\mathbf{x} = \mathbf{y}^T \mathbf{A}^T \mathbf{G}\mathbf{x} \tag{11.3}$$

Since this must be true for all $\mathbf{x}$ and $\mathbf{y}$ in $\mathbf{V}$,

$\Rightarrow$

$$\begin{aligned} \mathbf{G}\mathbf{A} &= \mathbf{A}^T \mathbf{G} \\ &= \mathbf{A}^T \mathbf{G}^T \quad (\text{since } \mathbf{G} = \mathbf{G}^T) \end{aligned}$$

$\Rightarrow$

$$\mathbf{G}\mathbf{A} = (\mathbf{G}\mathbf{A})^T \tag{11.4}$$

Thus, the matrix $\mathbf{A}$ is symmetric (or $\mathbf{T}$ is self-adjoint) with respect to the inner product (11.2) if the matrix $(\mathbf{G}\mathbf{A})$ is symmetric with respect to the standard inner product (11.1). Now consider the special case in which $\mathbf{G}$ is a diagonal matrix

https://doi.org/10.1515/9783111598055-013

$$\mathbf{G} = \begin{pmatrix} g_1 & 0 & . & . & 0 \\ 0 & g_2 & . & . & 0 \\ . & & & & . \\ . & & & & . \\ 0 & 0 & . & . & g_n \end{pmatrix}, \quad g_i > 0 \quad i = 1, 2, \ldots, n$$

Then

$$(\mathbf{GA}) = \{g_i a_{ij}\} \Rightarrow (\mathbf{GA}) = (\mathbf{GA})^T \Rightarrow a_{ij} g_i = a_{ji} g_j.$$

$\mathbf{G}$ is positive definite $\Rightarrow g_i > 0$ for all i. Thus, if we can choose g_i such that $g_i > 0$ and

$$a_{ij} g_i = a_{ji} g_j$$

then the operator $\mathbf{T} : \mathbf{V} \to \mathbf{V}$ (or equivalently the matrix $\mathbf{A}$) is symmetric (self-adjoint) w. r. t. the weighted inner product

$$\langle \mathbf{x}, \mathbf{y} \rangle = \sum_{i=1}^{n} g_i x_i y_i \tag{11.5}$$

Note that we can do so only if a_{ij} and a_{ji} are of the same sign. The above generalization of the inner product to make a nonsymmetric matrix into a symmetric matrix has many applications in chemical engineering. We illustrate here the use of weighted dot product with examples.

Example 11.1. Let

$$\mathbf{A} = \begin{pmatrix} 1 & 2 \\ 3 & 2 \end{pmatrix}$$

We note that $\mathbf{A}$ is not symmetric w. r. t. the usual inner product. Define

$$\langle \mathbf{x}, \mathbf{y} \rangle = \mathbf{y}^T \mathbf{G} \mathbf{x} = \mathbf{y}^T \begin{pmatrix} g_1 & 0 \\ 0 & g_2 \end{pmatrix} \mathbf{x}; \quad g_1, g_2 > 0$$

$\Rightarrow$

$$\mathbf{GA} = \begin{pmatrix} g_1 & 0 \\ 0 & g_2 \end{pmatrix} \begin{pmatrix} 1 & 2 \\ 3 & 2 \end{pmatrix} = \begin{pmatrix} g_1 & 2g_1 \\ 3g_2 & 2g_2 \end{pmatrix}$$

$\mathbf{GA}$ is symmetric if

$$2g_1 = 3g_2$$

Take $g_1 = 1 \Rightarrow g_2 = \frac{2}{3}$, and define

$$\langle \mathbf{x}, \mathbf{y} \rangle = x_1 y_1 + \frac{2}{3} x_2 y_2.$$

With respect to this inner product, **A** is symmetric. Note that this definition satisfies all the rules of inner product.

Eigenvalues and vectors of **A**:

$$\begin{aligned}
&(1-\lambda)(2-\lambda) - 6 = 0 \\
&\quad \Rightarrow \lambda^2 - 3\lambda - 4 = 0 \Rightarrow (\lambda - 4)(\lambda + 1) = 0 \Rightarrow \lambda = 4, -1 \\
&\qquad \lambda_1 = -1 \Rightarrow \begin{pmatrix} 2 & 2 \\ 3 & 3 \end{pmatrix} \begin{pmatrix} x_{11} \\ x_{12} \end{pmatrix} = \mathbf{0} \Rightarrow \mathbf{x}_1 = \begin{pmatrix} 1 \\ -1 \end{pmatrix} \\
&\qquad \lambda_2 = 4 \Rightarrow \begin{pmatrix} -3 & 2 \\ 3 & -2 \end{pmatrix} \begin{pmatrix} x_1 \\ x_2 \end{pmatrix} = \mathbf{0} \Rightarrow \mathbf{x}_2 = \begin{pmatrix} 2 \\ 3 \end{pmatrix}
\end{aligned}$$

It can be seen that $\langle \mathbf{x}_1, \mathbf{x}_2 \rangle = 1.2 + \frac{2}{3}(-1)(3) = 0$. Thus, $\mathbf{x}_1$ and $\mathbf{x}_2$ are orthogonal w. r. t. the new inner product.

In fact, it may be shown that any real or complex $n \times n$ matrix **A** that has real eigenvalues and a complete set of eigenvectors is self-adjoint (symmetric) with respect to some inner product.

Suppose that **A** has real eigenvalues and a complete set of eigenvectors. Then, ∃ a nonsingular matrix **T** such that

$$\mathbf{A} = \mathbf{T}\mathbf{\Lambda}\mathbf{T}^{-1} \tag{11.6}$$

where

$$\mathbf{\Lambda} = \text{spectral matrix} = \begin{pmatrix} \lambda_1 & & 0 \\ & \ddots & \\ 0 & & \lambda_n \end{pmatrix}$$

(11.6) ⇒

$$\mathbf{A}^T = (\mathbf{T}\mathbf{\Lambda}\mathbf{T}^{-1})^T = (\mathbf{T}^{-1})^T \mathbf{\Lambda}^T \mathbf{T}^T \tag{11.7}$$

⇒ If **T** diagonalizes **A** in a similarity transform, $(\mathbf{T}^{-1})^T$ or $(\mathbf{T}^T)^{-1}$ diagonalizes $\mathbf{A}^T$ in a similarity transform. For **A** to be symmetric w. r. t the inner product,

$$\langle \mathbf{x}, \mathbf{y} \rangle = \mathbf{y}^T \mathbf{G} \mathbf{x} \tag{11.8}$$

we should have

$$\mathbf{G}\mathbf{A} = (\mathbf{G}\mathbf{A})^T = \mathbf{A}^T \mathbf{G} \tag{11.9}$$

⇒

$$\mathbf{G}\mathbf{A}\mathbf{G}^{-1} = \mathbf{A}^T \tag{11.10}$$

(11.6) and (11.10) ⇒

$$\mathbf{G}\mathbf{T}\boldsymbol{\Lambda}\mathbf{T}^{-1}\mathbf{G}^{-1} = \mathbf{A}^T \tag{11.11}$$

(11.11) ⇒

$$\mathbf{G}\mathbf{T}\boldsymbol{\Lambda}(\mathbf{G}\mathbf{T})^{-1} = \mathbf{A}^T$$

Comparing (11.7) and (11.11) ⇒

$$(\mathbf{T}^{-1})^T = \mathbf{G}\mathbf{T} \Rightarrow \mathbf{G} = (\mathbf{T}^{-1})^T\mathbf{T}^{-1} = (\mathbf{T}^T)^{-1}\mathbf{T}^{-1}$$

or

$$\mathbf{G} = (\mathbf{T}\mathbf{T}^T)^{-1}.$$

Thus, if $\mathbf{T}$ is the modal matrix of $\mathbf{A}$ and we define

$$\langle \mathbf{x}, \mathbf{y} \rangle = \mathbf{y}^T(\mathbf{T}\mathbf{T}^T)^{-1}\mathbf{x}, \tag{11.12}$$

then $\mathbf{A}$ is symmetric w. r. t this inner product. Since each column of $\mathbf{T}$ is determined up to a constant there are n arbitrary (+ve) constants in (11.12).

Alternatively, if $\mathbf{W}$ is a nonsingular real matrix, then $\mathbf{W}^T\mathbf{W}$ is a symmetric positive definite matrix. We can define a weighted inner product

$$\langle \mathbf{x}, \mathbf{y} \rangle_{\mathbf{W}} = \langle \mathbf{W}\mathbf{x}, \mathbf{W}\mathbf{y} \rangle = \mathbf{y}^T\mathbf{W}^T\mathbf{W}\mathbf{x} \tag{11.13}$$

For the special case in which $\mathbf{W}$ is a diagonal matrix, equation (11.13) reduces to equation (11.5).

Example 11.2.

$$\mathbf{A} = \begin{pmatrix} 1 & 2 \\ 3 & 2 \end{pmatrix}$$

⇒

$$\mathbf{T} = \begin{pmatrix} c_1 & 2c_2 \\ -c_1 & 3c_2 \end{pmatrix}, \quad \mathbf{T}^T = \begin{pmatrix} c_1 & -c_1 \\ 2c_2 & 3c_2 \end{pmatrix}$$

⇒

$$\mathbf{T}\mathbf{T}^T = \begin{pmatrix} c_1 & 2c_2 \\ -c_1 & 3c_2 \end{pmatrix}\begin{pmatrix} c_1 & -c_1 \\ 2c_2 & 3c_2 \end{pmatrix} = \begin{pmatrix} c_1^2 + 4c_2^2 & -c_1^2 + 6c_2^2 \\ -c_1^2 + 6c_2^2 & c_1^2 + 9c_2^2 \end{pmatrix}$$

$$(\mathbf{T}\mathbf{T}^T)^{-1} = \frac{1}{25c_1^2c_2^2}\begin{pmatrix} c_1^2 + 9c_2^2 & c_1^2 - 6c_2^2 \\ c_1^2 - 6c_2^2 & c_1^2 + 4c_2^2 \end{pmatrix}$$

Take $c_1 = c_2 \Rightarrow$

$$\mathbf{G} = \frac{1}{25c^4}\begin{pmatrix} 10c^2 & -5c^2 \\ -5c^2 & 5c^2 \end{pmatrix} = \frac{2}{5c^2}\begin{pmatrix} 1 & -\frac{1}{2} \\ -\frac{1}{2} & \frac{1}{2} \end{pmatrix}$$

Take $c^2 = \frac{2}{5} \Rightarrow$

$$\mathbf{G} = \begin{pmatrix} 1 & -\frac{1}{2} \\ -\frac{1}{2} & \frac{1}{2} \end{pmatrix}$$

Then $\mathbf{A}$ is self-adjoint w. r. t the inner product

$$\begin{aligned} \langle \mathbf{x}, \mathbf{y} \rangle &= \mathbf{y}^T \begin{pmatrix} 1 & -\frac{1}{2} \\ -\frac{1}{2} & \frac{1}{2} \end{pmatrix} \mathbf{x} \\ &= (y_1 \quad y_2)\begin{pmatrix} 1 & -\frac{1}{2} \\ -\frac{1}{2} & \frac{1}{2} \end{pmatrix}\begin{pmatrix} x_1 \\ x_2 \end{pmatrix} \\ &= \left(y_1 - \frac{1}{2}y_2 \quad \frac{-y_1 + y_2}{2}\right)\begin{pmatrix} x_1 \\ x_2 \end{pmatrix} \\ &= x_1y_1 - \frac{1}{2}x_1y_2 - \frac{1}{2}x_2y_1 + \frac{1}{2}x_2y_2 \end{aligned}$$

Take

$$\mathbf{x} = \begin{pmatrix} 1 \\ 0 \end{pmatrix}$$

$\Rightarrow$

$$\begin{aligned} \langle \mathbf{x}, \mathbf{x} \rangle &= x_1^2 - x_1x_2 + \frac{1}{2}x_2^2 = 1 \\ \langle \mathbf{x}, \mathbf{y} \rangle &= \mathbf{0} \Rightarrow y_1 - \frac{1}{2}y_2 = 0 \Rightarrow y_2 = 2y_1 \end{aligned}$$

Take $y_2 = 2 \Rightarrow y_1 = 1$.
$\therefore$ $\mathbf{e}_1 = (1, 0)$, $\mathbf{e}_2 = (1, 2)$ is an orthonormal basis for $\mathbb{R}^2$. In this inner product space $\mathbf{e}_1, \mathbf{e}_2$ are orthonormal, i. e., they have unit length and orthogonal to each other. Clearly, the geometry of this space is quite different from what it is with the standard inner product (see Figure 11.1).

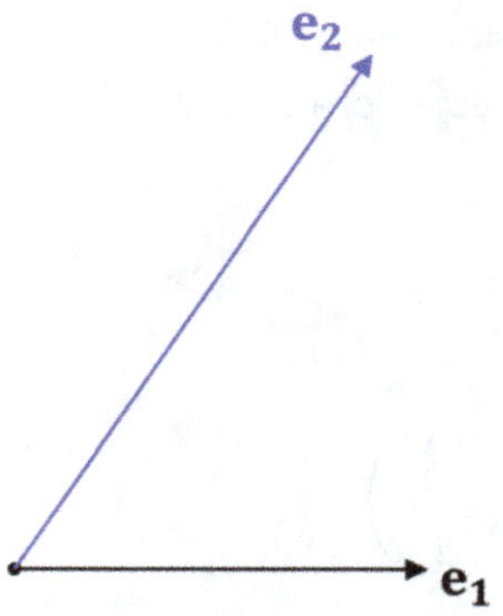

Figure 11.1: Schematic diagram of orthonormal basis vectors w. r. t. weighted inner product defined by equation (11.12).

11.2 Application of weighted inner product to interacting tank systems

The general form of the model for interacting tank systems (or discretized transient diffusion model) can be expressed as

$$\mathbf{C}\frac{d\mathbf{u}}{dt} = \mathbf{Q}\mathbf{u}, \quad \mathbf{u} = \mathbf{u}_0 \ @\ t = 0, \tag{11.14}$$

where $\mathbf{Q}$ is a symmetric exchange matrix, $\mathbf{C}$ is a diagonal capacitance matrix, with all positive diagonal elements, i. e.,

$$\mathbf{C} = \begin{pmatrix} a_1 & 0 & \cdots & 0 \\ 0 & a_2 & \cdots & 0 \\ \vdots & & & \vdots \\ 0 & 0 & \cdots & a_n \end{pmatrix}, \quad a_i > 0. \tag{11.15}$$

The model equation (11.14) can also be rewritten as

$$\frac{d\mathbf{u}}{dt} = \mathbf{C}^{-1}\mathbf{Q}\mathbf{u} = \mathbf{A}\mathbf{u}, \quad \mathbf{u} = \mathbf{u}_0 \ @\ t = 0. \tag{11.16}$$

Here, the matrix $\mathbf{A} = \mathbf{C}^{-1}\mathbf{Q}$ is not symmetric w. r. t. the usual inner product. However, if we define the capacitance weighted inner product

$$\langle \mathbf{u}, \mathbf{v} \rangle = \mathbf{v}^T\mathbf{C}\mathbf{u} = \sum_{i=1}^{n} a_i u_i v_i, \quad \text{for all real vectors } \mathbf{u} \text{ and } \mathbf{v} \tag{11.17}$$

then $\mathbf{A}$ is symmetric. This can be verified as follows:

$$\begin{aligned}\langle \mathbf{Au}, \mathbf{v}\rangle &= \mathbf{v}^T\mathbf{CAu} = \mathbf{v}^T\mathbf{CC}^{-1}\mathbf{Qu}\\ &= \mathbf{v}^T\mathbf{Qu}\end{aligned}$$

and

$$\begin{aligned}\langle \mathbf{u}, \mathbf{Av}\rangle &= (\mathbf{Av})^T\mathbf{Cu} = \mathbf{v}^T\mathbf{A}^T\mathbf{Cu}\\ &= \mathbf{v}^T(\mathbf{C}^{-1}\mathbf{Q})^T\mathbf{Cu}\\ &= \mathbf{v}^T\mathbf{Q}^T(\mathbf{C}^{-1})^T\mathbf{Cu}\\ &= \mathbf{v}^T\mathbf{QC}^{-1}\mathbf{Cu} \quad (\because \mathbf{C}^T = \mathbf{C} \text{ and } \mathbf{Q}^T = \mathbf{Q})\\ &= \mathbf{v}^T\mathbf{Qu}\end{aligned}$$

$\Longrightarrow$

$$\langle \mathbf{Au}, \mathbf{v}\rangle = \langle \mathbf{u}, \mathbf{Av}\rangle. \tag{11.18}$$

Thus, the solution of equations (11.14) or (11.16) can be obtained by taking the inner product (defined in equation (11.17)) with the eigenvectors $\mathbf{x}_j$ of $\mathbf{A}$:

$$\begin{aligned}\frac{d}{dt}\langle \mathbf{u}, \mathbf{x}_j\rangle &= \langle \mathbf{Au}, \mathbf{x}_j\rangle = \langle \mathbf{u}, \mathbf{Ax}_j\rangle\\ &= \lambda_j\langle \mathbf{u}, \mathbf{x}_j\rangle \quad (\because \mathbf{A} \text{ is symmetric, } \lambda_j \text{ are real})\end{aligned}$$

$\Longrightarrow$

$$\langle \mathbf{u}, \mathbf{x}_j\rangle = \langle \mathbf{u}_0, \mathbf{x}_j\rangle e^{\lambda_j t} \tag{11.19}$$

$\Longrightarrow$

$$\mathbf{u}(t) = \sum_{j=1}^{n} \frac{\langle \mathbf{u}, \mathbf{x}_j\rangle}{\langle \mathbf{x}_j, \mathbf{x}_j\rangle}\mathbf{x}_j = \sum_{j=1}^{n} \frac{\langle \mathbf{u}_0, \mathbf{x}_j\rangle}{\langle \mathbf{x}_j, \mathbf{x}_j\rangle} e^{\lambda_j t}\mathbf{x}_j. \tag{11.20}$$

Example 11.3 (Interacting two tank system). As an example, consider the two interacting tank system shown in Figure 11.2.

The model describing this system is given by

$$\begin{aligned}V_{R1}\frac{dc_1}{dt} &= -q_e c_1 + q_e c_2\\ V_{R2}\frac{dc_2}{dt} &= q_e c_1 - q_e c_2\end{aligned}$$

with initial condition

$$c_1 = c_{10} \quad \text{and} \quad c_2 = c_{20} \text{ @ } t = 0,$$

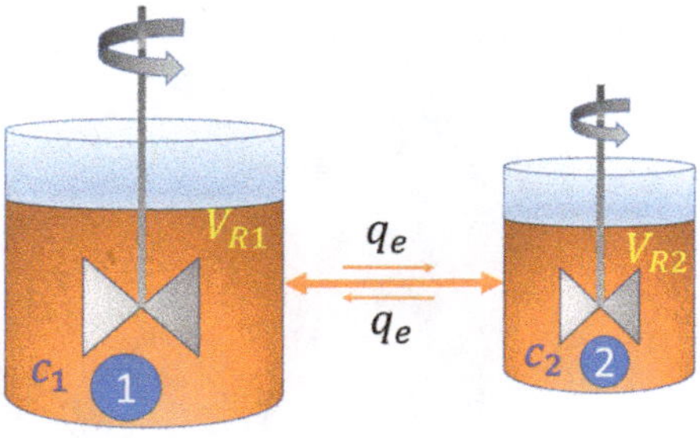

Figure 11.2: Interacting two tank system.

where V_{R1} and V_{R2} are the volumes of the tanks and q_e is the exchange flow rate. The model can be written in matrix-vector form:

$$\frac{d\mathbf{c}}{dt} = \mathbf{A}\mathbf{c}, \quad \mathbf{c}(t=0) = \mathbf{c}_0 = \begin{pmatrix} c_{10} \\ c_{20} \end{pmatrix}, \tag{11.21}$$

$$\mathbf{A} = \begin{pmatrix} -\alpha & \alpha \\ \beta & -\beta \end{pmatrix}; \quad \alpha = \frac{q_e}{V_{R1}}; \quad \beta = \frac{q_e}{V_{R2}}. \tag{11.22}$$

The matrix $\mathbf{A}$ is not symmetric w. r. t. the usual inner product unless $\alpha = \beta$ or $V_{R1} = V_{R2}$.

Standard solution

The eigenvalues of matrix $\mathbf{A}$ are

$$\lambda_1 = 0 \quad \text{and} \quad \lambda_2 = -(\alpha + \beta) \tag{11.23}$$

with corresponding eigenvectors

$$\mathbf{x}_1 = \begin{pmatrix} 1 \\ 1 \end{pmatrix}; \quad \mathbf{x}_2 = \begin{pmatrix} -\alpha \\ \beta \end{pmatrix} \tag{11.24}$$

and left eigenvectors

$$\mathbf{y}_1^T = \begin{pmatrix} \beta & \alpha \end{pmatrix}; \quad \mathbf{y}_2^T = \begin{pmatrix} -1 & 1 \end{pmatrix}. \tag{11.25}$$

Thus, the solution can be expressed using standard inner product as

$$\mathbf{c}(t) = \sum_{j=1}^{2} \frac{\mathbf{y}_j^T \mathbf{c}_0}{\mathbf{y}_j^T \mathbf{x}_j} e^{\lambda_j t} \mathbf{x}_j \tag{11.26}$$

$$= \frac{\beta c_{10} + \alpha c_{20}}{(\beta + \alpha)} \begin{pmatrix} 1 \\ 1 \end{pmatrix} + \frac{c_{20} - c_{10}}{(\beta + \alpha)} e^{-(\alpha+\beta)t} \begin{pmatrix} -\alpha \\ \beta \end{pmatrix}. \tag{11.27}$$

Solution using capacitance weighted inner product

Suppose that we define

$$\langle \mathbf{u}, \mathbf{v} \rangle = \mathbf{v}^T \mathbf{G} \mathbf{u}, \quad \mathbf{G} = \begin{pmatrix} g_1 & 0 \\ 0 & g_2 \end{pmatrix} \quad \text{with } g_i > 0 \tag{11.28}$$

$$= g_1 u_1 v_1 + g_2 u_2 v_2. \tag{11.29}$$

Then **A** is symmetric if

$$\mathbf{GA} = (\mathbf{GA})^T.$$

But

$$\mathbf{GA} = \begin{pmatrix} g_1 & 0 \\ 0 & g_2 \end{pmatrix} \begin{pmatrix} -\alpha & \alpha \\ \beta & -\beta \end{pmatrix}$$

$$= \begin{pmatrix} -\alpha g_1 & \alpha g_1 \\ \beta g_2 & -\beta g_2 \end{pmatrix}$$

$\implies$ **GA** is symmetric (in standard sense) if

$$\alpha g_1 = \beta g_2.$$

Thus, if we take

$$g_1 = \beta \quad \text{and} \quad g_2 = \alpha \tag{11.30}$$

then **A** is symmetric with weighted inner product:

$$\mathbf{G} = \begin{pmatrix} \beta & 0 \\ 0 & \alpha \end{pmatrix} \quad \text{and} \quad \langle \mathbf{u}, \mathbf{v} \rangle = \mathbf{v}^T \mathbf{G} \mathbf{u} = \beta u_1 v_1 + \alpha u_2 v_2. \tag{11.31}$$

Note that

$$\langle \mathbf{x}_1, \mathbf{x}_2 \rangle = \left\langle \begin{pmatrix} 1 \\ 1 \end{pmatrix}, \begin{pmatrix} -\alpha \\ \beta \end{pmatrix} \right\rangle = \beta.1.(-\alpha) + \alpha.1.\beta = 0.$$

Similarly,

$$\begin{aligned} \langle \mathbf{c}_0, \mathbf{x}_1 \rangle &= \beta c_{10} + \alpha c_{20} \\ \langle \mathbf{x}_1, \mathbf{x}_1 \rangle &= \beta + \alpha \\ \langle \mathbf{c}_0, \mathbf{x}_2 \rangle &= \alpha\beta(c_{20} - c_{10}) \\ \langle \mathbf{x}_2, \mathbf{x}_2 \rangle &= \alpha\beta(\alpha + \beta) \end{aligned}$$

Thus, the solution can be expressed as

$$\begin{aligned}\mathbf{c}(t) &= \sum_{j=1}^{2} \frac{\langle \mathbf{c}_0, \mathbf{x}_j \rangle}{\langle \mathbf{x}_j, \mathbf{x}_j \rangle} e^{\lambda_j t} \mathbf{x}_j \\ &= \frac{\langle \mathbf{c}_0, \mathbf{x}_1 \rangle}{\langle \mathbf{x}_1, \mathbf{x}_1 \rangle} \mathbf{x}_1 + \frac{\langle \mathbf{c}_0, \mathbf{x}_2 \rangle}{\langle \mathbf{x}_2, \mathbf{x}_2 \rangle} e^{-(\alpha+\beta)t} \mathbf{x}_2. \\ &= \frac{\beta c_{10} + \alpha c_{20}}{(\beta + \alpha)} \begin{pmatrix} 1 \\ 1 \end{pmatrix} + \frac{c_{20} - c_{10}}{(\beta + \alpha)} e^{-(\alpha+\beta)t} \begin{pmatrix} -\alpha \\ \beta \end{pmatrix} \end{aligned} \tag{11.32}$$

Equation (11.32) is the same solution as in equation (11.27) but we do not need to compute the eigenrows. Note that the vectors $\mathbf{x}_1 = \binom{1}{1}$ and $\mathbf{x}_2 = \binom{-\alpha}{\beta}$ are orthogonal w. r. t. the inner product defined in equation (11.31). The solution for $\alpha = 1$ and $\beta = 3$ is shown in Figure 11.3 for the initial condition of $\mathbf{c}_0 = \binom{1}{0}$.

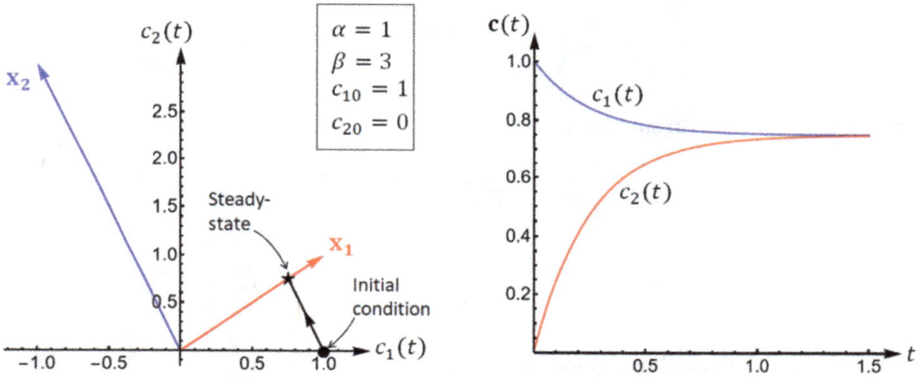

Figure 11.3: Solution diagram of interacting two tank system for $\alpha = 1, \beta = 3, c_{10} = 1$ and $c_{20} = 0$.

It can be seen from the plots that at steady-state (or after long time, $t \to \infty$), the solution is in the space spanned by the eigenvector $\mathbf{x}_1$ corresponding to zero eigenvalue (as expected).

11.3 Application of weighted inner product to monomolecular kinetics

Monomolecular kinetics (Wie–Prater scheme)

Consider the reaction scheme shown in Figure 11.4 between three species A_1, A_2, A_3, where k_{ij} = rate constant for the formation of species A_i from A_j.

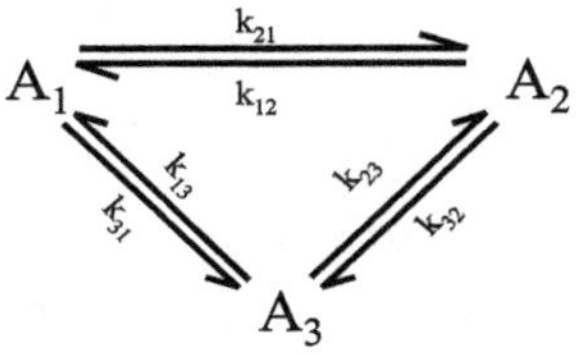

Figure 11.4: Reaction scheme between the species A_1, A_2 and A_3.

The conservation equations for a batch reactor are

$$\left.\begin{aligned}\frac{d[A_1]}{dt} &= -[k_{21}+k_{31}][A_1] + k_{12}[A_2] + k_{13}[A_3]\\ \frac{d[A_2]}{dt} &= k_{21}[A_1] - (k_{12}+k_{32})[A_2] + k_{23}[A_3]\\ \frac{d[A_3]}{dt} &= k_{31}[A_1] + k_{32}[A_2] - (k_{13}+k_{23})[A_3]\end{aligned}\right\} \quad \begin{array}{c} @\, t=0, \quad [A_i] = [A_{io}] \\ \text{(initial concentration of } A_i) \end{array}$$

Let

$$x_i = \frac{[A_i]}{\sum_{i=1}^{3}[A_i]} = \text{mole fraction of } A_i$$

then we have

$$\frac{d\mathbf{x}}{dt} = \mathbf{K}\mathbf{x}$$

where

$$\mathbf{x} = \begin{bmatrix} x_1 \\ x_2 \\ x_3 \end{bmatrix}, \quad \mathbf{K} = \begin{bmatrix} -(k_{21}+k_{31}) & k_{12} & k_{13} \\ k_{21} & -(k_{12}+k_{32}) & k_{23} \\ k_{31} & k_{32} & -(k_{13}+k_{23}) \end{bmatrix}$$

Generalizing this to n-species $(A_1, A_2, \ldots, A_n)$, we have

$$\frac{d\mathbf{x}}{dt} = \mathbf{K}\mathbf{x}, \quad \mathbf{x}(\mathbf{t}=0) = \mathbf{x}^0 \tag{11.33}$$

where

$$\mathbf{K} = \begin{bmatrix} k_{11} & k_{12} & k_{13} & . & . & k_{1n} \\ k_{21} & k_{22} & k_{23} & . & . & k_{2n} \\ . & & & & & \\ . & & & & & \\ . & & & & & \\ k_{n1} & k_{n2} & k_{n3} & . & . & k_{nn} \end{bmatrix}$$

and

$$k_{ii} = -\sum_{j=1,i}^{n} k_{ji}.$$

(The summation notation implies that j takes all values from 1 to n excluding $j = i$).

Note that each column of $\mathbf{K}$ sums to zero. Thus, rank of $\mathbf{K} < n$, and there is a nonzero solution of $\mathbf{Kx} = \mathbf{0}$. This is the equilibrium solution, denoted by $\mathbf{x}^*$. We assume that rank $\mathbf{K} = n-1$ so that there is no other equilibrium point. Clearly, the matrix $\mathbf{K}$ is nonsymmetric with respect to the usual inner product. A feature crucial to the following analysis is the *principle of microscopic reversibility* or principle of detailed balancing, which states that at equilibrium the rate of every reaction and its reverse must be equal, i. e.,

$$k_{ij}[A_j^*] = k_{ji}[A_i^*] \tag{11.34}$$

where $[A_j^*]$ is the equilibrium concentration of species A_j. Thus, we have (in terms of mole fractions)

$$k_{ij}x_j^* = k_{ji}x_i^* \Rightarrow \frac{k_{ij}}{x_i^*} = \frac{k_{ji}}{x_j^*}.$$

We define the weighted inner product by

$$\langle \mathbf{u}, \mathbf{v} \rangle = \sum_{j=1}^{n} \frac{u_j v_j}{x_j^*}, \tag{11.35}$$

and show that the matrix $\mathbf{K}$ is symmetric w. r. t. this inner product. The i-th element of the vector $\mathbf{Ku}$ is given by

$$(\mathbf{Ku})_i = \sum_{j=1}^{n} k_{ij}u_j$$

Thus,

$$\begin{aligned} \langle \mathbf{Ku}, \mathbf{v} \rangle &= \sum_{i=1}^{n} \left(\sum_{j=1}^{n} k_{ij}u_j \right) \frac{v_i}{x_i^*} \\ &= \sum_{i=1}^{n} \sum_{j=1}^{n} \frac{k_{ij}u_j v_i}{x_i^*} \end{aligned}$$

but

$$\frac{k_{ij}}{x_i^*} = \frac{k_{ji}}{x_j^*}$$

∴

$$\langle \mathbf{Ku}, \mathbf{v} \rangle = \sum_{j=1}^{n} \sum_{i=1}^{n} \frac{k_{ji} u_j v_i}{x_j^*}$$

Since the indices i and j are dummy we change $i \to j$ and $j \to i$ without changing the sum. Thus,

$$\begin{aligned}
\langle \mathbf{Ku}, \mathbf{v} \rangle &= \sum_{i=1}^{n} \sum_{j=1}^{n} k_{ij} v_j \frac{u_i}{x_i^*} \\
&= \sum_{i=1}^{n} \frac{(\sum_{j=1}^{n} k_{ij} v_j) u_i}{x_i^*} \\
&= \langle \mathbf{u}, \mathbf{Kv} \rangle
\end{aligned}$$

Therefore, $\mathbf{K}$ is self-adjoint w. r. t this inner product, which implies that all its eigenvalues are real and it has a set of n-eigenvectors. Let $\mathbf{z}_1 (= \mathbf{x}^*), \mathbf{z}_2, \mathbf{z}_3, \ldots, \mathbf{z}_n$ be the eigenvectors and $\lambda_1 (= 0), \lambda_2, \ldots, \lambda_n$ be the eigenvalues. Now, let us show that $\mathbf{K}$ is nonpositive, i. e., all the nonzero eigenvalues are strictly negative. We consider the quadratic form

$$\begin{aligned}
\langle \mathbf{Ku}, \mathbf{u} \rangle &= \sum_{i=1}^{n} \left(\frac{1}{x_i^*} u_i \right) \sum_{j=1}^{n} k_{ij} u_j \\
&= \sum_{i=1}^{n} \sum_{j=1}^{n} \frac{k_{ij} u_i u_j}{x_i^*} \\
&= \sum_{i=1}^{n} \left(\sum_{j=1,i}^{n} \frac{k_{ij} u_i u_j}{x_i^*} + \frac{k_{ii} u_i^2}{x_i^*} \right) \\
&= \sum_{i=1}^{n} \left(\sum_{j=1,i}^{n} \frac{k_{ij} u_i u_j}{x_i^*} - \sum_{j=1,i}^{n} \frac{k_{ji} u_i^2}{x_i^*} \right) \\
&= \sum\sum_{i \neq j} \left(\frac{k_{ij} u_i u_j}{x_i^*} \right) - \sum\sum_{i \neq j} \frac{k_{ji} u_i^2}{x_i^*} \\
&= \left[\sum\sum_{i \neq j} \sqrt{\frac{k_{ij}}{x_i^*}} \sqrt{\frac{k_{ji}}{x_j^*}} u_i u_j - \sum\sum_{i \neq j} \left(\sqrt{\frac{k_{ji}}{x_i^*}} u_i \right)^2 \right]
\end{aligned}$$

Interchanging indices gives

$$\langle \mathbf{Ku}, \mathbf{u} \rangle = \left[\sum\sum_{i \neq j} \sqrt{\frac{k_{ji}}{x_j^*}} \sqrt{\frac{k_{ij}}{x_i^*}} u_j u_i - \sum\sum_{i \neq j} \left(\sqrt{\frac{k_{ij}}{x_j^*}} u_j \right)^2 \right]$$

⇒

$$2\langle \mathbf{K}\mathbf{u}, \mathbf{u} \rangle = -\sum_{i \neq j} \sum \left(\sqrt{\frac{k_{ij}}{x_j^*}} u_j - \sqrt{\frac{k_{ji}}{x_i^*}} u_i \right)^2 \leq 0 \tag{11.36}$$

Thus, the nonzero eigenvalues are strictly negative.

The solution of

$$\frac{d\mathbf{x}}{dt} = \mathbf{K}\mathbf{x}, \quad \mathbf{x}(\mathbf{t} = 0) = \mathbf{x}^0$$

is given by

$$\mathbf{x} = \sum_{j=1}^{n} \langle \mathbf{x}^0, \mathbf{z}_j \rangle e^{\lambda_j t} \mathbf{z}_j$$

where $\mathbf{z}_j$ are orthonormal set of eigenvectors and the inner product is defined by equation (11.35). To obtain this form of the solution, take inner product with $\mathbf{z}_j$,

$\Longrightarrow$

$$\begin{aligned} \frac{d}{dt} \langle \mathbf{x}, \mathbf{z}_j \rangle &= \langle \mathbf{K}\mathbf{x}, \mathbf{z}_j \rangle \\ &= \langle \mathbf{x}, \mathbf{K}\mathbf{z}_j \rangle \\ &= \langle \mathbf{x}, \lambda_j \mathbf{z}_j \rangle \\ &= \lambda_j \langle \mathbf{x}, \mathbf{z}_j \rangle \end{aligned}$$

$\Rightarrow$

$$\langle \mathbf{x}, \mathbf{z}_j \rangle = \langle \mathbf{x}^0, \mathbf{z}_j \rangle e^{\lambda_j t}$$

$\Rightarrow$

$$\begin{aligned} \mathbf{x} &= \sum_{j=1}^{n} \langle \mathbf{x}^0, \mathbf{z}_j \rangle e^{\lambda_j t} \mathbf{z}_j \\ &= \langle \mathbf{x}^0, \mathbf{x}^* \rangle \mathbf{x}^* + \sum_{j=2}^{n} \langle \mathbf{x}^0, \mathbf{z}_j \rangle e^{\lambda_j t} \mathbf{z}_j. \end{aligned} \tag{11.37}$$

Using the fact that $\langle \mathbf{x}, \mathbf{x}^* \rangle = 1$, equation (11.37) simplifies to

$$\mathbf{x} = \mathbf{x}^* + \sum_{j=2}^{n} \langle \mathbf{x}^0, \mathbf{z}_j \rangle e^{\lambda_j t} \mathbf{z}_j. \tag{11.38}$$

The first term is the equilibrium solution, while the nonzero eigenvalues $(\lambda_j, j = 2, \ldots, n)$ determine the time scales associated with the transient process.

The Wie–Prater scheme is an experimental method for determining the eigenvalues and eigenvectors. From these experimental values, we can determine the rate constant matrix $\mathbf{K}$.

$$\text{Let } \mathbf{k}_j = \begin{pmatrix} k_{1j} \\ \cdot \\ \cdot \\ k_{nj} \end{pmatrix} = j\text{-th column of } \mathbf{K}.$$

Since $\mathbf{k}_j \in \mathbb{R}^n$, expand it in terms of the eigenvectors as

$$\mathbf{k}_j = \sum_{r=1}^{n} \langle \mathbf{k}_j, \mathbf{z}_r \rangle \mathbf{z}_r.$$

Now consider

$$\begin{aligned} \langle \mathbf{k}_j, \mathbf{z}_r \rangle &= \sum_{i=1}^{n} \frac{k_{ij} z_{ir}}{x_i^*} \\ &= \sum_{i=1}^{n} \frac{k_{ji}}{x_j^*} z_{ir} \\ &= \frac{1}{x_j^*} \sum_{i=1}^{n} k_{ji} z_{ir} \\ \mathbf{K}\mathbf{z}_r = \lambda_r \mathbf{z}_r &\Rightarrow \sum_{i=1}^{n} k_{ji} z_{ir} = \lambda_r z_{jr} \end{aligned}$$

$\because$

$$\langle \mathbf{k}_j, \mathbf{z}_r \rangle = \frac{1}{x_j^*} \lambda_r z_{jr}$$

$\Rightarrow$

$$\mathbf{k}_j = \sum_{r=1}^{n} \frac{\lambda_r z_{jr}}{x_j^*} \mathbf{z}_r = \frac{1}{x_j^*} \sum_{r=2}^{n} \lambda_r z_{jr} \mathbf{z}_r \quad (\text{since } \lambda_1 = 0)$$

$\Rightarrow$

$$k_{ij} = \frac{1}{x_j^*} \sum_{r=2}^{n} \lambda_r z_{jr} z_{ir} \tag{11.39}$$

All quantities on the RHS of equation (11.39) are experimentally determinable. Thus, we can determine all the rate constants from the eigenvectors and eigenvalues. An example in which this procedure was used by Wei and Prater [31] is the catalytic isomerization of butenes. The reaction network along with the relative values of the rate constants is shown in Figure 11.5.

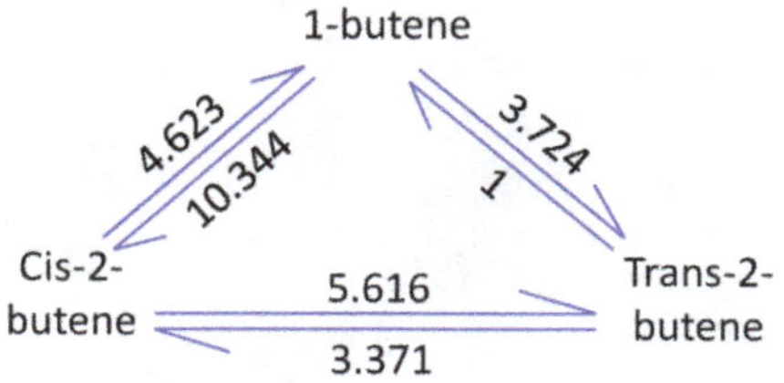

Figure 11.5: Reaction network for monomolecular kinetics for catalytic isomerization of butenes.

In this case, the eigenvalues and orthonormal set of eigenvectors are given by

$$\lambda_1 = 0, \quad \mathbf{z}_1 = \mathbf{x}^* = \begin{pmatrix} 0.1436 \\ 0.3213 \\ 0.5351 \end{pmatrix},$$

$$\lambda_2 = -9.2602, \quad \mathbf{z}_2 = \begin{pmatrix} 0.1903 \\ 0.3050 \\ 0.4953 \end{pmatrix},$$

$$\lambda_3 = -19.418, \quad \mathbf{z}_3 = \begin{pmatrix} 0.2946 \\ -0.3536 \\ 0.0590 \end{pmatrix}.$$

The experimentally observed reaction paths (which include straight line reaction paths) are shown in Figure 11.6.

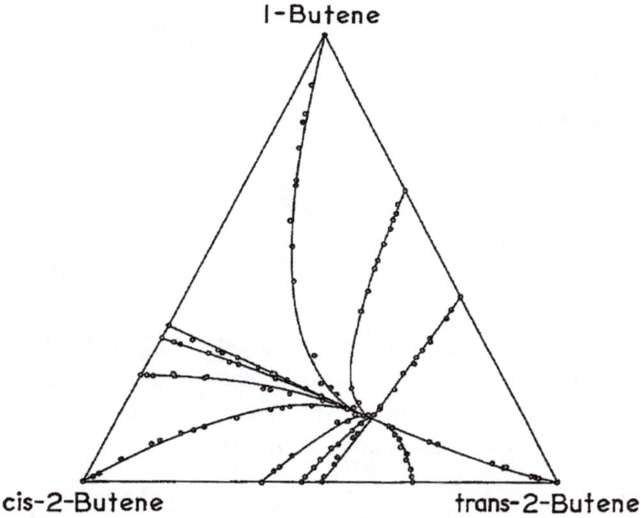

Figure 11.6: Experimentally observed reaction paths for catalytic isomerization of butenes, obtained from Wei and Prater [31].

Other applications of weighted inner product to stage operations, kinetics and weighted least squares are given in the exercises. Additional applications may also be found in the book by Ramkrishna and Amundson [25].

Problems

1. Consider the vector space of 2-tuples of real numbers over the real field, i. e., $\mathbb{R}^2$.
   (a) Find an inner product on $\mathbb{R}^2$ with respect to which the matrix $\mathbf{A} = \left(\begin{smallmatrix} -1 & 1 \\ 4 & -4 \end{smallmatrix}\right)$ is self-adjoint (symmetric).
   (b) Determine the eigenvalues and eigenvectors of $\mathbf{A}$ and verify that the eigenvectors are orthogonal w. r. t. the inner product defined in (a).
   (c) Determine the normalized eigenvectors and show a schematic plot of these eigenvectors.
   (d) Use the above results to obtain the solution of initial value problem

$$\frac{d\mathbf{u}}{dt} = \mathbf{A}\mathbf{u}; \quad \mathbf{u}(t=0) = \begin{pmatrix} 1 \\ 0 \end{pmatrix}.$$

2. Given the matrix

$$\mathbf{A} = \begin{pmatrix} -\gamma & \alpha & 0 \\ \beta & -\gamma & \alpha \\ 0 & \beta & -\gamma \end{pmatrix},$$

   where α, β and γ are positive and $\alpha \neq \beta$,
   (a) how would you define the inner product in $\mathbb{R}^3$ so that $\mathbf{A}$ is symmetric (self-adjoint)?
   (b) What additional condition need to be satisfied by α, β and γ so that $\mathbf{A}$ is negative definite?
3. The models used to describe the transient behavior of plate, gas absorbers, extraction units and rectifying and stripping and sections of a distillation column are essentially the same. As an illustration, consider a staged absorber containing N stages in which a heavy (liquid) phase and a light (gas) phase pass countercurrent to each other. The contacting may be assumed to be uniform so that equilibrium is attained in each stage and the equilibrium relationship is linear ($y = Kx$).
   (a) Show that the dynamic model of the system is of the form

$$\frac{dx_j}{dt} = \alpha x_{j-1} - (\alpha + \beta)x_j + \beta x_{j+1} \quad j = 1, 2, 3, \ldots, N$$

   where $\alpha = \frac{L}{h}$, $\beta = \frac{GK}{h}$, L is the liquid (heavy-phase) flow rate, h is the holdup (of the heavy phase), G, is the gas (light-phase) flow rate and x_j is the composition of the transferable component in the liquid stream leaving stage j. State any other assumptions involved.

(b) Assuming that the compositions $x_0(t)$ and $x_{N+1}(t) = \frac{y_{N+1}}{K}$, of the entering streams are known, show that the model may be written in the form

$$\frac{d\mathbf{x}}{dt} = \mathbf{A}\mathbf{x} + \mathbf{b}(t)$$

Identify the vectors $\mathbf{x}$, $\mathbf{b}(t)$ and the matrix $\mathbf{A}$.

(c) The matrix $\mathbf{A}$ is not self-adjoint with respect to the usual inner product in $\mathbb{R}^N$. However, show that if we define inner product as

$$\langle \mathbf{x}, \mathbf{y} \rangle = \sum_{i=1}^{N} \left(\frac{\beta}{\alpha}\right)^{i-1} x_i y_i$$

then $\mathbf{A}$ is self-adjoint with respect to this inner product.

(d) Using the above inner product, show that

$$\langle \mathbf{A}\mathbf{x}, \mathbf{x} \rangle = -\sum_{i=0}^{n} \alpha \left(\frac{\beta}{\alpha}\right)^{i-1} \left[x_i - \frac{\beta}{\alpha} x_{i+1}\right]^2 < 0$$

Thus, $\mathbf{A}$ is negative definite and all its eigenvalues are strictly negative.

(e) Compute the steady-state values for y_1 and x_3 when $N = 3$ (three stage process) and other parameters are given as follows:

$$L = 5, \quad G = 3, \quad h = 1, \quad K = 1, \quad x_0 = 0, \quad y_4 = 0.5$$

Compute and plot the transient response when there is a step change in y_4 from 0.5 to 0.7.

4. Consider a well-stirred batch reactor in which the following consecutive reversible first-order reactions occur:

$$A_1 \leftrightarrows A_2 \leftrightarrows \cdots \leftrightarrows A_n$$

(a) Denote the concentration vector by $\mathbf{c} = (c_1, c_2, \ldots, c_n)$ and identify the matrix $\mathbf{K}$, which yields the batch reactor equation

$$\frac{d\mathbf{c}}{dt} = \mathbf{K}\mathbf{c}.$$

(b) Find the inner product on $\mathbb{R}^n$ with respect to which $\mathbf{K}$ becomes self-adjoint. Is $\mathbf{K}$ negative or nonpositive? Obtain the solution to the differential equation above subject to the initial condition $\mathbf{c}(0) = \mathbf{c}_0$.

(c) Solve the transient continuous stirred tank reactor equation

$$\frac{d\mathbf{c}}{dt} = \frac{1}{\tau}(\mathbf{c_f} - \mathbf{c}) + \mathbf{Kc}, \quad \mathbf{c}(0) = \mathbf{c}_0$$

where $\mathbf{c_f}$ is the feed concentration vector and τ is the holding time.

5. Difference equations of the following type arise in the steady-state analysis of many stage operations:

$$\mathbf{c}_{k+1} = \mathbf{Ac}_k + \mathbf{b}_k; \quad \mathbf{c}_0 = \alpha,$$

where $\mathbf{A}$ is a constant $n \times n$ matrix and $\mathbf{c}_k, \mathbf{b}_k$ and α are $n \times 1$ vectors.

(a) Obtain a formal solution of the above set of difference equations.

(b) Show that the linear second-order difference equation of the form

$$c_{k+1} + ac_k + bc_{k-1} = b_k; \quad c_0 = \alpha_1, \quad c_1 = \alpha_2$$

can be written in the matrix form given in (a). Obtain an explicit solution to this equation.

6. Consider an extraction process in which G kg/s of a light phase containing Y_0 kg solute A per kg of carrier is fed to a cascade containing N ideal stages and contacted with L kg/s of solute-free heavy phase containing X_{N+1} kg solute A per kg of carrier.

(a) Formulate the relevant equations.

(b) If the concentration of A in the exit light and heavy streams are Y_N and X_1, respectively, show that the concentration of A in the light phase leaving stage i is given by

$$Y_i = \frac{Y_0 - KX_1}{1 - \frac{KG}{L}}\left(\frac{KG}{L}\right)^i + \frac{KX_{N+1} - \frac{KG}{L}Y_N}{1 - \frac{KG}{L}}$$

where K is the equilibrium constant defined by $Y_N = KX_N$.

7. A cascade of N stirred tank reactors is arranged to operate isothermally in series. Each reactor has a volume of $V\ m^3$ and is well stirred so that the composition of the reactor effluent is the same as the tank contents. If initially the tanks contain pure solvent only, and at a time designated $t_0, q\ m^3/h$ of a reactant A of concentration C_0 kg moles/m^3 are fed to the first tank, estimate the time required for the concentration of A leaving the N-th tank to be C_N kg moles/m^3. The reaction can be represented stoichiometrically by the equation

$$A \leftrightarrows B \longrightarrow C$$

and all the reactions are first order. There is no B or C in the feed but it may be assumed that the feed contains a catalyst that initiates the reaction as soon as the feed enters the first reactor.

8. Consider the three interacting tanks arranged in series as shown in Figure 11.7 below. The volume of tanks V_{R1}, V_{R2} and V_{R3} may not be identical. Similarly, the exchange flow rate q_1 between the tanks 1 and 2 may not be same as q_2 between tanks 2 and 3.

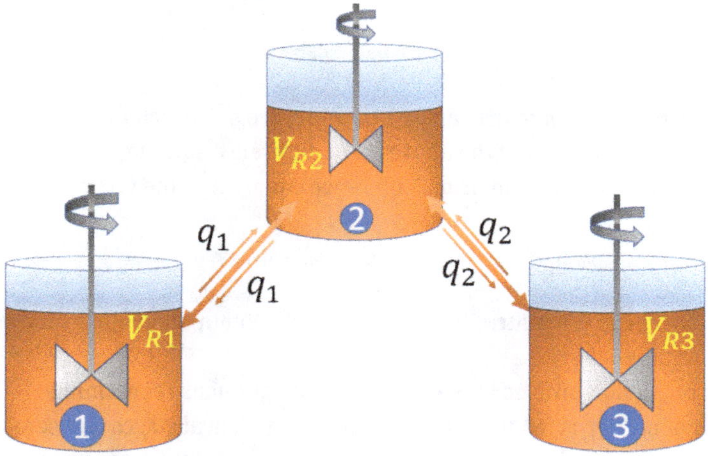

Figure 11.7: Schematic diagram of interacting tanks.

(a) Considering the transient process, formulate the model for concentrations c_i in each tank in the form:

$$\mathbf{V}_R \frac{d\mathbf{c}}{dt} = \mathbf{Q}\mathbf{c}$$

and show that (i) $\mathbf{V}_R$ is the diagonal positive definite matrix, and (ii) $\mathbf{Q}$ is a symmetric matrix with zero row and column sum (i. e., has a zero eigenvalue).

(b) Define a matrix $\mathbf{A} = \mathbf{V}_R^{-1}\mathbf{Q}$, show that (i) $\mathbf{A}$ is not symmetric w. r. t. usual inner product

$$\langle \mathbf{x}, \mathbf{y} \rangle = \mathbf{y}^T \mathbf{x} = \sum_{i=1}^{3} x_i y_i$$

(ii) $\mathbf{A}$ is symmetric w. r. t. weighted inner product

$$\langle \mathbf{x}, \mathbf{y} \rangle = \mathbf{y}^T \mathbf{V}_R \mathbf{x}$$

9. (*Weighted least squares*)
Consider a set of data points:

$$[(x_1, y_1), (x_2, y_2), \ldots, (x_N, y_N)]$$

and suppose that you want to fit a linear model

$$y = a_0 + a_1 x.$$

Suppose that the data point j, is given a weight w_j (> 0). Formulate the weighted least squares problem and determine the normal equations to be solved for a_0 and a_1.

Part III: **Linear ordinary differential equations-initial value problems, complex variables and Laplace transform**

12 The linear initial value problem

In earlier chapters, we have discussed the solution of linear initial value problems in which a square matrix of constant coefficients appeared. In this chapter, we consider the case of more general form of the linear initial value problem and discuss the relevant theory along with applications.

12.1 The vector initial value problem

Suppose that $\mathbf{A}(t)$ is an $n \times n$ matrix whose entries $\{a_{ij}(t)\}$ are continuous functions of t. Then, the most general linear initial value problem is defined by the equations

$$\frac{d\mathbf{u}}{dt} = \mathbf{A}(t)\mathbf{u} + \mathbf{b}(t), \quad 0 < t < a \tag{12.1}$$

$$\mathbf{u}(@\, t = 0) = \mathbf{u}^0 \tag{12.2}$$

where

$$\mathbf{u} = \begin{pmatrix} u_1(t) \\ u_2(t) \\ \cdot \\ \cdot \\ u_n(t) \end{pmatrix}$$

is an n-tuple of real (or complex) valued functions $u_i \in C^1[0, a]$, a is a positive constant and $\mathbf{u}^0$ is a constant vector determining the initial state of the system. The forcing vector $\mathbf{b}(t)$ is also an n-tuple of real (or complex) valued function of t. The following fundamental existence and uniqueness theorem may be stated for the initial value problem defined by equations (12.1) and (12.2).

Theorem. *Consider the IVP defined by equations* (12.1) *and* (12.2) *and suppose that*

$$a_{ij}(t) \in C[0, a], \quad b_j(t) \in C[0, a], \quad i, j = 1, 2, \ldots, n$$

and let $\mathbf{u}^0$ be any vector in $\mathbb{R}^n$ and t be a point in $[0, a]$. Then there exists one and only one solution $\mathbf{u}(t), 0 < t < a$, satisfying equations (12.1) *and* (12.2)*. This solution is a continuous function of the initial conditions $\mathbf{u}^0$ and if $a_{ij}(t)$ depend continuously on a parameter, so does the solution.*

[For proof of this theorem, see Coddington and Levinson [13].]

We now outline a method for obtaining the solution of the initial value problem defined by equations (12.1) and (12.2). As discussed previously, since equation (12.1) is linear, the principle of superposition may be used and the general solution may be written as

$$\mathbf{u}(t) = \mathbf{u}_h(t) + \mathbf{u}_p(t), \tag{12.3}$$

https://doi.org/10.1515/9783111598055-015

where $\mathbf{u}_h(t)$ is the solution to the homogeneous initial value problem

$$\frac{d\mathbf{u}_h}{dt} = \mathbf{A}(t)\mathbf{u}_h \tag{12.4}$$

$$\mathbf{u}_h(@\,t = 0) = \mathbf{u}^0 \tag{12.5}$$

and $\mathbf{u}_p(t)$ is a particular solution of the inhomogeneous system

$$\frac{d\mathbf{u}_p}{dt} = \mathbf{A}(t)\mathbf{u}_p + \mathbf{b}(t) \tag{12.6}$$

$$\mathbf{u}_p(@\,t = 0) = \mathbf{0}. \tag{12.7}$$

We now consider the homogeneous IVP defined by equation (12.4). The following properties may be easily established:

1. The set of all solutions to equation (12.4) forms a vector space.
2. There are n linearly independent solutions and every solution of equation (12.4) is expressible as a linear combination of these n solutions.
3. Let $\{\boldsymbol{\alpha}_1, \boldsymbol{\alpha}_2, \dots, \boldsymbol{\alpha}_n\}$ be any set of n linearly independent constant vectors in $\mathbb{R}^n/\mathbb{C}^n$ and $\mathbf{u}_j(t)$ be a solution of (12.4) satisfying the initial condition $\mathbf{u}_j(0) = \boldsymbol{\alpha}_j$. Then the set $\{\mathbf{u}_1(t), \mathbf{u}_2(t), \dots, \mathbf{u}_n(t)\}$ is linearly independent and forms a basis for the solution space.

Definition. A set of n linearly independent solutions of equation (12.4) is called a *fundamental set of solutions*. The matrix $\mathbf{U}(t) = [\mathbf{u}_1(t)\ \ \mathbf{u}_2(t) \dots \mathbf{u}_n(t)]$ whose columns form a fundamental set is called a *fundamental matrix*.

If $\mathbf{U}(t)$ is a fundamental matrix for equation (12.4), then every solution is of the form

$$\mathbf{u}_h(t) = \mathbf{U}(t)\mathbf{c} \tag{12.8}$$

where $\mathbf{c}$ is some constant vector.

Lemma.

(a) *The fundamental matrix satisfies the matrix differential equation:*

$$\frac{d\mathbf{U}}{dt} = \mathbf{A}(t)\mathbf{U}(t) \tag{12.9}$$

(b)

$$\det \mathbf{U}(t) = \det \mathbf{U}(\tau) \exp\left\{\int_\tau^t \operatorname{tr} \mathbf{A}(s)\, ds\right\} \tag{12.10}$$

and $\mathbf{U}(t)$ is nonsingular at every point in $[0, a]$. [Here, $\operatorname{tr} \mathbf{A}$ stands for the trace of the matrix $\mathbf{A}$].

(c) *If $\mathbf{B}$ is any constant nonsingular matrix, then $\mathbf{V}(\mathbf{t}) = \mathbf{U}(\mathbf{t})\mathbf{B}$ is also a fundamental matrix and every fundamental matrix may be written in this form.*

Note that the specific solution of equations (12.4) and (12.5) is given by

$$\mathbf{u}_h(t) = \mathbf{U}(t)\mathbf{U}(0)^{-1}\mathbf{u}^0. \tag{12.11}$$

The statements of the above lemma may be verified by direct calculation. We now consider the inhomogeneous equation (12.6) with homogeneous initial condition (equation (12.7)).

Theorem. *If* $\mathbf{U}(t)$ *is a fundamental matrix of the homogeneous system, a particular solution of equations* (12.6) *and* (12.7) *is given by*

$$\mathbf{u}_p(t) = \mathbf{U}(t)\int_0^t \mathbf{U}(s)^{-1}\mathbf{b}(s)\,ds \tag{12.12}$$

Proof. We use the variation of parameter technique. Writing

$$\mathbf{u}_p(t) = \mathbf{U}(t)\mathbf{h}(t) \tag{12.13}$$

⇒

$$\begin{aligned}\mathbf{u}_p' &= \mathbf{U}'(t)\mathbf{h}(t) + \mathbf{U}(t)\mathbf{h}'(t)\\ &= \mathbf{A}(t)\mathbf{U}(t)\mathbf{h}(t) + \mathbf{U}(t)\mathbf{h}'(t)\\ &= \mathbf{A}(t)\mathbf{U}(t)\mathbf{h}(t) + \mathbf{b}(t) \quad \text{(from Eq. (12.6))}\end{aligned}$$

⇒

$$\begin{aligned}\mathbf{U}(t)\mathbf{h}'(t) &= \mathbf{b}(t)\\ \mathbf{h}'(t) &= \mathbf{U}(t)^{-1}\mathbf{b}(t)\end{aligned}$$

Integrating and using the initial condition gives

$$\mathbf{h}(t) = \int_0^t \mathbf{U}(s)^{-1}\mathbf{b}(s)\,ds$$

and $\mathbf{u}_p(t)$ is given by equation (12.12).

Thus, the general solution of equation (12.1) is given by

$$\mathbf{u}(t) = \mathbf{U}(t)\mathbf{c} + \mathbf{U}(t)\int_0^t \mathbf{U}(s)^{-1}\mathbf{b}(s)\,ds \tag{12.14}$$

where $\mathbf{c}$ is a constant vector. The unique solution of equations (12.1) and (12.2) is given by

$$\mathbf{u}(t) = \mathbf{U}(t)\mathbf{U}(0)^{-1}\mathbf{u}^0 + \mathbf{U}(t)\int_0^t \mathbf{U}(s)^{-1}\mathbf{b}(s)\,ds. \tag{12.15}$$

We will be using equation (12.15) in many applications. □

12.2 The n-th order initial value problem

Consider the n-th order linear differential operator defined by

$$Lu = p_0(t)\frac{d^n u}{dt^n} + p_1(t)\frac{d^{n-1}u}{dt^{n-1}} + \cdots + p_{n-1}(t)\frac{du}{dt} + p_n(t)u, \quad 0 < t < a, \tag{12.16}$$

where $p_i(t), i = 0, 1, \ldots, n$ are real (or complex) valued functions of a real variable t. We assume that L is a regular differential operator, i. e., $p_o(t) \neq 0$ for $0 \leq t \leq a$ and $p_i(t) \in C[0, a]$. Suppose that $u(t) \in C^n[0, a]$, the class of n-times differentiable functions defined on the interval $[0, a]$. Then the most general form of a linear n-th order initial value problem is given by

$$Lu = f(t), \quad 0 < t < a \tag{12.17}$$

$$u(0) = \alpha_0, \quad u'(0) = \alpha_1, \quad \ldots, \quad u^{[n-1]}(0) = \alpha_{n-1}. \tag{12.18}$$

Defining

$$\begin{aligned} u_1(t) &= u(t) \\ u_2(t) &= \frac{du_1}{dt} = \frac{du}{dt} \\ &\cdot \\ &\cdot \\ u_n(t) &= \frac{du_{n-1}}{dt} = \frac{d^{n-1}u}{dt^{n-1}} \end{aligned} \tag{12.19}$$

Equations (12.17) and (12.18) may be written in the vector form of equations (12.1) and (12.2) with

$$\mathbf{u}^0 = \boldsymbol{\alpha}$$

$$\mathbf{u} = \begin{pmatrix} u(t) \\ u'(t) \\ \cdot \\ \cdot \\ \cdot \\ u^{[n-1]}(t) \end{pmatrix}, \quad \mathbf{A}(t) = \begin{pmatrix} 0 & 1 & 0 & 0 & . & . & . & 0 \\ 0 & 0 & 1 & 0 & . & . & . & 0 \\ 0 & 0 & 0 & 1 & . & . & . & 0 \\ . & . & . & . & . & . & . & . \\ -\frac{p_n(t)}{p_0(t)} & -\frac{p_{n-1}(t)}{p_0(t)} & . & . & . & . & . & -\frac{p_1(t)}{p_0(t)} \end{pmatrix}$$

$$\mathbf{b}(t) = \frac{f(t)}{p_0(t)} \begin{pmatrix} 0 \\ 0 \\ \cdot \\ \cdot \\ 0 \\ 1 \end{pmatrix} = \frac{f(t)}{p_0(t)}\mathbf{e}_n \tag{12.20}$$

Thus, the n-th order IVP is a special case of the vector initial value problem. In fact, every linear IVP (e. g., coupled higher-order scalar equations) may be written in the form given by equation (12.1).

The solutions of $Lu = 0$ are precisely those vectors (functions) whose image under L is zero, i. e., the kernel of L. It is easily shown that $\ker L$ is of dimension n and a basis for $\ker L$ consists of n-linearly independent solutions.

Definition. Any vector whose components form a basis for $\ker L$ is called a *fundamental vector* for $Lu = 0$. If $\psi^T = [\psi_1(t) \ \ \psi_2(t) \ \ \dots \ \ \psi_n(t)]$ is a fundamental vector, the general solution of $Lu = 0$ is of the form $u = \psi^T \mathbf{c}$, where $\mathbf{c}$ is a constant vector.

The fundamental matrix of the companion vector equation is called the *Wronskian matrix* of equation (12.17). The Wronskian matrix is denoted by

$$\mathbf{K}(\psi(t)) = \begin{pmatrix} \psi_1(t) & . & . & . & . & \psi_n(t) \\ \psi_1'(t) & . & . & . & . & \psi_n'(t) \\ . & . & . & . & . & . \\ . & . & . & . & . & . \\ . & . & . & . & . & . \\ \psi_1^{[n-1]}(t) & . & . & . & . & \psi_n^{[n-1]}(t) \end{pmatrix}$$

Similarly, the *Wronskian vector* of a solution $\psi(t)$ is defined by

$$\mathbf{k}(\psi(t)) = \begin{pmatrix} \psi(t) \\ \psi'(t) \\ . \\ . \\ . \\ \psi^{[n-1]}(t) \end{pmatrix}.$$

Theorem. *Given a linear n-th order homogeneous equation,*

$$Lu = \sum_{j=0}^{n} p_j(t) \frac{d^{n-j}u}{dt^{n-j}} = 0, \tag{12.21}$$

let

$$\mathbf{u}_1 = \mathbf{k}(\psi_1(t)) = \begin{pmatrix} \psi_1 \\ \frac{d\psi_1}{dt} \\ . \\ . \\ \frac{d^{n-1}\psi_1}{dt^{n-1}} \end{pmatrix}, \quad \dots, \quad \mathbf{u}_n = \mathbf{k}(\psi_n(t)) = \begin{pmatrix} \psi_n \\ \frac{d\psi_n}{dt} \\ . \\ . \\ \frac{d^{n-1}\psi_n}{dt^{n-1}} \end{pmatrix}$$

be a set of solutions to the associated vector equation, i. e., $\psi_1, \psi_2, \dots, \psi_n$ are solutions to equation (12.21). Then a necessary and sufficient condition that these solutions be linearly independent is that the Wronskian (defined as the determinant of the Wronskian matrix) be nonzero.

Proof. Suppose that $\psi_1, \psi_2, \dots, \psi_n$ are linearly independent, i. e., the only solution of

$$c_1\psi_1(t) + c_2\psi_2(t) + \cdots + c_n\psi_n(t) = 0$$

is $c_1 = c_2 = \cdots = c_n = 0$. Differentiating on both sides i times ($i = 1, 2, 3, \dots, n-1$), we get

$$\begin{gathered} c_1\psi_1'(t) + c_2\psi_2'(t) + \cdots + c_n\psi_n'(t) = 0 \\ \vdots \\ c_1\psi_1^{[n-1]} + c_2\psi_2^{[n-1]} + \cdots + c_n\psi_n^{[n-1]} = 0 \end{gathered}$$

Since the only solution to the above system of n equations is the trivial one, we have

$$W(t) = \begin{vmatrix} \psi_1(t) & \psi_2(t) & \cdots & \psi_n(t) \\ \psi_1'(t) & \psi_2'(t) & \cdots & \psi_n'(t) \\ \vdots & & & \\ \psi_1^{[n-1]}(t) & \psi_2^{[n-1]}(t) & \cdots & \psi_n^{[n-1]}(t) \end{vmatrix} \neq 0$$

Conversely, if $W(t) = 0$, then there is a nontrivial solution and $\psi_1, \psi_2, \dots, \psi_n$ are linearly dependent. □

Theorem. *The Wronskian $W(t)$ is either identically zero or is never zero, i. e., it never changes sign.*

Proof. By definition,

$$W(t) = \begin{vmatrix} \psi_1(t) & \psi_2(t) & \cdots & \psi_n(t) \\ \psi_1'(t) & \psi_2'(t) & \cdots & \psi_n'(t) \\ \psi_1''(t) & \psi_2''(t) & & \psi_n''(t) \\ \vdots & & & \\ \psi_1^{[n-1]}(t) & \psi_2^{[n-1]}(t) & \cdots & \psi_n^{[n-1]}(t) \end{vmatrix}$$

Using the rule for differentiation of a determinant, we get

$$\frac{dW(t)}{dt} = \begin{vmatrix} \psi_1' & \psi_2' & \cdots & \psi_n' \\ \psi_1' & \psi_2' & \cdots & \psi_n' \\ \psi_1'' & \psi_2'' & & \psi_n'' \\ \vdots & & & \vdots \\ \psi_1^{[n-1]} & \psi_2^{[n-1]} & \cdots & \psi_n^{[n-1]} \end{vmatrix}$$

$$+ \begin{vmatrix} \psi_1 & \psi_2 & \cdots & \psi_n \\ \psi_1'' & \psi_2'' & \cdots & \psi_n'' \\ \vdots & & & \vdots \\ \psi_1^{[n-1]} & \psi_2^{[n-1]} & \cdots & \psi_n^{[n-1]} \end{vmatrix} + \cdots + \begin{vmatrix} \psi_1 & \psi_2 & \cdots & \psi_n \\ \psi_1' & \psi_2' & \cdots & \psi_n' \\ \vdots & & & \vdots \\ \psi_1^{[n]} & \psi_2^{[n]} & \cdots & \psi_n^{[n]} \end{vmatrix}$$

$$= 0 + 0 + \cdots + 0 + \begin{vmatrix} \psi_1(t) & \psi_2(t) & \cdots & \psi_n(t) \\ \psi_1'(t) & \psi_2'(t) & \cdots & \psi_n'(t) \\ \psi_1''(t) & \psi_2''(t) & & \psi_n''(t) \\ \psi_1^{[n-2]}. & \psi_2^{[n-2]} & & \psi_n^{[n-2]} \\ \psi_1^{[n]}(t) & \psi_2^{[n]}(t) & \cdots & \psi_n^{[n]}(t) \end{vmatrix}$$

Now,

$$L\psi_j = 0$$

⇒

$$\psi_j^{[n]} = -\frac{p_1}{p_0}\psi_j^{[n-1]} - \frac{p_2}{p_0}\psi_j^{[n-2]} - \cdots - \frac{p_{n-1}}{p_0}\psi_j' - \frac{p_n}{p_0}\psi_j$$

Substitute this in the last row of the above determinant and perform the following $(n-1)$ row operations:

(i) Multiply row 1 by $\frac{p_n}{p_0}$ and add to the last row

(ii) Multiply row 2 by p_{n-1}/p_0 and add to the last row

⋮

Multiply row $(n-1)$ by p_2/p_0 and add to the last row

$\Rightarrow$

$$\frac{dW}{dt} = \begin{vmatrix} \psi_1 & \psi_2 & \cdots & \psi_n \\ \psi_1' & \psi_2' & \cdots & \psi_n' \\ \vdots & & & \vdots \\ -\frac{p_1}{p_0}\psi_1^{[n-1]} & -\frac{p_1}{p_0}\psi_2^{[n-1]} & \cdots & -\frac{p_1}{p_0}\psi_n^{[n-1]} \end{vmatrix} = -\frac{p_1}{p_0}W(t)$$

$\Rightarrow$

$$W(t) = W(t_0)\exp\left\{-\int_{t_0}^{t}\frac{p_1(s)}{p_0(s)}\,ds\right\}$$

If $W(t_0) = 0 \Rightarrow W(t) \equiv 0$ and if $W(t_0) \neq 0$, then $W(t)$ never vanishes
$\therefore$ The result. □

12.2.1 The *n*-th order inhomogeneous equation

Consider the inhomogeneous equation

$$Lu = f(t), \tag{12.22}$$

or its companion matrix equation

$$\frac{d\mathbf{u}}{dt} = \mathbf{A}(t)\mathbf{u} + \mathbf{b}(t). \tag{12.23}$$

Theorem. *Let L be regular and $f(t) \in C[t_0, b]$, $t_0 < b < \infty$. Then the solution of the IVP*

$$Lu = f(t)$$
$$\mathbf{k}(u(t_0)) = \boldsymbol{a}$$

is given by

$$u(t) = [\psi(t)]^T\mathbf{c} + [\psi(t)]^T\int_{t_0}^{t}\mathbf{K}(\psi(s))^{-1}\mathbf{e}_n\frac{f(s)}{p_0(s)}\,ds \tag{12.24}$$

where

$$\mathbf{e}_n = \begin{pmatrix} 0 \\ 0 \\ \cdot \\ \cdot \\ 0 \\ 1 \end{pmatrix}, \quad \psi(t) = \begin{pmatrix} \psi_1(t) \\ \psi_2(t) \\ \cdot \\ \cdot \\ \psi_n(t) \end{pmatrix}$$

is a fundamental vector of the homogeneous system and $\mathbf{K}(\psi(t))$ is the Wronskian matrix. The constant vector $\mathbf{c}$ is determined from the algebraic equations

$$\begin{aligned} a_0 &= [\psi(t_0)]^T \mathbf{c} \\ a_1 &= [\psi'(t_0)]^T \mathbf{c} \\ &\cdot \\ &\cdot \\ a_{n-1} &= [\psi^{[n-1]}(t_0)]^T \mathbf{c} \end{aligned}$$

or

$$\boldsymbol{a} = \mathbf{K}[\psi(t_0)]\mathbf{c} \quad or \quad \mathbf{c} = \mathbf{K}[\psi(t_0)]^{-1}\boldsymbol{a}.$$

Proof. The proof of equation (12.24) follows directly from that for the vector equation $\frac{d\mathbf{u}}{dt} = \mathbf{A}(t)\mathbf{u} + \mathbf{b}(t)$, whose solution is given by

$$\mathbf{u}(t) = \mathbf{U}(t)\mathbf{c} + \mathbf{U}(t) \int_{t_0}^{t} \mathbf{U}(s)^{-1}\mathbf{b}(s)\, ds.$$

Now, let

$$\mathbf{b}(s) = \begin{pmatrix} 0 \\ 0 \\ \cdot \\ \cdot \\ \frac{f(s)}{p_0(s)} \end{pmatrix} = \mathbf{e}_n \frac{f(s)}{p_0(s)}$$

and use the fact that the fundamental matrix for the scalar problem is the Wronskian matrix, i. e.,

$$\mathbf{U}(s) = \mathbf{K}(\psi(s)),$$

and take only the first component of the solution $\mathbf{u}(t)$.

A special case of equation (12.24) that is of interest in many applications is when $n = 2$ for which it may be written as

$$u(t) = c_1\psi_1(t) + c_2\psi_2(t) + \int_{t_0}^{t} \frac{\psi_1(s)\psi_2(t) - \psi_1(t)\psi_2(s)}{W(s)} \frac{f(s)}{p_0(s)}\, ds \tag{12.25}$$

where $W(s) = \psi_1(s)\psi_2'(s) - \psi_1'(s)\psi_2(s)$ is the Wronskian. □

Example 12.1. Consider the second-order equation

$$u'' + u = 2\sin t.$$

The two linearly independent solutions of the homogeneous equation are given by

$$\psi_1 = \sin t, \quad \psi_2 = \cos t$$

$$\mathbf{K}(\psi) = \begin{pmatrix} \sin t & \cos t \\ \cos t & -\sin t \end{pmatrix}, \quad W(t) = -1 \neq 0$$

∴

$$\begin{aligned} u_p(t) &= \text{particular solution} \\ &= \int_0^t (\sin t \cos s - \sin s \cos t) 2 \sin s\, ds \\ &= \sin t - t\cos t \end{aligned}$$

Thus, the general solution is given by

$$u(t) = c_1 \sin t + c_2 \cos t + \sin t - t\cos t.$$

12.3 Linear IVPs with constant coefficients

In this section, we consider some special cases of the general IVP.

(a) The first special case is of the vector IVP where the matrix $\mathbf{A}(t)$ is a constant matrix and the forcing vector $\mathbf{b}(t)$ is also a constant, i. e.,

$$\frac{d\mathbf{u}}{dt} = \mathbf{A}\mathbf{u} + \mathbf{b} \tag{12.26}$$

$$\mathbf{u} = \mathbf{u}^0 \ @\ t = 0 \tag{12.27}$$

In this case, $\mathbf{U} = \mathbf{e}^{\mathbf{A}t}$ is a fundamental matrix. Thus, $\mathbf{u}_h = e^{\mathbf{A}\mathbf{t}}\mathbf{c}$ and a particular solution is given by

$$\mathbf{u}_p = \mathbf{U}(t) \int_0^t \mathbf{U}(s)^{-1}\mathbf{b}(s)\, ds$$

$$= e^{\mathbf{A}t}\int_0^t e^{-\mathbf{A}s}\mathbf{b}\,ds = e^{\mathbf{A}t}[-\mathbf{A}^{-1}e^{-\mathbf{A}t}\mathbf{b} + \mathbf{A}^{-1}\mathbf{b}] \quad \text{(if } \mathbf{A} \text{ is invertible)}$$

Simplifying, we get

$$\mathbf{u}_p(t) = -\mathbf{A}^{-1}\mathbf{b} + \mathbf{A}^{-1}\mathbf{e}^{\mathbf{A}t}\mathbf{b} \tag{12.28}$$

The general solution in given by

$$\mathbf{u} = e^{\mathbf{A}t}\mathbf{c} + \mathbf{A}^{-1}e^{\mathbf{A}t}\mathbf{b} - \mathbf{A}^{-1}\mathbf{b}$$

Applying the initial condition,

$$\mathbf{u} = \mathbf{u}^0 \;@\; t = 0 \Rightarrow \mathbf{c} = \mathbf{u}^0 + \mathbf{0}$$

∴

$$\mathbf{u} = e^{\mathbf{A}t}\mathbf{u}^0 + \mathbf{A}^{-1}e^{\mathbf{A}t}\mathbf{b} - \mathbf{A}^{-1}\mathbf{b} \tag{12.29}$$

Another method: The solution to equation (12.26) may also be obtained by determining the steady-state solution and subtracting it to obtain a homogeneous equation. Let $\mathbf{z} = \mathbf{u} - \mathbf{u}_s$, $\mathbf{u}_s = -\mathbf{A}^{-1}\mathbf{b} \Rightarrow$

$$\begin{aligned}
\frac{d\mathbf{z}}{dt} &= \mathbf{A}\mathbf{z} \\
\mathbf{z} &= \mathbf{z}^0 = \mathbf{u}^0 - \mathbf{u}_s \;@\; t = 0 \\
\mathbf{z} &= e^{\mathbf{A}t}\mathbf{z}^0 \Rightarrow \mathbf{u} = -\mathbf{A}^{-1}\mathbf{b} + \mathbf{e}^{\mathbf{A}t}(\mathbf{u}^0 + \mathbf{A}^{-1}\mathbf{b}) \\
\mathbf{u} &= \mathbf{e}^{\mathbf{A}t}\mathbf{u}^0 + \mathbf{e}^{\mathbf{A}t}\mathbf{A}^{-1}\mathbf{b} - \mathbf{A}^{-1}\mathbf{b}
\end{aligned}$$

Thus, for the case of constant coefficients and constant forcing vector, the solution is given by

$$\mathbf{u}(t) = \mathbf{e}^{\mathbf{A}t}[\mathbf{u}^0 + \mathbf{A}^{-1}\mathbf{b}] - \mathbf{A}^{-1}\mathbf{b}. \tag{12.30}$$

[Remark: Since $\mathbf{A}^{-1}$ and $e^{\mathbf{A}t}$ commute, the solution given by equations (12.29) and (12.30) are identical].

(b) The second special case is that of scalar n-th order IVP with constant coefficients:

$$\begin{aligned}
p_0u^{[n]} + p_1u^{[n-1]} + \cdots + p_{n-1}u' + p_nu &= f(t). \\
u(0) &= \alpha_0, \\
u'(0) &= \alpha_1, \\
&\;\;\vdots \\
u^{[n-1]}(0) &= \alpha_{n-1}.
\end{aligned}$$

Let

$$u = e^{\lambda t}$$

⇒ The characteristic equation is given by

$$p_0\lambda^n + p_1\lambda^{n-1} + \cdots + p_{n-1}\lambda + p_n = 0$$

Denote the roots of the characteristic equation by

$$\lambda_1, \lambda_2, \ldots, \lambda_n,$$

and for simplicity assume that they are all distinct. Then

$$\psi_j(t) = e^{\lambda_j t}, \quad j = 1, 2, 3, \ldots, n$$

are linearly independent solutions. Therefore,

$$u_h = \sum_{j=1}^{n} c_j e^{\lambda_j t}$$

$$u_p(t) = \psi(t)^T \int_0^t [\mathbf{K}(\psi(s))]^{-1} \mathbf{e}_n \frac{f(s)}{p_0(s)}\, ds$$

where

$$\psi(t)^T = [\ \psi_1(t) \ \ . \ \ \psi_n(t)\ ]$$

$$\mathbf{K}(\psi(t)) = \begin{bmatrix} \psi_1 & \psi_2 & \cdot\ \cdot\ \cdot & \psi_n \\ \psi_1' & \psi_2' & \cdot\ \cdot\ \cdot & \psi_n' \\ \psi_1'' & \psi_2'' & & \psi_n'' \\ \cdot & & & \\ \cdot & & & \\ \psi_1^{[n-1]} & \psi_2^{[n-1]} & \cdot\ \cdot\ \cdot & \psi_n^{[n-1]} \end{bmatrix}$$

$$= \begin{bmatrix} e^{\lambda_1 t} & e^{\lambda_2 t} & \cdot\ \cdot\ \cdot & e^{\lambda_n t} \\ \lambda_1 e^{\lambda_1 t} & \lambda_2 e^{\lambda_2 t} & \cdot\ \cdot\ \cdot & \lambda_n e^{\lambda_n t} \\ \cdot & \cdot & & \cdot \\ \cdot & & & \\ \cdot & & & \\ \lambda_1^{n-1} e^{\lambda_1 t} & \lambda_2^{n-1} e^{\lambda_2 t} & \cdot\ \cdot\ \cdot & \lambda_n^{n-1} e^{\lambda_n t} \end{bmatrix}$$

Evaluation of $\mathbf{K}(\psi(t))$ at $t = 0$ leads to the Van der Monde matrix, whose inverse can be expressed analytically for the case of distinct eigenvalues.

Many examples illustrating the application of the above theory to first- and second-order scalar as well as vector equations are given as problems below and in Chapter 14.

Problems

1. The vibration of the spring-mass system (with identical springs and masses) may be described by the equations

$$\frac{m}{k}\frac{d^2u_1}{dt^2} = -2u_1 + u_2$$
$$\frac{m}{k}\frac{d^2u_2}{dt^2} = u_1 - 2u_2$$

where u_1 and u_2 are the displacements of the masses from their equilibrium positions. (a) Cast the above equations in dimensionless form, (b) Determine the natural frequencies of vibration and the modes (eigenvectors) of vibration and (c) What will be the form of the equations if damping is included? Write the equations in vector/matrix form.

2. Show that the solution of the inhomogeneous system

$$\frac{d\mathbf{u}}{dt} = \mathbf{A}\mathbf{u} + \mathbf{f}(t), \quad \mathbf{u}(t = 0) = \mathbf{u}_0,$$

where $\mathbf{A}$ is a constant coefficient matrix, is given by

$$\mathbf{u} = \mathbf{e}^{\mathbf{A}t}\mathbf{u}_0 + \int_0^t \mathbf{e}^{\mathbf{A}(t-t')}\mathbf{f}(t')\,dt'.$$

3. Determine a formula similar to that given in problem (2) for the solution of the inhomogeneous system

$$\frac{d^2\mathbf{u}}{dt^2} = -\mathbf{A}\mathbf{u} + \mathbf{f}(t), \quad \mathbf{u}(t = 0) = \mathbf{u}_0, \quad \frac{d\mathbf{u}}{dt}(t = 0) = \mathbf{v}_0.$$

4. Consider the flow system shown in Figure 12.1. Assume that each tank is well mixed and species A enters tank 1 at a concentration of $c_{\text{in}}(t)$ and leaves at $c_1(t)$. Assume further that $V_{R1} = 1\,\text{m}^3$, $V_{R2} = \frac{2}{3}\,\text{m}^3$ and $q_1 = q_2 = 2\,\text{m}^3/\text{min}$. (a) Formulate the differential equations describing the transient behavior of the system and put them in vector/matrix form. (b) Determine the response of the system (i. e., how the exit concentration $c_1(t)$ varies with time) for a unit step input ($c_{\text{in}}(t) = 1$ for $t > 0$ and 0 for $t < 0$). Assume that no A is present initially in either tank.
5. Solve problem (4) by converting the model to a scalar initial value problem for $c_2(t)$.

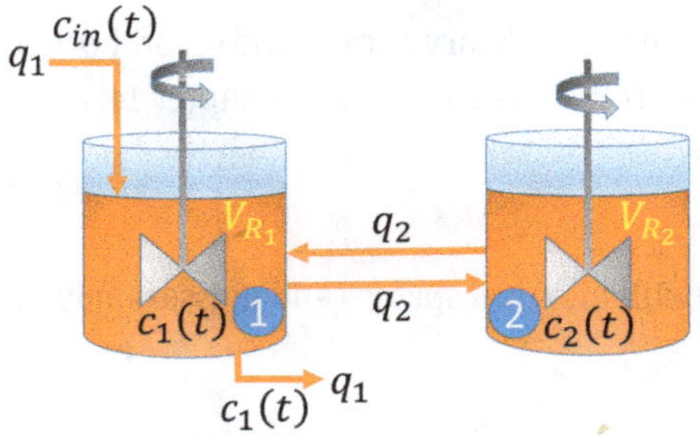

Figure 12.1: Schematic diagram of interacting tanks.

6. Consider the linear system

$$\frac{d\mathbf{u}}{dt} = \begin{pmatrix} 0 & 1 \\ -2 & 3 \end{pmatrix}\mathbf{u} + \begin{pmatrix} 1 \\ 1 \end{pmatrix}.$$

Determine (a) a fundamental matrix for the homogeneous system (b) a particular solution to the inhomogeneous system (c) the general form of the solution (d) the solution to the initial value problem $\mathbf{u}(0) = \binom{2}{1}$.

7. (a) Determine a fundamental matrix for the linear system

$$\frac{d\mathbf{u}}{dt} = \begin{pmatrix} 1 & 1 \\ 0 & 1 \end{pmatrix}\mathbf{u}.$$

(b) Determine the vector differential equations for which the following are fundamental matrices

$$\text{(i)}\quad \mathbf{U}(t) = \begin{pmatrix} e^t & te^t \\ e^t & (1+t)e^t \end{pmatrix} \qquad \text{(ii)}\quad \mathbf{U}(t) = \begin{pmatrix} 1 & t & t^2 \\ 0 & 1 & t \\ 0 & 0 & 1 \end{pmatrix}$$

(c) Verify that $\psi_1(t) = t^3$, $\psi_2(t) = t^{-2}$ are linearly independent solutionsof

$$Lu = u'' - 6t^{-2}u = 0.$$

Find a particular solution to the inhomogeneous equation $Lu = t\ln(t)$.

8. (a) Derive a formula for the solution of the initial value problem

$$Lu = f(t); \quad \mathbf{k}(u(0)) = \mathbf{b},$$

where L is an n-th order linear differential operator with constant coefficients, $\mathbf{k}$ is the Wronskian vector and $\mathbf{b}$ is a constant vector.

(b) Use the above formula to solve the initial value problem for forced oscillation of a second-order system

$$u'' + cu' + \alpha^2 u = a\cos\omega t$$

$$\mathbf{k}(u(0)) = \begin{pmatrix} 0 \\ 0 \end{pmatrix}$$

Here, c, α and a are positive constants. Plot the response for the overdamped ($c = 6, \alpha^2 = 5, \omega = 1$), critically damped ($c = 2, \alpha^2 = 1, \omega = 1$) and underdamped cases ($c = \alpha^2 = 2, \omega = 1$). Also, plot the amplitude of the asymptotic response as a function of the forcing frequency for a fixed α^2(say $\alpha^2 = 4$) and varying values of $c(\geq 0)$.

13 Linear systems with periodic coefficients

The theory of linear differential equations with periodic coefficients is encountered in applications such as the analysis of transport phenomena with spatially periodic properties, in the development of time averaged models of periodically forced systems, in the determination of the stability of periodic solutions of nonlinear systems, and so forth. In this chapter, we outline the theory briefly.

13.1 Scalar equation with a periodic coefficient

Consider the scalar initial value problem

$$\frac{du}{dt} = a(t)u, \quad u = u^0 \ @\ t = 0 \tag{13.1}$$

with the coefficient $a(t)$ periodic with period $T > 0$, i.e.,

$$a(t + T) = a(t).$$

Integrating equation (13.1), we get

$$\ln u = \int_0^t a(s)\, ds + c$$

$$t = 0, \quad u = u^0 \Rightarrow c = \ln u^0$$

⇒ The solution is given by

$$\ln\left(\frac{u}{u^0}\right) = \int_0^t a(s)\, ds. \tag{13.2}$$

To investigate the nature of this solution when $a(s)$ is periodic, we consider the different time intervals:

(i) $0 < t \leq T$

$$\frac{u(t)}{u^0} = \exp\left\{\int_0^t a(s)\, ds\right\} \tag{13.3}$$

(ii) $T < t < 2T$

$$\frac{u(t)}{u^0} = \exp\left\{\int_0^T a(s)\, ds + \int_T^t a(s)\, ds\right\}$$

https://doi.org/10.1515/9783111598055-016

$$= \exp\left\{\int_0^T a(s)\,ds\right\}.\exp\left\{\int_T^t a(s)\,ds\right\}$$

Let

$$m = \exp\left\{\int_0^T a(s)\,ds\right\} = \exp\left\{\int_{(k-1)T}^{kT} a(s)\,ds\right\}, \quad k = 1, 2, \ldots.$$

Then we get

$$\frac{u(t)}{u^0} = m\exp\left\{\int_T^t a(s)\,ds\right\}$$

We now make the substitution $\hat{s} = s - T$ in the integral inside the exponent to obtain

$$\frac{u(t)}{u^0} = m\exp\left\{\int_0^{t-T} a(\hat{s} + T)d\hat{s}\right\}$$

but $a(\hat{s} + T) = a(\hat{s}) \Rightarrow$

$$\frac{u(t)}{u^0} = m\exp\left\{\int_0^{t-T} a(\hat{s})d\hat{s}\right\}; \quad T < t < 2T \tag{13.4}$$

(iii) If $2T < t < 3T$, following the above procedure, it is easily seen that

$$\frac{u(t)}{u^0} = m^2\exp\left\{\int_0^{t-2T} a(s)\,ds\right\}; \quad 2T < t < 3T, \quad \text{etc.}$$

Thus, we can compute $u(t)$ for all t if we know $u(t)$ for $0 < t < T$.

Let

$$m = e^{\rho T} = \exp\left\{\int_0^T a(s)\,ds\right\}; \quad \rho = \frac{1}{T}\int_0^T a(s)\,ds$$

Claim: The solution to equation (13.1) may be expressed as

$$u(t) = e^{\rho t}p(t)$$

where

$$p(t+T) = p(t) = e^{-\rho t} \exp\left\{ \int_0^t a(s)\,ds \right\} u^0. \tag{13.5}$$

$$\begin{aligned} p(t+T) &= e^{-\rho(t+T)} \exp\left\{ \int_0^{t+T} a(s)\,ds \right\} u^0 \\ &= e^{-\rho t} . e^{-\rho T} . \exp\left\{ \int_0^T a(s)\,ds \right\} \exp\left\{ \int_T^{t+T} a(s)\,ds \right\} u^0 \\ &= e^{-\rho t} .1. \exp\left\{ \int_0^t a(s)\,ds \right\} u^0 \\ &= p(t) \end{aligned}$$

Thus, when $a(t)$ is periodic, the solution of equation (13.1) is of the form

$$u(t) = e^{\rho t} p(t)$$

where $p(t)$ is a periodic function. The solution is periodic iff $\rho = 0$ or $m = 1$
$\Rightarrow$

$$\exp\left\{ \int_0^T a(s)\,ds \right\} = 1$$

$\Rightarrow$

$$\int_0^T a(s)\,ds = 0.$$

Example 13.1. We consider the linear initial value problem

$$\frac{du}{dt} = (\cos t)u; \quad u = u^0.$$

Here,

$$\begin{aligned} a(s) &= \cos s \\ T &= 2\pi; \quad \int_0^{2\pi} \cos s\,ds = -\sin s|_0^{2\pi} = 0 \Longrightarrow m = 1 \text{ or } \rho = 0. \\ p(t) &= \exp\left\{ \int_0^t \cos s\,ds \right\} u^0 \\ &= \exp\{-\sin s|_0^t\} u^0 \\ &= \exp\{-\sin t\} u^0 \end{aligned}$$

Thus, the solution is given by

$$u(t) = p(t) = \exp\{-\sin t\}u^0.$$

A plot of this solution is shown in Figure 13.1 for $u^0 = 1$.

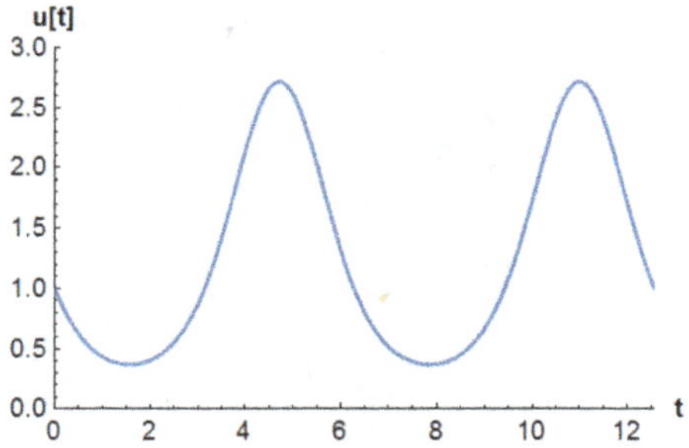

Figure 13.1: Plot of the periodic solution of the scalar linear equation with $\rho = 0$.

13.2 Vector equation with periodic coefficient matrix

We now consider the vector equation

$$\frac{d\mathbf{u}}{dt} = \mathbf{A}(t)\mathbf{u} \tag{13.6}$$

$$\mathbf{A}(t+T) = \mathbf{A}(t) \tag{13.7}$$

Here, **A** is a $n \times n$ matrix of real or complex valued continuous functions of real variable t. T is the period of $\mathbf{A}(t)$. The following theorem may be stated for the system defined by equations (13.6) and (13.7).

Theorem. *If* $\mathbf{U}(t)$ *is a fundamental matrix of equations* (13.6) *and* (13.7) *so is* $\mathbf{V}(t)$ *where*

$$\mathbf{V}(t) = \mathbf{U}(t+T)$$

Corresponding to every such $\mathbf{U}(t)$, $\exists$ *a nonsingular matrix* $\mathbf{P}(t)$, *which is periodic with period* T *and a constant* $n \times n$ *matrix* $\mathbf{C}$ *such that*

$$\mathbf{U}(t) = \mathbf{P}(t)e^{t\mathbf{C}}.$$

Proof. Since

$$\begin{aligned}\frac{d\mathbf{U}}{dt} = \mathbf{A}(t)\mathbf{U} \Rightarrow \mathbf{V}'(t) &= \mathbf{U}'(t+T).\mathbf{I}\\ &= \mathbf{A}(t+T)\mathbf{U}(t+T)\\ &= \mathbf{A}(t)\mathbf{V}(t)\\ \det \mathbf{V}(t) &= \det\{\mathbf{U}(t+T)\} \neq 0\end{aligned}$$

∴ $\mathbf{V}(t)$ is also a fundamental matrix.
∴ ∃ a constant nonsingular matrix $\mathbf{M}$ so that

$$\mathbf{U}(t+T) = \mathbf{U}(t)\mathbf{M}.$$

Since $\mathbf{M}$ is nonsingular, ∃ a constant matrix $\mathbf{C}$ such that

$$\mathbf{M} = e^{T\mathbf{C}} \quad \left(\text{i. e., } \ln \mathbf{M} = T\mathbf{C} \text{ or } \mathbf{C} = \frac{1}{T}\ln \mathbf{M}\right)$$

∴

$$\begin{aligned}\mathbf{U}(t+T) &= \mathbf{U}(t)e^{T\mathbf{C}}\\ &= \mathbf{U}(t)e^{-t\mathbf{C}}e^{(t+T)\mathbf{C}}\\ &= \mathbf{P}(t)e^{(t+T)\mathbf{C}},\end{aligned}$$

where the periodic matrix $\mathbf{P}(t)$ is defined by

$$\begin{aligned}\mathbf{P}(t) &= \mathbf{U}(t)e^{-t\mathbf{C}}\\ \mathbf{P}(t+T) &= \mathbf{U}(t+T)e^{-(t+T)\mathbf{C}}\\ &= \mathbf{P}(t)e^{(t+T)\mathbf{C}}.e^{-(t+T)\mathbf{C}}\\ &= \mathbf{P}(t)\end{aligned}$$

∴ The result. □

The significance of the above theorem is that the determination of the fundamental matrix $\mathbf{U}(t)$ over a finite interval of length T (e. g., $0 \le t \le T$) leads at once to the determination of $\mathbf{U}(t)$ over $(-\infty, \infty)$. This follows from the periodicity property:

$$\begin{aligned}\mathbf{U}(t+T) &= \mathbf{P}(t)e^{(t+T)\mathbf{C}}\\ &= \mathbf{P}(t)e^{t\mathbf{C}}.e^{T\mathbf{C}}\\ &= \mathbf{U}(t)e^{T\mathbf{C}} = \mathbf{U}(t)\mathbf{M}\end{aligned}$$

⇒

$$\begin{aligned}\mathbf{U}(t+2T) &= \mathbf{U}(t)e^{2T\mathbf{C}}\\ \mathbf{U}(t+nT) &= \mathbf{U}(t)e^{nT\mathbf{C}}, \quad n \text{ is a positive or negative integer}\\ &= \mathbf{U}(t)\mathbf{M}^n\end{aligned}$$

Remark. If we chose the initial condition

$$\mathbf{U}(0) = \mathbf{U}_0$$

Then

$$\mathbf{U}(T) = \mathbf{U}_0 e^{T\mathbf{C}} = \mathbf{U}_0\mathbf{M}$$

$\Rightarrow$

$$\mathbf{C} = \frac{1}{T}\ln[\mathbf{U}_0^{-1}\mathbf{U}(T)]$$

For the special case,

$$\mathbf{U}(0) = \mathbf{I}$$

$\Rightarrow$

$$\mathbf{C} = \frac{1}{T}\ln \mathbf{U}(T)$$

Thus, $\mathbf{C}$ can be determined from $\mathbf{U}(T)$. Each column of $\mathbf{U}(T)$ may be determined by integrating the IVP

$$\frac{d\mathbf{u}}{dt} = \mathbf{A}(t)\mathbf{u}; \quad \mathbf{u}(0) = \mathbf{e}_j; \quad j = 1, 2, \ldots, n$$

Definition. $\mathbf{M}$ is called the *monodromy* matrix. The eigenvalues of $\mathbf{M}$ are called the *characteristic multipliers* of the periodic initial value problem. The matrix $\mathbf{C}$ is called the Floquet matrix and the eigenvalues of $\mathbf{C}$ are called the *characteristic exponents* or *Floquet exponents.*

Remark. The characteristic multipliers $\mu_1, \mu_2, \ldots, \mu_n$ are uniquely defined. The characteristic exponents $\rho_1, \rho_2, \ldots, \rho_n$ are given by

$$T\rho_j = \ln \mu_j$$

If we take the principal value of the logarithm, then ρ_j is also uniquely defined.

Lemma 13.1. $\mathbf{M}$ *is not unique but any two are related by a similarity transform.*

Proof. Let $\mathbf{U}_i(t)$ be a fundamental matrix $(i = 1, 2)$

$$\mathbf{U}_1(t+T) = \mathbf{U}_1(t)\mathbf{M}_1$$
$$\mathbf{U}_2(t+T) = \mathbf{U}_2(t)\mathbf{M}_2$$

But $\mathbf{U}_2(t) = \mathbf{U}_1(t)\mathbf{D}$, $\mathbf{D}$ is nonsingular:

$$\begin{aligned}\mathbf{M}_2 &= \mathbf{U}_2^{-1}(t)\mathbf{U}_2(t+T)\\ &= \mathbf{D}^{-1}\mathbf{U}_1^{-1}(t)\mathbf{U}_1(t+T)\mathbf{D}\\ &= \mathbf{D}^{-1}\mathbf{M}_1\mathbf{D}\end{aligned}$$

Therefore, the eigenvalues of $\mathbf{M}_1$ and $\mathbf{M}_2$ are the same. □

In order to see the explicit form of the solution of the periodic system (equations (13.6) and (13.7)), we consider the simple case in which matrix $\mathbf{C}$ is diagonalizable, i. e., ∃ a nonsingular constant matrix $\mathbf{S}$ such that

$$\mathbf{C} = \mathbf{S}\boldsymbol{\rho}\mathbf{S}^{-1}$$

where

$$\boldsymbol{\rho} = \begin{pmatrix} \rho_1 & & 0 \\ & \cdot & \\ 0 & & \rho_n \end{pmatrix}$$

∴

$$e^{t\mathbf{C}} = \mathbf{S}\begin{pmatrix} e^{\rho_1 t} & & 0 \\ & \cdot & \\ 0 & & e^{\rho_n t} \end{pmatrix}\mathbf{S}^{-1}$$

∴

$$\mathbf{U}(t) = \mathbf{P}(t)\mathbf{S}\begin{pmatrix} e^{\rho_1 t} & & 0 \\ & \cdot & \\ 0 & & e^{\rho_n t} \end{pmatrix}\mathbf{S}^{-1}$$

From this, it is clear that $\mathbf{U}(t)$ consists of columns $\mathbf{u}_1(t), \dots, \mathbf{u}_n(t)$, which are of the form

$$\mathbf{u}_i(t) = \mathbf{p}_i(t)e^{\rho_i t}$$

where $\mathbf{p}_i(t + T) = \mathbf{p}_i(t)$ is a periodic n-vector.
⇒ The general form of the solution of equations (13.6) and (13.7) is

$$\mathbf{u} = \mathbf{p}_1(t)e^{\rho_1 t} + \cdots + \mathbf{p}_n(t)e^{\rho_n t}$$ □

Lemma 13.2. *A complex number ρ is a characteristic exponent of equations* (13.6) *if and only if there is a nontrivial solution of equations* (13.6) *of the form $e^{\rho t}\mathbf{p}(t)$ where $\mathbf{p}(t+T) = \mathbf{p}(T)$.*

Proof. This follows from above discussion. Note that from this lemma it follows that $\mathbf{u}(t)$ is a periodic solution iff ∃ a characteristic exponent of $\mathbf{C}$, which is zero, i. e., $\rho_i = 0$ for some i or equivalently $\mu_i = +1$ for some i. □

Lemma 13.3. *If $\mu_1, \mu_2, \dots, \mu_n$ are the characteristic multipliers of the periodic system (equations* (13.6) *and* (13.7)*), then*

$$\prod_{i=1}^{n} \mu_i = \exp\left\{\int_0^T \operatorname{trace} \mathbf{A}(s)\, ds\right\}$$

Proof. For the general linear system, we have seen that

$$\det \mathbf{U}(t) = \det \mathbf{U}(\tau) \exp\left\{ \int_{\tau}^{t} \operatorname{trace} \mathbf{A}(s)\, ds \right\}$$

Let $\tau = 0, t = T$

$\Rightarrow$

$$\det \mathbf{U}(T) = \det \mathbf{U}(0) \exp\left\{ \int_{0}^{T} \operatorname{trace} \mathbf{A}(s)\, ds \right\}$$

For the periodic system, we can take

$$\mathbf{U}(0) = \mathbf{I}$$
$$\mathbf{U}(T) = \mathbf{M}$$

$\Rightarrow$

$$\det \mathbf{M} = \exp\left\{ \int_{0}^{T} \operatorname{trace} \mathbf{A}(s)\, ds \right\}$$

but

$$\det \mathbf{M} = \mu_1 . \mu_2 \ldots \mu_n = \prod_{i=1}^{n} \mu_i$$

$\therefore$ The result. □

Thus, if $(n - 1)$ of the characteristic multipliers are known we can determine the remaining one using the above relation. For example, for the special case of a planar system $(n = 2)$, we have

$$\mu_1 \mu_2 = \exp\left\{ \int_{0}^{T} [a_{11}(s) + a_{22}(s)]\, ds \right\}$$

This result is useful in determining the stability of bifurcating periodic solutions of a planar system. In this case, one Floquet multiplier is unity and the magnitude of the second multiplier determines the stability of the bifurcating periodic solution.

Problems

1. Consider the linear system with periodic coefficients:

$$\mathbf{u}' = \mathbf{A}(t)\mathbf{u}; \quad \mathbf{A}(t) = \mathbf{A}(t + T) \tag{13.8}$$

(a) Show that $\mathbf{U}(t)$ is a fundamental matrix, it may be written as

$$\mathbf{U}(t) = \mathbf{P}(t)e^{t\mathbf{C}}, \quad \mathbf{P}(t) = \mathbf{P}(t+T)$$

where $\mathbf{P}(t)$ is a $n \times n$ periodic matrix and $\mathbf{C}$ is a constant matrix.

(b) Determine the asymptotic form of the solution to the initial value problem if $\mathbf{C}$ has (i) simple eigenvalues in the left half of the complex plane, (ii) a zero eigenvalue and all other in the left half-plane and (iii) a pair of purely imaginary eigenvalues and all others in the left half-plane.

(c) If $\mu_1, \mu_2, \ldots, \mu_n$ are the characteristic multipliers of the periodic system (i), show that

$$\prod_{i=1}^{n} \mu_i = \exp\left\{\int_0^T \operatorname{trace} \mathbf{A}(s)\, ds\right\}$$

2. If $\mu_1, \mu_2, \ldots, \mu_n$ are the characteristic multipliers of the periodic system,
   (a) Discuss the nature of solution when $\mu_i = -1$ for some i;
   (b) Discuss the nature of the solution when a pair of μ_i are on the unit circle.
3. Consider the linear system

$$\frac{du_1}{dt} = (-\sin 2t)u_1 + [(\cos 2t) - 1]u_2$$
$$\frac{du_2}{dt} = (1 + \cos 2t)u_1 + (\sin 2t)u_2$$

   (a) Verify that

$$\mathbf{U}(t) = \begin{pmatrix} e^t(\cos t - \sin t) & e^{-t}(\cos t + \sin t) \\ e^t(\cos t + \sin t) & e^{-t}(-\cos t + \sin t) \end{pmatrix}$$

   is a fundamental matrix.
   (b) Obtain the corresponding monodromy matrix and determine the characteristic multipliers and characteristic exponents.
4. Consider the nonlinear system

$$\frac{du_1}{dt} = u_1 - u_2 - u_1^3 - u_1u_2^2$$
$$\frac{du_2}{dt} = u_1 + u_2 - u_1^2u_2 - u_2^3 \tag{13.9}$$

   (a) Verify that $u_1^0 = \cos t$, $u_2^0 = \sin t$ is a (periodic) solution.
   (b) Show that the linearization of equation (13.9) around (u_1^0, u_2^0) gives the following linear system with period coefficients:

$$\frac{dz_1}{dt} = (-1 - \cos 2t)z_1 - [1 + \sin 2t]z_2$$

$$\frac{dz_2}{dt} = (1 - \sin 2t)z_1 + (\cos 2t - 1)z_2 \tag{13.10}$$

(c) Verify that

$$\mathbf{U}(t) = \begin{pmatrix} -\sin t & e^{-2t}\cos t \\ \cos t & e^{-2t}\sin t \end{pmatrix}$$

is fundamental matrix.

(d) Determine the monodromy matrix and the Floquent multipliers of the periodic system defined by equation (13.10).

5. Consider an ideal mixing tank with constant fluid density and tank volume but periodically varying flow rate. With appropriate notation,
   (a) Show that the concentration of a solute satisfies the scalar equation,

$$\frac{dc}{dt} = \frac{1 + \varepsilon \sin(\omega t)}{\tau}[c_{\text{in}}(t) - c], \quad t > 0; \quad c = c_0 \mathbin{@} t = 0$$

   (b) Obtain the solution of the above equation for $c_0 = 0$, and $c_{\text{in}} = H(t) =$ Heaviside's unit step function.
   (c) Obtain the solution for $c_{\text{in}}(t) = 0$ and $c_0 = 1$, and show a plot of the solution for $\varepsilon = 0$ and $\varepsilon > 0$.

14 Analytic solutions, adjoints and integrating factors

We have already shown that the solution to linear scalar and vector differential equations with constant coefficients can be expressed analytically in terms of the eigenvalues. This chapter deals with other cases in which it is possible to express the solutions in explicit form.

14.1 Analytic solutions

First consider the general nth order linear differential equation

$$p_0(t)\frac{d^n u}{dt^n} + p_1(t)\frac{d^{n-1}u}{dt^{n-1}} + \cdots + p_{n-1}\frac{du}{dt} + p_n(t)u = f(t)$$

or

$$Lu = f(t). \tag{14.1}$$

Consider only the homogeneous equation

$$Lu = 0 \tag{14.2}$$

Under certain special conditions, we can obtain the solution of equation (14.2) analytically. We discuss some of these special cases here.

(i) Scale Invariance in t

Suppose that equation (14.2) is invariant to the scaling $t \to at \Rightarrow Lu = 0$. Then the equation is called *equidimensional* or *Euler's equation*. To illustrate, consider the case of $n = 2$:

$$p_0(t)\frac{d^2u}{dt^2} + p_1(t)\frac{du}{dt} + p_2(t)u = 0 \tag{14.3}$$

and the special case of

$$p_0(t) = a_0 t^2, \quad p_1(t) = a_1 t, \quad p_2(t) = a_2$$

gives

$$a_0 t^2 u'' + a_1 t u' + a_2 u = 0.$$

Let $t' = at \Rightarrow \frac{d}{dt} = \frac{d}{dt'}.a$

https://doi.org/10.1515/9783111598055-017

$\Rightarrow$

$$\alpha_0\left(\frac{t'}{a}\right)^2 a^2\frac{d^2u}{dt'^2} + \alpha_1\left(\frac{t'}{a}\right).a\frac{du}{dt'} + \alpha_2 u = 0$$

$\Rightarrow$

$$\alpha_0 t'^2\frac{d^2u}{dt'^2} + \alpha_1 t'\frac{du}{dt'} + \alpha_2 u = 0.$$

Thus, the equation is invariant to the scaling $t \to at$. Scale invariant equations can be converted to constant coefficient equations by the transformation

$$t = e^x \quad \text{or} \quad x = \ln t \tag{14.4}$$

$\Rightarrow$

$$\frac{d}{dt} = \frac{d}{dx}.\frac{1}{t}$$

$\Rightarrow$

$$t\frac{d}{dt} = \frac{d}{dx}$$

$$\frac{d^2}{dt^2} = \frac{d^2}{dx^2}.\frac{1}{t^2} - \frac{1}{t^2}\frac{d}{dx} \Rightarrow t^2\frac{d^2}{dt^2} = \frac{d^2}{dx} - \frac{d}{dx}$$

Thus, using the transformation given by equation (14.4), equation (14.3) reduces to

$$\alpha_0\frac{d^2u}{dx^2} + (\alpha_1 - \alpha_0)\frac{du}{dx} + \alpha_2 u = 0$$

The solution of this equation is of the form

$$u(x) = c_1 e^{\lambda_1 x} + c_2 e^{\lambda_2 x}$$

where λ_1, λ_2 are roots of $\alpha_0\lambda^2 + (\alpha_1 - \alpha_0)\lambda + \alpha_2 = 0$,

$\Rightarrow$

$$u(t) = c_1 t^{\lambda_1} + c_2 t^{\lambda_2}$$

If $\lambda_1 = \lambda_2 \Rightarrow u(x) = c_1 e^{\lambda_1 x} + c_2 x e^{\lambda_2 x}$,

$\Rightarrow$

$$u(t) = c_1 t^{\lambda_1} + c_2(\ln t)t^{\lambda_1}.$$

The characteristic equation for equidimensional equations may also be obtained by the substitution

$$u = t^\lambda$$

⇒

$$u' = \frac{du}{dt} = \lambda t^{\lambda-1}$$
$$u'' = \lambda(\lambda - 1)t^{\lambda-2}$$

∴

$$\alpha_0\lambda(\lambda - 1) + \alpha_1\lambda + \alpha_2 = 0$$
$$\alpha_0\lambda^2 + (\alpha_1 - \alpha_0)\lambda + \alpha_2 = 0$$

[Remark: If we write $u(\lambda) = t^\lambda$, then $\frac{du}{d\lambda} = (\ln t)t^\lambda$. Hence, when λ is a double root, the two linearly independent solutions are t^λ and $(\ln t)t^\lambda$.]

Example 14.1. Consider the equation

$$u'' + \frac{u}{4t^2} = 0$$

⇒

$$\left(\lambda - \frac{1}{2}\right)^2 = 0 \Rightarrow \lambda = \frac{1}{2}, \frac{1}{2}$$

⇒

$$u(t) = c_1\sqrt{t} + c_2\sqrt{t}\ln t$$

Thus, for scale invariant equations in t, the linearly independent solutions are of the form t^λ or $t^\lambda \ln t$ is a solution for some λ determined by the characteristic equation.
(ii) Scale invariant equations in u

An ODE (linear or nonlinear) is called scale invariant w. r. t. u if the transformation $u \to au$ $(a \neq 0)$ leaves the equation invariant. All linear equations $Lu = 0$ are invariant w. r. t. u. If an equation is scale invariant w. r. t. u, we can reduce its order by one. To illustrate, consider the second-order equation

$$p_0(t)\frac{d^2u}{dt^2} + p_1(t)\frac{du}{dt} + p_2(t)u = 0, \tag{14.5}$$

which is scale invariant w. r. t. u. Define

$$u(t) = e^{w(t)} \tag{14.6}$$

⇒

$$u' = e^w.w' = uw'$$
$$u'' = e^w.(w')^2 + uw''$$

$\Rightarrow$

$$p_0(t).u[(w')^2 + w''] + p_1(t)uw' + p_2(t)u = 0$$

$\Rightarrow$

$$p_0[(w')^2 + w''] + p_1w' + p_2 = 0 \tag{14.7}$$

Let $w' = y$

$\Rightarrow$

$$p_0(t)\frac{dy}{dt} + p_0(t)y^2 + p_1(t)y + p_2(t) = 0 \tag{14.8}$$

This is a first-order (but nonlinear) equation in y.
(iii) Scale invariant equations in t and u

An ODE (linear or nonlinear) is called scale invariant in t and u if $\exists$ a transformation of the form $t \to at$, $u \to a^m u$(for some m and $a \neq 0$) that leaves the equation invariant.

It can be shown that all scale invariant equations can be transformed to equidimensional (Euler) equations in t.

Examples.
(1) $(u')^2 + uu'' + t = 0$ is scale invariant $t \to at$ and $u \to a^{3/2}u$
(2) $\frac{du}{dt} = F(\frac{u}{t})$ is scale invariant $t \to at$ and $u \to au$

First-order equations
(a) The linear equation

$$\frac{du}{dt} + p(t)u = q(t) \tag{14.9}$$

can be solved exactly in terms of quadratures

$$u \exp\left\{\int p(t')\,dt'\right\} = \int q(t) \exp\left\{\int p(t')\,dt'\right\} dt$$

(b) The Bernoulli's equation

$$\frac{du}{dt} + p(t)u = q(t)u^m \tag{14.10}$$

can be solved by the substitution

$$y = [u(t)]^{1-m}$$

to obtain a linear equation in y.

(c) The Riccati equation

$$\frac{du}{dt} = a(t)u^2 + b(t)u + c(t) \tag{14.11}$$

$$a(t) = 0 \Rightarrow \text{linear equation}$$

$$c(t) = 0 \Rightarrow \text{Bernoulli's equation}$$

When $a \neq 0$, let

$$u = \frac{-w'(t)}{a(t)w(t)}$$

$\Rightarrow$

$$w''(t) - \left[\frac{a'(t)}{a(t)} + b(t)\right]w'(t) + a(t)c(t)w = 0$$

This is a linear second-order equation.

Another special feature of the Riccati equation is that it can be solved exactly if we can find a (special) solution by inspection. Let

$$u = u_1(t) + v(t) \tag{14.12}$$

where u_1 is a special solution. Substitution of equation (14.12) in equation (14.11) gives

$$v'(t) = [b(t) + 2a(t)u_1]v(t) + a(t)v^2 \tag{14.13}$$

This is a Bernoulli equation for v.

(d) Separable and exact equations

$$a(u,t)\frac{du}{dt} + b(u,t) = 0$$

or

$$a(u,t)\,du + b(u,t)\,dt = 0 \tag{14.14}$$

If $a(u,t) = a_1(u)a_2(t)$ and $b(u,t) = b_1(u)b_2(t)$, then equation (14.14) is separable and can be solved by quadratures. If

$$\frac{\partial a}{\partial t} = \frac{\partial b}{\partial u}$$

then we have an *exact equation* that can be solved analytically.

A listing of all ODEs (linear and nonlinear) that have analytic solutions can be found in the book by E. Kamke [21].

14.2 Adjoints and integrating factors

The concept of adjoint plays an important role in the theory of differential equations. We discuss it here in the context of initial value problems and revisit it again in Chapter 18.

14.2.1 First-order equation

Consider the first-order homogeneous equation

$$Lu \equiv p_0(t)\frac{du}{dt} + p_1(t)u = 0 \tag{14.15}$$

Multiply both sides of equation (14.15) by $v(t)$,

$$vLu \equiv v(t)p_0(t)\frac{du}{dt} + v(t)p_1(t)u = 0 \tag{14.16}$$

Now using the chain rule:

$$\frac{d}{dt}(vp_0u) = vp_0\frac{du}{dt} + u\frac{d}{dt}(vp_0)$$

we can write

$$\begin{aligned} vLu &\equiv \frac{d}{dt}[p_0(t)u.v] - u\frac{d}{dt}(p_0(t)v) + v.p_1(t)u = 0 \\ \Longrightarrow vLu &\equiv (p_0uv)' + u[-(p_0v)' + p_1v] = 0 \end{aligned} \tag{14.17}$$

Suppose that $v(t)$ satisfies the equation

$$-(p_0v)' + p_1v = 0 \tag{14.18}$$

Then the LHS or RHS of equation (14.17) is an exact derivative. The function $v(t)$ is called the integrating factor of equation (14.15). To find the integrating factor of equation (14.15), we end up with equation (14.18) and to find the integrating factor of equation (14.18) we end up with equation (14.15). Hence, Lagrange called equation (14.18), the *adjoint equation.*

Thus, the adjoint operator of equation (14.15) is defined by

$$\begin{aligned} L^*v &= -(p_0v)' + p_1v \\ &= -p_0(t)\frac{dv}{dt} + [p_1(t) - p_0'(t)]v \end{aligned} \tag{14.19}$$

Also, we have

$$vLu - uL^*v = \frac{d}{dt}[p_0(t)u.v] \tag{14.20}$$

Equation (14.20) is known as the Lagrange identity.

Now we suppose that $v(t)$ satisfies the adjoint differential equation:

$$L^*v = -p_0(t)\frac{dv}{dt} + [p_1(t) - p_0'(t)]v = 0. \tag{14.21}$$

Then equation (14.20) becomes

$$vLu = \frac{d}{dt}[v(t)p_0(t)u(t)]. \tag{14.22}$$

Thus, the solution of the adjoint equation gives an integrating factor to equation (14.15). Now suppose that $u(t)$ satisfies $Lu = 0$. Then equation (14.20) gives

$$uL^*v = -\frac{d}{dt}[p_0(t)u.v] \tag{14.23}$$

Thus, $u(t)$ is an integrating factor for the adjoint equation. It is also seen that $(L^*)^* = L$, i. e., the adjoint of the adjoint equation is the original equation.

If $Lu = 0$ and $L^*v = 0$, equation (14.20) gives

$$\frac{d}{dt}(vp_0u) = 0. \tag{14.24}$$

or,

$$v(t)p_0(t)u(t) = c_0 \text{ (constant)} \tag{14.25}$$

Thus, if we have solution to $Lu = 0$, we can determine the solution to $L^*v = 0$ and vice versa. This idea extends to higher-order differential equations as shown below.

14.2.2 Second-order equation

Consider the second order differential operator

$$Lu = p_0(t)\frac{d^2u}{dt^2} + p_1(t)\frac{du}{dt} + p_2(t)u.$$

Multiplying by v and integrating by parts gives

$$\begin{aligned} vLu &= vp_0u'' + vp_1u' + vp_2u \\ &= (vp_0u')' - u'(p_0v)' + (vp_1u)' - u(p_1v)' + vp_2u \end{aligned}$$

$$\begin{aligned}
&= (vp_0u')' - [u(p_0v)']' + u(p_0v)'' + (vp_1u)' - u(p_1v)' + vp_2u \\
&= u[(p_0v)'' - (p_1v)' + vp_2] + (vp_0u')' - [u(p_0v)']' + (vp_1u)' \\
&= uL^*v + \frac{d}{dt}[vp_0u' - u(p_0v)' + vp_1u]
\end{aligned}$$

where

$$L^*v = (p_0v)'' - (p_1v)' + p_2v$$

$\Rightarrow$

$$vLu - uL^*v = \frac{d}{dt}[\pi(u,v)]$$

where the concomitant $\pi(u,v)$ is defined by

$$\begin{aligned}
\pi(u,v) &= vp_0u' - u(p_0v)' + p_1uv \\
&= p_0u'v - p_0uv' - p_0'uv + p_1u.v \\
&= p_0u'v - p_0uv' + (p_1 - p_0')uv \\
&= \begin{bmatrix} v & v' \end{bmatrix} \begin{bmatrix} p_1 - p_0' & p_0 \\ -p_0 & 0 \end{bmatrix} \begin{bmatrix} u \\ u' \end{bmatrix} \\
&= \mathbf{k}^t(v).\mathbf{P}.\mathbf{k}(u)
\end{aligned}$$

where

$$\mathbf{k}(u) = \begin{pmatrix} u \\ u' \end{pmatrix} = \text{Wronskian vector},$$

and the 2×2 matrix $\mathbf{P} = \begin{bmatrix} p_1-p_0' & p_0 \\ -p_0 & 0 \end{bmatrix}$ is called the *concomitant matrix.*

The bilinear form $\pi(u,v)$ is called the *bilinear concomitant.*

Suppose that $v(t)$ satisfies $L^*v = 0$ and $u(t)$ satisfies $Lu = 0$, then

$$\pi(u,v) = \text{constant}$$

$\Longrightarrow$

$$\begin{aligned}
&(p_0u' - p_0'u + p_1u)v - p_0uv' = \text{constant} \\
&\Longrightarrow p_0(u'v - uv') + (p_1 - p_0')uv = c_0. \qquad (14.26)
\end{aligned}$$

Thus, the solutions of $Lu = 0$ and $L^*v = 0$ are related by equation (14.26).

14.3 Relationship between solutions of $Lu = 0$ and $L^*v = 0$

When $v(t)$ satisfies the adjoint equation $L^*v = 0$, we have

$$vLu = \frac{d}{dt}[\pi(u, v)] = 0$$

$\Longrightarrow$

$$\pi(u, v) = \text{constant} \tag{14.27}$$

For the case of $n = 2$, equation (14.27) takes the form

$$p_0[u'(t)v(t) - u(t)v'(t)] + [p_1(t) - p_0'(t)]u(t)v(t) = c_0 \tag{14.28}$$

Thus, if ψ_1 and ψ_2 are two linearly independent solutions of $L^*v = 0$, we have

$$\begin{bmatrix} (p_1 - p_0')\psi_1 - p_0\psi_1' & p_0\psi_1 \\ (p_1 - p_0')\psi_2 - p_0\psi_2' & p_0\psi_2 \end{bmatrix} \begin{pmatrix} u \\ u' \end{pmatrix} = \begin{pmatrix} \pi(u, \psi_1) \\ \pi(u, \psi_2) \end{pmatrix} = \begin{pmatrix} c_{01} \\ c_{02} \end{pmatrix} \tag{14.29}$$

Solving equation (14.29) for $u(t)$ gives

$$u(t) = \frac{\psi_2(t)c_{01} - \psi_1(t)c_{02}}{p_0(t)W(t)} \tag{14.30}$$

where

$$W(t) = \begin{vmatrix} \psi_1 & \psi_2 \\ \psi_1' & \psi_2' \end{vmatrix} = \psi_1\psi_2' - \psi_1'\psi_2 \neq 0$$

Equation (14.30) gives the general solution to $Lu = 0$. In the general case of an n-th order equation, we have n relations

$$\pi(u, \psi_i) = c_{0i}, \quad i = 1, 2, \ldots n$$

Since π is linear in $u, u', u'', \ldots, u^{[n-1]}(t)$, we can eliminate (or solve) for these variables. The same reasoning also applies to the vector equation.

14.4 Vector initial value problem

Consider the vector form of initial value problem:

$$\frac{d\mathbf{u}}{dt} = \mathbf{A}(t)\mathbf{u} \tag{14.31}$$

Define linear operator L as

$$Lu = \frac{d\mathbf{u}}{dt} - \mathbf{A}(t)\mathbf{u} \tag{14.32}$$

then

$$\begin{aligned}\mathbf{v}^T L\mathbf{u} &= \mathbf{v}(t)^T \frac{d\mathbf{u}}{dt} - \mathbf{v}(t)^T \mathbf{A}(t)\mathbf{u} \\ &= \frac{d}{dt}(\mathbf{v}^T\mathbf{u}) - \frac{d\mathbf{v}^T}{dt}\mathbf{u} - \mathbf{v}(t)^T\mathbf{A}(t)\mathbf{u}\end{aligned} \tag{14.33}$$

is an exact derivative if

$$\frac{d\mathbf{v}^T}{dt}\mathbf{u} + \mathbf{v}^T\mathbf{A}(t)\mathbf{u} = \mathbf{0}$$

or

$$\frac{d\mathbf{v}^T}{dt} = -\mathbf{v}^T\mathbf{A}(t)$$

or

$$\frac{d\mathbf{v}}{dt} = -\mathbf{A}(t)^T\mathbf{v} \tag{14.34}$$

Thus, the adjoint operator L^* can be defined as

$$L^*\mathbf{v} = -\frac{d\mathbf{v}}{dt} - \mathbf{A}(t)^T\mathbf{v} \tag{14.35}$$

which rewrites equation (14.33) as

$$\mathbf{v}^T L\mathbf{u} = \frac{d}{dt}(\mathbf{v}^T\mathbf{u}) + (L^*\mathbf{v})^T\mathbf{u}$$

$\Longrightarrow$

$$\mathbf{v}^T L\mathbf{u} - (\mathbf{L}^*\mathbf{v})^T\mathbf{u} = \frac{d}{dt}(\mathbf{v}^T\mathbf{u}) \tag{14.36}$$

Integrating equation (14.36) from $t = 0$ to $t = a$ gives

$$\langle L\mathbf{u}, \mathbf{v}\rangle - \langle \mathbf{u}, L^*\mathbf{v}\rangle = \mathbf{v}^T(a)\mathbf{u}(a) - \mathbf{v}^T(0)\mathbf{u}(0) \tag{14.37}$$

where

$$\langle \mathbf{u}, \mathbf{v}\rangle = \int_0^a (\mathbf{v}^T\mathbf{u})dt.$$

Thus, if $\mathbf{u}(0) = \boldsymbol{\alpha}$, the Lagrange condition: $\langle L\mathbf{u}, \mathbf{v}\rangle = \langle \mathbf{u}, L^*\mathbf{v}\rangle$ can be satisfied when we have $\mathbf{v}^T(a)\mathbf{u}(a) = \mathbf{v}^T(0)\boldsymbol{\alpha}$. If we choose $\mathbf{v}(a) = \boldsymbol{\alpha}$, then $\boldsymbol{\alpha}^T\mathbf{u}(a) = \mathbf{v}^T(0)\boldsymbol{\alpha} = \boldsymbol{\alpha}^T\mathbf{v}(0) \Longrightarrow \mathbf{v}(0) = \mathbf{u}(a)$, i. e., the final condition of original IVP is the initial condition of the adjoint problem.

Thus, if we integrate the IVP,

$$\frac{d\mathbf{u}}{dt} = \mathbf{A}(t)\mathbf{u}, \quad 0 < t \leq a \quad \text{with } \mathbf{u}(t = 0) = \boldsymbol{\alpha}, \tag{14.38}$$

to obtain $\mathbf{u}(a)$, it is equivalent to integrating the adjoint IVP,

$$\frac{d\mathbf{v}}{dt} = -\mathbf{A}^T(t)\mathbf{v}, \quad 0 \leq t < a \quad \text{with } \mathbf{v}(t = a) = \boldsymbol{\alpha} \tag{14.39}$$

in backward direction to get $\mathbf{v}(0)$. Thus, forward integration of equation (14.38) and the backward integration of equation (14.39) are coupled. Further, if the final condition of original IVP is $\mathbf{u}(a) = \boldsymbol{\beta}$, the initial condition of the adjoint problem is $\mathbf{v}(0) = \mathbf{u}(a) = \boldsymbol{\beta}$. In other words, the adjoint IVP can be integrated in forward direction with the initial condition as $\mathbf{v}(0) = \boldsymbol{\beta} = \mathbf{u}(a)$ to get $\mathbf{v}(a) = \boldsymbol{\alpha} = \mathbf{u}(0)$. Thus, the adjoint problem may be used to determine what initial condition on equation (14.31) may lead to a given final state at $t = a$. This observation is useful in many control and optimization applications.

Let $\mathbf{u}_1, \mathbf{u}_2, \ldots, \mathbf{u}_n$ be n linearly independent solutions of

$$L\mathbf{u} \equiv \frac{d\mathbf{u}}{dt} - \mathbf{A}(t)\mathbf{u} = \mathbf{0}$$

and $\mathbf{v}(t)$ is a solution of the adjoint equation

$$L^*\mathbf{v} = -\frac{d\mathbf{v}}{dt} - \mathbf{A}^T(t)\mathbf{v} = \mathbf{0},$$

then we have

$$\frac{d}{dt}(\mathbf{v}^T\mathbf{u}) = 0 \Longrightarrow \mathbf{v}^T\mathbf{u} = \text{constant}$$

$\Longrightarrow$

$$\mathbf{v}^T\mathbf{u}_i = c_i, \quad i = 1, 2, \ldots, n$$

$\Longrightarrow$

$$\mathbf{v}^T\mathbf{U} = \mathbf{c}^T \tag{14.40}$$

where $\mathbf{U}(t)$ is a fundamental matrix. Equation (14.40) can also be rewritten as

$$\mathbf{U}^T\mathbf{v} = \mathbf{c} \Longrightarrow \mathbf{v} = [\mathbf{U}^T]^{-1}\mathbf{c} \tag{14.41}$$

Thus, we can find solutions of the adjoint equation if we know the solution of $L\mathbf{u} = \mathbf{0}$. It can also be shown that the fundamental matrices are related by

$$\mathbf{V}^T\mathbf{U} = \mathbf{B} = \text{a constant nonsingular matrix}$$

We return to the concept of adjoint again when we deal with boundary value problems in Chapter 18.

Problems

1. (*Linear first-order equation*): Find the general solution (or a particular solution if the initial condition is given) of the following first-order differential equations: (a) $\frac{dy}{dx} + y\cos x = \frac{1}{2}\sin 2x$, (b) $(1-x^2)\frac{dy}{dx} + 2xy = \frac{x}{\sqrt{1-x^2}}; y(0) = 0$, (c) $\frac{dy}{dx} - y\tan x = e^x \sec x$ and (d) $(1+x^3)\frac{dy}{dx} + 2xy = 4x^2$
2. (*Bernoulli's equation*): Find the general solution of the following first-order differential equations: (a) $x\frac{dy}{dx} + y = y^2 \ln x$, (b) $\frac{dy}{dx} + \frac{1}{x} = \frac{e^y}{x^2}$, (c) $\frac{1}{y^2}\frac{dy}{dx} + \frac{1}{xy} = 1$ and (d) $\frac{dy}{dx} = x^2y^3 - xy$
3. *(Transient behavior of a mixing tank):* A tank initially holds 80 gal of a brine solution containing 0.125 lb of salt per gallon. At $t = 0$, another brine solution containing 1 lb of salt per gallon is poured into the tank at a rate of 4 gal/min, while the well-stirred mixture leaves the tank at a rate of 8 gal/min. (a) Formulate a model for describing the transient behavior of the tank and (b) Find the amount of salt in the tank when the tank contains exactly 40 gal of the solution.
4. *(Compound interest modeling):* A depositor currently has $6,000 and plans to invest it in an account that accrues interest continuously. What is the required interest rate if the depositor needs to have $10,000 in 4 years? Formulate the model, solve it and use the solution to calculate the required rate of interest.
5. *(Application of Newton's second law for a free falling body):* A body weighing m kg is dropped from a height H with zero initial velocity. As it falls, the body encounters a force due to air resistance, which may be assumed to be proportional to its velocity. If the limiting/terminal velocity of this body is v_0 m/s, determine (a) an expression for the velocity of the body at any time t and (b) an expression for the position of the body at any time t.
6. *(Cooling of a pie):* A hot pie that was cooked at 325 °F is taken directly from an oven and placed outdoors in the shade to cool on a day when the air temperature in the shade is 85 °F. After 5 minutes in the shade, the temperature of the pie had been reduced to 250 °F. Determine (a) the temperature of the pie after 20 minutes and (b) the time at which the pie cools to a temperature of 90 °F.
7. *(Population growth):* The population of a certain state is known to grow at a rate proportional to the number of people presently living in the state. If after 10 years the population has trebled and if after 20 years the population is 200,000, find the number of people initially living in the state.
8. *(First-order process model)*: The process model for a first-order system is given by

$$\tau\frac{dy}{dt} + y = f(t); \quad y(0) = 0.$$

Here, $\tau > 0$ is the first-order time constant. (a) Determine and plot the response of the system for a unit step input and a unit impulse input and (b) Determine and plot the response of the system (amplitude and phase lag) when $f(t) = A\sin\omega t$.

9. *(Second-order irreversible reaction in a batch reactor)*: Consider the second-order irreversible reaction $2A \rightleftharpoons B + C$ occurring in a constant density batch reactor: (a) Formulate the differential equation for the concentration of A and cast it in dimensionless form and (b) Solve the equation in (a) for the initial condition corresponding to only A present at $t = 0$. Discuss the special cases of the solution corresponding to the equilibrium constant being infinity and a value of four.
10. (*Terminal velocity of a particle*): Consider the motion of a small particle falling through a fluid (such as a dust particle falling in air or very small solid particle of size smaller than 20 μm falling in water). Assume that the initial velocity of the particle is zero: (a) With appropriate notation, show that Newton's second law may be written as

$$\frac{\pi}{6}d_p^3\rho_s\frac{dv}{dt} = \frac{\pi}{6}d_p^3\rho_s g - \frac{\pi}{6}d_p^3\rho_f g - 3\pi\mu d_p v$$

Explain the meaning of each term in the above equation and (b) Solve the above model with the initial condition specified and show that the velocity of the particle at any time may be expressed as

$$v(t) = v_\infty\left[1 - \exp\left(-\frac{18\mu t}{\rho_s d_p^2}\right)\right]$$

where v_∞ is the terminal velocity (for $t \to \infty$). (c) Determine an expression for v_∞.
11. (*Modified population balance equation*): (a) Consider a population balance model in which the birth rate is proportional to the square of the population size while the death rate varies linearly: (i) With appropriate assumptions, show that the evolution equation is of the form

$$\frac{dN}{dt} = bN^2 - aN$$

where a and b are positive constants and $N(t)$ is the population size at time t, (ii) Solve the above equation and plot the solution for the two cases of $N(t = 0) < \frac{a}{b}$ and $N(t = 0) > \frac{a}{b}$.
12. (*Consecutive first-order reactions in a batch reactor*): Consider a well-stirred batch reactor in which the consecutive reactions $A \to B \to C$ occur. Assume that the density of the reaction mixture (and the volume of the reactor) remains constant and at time zero the reactor is charged with a solution containing only reactant A at a concentration of C_0. Further assume that the rate of the first reaction is given by $r_1 = -r_A = k_1C_A$ and that of the second reaction by $r_2 = r_C = k_2C_B$: (a) Formulate the differential equations (and the initial conditions) describing the concentrations of all the species as a function of time, (b) Solve the equations in (a) and determine the concentrations and (c) Determine the time at which the concentration of species B is maximum.

13. The process model for a second order system is given by

$$\tau^2 \frac{d^2u}{dt^2} + 2\gamma\tau \frac{du}{dt} + u = f(t); \quad u(0) = 0; \quad u'(0) = 0$$

Here, $\tau > 0$ is the system time constant, while $\gamma > 0$ is called the damping constant. (a) Determine and plot the response of the system for a unit step input and a unit impulse input and (b) Determine and plot the response of the system (amplitude and phase lag) when $f(t) = A \sin \omega t$ for three cases: $\gamma = 0.1, 1, 2$.

14. The transient response of a U-tube manometer to changes in pressure is described by the initial value problem

$$\frac{d^2h}{dt^2} + \frac{24\mu}{\rho d^2}\frac{dh}{dt} + \frac{3g}{2L}h = \frac{3}{4}\frac{\Delta p}{\rho L}; \quad t > 0, \quad h(0) = 0; \quad h'(0) = 0$$

where h is the deviation of the manometer liquid level from the equilibrium position, t is time, g is the gravitational acceleration, d is the tube diameter, ρ and μ are the density and viscosity of the manometer fluid, L is the total length of the fluid column and Δp is the change in pressure (a) Cast the above model in dimensionless form (b) Determine the critical tube radius above which the response is oscillatory (c) Determine and plot the transient response of the manometer for a step change in the pressure when the tube diameter is twice the critical value determined in (b).

15. The motion of a periodically driven pendulum (for small amplitudes), the classic mechanical spring-mass-dashpot oscillator and the RLC (resistor-inductor-capacitor) electric circuit may be described by the second- order IVP:

$$\frac{d^2u}{dt^2} + 2\gamma \frac{du}{dt} + \omega_0^2 u = f(t); \quad u(0) = 0; \quad u'(0) = 0$$

where γ is the damping constant and ω_0 is the natural frequency of oscillation in the absence of damping. (a) For each case, formulate the model, cast it into dimensionless form and identify/relate the constants γ and ω_0 to the physical quantities that appear in the model (b) Determine and plot the solution when $f(t) = a \cos \omega t$ with special attention to cases in which ω is close to ω_0.

16. Find a general solution of the following ODEs: (i) $y'' + 3y' + 2y = 0$, (ii) $y'' + 2y' + y = 0$ and (iii) $y'' + 4y = 0$ (iv) $y'' + 2y' + 5y = 0$

17. Determine whether or not the following functions are linearly independent: (i) e^{-x}, e^{-2x}, e^{-4x}, (ii) $x^2, x^2 \log x$, (iii) $\log x, \log(x^3)$, (iv) $\sin x, \sin 2x$ and (v) $e^{-x}, \cos x, 0$

18. Consider the second-order process model described by the following IVP:

$$\tau^2 \frac{d^2u}{dt^2} + 2\gamma\tau \frac{du}{dt} + u = f(t); \quad u(0) = 0; \quad u'(0) = 0$$

Determine and plot the response of the system for a unit step input when $\tau = 1$ and three values of the damping constant: $\gamma = 0.1, 1, 2$

19. The motion of a periodically driven pendulum (for small amplitudes) may be described by the second-order IVP:

$$\frac{d^2u}{dt^2} + 2\gamma\frac{du}{dt} + \omega_0^2 u = f(t); \quad u(0) = 0; \quad u'(0) = 0$$

where γ is the damping constant and ω_0 is the natural frequency of oscillation in the absence of damping. Consider the case of $\omega_0 = 1$ and $\gamma = 0.1$. Determine and plot the amplitude of the solution when $f(t) = \cos\omega t$ and ω is varied in the range (0.1, 2).
20. Consider the U-tube manometer described by the following IVP:

$$\frac{d^2h}{dt^2} + \frac{24\mu}{\rho d^2}\frac{dh}{dt} + \frac{3g}{2L}h = \frac{3}{4}\frac{\Delta p}{\rho L}; \quad t > 0, \quad h(0) = 0; \quad h'(0) = 0$$

If the viscous damping term can be neglected, determine the period of oscillation (or the natural frequency) for the case where the total length of the liquid column is 2 meters (b) Examine the eigenvalues of the homogeneous equation with the damping term and derive a formula for the critical diameter of the tube below which the system does not oscillate. Calculate this value when the manometer fluid is water at $20^\circ C$ with density $\approx 1.0\,\mathrm{g/cm^3}$ and viscosity ≈ 1 centipoise ($= 0.01\,\mathrm{g.cm^{-1}\,s^{-1}}$).
21. Find a general solution of the inhomogeneous equation $y'' + 3y' + 2y = f(x)$ for the following cases: (i) $f(x) = 1$, (ii) $f(x) = x^2$, (iii) $f(x) = e^{-4x}$, (iv) $f(x) = e^{-x}$ and (v) $\cos x$
22. Find a general solution of the inhomogeneous equation $y'' + 2y' + y = f(x)$ for the following cases: (i) $f(x) = 1$, (ii) $f(x) = x^2$, (iii) $f(x) = e^{-4x}$, (iv) $f(x) = e^{-x}$ and (v) $\sin x$
23. Find a general solution of the inhomogeneous equation $y'' + 4y = f(x)$ for the following cases: (i) $f(x) = 1$, (ii) $f(x) = x^2$, (iii) $f(x) = e^{-4x}$, (iv) $f(x) = e^{-x}$ and (v) $\sin 2x$
24. Find a general solution of the inhomogeneous equation $y'' + 2y' + 5y = f(x)$ for the following cases: (i) $f(x) = 1$, (ii) $f(x) = x^2$, (iii) $f(x) = e^{-x}\sin 2x$, (iv) $f(x) = e^{-x}$ and (v) $\sin 2x$
25. Use the method of variation of parameter to obtain a general formula for the solution of the equation in problem (21) and then solve the specific cases.
26. Find a general solution of the following ODEs: (i) $y^{iv} + 2y'' + y = 0$, (ii) $y^{iv} + 4y'' = 0$ and (iii) $y''' + 4y'' + 13y' = 0$
27. Determine the solution of the initial value problem:

$$t^3\frac{d^3y}{dt^3} + t\frac{dy}{dt} - y = t^2; \quad y(1) = 1; \quad y'(1) = 3; \quad y''(1) = 14$$

28. Let L be a linear differential operator defined by $L\mathbf{u} = \mathbf{u}' - \mathbf{A}(t)\mathbf{u}$ where $\mathbf{A}$ is a $n \times n$ matrix with components in $C[0, a]$. The equation $L\mathbf{u} = \mathbf{0}$ is equivalent to the linear system

$$\mathbf{u}' = \mathbf{A}(t)\mathbf{u} \tag{14.42}$$

(a) Show that the expression $\mathbf{v}^T L\mathbf{u}$ is a exact derivative if $\mathbf{v}$ satisfies the adjoint equation

$$\mathbf{v}' = -\mathbf{A}(t)^T\mathbf{v} \tag{14.43}$$

What is the adjoint operator?

(b) Show that

$$\mathbf{v}^T L\mathbf{u} - (L^*\mathbf{v})^T\mathbf{u} = (\mathbf{v}^T\mathbf{u})'$$

This is known as the Lagrange's identity and by integrating it we get the so-called Green's formula

$$\langle L\mathbf{u}, \mathbf{v}\rangle - \langle \mathbf{u}, L^*\mathbf{v}\rangle = \mathbf{v}(a)^T\mathbf{u}(a) - \mathbf{v}(0)^T\mathbf{u}(0)$$

(c) If $\mathbf{U}$ is a fundamental matrix for equation (14.42) and $\mathbf{V}$ for equation (14.43) show that $\mathbf{U}^T\mathbf{V} = \mathbf{C}$, where $\mathbf{C}$ is a nonsingular constant matrix.

(d) Discuss the relationship between scalar and vector adjoints for the case $n = 2$.

15 Introduction to the theory of functions of a complex variable

The theory of functions of a complex variable is helpful in determining and understanding the solutions (and their properties) of linear equations. Specifically, it is useful for (i) inverting the Laplace transformation, (ii) evaluation and inversion of Fourier transforms, (iii) series solutions of ordinary differential equations, (iv) solution of linear partial differential equations (conformal mapping and solution of Laplace's equation in the plane) and (v) evaluation of certain definite integrals. This chapter provides an introduction to the theory of functions of a complex variable with examples selected from applications.

15.1 Complex valued functions

15.1.1 Algebraic operations with complex numbers

We have already seen that the set of all complex numbers forms a field, which we shall denote by $\mathbb{C}$. The symbol $z = x + iy$, which can stand for any complex number in the set $\mathbb{C}$ is called a complex variable. We shall use the notation $\text{Re}\{z\}$ = real part of $z = x$ and $\text{Im}\{z\}$ = imaginary part of $z = y$. The complex conjugate of z is a complex number $x - iy$ and will be denoted by $\bar{z}$ or z^*.

If $z_1 = x_1 + iy_1 \in \mathbb{C}$ and $z_2 = x_2 + iy_2 \in \mathbb{C}$, then the usual algebraic operations (addition/subtraction and multiplication/division) are defined by

$$\begin{aligned}
z_1 \pm z_2 &= (x_1 \pm x_2) + i(y_1 \pm y_2) \\
z_1 z_2 &= (x_1 + iy_1)(x_2 + iy_2) \\
&= (x_1 x_2 - y_1 y_2) + i(x_1 y_2 + x_2 y_1) \\
\frac{z_1}{z_2} &= \frac{x_1 + iy_1}{x_2 + iy_2} = \frac{(x_1 + iy_1)(x_2 - iy_2)}{x_2^2 + y_2^2} \\
&= \frac{x_1 x_2 + y_1 y_2}{x_2^2 + y_2^2} + i\frac{(x_2 y_1 - x_1 y_2)}{x_2^2 + y_2^2} \quad (\text{for } z_2 \neq 0)
\end{aligned}$$

The *modulus* or *absolute value* of $z = x + iy$ is defined as $|z| = \sqrt{x^2 + y^2}$.

15.1.2 Polar form of complex numbers

Since a complex number $z = x + iy$ can be identified with the ordered pair (x, y), it can be represented as a point in the $x - y$ plane, called the complex plane or *Argand diagram*. This is shown in Figure 15.1.

https://doi.org/10.1515/9783111598055-018

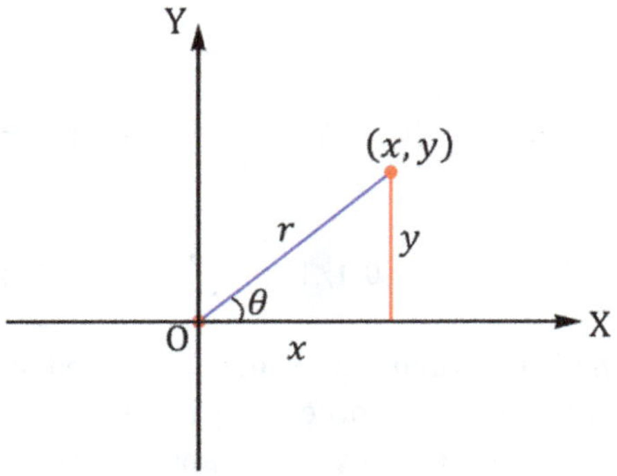

Figure 15.1: Representation of a complex number in the plane (Argand diagram).

Referring to this figure, we have

$$x = r\cos\theta; \quad y = r\sin\theta; \quad r = \sqrt{x^2 + y^2},$$

where r is the modulus and the angle θ with the positive x-axis (in the counterclockwise direction) is called the *argument*. In polar form, the complex number is written as

$$z = x + iy = r(\cos\theta + i\sin\theta) = re^{i\theta}.$$

The last equality follows from the Euler's formula

$$e^{i\theta} = \cos\theta + i\sin\theta. \tag{15.1}$$

The polar form of the complex number is convenient for some operations such as multiplication or division. For example, if we denote $z_k = r_k e^{i\theta_k}$ $(k = 1, 2, \ldots, n)$, then

$$z_1 z_2 = r_1 e^{i\theta_1} r_2 e^{i\theta_2} = r_1 r_2 e^{i(\theta_1+\theta_2)}.$$

Applying this to the n-th power of a complex number, we get *De Moivre's* theorem

$$z^n = (re^{i\theta})^n = r^n(\cos\theta + i\sin\theta)^n = r^n(\cos n\theta + i\sin n\theta) \tag{15.2}$$

or

$$\cos n\theta + i\sin n\theta = (\cos\theta + i\sin\theta)^n. \tag{15.3}$$

By expanding the RHS of equation (15.3) and equating the real and imaginary parts, we can express $\cos n\theta$ or $\sin n\theta$ (for $n = 1, 2, 3, \ldots$) in terms of $\cos\theta$ and $\sin\theta$.

15.1.3 Roots of complex numbers

If n is a positive integer, using *De Moivre's* theorem and the relation $e^{2k\pi i} = 1$ for $k = 0, 1, 2, \ldots$, we can write

$$z^{\frac{1}{n}} = (re^{i\theta})^{\frac{1}{n}} = r^{\frac{1}{n}}(e^{i\theta+2k\pi i})^{\frac{1}{n}} = r^{\frac{1}{n}} e^{i\frac{(\theta+2k\pi)}{n}}, \quad k = 0, 1, \ldots n-1 \tag{15.4}$$

It follows from equation (15.4) that there are n distinct values of $z^{\frac{1}{n}}$ that are located on a circle of radius $r^{\frac{1}{n}}$ and with argument differing by $\frac{2\pi}{n}$ (when ordered by increasing k). For example, for $n = 4$ and $r = 1$, we get the fourth roots of unity: 1, i, -1 and $-i$, located on a circle of unit radius as shown in Figure 15.2.

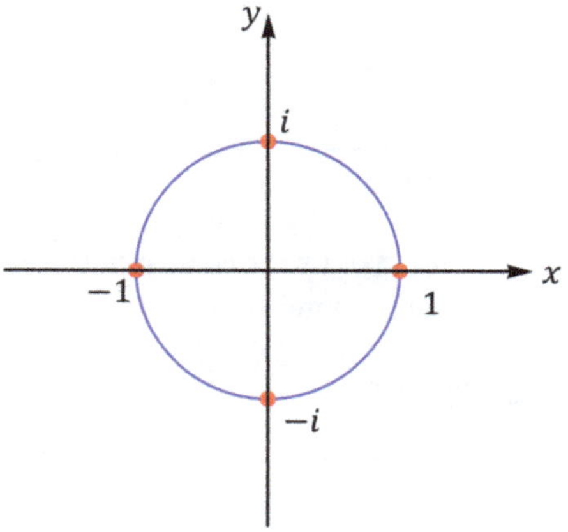

Figure 15.2: Fourth roots of unity on the unit circle.

Similarly, the nth roots of unity ($z = 1$) can be expressed as $1^{1/n} = 1, \omega, \omega^2, \ldots, \omega^{n-1}$ where $\omega = e^{2\pi i/n}$. For example, for $n = 4$, $\omega = e^{\frac{\pi i}{2}} = i$.

15.1.4 Complex-valued functions

If to each value of the complex number z in a set, we assign another complex number w such that $w = f(z)$, where f is a function, then w is called a function of a complex variable. The function can be single-valued or multivalued. In general, we write

$$\begin{aligned} w &= f(z) = f(x + iy) \\ &= u(x, y) + iv(x, y) \end{aligned} \tag{15.5}$$

where $u(x, y)$ and $v(x, y)$ are the real and imaginary part of $f(z)$. Unless otherwise specified, we assume $f(z)$ is single-valued.

Example 15.1. We consider some elementary functions and determine their real and imaginary parts.

1. Some elementary functions of a complex variables:
   (a) $w = z^2 = (x + iy)^2 = (x^2 - y^2) + i(2xy) \triangleq u(x,y) + iv(x,y)$.
   (b) $w = e^z = e^{x+iy} = e^x(\cos y + i\sin y) = e^x \cos y + ie^x \sin y \triangleq u(x,y) + iv(x,y)$. The function e^z is periodic with complex period $2\pi i$, since $e^z = e^{z+2k\pi i}$, $k = 0, \pm 1, \pm 2, \ldots$
   (c) $w = \sin z = \sin(x + iy) = \sin x \cos iy + \cos x \sin iy = \sin x \cosh y + i \cos x \sinh y \triangleq u(x,y) + iv(x,y)$. The function $\sin z$ is periodic with real period 2π.
2. Some functions in polar coordinates:
   (a) $w = z^2 = (re^{i\theta})^2 = r^2 e^{2i\theta} = r^2 \cos 2\theta + ir^2 \sin 2\theta \triangleq u(r,\theta) + iv(r,\theta)$.
   (b) $w = \ln z$. Writing $z = re^{i\theta} = re^{i\theta+2k\pi i}$, $k = 0, \pm 1, \pm 2, \ldots$ we get $w = \ln(re^{i\theta+2k\pi i}) = \ln r + i(\theta + 2k\pi) \triangleq u(r,\theta) + iv(r,\theta)$, $k = 0, \pm 1, \pm 2, \ldots$. The function $\ln z$ is infinitely many valued. The principal or main branch is given by $\ln r + i\theta$ and is denoted by $\operatorname{Ln} z$.

15.2 Limits, continuity and differentiation

15.2.1 Limits

Let $f(z)$ be a function defined in some neighborhood of a point z_0, with the possible exception of z_0 itself. We say that the limit of $f(z)$ as z approaches z_0 is w_0 and write

$$\lim_{z \to z_0} f(z) = w_0 \tag{15.6}$$

if for any $\epsilon > 0$, there exists a $\delta > 0$ such that

$$|f(z) - w_0| < \epsilon \tag{15.7}$$

whenever $0 < |z - z_0| < \delta$.

15.2.2 Continuity

Let $f(z)$ be a function defined in a neighborhood of z_0. Then we say that $f(z)$ is continuous at z_0 if (i) $\lim_{z \to z_0} f(z) = w_0$ exists, (ii) $f(z_0)$ is defined and (iii) $w_0 = f(z_0)$.

A function $f(z)$ is said to be continuous in a region if it is continuous at all points of the region.

15.2.3 Derivative

Let $f(z)$ be a complex-valued function defined in a neighborhood of z_0. Then the derivative of $f(z)$ at z_0 is defined by

$$f'(z_0) = \lim_{\Delta z \to 0} \frac{f(z_0 + \Delta z) - f(z_0)}{\Delta z} \tag{15.8}$$

provided the limit exists and is independent of the manner in which $\Delta z \to 0$.

Definition. A complex-valued function $f(z)$ is said to be *analytic* (*holomorphic* or *regular*) at a point z_0 if there exists a neighborhood $|z - z_0| < \delta$ at all points of which $f'(z)$ exists. The function $f(z)$ is analytic in an open region $\mathcal{R}$ if it is analytic at all points of $\mathcal{R}$.

Remarks.
(i) analyticity is a property defined over open sets while differentiability could conceivably hold at one point only. For example, for the function

$$f(z) = |z^2|,$$

the derivative exists at $z_0 = 0$ but does not exist at any other point.
(ii) As in the case of function of a real variable, differentiability implies continuity but the converse is not true. For example, the function $f(z) = |z^2|$ is continuous everywhere but is differentiable only at $z_0 = 0$. Similarly, the function $f(z) = \bar{z}$ is continuous everywhere but is not differentiable at any point.

15.2.4 The Cauchy–Riemann equations

The property of analyticity of a function of complex variable z dictates a relationship between the derivatives of its real and imaginary parts. If we write

$$f(z) = u(x,y) + iv(x,y) \tag{15.9}$$

and if $f(z)$ is differentiable at z_0, then $f'(z_0)$ is independent of how Δz approaches zero.

If $\Delta z = \Delta x$, then from the definition of the derivative, it follows that

$$f'(z_0) = \frac{\partial u}{\partial x}(x_0,y_0) + i\frac{\partial v}{\partial x}(x_0,y_0) \tag{15.10}$$

However, if $\Delta z = i\Delta y$, then we have

$$\begin{aligned} f'(z_0) &= \frac{1}{i}\frac{\partial u}{\partial y}(x_0,y_0) + \frac{\partial v}{\partial y}(x_0,y_0) \\ &= \frac{\partial v}{\partial y}(x_0,y_0) - i\frac{\partial u}{\partial y}(x_0,y_0) \end{aligned} \tag{15.11}$$

Thus, if $f'(z_0)$ exists, a necessary condition from equation (15.10) and equation (15.11) is

$$\frac{\partial u}{\partial x} = \frac{\partial v}{\partial y}, \quad \frac{\partial u}{\partial y} = -\frac{\partial v}{\partial x} \quad @\,(x_0,y_0) \tag{15.12}$$

Equations (15.12) are referred to as the Cauchy–Riemann (C–R) equations. It may be shown that a necessary and sufficient condition for $f(z)$ to be analytic in a region $\mathcal{R}$ is that the C–R equations hold and the first partial derivatives of $u(x,y)$ and $v(x,y)$ are continuous. In polar coordinates, the C–R equations become

$$\frac{\partial u}{\partial r} = \frac{1}{r}\frac{\partial v}{\partial \theta}, \quad \frac{\partial v}{\partial r} = -\frac{1}{r}\frac{\partial u}{\partial \theta} \tag{15.13}$$

Definition. A real-valued function $\phi(x,y)$ is said to be harmonic in a region $\mathcal{R}$ if all its second partial derivatives are continuous and

$$\frac{\partial^2 \phi}{\partial x^2} + \frac{\partial^2 \phi}{\partial y^2} = 0 \quad \text{(Laplace's eq.)} \tag{15.14}$$

at each point of $\mathcal{R}$.

Theorem (Analytic functions). *If $f(z) = u(x,y) + iv(x,y)$ is analytic in a region $\mathcal{R}$, then the functions $u(x,y)$ and $v(x,y)$ are harmonic in $\mathcal{R}$.*

The proof of this theorem follows from the C–R equations. $u(x,y)$ and $v(x,y)$ are called the *conjugate harmonic functions*.

15.2.5 Some elementary functions of a complex variable

Here, we give some examples of some elementary functions of a complex variable:

1. Polynomial functions
   Let

$$\begin{aligned} w &= P_n(z) \\ &= a_0 + a_1 z + a_2 z^2 + \cdots + a_n z^n, \end{aligned}$$

   where n is a nonnegative integer and $a_i \in \mathbb{C}$.
2. Algebraic functions
   These functions are defined by solutions to equations of the form:

$$P_0(z)w^n + P_1(z)w^{n-1} + \cdots + P_n(z) = 0.$$

   For example,

$$\begin{aligned} w_1(z) &= z^{\frac{1}{2}} \\ w_2(z) &= z^{\frac{1}{2}} + z^{\frac{1}{3}} \end{aligned}$$

   are algebraic functions. [$w_1(z)$ is two-valued function while $w_3(z)$ is six-valued.]

3. Rational algebraic functions
   These functions are of the form:

$$w(z) = \frac{P_n(z)}{P_m(z)}.$$

   For example, for $n = m = 1$, we obtain the bilinear transformation

$$w(z) = \frac{az + b}{cz + d}; \quad a, b, c, d \in \mathbb{C}.$$

4. Exponential functions

$$w = e^z = e^x \cos y + ie^x \sin y.$$

   This function is periodic with period $2\pi i$, i. e.,

$$w(z) = w(z + 2\pi i)$$

5. Logarithmic functions

$$\begin{aligned} w = \ln z = \ln(re^{i\theta}) &= \ln(re^{i\theta + 2k\pi i}), \quad k = 0, 1, 2, \dots \\ &= \ln r + i(\theta + 2k\pi), \quad k = 0, 1, 2, \dots \end{aligned}$$

   As stated earlier, this function (which is the inverse of the exponential function) is infinite-valued. The primary branch corresponding to $k = 0$ is denoted by

$$\operatorname{Ln} z = \ln r + i\theta.$$

   More generally, the function $f(z)^{g(z)}$ is defined by

$$f(z)^{g(z)} = e^{g(z) \ln f(z)}$$

   and may be infinite-valued. For example,

$$z^\alpha = e^{\alpha \ln z}$$

   is multi-valued for $\alpha = \frac{1}{n} (n = 2, 3, \dots)$ and infinite-valued for general (and complex values of) α.
6. Trigonometric functions

$$\sin z = \frac{e^{iz} - e^{-iz}}{2i}, \quad \cos z = \frac{e^{iz} + e^{-iz}}{2}, \quad \tan z = \frac{\sin z}{\cos z}$$

   are trigonometric functions and periodic with period 2π. The inverse function such as

$$\sin^{-1} z = \frac{1}{i} \ln(iz + \sqrt{1 - z^2})$$

is infinite-valued.

7. Hyperbolic functions

$$\sinh z = \frac{e^z - e^{-z}}{2}, \quad \cosh z = \frac{e^z + e^{-z}}{2}, \quad \tanh z = \frac{\sinh z}{\cosh z}$$

are periodic with period $2\pi i$. The inverse function such as

$$\sinh^{-1} z = \ln(z + \sqrt{z^2 - 1})$$

is infinite-valued.

15.2.6 Zeros and singular points of complex-valued functions

Definition (Zeros). If $f(z)$ is analytic at $z = a$ and $f(a) = f'(a) = f''(a) = \cdots = f^{[n-1]}(a) = 0$ but $f^{[n]}(a) \neq 0$, then we say that $f(z)$ has a zero at $z = a$ of order n.

Definition (Singular point). The point $z = z_0$ is a *singular point* of $f(z)$ if it is not analytic at z_0. It is called an *isolated singular point* if there is no other singular point in the neighborhood, i. e., $\exists \delta > 0 \ni$ there is no other singular point in $0 < |z - z_0| < \delta$.

Singular points may be classified as follows:

(a) Poles

If z_0 is a singular point of $f(z)$ such that

$$\lim_{z \to z_0} (z - z_0)^n f(z) = A \neq 0,$$

then z_0 is called a pole of order n. For $n = 1$, it is called a *simple pole*.

Example 15.2.

(i) $f_1(z) = \frac{e^z}{z-1}$ has a simple pole at $z = 1$.
(ii) $f_2(z) = \frac{1}{\sin z}$ has simple poles at $z = \pm k\pi$, $k = 0, 1, 2, \ldots$
(iii) $f_3(z) = \frac{1}{z^3}$ has a pole of order 3 at $z = 0$.

(b) Branch point

If $f(z)$ is a multivalued function centered at $z = z_0$ or equivalently, $f(z)$ takes multiple values on the circle $|z - z_0| = \varepsilon > 0$, then z_0 is called a branch point.

Example 15.3.

(i) $f_1(z) = \sqrt{z + 1}$ has a branch point at $z = -1$.
(ii) $f_2(z) = \ln z$ has a branch point at $z = 0$.
(iii) $f_3(z) = \cos \sqrt{z}$ has no branch points and is analytic for all z.

(c) Removable singularity
$z = z_0$ is called a removable singularity of $f(z)$ if

$$\lim_{z \to z_0} f(z) \text{ exists.}$$

For example, $f(z) = \frac{\sin\sqrt{z}}{\sqrt{z}}$ has a removable singularity at $z = 0$.

(d) Essential singularity
If $z = z_0$ is a singularity of $f(z)$ that is not a pole, branch point or removable singularity, then z_0 is an essential singularity. For example, the function $e^{\frac{1}{z}}$ has an essential singularity at $z = 0$.

(e) Singularity at infinity
Let

$$g(z) = f\left(\frac{1}{z}\right).$$

If $z = 0$ is a singularity of $g(z)$, then $z = \infty$ is a singularity of $f(z)$. For example, e^z has an essential singularity at $z = \infty$.

Definition. If $f(z)$ is analytic for all z(with $|z| < \infty$), it is called an *entire function*.

For example, $\sin z$, $\cos\sqrt{z}$, e^z and $J_0(z)$ are entire functions.

15.3 Complex integration, Cauchy's theorem and integral formulas

Let $f(z)$ be continuous at all points of a curve C (Figure 15.3), which is assumed to be of finite length (also called rectifiable). Let

$$S_n = \sum_{k=1}^{n} f(\xi_k)(z_k - z_{k-1}) = \sum_{k=1}^{n} f(\xi_k)\Delta z_k; \quad \Delta z_k = z_k - z_{k-1} \tag{15.15}$$

Define the complex line integral

$$\begin{aligned}\int_a^b f(z)\,dz &= \int_C f(z)\,dz \\ &= \lim_{\substack{n\to\infty \\ \max \Delta z_k \to 0}} S_n \end{aligned} \tag{15.16}$$

If this definite integral exists, $f(z)$ is said to be integrable along curve C.

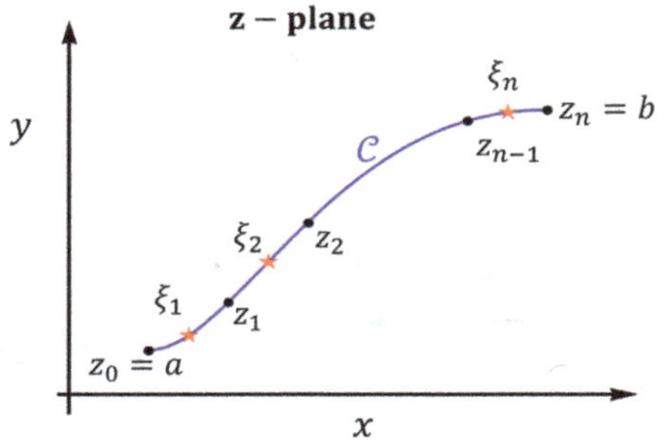

Figure 15.3: Schematic diagram illustrating integration in z-plane along a curve C.

From the above definition, it is seen that the complex line integral of $f(z) = u(x,y) + iv(x,y)$ can be expressed in terms of two real line integrals:

$$\begin{aligned}\int_C f(z)\,dz &= \int_C (u+iv)(dx + i\,dy)\\ &= \int_C u(x,y)\,dx - v(x,y)\,dy + i\int_C v(x,y)\,dx + u(x,y)\,dy\end{aligned} \tag{15.17}$$

Further, many properties of the real definite/line integrals also apply to complex integrals.

15.3.1 Simply and multiply connected domains

A region $\mathcal{R}$ is called *simply connected* if any simple closed curve, which lies in $\mathcal{R}$ can be shrunk to a point without leaving $\mathcal{R}$ (equivalently, the region $\mathcal{R}$ has no holes). In Figure 15.4, $\mathcal{R}_1$ and $\mathcal{R}_2$ are simply-connected while $\mathcal{R}_3$ and $\mathcal{R}_4$ are not. The regions $\mathcal{R}_3$ and $\mathcal{R}_4$ are multiply-connected (or they have one or more holes).

15.3.2 Contour integrals and traversal of a closed path

Let $\mathcal{R}$ be a region in the complex plane and C be its boundary. The boundary is said to be traversed in the positive direction if an observer moving on C has the region to the left. We use the notation

$$\oint_C f(z)\,dz \tag{15.18}$$

to denote the integration of $f(z)$ along the closed curve C, traversed in the positive direction.

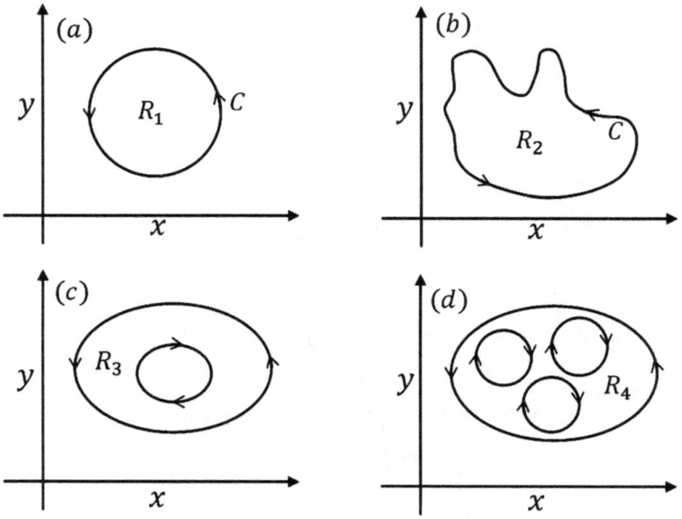

Figure 15.4: Schematic diagram illustrating simply and multiply connected domains.

Example 15.4. Evaluate $\oint_C z^3\,dz$ where C is the circle $|z| = 1$.

Along C, we have $z = e^{i\theta}$, and $dz = ie^{i\theta}\,d\theta$. Thus,

$$\oint_C z^3\,dz = \int_0^{2\pi} e^{i3\theta} ie^{i\theta}\,d\theta$$

$$= i\int_0^{2\pi} e^{i4\theta}\,d\theta = i\int_0^{2\pi} (\cos 4\theta + i\sin 4\theta)\,d\theta = 0$$

15.3.3 Cauchy's theorem

Suppose that $f(z)$ is analytic in a region $\mathcal{R}$ and on its boundary C. Then

$$\oint_C f(z)\,dz = 0. \tag{15.19}$$

This fundamental theorem may be shown to be valid for both simply and multiply connected domains. It was proved by Cauchy with the further assumption of $f'(z)$ to be continuous. However, Goursat removed the restriction and for this reason, it is also referred to as the *Cauchy–Goursat theorem*. Cauchy's proof utilizes Green's theorem in the plane.

Green's theorem. *Let $P(x,y)$ and $Q(x,y)$ be continuous and have continuous partial derivatives in a region $\mathcal{R}$ and on its boundary C. Then*

$$\oint_C P(x,y)\,dx + Q(x,y)\,dy = \iint_{\mathcal{R}} \left(\frac{\partial Q}{\partial x} - \frac{\partial P}{\partial y}\right) dx\,dy \tag{15.20}$$

This theorem is valid for simply as well as multiply connected domains.

If $f'(z)$ is continuous, the Cauchy–Riemann equations are valid and

$$\begin{aligned}\oint_C f(z)\,dz &= \int_C u(x,y)\,dx - v(x,y)\,dy + i\int_C v(x,y)\,dx + u(x,y)\,dy \\ &= \iint_R \left(\frac{-\partial v}{\partial x} - \frac{\partial u}{\partial y}\right) dx\,dy + i\iint_{\mathcal{R}} \left(\frac{\partial u}{\partial x} - \frac{\partial v}{\partial y}\right) dx\,dy \\ &= 0\end{aligned} \tag{15.21}$$

The consequences listed below follow from Cauchy's theorem:

1. If $f(z)$ is analytic in a simply connected region $\mathcal{R}$, the integral $\int_a^b f(z)\,dz$ is independent of the path in $\mathcal{R}$ joining any two points a and b in $\mathcal{R}$.
2. For a and z in $\mathcal{R}$, $F(z) = \int_a^z f(z)\,dz$ is analytic and $F'(z) = f(z)$.
3. If C_1 and C_2 are any closed curves in $\mathcal{R}$, and $f(z)$ is analytic on these curves and in the annular region between them, then

$$\oint_{C_1} f(z)\,dz = \oint_{C_2} f(z)\,dz,$$

where C_1 and C_2 are traversed in the positive sense relative to their interiors (see Figure 15.5).

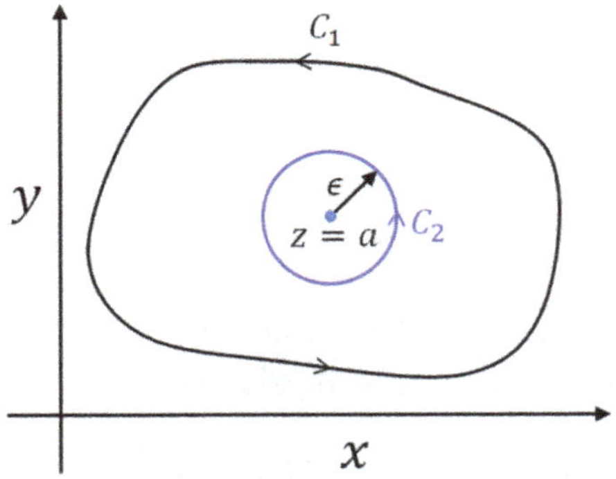

Figure 15.5: Schematic diagram illustrating positive traversal and Cauchy's theorem.

Example 15.5. Evaluate $\oint_C \frac{1}{(z-a)}\,dz$, where C is any simple closed curve and $z = a$ is (i) outside of C and (ii) inside C.

(i) If $z = a$ is outside of C, by Cauchy's theorem $\oint_C \frac{1}{(z-a)}\,dz = 0$.

(ii) If $z = a$ is inside C, let C_2 be a circle of radius ϵ centered at $z = a$. Then

$$I_1 = \oint_C \frac{1}{(z-a)}\,dz = \oint_{C_2} \frac{1}{(z-a)}\,dz$$

But on C_2,

$$z - a = \epsilon e^{i\theta}; \quad dz = i\epsilon e^{i\theta}\,d\theta$$

Thus,

$$I_1 = \oint_{C_2} \frac{1}{(z-a)}\,dz = \int_0^{2\pi} \frac{i\epsilon e^{i\theta}\,d\theta}{\epsilon e^{i\theta}} = 2\pi i$$

Example 15.6. Evaluate $I_n = \oint_C \frac{1}{(z-a)^n}\,dz$, $n = 2, 3, 4, \ldots$ where C is any simple closed curve.

Using the same procedure as above, it is easily seen that $I_n = 0$ for both cases, i. e., when $z = a$ is inside or outside of C.

15.3.4 Cauchy's integral formulas

Let $f(z)$ be analytic inside and on a simple closed curve C and let a be any point inside C. Then

$$f(a) = \frac{1}{2\pi i} \oint_C \frac{f(z)}{(z-a)}\,dz \tag{15.22}$$

$$f^{[n]}(a) = \frac{n!}{2\pi i} \oint_C \frac{f(z)}{(z-a)^{n+1}}\,dz, \quad n = 1, 2, 3, \ldots \tag{15.23}$$

These formulas follow from Cauchy's theorem and are remarkable as they imply that if $f(z)$ is known on a simple closed curve C, then the values of the function and all its derivatives can be found at all points inside C. An extended form of equation (15.22) is also useful for developing an inversion formula for the Laplace transform. It is also useful for solving Laplace's equation in two spatial dimensions with Dirichlet boundary conditions.

The following are some consequences of Cauchy's integral formulas:

1. Every polynomial of degree n,

$$P_n(z) \equiv a_0 + a_1 z + \cdots + a_n z^n = 0$$

with $n \geq 1$ and $a_n \neq 0$, has exactly n roots, counting multiplicity [also known as the fundamental theorem of algebra].

2. If $f(z)$ is analytic inside and on a circle C with center at a and radius r, then $f(a)$ is the mean value of $f(z)$ on C, i. e.,

$$f(a) = \frac{1}{2\pi} \int_0^{2\pi} f(a + re^{i\theta})\, d\theta$$

[This is also known as Gauss' mean value theorem.]
3. If $f(z)$ is analytic inside and on a simple closed curve C, except for a finite number of poles inside C, then

$$\frac{1}{2\pi i} \oint_C \frac{f'(z)}{f(z)}\, dz = N - P$$

where N and P are, respectively, the number of zeros and poles of $f(z)$ inside C.
4. Suppose that $f(z)$ and $g(z)$ are analytic inside and on a simple closed curve C and suppose that $|g(z)| < |f(z)|$ on C, then $f(z) + g(z)$ and $f(z)$ have the same number of zeros inside C. [This is also known as Rouche's theorem.]

For a proof of these and many other related theorems, we refer to the book by Spiegel [29].

Example 15.7 (Poisson's integral formula for a circle). Let $f(z)$ be analytic inside and on a circle C defined by $|z| = R$ and let $z = re^{i\theta}$ be any point inside C. We have by Cauchy's integral formula

$$f(z) = f(re^{i\theta}) = \frac{1}{2\pi i} \oint_C \frac{f(w)}{(w - z)}\, dw. \tag{15.24}$$

The inverse of the point z with respect to the circle C lies outside of C and is given by $\frac{R^2}{\bar{z}}$. By Cauchy's theorem,

$$0 = \frac{1}{2\pi i} \oint_C \frac{f(w)}{(w - \frac{R^2}{\bar{z}})}\, dw. \tag{15.25}$$

Subtraction of equation (15.25) from equation (15.24) gives

$$f(z) = \frac{1}{2\pi i} \oint_C \frac{(z - \frac{R^2}{\bar{z}})}{(w - z)(w - \frac{R^2}{\bar{z}})} f(w)\, dw \tag{15.26}$$

Let $z = re^{i\theta}$ and $w = \mathrm{R}e^{i\phi}$ in equation (15.26) to obtain

$$f(re^{i\theta}) = \frac{1}{2\pi} \int_0^{2\pi} \frac{(R^2 - r^2) f(\mathrm{R}e^{i\phi})}{R^2 - 2Rr\cos(\theta - \phi) + r^2} \, d\phi \tag{15.27}$$

Writing

$$f(re^{i\theta}) = u(r, \theta) + iv(r, \theta)$$
$$\text{and} \quad f(\mathrm{R}e^{i\phi}) = u(R, \phi) + iv(R, \phi)$$

and separating the real and imaginary parts of equation (15.27) gives

$$u(r, \theta) = \frac{1}{2\pi} \int_0^{2\pi} \frac{(R^2 - r^2) u(R, \phi)}{R^2 - 2Rr\cos(\theta - \phi) + r^2} \, d\phi \tag{15.28}$$

and

$$v(r, \theta) = \frac{1}{2\pi} \int_0^{2\pi} \frac{(R^2 - r^2) v(R, \phi)}{R^2 - 2Rr\cos(\theta - \phi) + r^2} \, d\phi \tag{15.29}$$

We note that the Poisson's integral formula given by equation (15.28) is the solution of the Laplace's equation

$$\nabla^2 u = \frac{1}{r} \frac{\partial}{\partial r}\left(r \frac{\partial u}{\partial r} \right) + \frac{1}{r^2} \frac{\partial^2 u}{\partial \theta^2} = 0$$

in a circle with boundary value $u(R, \phi)$ specified [Dirichlet problem]. A similar formula may be obtained for the solution of $\nabla^2 u = 0$ in the upper half-plane ($y > 0$) with $u(x, 0)$ specified.

15.4 Infinite series: Taylor's and Laurent's series

Let $u_1(z), u_2(z), \ldots, u_n(z), \ldots$ be a sequence of functions of z and single valued in some region of the z-plane. We call $U(z)$ the limit of the sequence $\{u_n(z)\}$ as $n \to \infty$ and write

$$\lim_{n \to \infty} u_n(z) = U(z)$$

if given $\epsilon > 0$, we can find a number N (depending in general on both ϵ and z) such that

$$|u_n(z) - U(z)| < \epsilon \quad \text{for all } n > N$$

In such case, we say that the sequence converges to $U(z)$. If a sequence $\{u_n(z)\}$ converges for all z in a region $\mathcal{R}$, we call $\mathcal{R}$ the region of convergence of the sequence. A sequence which is not convergent is called divergent.

Example 15.8. Consider the sequence $\{u_n(z) = \frac{2n-1}{n} + i\frac{n+2}{n}\} = \{1+3i, \frac{3}{2}+2i, \frac{5}{3}+i\frac{5}{3}, \ldots\}$. We claim that

$$\lim_{n\to\infty} u_n(z) = 2 + i$$

To verify this, we note that

$$\begin{aligned} |u_n(z) - 2 - i| &= \left|\frac{2n-1}{n} + i\frac{n+2}{n} - 2 - i\right| \\ &= \left|-\frac{1}{n} + i\frac{2}{n}\right| \\ &= \frac{\sqrt{5}}{n} \end{aligned}$$

If $\frac{\sqrt{5}}{n} < \varepsilon \Rightarrow n > \frac{\sqrt{5}}{\varepsilon}$. Thus, if $\varepsilon = 0.01$, $n > 223 = N$ and all terms after the 223rd are inside a circle of radius 0.01 centered at $2 + i$.

15.4.1 Taylor's series

Taylor's theorem. *Let $f(z)$ be analytic in a region $\mathcal{R}$ and let $z = a$ be any point in $\mathcal{R}$. Then there exists precisely one power series with center at $z = a$ representing $f(z)$ and it is given by*

$$\begin{aligned} f(z) &= \sum_{n=0}^{\infty} \frac{f^{(n)}(a)}{n!}(z-a)^n \\ &= f(a) + \frac{f'(a)}{1!}(z-a) + \frac{f''(a)}{2!}(z-a)^2 + \cdots \end{aligned}$$

The above representation is referred to as Taylor's series expansion of $f(z)$ and is valid in the largest open disk with center at $z = a$ in $\mathcal{R}$, i. e., the radius of convergence of the above Taylor's series is equal to the distance of the point $z = a$ to the nearest singularity of $f(z)$, or to the boundary as shown in Figure 15.6 *schematically.*

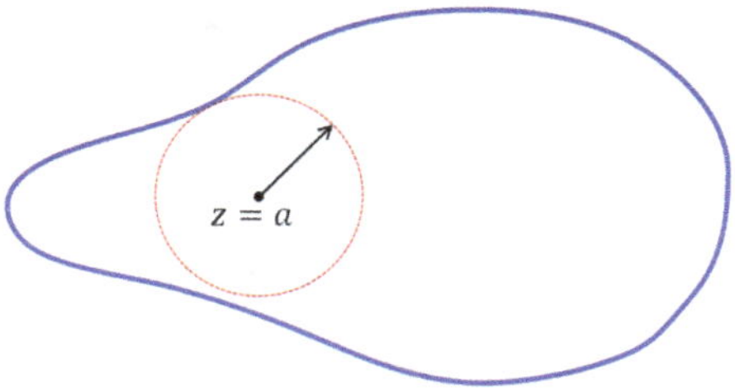

Figure 15.6: Schematic diagram illustrating the region of convergence of Taylor series.

15.4.2 Practical methods of obtaining power series

Let $f(z)$ be analytic in a region $\mathcal{R}$ and let $z = a$ be any point in $\mathcal{R}$. Then we have seen that $f(z)$ has a power series representation

$$f(z) = \sum_{n=0}^{\infty} \frac{f^{(n)}(a)}{n!}(z-a)^n$$

that converges uniformly in the disk $|z-a| < b\,(b > 0)$. The power series may be obtained either by evaluating the derivatives of $f(x)$ or by other methods.

Example 15.9.

(a)

$$\begin{aligned} f(z) &= \frac{1}{1+z} \\ &= 1 - z + z^2 - z^3 + \cdots + (-1)^n z^n + \cdots \qquad (|z| < 1) \end{aligned}$$

(b)

$$\begin{aligned} f(z) &= \ln(1+z) \\ &= z - \frac{z^2}{2} + \frac{z^3}{3} + \cdots + \frac{(-1)^{n-1} z^n}{n} + \cdots \qquad (|z| < 1) \end{aligned}$$

(c)

$$\begin{aligned} f(z) &= \tan^{-1} z \\ &= z - \frac{z^3}{3} + \frac{z^5}{5} + \cdots + \frac{(-1)^n z^{2n+1}}{2n+1} + \cdots \qquad (|z| < 1) \end{aligned}$$

In cases (b) and (c), the series represent the principal value of the function.

15.4.3 Laurent series

In many applications, it is necessary to expand a function around points where $f(z)$ is singular. In such cases, Taylor's theorem cannot be applied and a new type of series, called the Laurent series, is necessary.

Example 15.10. Consider the function $f(z) = \frac{1}{z^2(1-z)}$. Using the result,

$$\frac{1}{1-z} = 1 + z + z^2 + z^3 + \cdots; |z| < 1,$$

we obtain

$$f(z) = \frac{1}{z^2} + \frac{1}{z} + 1 + z + z^2 + z^3 + \cdots; \quad 0 < |z| \leq \gamma < 1$$

which is valid in the punctured disk $0 < |z| \leq \gamma$ (i. e., all points of the disk $|z| \leq \gamma$ excluding the center). This is the Laurent series of the function $f(z)$. The first part containing the reciprocal powers of z is called the *principal part* while the rest of the series containing the constant and positive power of z is called the *analytic part* of $f(z)$.

Theorem (Laurent). *If $f(z)$ is analytic and single-valued on two concentric circles C_1 and C_2 with center at $z = a$ and in the annulus between them, then $f(z)$ can be represented by the (Laurent) series*

$$f(z) = \sum_{n=-\infty}^{\infty} a_n (z-a)^n$$

where

$$a_n = \frac{1}{2\pi i} \oint_C \frac{f(w)}{(w-a)^{n+1}}\, dw$$

and C is a closed curve in the annulus that encircles C_1 (Figure 15.7). The series converges and represents $f(z)$ in the open annulus obtained from the given annulus by continuously increasing C_1 and decreasing C_2 until each of the two circles reaches a point where $f(z)$ is singular.

15.5 The residue theorem and integration by the method of residues

If $f(z)$ is analytic in a neighborhood of a point $z = a$, then by Cauchy's integral theorem, we can write

$$\int_C f(z)\, dz = 0 \tag{15.30}$$

for any contour C in that neighborhood. If, however, $f(z)$ has a pole or an isolated essential singularity at $z = a$, and $z = a$ lies in the interior of C, then the integral (equation (15.30)) will, in general, be different from zero. In this case, we may represent $f(z)$ by a Laurent series

$$f(z) = \sum_{n=0}^{\infty} a_n (z-a)^n + \sum_{n=1}^{\infty} \frac{b_n}{(z-a)^n}, \tag{15.31}$$

which converges in the annulus $0 < |z-a| < R$. Here, R is the distance from $z = a$ to the nearest singularity of $f(z)$ and

$$a_n = \frac{1}{2\pi i} \oint_{C_2} \frac{f(w)}{(w-a)^{n+1}}\, dw \tag{15.32}$$

$$b_n = \frac{1}{2\pi i}\oint_{C_1}(w-a)^{n-1}f(w)\,dw \tag{15.33}$$

where C_1 and C_2 are the contours enclosing the point $z = a$ as shown in Figure 15.7.

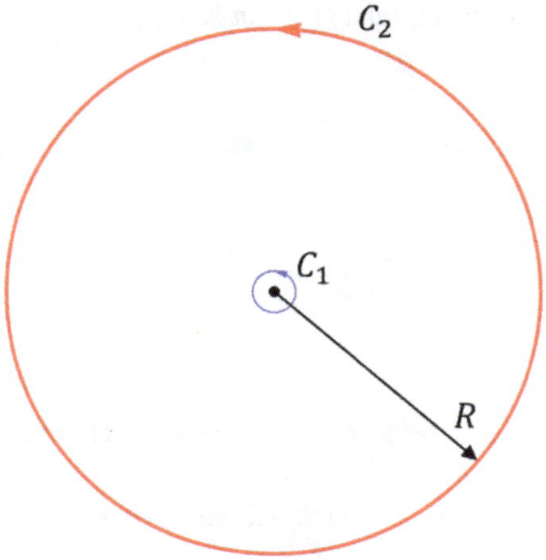

Figure 15.7: Schematic diagram illustrating Laurant's theorem.

From equations (15.32)–(15.33), we get

$$b_1 = \frac{1}{2\pi i}\oint_{C_1} f(w)\,dw$$
$$\Longrightarrow \oint_{C_1} f(z)\,dz = 2\pi i b_1 \tag{15.34}$$

The coefficient b_1 in the expansion (equation (15.31)) is called the residue of $f(z)$ at $z = a$.

Formally, we may obtain equation (15.34) from equation (15.31) by integrating term by term as follows:

$$\begin{aligned}\oint_{C_1} f(z)\,dz &= \oint_{C_1}\left[\sum_{n=0}^{\infty} a_n(z-a)^n + \sum_{n=1}^{\infty}\frac{b_n}{(z-a)^n}\right]dz\\ &= \sum_{n=0}^{\infty} a_n\oint_{C_1}(z-a)^n\,dz + \sum_{n=1}^{\infty} b_n\oint_{C_1}\frac{dz}{(z-a)^n}\\ &= 0 + \sum_{n=1}^{\infty} b_n\oint_{C_1}\frac{dz}{(z-a)^n}\end{aligned}$$

But on contour C_1, we can write

$$z - a = \varepsilon e^{i\theta} \quad \text{and} \quad dz = i\varepsilon e^{i\theta}\, d\theta, \quad \text{with } |\varepsilon| < R$$

$\Longrightarrow$

$$\oint_{C_1} \frac{dz}{(z-a)^n} = \int_0^{2\pi} \frac{i\varepsilon e^{i\theta}\, d\theta}{\varepsilon^n e^{in\theta}} = i\varepsilon^{1-n} \int_0^{2\pi} e^{i(1-n)\theta}\, d\theta$$

$$= \begin{cases} 0, & n \neq 1 \\ 2\pi i, & n = 1 \end{cases}$$

Thus,

$$\oint_{C_1} f(z)\, dz = \sum_{n=1}^{\infty} b_n \oint_{C_1} \frac{dz}{(z-a)^n} = 2\pi i b_1$$
$$= 2\pi i.\, \mathrm{Res} f(z)|_{z=a} \tag{15.35}$$

Since b_1 is the only term that contributes to the integral, it is called the *residue*.

Note that the Laurent expansion may be obtained by various methods without using the formula (equations (15.32)–(15.33)). Hence, we may determine the residue by one of these methods and then use the formula (equation (15.35)) for evaluating the contour integral.

Example 15.11.

1. Evaluate

$$\oint_C e^z\, dz, \quad \text{where } C : |z| = 1$$

Since e^z is analytic function inside C, we have

$$\oint_C e^z\, dz = 0$$

2. Evaluate

$$\oint_C z^2 e^{\frac{1}{z}}\, dz, \quad \text{where } C : |z| = 5$$

Note that the point $z = 0$ is an essential singularity of $e^{\frac{1}{z}}$. The expansion of $e^{\frac{1}{z}}$ can be expressed as

$$e^{\frac{1}{z}} = 1 + \frac{1}{z} + \frac{1}{2!}\frac{1}{z^2} + \frac{1}{3!}\frac{1}{z^3} + \frac{1}{4!}\frac{1}{z^4} \cdots$$

$\Longrightarrow$

$$z^2 e^{\frac{1}{z}} = z^2 + z + \frac{1}{2!} + \frac{1}{3!}\frac{1}{z} + \frac{1}{4!}\frac{1}{z^2} \cdots$$

$\Longrightarrow$

$$b_1 = \text{Res}(z^2 e^{\frac{1}{z}})\big|_{z=0} = \frac{1}{3!}$$

Thus,

$$\oint_C z^2 e^{\frac{1}{z}}\, dz = 2\pi i b_1 = \frac{\pi i}{3}$$

3. Evaluate

$$\oint_C \frac{\sin z}{z^3}\, dz, \quad \text{where } C : |z| = 1$$

The expansion of $\frac{\sin z}{z^3}$ can be expressed as

$$\begin{aligned} \frac{\sin z}{z^3} &= \frac{1}{z^3}\left(z - \frac{z^3}{3!} + \frac{z^5}{5!} - \cdots\right) \\ &= \frac{1}{z^2} - \frac{1}{3!} + \frac{z^2}{5!} - \cdots \end{aligned}$$

Thus, $\frac{\sin z}{z^3}$ has a second-order pole at $z = 0$ with residue as

$$b_1 = \text{Res}\left(\frac{\sin z}{z^3}\right)\bigg|_{z=0} = 0$$

Thus,

$$\oint_C \frac{\sin z}{z^3}\, dz = 2\pi i b_1 = 0.$$

15.5.1 Other methods for evaluating residues

(a) Simple pole

Suppose that $f(z)$ has a simple pole at $z = a$. Then, near $z = a, f(z)$ has a representation

$$f(z) = \frac{p(z)}{q(z)}$$

where $p(z)$ and $q(z)$ are analytic at $z = a$,

$$q(z) = (z-a)\left[q'(a) + \frac{(z-a)}{2!}q''(a) + \frac{(z-a)^2}{3!}q'''(a) + \cdots\right],$$

and

$$q'(a) \neq 0, \quad p(a) \neq 0,$$

i. e., $q(z)$ has a simple zero at $z = a$. Thus, the residue of $f(z)$ is given by

$$\begin{aligned}\operatorname{Res} f(z)|_{z=a} &= \lim_{z\to a} \frac{(z-a)p(z)}{q(z)} \\ &= \frac{p(a)}{q'(a)}\end{aligned} \tag{15.36}$$

(using L'Hospital's rule)

Example 15.12.
(i) Consider $f(z) = \frac{4-3z}{z(z-1)}$, which has simple poles at $z = 0$ and $z = 1$. Thus,

$$\operatorname{Res} f(z)|_{z=0} = \lim_{z\to 0} \frac{4-3z}{(z-1)} = -4$$
$$\operatorname{Res} f(z)|_{z=1} = \lim_{z\to 0} \frac{4-3z}{z} = 1$$

(ii) Consider $f(z) = \tan z = \frac{\sin z}{\cos z}$, which has simple poles at

$$a_k = (2k-1)\frac{\pi}{2}, \quad k = 0, \pm 1, \pm 2, \ldots$$

with residue given by

$$\begin{aligned}\operatorname{Res} f(z)|_{z=a_k} &= \lim_{z\to a_k} \frac{(z-a_k)\sin z}{\cos z} = \sin a_k \lim_{z\to a_k} \frac{(z-a_k)}{\cos z} \\ &= \sin a_k \lim_{z\to a_k} \frac{1}{\sin z} = \frac{\sin a_k}{\sin a_k} = 1\end{aligned}$$

(b) Poles of higher order
If $f(z)$ has a pole of order $m(> 1)$ at $z = a$, then the Laurent series expansion is of the form:

$$f(z) = \sum_{n=0}^{\infty} a_n (z-a)^n + \frac{b_1}{z-a} + \frac{b_2}{(z-a)^2} + \cdots + \frac{b_m}{(z-a)^m}$$

$\Longrightarrow$

$$(z-a)^m f(z) = b_m + b_{m-1}(z-a) + \cdots + b_1(z-a)^{m-1} + \sum_{n=0}^{\infty} a_n (z-a)^{n+m}$$

$\Longrightarrow$

$$\frac{d^{m-1}}{dz^{m-1}}[(z-a)^m f(z)]\big|_{z=a} = (m-1)!b_1$$

Thus,

$$\operatorname{Res} f(z)|_{z=a} = b_1 = \frac{1}{(m-1)!}\lim_{z\to a}\left\{\frac{d^{m-1}}{dz^{m-1}}[(z-a)^m f(z)]\right\} \tag{15.37}$$

Example 15.13. Consider $f(z) = \frac{e^z}{(z-1)^3}$, which has pole of order 3 at $z = 1$. Thus,

$$\operatorname{Res} f(z)|_{z=1} = \lim_{z\to 1}\frac{1}{2!}\lim_{z\to 1}\left\{\frac{d^2}{dz^2}[e^z]\right\} = \frac{e}{2}$$

Alternatively, we can rewrite the function $f(z)$ by transforming $z = 1 + u$, which simplifies $f(z)$ as

$$\begin{aligned} f &= \frac{e^{1+u}}{u^3} = \frac{e}{u^3}\left(1 + u + \frac{u^2}{2!} + \frac{u^3}{3!} + \frac{u^4}{4!} + \cdots\right) \\ &= \frac{e}{u^3} + \frac{e}{u^2} + \frac{e}{2!u} + \frac{1}{3!} + \frac{u}{4!} + \cdots \end{aligned}$$

$\Longrightarrow$

$$\mathrm{b}_1 = \operatorname{Res} f(z)|_{z=1} = \operatorname{Res} f(u)|_{u=0} = \frac{e}{2!} = \frac{e}{2}$$

15.5.2 Residue theorem

Let $f(z)$ be a single-valued function, which is analytic inside a simple closed path C and on C, except for finitely many singular points at $a_1, a_2, \ldots a_m$ inside C. Then

$$\oint_C f(z)\,dz = 2\pi i\sum_{j=1}^{m}\operatorname{Res} f(z)|_{z=a_j}$$

Proof. Consider the schematic of a simply connected contour C and positively oriented circles C_k (interior to C and centered at $z = a_k$) as shown in Figure 15.8.

It follows from Cauchy's theorem that

$$\begin{aligned} \oint_C f(z)\,dz &= \oint_{C_1} f(z)\,dz + \oint_{C_2} f(z)\,dz + \cdots + \oint_{C_m} f(z)\,dz \\ &= 2\pi i \operatorname{Res} f(z)|_{z=a_1} + 2\pi i \operatorname{Res} f(z)|_{z=a_2} + \cdots + 2\pi i \operatorname{Res} f(z)|_{z=a_m} \\ &= 2\pi i\sum_{j=1}^{m}\operatorname{Res} f(z)|_{z=a_j} \end{aligned} \tag{15.38}$$

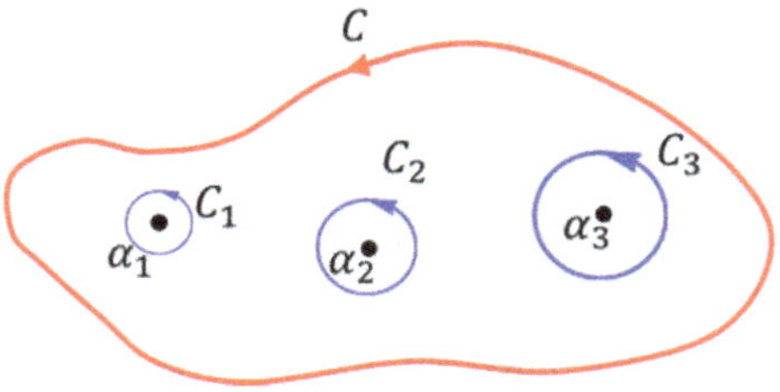

Figure 15.8: Simply connected contour C containing positively oriented circles C_k centered at $z = \alpha_k$.

This important theorem has various applications in connection with complex and real integrals that appears in the inversion of Laplace and Fourier transforms. This will be illustrated in later chapters. □

Problems

1. (*Complex numbers and functions*):
   (a) Evaluate the following:
   (i) $\sin^{-1} 2$, (ii) $\cos(1+i)$, (iii) i^i, (iv) $\ln(-4)$ and (v) $\sinh^{-1} z$
   (b) Use the definition of elementary functions to prove the following:
   (i) $\sin iz = i \sinh z$, (ii) $\cos iz = \cosh z$ and (iii) $\tan z = \frac{\sin 2x}{\cos 2x + \cosh 2y} + i\frac{\sinh 2y}{\cos 2x + \cosh 2y}$
2. (*Real and complex roots of nonlinear equations*): Determine all the roots (real and complex) of the following equations:
   (a) $z \sin z = \gamma \cos z \quad (\gamma = 2)$
   (b) $1 + \frac{k_p e^{-zD}}{1+\tau z} = 0 \; (k_p = 1, \tau = 1, D = 1)$

   Remark. Equation (a) appears in the solution of unsteady state heat/mass transform problems while (b) appears in the stability analysis of a closed loop control system with delay.

3. Determine which of the following functions are analytic:
   (a) $z^3 + z$
   (b) $\arg z$
   (c) $\frac{1}{1-z}$
   (d) $\frac{z}{\bar{z}+2}$.
4. Determine the six branches of the function $f(z) = z^{\frac{1}{2}} + z^{\frac{1}{3}}$ in terms of polar coordinates.
5. Show that the following functions are harmonic and find the analytic function $f(z) = u + iv$:
   (a) $u = \ln(x^2 + y^2)$ and
   (b) $v = -\sin x \sinh y$
   (c) $u = xe^x \cos y - ye^x \sin y$
6. Verify Cauchy's theorem for the function $f(z) = 3z^2 + iz - 4$ if C is the unit circle.

7. Show that

$$\frac{1}{2\pi i}\oint_C \frac{e^{zt}}{z^2+1}\,dz = \sin t, \quad t > 0$$

where C is the circle $|z| = 3$.

8. Find the steady-state temperature T at each point inside the unit disk if the temperature of the rim is at the levels shown in Figure 15.9. What is the temperature at the center?

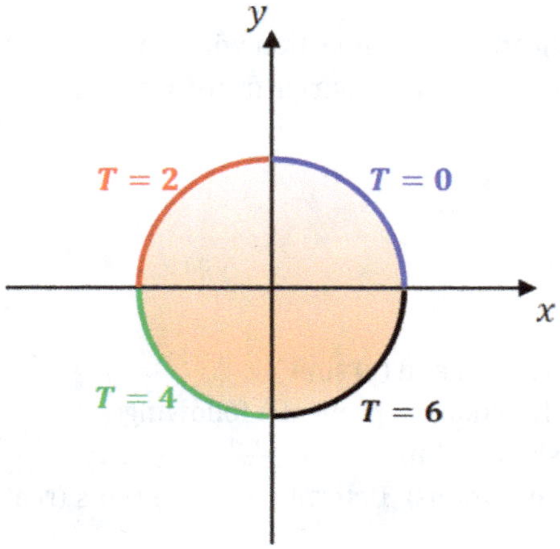

Figure 15.9: Schematic of a disc subjected to a temperature distribution at the rim.

9. (*Singularity and convergence*)
   (a) Locate and name the singularities of the following functions:
   (i) $\frac{\sqrt{1+z^2}}{(z^3-1)}\sin(\frac{1}{1+z})$, (ii) $\operatorname{sech} z$, (iii) $\tan^{-1}(z^2+2z+2)$ and (iv) $\exp(\tan\frac{1}{z})$ (v) $\frac{\sinh z\sqrt{a}}{z\sinh\sqrt{z}}$
   (b) Determine the region of convergence for:
   (i) $\sum_{n=1}^{\infty}\frac{e^{2\pi inz}}{(n+1)^{3/2}}$, (ii) $\sum_{n=1}^{\infty}\frac{(z+i)^n}{(n+1)(n+2)}$ and (iii) $\sum_{n=1}^{\infty}\frac{n!}{(n+1)^{n+1}}z^n$
   (c) Expand each of the following in a Laurent series about $z = 0$, naming the type of singularity in each case:
   (i) $\frac{1-\cos z}{z}$, (ii) $\frac{e^{z^2}}{z^3}$, (iii) $z^2e^{-z^4}$, (iv) $\frac{\cosh z^{-1}}{z}$ and (v) $z\sinh\sqrt{z}$
   (d) State whether each of the following functions are entire, meromorphic or neither:
   (i) z^2e^{-z}, (ii) $\cot 2z$, (iii) $\frac{1-\cos z}{z}$, (iv) $z\sin\frac{1}{z}$ and (v) $\frac{\sin\sqrt{z}}{\sqrt{z}}$ (vi) $z+\frac{1}{z}$

10. Show that

$$\int_0^{\infty}\frac{\sin ax}{e^{2\pi x}-1}\,dx = \frac{1}{4}\coth\frac{a}{2} - \frac{1}{2a}$$

Hint: Use $\frac{e^{aiz}}{e^{2\pi z}-1}$ around a rectangle with vertices at $0, R, R+i, i$ and let $R \to \infty$.

11. Nyquist stability criterion: Let $f(z) = P_n(z) + Q_m(z)e^{-zD}$, where $D > 0$; $P_n(z)$ and $Q_m(z)$ are relative prime polynomials, $n > m$ and $f(z)$ has no zeros on the imaginary axis while $P_n(z)$ has N zeros in the right half-plane. Prove that for the function $f(z)$ to have no zeros in the right half-plane, it is necessary and sufficient that the point

$$w = -\frac{Q_m(z)}{P_n(z)}e^{-zD}$$

wind around the point $w = 1$, N times in the positive direction while the point z traverses the entire imaginary axis upwards.

(a) Apply the theorem to the following function to determine the domain of the real numbers k_p and τ for which all the zeros of the function lie in the left half-plane:

$$f(z) = k_p e^{-zD} + \tau z + 1$$

(b) Extend the results in (a) to second- and higher-order systems.

12. Use the contour integration and residue theorem to show that

$$\int_0^\infty \frac{x^{b-1}}{1+x} = \frac{\pi}{\sin b\pi} \qquad (0 < b < 1).$$

16 Series solutions and special functions

In this chapter, we illustrate ordinary, regular and irregular singular points of first- and second-order differential equations and method of obtaining series solutions. We also introduce various special functions that arise in applications.

16.1 Series solution of a first-order ODE

To illustrate the types of solutions, we consider a linear first-order equation:

$$\frac{dw}{dz} + p(z)w = 0 \tag{16.1}$$

and four special cases of the coefficient function $p(z)$ as discussed below.

Case 1: $p(z) = 1$
The exact solution is $w(z) = c\exp(-z)$, where c is a constant.
$z = 0$ is an ordinary point and the solution is an *entire function.*

Case 2: $p(z) = \frac{1}{2z}$
The exact solution is $w(z) = \frac{c}{\sqrt{z}}$.
$z = 0$ is a regular singular point and the solution has a *branch point* at $z = 0$.

Case 3: $p(z) = \frac{3}{z}$
The exact solution is $w(z) = \frac{c}{z^3}$.
$z = 0$ is a regular singular point and the solution has a *pole of order 3* at $z = 0$.

Case 4: $p(z) = \frac{1}{z^2}$
The exact solution is $w(z) = c\exp(\frac{1}{z})$.
$z = 0$ is an irregular singular point and the solution has an *essential singularity* at $z = 0$.

More generally, when $p(z) = kz^{n-1}$ with $k \neq 0$, the exact solution is given by

$$w(z) = \begin{cases} c\exp(-\frac{k}{n}z^n), & n \neq 0 \\ cz^{-k}, & n = 0 \end{cases} \tag{16.2}$$

where c is a constant. Thus, when $n < 0$ or when n is not an integer, $z = 0$ is an irregular singular point and the solution has an essential singularity at $z = 0$. When n is a natural number, $z = 0$ is an ordinary point and the solution is an entire function. But when $n = 0$, (i) for all noninteger k, $z = 0$ is a regular singular point and the solution has a branch

https://doi.org/10.1515/9783111598055-019

point at $z = 0$; (ii) for any positive integer k, $z = 0$ is a regular singular point and the solution has a pole of order k and (iii) for any negative integer k, $z = 0$ is an ordinary point and the solution is an entire function.

16.2 Ordinary and regular singular points

Consider the homogeneous linear ODE of order n,

$$w^{[n]} + p_{n-1}(z)w^{[n-1]} + \cdots + p_1(z)w'(z) + p_0(z)w(z) = 0 \tag{16.3}$$

where

$$w^{[i]} = \frac{d^i w}{dz^i}$$

Definition. $z_0(\neq \infty)$ is called an *ordinary point* of equation (16.3) if $p_i(z)$, $i = 0, 1, \ldots, n-1$ are analytic at z_0.

Examples. (a)

$$w''(z) + z^2 w'(z) + (z-3)w = 0 \tag{16.4}$$

All points are ordinary points

(b)

$$w''(z) + \frac{1}{(1+z^2)} w'(z) + e^z w = 0 \tag{16.5}$$

All points except $z = \pm i$ are ordinary points.

Theorem. *Suppose that z_0 is an ordinary point of* (16.3). *Then* (16.3) *has n linearly independent solutions that are analytic at z_0. Each solution may be expanded in a Taylor series*

$$w(z) = \sum_{n=0}^{\infty} a_n (z - z_0)^n \tag{16.6}$$

and the radius of convergence of this series is at least as large as the distance to the nearest singularity of the coefficient functions $p_i(z)$.

Example 16.1. Consider the second-order equation

$$w''(z) + \frac{w'(z)}{1+z^2} + e^z w = 0, \tag{16.7}$$

and note that $z_0 = 0$ is an ordinary point of equation (16.7). To determine solutions of the form,

$$w(z) = \sum_{n=0}^{\infty} a_n z^n \tag{16.8}$$

We substitute equation (16.8) into equation (16.7) to obtain

$$(1+z^2)\sum_{n=0}^{\infty} n(n-1)a_n z^{n-2} + \sum_{n=0}^{\infty} n a_n z^{n-1} + (1+z^2)\left(\sum_{n=0}^{\infty} \frac{z^n}{n!}\right)\left(\sum_{n=0}^{\infty} a_n z^n\right) = 0$$

We solve for a_n by equating the coefficients of various powers of z to zero.

z^0: $2a_2 + a_1 + a_0 = 0$

z^1: $6a_3 + 2a_2 + a_1 + a_0 = 0$

z^2: $12a_4 + 3a_3 + 3a_2 + a_1 + \frac{3}{2}a_0 = 0,$

etc.

$\Rightarrow$

$$\begin{aligned} w(z) &= a_0\left(1 - \frac{z^2}{2} + \cdots\right) + a_1\left(z - \frac{z^2}{2} + \frac{z^4}{24} + \cdots\right) \\ &= a_0 w_0(z) + a_1 w_1(z) \end{aligned}$$

Here, $w_0(z)$ and $w_1(z)$ are the two linearly independent solutions.

Definition (Regular singular point, r. s. p.). The point z_0 is a regular singular point of equation (16.3) if not all $p_i(z)$, $i = 0, 1, 2, \ldots, (n-1)$ are analytic at z_0 but if $(z-z_0)^n p_0(z)$, $(z-z_0)^{n-1} p_1(z), \ldots, (z-z_0)p_{n-1}(z)$, are analytic at z_0.

Example 16.2. Consider the second-order equation

$$z^2 w'' + z w' - w = 0$$

$$w'' + \frac{w'}{z} - \frac{w}{z^2} = 0$$

$$p_0(z) = -\frac{1}{z^2}, \quad p_1(z) = \frac{1}{z}$$

$z_0 = 0$ is a regular singular point.

Theorem. *Suppose that z_0 is regular singular point of* (16.3). *Then the n linearly independent solutions of* (16.3) *have one of the following forms:*

$$\begin{aligned} w_1(z) &= (z-z_0)^{\alpha_1} f_1(z) \\ w_2(z) &= (z-z_0)^{\alpha_i} f_i(z), \quad \alpha_i \neq \alpha_1 \\ w_3(z) &= (z-z_0)^{\alpha_1} f_1(z)\ln(z-z_0) + (z-z_0)^{\beta} g_1(z) \\ w_4(z) &= (z-z_0)^{\gamma} \sum_{i=0}^{k} [\ln(z-z_0)]^i f_i(z), \quad \textit{for some } k = 1, 2, \ldots, n-1 \end{aligned}$$

where $f_i(z)$, $g_i(z)$ are analytic at z_0 and have a Taylor series, which converges in a disk extending to the nearest singularity of $p_i(z)$. α_i, β and γ are called indicial exponents. A solution of equation (16.3) is either analytic at z_0 or if it is not analytic, the singularity must be either a pole or an algebraic or logarithmic branch point. The indicial exponents can be obtained by solving a polynomial of degree n whose coefficients depend on $p_i(z)$.

Frobenius method

To determine the solution(s) of the form $w_1(z)$, we write

$$w_1(z) = (z - z_0)^{\alpha_1} \sum_{i=0}^{\infty} a_i (z - z_0)^i$$

$$p_k(z) = \sum_{i=0}^{\infty} p_{ki}(z - z_0)^i, \quad k = 0, 1, \ldots, n - 1$$

To find other forms of solutions, expand $f_i(z)$ and $g_i(z)$ in a Taylor series around z_0. We now consider the special case $n = 2$ (second-order differential equation). Let α_1 and α_2 be the two indicial exponents. Then the different forms of the solutions are

$$w_1(z) = (z - z_0)^{\alpha_1} f_1(z) \tag{16.9}$$

$$w_2(z) = (z - z_0)^{\alpha_2} f_2(z) \tag{16.10}$$

or

$$w_2(z) = (z - z_0)^{\alpha_2} g_1(z) + (z - z_0)^{\alpha_1} \ln(z - z_0) g_2(z) \tag{16.11}$$

The indicial equation is given by

$$\alpha^2 + [p_{10} - 1]\alpha + p_{00} = 0$$

where $p_{10} = \lim_{z \to 0}(z - z_0)p_1(z_0)$, $p_{00} = \lim_{z \to 0}(z - z_0)^2 p_0(z_0)$.

Case 1: $\alpha_1 \neq \alpha_2$; $\alpha_1 - \alpha_2 \neq$ integer

Two Frobenius solutions of the form given in equation (16.9) and equation (16.10)

Case 2: $\alpha_1 - \alpha_2 = 0, 1, 2, \ldots$

Either two Frobenius solutions or one Frobenius solution and one with logarithm

Example 16.3.

$$zw'' + w' + zw = 0$$

(This equation is a special case of $z^2 w'' + zw' + (z^2 - n^2)w = 0$, which is Bessel's equation of order n)

$$w'' + \frac{w'}{z} + w = 0$$

$z = 0$ is regular singular point.

$$p_0(z) = 1, \quad p_1(z) = \frac{1}{z}$$

Indicial equation

$$\alpha^2 + [p_{10} - 1]\alpha + p_{00} = 0$$

$$p_{10} = \lim_{z\to 0} z.\frac{1}{z} = 1$$

$$p_{00} = \lim_{z\to 0} z^2.1 = 0$$

⇒

$$\alpha^2 = 0$$

⇒

$$\alpha = 0, 0$$

Substituting

$$w = \sum_{n=0}^{\infty} a_n z^n$$

gives

$$zw'' + w' + zw = \sum_{n=0}^{\infty} \{a_n n^2 + a_{n-2}\} z^{n-1}$$

⇒

$$a_n = -\frac{a_{n-2}}{n^2}$$

$$a_{2n+1} = 0$$

$$a_2 = -\frac{a_0}{2^2}, \quad a_4 = -\frac{a_2}{4^2}, \quad a_6 = -\frac{a_4}{6^2}, \quad \text{etc.}$$

⇒

$$w_0(z) = a_0 \sum_{k=0}^{\infty} \frac{(-1)^k z^{2k}}{2^{2k}(k!)^2} = a_0 J_0(z)$$

where

$$J_0(z) = \sum_{k=0}^{\infty} \frac{(-1)^k (\frac{z}{2})^{2k}}{(k!)^2} = \text{Bessel function of order zero}$$

$I_0(z) = J_0(iz)$ = Bessel function of second kind of order zero (also known as modified Bessel function). For $|z| \gg 1, J_0(z) = \sqrt{\frac{2}{\pi z}} \cos(z - \frac{\pi}{4})$. A graph of $J_0(z)$ is shown in Figure 16.1.

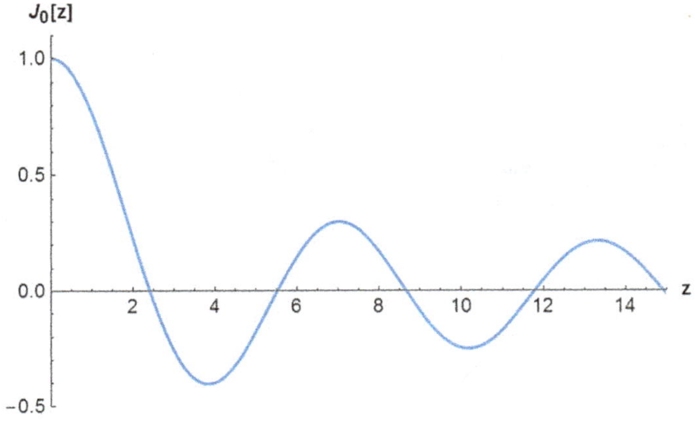

Figure 16.1: Zeroth-order Bessel function of the first kind.

To determine the second solution, we write

$$w_1(z) = g_1(z) + (\ln z) g_2(z),$$

where g_1 and g_2 are analytic functions:

$$g_1(z) = \sum_{n=0}^{\infty} a_n z^n$$

$$g_2(z) = \sum_{n=0}^{\infty} b_n z^n$$

$$w_1' = g_1' + (\ln z) g_2' + \frac{g_2}{z}$$

$$w_1'' = g_1'' + g_2'' \ln z + \frac{2g_2'}{z} - \frac{g_2}{z^2}$$

Substitute these in the differential equation and compare coefficients. After some algebra, we get the second solution

$$w_1(z) = cY_0(z), \quad c = \text{constant}$$

where

$$Y_0(z) = \frac{2}{\pi}\left\{\ln\left(\frac{z}{2}\right) + \gamma\right\}J_0(z) + \frac{2}{\pi}\left\{\frac{z^2}{2^2} - \frac{z^4}{2^2.4^2}\left(1+\frac{1}{2}\right) + \frac{z^6}{2^2.4^2.6^2}\left(1+\frac{1}{2}+\frac{1}{3}\right) - \cdots\right\}$$

and

$$\gamma = \lim_{n\to\infty}\left\{\left(1+\frac{1}{2}+\frac{1}{3}+\cdots+\frac{1}{n}\right) - \ln n\right\} = 0.5772 \quad \text{(Euler's constant)}$$

For $|z| \gg 1$, $Y_0(z) \approx \sqrt{\frac{2}{\pi z}}\sin(z - \frac{\pi}{4})$. A graph of $Y_0(z)$ for real z is shown below (see Figure 16.2). [Remark: This second linearly independent solution may also be obtained by using the formula in the section below.]

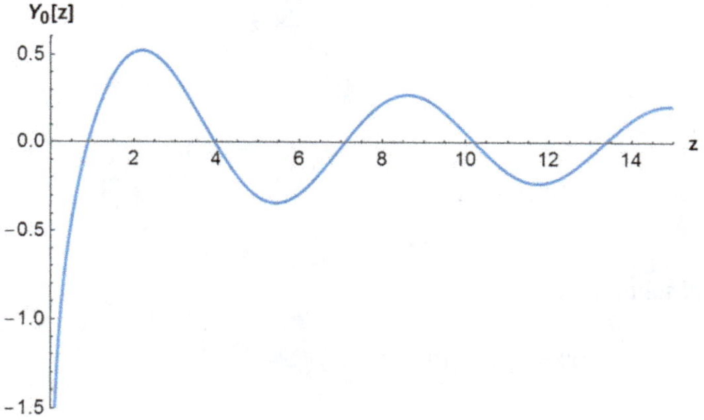

Figure 16.2: Zeroth-order Bessel function of the second kind.

16.3 Series solutions of second-order ODEs

Consider a second-order homogeneous ODE

$$p_0(z)\frac{d^2u}{dz^2} + p_1(z)\frac{du}{dz} + p_2(z)u = 0 \tag{16.12}$$

Let

$$p(z) = \frac{p_1(z)}{p_0(z)}, \quad q(z) = \frac{p_2(z)}{p_0(z)}$$

and write equation (16.12) as

$$\frac{d^2u}{dz^2} + p(z)\frac{du}{dz} + q(z)u = 0 \tag{16.13}$$

Let $\psi_1(z)$ and $\psi_2(z)$ be the two linearly independent solutions of equation (16.13), and

$$W(z) = \begin{vmatrix} \psi_1(z) & \psi_2(z) \\ \psi_1'(z) & \psi_2'(z) \end{vmatrix} = \psi_1\psi_2' - \psi_1'\psi_2 = \text{Wronskian}$$

Now,

$$\begin{aligned} \frac{dW}{dz} &= \begin{vmatrix} \psi_1(z) & \psi_2(z) \\ \psi_1''(z) & \psi_2''(z) \end{vmatrix} = \begin{vmatrix} \psi_1 & \psi_2 \\ -p\psi_1' - q\psi_1 & -p\psi_3' - q\psi_2 \end{vmatrix} \\ &= \begin{vmatrix} \psi_1 & \psi_2 \\ -p\psi_1' & -p\psi_3' \end{vmatrix} = -p(z)\begin{vmatrix} \psi_1 & \psi_2 \\ \psi_1' & \psi_3' \end{vmatrix} \end{aligned}$$

$\Longrightarrow$

$$\frac{dW}{dz} = -p(z)W(z)$$

or

$$W(z) = W(z_0)\exp\left[-\int_{z_0}^{z} p(t')\,dt'\right]$$

$\Longrightarrow$

$$\frac{\psi_1\psi_2' - \psi_1'\psi_2}{\psi_1^2} = \frac{W(z)}{\psi_1^2} = \frac{W(z_0)}{\psi_1^2}\exp\left[-\int_{z_0}^{z} p(t')\,dt'\right]$$

$\Longrightarrow$

$$\frac{d}{dz}\left(\frac{\psi_2}{\psi_1}\right) = \frac{W(z)}{\psi_1^2} = \frac{W(z_0)}{\psi_1^2(z)}\exp\left[-\int_{z_0}^{z} p(t')\,dt'\right]$$

$\Longrightarrow$

$$\psi_2(z) = \psi_1(z)\int_{z_0}^{z} \frac{1}{\psi_1^2(t')}\exp\left[-\int_{z_0}^{t'} p(y)\,dy\right]dt' \tag{16.14}$$

up to a constant. Thus, if one solution of equation (16.13) is known, a second linearly independent solution can be determined using equation (16.14). This result is often used in defining special functions.

Example 16.4. Consider the second-order ODE

$$u'' + \frac{u}{4z^2} = 0$$

Here, $p(z) = 0$ and $q(z) = \frac{1}{4z^2}$. It can be seen that $\psi_1(z) = \sqrt{z}$ is a solution, as $\psi_1' = \frac{1}{2\sqrt{z}}$, $\psi_1'' = -\frac{1}{4z\sqrt{z}} \Longrightarrow 4z^2\psi_1'' + \psi_1 = 0$. Thus, from equation (16.7), we get

$$\psi_2 = \sqrt{z}\int_{z_0}^{z} \frac{1}{t'}\,dt' = \sqrt{z}(\ln z - \ln z_0) = \sqrt{z}\ln z + c_1\psi_1$$

Thus, we can take $\psi_2 = \sqrt{z}\ln z$ as the second linearly independent solution.

Example 16.5 (Legendre equation). Consider the second-order Legendre equation

$$(1 - z^2)w'' - 2zw' + n(n + 1)w = 0.$$

Here, $p(z) = -\frac{2z}{1-z^2} \Longrightarrow$

$$\exp\left[-\int^{z} p(t')\,dt'\right] = \frac{1}{1 - z^2}$$

For $n = 0$, $w = P_0(z) = 1$ is a solution (zeroth-order Legendre function of the first kind). Using equation (16.14), the second solution is given by

$$Q_0(z) = 1.\int \frac{1}{(1 - z^2).1^2}\,dz = \frac{1}{2}\ln\left(\frac{1 + z}{1 - z}\right)$$

which is zeroth-order Legendre function of the second kind.

For $n = 1$, the solution is same as in the above example where first-order Legendre function of first kind is $P_1(z) = \psi_1(z) = z$ and first-order Legendre function of second kind is $Q_1(z) = \psi_2(z) = \frac{z}{2}\ln(\frac{1+z}{1-z}) - 1$.

For $n = 2$, the solution is given by second-order Legendre function of first and second kind as given by

$$P_2(z) = \frac{1}{2}(3z^2 - 1)$$

$$Q_2(z) = \frac{3z^2 - 1}{4}\ln\left(\frac{1 + z}{1 - z}\right) - \frac{3z}{2}$$

Similarly, higher-order Legendre functions of both kinds can be obtained.

16.4 Special functions defined by second-order ODEs

In this section, we summarize some second-order equations and their solutions in terms of special functions. These appear in many of our applications.

16.4.1 Airy equation

The Airy equation is given by

$$w'' = zw$$

The two linearly independent solutions are expressed in powers of z (as $z = 0$ is an ordinary point) and are denoted by Ai(z) and $Bi(z)$. These are defined by

$$\text{Ai}(z) = \frac{1}{\pi 3^{2/3}} \sum_{n=0}^{\infty} \frac{\Gamma(\frac{n+1}{3})}{n!} (3^{1/3}z)^n \sin\left[\frac{2(n+1)\pi}{3}\right]$$

$$Bi(z) = \frac{1}{\pi 3^{1/6}} \sum_{n=0}^{\infty} \frac{\Gamma(\frac{n+1}{3})}{n!} (3^{1/3}z)^n \left|\sin\left[\frac{2(n+1)\pi}{3}\right]\right|$$

where Γ is the Gamma function. For a plot of the Airy Ai(z) and Bairy Bi(z) functions, see Figure 16.3. Note that $\text{Ai}(0) = \frac{1}{3^{2/3}\Gamma(\frac{2}{3})} = 0.3550$ and $\text{Bi}(0) = \frac{1}{3^{1/6}\Gamma(\frac{2}{3})} = 0.6149$.

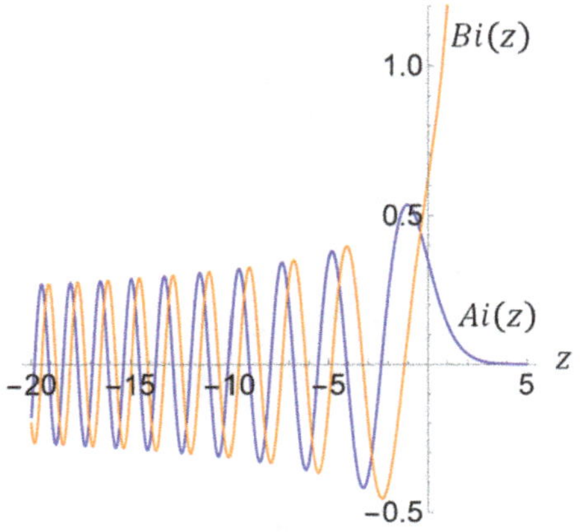

Figure 16.3: Airy functions Ai(z) and Bi(z).

16.4.2 Bessel equation

The Bessel equation is given by

$$z^2w'' + zw' + (z^2 - \nu^2)w = 0$$

where ν may not have to be an integer. Note that $z = 0$ is a regular singular point. The two linearly independent solutions are denoted by $J_\nu(z)$ and $Y_\nu(z)$ and referred to as Bessel functions of first and second kind, respectively. These are defined by

$$J_\nu(z) = \sum_{m=0}^{\infty} \frac{(-1)^m}{m!\Gamma(m+\nu+1)}\left(\frac{z}{2}\right)^{2m+\nu}, \quad \forall\nu \text{ (including nonintegers)}$$

$$Y_\nu(z) = \frac{J_\nu(z)\cos(\nu\pi) - J_{-\nu}(z)}{\sin(\nu\pi)}, \quad \nu \text{ is not an integer}$$

For $\nu = 0$, the Bessel equation simplifies to

$$zw'' + w' + zw = 0$$

with the two linearly independent solutions $J_0(z)$ and $Y_0(z)$.

For $\nu = n$ (integer), the Bessel function of second kind is defined by

$$Y_n(z) = \frac{2}{\pi}\left[\ln\left(\frac{z}{2}\right) + \gamma\right]J_n(z) - \frac{1}{\pi}\sum_{k=0}^{n-1}\frac{(n-k-1)!}{k!}\left(\frac{z}{2}\right)^{2k-n}$$
$$- \frac{1}{\pi}\sum_{k=0}^{n-1}(-1)^k\frac{[\phi(k)+\phi(n+k)]}{(n+k)!k!}\left(\frac{z}{2}\right)^{2k+n}$$

where γ = Euler's constant and

$$\phi(p) = 1 + \frac{1}{2} + \frac{1}{3} + \cdots + \frac{1}{p}; \quad \phi(0) = 0.$$

A plot of the Bessel functions is shown in Figure 16.4.

16.4.3 Modified Bessel equation

The modified Bessel equation is given by

$$z^2w'' + zw' - (z^2 + \nu^2)w = 0$$

where ν is a real constant. When ν is not an integer, the two linearly independent solutions are denoted by $I_\nu(z)$ and $K_\nu(z)$, called modified Bessel functions of first and second kind respectively, where

$$I_\nu(z) = i^{-\nu}J_\nu(iz) = \sum_{m=0}^{\infty}\frac{1}{m!\Gamma(m+\nu+1)}\left(\frac{z}{2}\right)^{2m+\nu} \quad \forall\nu \text{ (including nonintegers)}$$

$$K_\nu(z) = \frac{\pi}{2}\frac{I_{-\nu}(z) - I_\nu(z)}{\sin(\nu\pi)}, \quad \nu \text{ is not an integer}$$

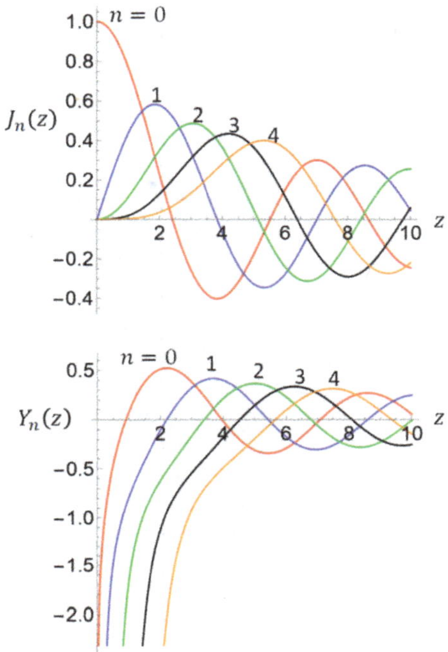

Figure 16.4: Bessel functions $J_n(z)$ and $Y_n(z)$ for $n = 0, 1, 2, 3, 4$.

For $\nu = 0$, the modified Bessel equation simplifies to

$$zw'' + w' - zw = 0$$

with the two linearly independent solutions $I_0(z)$ and $K_0(z)$, where

$$I_0(z) = 1 + \frac{z^2}{2^2} + \frac{z^4}{2^2.4^2} + \frac{z^2}{2^2.4^2.6^2} + \cdots$$

$$K_0(z) = -\left[\ln\left(\frac{z}{2}\right) + \gamma\right] I_0(z) + \frac{z^2}{2^2} + \frac{z^4}{2^2.4^2}\left(1 + \frac{1}{2}\right) + \frac{z^6}{2^2.4^2.6^2}\left(1 + \frac{1}{2} + \frac{1}{3}\right) + \cdots$$

For $\nu = \pm\frac{1}{2}$, $I_{\frac{1}{2}}(z) = \sqrt{\frac{2}{\pi z}} \sinh z$ and $I_{-\frac{1}{2}}(z) = \sqrt{\frac{2}{\pi z}} \cosh z$. A plot of modified Bessel functions for real z is shown in Figure 16.5. Note that $z = 0$ is a regular singular point of modified Bessel equation.

16.4.4 Spherical Bessel equation

The spherical Bessel equation is given by

$$z^2 w'' + 2zw' + [z^2 - n(n+1)]w = 0; \quad n = 0, 1, 2, \ldots$$

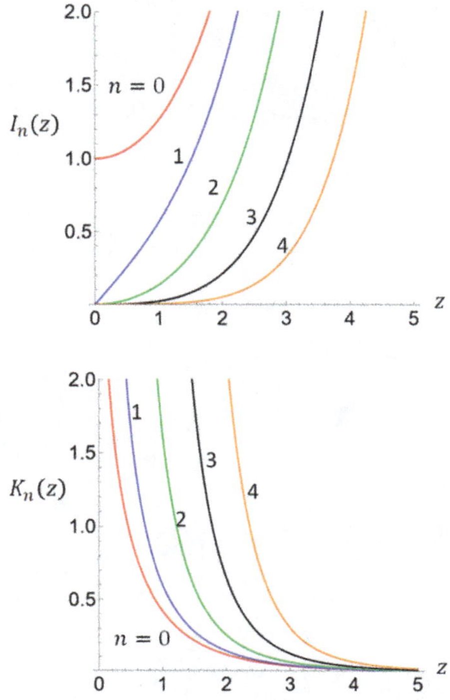

Figure 16.5: Modified Bessel function $I_n(z)$ and $K_n(z)$ for $n = 0, 1, 2, 3, 4$.

For $n = 0$, the spherical Bessel equation simplifies to

$$zw'' + 2w' + zw = 0$$

with the two linearly independent solutions $j_0(z)$ and $y_0(z)$, called as spherical Bessel functions of first and second kind, respectively, where

$$j_0(z) = \frac{\sin z}{z}; \quad \text{and} \quad y_0(z) = -\frac{\cos z}{z}$$

Note that the negative sign in $y_0(z)$ is used as convention so that these functions are similar to $J_0(z)$ and $Y_0(z)$.

For $n > 0$, the two linearly independent solutions of the spherical Bessel equation are

$$j_n(z) = (-1)^{n+1}\sqrt{\frac{\pi}{2}}\frac{J_{n+\frac{1}{2}}(z)}{\sqrt{z}}$$
$$y_n(z) = \sqrt{\frac{\pi}{2}}\frac{Y_{n+\frac{1}{2}}(z)}{\sqrt{z}} = (-1)^{n+1}\sqrt{\frac{\pi}{2}}\frac{J_{-(n+\frac{1}{2})}(z)}{\sqrt{z}}$$

For example, for $\nu = 1$, $j_1(z) = \frac{\sin z}{z^2} - \frac{\cos z}{z}$ and $y_1(z) = -\frac{\cos z}{z^2} - \frac{\sin z}{z}$. A plot of the spherical Bessel functions is shown in Figure 16.6. Note that $z = 0$ is a regular singular point of the spherical Bessel equation.

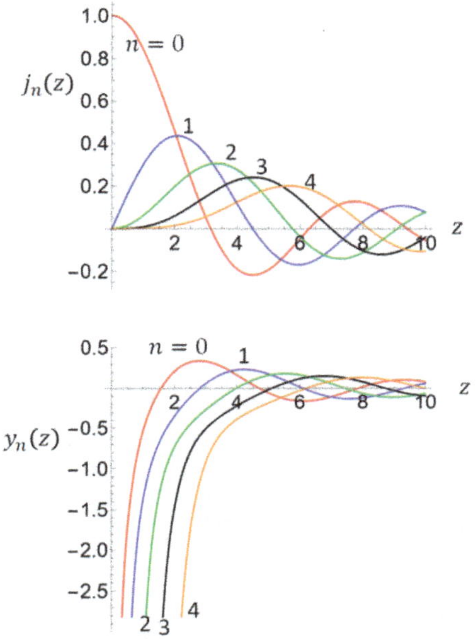

Figure 16.6: Spherical Bessel function $j_n(z)$ and $y_n(z)$ for $n = 0, 1, 2, 3, 4$.

16.4.5 Legendre equation

The Legendre equation is given by

$$(1 - z^2)w'' - 2zw' + n(n + 1)w = 0; \quad n = 0, 1, 2, \ldots$$

Note that $z = \pm 1$ is a regular singular point. The two linearly independent solutions are denoted by $P_n(z)$ and $Q_n(z)$ and are called Legendre functions of first kind and of second kind, respectively, and are related as

$$\begin{aligned}(n + 1)P_{n+1}(z) &= (2n + 1)zP_n(z) - nP_{n-1}(z)\\ (n + 1)Q_{n+1}(z) &= (2n + 1)zQ_n(z) - nQ_{n-1}(z), \quad n \geq 1\end{aligned}$$

where

$$P_0(z) = 1; \quad Q_0(z) = \frac{1}{2}\ln\left(\frac{1 + z}{1 - z}\right)$$

$$P_1(z) = z; \quad Q_1(z) = \frac{z}{2}\ln\left(\frac{1+z}{1-z}\right) - 1$$

These functions are plotted in Figure 16.7.

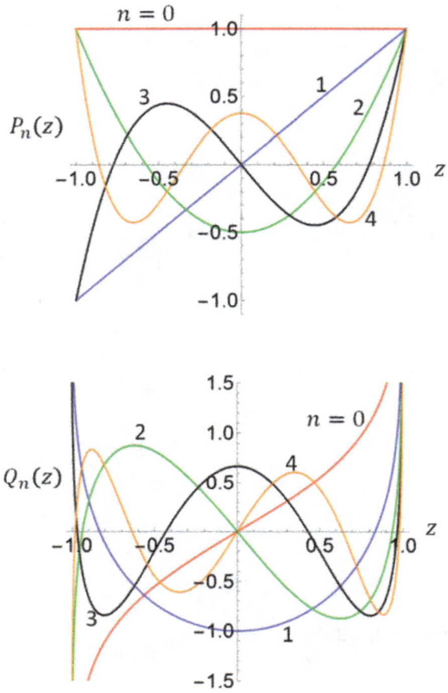

Figure 16.7: Legendre polynomial $P_n(z)$ and Legendre's function of second kind $Q_n(z)$ for $n = 0, 1, 2, 3, 4$.

16.4.6 Associated Legendre equation

The Associated Legendre equation is given by

$$(1-z^2)w'' - 2zw' + \left[n(n+1) - \frac{m^2}{1-z^2}\right]w = 0,$$

where m and n are nonnegative integers. Note that $z = \pm 1$ is a regular singular point. The two linearly independent solutions are denoted by $P_n^m(z)$ and $Q_n^m(z)$ and are called the associated Legendre functions of first kind and of second kind, respectively. For $m = 0$, P_n^0 reduces to Legendre polynomials. Figure 16.8 shows the plot of P_n^1 and Q_n^1 for $n = 1, 2, 3, 4, 5$.

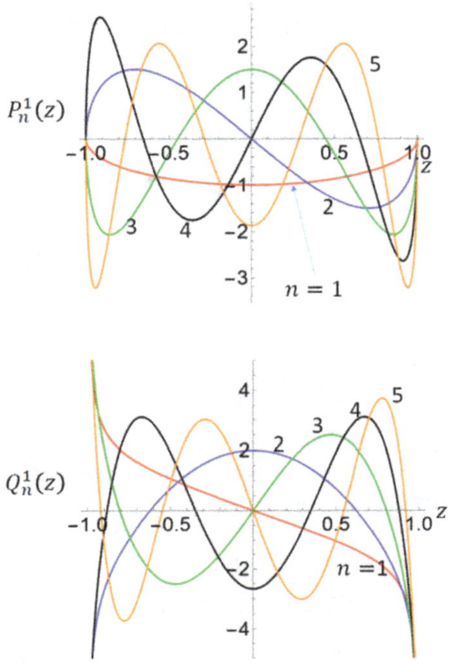

Figure 16.8: Associated Legendre polynomial $P_n^1(z)$ and Legendre's function of second kind $Q_n^1(z)$ for $n = 1, 2, 3, 4, 5$.

16.4.7 Hermite's equation

The Hermite's equation is given by

$$w'' - 2zw' + 2nw = 0; \quad n = 0, 1, 2, \ldots,$$

which has an irregular singular point at infinity. The bounded solutions are the Hermite polynomials $H_n(z)$ that are given as

$$H_0(z) = 1; \quad H_1(z) = 2z; \quad H_2(z) = 4z^2 - 2$$
$$H_n(z) = (-1)^n \exp(z^2) \frac{d^n}{dz^n} \exp(-z^2)$$

16.4.8 Laguerre's equation

The Laguerre's differential equation is given by

$$w'' + (1 - z)w' + nw = 0; \quad n = 0, 1, 2, \ldots$$

where the bounded solutions are the Laguerre polynomials $L_n(z)$ that are given as

$$L_0(z) = 1; \quad L_1(z) = 1 - z; \quad L_2(z) = z^2 - 4z + 2$$
$$L_n(z) = \exp(z)\frac{d^n}{dz^n}[z^n \exp(-z)]$$

16.4.9 Chebyshev's equation

The Chebyshev's differential equation is given by

$$(1 - z^2)w'' - zw' + n^2 w = 0; \quad n = 0, 1, 2, \ldots$$

where the two linearly independent solutions are $T_n(z)$ and $\sqrt{1 - z^2}U_n$ with $T_n(z)$ and $U_n(z)$ being Chebyshev polynomials of degree n of first and second kind, respectively. These polynomials are given by following recurring relations:

$$T_0(z) = 1; \quad T_1(z) = z; \quad T_{n+1}(z) = 2zT_n(z) - T_{n-1}(z)$$
$$U_0(z) = 1; \quad U_1(z) = 2z; \quad U_{n+1}(z) = 2zU_n(z) - U_{n-1}(z).$$

For additional discussion on series solutions and special functions, we refer to the book by Bender and Orszag [7].

Problems

1. Determine the Taylor series expansion about the point $z = 0$ of the solution to the following initial value problems:
   (a) $w'' = (z - 1)w$; $w(0) = 1$, $w'(0) = 0$
   (b) $w''' = z^3 w$; $w(0) = 1$, $w'(0) = 0$, $w''(0) = 0$
2. Determine the series expansions of all solutions of the following differential equations (about the point $z = 0$) and identify the functions that appear:
   (a) $zw'' + w = 0$
   (b) $w'' - z^2 w = 0$
3. Determine the linearly independent (series) solutions of the following differential equations and identify the functions that appear:
   (a) $zw'' + w' - zw = 0$
   (b) $w'' + \lambda z^2 w = 0$; λ is a positive constant.
   (c) $(1 - z^2)w'' - 2zw' + n(n + 1)w = 0$; $n = 0, 1, 2, \ldots$.
   (d) $z^2 w'' + 2zw' - [z^2 + n(n + 1)]w = 0$; $n = 0, 1, 2, \ldots$

17 Laplace transforms

The Laplace transform is a special case of general linear integral transformation of a function of variable t and parameter s. The general transformation with kernel $K(t,s)$ is of the form

$$T\{f(t)\} = F(s) = \int_a^b K(t,s)f(t)\,dt \tag{17.1}$$

where $F(s)$ is called the image or transform of $f(t)$. Integral transforms of the above form have been studied by Laplace (1749–1827) and Cauchy (1789–1857), and hence the name. When $a = 0$, $b \to \infty$ and $K(t,s) = e^{-st}$, the general integral transform becomes the Laplace transform.

The Laplace transform technique is useful for solving (i) linear differential equations (initial value problems and boundary value problems), (ii) linear difference equations, (iii) linear integral equations, (iv) linear ordinary differential equations with time delay, (v) linear integro-differential equations and (vi) linear partial differential equations that arise in many applications. We review first the theory of Laplace transform and then illustrate its usefulness with some chemical engineering applications. Further applications are given in the last section.

17.1 Definition of Laplace transform

Definition. Let $f(t)$ be a real or complex valued function of a real variable t satisfying the following conditions:

(i) $f(t) \equiv 0, t < 0$

(ii) $f(t)$ has at most a finite number of discontinuities in $0 \le t \le \lambda < \infty$ for any finite λ, i. e., $f(t)$ is sectionally continuous

(iii) $f(t)$ has a bounded order of growth for $t \to \infty$, i. e., $\exists$ positive constants M and γ such that for all $t > 0$

$$|f(t)| \le Me^{\gamma t} \tag{17.2}$$

(This last condition implies that the function $f(t)$ does not grow faster than an exponential function. Thus, for functions such as e^{t^2}, the Laplace transform does not exist. There are also other classes of functions such as $\frac{1}{t^n}; n \ge 1$, which are unbounded at $t = 0$ and for which the Laplace transform does not exist.)

The Laplace transform associates with $f(t)$ a function $F(s)$ of the complex variable $s = x + iy$ defined by the integral

https://doi.org/10.1515/9783111598055-020

$$F(s) = \int_0^\infty e^{-st} f(t)\, dt = \ell\{f(t)\} \tag{17.3}$$

Theorem 17.1. *If $f(t)$ satisfies conditions* (i)–(iii) *above, the Laplace transform exists for* Re $s > \gamma$.

Proof. From the definition, we have

$$\begin{aligned} |F(s)| &= \left| \int_0^\infty e^{-st} f(t)\, dt \right| \\ &\le M \int_0^\infty |e^{-st}| e^{\gamma t}\, dt \end{aligned}$$

Now,

$$|e^{-st}| = |e^{-(x+iy)t}| = e^{-xt}$$

∴

$$|F(s)| \le M \int_0^\infty e^{(\gamma - x)t}\, dt = \frac{M}{(x-\gamma)}, \quad x > \gamma$$

Thus, if Re $s = x > x_0 > \gamma$,

$$|F(s)| \le \frac{M}{(x_0 - \gamma)} \tag{17.4}$$

and the integral converges uniformly in the domain Re $s > x_0 > \gamma$. □

Theorem 17.2. *The Laplace transform $F(s)$ of $f(t)$ is an analytic function of the complex variable s in the domain* Re $s > \gamma$

Proof. From the definition,

$$\begin{aligned} F(s) &= \int_0^\infty e^{-(x+iy)t} f(t)\, dt \\ &= \int_0^\infty e^{-xt} (\cos yt) f(t)\, dt - i \int_0^\infty e^{-xt} (\sin yt) f(t)\, dt \\ &\triangleq u(x,y) + iv(x,y) \end{aligned} \tag{17.5}$$

Now,

$$\frac{\partial u}{\partial x} = -\int_0^\infty t e^{-xt} (\cos yt) f(t)\, dt \tag{17.6}$$

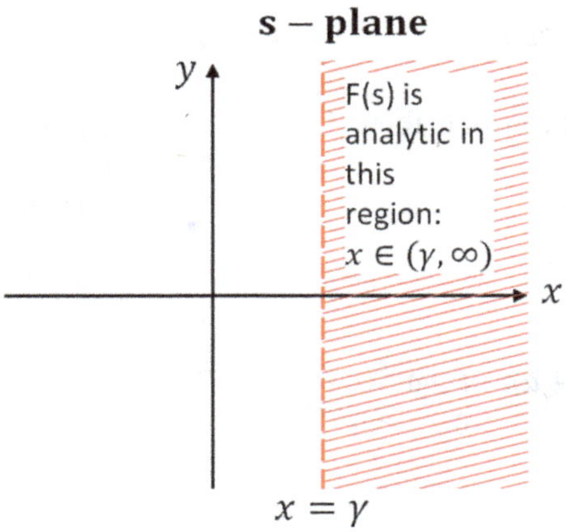

Figure 17.1: Schematic diagram illustrating the region $x > \gamma$ in which $F(s)$ is analytic.

and

$$\left|\frac{\partial u}{\partial x}\right| \leq \int_0^\infty e^{-xt} t M e^{\gamma t}\, dt = \frac{M}{(x-\gamma)^2}. \tag{17.7}$$

Thus, $\frac{\partial u}{\partial x}$ exists and it is continuous in the domain $x > \gamma$. Also, from the definition,

$$\frac{\partial v}{\partial y} = -\int_0^\infty t e^{-xt} (\cos yt) f(t)\, dt = \frac{\partial u}{\partial x}. \tag{17.8}$$

Similarly, it can be shown that

$$\frac{\partial u}{\partial y} = -\frac{\partial v}{\partial x}.$$

Thus, the Cauchy–Riemann equations are satisfied and the first partial derivatives are continuous $\Rightarrow F(s)$ is analytic for $\operatorname{Re} s > \gamma$. This theorem also implies that any singularities of $F(s)$ must lie to the left of the line $\operatorname{Re} s = \gamma$. Figure 17.1 shows a schematic diagram illustrating the region in Laplace domain where $F(s)$ is analytic. □

17.2 Properties of Laplace transform

The following properties of Laplace transform, which can be established from the definition, are useful in the solution of linear equations:

1. Linearity
   If $\ell\{f_1(t)\} = F_1(s)$ and $\ell\{f_2(t)\} = F_2(s)$, then

$$\ell\{c_1 f_1(t) + c_2 f_2(t)\} = c_1 F_1(s) + c_2 F_2(s), \tag{17.9}$$

where c_1 and c_2 are real or complex constants.

2. Shifting or Translation
   (a) If $\ell\{f(t)\} = F(s)$ and a is any complex constant, then

$$\ell\{e^{at}f(t)\} = F(s-a) \tag{17.10}$$

   (b) If $\ell\{f(t)\} = F(s)$ and

$$g(t) = \begin{cases} f(t-a), & t \geq a \\ 0, & 0 < t < a, \end{cases}$$

   where a is a real constant ($a > 0$),

$$\ell\{g(t)\} = e^{-as}F(s) \tag{17.11}$$

3. Scaling property
   If $\ell\{f(t)\} = F(s)$

$$\ell\{f(at)\} = \frac{1}{a}F\left(\frac{s}{a}\right), \quad \text{where } a \neq 0 \text{ is any complex number.} \tag{17.12}$$

4. Transforms of derivatives
   If $\ell\{f(t)\} = F(s)$ and $f(t)$ has continuous derivatives,

$$\ell\{f'(t)\} = sF(s) - f(0).$$

   This formula can be established from the definition using integration by parts. Repeated application of the above formula gives

$$\ell\{f''(t)\} = s^2F(s) - sf(0) - f'(0), \tag{17.13}$$

$$\ell\{f'''(t)\} = s^3F(s) - s^2f(0) - sf'(0) - f''(0), \quad \text{and so forth.} \tag{17.14}$$

   If $f(t)$ has discontinuity at $t = a (a > 0)$,

$$\ell\{f'(t)\} = sF(s) - f(0) - e^{-as}[f(a^+) - f(a^-)] \tag{17.15}$$

5. Transforms of integrals
   If $\ell\{f(t)\} = F(s) = \int_0^\infty e^{-st}f(t)\,dt$, then

$$\ell\left\{\int_0^t f(t')\,dt'\right\} = \frac{F(s)}{s}, \tag{17.16}$$

6. Differentiation and integration of transforms

$$\mathcal{L}\{(-1)^n t^n f(t)\} = \frac{d^n F(s)}{ds^n} \tag{17.17}$$

$$\mathcal{L}\left\{\frac{f(t)}{t}\right\} = \int_s^\infty F(s')\,ds' \tag{17.18}$$

7. The transform of a convolution
The *convolution* of two functions $f_1(t)$ and $f_2(t)$ is defined by

$$\begin{aligned}\phi(t) &= f_1(t) * f_2(t)\\ &= \int_0^t f_1(t')f_2(t-t')\,dt' = \int_0^t f_1(t-t')f_2(t')\,dt'\end{aligned} \tag{17.19}$$

If $\mathcal{L}\{f_i(t)\} = F_i(s)$, $i = 1, 2$, it may be shown that

$$\mathcal{L}\{\phi(t)\} = F_1(s)F_2(s) \tag{17.20}$$

8. Periodic functions
If $f(t+T) = f(t)$, $T > 0$

$$\mathcal{L}\{f(t)\} = \frac{1}{[1-e^{-sT}]}\int_0^T e^{-st} f(t)\,dt \tag{17.21}$$

9. Initial and final value theorems:
If the indicated limits exist, it may be shown that

$$\lim_{t\to 0} f(t) = \lim_{s\to\infty} sF(s) \quad \text{(Initial value theorem)} \tag{17.22}$$

$$\lim_{t\to\infty} f(t) = \lim_{s\to 0} sF(s) \quad \text{(Final value theorem)} \tag{17.23}$$

10. Moment theorem:
Expanding the exponential in the definition,

$$\begin{aligned}F(s) &= \int_0^\infty e^{-st} f(t)\,dt = \int_0^\infty \left[1 - st + \frac{s^2t^2}{2!} - \frac{s^3t^3}{3!} + \cdots\right] f(t)\,dt\\ &= M_0 - \frac{s}{1!}M_1 + \frac{s^2}{2!}M_2 - \frac{s^3}{3!}M_3 + \ldots\end{aligned}$$

where M_j is the jth moment of $f(t)$, which is defined by

$$M_j = \int_0^\infty t^j f(t)\,dt; \quad j = 0, 1, 2, \ldots. \tag{17.24}$$

Thus, it follows that

$$M_j = (-1)^j \frac{d^j F(s)}{ds^j}\bigg|_{s=0}. \tag{17.25}$$

Equation (17.25) shows that the jth moment of $f(t)$ can be obtained from $F(s)$ by expanding it in power of s.

17.2.1 Examples of Laplace transform

Example 17.1. Consider the exponential function $f(t) = e^{at}$, $t > 0$ (a is real or complex). We have

$$\ell\{e^{at}\} = \int_0^{\infty} e^{at} e^{-st}\, dt = \frac{1}{s-a} \tag{17.26}$$

From equation (17.26), we can obtain the Laplace transform of many elementary functions using the above properties:

(set $a = 0$)

$$\ell\{1\} = \frac{1}{s}$$

$$\ell\{t\} = \frac{1}{s^2}$$

$$\vdots$$

$$\ell\{t^n\} = \frac{n!}{s^{n+1}}$$

(set $a = \alpha + i\beta$)

$$\Rightarrow$$

$$\ell\{e^{\alpha t}(\cos\beta t + i\sin\beta t)\}$$

$$= \frac{1}{s - \alpha - i\beta}$$

$$\Rightarrow \ell\{e^{\alpha t}\cos\beta t\} = \frac{s-\alpha}{(s-\alpha)^2 + \beta^2}$$

(differentiation w. r. t. a)

$$\ell\{te^{at}\} = \frac{1}{(s-a)^2}$$

$$\vdots$$

$$\ell\{t^n e^{at}\} = \frac{n!}{(s-a)^{n+1}}$$

(set $a = iw$)

$$\ell\{e^{iwt}\} = \frac{1}{s-iw}$$

$$\ell\{e^{-iwt}\} = \frac{1}{s+iw}$$

$$\Rightarrow \ell\{\cos wt\} = \frac{s}{s^2+w^2}.$$

$$\ell\{\sin wt\} = \frac{w}{s^2+w^2}$$

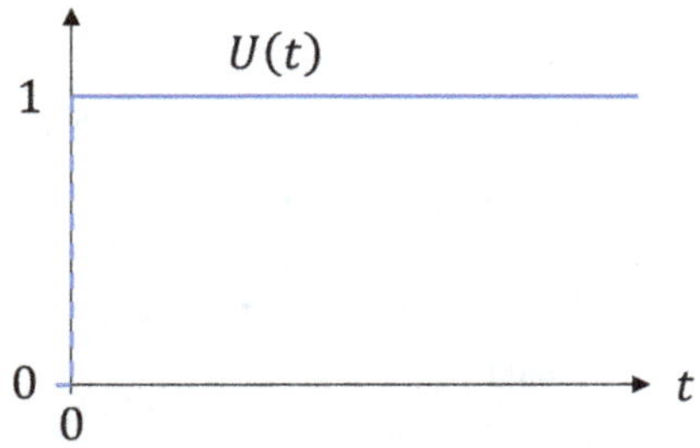

Figure 17.2: Schematic diagram of unit step input at $t = 0$.

Example 17.2 (Unit step function (Heaviside's function)). The Heaviside's function, also known as the unit-step function is shown in Figure 17.2.

$$H(t) = U(t) = \begin{cases} 1, & t \geq 0 \\ 0, & t < 0 \end{cases}$$

$$\ell\{U(t)\} = \frac{1}{s} \tag{17.27}$$

Similarly, if the unit step is located at $t = \alpha$ ($\alpha > 0$), then we denote the function by $U(t - \alpha)$ and

$$\ell\{U(t - \alpha)\} = \frac{1}{s}e^{-\alpha s} \tag{17.28}$$

Example 17.3. Consider the function

$$f(t) = \begin{cases} \frac{1}{\varepsilon}, & 0 < t < \varepsilon \\ 0, & t \geq \varepsilon \end{cases}$$

The Laplace transform is given by

$$F(s) = \int_0^{\varepsilon} \frac{1}{\varepsilon}e^{-st}dt = \frac{1 - e^{-\varepsilon s}}{\varepsilon s}$$

Taking the limit $\varepsilon \to 0$ (but keeping the area under the curve constant), we get the so-called unit impulse function (also known as the Dirac–delta function):

$$\begin{aligned} \delta(t) &= \text{Dirac–delta function (Unit impulse function)} \\ &= \lim_{\varepsilon \to 0} f(t, \varepsilon) \\ \ell\{\delta(t)\} &= 1 \end{aligned} \tag{17.29}$$

Figure 17.3a shows an approximation to unit impulse for small values of ε. Similarly, for the unit impulse function at time $t = t_0$, denoted $\delta(t - t_0)$, we have

$$\ell\{\delta(t - t_0)\} = e^{-st_0} \tag{17.30}$$

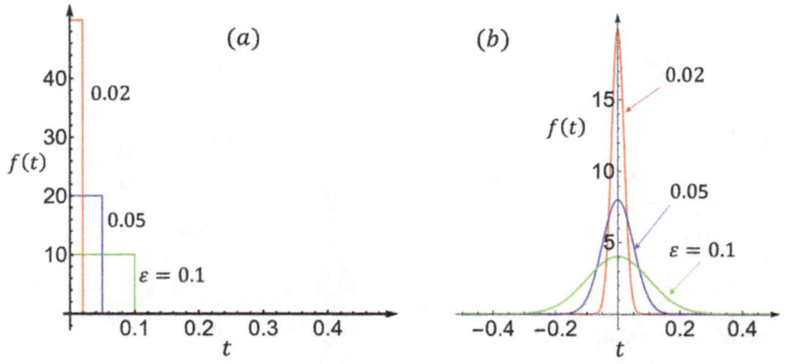

Figure 17.3: Approximation to (a) one-sided and (b) two-sided unit impulse (Dirac delta) input at $t = 0$.

Remark. The Dirac delta function may be approached by using many "test functions." The illustration shown in Figure 17.3a used one-sided test function. A test function using the two-sided or symmetric around the origin is the Gaussian function

$$f(x,\varepsilon) = \frac{1}{\sqrt{2\pi\varepsilon^2}} \exp\left(-\frac{x^2}{2\varepsilon^2}\right).$$

In the limit of $\varepsilon \to 0$, this function approaches $\delta(x)$ as shown in Figure 17.3b.

Example 17.4. Laplace transform of *Bessel functions*

$$f(t) = J_0(t) = \sum_{m=0}^{\infty} \frac{(-1)^m t^{2m}}{2^{2m}(m!)^2}$$

$$\ell\{J_0(t)\} = \sum_{m=0}^{\infty} \frac{(-1)^m (2m)!}{2^{2m}(m!)^2} \frac{1}{s^{2m+1}} = \frac{1}{\sqrt{s^2+1}}$$

$$\ell\{J_0(at)\} = \frac{1}{\sqrt{s^2+a^2}}$$

$$\ell\{I_0(at)\} = \frac{1}{\sqrt{s^2-a^2}}$$

Example 17.5. Laplace Transform of *Error and general power functions*

$$\operatorname{erf} t = \frac{2}{\sqrt{\pi}} \int_0^t e^{-u^2}\, du = \text{Error function (definition)}$$

$$\operatorname{erf} \sqrt{t} = \frac{2}{\sqrt{\pi}} \left\{ t^{\frac{1}{2}} - \frac{t^{3/2}}{3(1!)} + \frac{t^{5/2}}{5(2!)} - \frac{t^{\frac{7}{2}}}{7(3!)} + \cdots \right\}$$

$$\ell\{\operatorname{erf} \sqrt{t}\} = \frac{2}{\sqrt{\pi}} \left\{ \frac{\Gamma(3/2)}{s^{3/2}} - \frac{\Gamma(5/2)}{3s^{5/2}} + \cdots \right\} = \frac{1}{s\sqrt{s+1}}$$

$$\ell\{t^a\} = \int_0^\infty e^{-st} t^a\, dt; \Leftarrow st = u$$

$$= \int_0^\infty \frac{e^{-u}u^\alpha}{s^{\alpha+1}}\,du = \frac{\Gamma(\alpha+1)}{s^{\alpha+1}}$$

where Γ is the *Gamma (factorial) function* defined by

$$\Gamma(\alpha+1) = \int_0^\infty u^\alpha e^{-u}\,du = \alpha!$$

$$\Gamma(\alpha+1) = \alpha\Gamma(\alpha)$$

$$\Gamma\left(\frac{1}{2}\right) = \int_0^\infty u^{-\frac{1}{2}} e^{-u}\,du, \quad u = v^2 \Rightarrow \Gamma\left(\frac{1}{2}\right) = 2\int_0^\infty e^{-v^2}\,dv = \sqrt{\pi}.$$

Example 17.6 (Laplace transform directly from the D. E.). Consider the Bessel equation of order zero:

$$tJ_0'' + J_0' + tJ_0 = 0,$$

with initial conditions:

$$J_0(0) = 1, \quad J_0'(0) = 0.$$

Let

$$\mathcal{L}\{J_0(t)\} = F(s)$$

$\Rightarrow$

$$-\frac{d}{ds}\{s^2F(s) - s\} + sF - 1 - \frac{dF}{ds} = 0$$

$$\Rightarrow \frac{dF}{ds} = -\frac{sF}{1+s^2} \Rightarrow F = \frac{c}{\sqrt{1+s^2}},$$

$$\text{and} \quad \lim_{s\to\infty} sF = c = 1$$

$\therefore$

$$\mathcal{L}\{J_0(t)\} = \frac{1}{\sqrt{1+s^2}}.$$

17.3 Inversion of Laplace transform

We have shown that if $f(t)$ is sectionally continuous and is of exponential order, i. e.,

$$|f(t)| < Me^{\gamma_0 t} \tag{17.31}$$

then the Laplace transform of $f(t)$,

$$F(s) = \int_0^\infty e^{-st} f(t)\, dt$$

is analytical in the region $\operatorname{Re} s > \gamma_0$. Assume that $F(s)$ is real for s real. Suppose that s is any point on the real axis and $\gamma > \gamma_0$ as shown in Figure 17.4.

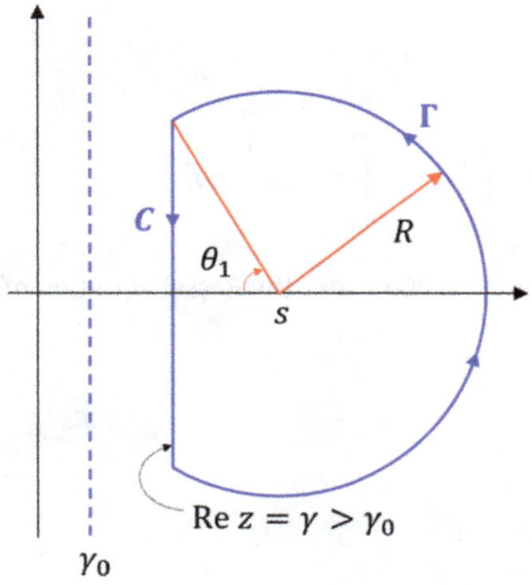

Figure 17.4: Schematic diagram of the contours for Cauchy's integral formula.

If C is the vertical line and Γ is the curved contour as shown in Figure 17.4, Cauchy's integral formula gives

$$\begin{aligned} F(s) &= \frac{1}{2\pi i}\left[\int_C \frac{F(z)}{(z-s)}\, dz + \int_\Gamma \frac{F(z)}{(z-s)}\, dz\right] \\ &= \frac{1}{2\pi i}\int_{\gamma+iR\sin\theta_1}^{\gamma-iR\sin\theta_1} \frac{F(z)}{(z-s)}\, dz + \frac{1}{2\pi i}\int_\Gamma \frac{F(z)}{(z-s)}\, dz \end{aligned} \tag{17.32}$$

where Γ = part of circular arc with radius R as shown in Figure 17.4. Consider the integral

$$J = \frac{1}{2\pi i}\int_\Gamma \frac{F(z)}{(z-s)}\, dz \tag{17.33}$$

on Γ, $z - s = \mathrm{R}e^{i\theta} \Rightarrow dz = i\,\mathrm{R}e^{i\theta}\, d\theta$.

$\therefore$

$$|J| \le \left(\frac{1}{2\pi}\right) \cdot \int_{\pi-\theta_1}^{\pi+\theta_1} |F(z)|\, d\theta. \tag{17.34}$$

If we assume that for sufficiently large R,

$$|F(z)| \leq \frac{M}{R^k} \tag{17.35}$$

⇒

$$|J| \leq \frac{1}{2\pi} \cdot \frac{M}{R^k} 2\theta_1 = \frac{M\theta_1}{\pi R^k}. \tag{17.36}$$

Thus, $\lim_{R\to\infty} |J| = 0$ if $k > 0$.

⇒

$$F(s) = \frac{1}{2\pi i} \int_{\gamma-i\infty}^{\gamma+i\infty} \frac{F(z)}{(s-z)}\, dz \tag{17.37}$$

where s is real. The above formula is an extended form of Cauchy's integral formula for $F(z)$ satisfying equation (17.35) with $k > 0$.

17.3.1 Bromwich's complex inversion formula

Now, if $\ell\{f(t)\} = F(s) \Rightarrow \ell^{-1}\{F(s)\} = f(t)$, where ℓ^{-1} is the inverse operator. Applying ℓ^{-1} on both sides of equation (17.37) ⇒

$$f(t) = \frac{1}{2\pi i} \ell^{-1} \int_{\gamma-i\infty}^{\gamma+i\infty} \frac{F(z)}{(s-z)}\, dz \tag{17.38}$$

ℓ^{-1} is w. r. t. s while integration is w. r. t. z. Thus, we can take ℓ^{-1} inside the integral

∴

$$f(t) = \frac{1}{2\pi i} \int_{\gamma-i\infty}^{\gamma+i\infty} F(z) \ell^{-1}\left(\frac{1}{s-z}\right) dz \tag{17.39}$$

⇒

$$f(t) = \frac{1}{2\pi i} \int_{\gamma-i\infty}^{\gamma+i\infty} e^{zt} F(z)\, dz. \tag{17.40}$$

This is the desired inversion formula. Note that to compute/evaluate $f(t)$, we have to evaluate the complex line integral given by equation (17.40).

Theorem. *Suppose that $F(s)$ is analytic in some right half-plane* $\operatorname{Re} s > \gamma$, $|F(s)| < \frac{M}{s^k}$ *for some $k > 0$ and $F(s)$ is real for s real. Then the integral*

$$f(t) = \frac{1}{2\pi i} \int_{\gamma-i\infty}^{\gamma+i\infty} e^{st} F(s)\, ds \tag{17.41}$$

is independent of γ and converges to a real valued function $f(t)$ whose Laplace transform is $F(s)$.

A proof of this theorem may be found in the literature (Churchill [12]).

The Laplace transform pair may be summarized as

$$F(s) = \int_0^\infty e^{-st} f(t)\, dt, \quad f(t) = \frac{1}{2\pi i} \int_{\gamma-i\infty}^{\gamma+i\infty} e^{st} F(s)\, ds. \tag{17.42}$$

In Part V, we compare it with the Fourier transform pair:

$$F(\alpha) = \int_{-\infty}^\infty e^{-i\alpha t} f(t)\, dt, \quad f(t) = \frac{1}{2\pi} \int_{-\infty}^\infty e^{i\alpha t} F(\alpha)\, d\alpha. \tag{17.43}$$

17.3.2 Computing the Bromwich's integral

Case A: only singularities of $F(s)$ for $\operatorname{Re} s < \gamma$ are poles

Choose R large enough so that the contour $ABCEFA$ shown in Figure 17.5 encloses all the poles $F(s)$.

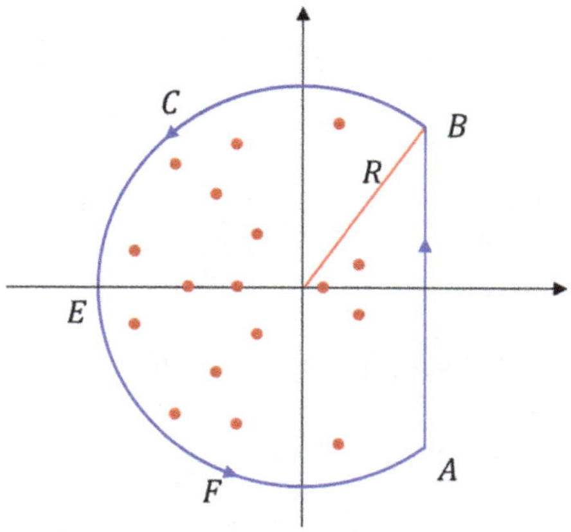

Figure 17.5: Schematic diagram illustrating the contour enclosing all the poles for Bromwich's integral formula.

By the residue theorem,

$$\frac{1}{2\pi i}\oint e^{st}F(s)\,ds = \sum \text{Residues of } e^{st}F(s) \text{ at poles of } F(s)$$

$$\text{LHS} = \frac{1}{2\pi i}\int_{\gamma-i\infty}^{\gamma+i\infty} e^{st}F(s)\,ds + \frac{1}{2\pi i}\int_{\Gamma} e^{st}F(s)\,ds,$$

where Γ is the curve BCEFA. If $F(s)$ is such that on Γ, $|F(s)| < \frac{M}{R^k}, k > 0$, then it can be shown that for $R \to \infty$, the second integral goes to zero and

$$\text{LHS} = \frac{1}{2\pi i}\int_{\gamma-iR}^{\gamma+iR} e^{st}F(s)\,ds$$

Taking the limit as $R \to \infty$, gives

$$f(t) = \sum \text{Residues of } e^{st}F(s) \text{ at poles of } F(s)$$

Case B: $F(s)$ has a branch point (say at $s = 0$)

As an example, we consider the function

$$F(s) = \frac{e^{-a\sqrt{s}}}{s} \qquad (a > 0),$$

which has a branch point at $s = 0$. The appropriate contour to consider is shown in Figure 17.6.

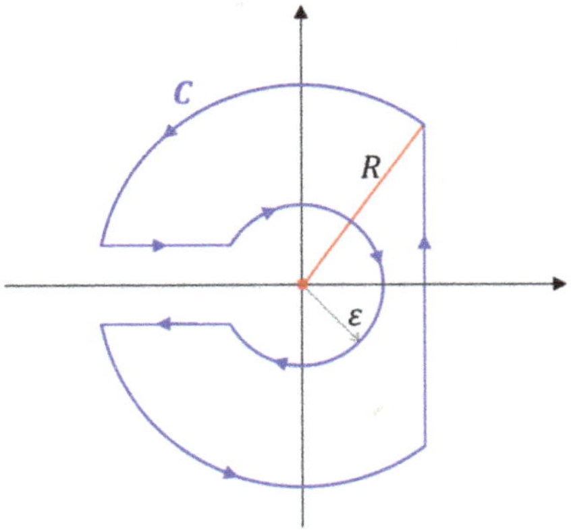

Figure 17.6: Schematic diagram illustrating the contour enclosing the branch point at $s = 0$ for Bromwich's integral formula.

Thus,

$$\mathcal{L}^{-1}\left\{\frac{e^{-a\sqrt{s}}}{s}\right\} = \frac{1}{2\pi i}\int_{\gamma-i\infty}^{\gamma+i\infty}\left\{\frac{e^{-st-a\sqrt{s}}}{s}\right\} ds.$$

The function $f(t)$ can be obtained by evaluating the integral along the contour shown in Figure 17.6 and taking the limits $\varepsilon \to 0$ and $R \to \infty$.

Case C: $F(s)$ has poles as well as one or more branch points

(i) If poles are not on branch lines, modify the contour C as in case B.

(ii) If poles are on branch lines, then modify the contour C to go around the poles that are on branch line.

Case D: (numerical inversion)

It follows from equation (17.41) that the evaluation of $f(t)$ for a fixed t requires the evaluation of a complex line integral along the line $\operatorname{Re} s = \gamma$. This can be done numerically in various ways such as using (i) Fourier transform (or fast Fourier transform) methods, (ii) Gaver functionals, (iii) Laguerre functions or (iv) deformation of the Bromwich contour. We refer to the articles by Abate and Valko [1] and Defreitas and Kane [17] for further details. Examples illustrating numerical inversion are given in Sections 17.5, 17.6 and Chapter 29.

17.4 Solution of linear differential equations by Laplace transform

In this section, we illustrate the solution of various linear differential equations (in one independent variable) using the Laplace transform technique.

17.4.1 Initial value problems with constant coefficients

Example 17.7 (Homogeneous equation). Solve

$$a_0\frac{d^2u}{dt^2} + a_1\frac{du}{dt} + a_2u = 0 \tag{17.44}$$

$$u(0) = d_0 \tag{17.45}$$

$$u'(0) = d_1 \tag{17.46}$$

where d_0 and d_1 are constants. Let

$$U(s) = \mathcal{L}\{u(t)\} = \int_0^\infty e^{-st} u(t)\, dt \tag{17.47}$$

$$\mathcal{L}\left\{\frac{du}{dt}\right\} = sU(s) - u(0) \tag{17.48}$$

$$\mathcal{L}\left\{\frac{d^2u}{dt^2}\right\} = s^2U(s) - su(0) - u'(0) \tag{17.49}$$

Taking the Laplace transform of equation (17.44) gives

$$a_0\left[s^2U(s) - sd_0 - d_1\right] + a_1\left[sU(s) - d_0\right] + a_2U(s) = 0$$

⇒

$$U(s) = \frac{a_0(sd_0 + d_1) + a_1 d_0}{a_0 s^2 + a_1 s + a_2} \tag{17.50}$$

Now, consider the n-th order equation with constant coefficients

$$a_0 \frac{d^n u}{dt^n} + a_1 \frac{d^{n-1} u}{dt^{n-1}} + \cdots + a_{n-1} \frac{du}{dt} + a_n u = 0 \tag{17.51}$$

and initial conditions

$$u(0) = d_0, \quad u'(0) = d_1, \ldots, u^{[n-1]}(0) = d_{n-1}. \tag{17.52}$$

Then, taking LT of equation (17.51) gives

$$U(s) = \frac{Q_m(s)}{P_n(s)}, \quad m \le n - 1 \tag{17.53}$$

where P_n and Q_m are polynomials in s of degree n and m, respectively. To determine $u(t)$, let $s_1, s_2, \ldots, s_n$ denote the zeros of the polynomial $P_n(s)$ and assume (for simplicity) these are simple and $Q_m(s_i) \neq 0$. Then

$$\begin{aligned} u(t) &= \mathcal{L}^{-1}\{U(s)\} \\ &= \sum_{i=1}^{n} \text{Residue } e^{st} \frac{Q_m(s)}{P_n(s)} \ @\, s = s_i \end{aligned} \tag{17.54}$$

⇒

$$u(t) = \sum_{i=1}^{n} e^{s_i t} \frac{Q_m(s_i)}{P_n'(s_i)} \tag{17.55}$$

Equation (17.55) is called the Heaviside's expansion formula. An elementary derivation of this equation is given below.

Example 17.8 (Inhomogeneous equation). Now consider the inhomogeneous equation:

$$a_0 \frac{d^n u}{dt^n} + a_1 \frac{d^{n-1}u}{dt^{n-1}} + \cdots + a_{n-1}\frac{du}{dt} + a_n u = g(t) \tag{17.56}$$

with initial conditions

$$u(0) = d_0, \quad u'(0) = d_1, \quad \ldots, \quad u^{[n-1]}(0) = d_{n-1} \tag{17.57}$$

Let $\ell\{u(t)\} = U(s)$ and $\ell\{g(t) = \hat{g}(s)\}$. Taking $L.T$ of equations (17.56)–(17.57) gives

$$P_n(s)U(s) - Q_m(s) = \hat{g}(s)$$

$\Rightarrow$

$$U(s) = \frac{Q_m(s)}{P_n(s)} + \frac{\hat{g}(s)}{P_n(s)} \tag{17.58}$$

$$\begin{aligned} u(t) &= \ell^{-1}\left\{\frac{Q_m(s)}{P_n(s)}\right\} + \ell^{-1}\left\{\frac{\hat{g}(s)}{P_n(s)}\right\} \\ &= u_1(t) + u_2(t), \end{aligned} \tag{17.59}$$

where

$$u_1(t) = \text{solution to the homogeneous equation} \tag{17.60}$$

$$u_2(t) = \text{particular solution} \tag{17.61}$$

$u_1(t)$ is obtained by setting $g(t) = 0$, while $u_2(t)$ is obtained by setting $d_i = 0$ $(i = 0, 1, 2, \ldots, n-1)$.

$$u_2(t) = \ell^{-1}\left\{\frac{\hat{g}(s)}{P_n(s)}\right\} \tag{17.62}$$

Let

$$\ell^{-1}\left\{\frac{1}{P_n(s)}\right\} = \sum_{i=1}^{n} \frac{e^{s_i t}}{P_n'(s_i)} \equiv G(t) \tag{17.63}$$

Using the convolution theorem, we get

$$u_2(t) = \int_0^t G(t')g(t-t')\,dt' \tag{17.64}$$

$$= \int_0^t G(t-t')g(t')\,dt' \tag{17.65}$$

The function $G(t-t')$ is also called the Green's function for the initial value problem. We note that when $g(t) = \delta(t-t_0)$, equation (17.65) reduces to $u_2(t) = G(t-t_0)$. Thus, $G(t-t_0)$ is the response of the system with homogeneous (zero) initial condition for a unit impulse at $t = t_0$.

17.4.2 Elementary derivation of Heaviside's formula

Consider the scalar nth order IVP

$$Lu = f(t), \quad t > 0 \tag{17.66}$$

$$\mathbf{k}(u(0)) = \mathbf{d} \tag{17.67}$$

where L has constant coefficients. Taking Laplace transform and solving for the transform, we get

$$U(s) = \frac{Q_m(s)}{P_n(s)} + \frac{F(s)}{P_n(s)} \tag{17.68}$$

where $Q_m(s)$ is a polynomial of degree $m(< n)$ depending only on the initial conditions while $P_n(s)$ is a polynomial depending on the coefficients of L. Equation (17.68) is written as

$$U(s) = U_h(s) + U_p(s) \tag{17.69}$$

or

$$u(t) = u_h(t) + u_p(t) \tag{17.70}$$

Homogeneous part

We now derive the Heaviside's formula for $u_h(t)$.

Let

$$U(s) = \frac{Q_m(s)}{P_n(s)} \tag{17.71}$$

Since $m \leq (n-1)$, we can express equation (17.71) in partial fractions

$$\frac{Q_m(s)}{P_n(s)} = \frac{A_1}{(s-\lambda_1)} + \frac{A_2}{(s-\lambda_2)} + \cdots + \frac{A_n}{(s-\lambda_n)} \tag{17.72}$$

[Assuming λ_i, $i = 1, 2, \ldots, n$ is a simple root of $P_n(s) = 0$.] Multiply equation (17.72) by $(s-\lambda_i)$ and let $s \to \lambda_i \Longrightarrow$

$$A_i = \lim_{s \to \lambda_i} \frac{(s - \lambda_i)}{P_n(s)} Q_m(s)$$
$$= \frac{Q_m(\lambda_i)}{P_n'(\lambda_i)} \tag{17.73}$$

Taking inverse Laplace transform of equation (17.72) gives

$$u_h(t) = \sum_{j=1}^{n} A_j e^{\lambda_j t}$$
$$= \sum_{j=1}^{n} e^{\lambda_j t} \frac{Q_m(\lambda_j)}{P_n'(\lambda_j)}, \tag{17.74}$$

which is the Heaviside's formula.

Heaviside's formula for repeated roots

Consider again

$$U(s) = \frac{Q_m(s)}{P_n(s)} \tag{17.75}$$

Pole of order 2

Suppose that

$$P_n(s) = (s - a)^2 R_{n-2}(s), \quad \text{where } R_{n-2}(a) \neq 0 \tag{17.76}$$

Since $s = a$ is a pole of order 2, we can write

$$U(s) = \frac{\alpha_1}{s - a} + \frac{\alpha_2}{(s - a)^2} + \sum_{j=3}^{n} \frac{\alpha_j}{(s - s_j)} \tag{17.77}$$

$\Longrightarrow$

$$(s - a)^2 U(s) = \alpha_1 (s - a) + \alpha_2 + (s - a)^2 \sum_{j=3}^{n} \frac{\alpha_j}{(s - s_j)}$$

$\Longrightarrow$

$$\alpha_2 = \lim_{s \to a} [(s - a)^2 U(s)] \tag{17.78}$$

$$\alpha_1 = \lim_{s \to a} \frac{d}{ds} [(s - a)^2 U(s)] \tag{17.79}$$

$$\alpha_j = \frac{Q_m(s_j)}{P_n'(s_j)}, \quad j \geq 3 \tag{17.80}$$

and

$$u(t) = \alpha_1 e^{at} + \alpha_2 t e^{at} + \sum_{j=3}^{n} \alpha_j e^{s_j t} \tag{17.81}$$

Pole of order 3

let $s = a$ is a pole of order 3, we can write

$$U(s) = \frac{\alpha_1}{s-a} + \frac{\alpha_2}{(s-a)^2} + \frac{\alpha_3}{(s-a)^3} + \sum_{j=4}^{n} \frac{\alpha_j}{(s-s_j)} \tag{17.82}$$

where

$$\alpha_3 = \lim_{s\to a}[(s-a)^3 U(s)] \tag{17.83}$$

$$\alpha_2 = \lim_{s\to a} \frac{d}{ds}[(s-a)^3 U(s)] \tag{17.84}$$

$$\alpha_1 = \frac{1}{2!} \lim_{s\to a} \frac{d^2}{ds^2}[(s-a)^2 U(s)] \tag{17.85}$$

$$\alpha_j = \frac{Q_m(s_j)}{P_n'(s_j)}, \quad j \geq 4 \tag{17.86}$$

and

$$u(t) = \alpha_1 e^{at} + \alpha_2 t e^{at} + \alpha_3 \frac{t^2}{2!} e^{at} + \sum_{j=4}^{n} \alpha_j e^{s_j t} \tag{17.87}$$

and so on for higher-order poles.

Similar results may also be derived using the residue theorem. For example, consider the case with pole of order 2. In this case, we can write

$$U(s) = \frac{Q_m(s)}{(s-a)^2 R_{n-2}(s)}, \quad \text{where } R_{n-2}(a) \neq 0 \tag{17.88}$$

Thus, the residue theorem leads to

$$u(t) = \lim_{s\to a} \frac{d}{ds}\left[\frac{(s-a)^2 Q_m(s)}{P_n(s)} e^{st}\right] + \sum_{j=3}^{n} \lim_{s\to s_j}\left[\frac{(s-s_j)Q_m(s)}{P_n(s)} e^{st}\right]$$

$\Longrightarrow$

$$\begin{aligned} u(t) &= \lim_{s\to a} \frac{d}{ds}\left[\frac{Q_m(s)}{R_{n-2}(s)} e^{st}\right] + \sum_{j=3}^{n} \frac{Q_m(s_j)}{P_n'(s_j)} e^{s_j t} \\ &= \left(\lim_{s\to a} \frac{d}{ds}\left[\frac{Q_m(s)}{R_{n-2}(s)}\right]\right) e^{at} + t \frac{Q_m(a)}{R_{n-2}(a)} e^{at} + \sum_{j=3}^{n} \frac{Q_m(s_j)}{P_n'(s_j)} e^{s_j t} \end{aligned}$$

$$= \alpha_1 e^{at} + \alpha_2 t e^{at} + \sum_{j=3}^{n} \alpha_j e^{s_j t} \tag{17.89}$$

where

$$\alpha_1 = \left(\lim_{s\to a} \frac{d}{ds}\left[\frac{Q_m(s)}{R_{n-2}(s)}\right]\right) = \left(\lim_{s\to a} \frac{d}{ds}[(s-a)^2 U(s)]\right) \tag{17.90}$$

$$\alpha_2 = \frac{Q_m(a)}{R_{n-2}(a)} = \lim_{s\to a}[(s-a)^2 U(s)] \tag{17.91}$$

$$\alpha_j = \frac{Q_m(s_j)}{P_n'(s_j)}, \quad j \geq 3 \tag{17.92}$$

As can be expected, the results from residue theorem (equations (17.89)–(17.92)) are identical to those obtained from Heaviside's formula (equations (17.78)–(17.81)).

Inhomogeneous part

We now derive the Heaviside's formula for $u_p(t)$.

Let

$$U_p(s) = \frac{F(s)}{P_n(s)} = \frac{F(s)}{P_n(s)} \tag{17.93}$$

Writing

$$P_n(s) = a_0(s-\lambda_1)(s-\lambda_2)\dots(s-\lambda_n) \tag{17.94}$$

we get

$$\ell^{-1}\left[\frac{1}{P_n(s)}\right] = \sum_{j=1}^{n} \frac{e^{\lambda_j t}}{P_n'(\lambda_j)} \equiv G(t) \tag{17.95}$$

Thus, by convolution,

$$\begin{aligned} u_p(t) &= \ell^{-1}\left[\frac{1}{P_n(s)}.F(s)\right] \\ &= \int_0^t G(t-t')f(t')\,dt' \\ &= \int_0^t G(t')f(t-t')\,dt'. \end{aligned}$$

Here, $G(t)$ is the Green's function of the IVP and is defined by equation (17.63).

Generalized Heaviside's formula

Consider the inversion of

$$U(s) = \frac{Q(s)}{P(s)}, \tag{17.96}$$

where $P(s)$ and $Q(s)$ have no common roots, $Q(s)$ is analytic and the only singularities of $U(s)$ are simple poles at $s = s_j, j = 1, 2, 3, \ldots$. Then using the residue theorem, it may be shown that

$$u(t) = \sum_j \frac{Q(s_j)}{P'(s_j)} e^{s_j t}. \tag{17.97}$$

The main application of this extended formula is for cases in which $P(s)$ is not a polynomial and $P(s) = 0$ has infinite number of roots.

17.4.3 Two-point boundary value problems

Consider solving

$$\frac{d^2u}{dx^2} + a_1 \frac{du}{dx} + a_0 u = f(x), \quad a < x < b \tag{17.98}$$

$$u(a) = \alpha_1, \quad u(b) = \alpha_2 \tag{17.99}$$

We may assume $a = 0$. (If $a \neq 0$, we can define $x' = x - a$, so that the new domain is 0, $b - a$). To solve equations (17.98)–(17.99), we replace it by the modified problem

$$u'' + a_1 u' + a_0 u = f(x), \quad 0 < x < b \tag{17.100}$$

$$u(0) = \alpha_1, \quad u'(0) = \beta_1, \tag{17.101}$$

where the unknown constant β_1 is to be determined.

Taking LT. of equations (17.100)–(17.101) gives

$$U(s) = \frac{s\alpha_1 + \beta_1 + a_1\alpha_1}{s^2 + a_1 s + a_0} + \frac{\hat{f}(s)}{s^2 + a_1 s + a_0} \tag{17.102}$$

$\Rightarrow$

$$u(x) = e^{s_1 x} \frac{(s_1\alpha_1 + \beta_1 + a_1\alpha_1)}{(s_1 - s_2)} + e^{s_2 x} \frac{(s_2\alpha_1 + \beta_1 + a_1\alpha_1)}{(s_2 - s_1)} + \int_0^x \frac{e^{s_1 x'} - e^{s_2 x'}}{(s_1 - s_2)} f(x - x')\, dx' \tag{17.103}$$

Determine β_1 so that $u(b) = \alpha_2$. When $\alpha_1 = \alpha_2 = 0$, we get a linear equation for determining β_1 as follows:

$$\beta_1\left(\frac{e^{s_1 b} - e^{s_2 b}}{s_1 - s_2}\right) + \int_0^b G(x, x')f(x')\,dx' = 0.$$

Also, in this case, the solution of the BVP with homogeneous BCs may be expressed as

$$u(x) = \beta_1\left(\frac{e^{s_1 x} - e^{s_2 x}}{s_1 - s_2}\right) + \int_0^x G(x, x')f(x')\,dx' \tag{17.104}$$

where

$$G(x, x') = \frac{e^{s_1(x-x')} - e^{s_2(x-x')}}{(s_1 - s_2)}. \tag{17.105}$$

Equations (17.104)–(17.105) can be simplified further to determine the Green's function for the BVP. This is discussed in more detail and generality in Part IV.

17.4.4 Linear ODEs with variable coefficients:

Example 17.9. Solve

$$t\frac{d^2u}{dt^2} - \frac{du}{dt} - t.u = 0$$

$$u(0) = 0$$

$$U(s) = \mathcal{L}\{u(t)\}$$

$$\mathcal{L}\{u''(t)\} = s^2U(s) - 0 - u'(0), \quad \mathcal{L}(u'(t)) = sU(s)$$

$$\mathcal{L}\{tu''(t)\} = -\frac{d}{ds}\{s^2U(s) - u'(0)\}$$

⇒

$$-\frac{d}{ds}\{s^2U(s)\} - sU(s) + \frac{dU}{ds} = 0$$

$$(s^2 - 1)\frac{dU}{ds} + 3sU = 0$$

⇒

$$U(s) = \frac{c}{(s^2 - 1)^{3/2}}$$

$$u(t) = c\ \mathcal{L}^{-1}\{(s^2 - 1)^{-3/2}\} = c\ t\ I_1(t),$$

where I_1 is the modified Bessel function of order one. Note that

$$
\begin{aligned}
(s^2-1)^{-3/2} &= s^{-3}\left(1-\frac{1}{s^2}\right)^{-3/2} \\
&= s^{-3}\left[1+\frac{3}{2}.\frac{1}{s^2}+\frac{(-\frac{3}{2})(-\frac{5}{2})}{2!}\frac{1}{s^4}+\cdots\right] \\
&= \frac{1}{s^3}+\frac{3}{2.s^5}+\frac{3.5}{2!.2^2s^7}+\frac{3.5.7}{3!2^3s^9}+\cdots
\end{aligned}
$$

Taking the inverse transform term-by-term gives

$$u(t) = tI_1(t),$$

where

$$I_\nu(t) = \sum_{m=0}^{\infty}\frac{(t/2)^{2m+\nu}}{m!\Gamma(m+\nu+1)}$$

Other Bessel equations

(i)

$$t\frac{d^2u}{dt^2}+\frac{du}{dt}+tu=0, \quad u(0)=1$$

$\Rightarrow$

$$U(s)=\frac{c}{\sqrt{1+s^2}}, \quad c=u(0)=\lim_{s\to\infty} s.U(s)=1$$

$\Rightarrow$

$$u(t)=J_0(t)$$

(ii)

$$
\begin{aligned}
&t\frac{d^2u}{dt^2}+\frac{du}{dt}-tu=0 \\
&u(0)=1 \Rightarrow U(s)=\frac{1}{\sqrt{s^2-1}}, \quad u(t) = I_0(t)
\end{aligned}
$$

17.4.5 Simultaneous ODEs with constant coefficients

Example 17.10. Consider the coupled initial value problem

$$\ddot{x}-x+\dot{y}-y=0 \tag{17.106}$$

$$-2\dot{x}-2x+\ddot{y}-y=e^{-t} \tag{17.107}$$

$$x(0) = 0, \quad y(0) = 1 \tag{17.108}$$

$$\dot{x}(0) = -1, \quad \dot{y}(0) = 1 \tag{17.109}$$

Taking L.T. gives

$$s^2X(s) + 1 - X(s) + sY(s) - 1 - Y(s) = 0 \tag{17.110}$$

$$-2[sX(s)] - 2X(s) + s^2Y(s) - s - 1 - Y(s) = \frac{1}{s+1} \tag{17.111}$$

$\Rightarrow$

$$X(s) = -\frac{(s^2 + 2s + 2)}{(s+1)^2(s^2+1)} \tag{17.112}$$

$$Y(s) = \frac{s^2 + 2s + 2}{(s^2+1)(s+1)} \tag{17.113}$$

Using Heaviside's formula gives

$$X(t) = \frac{1}{2}\cos t - \sin t - \frac{1}{2}(1+t)e^{-t} \tag{17.114}$$

$$Y(t) = \frac{1}{2}\cos t + \frac{3}{4}\sin t + \frac{1}{2}e^{-t}. \tag{17.115}$$

Example 17.11 (Autonomous linear systems). Consider the vector initial value problem

$$\frac{d\mathbf{u}}{dt} = \mathbf{A}\mathbf{u}, \mathbf{u}(@\, t = 0) = \mathbf{u}^0, \quad (\mathbf{u} \text{ is a n-vector, } \mathbf{A} \text{ is a } n \times n \text{ matrix}) \tag{17.116}$$

$\ell\{\mathbf{u}\} = \hat{\boldsymbol{u}} \Rightarrow$

$$s\hat{\mathbf{u}}(s) - \mathbf{u}^0 = \mathbf{A}\hat{\mathbf{u}}(s) \Rightarrow \hat{\mathbf{u}}(s) = (s\mathbf{I} - \mathbf{A})^{-1}\mathbf{u}^0 \tag{17.117}$$

Complex inversion formula gives

$$\mathbf{u}(t) = \frac{1}{2\pi i}\int_{\gamma-i\infty}^{\gamma+i\infty} e^{st}(s\mathbf{I} - \mathbf{A})^{-1}\mathbf{u}^0\, ds \tag{17.118}$$

$$(s\mathbf{I} - \mathbf{A})^{-1} = \frac{\operatorname{adj}(s\mathbf{I} - \mathbf{A})}{\det(s\mathbf{I} - \mathbf{A})} = \frac{\operatorname{adj}(s\mathbf{I} - \mathbf{A})}{P_n(s)} \tag{17.119}$$

Let $s_1, s_2, \ldots, s_n$ be roots of $P_n(s) = 0 \Rightarrow$

$$\begin{aligned} u(t) &= \sum_{i=1}^{n} \text{Residue } e^{st}(s\mathbf{I} - \mathbf{A})^{-1}\mathbf{u}^0|_{s=s_i} \\ &= \sum_{i=1}^{n} e^{s_i t}\frac{\operatorname{adj}(s_i\mathbf{I} - \mathbf{A})}{P_n'(s_i)}\mathbf{u}^0 \end{aligned}$$

$$= \sum_{i=1}^{n} e^{s_i t} \mathbf{E}_i \mathbf{u}^0 = e^{\mathbf{A}t} \mathbf{u}^0. \qquad (17.120)$$

This example shows the connection between the spectral theorem and residue theorem.

17.5 Solution of linear partial differential equations by Laplace transform

In this section, we illustrate the application of the Laplace transform method to solve linear differential equations in two independent variables.

17.5.1 Heat transfer in a finite slab

Consider the heat/diffusion equation in dimensionless form

$$\frac{\partial \theta}{\partial t} = \frac{\partial^2 \theta}{\partial \xi^2}; \quad 0 < \xi < 1, \ t > 0 \qquad (17.121)$$

$$\theta(\xi, 0) = 0 \quad \text{(initial condition)} \qquad (17.122)$$

$$\theta(0, t) = 0; \quad \theta(1, t) = 1 \quad \text{(boundary conditions)} \qquad (17.123)$$

Let $\bar{\theta}(\xi, s) = \ell\{\theta(\xi, t)\}$. Taking LT of the equation and BCs gives

$$\frac{d^2\bar{\theta}}{d\xi^2} - s\bar{\theta} = 0$$

$$\bar{\theta}(0, s) = 0 \text{ BC1}$$

$$\bar{\theta}(1, s) = \frac{1}{s} \text{ BC2}$$

$\Rightarrow$

$$\bar{\theta} = c_1 \cosh \sqrt{s}\xi + c_2 \sinh \sqrt{s}\xi$$

BC1$\Rightarrow$ $c_1 = 0$, BC2 $\Rightarrow c_2 = \frac{1}{s.\sinh\sqrt{s}} \Rightarrow$

$$\bar{\theta} = \frac{\sinh \sqrt{s}\xi}{s \sinh \sqrt{s}} \qquad (17.124)$$

$$\theta(\xi, t) = \frac{1}{2\pi i} \int_{\gamma - i\infty}^{\gamma + i\infty} e^{st}.\bar{\theta}(\xi, s)\, ds \qquad (17.125)$$

We note that $s = 0$ is a simple pole (not a branch point) and simple poles at $\sinh \sqrt{s} = 0 \Rightarrow \sqrt{s} = \pm n\pi i$, $s_n = -n^2\pi^2$, $n = 1, 2, 3, \ldots$. Now,

$$\text{Residue } e^{st}\frac{\sinh\sqrt{s}\xi}{s\sinh\sqrt{s}}\bigg|_{s=0} = \xi$$

$$\text{Residue } e^{st}\frac{\sinh\sqrt{s}\xi}{s\sinh\sqrt{s}}\bigg|_{s=s_n} = (-1)^n\frac{2}{n\pi}e^{-n^2\pi^2 t}\sin(n\pi\xi).$$

Thus,

$$\theta(\xi,t) = \xi + \frac{2}{\pi}\sum_{n=1}^{\infty}\frac{(-1)^n\sin(n\pi\xi)}{n}e^{-n^2\pi^2 t} \tag{17.126}$$

Note that for $t \to \infty$, $\theta(\xi,t) \to \xi$, i. e. the steady-state profile is linear. The solution is plotted in Figure 17.7 using 1000 terms in the summation using Mathematica®. [Remark: The curves for $t = 0.5$ and 1 are indistinguishable in this plot.]

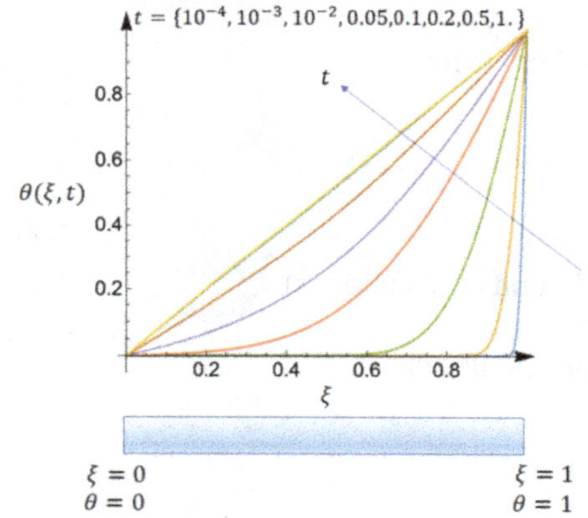

Figure 17.7: Temperature profile at various times in a slab from one-dimensional heat conduction model given in equations (17.121)–(17.123).

17.5.2 TAP reactor model

A temporal analysis of products (TAP) reactor is a packed-bed of catalyst particles operated under vacuum conditions. In a typical experiment, a pulse of molecules is injected at the inlet to the TAP reactor and the product composition at the exit is measured as a function of time. The response curves are analyzed to determine the kinetic constants and species diffusivities (in the Knudsen regime). A schematic of the TAP reactor is shown in Figure 17.8.

The mathematical model that describes the concentration of an inert molecule in a TAP (temporal analysis of products) reactor in dimensionless form is given by

$$\frac{\partial c}{\partial t} = \frac{\partial^2 c}{\partial z^2}; \quad 0 < z < 1, \quad t > 0 \tag{17.127}$$

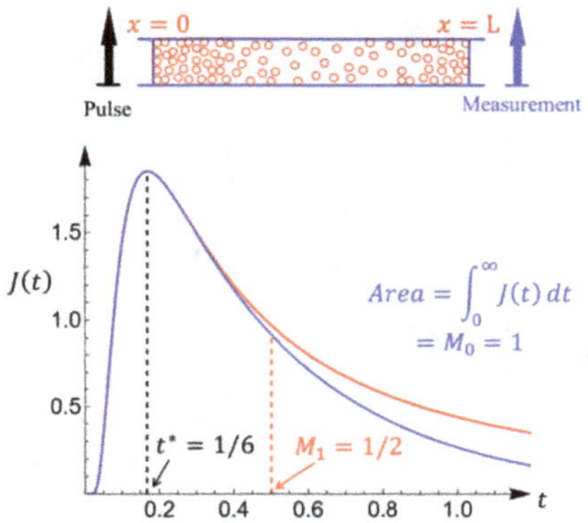

Figure 17.8: Temporal profile of flux for TAP reactor (exact solution in blue and short time solution in red curves).

with BCs:

$$-\frac{\partial c}{\partial z}(0,t) = \delta(t); \quad c(1,t) = 0 \tag{17.128}$$

and IC:

$$c(z,0) = 0 \tag{17.129}$$

The quantity of interest is the dimensionless exit flux J defined by

$$J(t) = -\frac{\partial c}{\partial z}(1,t). \tag{17.130}$$

The Laplace transform method can be used to solve for dimensionless concentration c and flux J.

Let

$$\widehat{c}(z,s) = \ell[c(z,t)] = \text{LT of } c(z,t) \tag{17.131}$$

Taking LT of governing equation and boundary conditions (equations (17.127)–(17.129)) gives

$$s\widehat{c} = \frac{d^2\widehat{c}}{dz^2}; \quad -\frac{d\widehat{c}}{dz}(z=0,s) = 1; \quad \widehat{c}(z=1,s) = 0 \tag{17.132}$$

$\Longrightarrow$

$$\widehat{c}(z,s) = \frac{\sinh[\sqrt{s}(1-z)]}{\sqrt{s}\cosh[\sqrt{s}]} \tag{17.133}$$

$\Longrightarrow$ LT of exit flux is

$$\widehat{J}(s) = -\frac{d\widehat{c}}{dz}(z=1,s) = \frac{1}{\cosh[\sqrt{s}]} \tag{17.134}$$

Temporal moments of flux $J(t)$

From the definition of LT of flux,

$$\hat{J}(s) = \int_0^{\infty} e^{-st} J(t)\, dt$$
$$= \int_0^{\infty} \left(1 - st + \frac{s^2 t^2}{2!} \dots\right) J(t)\, dt$$
$$= M_0 - sM_1 + \frac{s^2}{2!} M_2 - \cdots \tag{17.135}$$

where M_0, M_1 and M_2 are the zeroth, first and second moments of $J(t)$. Thus,

$$M_0 = \hat{J}|_{s=0} = 1 \tag{17.136}$$
$$M_1 = -\left.\frac{d\hat{J}}{dz}\right|_{s=0} = \left.\frac{\sinh\sqrt{s}}{2\sqrt{s}(\cosh\sqrt{s})^2}\right|_{s=0} = \frac{1}{2} \tag{17.137}$$
$$M_2 = \left.\frac{d^2\hat{J}}{dz^2}\right|_{s=0} = \frac{5}{12} \tag{17.138}$$

Thus, the dimensionless variance (second central moment) of the response is given by

$$\sigma^2 = \frac{M_2}{M_1^2} - 1 = \frac{2}{3}. \tag{17.139}$$

The moments can also be found by using the expansion

$$\frac{1}{\cosh\sqrt{s}} = \frac{1}{1 + \frac{s}{2!} + \frac{s^2}{4!} + \cdots + \frac{s^n}{(2n)!} + \cdots} = 1 - \frac{s}{2} + \frac{s^2}{2!}\frac{5}{12} + \cdots$$

Laplace inverse of flux and solution in time domain

Equation (17.134) suggests that $\hat{J}$ has poles at

$$\cosh\sqrt{s} = 0 \Longrightarrow e^{2\sqrt{s}} + 1 = 0$$

or

$$s_k = -(2k-1)^2 \frac{\pi^2}{4}; \quad k = 1, 2, 3, \dots \tag{17.140}$$

All these poles are simple. Thus, using Heaviside's formula or residue theorem, we get

$$J(t) = \sum_{k=1}^{\infty} e^{s_k t} \frac{2\sqrt{s_k}}{\sinh\sqrt{s_k}}$$

$$= \sum_{k=1}^{\infty} (-1)^{k-1} (2k-1)\pi \exp\left[-(2k-1)^2 \frac{\pi^2}{4} t\right] \tag{17.141}$$

$$= \pi[e^{-\frac{\pi^2}{4}t} - 3e^{-\frac{9\pi^2}{4}t} + 5e^{-\frac{25\pi^2}{4}t} - \cdots] \tag{17.142}$$

This solution is valid for all $t > 0$, though the convergence may be slow for short times. To simplify the solution for short times, we note that for large s (or small t),

$$\hat{J}(s) = 2e^{-\sqrt{s}}, \quad \text{large } s$$

$$\Longrightarrow J(t) = \frac{1}{\sqrt{\pi t^3}} e^{-\frac{1}{4t}}, \quad \text{small } t \tag{17.143}$$

The general solution from equation (17.141) with 100 terms in the summation and the short time solution from equation (17.143) are shown in Figure 17.8 as blue and red curves, respectively.

It can be seen from this figure that the flux attains a maximum when

$$\left.\frac{dJ}{dt}\right|_{t=t^*} = 0 \Longrightarrow t^* = \frac{1}{6} \quad \text{and} \quad J_{\max} = 1.85 \tag{17.144}$$

We can also show from initial and final value theorem that

$$\lim_{t\to 0} J(t) = \lim_{s\to\infty} s\hat{J}(s) = 0 \tag{17.145}$$

$$\lim_{t\to\infty} J(t) = \lim_{s\to 0} s\hat{J}(s) = 0. \tag{17.146}$$

17.5.3 Dispersion of tracers in unidirectional flow

Tracer tests are used to determine the flow maldistributions, leaks and performance of several types of process equipment such as reactors, separation/adsorption and distillation columns. In a typical tracer test, a pulse of (inert) tracer is injected at the inlet to the device and the exit concentration of the tracer is recorded. The response of the equipment for a unit impulse input is known as the "*residence time distribution*" or RTD curve and is denoted by $E(t)$. The exit response to the unit step input is referred to as the "*breakthrough curve*" or cumulative RTD curve, and is denoted by $F(t)$.

Figure 17.9 shows some schematic RTD curves corresponding to unit-step and impulse inputs to a tubular reactor.

Let $E(t)$ = RTD curve = response to a unit impulse function and $F(t)$ = response to a unit step function:

$$E(t) = \frac{dF}{dt} \quad \text{or} \quad F(t) = \int_0^t E(t')\, dt'.$$

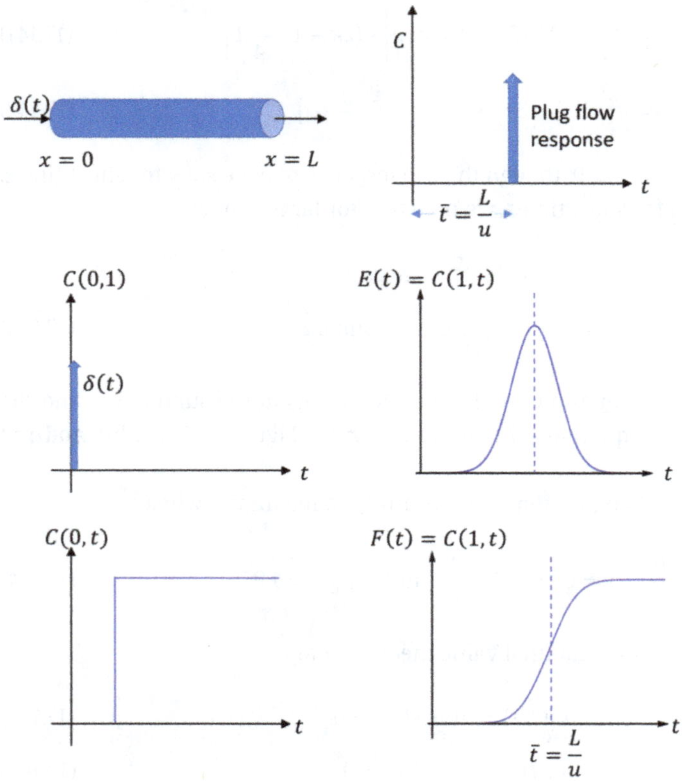

Figure 17.9: Schematic of RTD curves corresponding to unit-pulse and unit-step input in a tubular reactor.

Axial dispersion model

The axial dispersion model describes the dispersion of a tracer in a tube. If the tube is empty (i. e., no packing), the dispersion is due to the combined effect of velocity profile and molecular diffusion. If the tube is packed, the dispersion may be due to the combined effect of molecular diffusion, interstitial mixing and intraparticle diffusion. The simplest model that describes the tracer dispersion in both cases is the *axial dispersion model*. The model for a tube of length L is defined by

$$\frac{\partial c}{\partial t'} + \langle u\rangle \frac{\partial c}{\partial x} = D\frac{\partial^2 c}{\partial x^2}, \quad 0 < x < L \tag{17.147}$$

with BCs:

$$\left.\begin{array}{ll} \langle u\rangle C_{in}(t') = \langle u\rangle c - D\frac{\partial c}{\partial x} & @\, x = 0 \\ \frac{\partial c}{\partial x} = 0 & @\, ;x = L \end{array}\right\} \quad \text{Danckwert's BCs,} \tag{17.148}$$

and IC.:

$$c(x,0) = f(x) \tag{17.149}$$

To cast the model in dimensionless form, define

$$t = \frac{t'\langle u\rangle}{L}; \quad z = \frac{x}{L}; \quad \mathrm{Pe} = \frac{\langle u\rangle L}{D}; \quad g(z) = f(Lz); \quad c_{in}(t) = C_{in}\left(\frac{Lt}{\langle u\rangle}\right) \tag{17.150}$$

Then the differential equation and the boundary and initial conditions may be written as

$$\begin{aligned} \frac{\partial c}{\partial t} + \frac{\partial c}{\partial z} &= \frac{1}{\mathrm{Pe}}\frac{\partial^2 c}{\partial z^2}, \quad 0 < z < 1, \quad t > 0 \\ \text{BCs:} \quad c_{in}(t) &= c - \frac{1}{\mathrm{Pe}}\frac{\partial c}{\partial z} \ @\, z = 0 \\ \frac{\partial c}{\partial z} &= 0 \ @\, z = 1 \\ \text{IC:} \quad c(z,0) &= g(z) \end{aligned} \tag{17.151}$$

The solution of this more general model will be considered in part V using the Fourier transform method. Here, we consider the special case in which

$$g(z) = 0 \quad \text{and} \quad c_{in}(t) = \delta(t) \tag{17.152}$$

Let

$$\hat{c}(z,s) = \ell\{c(z,t)\} \tag{17.153}$$

⇒

$$\begin{aligned} \frac{1}{\mathrm{Pe}}\frac{d^2\hat{c}}{dz^2} - \frac{d\hat{c}}{dz} - s\hat{c} &= 0, \quad 0 < z < 1, \quad t > 0 \\ \text{BCs:} \quad 1 &= \hat{c} - \frac{1}{\mathrm{Pe}}\frac{d\hat{c}}{dz} \ @\, z = 0 \\ \frac{d\hat{c}}{dz} &= 0 \ @\, z = 1 \end{aligned} \tag{17.154}$$

⇒

$$\hat{c} = \alpha_1 e^{\lambda_1 z} + \alpha_2 e^{\lambda_2 z} \tag{17.155}$$

where

$$\begin{aligned} \frac{\lambda^2}{\mathrm{Pe}} - \lambda - s &= 0 \\ \lambda_{1,2} &= \frac{1 \pm \sqrt{1 + \frac{4s}{\mathrm{Pe}}}}{\frac{2}{\mathrm{Pe}}} \end{aligned} \tag{17.156}$$

BCs ⇒

$$1 = \alpha_1 + \alpha_2 - \frac{1}{\mathrm{Pe}}[\alpha_1\lambda_1 + \alpha_2\lambda_2]$$
$$0 = \alpha_1\lambda_1 e^{\lambda_1} + \alpha_2\lambda_2 e^{\lambda_2}$$

solving ⇒

$$\alpha_1 = \frac{\mathrm{Pe}\,\lambda_2 e^{\lambda_2}}{\lambda_2^2 e^{\lambda_2} - \lambda_1^2 e^{\lambda_1}}, \quad \alpha_2 = -\frac{\mathrm{Pe}\,\lambda_1 e^{\lambda_1}}{\lambda_2^2 e^{\lambda_2} - \lambda_1^2 e^{\lambda_1}} \tag{17.157}$$

∴

$$\hat{c}(z,s) = \frac{\mathrm{Pe}}{(\lambda_2^2 e^{\lambda_2} - \lambda_1^2 e^{\lambda_1})}[\lambda_2 e^{\lambda_2+\lambda_1 z} - \lambda_1 e^{\lambda_1+\lambda_2 z}] \tag{17.158}$$

where

$$\lambda_1 = \frac{\mathrm{Pe}}{2}\left[1 + \sqrt{1 + \frac{4s}{\mathrm{Pe}}}\right] \tag{17.159}$$
$$\lambda_2 = \frac{\mathrm{Pe}}{2}\left[1 - \sqrt{1 + \frac{4s}{\mathrm{Pe}}}\right] \tag{17.160}$$

Since

$$E(t) = c(1,t) \tag{17.161}$$

⇒

$$\hat{E}(s) = \hat{c}(1,s) = \text{ Laplace transform of the RTD curve} \tag{17.162}$$

∴

$$\hat{E}(s) = \frac{\mathrm{Pe}\ e^{\mathrm{Pe}}[\lambda_2 - \lambda_1]}{\lambda_2^2 e^{\lambda_2} - \lambda_1^2 e^{\lambda_1}} \tag{17.163}$$

Write

$$\lambda_1 = \frac{\mathrm{Pe}}{2}(1+q); \quad q = \sqrt{1 + \frac{4s}{\mathrm{Pe}}} \tag{17.164}$$
$$\lambda_2 = \frac{\mathrm{Pe}}{2}(1-q) \Rightarrow \lambda_2 - \lambda_1 = -\,\mathrm{Pe}\ q \tag{17.165}$$

⇒

$$\hat{E}(s) = \frac{4qe^{\frac{\mathrm{Pe}}{2}}}{(1+q)^2 e^{\frac{\mathrm{Pe}\,q}{2}} - (1-q)^2 e^{-\frac{\mathrm{Pe}\,q}{2}}} = \frac{4qe^{\frac{\mathrm{Pe}}{2}}}{H(q)} \tag{17.166}$$

where

$$H(q) = (1+q)^2 e^{\frac{\mathrm{Pe}\, q}{2}} - (1-q)^2 e^{-\frac{\mathrm{Pe}\, q}{2}} \tag{17.167}$$

and $H(-q) = -H(q)$. Thus, $H(q)$ is an odd function and may be written as

$$H(q) = q h(q^2) \tag{17.168}$$

$\Rightarrow \hat{E}(s)$ contains only even power of q, and hence $q = 0$, or equivalently, $s = \frac{-\mathrm{Pe}}{4}$ is not a branch point. The poles of $\hat{E}(s)$ are given by

$$H(q) = 0 \quad \left(q = 0 \text{ is not a branch point since } \hat{E}(s) = \frac{4e^{\frac{\mathrm{Pe}}{2}}}{h(q^2)} \text{ for } q \to 0 \right)$$

$\Rightarrow$

$$\frac{(1-q)^2}{(1+q)^2} = e^{\mathrm{Pe}\, q}$$
$$\frac{1+q^2-2q}{1+q^2+2q} = e^{\mathrm{Pe}\, q}$$

$\Rightarrow$

$$\frac{2q}{1+q^2} = \frac{1-e^{\mathrm{Pe}\, q}}{1+e^{\mathrm{Pe}\, q}} = \tanh\left(-\frac{\mathrm{Pe}\, q}{2}\right) \tag{17.169}$$

Let

$$q = iQ \tag{17.170}$$

$\Rightarrow$

$$\frac{2Q}{1-Q^2} + \tan\left(\frac{\mathrm{Pe}\, Q}{2}\right) = 0 \tag{17.171}$$

$\Rightarrow$ The poles are determined by the equation

$$\tan\Lambda = \frac{\Lambda\, \mathrm{Pe}}{\Lambda^2 - \frac{\mathrm{Pe}^2}{4}} \tag{17.172}$$

$\Rightarrow$

$$\Lambda = \frac{\mathrm{Pe}\, Q}{2} = \frac{\mathrm{Pe}}{2} iq = i\frac{\mathrm{Pe}}{2}\sqrt{1+\frac{4s}{\mathrm{Pe}}} \tag{17.173}$$

$$\Lambda^2 = -\frac{\mathrm{Pe}^2}{4}\left(1+\frac{4s}{\mathrm{Pe}}\right) = -\frac{\mathrm{Pe}^2}{4} - s\,\mathrm{Pe} \tag{17.174}$$

Denote the roots of the characteristic equation (17.172) as $\Lambda_1, \Lambda_2 \ldots$. Each of these roots corresponds to a pole of $\hat{E}(s)$, and are located on the negative real axis:

$$s = s_j = -\frac{1}{\mathrm{Pe}}\left(\Lambda_j^2 + \frac{\mathrm{Pe}^2}{4}\right), \quad j = 1, 2 \ldots (\Lambda_i \neq 0) \tag{17.175}$$

These roots are shown in Figure 17.10 for Pe = 5.

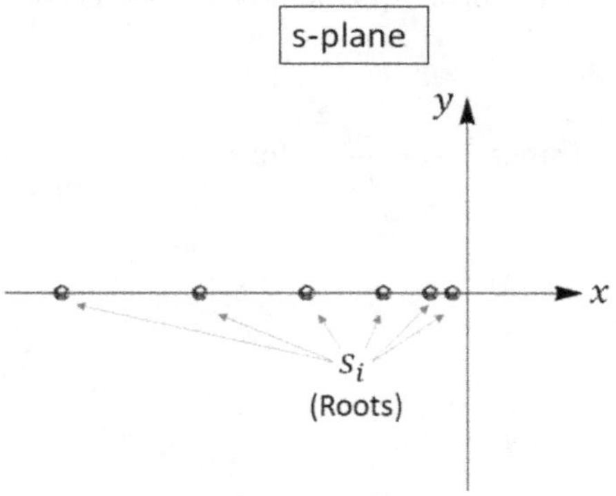

Figure 17.10: Schematic diagram showing the location of the roots of characteristic equation in Laplace domain.

A detailed discussion of the nature of the roots of equations (17.172) or (17.173)–(17.174) is given in Chapter 23.

RTD curve

The RTD curve can be obtained from the residue theorem as

$$E(t) = \frac{1}{2\pi i}\int_{\gamma - i\infty}^{\gamma + i\infty} e^{st}\hat{E}(s)\,ds = \sum_j \text{Residue}\; e^{st}\hat{E}(s)|_{s=s_j} \tag{17.176}$$

Let

$$\text{Residue}\; e^{st}\hat{E}(s)|_{s=s_j} = R_j = e^{s_j t}\lim_{s\to s_j}(s - s_j)\hat{E}(s). \tag{17.177}$$

Evaluating the limit and simplification gives

$$R_j = e^{s_j t}A_j \tag{17.178}$$

Thus, we obtain the solution as an infinite sum

$$E(t) = \sum_{j=1}^{\infty} e^{s_j t} A_j \tag{17.179}$$

where

$$s_j = -\frac{1}{\mathrm{Pe}}\left(\Lambda_j^2 + \frac{\mathrm{Pe}^2}{4}\right) \tag{17.180}$$

and

$$A_j = \frac{2\,\mathrm{Pe}(1 + \frac{4s_j}{\mathrm{Pe}})e^{\frac{\mathrm{Pe}}{2}}}{H'(q_j)} \tag{17.181}$$

⇒

$$E(t) = e^{\frac{\mathrm{Pe}}{2}} \sum_{j=1}^{\infty} \frac{(-1)^{j+1} 8\Lambda_j^2}{4\Lambda_j^2 + 4\,\mathrm{Pe} + \mathrm{Pe}^2} e^{s_j t} \tag{17.182}$$

A plot of the RTD curve is shown in Figure 17.11 for three different values of the Peclet number.

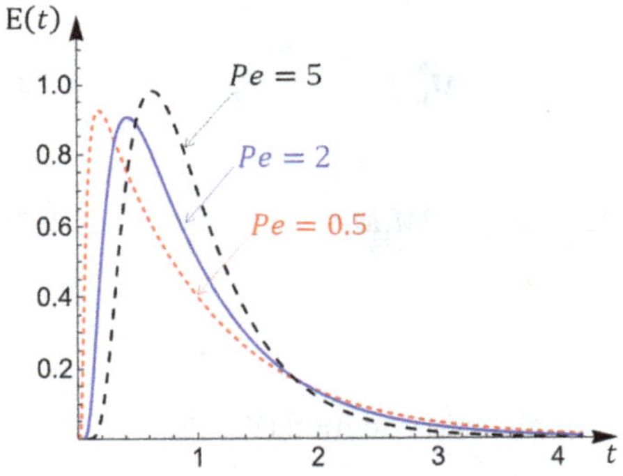

Figure 17.11: RTD curves for Pe = 0.5, 2.0 and 5.0.

Similarly, the results from numerical inversion is shown in Figure 17.12 for Pe = 5.0, along with the solution from the Residue theorem (equation (17.182)) with 10 terms in the summation.

Alternate method of analysis of the RTD curve

Moments

Recall the property (equations (17.24)–(17.25)) of Laplace transform (discussed in earlier section), the nth moment of a function $f(t)$ can be obtained from its Laplace transform

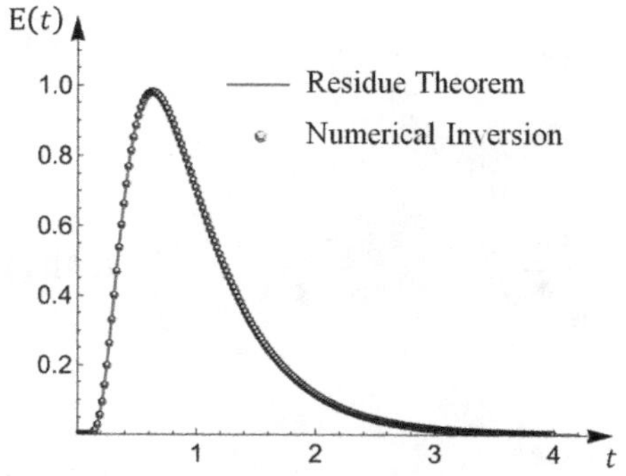

Figure 17.12: RTD curves for Pe = 5.0 from residue theorem (solid line) and numerical inversion of Laplace transform (marker points).

$F(s) = \ell[f(t)]$ as follows:

$$M_n = (-1)^n \frac{d^n F(s)}{ds^n}\bigg|_{s=0} \tag{17.183}$$

Central moments

$$m_2 = \sigma^2 = \int_0^\infty (t-\bar{t})^2 f(t)\,dt = M_2 - M_1^2 \tag{17.184}$$

$$m_3 = \int_0^\infty (t-\bar{t})^3 f(t)\,dt = M_3 + 2M_1^3 - 3M_1M_2 \tag{17.185}$$

Solution of axial dispersion model

Consider the solution of axial dispersion model as given in equation (17.166):

$\Rightarrow$

$$\hat{E}(s) = \frac{4qe^{\frac{\mathrm{Pe}}{2}}}{(1+q)^2 e^{\frac{\mathrm{Pe}\,q}{2}} - (1-q)^2 e^{-\frac{\mathrm{Pe}\,q}{2}}} \tag{17.186}$$

where q is given from equation (17.164) as

$$q = \sqrt{1 + \frac{4s}{\mathrm{Pe}}}$$

$\Longrightarrow$

$$q = \sqrt{1 + \frac{4s}{\mathrm{Pe}}} = 1 + \frac{2s}{\mathrm{Pe}} - \frac{2s^2}{\mathrm{Pe}^2} + \cdots$$

and

$$\exp\left(\frac{\mathrm{Pe}}{2}q\right) = e^{\frac{\mathrm{Pe}}{2}} e^{s-\frac{s^2}{\mathrm{Pe}}+\cdots}$$

Thus,

$$\begin{aligned}\hat{E}(s) &= \frac{4(1+\frac{2s}{\mathrm{Pe}}-\frac{2s^2}{\mathrm{Pe}^2}+\cdots)e^{\frac{\mathrm{Pe}}{2}}}{(4+\frac{8s}{\mathrm{Pe}}-\frac{4s^2}{\mathrm{Pe}^2})e^{\frac{\mathrm{Pe}}{2}}e^{s-\frac{s^2}{\mathrm{Pe}}+\cdots}+\frac{4s^2}{\mathrm{Pe}^2}e^{-\frac{\mathrm{Pe}}{2}}e^{-s+\frac{s^2}{\mathrm{Pe}}+\cdots}} \\ &= 1 - s + \left(\frac{1}{2}+\frac{1}{\mathrm{Pe}}-\frac{1}{\mathrm{Pe}^2}+\frac{e^{-\mathrm{Pe}}}{\mathrm{Pe}^2}\right)s^2 + O(s^3) \end{aligned} \tag{17.187}$$

But

$$\hat{E}(s) = \ell[E(t)] = \int_0^\infty e^{-st}E(t)\,dt$$

$\Longrightarrow$ from equation (17.183)

$$\begin{aligned}\hat{E}(s) &= \sum_{n=0}^{\infty} \left.\frac{d^n\hat{E}(s)}{ds^n}\right|_{s=0}\frac{s^n}{n!} = \sum_{n=0}^{\infty}(-1)^n\frac{M_n}{n!}s^n \\ &= M_0 - M_1 s + \frac{M_2}{2!}s^2\ldots \end{aligned} \tag{17.188}$$

Comparing (17.187) and (17.188), we get

$$M_0 \text{ (zeroth moment)} = 1 \tag{17.189}$$

$$M_1 \text{ (first moment)} = 1 \tag{17.190}$$

$$M_2 \text{ (second moment)} = 1 + \frac{2}{\mathrm{Pe}} - \frac{2}{\mathrm{Pe}^2} + \frac{2e^{-\mathrm{Pe}}}{\mathrm{Pe}^2} \tag{17.191}$$

$\Longrightarrow$ from equation (17.184)

$$\begin{aligned}\sigma^2 &= M_2 - M_1^2 \\ &= \frac{2}{\mathrm{Pe}} - \frac{2}{\mathrm{Pe}^2}(1-e^{-\mathrm{Pe}}) \end{aligned} \tag{17.192}$$

Note that for $\mathrm{Pe} \gg 1$, $\sigma^2 \approx \frac{2}{\mathrm{Pe}}$. The higher-order moments can be obtained similarly by Taylor series expansion (using Mathematica®) as

$$M_3 = 1 + \frac{6}{\mathrm{Pe}} - \frac{6}{\mathrm{Pe}^2} + \frac{18e^{-\mathrm{Pe}}}{\mathrm{Pe}^2} - \frac{24}{\mathrm{Pe}^3} + \frac{24e^{-\mathrm{Pe}}}{\mathrm{Pe}^3} \tag{17.193}$$

$$M_4 = 1 + \frac{12}{\mathrm{Pe}} - \frac{48}{\mathrm{Pe}^2} + \frac{108e^{-\mathrm{Pe}}}{\mathrm{Pe}^2} + \frac{360e^{-\mathrm{Pe}}}{\mathrm{Pe}^3} - \frac{336e^{-\mathrm{Pe}}}{\mathrm{Pe}^4} + \frac{312e^{-\mathrm{Pe}}}{\mathrm{Pe}^4} + \frac{24e^{-2\,\mathrm{Pe}}}{\mathrm{Pe}^4}. \tag{17.194}$$

The third central moment is given by

$$m_3 = M_3 + 2M_1^3 - 3M_1M_2$$
$$= \frac{12}{\mathrm{Pe}^2} + \frac{12e^{-\mathrm{Pe}}}{\mathrm{Pe}^2} - \frac{24}{\mathrm{Pe}^3} + \frac{24e^{-\mathrm{Pe}}}{\mathrm{Pe}^3} \tag{17.195}$$

For Pe ≫ 1,

$$m_3 \approx \frac{12}{\mathrm{Pe}^2} \tag{17.196}$$

which means that the RTD curve is positively skewed (i. e., longer tail on right side). This can be observed in the plots shown in Figures 17.11 and 17.12.

17.5.4 Unsteady-state operation of a packed-bed

A number of chemical engineering processes utilize the unsteady-state operation of a packed-bed. These include a heat storage or regenerator, an adsorption column, an ion-exchange column and a chromatographic column. Typically, the equipment consists of a column filled with a loosely packed solid as shown schematically in Figure 17.13. A fluid is passed through the column, and the fluid and solid exchange heat or mass or both. Since the solid remains fixed in place, the temperature and/or composition of the solid

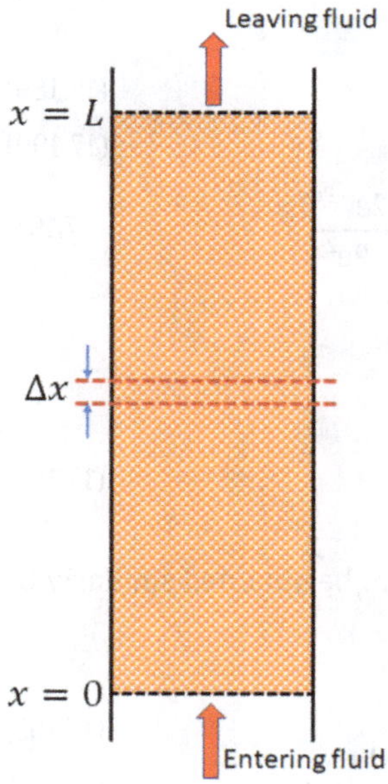

Figure 17.13: Schematic diagram of a packed-bed.

changes with time, and the operation of the column is an unsteady-state process. We consider here the heat transfer problem. A closely related problem of chromatography is discussed in Chapter 29.

Formulation of the heat transfer problem

Consider a packed bed in which heat is transferred from the fluid phase to the solid phase. Neglect conduction in both solid and fluid phases. Let

$$\rho_s, C_{ps} = \text{solid density, heat capacity}$$

$$\rho_f, C_{pf} = \text{fluid density, heat capacity}$$

$$\varepsilon = \text{fractional volume of the bed available for flow(void fraction)}$$

$$h = \text{overall heat transfer coefficient between solid and fluid}$$

$$a_v = \text{specific surface area (area of transfer per unit volume of bed)}$$

$$u_0 = \text{interstitial fluid velocity} \left(= \frac{\text{superficial velocity}}{\varepsilon} \right)$$

$$T_s = \text{solid temperature}$$

$$T_f = \text{fluid temperature}$$

$$T_r = \text{reference temperature}$$

$$A_c = \text{area of cross-section of the bed}$$

$$L = \text{Length of bed}$$

$$x = \text{distance from the bed entrance}$$

Energy balance for the fluid phase

$$\underset{\text{(inflow)}}{A_c\varepsilon u_0\rho_f C_{pf}(T_f - T_r)|_x} - \underset{\text{(outflow)}}{A_c\varepsilon u_0\rho_f C_{pf}(T_f - T_r)|_{x+\Delta x}} - \underset{\text{(transfer to solid phase)}}{A_c\Delta x a_v h(T_f - T_s)} = \underset{\text{(accumulation)}}{\frac{\partial}{\partial t}[A_c\Delta x\varepsilon\rho_f C_{pf}(T_f - T_r)]} \tag{17.197}$$

Assuming constant (and average) physical properties, and taking the limit $\triangle x \to 0$, we get

$$\varepsilon\rho_f C_{pf}\frac{\partial T_f}{\partial t} = -ha_v(T_f - T_s) - \varepsilon u_0\rho_f C_{pf}\frac{\partial T_f}{\partial x} \tag{17.198}$$

Energy balance for the solid phase

$$ha_vA_c\Delta x(T_f - T_s) = \frac{\partial}{\partial t}[A_c\Delta x(1-\varepsilon)\rho_s C_{ps}(T_S - T_r)]$$

⇒

$$(1-\varepsilon)\rho_s C_{ps}\frac{\partial T_s}{\partial t} = ha_v(T_f - T_s) \tag{17.199}$$

Assume that at time $t = 0$, the fluid and solid in the bed are at $T = T_0$ and for $t > 0$ the fluid enters the bed at a temperature $T = T_{\text{in}}(t)$, i. e., the initial and boundary conditions can be expressed as

$$T_s(x,0) = T_0; \quad T_f(x,0) = T_0; \quad T_f(0,t) = T_{\text{in}}(t) \tag{17.200}$$

Defining dimensionless variables

$$z = \frac{x}{L}; \quad \tau = t\frac{u_0}{L}; \quad \theta_f = \frac{T_f - T_0}{T_0}; \quad \theta_s = \frac{T_s - T_0}{T_0}, \tag{17.201}$$

Equations (17.198) and (17.199) may be written as

$$\frac{\partial \theta_s}{\partial \tau} = \frac{ha_vL}{u_0(1-\varepsilon)\rho_s C_{ps}}(\theta_f - \theta_s) \tag{17.202}$$

$$\frac{\partial \theta_f}{\partial \tau} = -\frac{\partial \theta_f}{\partial z} - \frac{ha_vL}{u_0\varepsilon\rho_f C_{pf}}(\theta_f - \theta_s) \tag{17.203}$$

with initial and boundary conditions (equation (17.200)) simplifying to

$$\theta_s(z,0) = \theta_f(z,0) = 0 \tag{17.204}$$

$$\theta_f(0,\tau) = \theta_{\text{in}}(\tau) = \frac{T_{\text{in}}(t) - T_0}{T_0}. \tag{17.205}$$

The dimensionless equations (17.202) and (17.203) contain two dimensionless groups, namely

$$p_h = \frac{u_0\varepsilon\rho_f C_{pf}}{ha_vL} = \frac{t_h}{t_c} = \text{local (or transverse) heat Peclet number;} \tag{17.206}$$

$$a_h = \frac{(1-\varepsilon)\rho_s C_{ps}}{\varepsilon\rho_f C_{pf}} = \text{heat capacitance ratio of solid to fluid in the bed.} \tag{17.207}$$

Here, t_c (= $\frac{L}{u_0\varepsilon}$) is the convection time while t_h (= $\frac{\rho_f C_{pf}}{ha_v}$) is the heat exchange time between solid and fluid. The model in dimensionless form can be expressed as

$$a_h p_h\frac{\partial \theta_s}{\partial \tau} = (\theta_f - \theta_s) \tag{17.208}$$

$$p_h\left(\frac{\partial \theta_f}{\partial \tau}+\frac{\partial \theta_f}{\partial z}\right)=-(\theta_f-\theta_s) \tag{17.209}$$

with ICs

$$\theta_s(z,0)=\theta_f(z,0)=0 \tag{17.210}$$

and boundary (or inlet) condition

$$\theta_f(0,\tau)=\theta_{\text{in}}(\tau). \tag{17.211}$$

This model ignores heat conduction in the solid and fluid phases. This is the simplest nontrivial model for unsteady-state heat transfer in a packed-bed. As shown, later, the same model appears in many other packed-bed operations such as chromatography and mass transfer operations. We consider here only the special case of a unit-step input, i. e., $\theta_{\text{in}}(\tau)=H(\tau)=$ Heaviside's function.

Let

$$\Theta_s(z,s)=\ell\{\theta_s(z,\tau)\}=\int_0^{\infty} e^{-st}\theta_s(z,t)\,dt \quad \text{and} \quad \Theta_f(z,s)=\ell\{\theta_f(z,\tau)\}$$

Taking Laplace transformations on both sides, equations (17.208)–(17.211) ⇒

$$\alpha_h p_h s\Theta_s=(\Theta_f-\Theta_s) \tag{17.212}$$

$$p_h\left(s\Theta_f+\frac{d\Theta_f}{dz}\right)=-(\Theta_f-\Theta_s) \tag{17.213}$$

$$\Theta_f(0,s)=\frac{1}{s} \tag{17.214}$$

Equation (17.212) ⇒

$$\Theta_s=\frac{1}{1+\alpha_h p_h s}\Theta_f \tag{17.215}$$

substituting equation (17.215) in equation (17.213), we get

$$\frac{d\Theta_f}{dz}+\Theta_f\left[1+\frac{\alpha_h}{1+\alpha_h p_h s}\right]s=0 \tag{17.216}$$

⇒

$$\begin{aligned}\Theta_f&=\frac{1}{s}\exp\left[-s\left(1+\frac{\alpha_h}{1+\alpha_h p_h s}\right)z\right]\\ &=\frac{1}{s}\exp[-sz]\exp\left[-\frac{\alpha_h s}{1+\alpha_h p_h s}z\right]\\ &=\frac{1}{s}\exp[-sz]\exp\left[-\frac{z}{p_h}\left(1-\frac{1}{1+\alpha_h p_h s}\right)\right]\end{aligned}$$

$\Longrightarrow$

$$\Theta_f = \exp\left[-\frac{z}{p_h}\right]\frac{\exp[-sz]}{s}\exp\left[\frac{1}{(s+\frac{1}{a_h p_h})}\frac{z}{a_h p_h^2}\right] \tag{17.217}$$

Equation (17.215) $\Rightarrow$

$$\Theta_s = \frac{1}{a_h p_h}\exp\left[-\frac{z}{p_h}\right]\frac{\exp[-sz]}{s}\left(\frac{1}{s+\frac{1}{a_h p_h}}\exp\left[\frac{1}{(s+\frac{1}{a_h p_h})}\frac{z}{a_h p_h^2}\right]\right) \tag{17.218}$$

Equations (17.217) and (17.218) are the Laplace transformation of the fluid and solid temperatures. By inverting them, we get the temperature profiles in the time domain.

Using the formulas,

$$\ell^{-1}\left\{\frac{1}{s}\right\} = 1,$$

$\Longrightarrow$

$$\ell^{-1}\left\{\frac{e^{-sz}}{s}\right\} = H(\tau - z) = \begin{cases} 1, & \tau > z \\ 0, & \tau < z, \end{cases} \tag{17.219}$$

and

$$\ell\{f(t)\} = F(s), \quad \text{then } \ell\{e^{-at}f(t)\} = F(s+a) \text{ (Shift theorem)}, \tag{17.220}$$

we have

$$\ell^{-1}\left[\frac{1}{(s+\beta)}\exp\left(\frac{\lambda}{s+\beta}\right)\right] = e^{-\beta\tau}\ell^{-1}\left[\frac{1}{s}e^{\frac{\lambda}{s}}\right] \tag{17.221}$$

Since

$$\frac{1}{s}\exp\left(\frac{\lambda}{s}\right) = \frac{1}{s}\sum_{n=0}^{\infty}\frac{\lambda^n}{n!s^n} = \sum_{n=0}^{\infty}\frac{\lambda^n}{n!s^{n+1}}$$

and

$$\ell^{-1}\left\{\frac{1}{s^{n+1}}\right\} = \frac{\tau^n}{n!}$$

$\Rightarrow$

$$\ell^{-1}\left\{\frac{1}{s}\exp\left(\frac{\lambda}{s}\right)\right\} = \sum_{n=0}^{\infty}\frac{(\lambda\tau)^n}{(n!)^2} = \sum_{n=0}^{\infty}\frac{(2\sqrt{\lambda\tau})^{2n}}{2^{2n}(n!)^2} = I_0(2\sqrt{\lambda\tau}) \tag{17.222}$$

where I_0 is the modified Bessel function of order zero, which is defined by

$$I_0(\xi) = \sum_{n=0}^{\infty} \frac{(\frac{\xi}{2})^{2k}}{(k!)^2} = \sum_{k=0}^{\infty} \frac{\xi^{2k}}{2^{2k}(k!)^2}.$$

Thus, by replacing $\beta = \frac{1}{\alpha_h p_h}$ and $\lambda = \frac{z}{\alpha_h p_h^2}$, equations (17.221) and (17.222) $\Longrightarrow$

$$\begin{aligned}
&\ell^{-1}\left[\frac{1}{s+\frac{1}{\alpha_h p_h}} \exp\left(\frac{1}{s+\frac{1}{\alpha_h p_h}} \frac{z}{\alpha_h p_h^2}\right)\right] \\
&= \exp\left(\frac{-\tau}{\alpha_h p_h}\right)\ell^{-1}\left[\frac{1}{s}\exp\left(\frac{z}{\alpha_h p_h^2}\frac{1}{s}\right)\right] \\
&= \exp\left(\frac{-\tau}{\alpha_h p_h}\right) I_0\left(2\sqrt{\frac{z\tau}{\alpha_h p_h^2}}\right) \qquad (17.223)
\end{aligned}$$

Thus, using the convolution theorem (17.19)–(17.20) and equations (17.219) and (17.223), we can express inverse Laplace transform of equation (17.218) as

$$\begin{aligned}
\theta_s(z,\tau) &= \ell^{-1}[\Theta_s(z,s)] \\
&= \frac{1}{\alpha_h p_h} \exp\left[-\frac{z}{p_h}\right] \int_0^{\tau} H(\tau - t - z) \exp\left(\frac{-t}{\alpha_h p_h}\right) I_0\left(2\sqrt{\frac{zt}{\alpha_h p_h^2}}\right) dt \qquad (17.224)
\end{aligned}$$

Since

$$H(\tau - t - z) = \begin{cases} 1, & 0 < t < \tau - z \\ 0, & t > \tau - z \end{cases}$$

Equation (17.224) simplifies to

$$\theta_s(z,\tau) = \begin{cases} \frac{1}{\alpha_h p_h} \exp[-\frac{z}{p_h}] \int_0^{\tau - z} \exp(\frac{-t}{\alpha_h p_h}) I_0(2\sqrt{\frac{zt}{\alpha_h p_h^2}})\, dt, & \tau > z \\ 0, & \tau < z \end{cases} \qquad (17.225)$$

Now we can obtain $\theta_f(z,\tau)$ by taking inverse Laplace transform of equation (17.217). Alternatively, we can use equations (17.208) and (17.225) to obtain $\theta_f(z,\tau)$, which is given as follows:

$$\begin{aligned}
\theta_f &= \theta_s + \alpha_h p_h \frac{\partial \theta_s}{\partial \tau} \\
&= \begin{cases} \frac{1}{\alpha_h p_h} \exp[-\frac{z}{p_h}] \int_0^{\tau - z} \exp(\frac{-t}{\alpha_h p_h}) I_0(2\sqrt{\frac{zt}{\alpha_h p_h^2}})\, dt \\ \quad + \exp[-\frac{z}{p_h}] \exp(\frac{z-\tau}{\alpha_h p_h}) I_0(2\sqrt{\frac{z(\tau - z)}{\alpha_h p_h^2}}), \quad \tau > z \\ 0, \quad \tau < z \end{cases} \qquad (17.226)
\end{aligned}$$

From equation (17.225), we obtain

$$\theta_s(0,\tau) = 1 - e^{-\beta\tau}$$

Figure 17.14 shows the breakthrough curve, fluid temperature at the exit, i. e., $\theta_f(1,\tau)$, when $\alpha_h = 10$ and $p_h = 0.01$, 0.05 and 0.1. The 3D and density plots of solid phase temperatures (θ_s) are also shown in this figure (on right) with $\alpha_h = 10$ and $p_h = 0.1$.

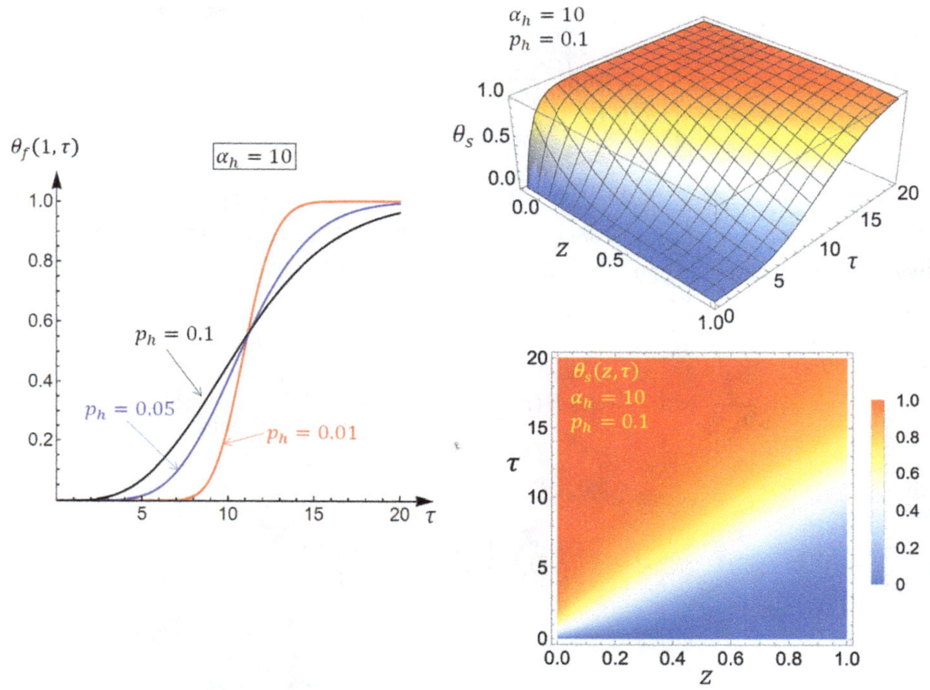

Figure 17.14: Breakthrough curves $\theta_f(1,\tau)$ for $\alpha_h = 10$ and different Peclet numbers p_h between 0.01 to 0.1 (left); and 3D and density plots of dimensionless temperatures of solid $\theta_s(z,\tau)$ for $\alpha_h = 10$ and $p_h = 0.1$ (right).

Note that for small values of p_h, the temperature front moves with a speed of $(\frac{1}{1+\alpha_h})$ or the breakthrough time is $\tau \approx (1+\alpha_h)$, and the spread (dispersion) is symmetric. In the limit as $p_h \to 0$, the breakthrough curve is a step function with a jump at $\tau = 1 + \alpha_h$.

17.6 Control system with delayed feedback

In this section, we illustrate the solution of linear initial value problems with time delay using the Laplace transform technique.

17.6.1 PI control with delayed feedback

Consider a single input-single output (SISO) first-order system with PI control. The dynamics of such control system can be described by the linear equations:

$$\tau\frac{dx}{dt} + x = u + f(t) \tag{17.227}$$

$$u = -k_P\left[x(t-\tau_D) + k_I\int_0^t x(t')\,dt'\right] \tag{17.228}$$

$$x(t) = 0, \quad -\tau_D < t \leq 0 \tag{17.229}$$

where x and u are the state and the control variables, τ is the process time constant, τ_D is the delay time, k_P and k_I are proportional and integral gains, and $f(t)$ is the disturbance function. Let $X(s)$ and $F(s)$ be the Laplace transform of $x(t)$ and $f(t)$, i. e.,

$$\ell[x(t)] = X(s); \quad \ell[f(t)] = F(s) \tag{17.230}$$

$\Longrightarrow$

$$\ell\left[\int_0^t x(t')\,dt'\right] = \frac{X(s)}{s}; \tag{17.231}$$

$$\ell\left[\frac{dx}{dt}\right] = sX(s) - x(0) = sX(s); \tag{17.232}$$

and

$$\begin{aligned}
\ell[x(t-\tau_D)] &= \int_0^\infty \exp(-st')x(t'-\tau_D)\,dt' \\
&= \int_{-\tau_D}^\infty \exp(-s(t_1+\tau_D))x(t_1)\,dt_1 \quad (\text{taking } t_1 = t'-\tau_D) \\
&= \exp(-s\tau_D)\left[\int_{-\tau_D}^0 \exp(-s(t_1))x(t_1)\,dt_1 + \int_0^\infty \exp(-st_1)x(t_1)\,dt_1\right] \\
&= \exp(-s\tau_D)[0 + X(s)]
\end{aligned}$$

$\Longrightarrow$

$$\ell[x(t-\tau_D)] = \exp(-s\tau_D)X(s) \tag{17.233}$$

Thus, the Laplace transform of the control variable u may be obtained as

$$\ell[u] = U(s) = -k_P\left[\exp(-s\tau_D) + \frac{k_I}{s}\right]X(s) \tag{17.234}$$

Thus, taking the Laplace transform of equations (17.227)–(17.229) on both sides, we get

$$(s\tau + 1)X(s) = U(s) + \ell[f(t)]$$
$$= -k_P\left[\exp(-s\tau_D) + \frac{k_I}{s}\right]X(s) + F(s)$$

$\Longrightarrow$

$$X(s) = \frac{F(s)}{(s\tau + 1) + k_P\exp(-s\tau_D) + \frac{k_P k_I}{s}} \tag{17.235}$$

For a unit step disturbance, i. e.,

$$f(t) = H(t) = \begin{cases} 1, & t \geq 0 \\ 0, & t < 0, \end{cases} \quad \text{or} \quad F(s) = \frac{1}{s},$$

the response to the unit step disturbance can be expressed in Laplace domain as

$$X(s) = \frac{1}{\tau s^2 + [1 + k_P\exp(-s\tau_D)]s + k_P k_I}. \tag{17.236}$$

We consider the inversion of equation (17.236) for various cases.

Case 1: PI control with no delay

Consider the PI control system with no delay (i. e., $\tau_D = 0$ and $k_I \neq 0$). In this case, the response to a unit-step disturbance can be simplified in the Laplace domain from equation (17.236) as

$$X(s) = \frac{1}{\tau s^2 + (1 + k_P)s + k_P k_I} \tag{17.237}$$

which can be further simplified as

$$X(s) = \frac{1}{\tau(s - s_1)(s - s_2)} \tag{17.238}$$

where s_1 and s_2 are the roots of the denominator function in equation (17.237), which can be expressed as

$$s_{1,2} = \frac{-(1 + k_P) \pm \sqrt{(1 + k_P)^2 - 4\tau k_P k_I}}{2\tau} \tag{17.239}$$

The response function $X(s)$ can be simplified further from equation (17.238) as

$$X(s) = \frac{1}{\tau(s_1 - s_2)}\left[\frac{1}{s - s_1} - \frac{1}{s - s_2}\right] \tag{17.240}$$

$\Longrightarrow$

$$x(t) = \frac{1}{\tau}\left[\frac{\exp(s_1 t) - \exp(s_2 t)}{(s_1 - s_2)}\right] \tag{17.241}$$

Note from equation (17.239) since $k_P > 0$ and $\tau > 0$ for a physical system and PI control, the real part of s_1 and s_2 is always negative, i. e.,

$$\operatorname{Re}(s_1) < 0 \quad \text{and} \quad \operatorname{Re}(s_2) < 0$$

therefore, $x(t)$ goes to zero for $t \to \infty$, i. e.,

$$\lim_{t\to\infty} x(t) = 0.$$

In other words, such control system leads to no offset. In addition, when s_1 and s_2 are real (i. e., $(1+k_P)^2 \geq 4\tau k_P k_I$), then $x(t)$ goes through a maximum at $t = \frac{\ln(\frac{s_1}{s_2})}{s_1 - s_2}$. But when s_1 and s_2 are complex, $x(t)$ goes to zero in an oscillatory manner. For example, when s_1 and s_2 are complex, i. e., $s_{1,2} = -a \pm ib$, the response to unit step can be given from equation (17.241) as

$$\begin{aligned} x(t) &= \frac{\exp(-at)}{b\tau}\left[\frac{\exp(ibt) - \exp(-ibt)}{2i}\right] \\ &= \frac{\exp(-at)}{b\tau}\sin(bt) \end{aligned} \tag{17.242}$$

Further, when $k_I = 0$ (i. e., proportional control only), the roots are $s_1 = 0$ and $s_2 = -(1 + k_P)$, and in this case the response to a unit disturbance can be expressed as

$$x(t) = \frac{1}{\tau}\left[\frac{1 - \exp[-t(1 + k_P)]}{1 + k_P}\right]. \tag{17.243}$$

Similarly, when $(1 + k_P)^2 = 4\tau k_P k_I$, then such system has repeated roots, i. e., $s_1 = s_2 = \frac{-(1+k_P)}{2\tau}$. In this case, the response function simplifies from equation (17.241) to

$$\begin{aligned} x(t) &= \lim_{s_1 \longrightarrow s_2} \frac{1}{\tau}\left[\frac{\exp(s_1 t) - \exp(s_2 t)}{(s_1 - s_2)}\right] \\ &= \frac{t}{\tau}\exp(s_2 t) \\ &= \frac{t}{\tau}\exp\left[\frac{-(1 + k_P)t}{2\tau}\right] \end{aligned} \tag{17.244}$$

Figure 17.15 shows the response of PI control systems for few cases discussed below:

1. $\tau = 1, k_I = 0, k_P = 1$. In this case, the roots are given from equation (17.239) as

$$s_{1,2} = 0, -2$$

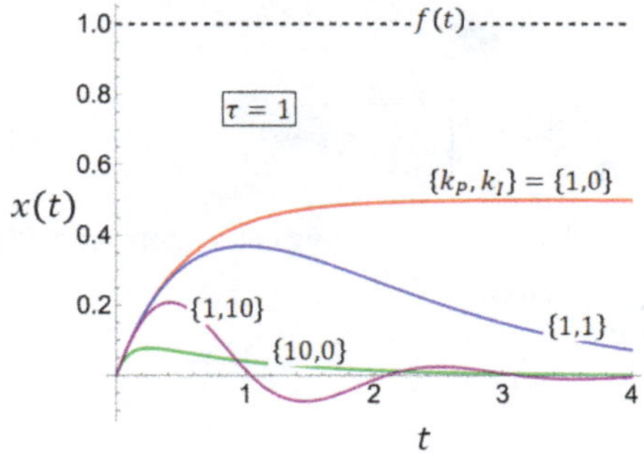

Figure 17.15: Response to a unit step disturbance for a PI control system.

⟹ from equation (17.243) as

$$x(t) = \left[\frac{1 - \exp(-2t)}{2}\right]$$

2. $\tau = 1$, $k_I = 1$, $k_P = 1$. In this case, the roots are given from equation (17.239) as

$$s_{1,2} = -1, -1$$

⟹ from equation (17.241) as

$$x(t) = t\exp(-t)$$

3. $\tau = 1$, $k_I = 1$, $k_P = 10$. In this case, the roots are given from equation (17.239) as

$$s_{1,2} = \frac{-11 \pm \sqrt{121 - 40}}{2} = \frac{-11 \pm 9}{2} = -1, -10$$

⟹ from equation (17.241) as

$$x(t) = \frac{\exp(-t) - \exp(-10t)}{9}$$

4. $\tau = 1$, $k_I = 10$, $k_P = 1$. In this case, the roots are given from equation (17.239) as

$$s_{1,2} = \frac{-2 \pm \sqrt{4 - 40}}{2} = \frac{-2 \pm 6i}{2} = -1 + 3i, -1 - 3i$$

⟹ from equation (17.242) as

$$x(t) = \frac{\exp(-t)}{3}\sin(3t)$$

Case 2: proportional control with delay

Consider the proportional control system with delay (i. e., $k_I = 0$ and $\tau_D \neq 0$). In this case, the response to a unit-step disturbance can be simplified in Laplace domain from equation (17.236) as

$$X(s) = \frac{1}{s[\tau s + 1 + k_P \exp(-s\tau_D)]}; \tag{17.245}$$

Let the function in the denominator be written as $sG(s)$. Then we have

$$X(s) = \frac{1}{sG(s)} \tag{17.246}$$

$$G(s) = \tau s + 1 + k_P \exp(-s\tau_D) \tag{17.247}$$

$$G'(s) = \tau - k_P\tau_D \exp(-s\tau_D) \tag{17.248}$$

$$G''(s) = k_P\tau_D^2 \exp(-s\tau_D) > 0 \tag{17.249}$$

Thus, it can be seen that $G''(s)$ is always positive, and hence $G(s) = 0$ cannot have a root repeated more than twice. The roots that are repeated twice, can be obtained by solving $G(s) = G'(s) = 0$, simultaneously, which leads to

$$\frac{\tau}{\tau_D} = k_P \exp(-s_j\tau_D) = -(1 + \tau s_j) \tag{17.250}$$

This results into

$$s_j = \frac{-(1+\alpha)}{\tau_D}; \; \alpha = \frac{\tau_D}{\tau}; \; \frac{1}{k_P} = \alpha \exp(1+\alpha) \text{ and } G''(s_j) = \tau_D\tau \tag{17.251}$$

Thus, if k_p, τ and τ_D satisfy the constraint given in equation (17.251), then $s_j = -\frac{1+\alpha}{\tau_D}$ is a repeated root of $G(s)$ or a second-order pole of $X(s)$, otherwise s_j given by

$$G(s_j) = \tau s_j + 1 + k_P \exp(-s_j\tau_D) = 0$$

is a simple pole. Also note that since $\alpha \exp(1+\alpha)$ is an increasing function of α, the relation given by $\frac{1}{k_P} = \alpha \exp(1+\alpha)$ leads to unique solution in α for a given k_p. Thus, for a given set of k_p, τ and τ_D, that satisfy the relation given by equation (17.251), only one repeated root exits.

Note that $s = 0$ is a simple pole of $X(s)$ and the residue at $s_0 = 0$ is given by

$$\operatorname{Res}[\exp(st)X(s)]\Big|_{s=s_i} = \lim_{s\to 0} \frac{1}{G(s)} = \frac{1}{1+k_P} \tag{17.252}$$

If s^* is the root repeated twice and s_j are the other nonzero simple roots of $G(s)$, the response to unit disturbance can be obtained from residue theorem as

$$x(t) = \sum \operatorname{Res}[\exp(st)X(s)]$$

$$= \text{Res}[\exp(st)X(s)]\Big|_{s=0} + \sum_{i=1}^{\infty} \text{Res}[\exp(st)X(s)]\Big|_{s=s_i} + \text{Res}[\exp(st)X(s)]\Big|_{s=s^*}$$

$$= \frac{1}{1+k_P} + \sum_{i=1}^{\infty} \frac{\exp(s_i t)}{s_i} \frac{1}{G'(s_i)} + \frac{\exp(s^* t)}{s^*} \frac{2}{G''(s^*)}$$

$\Longrightarrow$

$$x(t) = \frac{1}{1+k_P} + \sum_{i=1}^{\infty} \frac{\exp(s_i t)}{s_i[\tau - k_P \tau_D \exp(-s_i \tau_D)]} + \frac{2\exp(s^* t)}{s^* \tau_D \tau} \tag{17.253}$$

When all the roots of $G(s) = 0$ are simple, the last term in the above equation (17.253) can be dropped.

Stability analysis

Consider again case 2 (i. e., proportional control with delay). The control system is stable when all the roots of denominator in $X(s)$ lie in the left half-plane (i. e., $\text{Re}(s_j) < 0$). We can determine the criteria at which these roots cross the y-axis (imaginary axis) and go from left to right half-plane. For this, we can substitute $s = j\omega$, $j = \sqrt{-1}$ (i. e., the roots lying on imaginary axis) in the expression of $G(s)$ given in equation (17.247), which leads to

$$\begin{aligned} G(j\omega) &= j\omega\tau + 1 + k_P \exp(-j\omega\tau_D) \\ &= j\omega\tau + 1 + k_P[\cos(\omega\tau_D) - j\sin(\omega\tau_D)] \\ &= [1 + k_P\cos(\omega\tau_D)] + j[\omega\tau - k_P\sin(\omega\tau_D)] = 0 \end{aligned}$$

$\Longrightarrow$

$$\cos(\omega\tau_D) = -\frac{1}{k_P} \tag{17.254}$$

$$\text{and} \quad \sin(\omega\tau_D) = \frac{\omega\tau}{k_P} \tag{17.255}$$

The relation (17.254)–(17.255) cannot be satisfied if $k_p < 1$. In other words, when $k_p < 1$ the roots never cross the imaginary axis and lie always in left half-plane, i. e., the control system is stable. When $k_P > 1$, equation (17.254) lead to

$$\omega\tau_D = \cos^{-1}\left(\frac{-1}{k_P}\right) = \sin^{-1}\left(\frac{\sqrt{k_P^2 - 1}}{k_P}\right). \tag{17.256}$$

Similarly, equations (17.254) and (17.255) give

$$\omega\tau = \sqrt{k_P^2 - 1} \tag{17.257}$$

Thus, ω can be eliminated from equations (17.256) and (17.257), which leads to

$$\frac{\tau_D}{\tau} = \frac{\cos^{-1}(\frac{-1}{k_P})}{\sqrt{k_P^2 - 1}} = \frac{\sin^{-1}(\frac{\sqrt{k_P^2-1}}{k_P})}{\sqrt{k_P^2 - 1}} \tag{17.258}$$

Thus, we summarize the analysis by the following two observations:

1. If $0 \le k_p \le 1$, then all the roots of $G(s) = 0$ are in the left half-plane and control system is stable.
2. If $k_p > 1$, then the locus separating the stable and unstable region is given by equation (17.258). This is also shown in Figure 17.16 schematically.

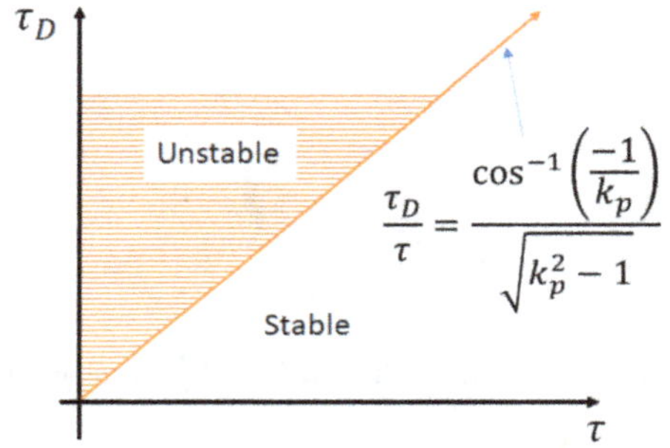

Figure 17.16: Stability region of linear control system with delayed proportional control.

The above results can be extended to systems described by second-, third- and higher-order transfer functions. To illustrate the above two points, we consider three examples with $k_p = 2$ and $\tau = 1$, i. e. $\tau_D^* = \frac{\cos^{-1}(\frac{-1}{k_p})}{\sqrt{k_p^2-1}} = \frac{2\pi}{3\sqrt{3}} = 1.2092$: (i) $\tau_D = \tau_D^* = 1.2092$, (ii) $\tau_D = 0.605 < \tau_D^*$ and (iii) $\tau_D = 2.418 > \tau_D^*$. The responses to unit-step disturbance corresponding to these cases are shown in Figure 17.17.

Determination of the poles of the transfer function in the complex plane

Express the root of $G(s)$ as

$$s = a + ib \tag{17.259}$$

$\Longrightarrow$ from equation (17.247)

$$G(s) = 0 \Longrightarrow \tau(a + ib) + 1 + k_P \exp(-\tau_D a) \exp(-i\tau_D b) = 0$$

$\Longrightarrow$

$$1 + \tau a + k_P \exp(-\tau_D a) \cos(\tau_D b) = 0$$
$$\text{and} \quad [\tau b - k_P \exp(-\tau_D a) \sin(\tau_D b)] = 0$$

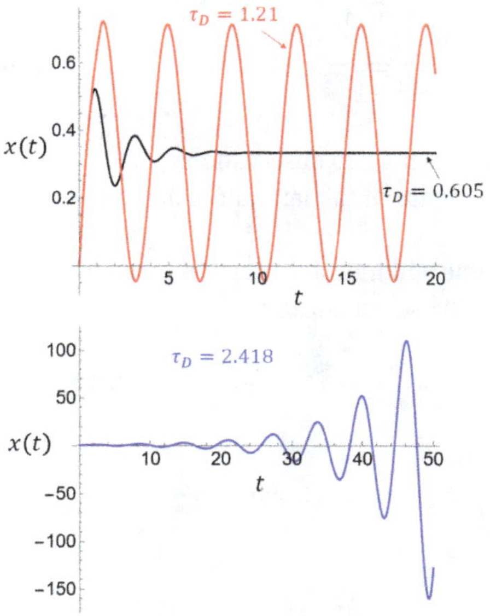

Figure 17.17: Response to unit step function for proportional control with delay for stable, unstable and oscillatory region.

$\Longrightarrow$

$$a = \frac{1}{\tau_D} \ln\left[\frac{k_P \sin(\tau_D b)}{\tau b}\right] \tag{17.260}$$

and

$$1 + \frac{\tau}{\tau_D} \ln\left[\frac{k_P \sin(\tau_D b)}{\tau b}\right] + \tau b \cot(\tau_D b) = 0 \tag{17.261}$$

Thus, for a given set of τ, τ_D and k_p, equations (17.260)–(17.261) can be solved for real values of b and a numerically, and hence the roots. Using these roots, the response curve can be obtained using equation (17.253) with $s_j = a_j \pm ib_j$.

Example: $\tau = \tau_D = k_P = 1$;

In this case, the response function can be given from equations (17.246)–(17.249) as

$$X(s) = \frac{1}{sG(s)}; \quad G(s) = 1 + s + \exp(-s)$$

The roots $s = a + ib$ can be obtained by solving equations (17.260)–(17.261):

$$1 + \ln\left[\frac{\sin(b)}{b}\right] + b \cot(b) = 0 \quad \text{and} \quad a = \ln\left[\frac{\sin(b)}{b}\right] \tag{17.262}$$

Table 17.1: First few roots for proportional control with delayed feedback ($\tau = \tau_D = k_p = 1$).

j	a_j	b_j	$s_j = a_j + ib_j$
1	−0.605021	±1.78819	−0.605021 ± 1.78819i
2	−2.05283	±7.71841	−2.05283 ± 7.71841i
3	−2.64736	±14.0202	−2.64736 ± 14.0202i
4	−3.01658	±20.3214	−3.01658 ± 20.3214i
5	−3.28526	±26.6179	−3.28526 ± 26.6179i
6	−3.49668	±32.911	−3.49668 ± 32.911i
7	−3.67104	±39.2019	−3.67104 ± 39.2019i
8	−3.81944	±45.4912	−3.81944 ± 45.4912i
9	−3.94861	±51.7794	−3.94861 ± 51.7794i
10	−4.06298	±58.0668	−4.06298 ± 58.0668i
11	−4.1656	±64.3535	−4.1656 ± 64.3535i
12	−4.25866	±70.6397	−4.25866 ± 70.6397i
13	−4.34378	±76.9256	−4.34378 ± 76.9256i
14	−4.42223	±83.2111	−4.42223 ± 83.2111i
15	−4.49496	±89.4964	−4.49496 ± 89.4964i
16	−4.56276	±95.7814	−4.56276 ± 95.7814i

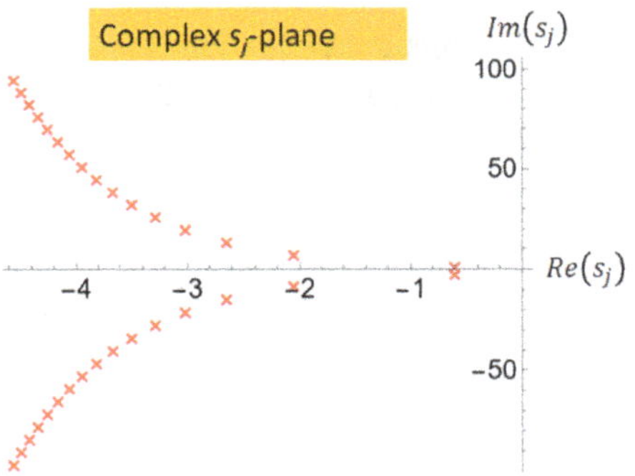

Figure 17.18: First few roots of $G(s) = 0$ for proportional control with delayed feedback ($\tau = \tau_D = k_p = 1$).

Table 17.1 lists first few roots of $G(s)$ in this example. These roots are also shown in Figure 17.18.

Using these roots, the response function can be expressed from equation (17.253) as

$$x(t) = \frac{1}{2} + \sum_{n=1}^{\infty}\left(\frac{\exp[(a_n + ib_n)t]}{(a_n + ib_n)[\tau - k_P\tau_D \exp[-(a_n + ib_n)\tau_D]]} + \frac{\exp[(a_n - ib_n)t]}{(a_n - ib_n)[\tau - k_P\tau_D \exp[-(a_n - ib_n)\tau_D]]}\right). \tag{17.263}$$

Using first 16 conjugate roots (listed in Table 17.1), the expression given in equation (17.263) is plotted in Figure 17.19 in red solid lines, along with the method of steps based numerical solution (green dashed line).

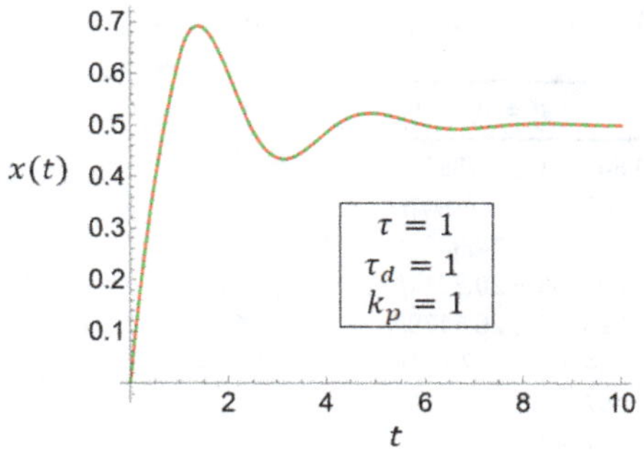

Figure 17.19: Response to unit step disturbance equipped with proportional control with delayed feedback for $\tau = \tau_D = k_p = 1$.

Both match very good with each other as expected. Note that since $\mathrm{Re}(s_j) < 0\ \forall j$, $\lim_{t\to\infty} x(t) = \frac{1}{1+k_p} = \frac{1}{2}$.

Numerical inversion of the Laplace transform

In addition to the residue theorem and other numerical methods, equation (17.245) can also be solved by inverting the Laplace transform numerically. For example, Figure 17.20 shows a comparison of the results from numerical inversion and residue theorem for

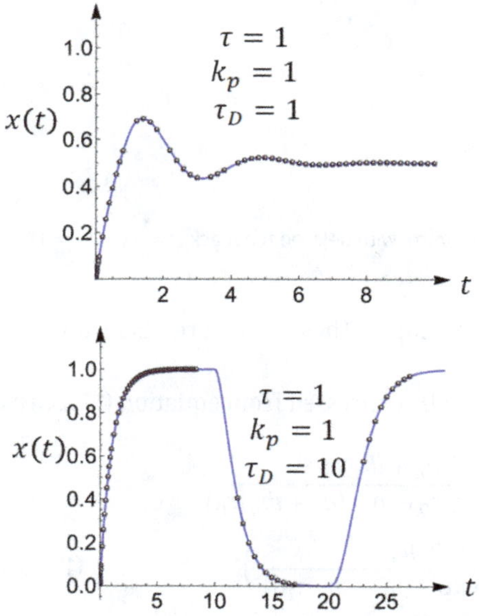

Figure 17.20: Response of a first-order system to unit-step disturbance equipped with proportional control with delayed feedback for $\tau = k_p = 1$ with $\tau_D = 1$ (top plot) and $\tau_D = 10$ (bottom plot) and from residue theorem (solid lines) and numerical inversion of Laplace transform (marker points).

$\tau = k_P = 1$ with small delay of $\tau_D = 1$ (top plot) as well as larger delay of $\tau_D = 10$ (bottom plot).

It can be seen that the numerical inversion leads to good match with the solution from residue theorem. However, numerical inversion sometimes fails at certain times where the plot is sharper or the slope is high, especially for larger delay [see the bottom plot ($\tau_D = 10$) in Figure 17.20 near time $t = 10$ or 20].

Problems

1. Use the complex inversion formula to evaluate the inverse Laplace transform of the following functions:
   (i) $\frac{1}{(s+1)(s^2+1)}$ (ii) $\frac{1}{(s^4+4)}$ (iii) $\frac{s}{(s^2+1)^4}$
2. Use the complex inversion formula to evaluate the inverse Laplace transform of the following functions:
   (i) $e^{-\sqrt{s}}$ (ii) $\frac{1}{s\sqrt{s+1}}$ (iii) $\frac{1}{s^2 \cosh \sqrt{s}}$ (iv) $\frac{\cosh x\sqrt{s}}{s \cosh a\sqrt{s}}$ $(0 < x < a)$
3. Solve the following linear initial value problems using the Laplace transformation method:
   (i)
$$\frac{d^4u}{dt^4} - 2\frac{d^2u}{dt^2} + u = 0$$
$$u(0) = 1, \quad u'(0) = 0, \quad u''(0) = 1, \quad u'''(0) = 0$$
   (ii)
$$Lu = f(t), \quad u^{[i]}(0) = b_i, \quad i = 0, 1, \dots, n-1$$
   where L is a linear differential operator with constant coefficients. Determine the asymptotic frequency response of the system for homogeneous initial conditions, i. e., the form of the solution for $t \to \infty$ when $f(t) = A \sin \omega t$ and $b_i = 0$ $(i = 0, 1, \dots, n-1)$,
   (iii)
$$t^2\frac{d^2u}{dt^2} + t\frac{du}{dt} + (t^2 - 1)u = 0, \quad u(1) = 2, \quad u(t) \text{ bounded for all } t$$
4. Solve the following linear integral, integrodifferential and delay (difference) equations using Laplace transformation
   (i)
$$\int_0^t u(t')u(t-t')\,dt' = 2u(t) + \frac{t^3}{6} - 2t$$

(ii)

$$u''(t) + au'(t) = \int_0^t g(t-t')u(t')\,dt' + bu(t) + f(t)$$

$$u(0) = c_0, \quad u'(0) = c_1$$

(iii)

$$u'(t) = au(t) + bu(t - \tau_D) + \alpha \sin \omega t; \quad u(0) = 0$$

(iv)

$$u(t) + au(t-1) + bu(t-2) = f(t)$$

where

$$u(t) = 0 \quad \text{for } t < 0 \quad \text{and} \quad f(t) = \begin{cases} e^{-t} & t > 0 \\ 0 & t < 0 \end{cases}$$

5. The dynamic model for a cascade of N perfect continuous-flow stirred tank reactors (CSTRs) is given by

$$\frac{\tau}{N}\frac{dc_1}{dt} = c_0(t) - c_1(t)$$
$$\frac{\tau}{N}\frac{dc_i}{dt} = c_{i-1}(t) - c_i(t), \quad i = 2, 3, \ldots, N$$

where τ is the mean residence (space) time in the cascade, N is the number of tanks, $c_0(t)$ is the inlet concentration and $c_{i-1}(t)$ and $c_i(t)$ are the concentrations of a tracer in the stream entering and leaving tank i, respectively. The response of the system to a unit impulse defined the residence time distribution (RTD) function, denoted as $E(t)$, i. e.,

$$E(t) = c_N(t)$$

for $c_0(t) = \delta(t)$ and the initial conditions $c_i(t = 0) = 0, i = 1, 2, \ldots, N$.

(a) Use the Laplace transform technique to show that $\hat{E}(s) = \frac{1}{(1+\frac{s\tau}{N})^N}$ where $\hat{E}(s)$ is the Laplace transform of $E(t)$. Determine the function $E(t)$.

(b) If the i-th moment of $E(t)$ is defined as

$$M_i = \int_0^\infty t^i E(t)\,dt,$$

use the function $\hat{E}(s)$ to determine the first three moments. If the i-th central moment is defined by

$$m_i = \int_0^\infty (t - M_1)^i E(t)\,dt; \quad i \geq 2$$

show that

$$m_2 = M_2 - M_1^2 = \frac{\tau^2}{N}$$

$$m_3 = M_3 - 3M_1M_2 + 2M_1^3 = \frac{2\tau^3}{N^2}$$

6. The dynamic model for a recycle reactor (tubular plug flow reactor with recycle) is given by

$$\frac{\partial c}{\partial t} + \frac{(1+R)}{\tau}\frac{\partial c}{\partial z} = 0 \quad 0 < z < 1, \quad t > 0$$

$$c(0,t) = \frac{R}{1+R}c(1,t) + \frac{c_{\text{in}}(t)}{1+R}$$

$$c(z,0) = 0, \quad 0 < z < 1,$$

where τ is the space time and R is the recycle ratio.

(a) Use the Laplace transform technique to show that the RTD function ($E(t) = c(1,t)$), which is the response of the system for a unit pulse input ($c_{\text{in}}(t) = \delta(t)$) is given by

$$E(t) = \sum_{i=0}^{\infty} \frac{R^i}{(1+R)^{1+i}}\delta\left(t - \frac{(i+1)\tau}{1+R}\right)$$

Plot a schematic diagram of $E(t)$ for varying values of R.

(b) If the moments of the RTD function are defined by

$$M_i = \int_0^\infty t^i E(t)\,dt, \quad m_i = \int_0^\infty (t - M_1)^i E(t)\,dt, \quad i = 1, 2, \ldots$$

show that

$$M_1 = \tau, \quad M_2 = \frac{(1+2R)}{(1+R)}\tau^2$$

$$m_1 = 0, m_2 = \frac{R}{1+R}\tau^2$$

7. The dynamics of a single input-single output (SISO) first-order system with PI control is described by the linear equations

$$\tau \frac{dx}{dt} + x = u + f(t)$$

$$u = -k_p \left[x(t - \tau_D) + k_1 \int_0^t x(t')\, dt' \right],$$

where x and u are state and control variables, τ is the process time constant, τ_D is the delay time, k_P and k_I are proportional and integral gains and $f(t)$ is the disturbance function. Use the Laplace transform method to determine the response of the system for a unit step disturbance for the following cases:

	τ	τ_D	k_I	k_P
(a)	1	0	0	0, 1, 10
(b)	1	0	$\frac{1}{2}$, 1, 2	1
(c)	1	1	0	1
(d)	1	0.5	1	1

8. Solve the following initial-boundary value problem using the Laplace transformation and make a schematic plot of the solution at any fixed position ($x \neq 0$) as a function of time

$$\frac{\partial^2 u}{\partial x^2} = \frac{\partial u}{\partial t}; \qquad 0 < x < 1, \quad t > 0$$

$$\text{I.C} \quad u(x,0) = 0; \quad \text{BCs} \quad u(0,t) = \delta(t), \quad u(1,t) = 0$$

9. Consider the flow system shown in Figure 17.21. Assume that each tank is well mixed and species A enters tank 1 at a concentration of $c_{\text{in}}(t)$ and leaves at $c_1(t)$. Assume further that $V_{R1} = 1\,\text{m}^3$, $V_{R2} = \frac{2}{3}\,\text{m}^3$ and $q_1 = q_2 = 2\,\text{m}^3/\text{min}$.

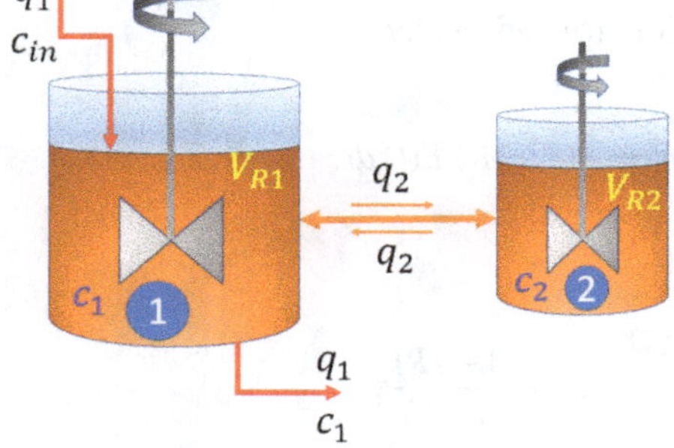

Figure 17.21: Schematic diagram of interacting tanks.

(a) Formulate the differential equations describing the system.

(b) Determine the response of the system (i. e., how the exit concentration c_i varies with time) for a unit impulse input $c_{in}(t) = \delta(t)$. Assume that no A is present initially in either tank.
(c) Show a schematic diagram of the response
(d) Determine the first two moments, and hence the mean and variance of the response curve.

10. The transient behavior of some biological systems with delayed feedback [as well as other control systems with delay] is described by the linear equation

$$\frac{du}{dt} = u(t) - \beta u(t-\tau) + f(t), \quad t > 0;$$
$$u(t) = 0, \quad -\tau \le t \le 0.$$

Here, $f(t)$ is the external input disturbance, $\tau > 0$ is the delay time and $\beta > 1$ is the strength of the delayed feedback.
(a) Determine the steady-state response for a unit-step input.
(b) Write the general form of the transient response for a unit-step input (no need to compute it).
(c) Can the system go unstable, i. e., any poles/eigenvalues cross the imaginary axis? If so, determine the smallest value of τ for which the system becomes unstable.
(d) Show a schematic diagram of the transient response for $\beta = 2$ and $\tau = 0.60$ [Hint: $\frac{\pi}{3\sqrt{3}} = 0.6046$].

11. [*Frequency response*]
Consider the inhomogeneous n-th order scalar differential equation with constant coefficients:

$$Lu = a_0 \frac{d^n u}{dt^n} + a_1 \frac{d^{n-1}u}{dt^{n-1}} + \cdots + a_{n-1}\frac{du}{dt} + a_n u = f(t), \quad t > 0$$
$$\text{I.Cs:} \quad u^{[k]}(t=0) = 0, \quad k = 0, 1, \ldots, (n-1).$$

(a) Show that the solution in Laplace domain can be expressed as

$$U(s) = \ell\{u(t)\} = \frac{F(s)}{P_n(s)}$$

where $F(s) = \ell\{f(t)\}$ and

$$P_n(\lambda) = a_0\lambda^n + a_1\lambda^{n-1} + \cdots + a_{n-1}\lambda + a_n$$

(b) Consider the frequency response, i. e., when

$$f(t) = A \sin \omega t \Longrightarrow F(s) = \frac{A\omega}{s^2 + \omega^2},$$

and assume that $\lambda_1, \lambda_2, \ldots, \lambda_n$ are n distinct roots of $P_n(\lambda) = 0$, then the general form of solution can be expressed as

$$u(t) = A\left[\sum_{j=1}^{j=n} \frac{\omega e^{\lambda_j t}}{\omega^2 + \lambda_j^2} \frac{1}{P_n'(\lambda_j)} + A_R \sin(\omega t - \phi)\right]$$

where

$$A_R = \text{amplitude ratio} = \frac{1}{|P_n(i\omega)|}$$

$$\phi = \text{phase lag} = \arg P_n(i\omega) = -\arg \frac{1}{P_n(i\omega)}$$

[Remark: When $\text{Re}(\lambda_j) > 0$ for all j, the steady-state frequency response is simplified to

$$u(t) = AA_R \sin(\omega t - \phi),$$

which can be obtained by taking the system transfer function $\frac{1}{P_n(s)}$ and replacing s by $i\omega$ to get the complex number $\frac{1}{P_n(i\omega)}$. If we write $\frac{1}{P_n(i\omega)} = \frac{1}{|P_n(i\omega)|e^{i\phi}}$, we obtain A_R and ϕ.]

Part IV: **Linear ordinary differential equations-boundary value problems**

18 Two-point boundary value problems

In this and the next chapter, we discuss the solution of linear differential equations with prescribed end or boundary conditions. Specifically, we discuss the problem of determining a function $u(x)$ satisfying an n-th order differential equation in the independent variable x ($a < x < b$) and n boundary conditions involving the function and its first $(n-1)$ derivatives at the end points $x = a$ and $x = b$.

18.1 The adjoint differential operator

Before we study the properties of linear boundary value problems, we revisit the concept of an adjoint to a linear differential operator. As shown in Chapter 14, the concept of an adjoint plays a very important role in the theory of differential equations and was first introduced by Lagrange in connection with the problem of finding the integrating factors. Consider the first-order linear differential operator:

$$Lu = p_0(x)\frac{du}{dx} + p_1(x)u \tag{18.1}$$

In general, the RHS of equation (18.1) is not an exact derivative. We would like to find a function $v(x)$ such that vLu is an exact derivative. Multiplying equation (18.1) by $v(x)$, we have (using $'$ to denote differentiation with respect to x),

$$\begin{aligned} vLu &= vp_0(x)\frac{du}{dx} + vp_1(x)u \\ &= \frac{d}{dx}[vp_0u] - u(vp_0)' + uvp_1 \\ &= \frac{d}{dx}[vp_0u] + [-(p_0v)' + p_1v]u \end{aligned} \tag{18.2}$$

Let us define

$$\begin{aligned} L^*v &= -(p_0v)' + p_1v \\ &= -p_0v' + (p_1 - p_0')v, \end{aligned}$$

where L^* is also a linear differential operator. Equation (18.2) may now be written as

$$vLu - uL^*v = \frac{d}{dx}[vp_0u] \tag{18.3}$$

Now suppose that v satisfies the homogeneous equation

$$L^*v = 0. \tag{18.4}$$

Then equation (18.3) reduces to

https://doi.org/10.1515/9783111598055-022

$$vLu = \frac{d}{dx}[vp_0u], \tag{18.5}$$

which is an exact derivative. Equation (18.4) is called the adjoint equation to $Lu = 0$ and equation (18.3) is called the Lagrange identity. We note that the adjoint equation

$$L^*v = 0$$

may be written as

$$v' = \frac{p_1 - p_0'}{p_0}v$$

Integrating once, we get

$$\begin{aligned}\ln v &= \int\left(\frac{-p_0'}{p_0} + \frac{p_1}{p_0}\right)dx\\ &= -\ln p_0(x) + \int \frac{p_1(x)}{p_0(x)}\,dx\end{aligned}$$

Thus, the integrating factor to equation (18.1) is given by

$$v(x) = \frac{1}{p_0(x)}\exp\left\{\int \frac{p_1(x)}{p_0(x)}\,dx\right\}. \tag{18.6}$$

Now, if we consider the homogeneous equations $Lu = 0$ and $L^*v = 0$, we have from equation (18.3),

$$\frac{d}{dx}[vp_0u] = 0.$$

Integrating, we obtain

$$v(x)p_0(x)u(x) = \text{constant}. \tag{18.7}$$

Thus, if we know either $u(x)$ or $v(x)$, we can determine the other, or the integrating factors as the solutions to the equations $Lu = 0$ and $L^*v = 0$ are intimately related.

Now consider the second-order operator

$$Lu = p_0(x)\frac{d^2u}{dx^2} + p_1(x)\frac{du}{dx} + p_2(x)u \tag{18.8}$$

Following the same procedure, we obtain

$$\begin{aligned}vLu &= vp_0\frac{d^2u}{dx^2} + vp_1(x)\frac{du}{dx} + vp_2(x)u\\ &= vp_0u'' + vp_1u' + vp_2u\end{aligned}$$

$$
\begin{aligned}
&= (vp_0u')' - u'(vp_0)' + (vp_1u)' - u(vp_1)' + vp_2u \\
&= (vp_0u')' - [u(vp_0)']' + u(p_0v)'' + (vp_1u)' - u(vp_1)' + vp_2u \\
&= u[(p_0v)'' - (p_1v)' + p_2v] + \frac{d}{dx}[vp_0u' - u(vp_0)' + vp_1u]
\end{aligned}
$$

$\Longrightarrow$

$$
\begin{aligned}
vLu - uL^*v &= \frac{d}{dx}[vp_0u' - u(vp_0)' + vp_1u] \\
&= \frac{d}{dx}[\pi(u,v)]
\end{aligned}
\tag{18.9}
$$

Thus, if v satisfies the adjoint equation, i. e., $L^*v = 0$, where

$$
\begin{aligned}
L^*v &= (p_0v)'' - (p_1v)' + p_2v \\
&= p_0''v + 2p_0'v' + p_0v'' - (p_1'v + p_1v') + p_2v \\
&= p_0v'' + (2p_0' - p_1)v' + (p_0'' - p_1' + p_2)v,
\end{aligned}
\tag{18.10}
$$

vLu is an exact derivative of

$$
\begin{aligned}
\pi(u,v) &= vp_0u' - uvp_0' - uv'p_0 + vp_1u \\
&= vu'p_0 - v'up_0 + (p_1 - p_0')vu \\
&= \begin{bmatrix} v & v' \end{bmatrix} \begin{bmatrix} p_1 - p_0' & p_0 \\ -p_0 & 0 \end{bmatrix} \begin{bmatrix} u \\ u' \end{bmatrix} \\
&= \mathbf{k}^T(v(x))\mathbf{P}(x)\mathbf{k}(u(x)),
\end{aligned}
\tag{18.11}
$$

where $\mathbf{k}$ is the Wronskian vector. The function $\pi(u,v)$ is called the bilinear concomitant and $\mathbf{P}(x)$ is called the concomitant matrix. Thus,

$$
vLu - uL^*v = \frac{d}{dx}[\mathbf{k}^T(v)\mathbf{P}(x)\mathbf{k}(u)] \tag{18.12}
$$

Equation (18.12) is again the *Lagrange identity* in terms of the Wronskian vectors and concomitant matrix. Note that if two linearly independent solutions of the adjoint equation $L^*v = 0$ are known, then we have

$$
\pi(u,v_1) = v_1(x)p_0(x)u'(x) - u(x)(v_1(x)p_0(x))' + v_1(x)p_1(x)u(x) = c_1 \tag{18.13}
$$

$$
\pi(u,v_2) = v_2(x)p_0(x)u'(x) - u(x)(v_2(x)p_0(x))' + v_2(x)p_1(x)u(x) = c_2 \tag{18.14}
$$

These two linear equations can be solved for $u(x)$ and $u'(x)$ to determine the two linearly independent solutions of the equation $Lu = 0$. Similarly, if two solutions $u_1(x)$ and $u_2(x)$ of $Lu = 0$ are known, we can determine the solutions of $L^*v = 0$. Thus, the solutions of the two homogeneous equations $Lu = 0$ and $L^*v = 0$ are closely related.

18.1.1 The Lagrange identity for an *n*-th order linear differential operator

Now, consider an n-th order linear differential operator:

$$\begin{aligned} Lu &\equiv p_0(x)\frac{d^n u}{dx^n} + p_1(x)\frac{d^{n-1}u}{dx^{n-1}} + \cdots + p_{n-1}(x)\frac{du}{dx} + p_n(x)u \\ &= \sum_{j=0}^{n} p_{n-j}(x)u^{[j]} \end{aligned} \tag{18.15}$$

where

$$u^{[j]} = \frac{d^j u}{dx^j} \tag{18.16}$$

It may be shown that the adjoint equation is given by

$$L^* v = \sum_{j=0}^{n} (-1)^j \left[p_{n-j}(x)v\right]^{[j]} \tag{18.17}$$

and the Lagrange identity is given by

$$vLu - uL^*v = \frac{d}{dx}\left[\mathbf{k}^T(v)\mathbf{P}\mathbf{k}(u)\right] \tag{18.18}$$

where

$$\mathbf{k}(u) = \begin{pmatrix} u(x) \\ u'(x) \\ u''(x) \\ \vdots \\ u^{[n-1]}(x) \end{pmatrix} \tag{18.19}$$

is the Wronskian vector of $u(x)$ and the elements of the concomitant matrix $\mathbf{P}$ are defined by

$$p^*_{ij}(x) = \begin{cases} \sum_{h=i}^{n-j+1}(-1)^{h-1}\begin{pmatrix} h-1 \\ i-1 \end{pmatrix} p^{[h-i]}_{n-h-j+1}(x), & i \le n-j+1 \\ = 0 & i > n-j+1 \end{cases} \tag{18.20}$$

The bilinear concomitant for the n-th order case may be expressed as

$$\pi(u,v) = \sum_{i=1}^{n}\sum_{l=1}^{n-i+1} p^*_{li}(x)v^{[l-1]}u^{[i-1]}. \tag{18.21}$$

It is clear that $\pi(u,v)$ defined by equation (18.21) is a bilinear form, and hence the name bilinear concomitant. For example, for n=3, we get

$$\mathbf{P} = \begin{pmatrix} p_0'' - p_1' + p_2 & p_1 - p_0' & p_0 \\ 2p_0' - p_1 & -p_0 & 0 \\ p_0 & 0 & 0 \end{pmatrix} \tag{18.22}$$

Using the Lagrange identity, we can prove the following theorems.

Theorem 18.1. *The operators L and L^* are adjoint to each other, i. e., $L^{**}y = Ly$ (the adjoint relationship is a reciprocal one).*

Theorem 18.2. *The concomitant matrix $\mathbf{P}$ is nonsingular and its determinant is given by* $\det \mathbf{P}(x) = \{p_0(x)\}^n$.

Proof. Observe that for an $n \times n$ matrix

$$\begin{pmatrix} a_{11} & a_{12} & \cdot & \cdot & a_{1n} \\ a_{21} & a_{22} & \cdot & \cdot & a_{2n} \\ \cdot & & & & \cdot \\ \cdot & & & & \cdot \\ a_{n1} & a_{n2} & \cdot & \cdot & a_{nn} \end{pmatrix}$$

the indices on the antidiagonal sum to $n+1$, i. e., if $i+j = n+1$, then a_{ij} is an antidiagonal element. [Remark: antidiagonal elements are those on the line connecting a_{1n} and a_{n1}.] Since $p_{ij}^* = 0$, for $i > n-j+1 \Longrightarrow \mathbf{P}$ is an upper triangular matrix with

$$\begin{aligned} p_{n-j+1,j}^* &= (-1)^{n-j} p_0(x) \\ \Longrightarrow \det \mathbf{P} &= \{p_0(x)\}^n \end{aligned}$$

□

Definition. If $L^* = L$, then we say that the differential operator L is formally self-adjoint.

Theorem 18.3. *A necessary and sufficient condition for L to be formally self-adjoint is that*

$$\mathbf{P} = -\mathbf{P}^T$$

i. e., $\mathbf{P}$ is a skew-symmetric matrix.

Theorem 18.4. *If $\mathbf{u}$ and $\mathbf{v}$ are fundamental vectors for $Lu = 0$ and $L^*v = 0$, respectively, then*

$$\mathbf{K}^T(\mathbf{v})\mathbf{P}\mathbf{K}(\mathbf{u}) = \mathbf{C}$$

where $\mathbf{C}$ is a nonsingular constant matrix. Further, we can choose $\mathbf{v}(x)$ and $\mathbf{u}(x)$ such that $\mathbf{C}$ is the identity matrix (here $\mathbf{K}$ is the Wronskian matrix).

Example 18.1. (a) Self-adjoint form of a second-order operator

$$\begin{aligned} Lu &= p_0(x)u'' + p_1(x)u' + p_2(x)u \\ L^*v &= p_0(x)v'' + (2p_0' - p_1)v' + (p_0'' - p_1' + p_2)v \\ L = L^* &\Longrightarrow p_0'' - p_1' = 0 \quad \text{and} \quad p_1 = 2p_0' - p_1 \Longrightarrow p_1 = p_0' \end{aligned}$$

$\therefore$

$$\begin{aligned} Lu &= p_0(x)u'' + p_0'(x)u' + p_2(x)u \\ &= (p_0(x)u')' + p_2(x)u \\ &= \frac{d}{dx}\left(p_0(x)\frac{du}{dx}\right) + p_2(x)u \end{aligned}$$

is formally self-adjoint.

(b) Self-adjoint form of a fourth-order operator

$$Lu = p_0(x)u'''' + p_1(x)u''' + p_2(x)u'' + p_3(x)u' + p_4(x)u$$

Skew-symmetry of the concomitant matrix leads to

$$\begin{aligned} p_1 &= 2p_0' \\ p_3 &= p_2' - p_0''' \end{aligned}$$

Thus, $L = L^*$ when L is of the form

$$Lu = (p_0(x)u'')'' + (q_2(x)u')' + p_4u,$$

where $q_2 = p_2 - p_0''$.

For algebraic details and proofs of the theorems stated above, we refer to the book by R. H. Cole [14].

18.2 Two-point boundary value problems

Let

$$L = p_0(x)\frac{d^n}{dx^n} + p_1(x)\frac{d^{n-1}}{dx^{n-1}} + \cdots + p_{n-1}(x)\frac{d}{dx} + p_n(x) \tag{18.23}$$

be an n-th order differential operator and $u(x) \in C^n[a, b]$, $p_0(x) \neq 0$ in $[a, b]$. Consider the problem of solving

$$Lu = -f(x), \quad a < x < b \tag{18.24}$$

subject to the boundary conditions

$$\begin{aligned} \alpha_{11}u(a) + \cdots + \alpha_{1n}u^{[n-1]}(a) + \beta_{11}u(b) + \cdots + \beta_{1n}u^{[n-1]}(b) &= d_1 \\ \alpha_{21}u(a) + \cdots + \alpha_{2n}u^{[n-1]}(a) + \beta_{21}u(b) + \cdots + \beta_{2n}u^{[n-1]}(b) &= d_2 \\ &\;\;\vdots \end{aligned}$$

$$\alpha_{m1}u(a) + \cdots + \alpha_{mn}u^{[n-1]}(a) + \beta_{m1}u(b) + \cdots + \beta_{mn}u^{[n-1]}(b) = d_m \tag{18.25}$$

Let

$$\mathbf{k}(u(x)) = \begin{pmatrix} u(x) \\ u'(x) \\ \cdot \\ \cdot \\ u^{[n-1]}(x) \end{pmatrix} \quad \text{be the Wronskian vector of } u(x)$$

and define the coefficient matrices

$$\mathbf{W}_a = \begin{pmatrix} \alpha_{11} & \alpha_{12} & \cdot & \cdot & \alpha_{1n} \\ \alpha_{21} & \alpha_{22} & \cdot & \cdot & \alpha_{2n} \\ \cdot & & & & \\ \cdot & & & & \\ \alpha_{m1} & \alpha_{m2} & \cdot & \cdot & \alpha_{mn} \end{pmatrix}, \quad \mathbf{W}_b = \begin{pmatrix} \beta_{11} & \beta_{12} & \cdot & \cdot & \beta_{1n} \\ \beta_{21} & \beta_{22} & \cdot & \cdot & \beta_{2n} \\ \cdot & & & & \\ \cdot & & & & \\ \beta_{m1} & \beta_{m2} & \cdot & \cdot & \beta_{mn} \end{pmatrix}$$

Then the boundary conditions (18.25) may be written as

$$\mathbf{W}_a\mathbf{k}(u(a)) + \mathbf{W}_b\mathbf{k}(u(b)) = \mathbf{d}, \tag{18.26}$$

where $\mathbf{d}$ is a m-vector with components d_i. Let

$$\mathbf{W} = (\mathbf{W}_a \quad \mathbf{W}_b) = m \quad \text{by } 2n \text{ matrix and write Eq. (18.26) as}$$

$$\mathbf{W}\begin{pmatrix} \mathbf{k}(u(a)) \\ \mathbf{k}(u(b)) \end{pmatrix} = \mathbf{d}. \tag{18.27}$$

In practice and most of our applications, $m = n$, but, for the present we do not impose this restriction. We assume that rank $\mathbf{W} = m$, i. e., the boundary conditions are independent. The n-th order two-point BVP is defined by equations (18.24) and (18.27). Its solution requires determining a function $u(x)$ satisfying the differential equation as well as the end conditions.

We can use the principle of superposition to write the solution of equations (18.24) and (18.27) as

$$u(x) = u_1(x) + u_2(x), \tag{18.28}$$

where $u_1(x)$ satisfies the inhomogeneous equation with homogeneous boundary conditions:

$$Lu_1(x) = -f(x) \tag{18.29}$$

$$\mathbf{W}\begin{pmatrix} \mathbf{k}(u_1(a)) \\ \mathbf{k}(u_1(b)) \end{pmatrix} = \mathbf{0} \tag{18.30}$$

while $u_2(x)$ satisfies the homogeneous equation with inhomogeneous BCs:

$$Lu_2(x) = 0 \tag{18.31}$$

$$\mathbf{W}\begin{pmatrix} \mathbf{k}(u_2(a)) \\ \mathbf{k}(u_2(b)) \end{pmatrix} = \mathbf{d}. \tag{18.32}$$

As in the case of linear algebraic equations, whether the boundary value problem defined by equations (18.29)–(18.30) has no solution, a unique solution or an infinite number of solutions, depends on the properties of the homogeneous problem. These properties are discussed now.

The two-point homogeneous BVP

Consider the two-point homogeneous BVP defined by

$$Lu = 0, \tag{18.33}$$

$$\mathbf{W}\begin{pmatrix} \mathbf{k}(u(a)) \\ \mathbf{k}(u(b)) \end{pmatrix} = \mathbf{0}. \tag{18.34}$$

Theorem 18.5. *The solutions of the two-point homogeneous boundary value problem defined by equations* (18.33)–(18.34) *form a vector space. Let $C^r[a,b]$ be the vector space of all functions that are r-times differentiable* ($r \geq n$). *We can think of L as a linear operator, i. e.,*

$$L : C^r[a,b] \longrightarrow C[a,b]$$

Then the solutions of $Lu = 0$ form the kernel of L, i. e., these are elements whose image under L is the zero image.

Figure 18.1 shows schematically the domain, codomain and transformation of kernel of the operator L. We have already shown that the kernel of L is a vector space V (which is a subspace of $C^r[a,b]$) of dimension n. Now let $\psi_1(x)$ be any element in $C^r[a,b]$.

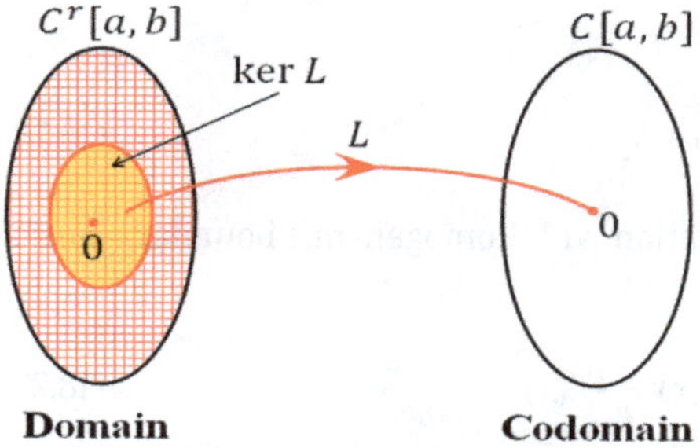

Figure 18.1: Schematic of domain, codomain and transformation of kernel of an operator L.

Define a linear transformation (mapping) from $C^r[a,b]$ to $\mathbb{R}^{2n}$ by the relation

$$\mathcal{B}(\psi_1(x)) = \begin{pmatrix} \psi_1(a) \\ \psi_1'(a) \\ \cdot \\ \psi_1^{[n-1]}(a) \\ \psi_1(b) \\ \cdot \\ \psi_1^{[n-1]}(b) \end{pmatrix} = \text{boundary vector} \tag{18.35}$$

where $\mathcal{B}$ is a linear mapping (see Figure 18.2). This mapping is not obviously one-to-one. But we can define inverse images of sets in R^{2n} in a familiar fashion. If S is any set of elements, then we use $\mathcal{B}^{-1}(S)$ to represent the set of all elements in V whose images are in S.

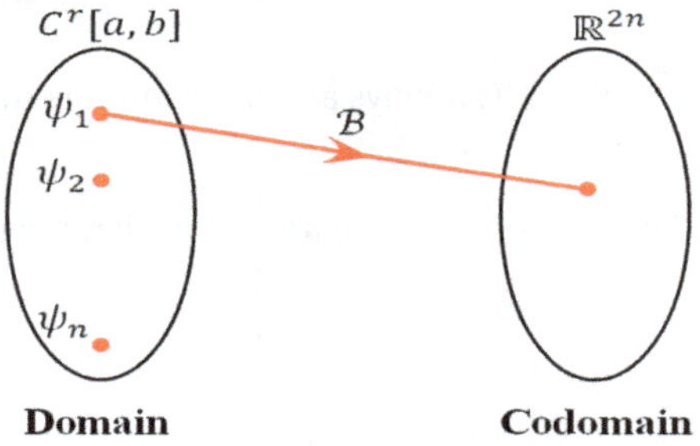

Figure 18.2: Schematic of domain and codomain of the boundary operator $\mathcal{B}$.

Lemma. *If S is a subspace of R^{2n}, then $\mathcal{B}^{-1}(S)$ is a subspace of $C^r[a,b]$.*

Proof. Let $\psi_1, \psi_2 \in B^{-1}(S)$ and α_1, α_2 be any constants. Then the boundary vector of $\alpha_1\psi_1 + \alpha_2\psi_2$ is given by

$$\mathcal{B}(\alpha_1\psi_1 + \alpha_2\psi_2) = \alpha_1\mathcal{B}\psi_1 + \alpha_2\mathcal{B}\psi_2 \tag{18.36}$$

Since S is a subspace, this vector must be in S,

$$\Longrightarrow \alpha_1\psi_1 + \alpha_2\psi_2 \quad \text{is in } \mathcal{B}^{-1}(S).$$

$\therefore$ The result. □

Proof of Theorem 18.5. Equation (18.34) requires the boundary vector to be orthogonal to the rows of **W**. Thus, it defines a subspace S of R^{2n}. Since $\mathcal{B}$ is linear, from the above lemma, the inverse image of this subspace is a subspace of $C^r[a,b]$. □

Definition. The dimension of the solution space of equations (18.33)–(18.34) is called the *index of compatibility* of the BVP. The BVP is said to be *incompatible* if its index of compatibility is zero. (In this case, $u(x) \equiv 0$ is the only solution to the BVP).

Definition. Let $\psi_1(x), \ldots, \psi_n(x)$ be a fundamental set of solutions to $Lu = 0$ and

$$\mathbf{K}(\psi(x)) = \text{Wronskian matrix of } \psi, \quad \text{where } \psi = \begin{pmatrix} \psi_1 \\ \psi_2 \\ \cdot \\ \cdot \\ \cdot \\ \cdot \\ \psi_n \end{pmatrix} \tag{18.37}$$

Define

$$\mathbf{D} = \mathbf{W}_a \mathbf{K}(\psi(a)) + \mathbf{W}_b \mathbf{K}(\psi(b)) \tag{18.38}$$

The matrix $\mathbf{D}$ is called the *characteristic matrix* of the BVP. It plays an important role in determining the properties of the BVP.

Theorem. *If the BVP defined by* (18.33)–(18.34) *has a characteristic matrix of rank r, then its index of compatibility is $n - r$.*

Proof. If ψ is a fundamental vector, then

$$u = \psi^T \mathbf{c} \tag{18.39}$$

is a solution. From equation (18.39), we evaluate the Wronskian matrix

$$\mathbf{K}(u(x)) = \mathbf{K}(\psi(x))\mathbf{c} \tag{18.40}$$

Thus, equation (18.34) $\Longrightarrow$

$$\mathbf{D}\mathbf{c} = \mathbf{0} \tag{18.41}$$

If rank of $\mathbf{D}$ is r then, equation (18.41) has $(n - r)$ linearly independent solutions. These determine precisely $(n - r)$ linearly independent solutions of the BVP. □

Example 18.2. Find the index of compatibility of the system:

$$u''' = 0, \quad 0 < x < 1$$
$$u(0) = 0, \quad u(1) = u'(0), \quad u'(1) = u''(0)$$

Solution:

$$\psi_1(x) = 1$$
$$\psi_2(x) = x$$
$$\psi_3(x) = x^2$$

are linearly independent solutions of the homogeneous system. We have

$$\mathbf{K}(\psi(x)) = \begin{pmatrix} 1 & x & x^2 \\ 0 & 1 & 2x \\ 0 & 0 & 2 \end{pmatrix}$$

⇒

$$\mathbf{K}(\psi(0)) = \begin{pmatrix} 1 & 0 & 0 \\ 0 & 1 & 0 \\ 0 & 0 & 2 \end{pmatrix}, \quad \mathbf{K}(\psi(1)) = \begin{pmatrix} 1 & 1 & 1 \\ 0 & 1 & 2 \\ 0 & 0 & 2 \end{pmatrix}$$
$$\mathbf{W}_0 = \begin{pmatrix} 1 & 0 & 0 \\ 0 & -1 & 0 \\ 0 & 0 & -1 \end{pmatrix}, \quad \mathbf{W}_1 = \begin{pmatrix} 0 & 0 & 0 \\ 1 & 0 & 0 \\ 0 & 1 & 0 \end{pmatrix}$$

∴

$$\begin{aligned}\mathbf{D} &= \begin{pmatrix} 1 & 0 & 0 \\ 0 & -1 & 0 \\ 0 & 0 & -1 \end{pmatrix}\begin{pmatrix} 1 & 0 & 0 \\ 0 & 1 & 0 \\ 0 & 0 & 2 \end{pmatrix} + \begin{pmatrix} 0 & 0 & 0 \\ 1 & 0 & 0 \\ 0 & 1 & 0 \end{pmatrix}\begin{pmatrix} 1 & 1 & 1 \\ 0 & 1 & 2 \\ 0 & 0 & 2 \end{pmatrix} \\ &= \begin{pmatrix} 1 & 0 & 0 \\ 0 & -1 & 0 \\ 0 & 0 & -2 \end{pmatrix} + \begin{pmatrix} 0 & 0 & 0 \\ 1 & 1 & 1 \\ 0 & 1 & 2 \end{pmatrix} \\ &= \begin{pmatrix} 1 & 0 & 0 \\ 1 & 0 & 1 \\ 0 & 1 & 0 \end{pmatrix}\end{aligned}$$

$$\text{rank}\,\mathbf{D} = 3$$

∴ index of compatibility = 0. Thus, the only solution to the homogeneous problem is the trivial one. [Remark: In this simpler case, the result can also be verified by direct integration and application of the boundary conditions.]

Example 18.3. Find the index of compatibility of the system:

$$\begin{aligned} u'' - 3u' + 2u &= 0, \quad 0 < x < 1 \\ u(0) - u'(0) &= 0, \quad -u'(1) + u(1) = 0 \end{aligned}$$

We note that the two linearly independent solutions of the homogeneous equation are

$$\psi_1(x) = e^x, \quad \psi_2(x) = e^{2x}$$

$\Rightarrow$

$$\mathbf{K}(\psi(x)) = \begin{pmatrix} e^x & e^{2x} \\ e^x & 2e^{2x} \end{pmatrix}$$

$$\mathbf{K}(\psi(0)) = \begin{pmatrix} 1 & 1 \\ 1 & 2 \end{pmatrix}, \quad \mathbf{K}(\psi(1)) = \begin{pmatrix} e & e^2 \\ e & 2e^2 \end{pmatrix}$$

$\therefore$

$$\begin{aligned} \mathbf{D} &= \begin{pmatrix} 1 & -1 \\ 0 & 0 \end{pmatrix}\begin{pmatrix} 1 & 1 \\ 1 & 2 \end{pmatrix} + \begin{pmatrix} 0 & 0 \\ 1 & -1 \end{pmatrix}\begin{pmatrix} e & e^2 \\ e & 2e^2 \end{pmatrix} \\ &= \begin{pmatrix} 0 & -1 \\ 0 & 0 \end{pmatrix} + \begin{pmatrix} 0 & 0 \\ 0 & -e^2 \end{pmatrix} \\ &= \begin{pmatrix} 0 & -1 \\ 0 & -e^2 \end{pmatrix} \\ \mathbf{Dc} = \mathbf{0} &\Longrightarrow \begin{pmatrix} 0 & -1 \\ 0 & -e^2 \end{pmatrix}\begin{pmatrix} c_1 \\ c_2 \end{pmatrix} = \mathbf{0} \Longrightarrow c_2 = 0, \quad c_1 = 1 \\ \text{Rank } \mathbf{D} &= 1 \end{aligned}$$

$\therefore$ index of compatibility is one, i. e., the solution space has dimension one.

$$\psi_1(x) = e^x \text{ is a basis for the solution space.}$$

18.3 The adjoint boundary value problem

Recall the standard inner product of two real valued functions of a real variable x, $u(x)$ and $v(x)$ in $C^r[a, b]$:

$$\langle u, v\rangle = \int_a^b u(x)v(x)\,dx \tag{18.42}$$

If the functions are complex-valued, the inner product is defined by

$$\langle u, v\rangle = \int_a^b u(x)\overline{v(x)}\,dx \tag{18.43}$$

where the over bar denotes complex conjugate.

Green's formula:
Consider the n-th order homogeneous equation

$$Lu \equiv p_0(x)u^{[n]} + p_1(x)u^{[n-1]} + \cdots + p_n(x)u = \sum_{j=0}^{n} p_{n-j}u^{[j]} = 0 \tag{18.44}$$

We have shown that the adjoint operator is defined by

$$L^*v = \sum_{j=0}^{n}(-1)^j [p_{n-j}(x)v]^{[j]} = 0 \tag{18.45}$$

and the Lagrange identity may be expressed as

$$vLu - uL^*v = \frac{d}{dx}[\pi(u,v)] = \frac{d}{dx}\{\mathbf{k}^T(v(x))\mathbf{P}(x)\mathbf{k}(u(x))\} \tag{18.46}$$

Integrating equation (18.46) both sides from $x = a$ to $x = b$, we get

$$\begin{aligned}\int_a^b (vLu - uL^*v)\,dx &= \langle Lu, v\rangle - \langle u, L^*v\rangle \\ &= \mathbf{k}^T(v(b))\mathbf{P}(b)\mathbf{k}(u(b)) - \mathbf{k}^T(v(a))\mathbf{P}(a)\mathbf{k}(u(a)) \\ &= \begin{bmatrix} \mathbf{k}^T(v(a)) & \mathbf{k}^T(v(b)) \end{bmatrix}\begin{bmatrix} -\mathbf{P}(a) & \mathbf{0} \\ \mathbf{0} & \mathbf{P}(b)\end{bmatrix}\begin{bmatrix} \mathbf{k}(u(a)) \\ \mathbf{k}(u(b))\end{bmatrix}\end{aligned} \tag{18.47}$$

Equation (18.47) is called the Green's formula. Now, consider the homogeneous two-point BVP,

$$Lu = 0 \tag{18.48}$$

$$\mathbf{W}_a\mathbf{k}(u(a)) + \mathbf{W}_b\mathbf{k}(u(b)) = \mathbf{0} \tag{18.49}$$

We want to impose boundary conditions on the function $v(x)$ of the adjoint problem such that when these conditions and the adjoint equation

$$L^*v = 0 \tag{18.50}$$

are satisfied, the right-hand side of Green's formula is zero, i. e., we want to find a set of boundary conditions (called the adjoint BCs) such that

$$\langle Lu, v\rangle = \langle u, L^*v\rangle.$$

Write Green's formula as

$$\underset{1\times 2n}{\begin{bmatrix} \mathbf{k}^T(v(a)) & \mathbf{k}^T(v(b)) \end{bmatrix}} \underset{2n\times 2n}{\begin{bmatrix} -\mathbf{P}(a) & \mathbf{0} \\ \mathbf{0} & \mathbf{P}(b) \end{bmatrix}} \underset{2n\times 1}{\begin{bmatrix} \mathbf{k}(u(a)) \\ \mathbf{k}(u(b)) \end{bmatrix}} = 0 \tag{18.51}$$

The BCs on u may be written as

$$\begin{bmatrix} \mathbf{W}_a & \mathbf{W}_b \end{bmatrix} \begin{bmatrix} \mathbf{k}(u(a)) \\ \mathbf{k}(u(b)) \end{bmatrix} = \mathbf{0} \Rightarrow \mathbf{W} \begin{bmatrix} \mathbf{k}(u(a)) \\ \mathbf{k}(u(b)) \end{bmatrix} = \mathbf{0} \tag{18.52}$$

i. e., the boundary vector is orthogonal to the rows of $\mathbf{W}$. Thus, we can satisfy (18.51) if we require the vector

$$\begin{bmatrix} \mathbf{k}^T(v(a)) & \mathbf{k}^T(v(b)) \end{bmatrix} \begin{bmatrix} -\mathbf{P}(a) & \mathbf{0} \\ \mathbf{0} & \mathbf{P}(b) \end{bmatrix}$$

to belong to the row space of $\mathbf{W}$, i. e.,

$$\begin{bmatrix} \mathbf{k}^T(v(a)) & \mathbf{k}^T(v(b)) \end{bmatrix} \begin{bmatrix} -\mathbf{P}(a) & \mathbf{0} \\ \mathbf{0} & \mathbf{P}(b) \end{bmatrix} = \mathbf{a}^T\mathbf{W} \tag{18.53}$$

$$= \begin{pmatrix} a_1 & a_2 & \dots & a_n \end{pmatrix} \begin{pmatrix} \mathbf{w}_1^T \\ \mathbf{w}_2^T \\ \cdot \\ \mathbf{w}_n^T \end{pmatrix} \tag{18.54}$$

where $\mathbf{a}$ is a vector in $\mathbf{R}^n$ and $\mathbf{w}_j^T$ is the j-th row of $\mathbf{W}$. These are called the adjoint boundary conditions. Taking the transpose of equation (18.54), we get

$$\mathbf{W}^T\mathbf{a} = \begin{bmatrix} -\mathbf{P}^T(a) & \mathbf{0} \\ \mathbf{0} & \mathbf{P}^T(b) \end{bmatrix} \begin{bmatrix} \mathbf{k}(v(a)) \\ \mathbf{k}(v(b)) \end{bmatrix} \tag{18.55}$$

Equation (18.55) defines a set of $2n$ relations in $\mathbf{k}(v(a))$ and $\mathbf{k}(v(b))$. However, these relations contain n unknown constants $\mathbf{a}$. By eliminating these constants, we obtain a set of n relations in terms of $\mathbf{k}(v(a))$ and $\mathbf{k}(v(b))$. These give the adjoint boundary conditions. Equation (18.55) may be written as

$$-\mathbf{P}^T(a)\mathbf{k}(v(a)) = \mathbf{W}_a^T\mathbf{a} \quad \text{and} \quad \mathbf{P}^T(b)\mathbf{k}(v(b)) = \mathbf{W}_b^T\mathbf{a} \tag{18.56}$$

This is another form of the adjoint BCs.

Remark: The above procedure can be modified for complex valued functions with inner product given by equation (18.43).

Example 18.4. Consider the BVP:

$$Lu \equiv u'' - 3u' + 2u = 0$$
$$u(0) - u'(0) = 0, \quad u(1) - u'(1) = 0$$

We have already determined the adjoint homogeneous equation

$$L^*v = v'' + 3v' + 2v = 0.$$

To find the adjoint BCs, use equation (18.55):

$$\begin{bmatrix} 1 & -1 & 0 & 0 \\ 0 & 0 & 1 & -1 \end{bmatrix}\begin{bmatrix} u(0) \\ u'(0) \\ u(1) \\ u'(1) \end{bmatrix} = \begin{pmatrix} 0 \\ 0 \end{pmatrix}$$

∴

$$\mathbf{W} = \begin{bmatrix} 1 & -1 & 0 & 0 \\ 0 & 0 & 1 & -1 \end{bmatrix}$$

and

$$\mathbf{P} = \begin{pmatrix} -3 & 1 \\ -1 & 0 \end{pmatrix}$$

∴

$$\begin{bmatrix} 3 & 1 & 0 & 0 \\ -1 & 0 & 0 & 0 \\ 0 & 0 & -3 & -1 \\ 0 & 0 & 1 & 0 \end{bmatrix}\begin{bmatrix} v(0) \\ v'(0) \\ v(1) \\ v'(1) \end{bmatrix} = \begin{bmatrix} 1 & 0 \\ -1 & 0 \\ 0 & 1 \\ 0 & -1 \end{bmatrix}\begin{bmatrix} a_1 \\ a_2 \end{bmatrix}$$
$$= \begin{bmatrix} a_1 \\ -a_1 \\ a_2 \\ -a_2 \end{bmatrix}$$

⟹

$$\begin{bmatrix} 3v(0) + v'(0) \\ -v(0) \\ -3v(1) - v'(1) \\ v(1) \end{bmatrix} = \begin{bmatrix} a_1 \\ -a_1 \\ a_2 \\ -a_2 \end{bmatrix}$$

Add rows (1) and (2), and rows (3) and (4),

⟹

$$\begin{bmatrix} 2v(0) + v'(0) \\ -v(0) \\ -2v(1) - v'(1) \\ +v(1) \end{bmatrix} = \begin{bmatrix} 0 \\ -a_1 \\ 0 \\ -a_2 \end{bmatrix}$$

Thus, the adjoint boundary conditions are

$$2v(0) + v'(0) = 0$$
$$-2v(1) - v'(1) = 0$$

The boundary value problems

$$\begin{aligned} u'' - 3u' + 2u &= 0 \quad & v'' + 3v' + 2v &= 0 \\ u(0) - u'(0) &= 0 \quad & 2v(0) + v'(0) &= 0 \\ u(1) - u'(0) &= 0 \quad & 2v(1) + v'(1) &= 0 \end{aligned}$$

are adjoint to each other. It may be verified that $\phi_1(x) = e^{-2x}$ is a basis for the solution space of the adjoint problem.

Remark. While the above general formalism is useful for higher order BVPs, for the case of second- and fourth- order BVPs where the concomitant is a simple expression, we can determine the adjoint BCs by simply using the relation

$$\pi\big(u(a), v(a)\big) - \pi\big(u(b), v(b)\big) = 0.$$

For example, in the above example,

$$\pi(u, v) = u'v - uv' - 3uv,$$

and equating it at the two end points leads to

$$u'(1)v(1) - u(1)v'(1) - 3u(1)v(1) - u'(0)v(0) + u(0)v'(0) + 3u(0)v(0) = 0.$$

Now, using the BCs on $u(x)$, this simplifies to

$$-u(1)\big[v'(1) + 2v(1)\big] + u(0)\big[v'(0) + 2v(0)\big] = 0.$$

The adjoint BCs are obtained by setting the quantities in the brackets to zero.

Example 18.5. Consider the fourth-order differential equation and boundary conditions

$$Lu = \frac{d^4u}{dx^4} + q(x)u = 0;$$
$$u(0) = u''(0) = 0, \quad u(1) = u''(1) = 0$$

Following the above procedure, it may be verified that this BVP is self-adjoint, i. e., the adjoint operator and BCs are the same.

18.3.1 Adjoint BCs and conditions for self-adjointness of the BVP

If the adjoint BCs may be written in the form

$$\mathbf{Q}\begin{bmatrix} \mathbf{k}(v(a)) \\ \mathbf{k}(v(b)) \end{bmatrix} = \mathbf{0}, \tag{18.57}$$

then the two systems

$$Lu = 0 \tag{18.58}$$

$$\mathbf{W}\begin{bmatrix} \mathbf{k}(u(a)) \\ \mathbf{k}(u(b)) \end{bmatrix} = \mathbf{0} \tag{18.59}$$

and

$$L^* v = 0 \tag{18.60}$$

$$\mathbf{Q}\begin{bmatrix} \mathbf{k}(v(a)) \\ \mathbf{k}(v(b)) \end{bmatrix} = \mathbf{0} \tag{18.61}$$

are adjoint to each other. We have seen that the set of solutions to equations (18.58)–(18.59) forms the vector space, which is a subspace of $C^n[a, b]$. The set of solutions to equations (18.60)–(18.61) also form a subspace of $C^n[a, b]$. If these two subspaces are identical, then we say that the BVP is self-adjoint. This is so if $L = L^*$ and $\mathbf{Q} = \mathbf{W}$ or $\mathbf{Q} = \mathbf{CW}$, where $\mathbf{C}$ is a nonsingular matrix. In terms of the coefficient and concomitant matrices, the condition for self-adjointness of the BVP may be expressed as

$$\mathbf{Q}\begin{bmatrix} \mathbf{P}^{-1}(a) & \mathbf{0} \\ \mathbf{0} & -\mathbf{P}^{-1}(b) \end{bmatrix}^T \mathbf{W}^T = \mathbf{0}. \tag{18.62}$$

Further, it may be shown that the index of compatibility of the adjoint system is the same as that of the original system. If we start with the Lagrange identity

$$vLu - uL^* v = \frac{d}{dx}[\mathbf{k}^T(v(x))\mathbf{P}(x)\mathbf{k}(u(x))]$$

and assume that u satisfies

$$Lu = 0$$

and v satisfies

$$L^* v = 0$$

$\Longrightarrow$

$$\frac{d}{dx}[\mathbf{k}^T(v(x))\mathbf{P}(x)\mathbf{k}(u(x))] = 0 \Longrightarrow \mathbf{k}^T(v(x))\mathbf{P}(x)\mathbf{k}(u(x)) = \text{constant} \tag{18.63}$$

Let

$$u(x) = \mathbf{u}^T\mathbf{c}$$

where $\mathbf{u}(x)$ is a fundamental vector. Similarly, let

$$v(x) = \mathbf{v}^T\mathbf{d}$$

$\Longrightarrow$

$$\mathbf{k}^T(\mathbf{v}^T\mathbf{d})\mathbf{P}(x)\mathbf{k}(\mathbf{u}^T\mathbf{c}) = \text{constant}$$

or

$$\mathbf{d}^T\mathbf{K}^T(\mathbf{v}(x))\mathbf{P}(x)\mathbf{K}(\mathbf{u}(x))\mathbf{c} = \text{constant}$$

Since $\mathbf{d}$ and $\mathbf{c}$ are arbitrary vectors in $\mathbb{R}^n \Longrightarrow$

$$\mathbf{K}^T(\mathbf{v}(x))\mathbf{P}(x)\mathbf{K}(\mathbf{u}(x)) = \mathbf{C} \quad \text{(constant matrix)}$$

We can choose $\mathbf{u}$ such that $\mathbf{C} = \mathbf{I}$,
$\therefore$

$$\mathbf{K}^T(\mathbf{v}(x))\mathbf{P}(x)\mathbf{K}(u(x)) = \mathbf{I} \tag{18.64}$$

Now, to determine the index of compatibility of the adjoint system, we use the adjoint BCs,

$$-\mathbf{k}^T(v(a))\mathbf{P}(a) = \mathbf{a}^T\mathbf{W}_a$$
$$\mathbf{k}^T(v(b))\mathbf{P}(b) = \mathbf{a}^T\mathbf{W}_b$$

If $\mathbf{v}$ is fundamental vector, then

$$v = \mathbf{c}^T\mathbf{v} \quad \text{for some } \mathbf{c} \in \mathbf{R}^n$$

$\Longrightarrow$

$$-\mathbf{c}^T\mathbf{K}^T(\mathbf{v}(a))\mathbf{P}(a) = \mathbf{a}^T\mathbf{W}_a \tag{18.65}$$
$$\mathbf{c}^T\mathbf{K}^T(\mathbf{v}(b))\mathbf{P}(b) = \mathbf{a}^T\mathbf{W}_b \tag{18.66}$$

Multiply equation (18.65) by $\mathbf{K}(\mathbf{u}(a))$ and multiply equation (18.66) by $\mathbf{K}(\mathbf{u}(b))$ and use equation (18.64),

$\Longrightarrow$

$$-\mathbf{c}^T = \mathbf{a}^T\mathbf{W}_a\mathbf{K}(\mathbf{u}(a))$$
$$\mathbf{c}^T = \mathbf{a}^T\mathbf{W}_b\mathbf{K}(\mathbf{u}(b))$$

that leads after adding to

$\Longrightarrow$

$$\mathbf{0} = \mathbf{a}^T\mathbf{D}$$

or

$$\mathbf{D}^T\mathbf{a} = \mathbf{0} \tag{18.67}$$

If rank $\mathbf{D} = r$, then there are $(n - r)$ solutions, $\mathbf{a}_j$, to these equations. From each of these solutions, we get a solution to the adjoint problem

$$v_j = -\mathbf{a}_j^T\mathbf{W}_a\mathbf{K}(\mathbf{u}(a))\mathbf{v}(x) \tag{18.68}$$

Thus, the index of compatibility of the adjoint system is also $(n - r)$.

Problems

1. (a) Show that the differential operator

$$Lu = -\frac{1}{w(x)}[(p(x)u')' + q(x)u], \quad a < x < b$$

is formally self-adjoint with respect to the inner product

$$\langle u, v\rangle = \int_a^b w(x)u(x)v(x)\,dx$$

(b) Show that any formally self-adjoint operator of order $2m$ with respect to the usual inner product may be written in the form $Lu = \frac{d^m}{dx^m}\{q_0(x)u^{[m]}\} + \frac{d^{m-1}}{dx^{m-1}}\{q_1(x)u^{[m-1]}\} + \cdots + q_m(x)u$

2. (a) Find the index of compatibility and the solution space of each of the following boundary value problems:
(i) $u''' = 0, 0 < x < 1; u(0) = 0, u'(1) = 0, u'(0) - 2u(1) = 0$ and (ii) $u'' - 3u' + 2u = 0$, $0 < x < 1; u(0) - u(1) = 0, u'(0) - u'(1) = 0$

(b) Determine the values of the parameter λ for which the following boundary value problems are compatible:

(i)

$$u'' + \lambda^2 u = 0, \quad 0 < x < \pi; \quad u(0) = u(\pi) = 0$$

(ii)

$$u^{[4]} - \lambda^4 u = 0, \quad -1 < x < 1; \quad u(-1) = u(1) = u''(-1) = u''(1) = 0$$

3. Given the fourth-order operator $Lu = \frac{d^4u}{dx^4}$, and boundary conditions as follows:
(a) $u(0) = 0, u''(0) = 0, u(1) = 0, u''(1) = 0$
(b) $u(0) = 0, u'(0) = 0, u(1) = 0, u'(1) = 0$
(c) $u(0) = 0, u'(0) = 0, u''(1) = 0, u'''(1) = 0$
(d) $u(0) = 0, u'''(0) = 0, u(1) = 0, u'(1) = 0$
(e) $u(0) = 0, u'''(0) = 0, u(1) = 0, u''(1) = 0$
Determine the adjoint boundary conditions.
4. The following boundary value problem is known as the Orr–Sommerfeld equation and arises in the stability analysis of parallel shear flows:

$$u^{[4]} - 2k^2u'' + k^4u - ik\,\mathrm{Re}\{[1 - x^2 - c][u'' - k^2u] + 2u\} = 0,$$
$$0 < x < 1, \quad i = \sqrt{-1}$$
$$u'(0) = u'''(0) = u(1) = u'(1) = 0$$

Here, k is the wave number ($k > 0$), Re is the Reynolds number (Re > 0) and c is a complex number (which is the dimensionless wave speed). Show that the adjoint system is given by

$$v^{[4]} - 2k^2v'' + k^4v + ik\,\mathrm{Re}\{[1 - x^2 - \overline{c}][v'' - k^2v] - 4xv'\} = 0,$$
$$v'(0) = v'''(0) = v(1) = v'(1) = 0; \quad \overline{c} = \text{complex conjugate of } c$$

5. Let $V = C[a, b]$, the vector space of complex valued continuous functions defined over the real interval $[a, b]$ and T be the linear operator on V defined by

$$Tu(x) = \int_a^b K(s, x)u(s)\,ds$$

where $u \in C[a, b]$ and $K(s, x)$ is continuous in $[a, b] \times [a, b]$. Determine the adjoint operator with respect to the usual inner product.
6. (a) Given the linear operator

$$Lu = \frac{\partial^2 u}{\partial x^2} + \frac{\partial^2 u}{\partial y^2} + \lambda u, \quad 0 < x < a, \quad 0 < y < b$$

and boundary conditions

$$u(0,y) = 0, \quad \frac{\partial u}{\partial x}(a,y) = 0,$$
$$u(x,0) = 0, \quad \frac{\partial u}{\partial y}(x,b) + au(x,b) = 0$$

Determine the adjoint system.

7. Given the linear operator

$$Lu = \frac{\partial^2 u}{\partial x^2} - \frac{\partial u}{\partial t}, \quad 0 < x < 1, \quad t > 0$$

and boundary and initial conditions

$$\frac{\partial u}{\partial x} - au = 0, \quad @\, x = 0; \quad u = 0, \quad @\, x = 1$$
$$u = 0, \quad @\, t = 0$$

Determine the adjoint system.

8. Consider the linear partial differential equation

$$Lu = \sum_{i=1}^{n}\sum_{j=1}^{n} a_{ij}\frac{\partial^2 u}{\partial x_i \partial x_j} + \sum_{i=1}^{n} b_i \frac{\partial u}{\partial x_i} + cu$$

where the coefficients a_{ij}, b_i and c are real analytic functions of real variables $x_1, \ldots, x_n$ and where the matrix of elements a_{ij} is symmetric. (a) Determine the adjoint operator and the form of the Lagrange identity (b) Use the divergence theorem to obtain a formula analogous to Green's formula.

9. The steady-state conversion (u) in a radial flow reactor (see Figure 18.3) is given by the boundary value problem

$$\frac{1}{\mathrm{Pe}}\frac{1}{r}\frac{d}{dr}\left(r\frac{du}{dr}\right) - \frac{r_0}{r}\frac{du}{dr} - \mathrm{Da}\, u = -\mathrm{Da}, \quad r_0 < r < 1$$
$$\frac{du}{dr}(1) = 0; \quad \frac{1}{\mathrm{Pe}}\frac{du}{dr}(r_o) - u(r_0) = 0$$

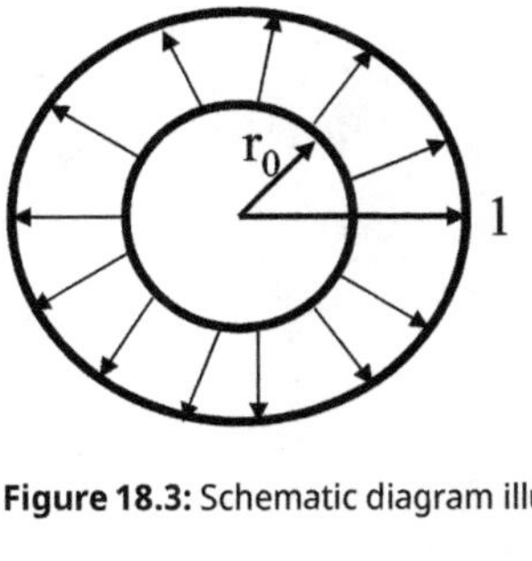

Figure 18.3: Schematic diagram illustrating a radial flow reactor.

where Pe is the Peclet number, Da is the Damköhler number and r_0 is the dimensionless inner radius.

(a) Determine the adjoint homogeneous problem w. r. t. the inner product

$$\langle u, v \rangle = \int_{r_0}^{1} ruv\, dr$$

(b) Give a physical interpretation of the adjoint problem

(c) Write down the BVP that describes the conversion $v(r)$ when the flow direction is reversed.

(d) Use the Lagrange identity and the above results to show that $u(1) = v(r_0)$, i. e., the exit conversion is independent of the flow direction.

10. Let V be the vector space of complex valued functions $u(x)$ defined on the interval $(0, a)$ satisfying the periodicity condition $u(0) = u(a)$. Let L be a linear operator on V defined by

$$Lu(x) = -i\frac{du}{dx}; \quad i = \sqrt{-1}$$

Show that this operator is self-adjoint with respect to the usual inner product on V, i. e.,

$$\langle u, v \rangle = \int_{0}^{a} u(x)\overline{v(x)}\, dx.$$

19 The nonhomogeneous BVP and Green's function

19.1 Introduction to Green's function

Consider the nonhomogeneous BVP

$$Lu = -f(x), \quad a < x < b \tag{19.1}$$

with homogeneous BCs

$$\mathbf{W}_a \mathbf{k}(u(a)) + \mathbf{W}_b \mathbf{k}(u(b)) = \mathbf{0}. \tag{19.2}$$

Suppose that we can express the solution of equations (19.1) and (19.2) as

$$u(x) = \int_a^b G(x, \xi) f(\xi)\, d\xi, \tag{19.3}$$

where $G(x, \xi)$ is called the Green's function of the linear operator L with homogeneous BCs

$$\mathbf{W}_a \mathbf{k}(G(a, \xi)) + \mathbf{W}_b \mathbf{k}(G(b, \xi)) = \mathbf{0}. \tag{19.4}$$

Remark. Some authors define the Green's function with a positive sign on the RHS of equation (19.1). However, we shall use the definition above to be consistent with the engineering literature. With this notation, positive (negative) values of $f(x)$ correspond to source (sink), when L is the diffusion/conduction operator.

We first discuss an example in which the Green's function is obtained by direct integration. We then obtain a formula for the Green's function of a second-order self-adjoint BVP. Lastly, we discuss the general case of inhomogeneous n-th order BVP.

Example 19.1. Consider the second-order BVP defined by

$$\frac{d^2u}{dx^2} = -f(x), \quad 0 < x < 1$$
$$u'(0) = 0, \quad u(1) = 0.$$

Integrating once from 0 to x and using the first boundary condition, we get

$$\frac{du}{dx} = -\int_0^x f(\eta)\, d\eta$$

Integrating again,

https://doi.org/10.1515/9783111598055-023

$$u(x) = u(0) - \int_0^x \left[\int_0^\xi f(\eta)\, d\eta \right] d\xi,$$

$$\Longrightarrow u(1) = u(0) - \int_0^1 \int_0^\xi f(\eta)\, d\eta\, d\xi$$

$\Longrightarrow$

$$u(x) - u(1) = \int_0^1 \left[\int_0^\xi f(\eta)\, d\eta \right] d\xi - \int_0^x \left[\int_0^\xi f(\eta)\, d\eta \right] d\xi$$

$u(1) = 0 \Longrightarrow$

$$u(x) = \int_0^1 \left[\int_0^\xi f(\eta)\, d\eta \right] d\xi + \int_x^0 \left[\int_0^\xi f(\eta)\, d\eta \right] d\xi,$$

which is simplified to

$$u(x) = \int_x^1 \left[\int_0^\xi f(\eta)\, d\eta \right] d\xi$$

$$= \int_x^1 \int_0^x f(\eta)\, d\eta\, d\xi + \int_x^1 \int_x^\xi f(\eta)\, d\eta\, d\xi$$

To evaluate this double integral, we change the order of integration. Figure 19.1 shows the schematic of the domain and change in order of variables in the second double integration.

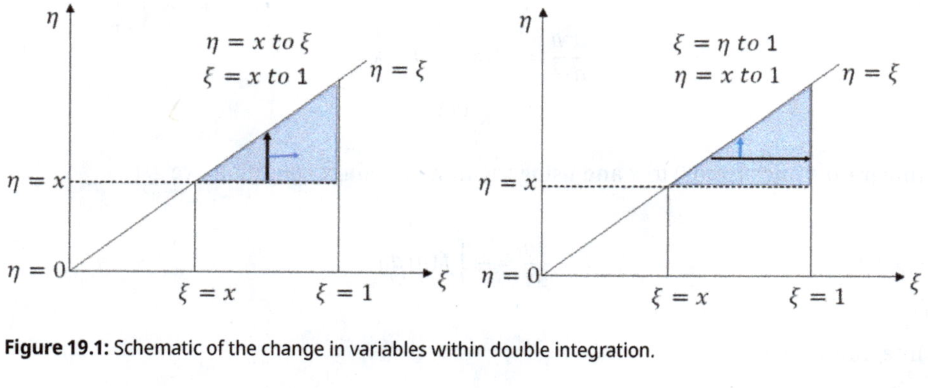

Figure 19.1: Schematic of the change in variables within double integration.

Thus,

$$u(x) = \int_0^x \int_x^1 f(\eta)\,d\xi\,d\eta + \int_{\eta=x}^1 \int_{\xi=\eta}^{\xi=1} f(\eta)\,d\xi\,d\eta$$

$$= (1-x)\int_0^x f(\eta)\,d\eta + \int_x^1 (1-\eta)f(\eta)\,d\eta$$

$$= \int_0^1 G(x,\eta)f(\eta)\,d\eta$$

where

$$G(x,\eta) = \begin{cases} (1-x), & 0 \le \eta \le x \\ (1-\eta), & x < \eta \le 1 \end{cases}$$

is the Green's function.

[Remark: The above procedure does not change if we replace $f(x)$ by a nonlinear function $f(x,u(x))$. In this case, we convert the nonlinear BVP to a nonlinear integral equation

$$u(x) = \int_0^1 G(x,\eta)f(\eta,u(\eta))\,d\eta,$$

where the Green's function remains the same].

19.2 Green's function for second-order self-adjoint TPBVP

We now generalize the above procedure to the more general second-order BVP,

$$Lu \equiv \frac{d}{dx}\left(p(x)\frac{du}{dx}\right) - q(x)u = -f(x), \quad a < x < b, \tag{19.5}$$

with one of the following three types of BCs:

$$\left.\begin{aligned} \alpha_1 u(a) + \alpha_2 u'(a) = 0 \\ \beta_1 u(b) + \beta_2 u'(b) = 0 \end{aligned}\right\} \quad \text{Robin/mixed/radiation boundary conditions} \tag{19.6}$$

$$\left.\begin{aligned} u(a) = 0 \\ u(b) = 0 \end{aligned}\right\} \quad \text{Dirichlet boundary conditions} \tag{19.7}$$

$$\left.\begin{aligned} u(a)p(a) = u(b)p(b) \\ u'(a) = u'(b) \end{aligned}\right\} \quad \text{Periodic boundary conditions} \tag{19.8}$$

It may be easily verified that the TPBVP defined by equation (19.5) with one of the above set of BCs (19.6)–(19.8) is self-adjoint. We want to write the solution of equation (19.5) with Robin/Dirichlet/Periodic BCs in the form

$$u(x) = \int_a^b G(x,s)f(s)\,ds \tag{19.9}$$

where $G(x,s)$ is the Green's function. The method for developing Green's functions is based on the method of variation of parameters. Suppose that $u_1(x)$ and $u_2(x)$ are two linearly independent solutions of the homogeneous equation

$$Lu = (pu')' - qu = 0$$

Assume that $u_1(x)$ satisfies the boundary condition at $x = a$ and $u_2(x)$ satisfies the boundary condition at $x = b$, i. e.,

$$\alpha_1 u_1(a) + \alpha_2 u_1'(a) = 0$$
$$\beta_1 u_2(b) + \beta_2 u_2'(b) = 0$$

Express the solution to equations (19.5) and (19.6) as

$$u = c_1(x)u_1(x) + c_2(x)u_2(x) \tag{19.10}$$

where the functions $c_j(x), j = 1, 2$, are yet to be determined.

$\Longrightarrow$

$$u' = c_1 u_1' + c_2 u_2' + c_1' u_1 + c_2' u_2 \tag{19.11}$$

Choose c_1 and c_2 such that

$$c_1' u_1 + c_2' u_2 = 0 \tag{19.12}$$

$\Longrightarrow$

$$u' = c_1 u_1' + c_2 u_2'$$

$\Longrightarrow$

$$pu' = c_1 pu_1' + c_2 pu_2'$$

$\Longrightarrow$

$$(pu')' = c_1(pu_1')' + c_2(pu_2')' + c_1' pu_1' + c_2' pu_2'$$

and

$$qu = qc_1u_1 + qc_2u_2$$

From the above two equations, we get

$$-f(x) = c_1[(pu_1')' - qu_1] + c_2[(pu_2')' - qu_2] + c_1'pu_1' + c_2'pu_2'$$

$\Longrightarrow$

$$c_1'u_1' + c_2'u_2' = -\frac{f}{p} \tag{19.13}$$

Equations (19.12) and (19.13) $\Longrightarrow$

$$\begin{bmatrix} u_1 & u_2 \\ u_1' & u_2' \end{bmatrix}\begin{bmatrix} c_1' \\ c_2' \end{bmatrix} = \begin{bmatrix} 0 \\ -\frac{f}{p} \end{bmatrix} = -\mathbf{e}_2\frac{f}{p}$$

$\Longrightarrow$

$$\mathbf{K}(\mathbf{u}(x))\mathbf{c}' = -\mathbf{e}_2\frac{f}{p}$$
$$\mathbf{c}' = -\mathbf{K}^{-1}(\mathbf{u}(x))\mathbf{e}_2\frac{f}{p}$$

$\Longrightarrow$

$$\mathbf{c}(x) = \mathbf{c}_0 - \int_a^x \mathbf{K}^{-1}(\mathbf{u}(s))\mathbf{e}_2\frac{f(s)}{p(s)}\,ds, \tag{19.14}$$

where $\mathbf{c}_0$ is a constant vector.

Substituting equation (19.14) in equation (19.10) gives

$$u(x) = \begin{bmatrix} u_1(x) & u_2(x) \end{bmatrix}\begin{bmatrix} c_{10} \\ c_{20} \end{bmatrix} - \begin{bmatrix} u_1(x) & u_2(x) \end{bmatrix}\int_a^x \mathbf{K}^{-1}(\mathbf{u}(s))\mathbf{e}_2\frac{f(s)}{p(s)}\,ds$$

$$u'(x) = \begin{bmatrix} u_1'(x) & u_2'(x) \end{bmatrix}\begin{bmatrix} c_{10} \\ c_{20} \end{bmatrix} - \begin{bmatrix} u_1'(x) & u_2'(x) \end{bmatrix}\int_a^x \mathbf{K}^{-1}(\mathbf{u}(s))\mathbf{e}_2\frac{f(s)}{p(s)}\,ds$$
$$- \begin{bmatrix} u_1(x) & u_2(x) \end{bmatrix}\mathbf{K}^{-1}(\mathbf{u}(x))\mathbf{e}_2\frac{f(x)}{p(x)} \tag{19.15}$$

The last term in equation (19.15) is identically zero as can be seen from the following calculation:

$$\mathbf{K}(\mathbf{u}(x)) = \begin{bmatrix} u_1 & u_2 \\ u_1' & u_2' \end{bmatrix}$$

$$\det \mathbf{K} \equiv W(x) = u_1 u_2' - u_2 u_1'$$

∴

$$\mathbf{K}^{-1}(\mathbf{u}(x)) = \begin{bmatrix} u_2' & -u_2 \\ -u_1' & u_1 \end{bmatrix} \frac{1}{W}$$

$$\begin{aligned}\mathbf{u}^T\mathbf{K}^{-1}(\mathbf{u}(x)) &= \begin{bmatrix} u_1 & u_2 \end{bmatrix} \begin{bmatrix} u_2' & -u_2 \\ -u_1' & u_1 \end{bmatrix} \frac{1}{W} \\ &= \begin{bmatrix} u_1u_2' - u_2u_1' & -u_1u_2 + u_1u_2 \end{bmatrix} \frac{1}{W} \\ &= \begin{bmatrix} 1 & 0 \end{bmatrix}\end{aligned}$$

∴

$$\begin{aligned}\mathbf{u}^T\mathbf{K}^{-1}(\mathbf{u})\mathbf{e}_2 \frac{f(x)}{p(x)} &= \begin{bmatrix} 1 & 0 \end{bmatrix} \begin{bmatrix} 0 \\ 1 \end{bmatrix} \frac{f(x)}{p(x)} \\ &= 0\end{aligned}$$

⟹

$$u'(x) = \begin{bmatrix} u_1' & u_2' \end{bmatrix} \begin{bmatrix} c_{10} \\ c_{20} \end{bmatrix} - \begin{bmatrix} u_1' & u_2' \end{bmatrix} \int_a^x \mathbf{K}^{-1}(\mathbf{u}(s))\mathbf{e}_2 \frac{f(s)}{p(s)}\, ds \tag{19.16}$$

Thus,

$$\begin{aligned}u(a) &= \begin{bmatrix} u_1(a) & u_2(a) \end{bmatrix} \begin{bmatrix} c_{10} \\ c_{20} \end{bmatrix} \\ u'(a) &= \begin{bmatrix} u_1'(a) & u_2'(a) \end{bmatrix} \begin{bmatrix} c_{10} \\ c_{20} \end{bmatrix} \quad \text{and} \\ \alpha_1 u(a) + \alpha_2 u'(a) &= \alpha_1[c_{10}u_1(a) + c_{20}u_2(a)] + \alpha_2[c_{10}u_1'(a) + c_{20}u_2'(a)] \\ &= c_{10}[\alpha_1 u_1(a) + \alpha_2 u_1'(a)] + c_{20}[\alpha_1 u_2(a) + \alpha_2 u_2'(a)] \\ &= c_{20}[\alpha_1 u_2(a) + \alpha_2 u_2'(a)] = 0\end{aligned}$$

⟹

$$\begin{aligned}c_{20} &= 0 \\ u(b) &= \mathbf{u}^T(b)\mathbf{c}_0 - \mathbf{u}^T(b) \int_a^b \mathbf{K}^{-1}(\mathbf{u}(s))\mathbf{e}_2 \frac{f(s)}{p(s)}\, ds\end{aligned}$$

$$u'(b) = \mathbf{u}'^T(b)\mathbf{c}_0 - \mathbf{u}'^T(b)\int_a^b \mathbf{K}^{-1}(\mathbf{u}(s))\mathbf{e}_2\frac{f(s)}{p(s)}\,ds$$

$\beta_1 u(b) + \beta_2 u'(b) = 0 \Longrightarrow$

$$\beta_1 c_{10}u_1(b) + \beta_2 c_{10}u_1'(b) = \begin{bmatrix} \beta_1 u_1(b) + \beta_2 u_1'(b) & \beta_1 u_2(b) + \beta_2 u_2'(b) \end{bmatrix}\int_a^b \mathbf{K}^{-1}\mathbf{e}_2\frac{f(s)}{p(s)}\,ds$$

$\Longrightarrow$

$$c_{10} = \begin{bmatrix} 1 & 0 \end{bmatrix}\int_a^b \mathbf{K}^{-1}(\mathbf{u}(s))\mathbf{e}_2\frac{f}{p}\,ds$$

$\therefore$

$$\begin{aligned} u(x) &= [u_1(x) \quad 0]\int_a^b \mathbf{K}^{-1}(\mathbf{u}(s))\mathbf{e}_2\frac{f(s)}{p(s)}\,ds - \begin{bmatrix} u_1 & u_2 \end{bmatrix}\int_a^x \mathbf{K}^{-1}(\mathbf{u}(s))\mathbf{e}_2\frac{f}{p}\,ds \\ &= [u_1 \quad 0]\left\{\int_a^x \{.\}\,ds + \int_x^b \{.\}\,ds\right\} - [u_1 \quad 0]\int_a^x \{.\}\,ds - \begin{bmatrix} 0 & u_2 \end{bmatrix}\int_a^x \{.\}\,ds \\ &= [u_1 \quad 0]\int_x^b \{.\}\,ds - \begin{bmatrix} 0 & u_2 \end{bmatrix}\int_a^x \{.\}\,ds \end{aligned}$$

Now,

$$\begin{aligned} \mathbf{K}^{-1}(\mathbf{u}(s))\mathbf{e}_2 &= \frac{1}{W(s)}\begin{bmatrix} u_2'(s) & -u_2(s) \\ -u_1'(s) & u_1(s) \end{bmatrix}\begin{bmatrix} 0 \\ 1 \end{bmatrix} \\ &= \frac{1}{W(s)}\begin{bmatrix} -u_2(s) \\ u_1(s) \end{bmatrix} \end{aligned}$$

$\therefore$

$$\begin{aligned} u(x) &= \int_x^b \frac{-u_1(x)u_2(s)f(s)}{p(s)W(s)}\,ds + \int_a^x \frac{-u_2(x)u_1(s)f(s)}{p(s)W(s)}\,ds \\ &= \int_a^b G(x,s)f(s)\,ds \end{aligned}$$

where

$$G(x,s) = \begin{cases} \frac{-u_1(s)u_2(x)}{p(s)W(s)}, & a \le s \le x \\ \frac{-u_1(x)u_2(s)}{p(s)W(s)}, & x \le s \le b \end{cases} \tag{19.17}$$

is the Green's function. To simplify the expression for the Green's function further, we make the following observation:

$$\frac{d}{dx}\{p(x)W(x)\} = 0$$

To prove,

$$\begin{aligned} \text{LHS} &= \frac{d}{dx}\{p(x)(u_1u_2' - u_2u_1')\} \\ &= u_1(pu_2')' + u_1'pu_2' - u_2(pu_1')' - pu_1'u_2' \\ &= qu_1u_2 - u_2[qu_1] = 0 \end{aligned}$$

Therefore, $p(x)W(x) = \text{constant}$. We may choose u_1 and u_2 such that the constant is equal to minus one (−1). Then

$$G(x,s) = \begin{cases} u_1(s)u_2(x), & a < s < x \\ u_1(x)u_2(s), & x < s < b. \end{cases} \tag{19.18}$$

Figure 19.2 shows an interpretation of the Green's function considered as a function of x (with s fixed) or function of s (with x fixed).

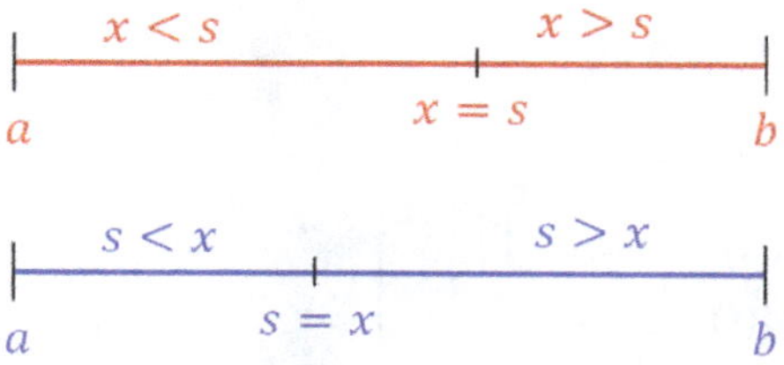

Figure 19.2: Geometric interpretation of Green's function.

It may be shown that the above formula for the Green's function, though derived for the case of Robin BCs is also valid for Dirichlet boundary conditions. The same procedure can be applied for the case of periodic boundary conditions.

Example 19.2. Consider the boundary value problem

$$\frac{d^2u}{dx^2} = -f(x), \quad u'(0) = 0, \quad u(1) = 0$$

⇒

$$u_1(x) = 1, \quad u_2(x) = 1 - x, \quad p(x) = 1, \quad q(x) = 0$$

$$W = \begin{vmatrix} 1 & 1-x \\ 0 & -1 \end{vmatrix} = -1$$

$\therefore$

$$G(x,s) = \begin{cases} 1-x, & a < s < x \\ 1-s, & x < s < 1 \end{cases}$$

Example 19.3. Consider

$$\frac{d^2u}{dx^2} = -f(x)$$
$$u(0) = 0, \quad u(1) = 0$$
$$u_1(x) = x, \quad u_2(x) = 1-x, \quad p(x) = 1, \quad q(x) = 0$$
$$W = \begin{vmatrix} x & 1-x \\ 1 & -1 \end{vmatrix} = -x - 1 + x = -1$$

Therefore,

$$G(x,s) = \begin{cases} s(1-x), & 0 < s < x \\ x(1-s), & x < s < 1 \end{cases}$$

$$u(x) = \int_0^x s(1-x)f(s)\,ds + \int_x^1 x(1-s)f(s)\,ds$$
$$= (1-x)\int_0^x sf(s)\,ds + x\int_x^1 (1-s)f(s)\,ds$$

Example 19.4. Consider the boundary value problem

$$Lu \equiv \frac{d}{dx}\left(x\frac{du}{dx}\right) - \frac{n^2}{x}u = -f(x), \quad n \text{ is a positive integer.}$$
$$u'(0) = 0, \quad u(1) = 0$$

This operator arises in the solution of inhomogeneous problems in cylindrical geometry. We note that the homogeneous equation is a Euler equation and

$$u_1(x) = x^n, \quad u_2(x) = x^{-n} - x^n$$

are linearly independent solutions satisfying the boundary conditions at the ends:

$$W(x) = \begin{vmatrix} x^n & x^{-n} - x^n \\ nx^{n-1} & -nx^{-n-1} - nx^{n-1} \end{vmatrix} = \frac{-2n}{x}$$

$\therefore$

$$p(x)W(x) = x\left(\frac{-2n}{x}\right) = -2n$$

∴

$$G(x,s) = \begin{cases} \frac{s^n(x^{-n}-x^n)}{2n}, & 0 < s < x \\ \frac{x^n(s^{-n}-s^n)}{2n}, & x < s < 1 \end{cases}$$
$$= \begin{cases} [(\frac{s}{x})^n - (sx)^n]/2n, & 0 < s < x \\ [(\frac{x}{s})^n - (sx)^n]/2n, & x < s < 1 \end{cases}$$

19.3 Properties of the Green's function for the second-order self-adjoint BVP

We now examine the properties of the Green's function for the second-order self-adjoint problem:

$$Lu \equiv \frac{d}{dx}\left(p(x)\frac{du}{dx}\right) - q(x)u = -f(x), \quad a < x < b \tag{19.19}$$

$$G(x,s) = \begin{cases} -\frac{u_1(s)u_2(x)}{p(s)W(s)}, & a < s < x \\ -\frac{u_1(x)u_2(s)}{p(s)W(s)}, & x < s < b \end{cases} \tag{19.20}$$

Properties of $G(x,s)$:

1. $G(x,s)$ is a continuous function in $[a,b] \times [a,b]$, even at $x = s$.
2. $G(x,s)$ is symmetric, i. e.,

$$G(x,s) = G(s,x) \tag{19.21}$$

3. $G(x,s)$ satisfies the differential equation (in both variables x and s)

$$Lu = 0$$

except perhaps at $x = s$, i. e.,

$$\frac{d}{dx}\left(p(x)\frac{dG}{dx}\right) - q(x)G = 0,\ x \neq s \tag{19.22}$$
$$\frac{d}{ds}\left(p(s)\frac{dG}{ds}\right) - q(s)G = 0,\ s \neq x$$

This is a simple and straightforward calculation.

4. The derivative of $G(x,s)$ has a jump discontinuity at $x = s$. Treating $G(x,s)$ as a function of x, and denoting $p(x)W(x) = p(s)W(s) = C$, we have

$$G(x,s) = \begin{cases} -\frac{u_1(x)u_2(s)}{C}, & a < x < s \\ -\frac{u_2(x)u_1(s)}{C}, & s < x < b \end{cases}$$

$$\left.\frac{\partial G}{\partial x}\right|_{x=s^+} = \left.\frac{-u_1(s)u_2'(x)}{C}\right|_{x=s} \quad (x \text{ approaching } s \text{ from the RHS})$$

$$= \frac{-u_1(s)u_2'(s)}{C}$$

$$\left.\frac{\partial G}{\partial x}\right|_{x=s^-} = \left.\frac{-u_1'(x)u_2(s)}{C}\right|_{x=s} \quad (x \text{ approaching } s \text{ from the LHS})$$

$$= \frac{-u_1'(s)u_2(s)}{C}$$

∴

$$\left.\frac{\partial G}{\partial x}\right|_{x=s^+} - \left.\frac{\partial G}{\partial x}\right|_{x=s^-} = \text{Jump at } x = s$$

$$= \frac{u_1'u_2 - u_1u_2'}{C} = \frac{-W(s)}{p(s)W(s)}$$

$$\frac{\partial G}{\partial x}(s^+, s) - \frac{\partial G}{\partial x}(s^-, s) = \frac{-1}{p(s)} \tag{19.23}$$

Similarly, treating $G(x,s)$ as a function of s,

$$\left.\frac{\partial G}{\partial s}\right|_{s=x^+} = \left.\frac{-u_1(x)u_2'(s)}{C}\right|_{s=x} = \frac{-u_1(x)u_2'(x)}{C}$$

and

$$\left.\frac{\partial G}{\partial s}\right|_{s=x^-} = \left.\frac{-u_1'(s)u_2(x)}{C}\right|_{s=x} = \frac{-u_1'(x)u_2(x)}{C}$$

$$\text{Jump} = \left.\frac{\partial G}{\partial s}\right|_{s=x^+} - \left.\frac{\partial G}{\partial s}\right|_{s=x^-} = \frac{u_1'u_2 - u_1u_2'}{C} = \frac{-W(x)}{W(x)p(x)}$$

$$\Rightarrow \frac{\partial G}{\partial s}(x^+, x) - \frac{\partial G}{\partial s}(x^-, x) = -\frac{1}{p(x)} \tag{19.24}$$

5. $G(x,s)$ satisfies the boundary conditions in both variables.

Proof. Consider $G(x,s)$ as a function of x

BC1

$$\alpha_1 G(a,s) + \alpha_2 G'(a,s) = \alpha_1 \frac{-u_1(a)u_2(s)}{C} + \alpha_2 \frac{-u_1'(a)u_2(s)}{C}$$

$$= -\frac{u_2(s)}{C}[\alpha_1 u_1(a) + \alpha_2 u_1'(a)] = 0 \tag{19.25}$$

BC2

$$\begin{aligned}\beta_1 G(b,s) + \beta_2 G'(b,s) &= \beta_1 \frac{-u_1(s)u_2(b)}{C} + \beta_2 \frac{-u_1(s)u_2'(b)}{C}\\ &= \frac{-u_1(s)}{C}[\beta_1 u_2(b) + \beta_2 u_2'(b)] = 0\end{aligned} \tag{19.26}$$

Consider $G(x,s)$ as a function of s

BC1

$$\begin{aligned}\alpha_1 G(x,a) + \alpha_2 G'(x,a) &= \alpha_1 \left[\frac{-u_1(a)u_2(x)}{C}\right] + \alpha_2 \left[\frac{-u_1'(a)u_2(x)}{C}\right]\\ &= \frac{-u_2(x)}{C}[\alpha_1 u_1(a) + \alpha_2 u_1'(a)] = 0\end{aligned} \tag{19.27}$$

BC2

$$\begin{aligned}\beta_1 G(x,b) + \beta_2 G'(x,b) &= \beta_1 \left[\frac{-u_1(x)u_2(b)}{C}\right] + \beta_2 \left[\frac{-u_1(x)u_2'(b)}{C}\right]\\ &= \frac{-u_1(x)}{C}[\beta_1 u_2(b) + \beta_2 u_2'(b)] = 0\end{aligned} \tag{19.28}$$

6. $u(x) = \int_a^b G(x,s)f(s)\,ds$ is indeed the solution to $Lu = -f(x)$ with the homogeneous boundary conditions.

$$\begin{aligned}\alpha_1 u(a) + \alpha_2 u'(a) &= 0\\ \beta_1 u(b) + \beta_2 u'(b) &= 0\end{aligned}$$

□

Proof.

$$\begin{aligned}u(x) &= \int_a^b G(x,s)f(s)\,ds \qquad (19.29)\\ &= \int_a^x G(x,s)f(s)\,ds + \int_x^b G(x,s)f(s)\,ds\end{aligned}$$

$\Longrightarrow$

$$\begin{aligned}u'(x) = &\int_a^x \frac{\partial G(x,s)}{\partial x} f(s)\,ds + \int_x^b \frac{\partial G(x,s)}{\partial x} f(s)\,ds\\ &+ G(x,x^-)f(x^-) - G(x,x^+)f(x^+)\end{aligned} \tag{19.30}$$

Since G and f are continuous, the last two terms cancel and we get

$$u'(x) = \int_a^x \frac{\partial G(x,s)}{\partial x} f(s)\,ds + \int_x^b \frac{\partial G(x,s)}{\partial x} f(s)\,ds \tag{19.31}$$

Now, using equations (19.29) and (19.31), we get

$$u(a) = \int_a^b G(a,s) f(s)\, ds$$

$$u'(a) = \int_a^b \frac{\partial G}{\partial x}(a,s) f(s)\, ds$$

$\Longrightarrow$

$$\alpha_1 u(a) + \alpha_2 u'(a) = \int_a^b \left[\alpha_1 G(a,s) + \alpha_2 \frac{\partial G}{\partial x}(a,s) \right] f(s)\, ds = 0$$

since G satisfies the BC at $x = a$. Similarly,

$$u(b) = \int_a^b G(b,s) f(s)\, ds$$

$$u'(b) = \int_a^b \frac{\partial G}{\partial x}(b,s) f(s)\, ds$$

$$\beta_1 u(b) + \beta_2 u'(b) = \int_a^b \left[\beta_1 G(b,s) + \beta_2 \frac{\partial G}{\partial x}(b,s) \right] f(s)\, ds = 0$$

since G satisfies the BC at $x = b$. Thus, the solution given by equation (19.29) satisfies the BCs.

Now, equation (19.31) $\Longrightarrow$

$$\begin{aligned}(pu')' &= \frac{\partial}{\partial x}\left[p(x) \int_a^x \frac{\partial G}{\partial x}(x,s) f(s)\, ds + p(x) \int_x^b \frac{\partial G}{\partial x}(x,s) f(s)\, ds \right] \\ &= \int_a^x (pG')' f(s)\, ds + \int_x^b (pG')' f(s)\, ds + p(x) f(x) \left(\frac{\partial G}{\partial x}(x, x^-) - \frac{\partial G}{\partial x}(x, x^+) \right)\end{aligned}$$

$\Longrightarrow$

$$\begin{aligned}(pu')' + qu &= \int_a^x [(pG')' + qG] f(s)\, ds + \int_x^b [(pG')' + qG] f(s)\, ds \\ &\quad + p(x) f(x) \left[\frac{\partial G}{\partial x}(x, x_-) - \frac{\partial G}{\partial x}(x, x_+) \right] \\ &= 0 + 0 + p(x) f(x) \left[\frac{-1}{p(x)} \right] \\ &= -f(x)\end{aligned}$$

Thus, $G(x,s)$ is the solution of $Lu = -f(x)$.

(7) $G(x,s)$ is a unique function. □

Proof. Suppose that there are two, say $G_1(x,s)$ and $G_2(x,s)$ and form

$$\bar{G} = G_1 - G_2$$

(i) $\bar{G}$ satisfies the BCs
(ii) $\bar{G}$ satisfies the homogeneous equation even at $x = s$ because the jump disappears.

Thus, $\bar{G}$ is a regular solution of the homogeneous system which by assumption is incompatible. Thus, $\bar{G} \equiv 0$ is the only solution $\Longrightarrow G_1 = G_2$ or $G(x,s)$ is a unique function.

(8) The solution

$$u(x) = \int_a^b G(x,s)f(s)\,ds$$

is unique. The proof is similar to (7). Suppose that there are two solutions $\hat{u}_1(x)$ and $\hat{u}_2(x)$. Then,

$$\hat{u}_1(x) - \hat{u}_2(x) = \int_a^b G(x,s)f(s)\,ds - \int_a^b G(x,s)f(s)\,ds \equiv 0,$$

since $G(x,s)$ is continuous.

$\Longrightarrow$

$$\hat{u}_1(x) = \hat{u}_2(x).$$

□

19.4 Green's function for the *n*-th order TPBVP

Consider *n*-th order TPBVP:

$$Lu = -f(x), \quad a < x < b \tag{19.32}$$

$$\mathbf{W}_a\mathbf{k}(u(a)) + \mathbf{W}_b\mathbf{k}(u(b)) = \mathbf{0} \tag{19.33}$$

Let

$$\psi(x)^T = [\ \psi_1(x) \quad \psi_2(x) \quad . \quad . \quad \psi_n(x)\ ]$$

be a fundamental vector for $Lu = 0$. Then the general solution of equations (19.32)–(19.33) is given by

$$u(x) = \psi(x)^T\mathbf{c} + \psi(x)^T \int_a^x \mathbf{K}(\psi(s))^{-1} \cdot \mathbf{e}_n\left[\frac{-f(s)}{p_0(s)}\right] ds \tag{19.34}$$

Now,

$$\psi(x)^T = [\psi_1(x) \quad \psi_2(x) \quad \cdots \quad \psi_n(x)]$$
$$= \mathbf{e}_1^T \begin{bmatrix} \psi_1 & \psi_2 & . & \psi_n \\ \psi_2' & \psi_2' & . & \psi_n' \\ . & & & \\ \psi_1^{[n-1]} & \psi_1^{[n-2]} & . & \psi_n^{[n-1]} \end{bmatrix} = \mathbf{e}_1^T \mathbf{K}(\boldsymbol{\psi}(x)) \tag{19.35}$$

$\therefore$

$$u(x) = \boldsymbol{\psi}(x)^T \mathbf{c} + \mathbf{e}_1^T \mathbf{K}(\boldsymbol{\psi}(x)) \int_a^x \mathbf{K}(\boldsymbol{\psi}(s))^{-1} \cdot \mathbf{e}_n \left[\frac{-f(s)}{p_0(s)}\right] ds \tag{19.36}$$

$$= u_h + u_p$$
$$u_h = \boldsymbol{\psi}(x)^T \mathbf{c} = c_1\psi_1 + c_2\psi_2 + \cdots + c_n\psi_n$$
$$u_h' = c_1\psi_1' + c_1\psi_2' + \cdots + c_n\psi_n'$$
$$.$$
$$.$$
$$u_h^{[n-1]} = c_1\psi_1^{[n-1]} + c_2\psi_2^{[n-1]} + \cdots + c_n\psi_n^{[n-1]}$$

$\therefore$

$$\mathbf{k}(u_h(x)) = \mathbf{K}(\boldsymbol{\psi}(x))\mathbf{c} \tag{19.37}$$

$$u_p(x) = \mathbf{e}_1^T \mathbf{K}(\boldsymbol{\psi}(x)) \int_a^x \mathbf{K}(\boldsymbol{\psi}(s))^{-1} \cdot \mathbf{e}_n \left[\frac{-f(s)}{p_0(s)}\right] ds$$

$$u_p(x) = [\psi_1(x) \quad \psi_2(x) \quad \cdots \quad \psi_n(x)] \int_a^x \mathbf{K}(\boldsymbol{\psi}(s))^{-1} \cdot \mathbf{e}_n \left[\frac{-f(s)}{p_0(s)}\right] ds$$

$$u_p' = [\psi_1' \quad \psi_2' \quad \ldots \quad \psi_n'] \int_a^x \mathbf{K}(\boldsymbol{\psi}(s))^{-1} \cdot \mathbf{e}_n \left[\frac{-f(s)}{p_0(s)}\right] ds$$
$$+ [\psi_1(x) \quad \psi_2(x) \quad \cdots \quad \psi_n(x)] \mathbf{K}(\boldsymbol{\psi}(x))^{-1} \cdot \mathbf{e}_n \left[\frac{-f(x)}{p_0(x)}\right]$$

$$\text{Second term} = \mathbf{e}_1^T \mathbf{K}(\boldsymbol{\psi}(x)) \mathbf{K}(\boldsymbol{\psi}(x))^{-1} \cdot \mathbf{e}_n \frac{-f(x)}{p_0(x)}$$
$$= \mathbf{e}_1^T . \mathbf{e}_n \frac{-f(x)}{p_0(x)} = 0$$

$\therefore$

$$u_p' = [\psi_1' \quad \psi_2' \quad \ldots \quad \psi_n'] \int_a^x \mathbf{K}(\boldsymbol{\psi}(s))^{-1} \cdot \mathbf{e}_n \frac{-f(s)}{p_0(s)} ds$$

$$= \mathbf{e}_2^T \mathbf{K}(\psi(x)) \int_a^x \mathbf{K}(\psi(s))^{-1} \cdot \mathbf{e}_n \left[\frac{-f(s)}{p_0(s)} \right] ds$$

Similarly,

$$u_p'' = \mathbf{e}_3^T \mathbf{K}(\psi(x)) \int_a^x \mathbf{K}(\psi(s))^{-1} \cdot \mathbf{e}_n \left[\frac{-f(s)}{p_0(s)} \right] ds$$

$$\vdots$$

$$u_p^{[n-1]} = \mathbf{e}_n^T \mathbf{K}(\psi(x)) \int_a^x \mathbf{K}(\psi(s))^{-1} \cdot \mathbf{e}_n \left[\frac{-f(s)}{p_0(s)} \right] ds$$

$\Longrightarrow$

$$\mathbf{k}(u_p(x)) = \mathbf{K}(\psi(x)) \int_a^x \mathbf{K}(\psi(s))^{-1} \cdot \mathbf{e}_n \left[\frac{-f(s)}{p_0(s)} \right] ds \tag{19.38}$$

Substitute equation (19.38) in the boundary conditions, equation (19.33): $\Longrightarrow$

$$\mathbf{W}_a \mathbf{K}(\psi(a))\mathbf{c} + \mathbf{W}_b \left[\mathbf{K}(\psi(b))\mathbf{c} + \mathbf{K}(\psi(b)) \int_a^b \mathbf{K}(\psi(s))^{-1} \cdot \mathbf{e}_n \left[\frac{-f(s)}{p_0(s)} \right] ds \right] = \mathbf{0}$$

$\Longrightarrow$

$$\mathbf{D}\mathbf{c} = -\mathbf{W}_b \mathbf{K}(\psi(b)) \int_a^b \mathbf{K}(\psi(s))^{-1} \cdot \mathbf{e}_n \frac{-f(s)}{p_0(s)} ds$$

Assuming $\mathbf{D}$ is not singular $\Longrightarrow$

$$\mathbf{c} = -\mathbf{D}^{-1} \mathbf{W}_b \mathbf{K}(\psi(b)) \int_a^b \mathbf{K}(\psi(s))^{-1} \cdot \mathbf{e}_n \left[\frac{-f(s)}{p_0(s)} \right] ds$$

$\Longrightarrow$

$$\begin{aligned} u(x) = & -\psi(x)^T \mathbf{D}^{-1} \mathbf{W}_b \mathbf{K}(\psi(b)) \int_a^b \mathbf{K}(\psi(s))^{-1} \cdot \mathbf{e}_n \left[\frac{-f(s)}{p_0(s)} \right] ds \\ & + \psi(x)^T \int_a^x \mathbf{K}(\psi(s))^{-1} \cdot \mathbf{e}_n \left[\frac{-f(s)}{p_0(s)} \right] ds \end{aligned} \tag{19.39}$$

Insert the identity

$$\mathbf{D}^{-1}[\mathbf{W}_a\mathbf{K}(\psi(a)) + \mathbf{W}_b\mathbf{K}(\psi(a))] = \mathbf{I} \tag{19.40}$$

in the second term of equation (19.39) before the integral sign. Also, split the first term of equation (19.39) into two terms.

$\Longrightarrow$

$$\begin{aligned} u(x) = &- \psi(x)^T\mathbf{D}^{-1}\mathbf{W}_b\mathbf{K}(\psi(b))\left(\int_a^x + \int_x^b\right)\mathbf{K}(\psi(s))^{-1}\cdot\mathbf{e}_n\left[\frac{-f(s)}{p_0(s)}\right]ds \\ &+ \psi(x)^T\int_a^x \mathbf{K}(\psi(s))^{-1}\cdot\mathbf{e}_n\left[\frac{-f(s)}{p_0(s)}\right]ds \end{aligned}$$

$\Longrightarrow$

$$\begin{aligned} u(x) = &- \psi(x)^T\mathbf{D}^{-1}\mathbf{W}_b\mathbf{K}(\psi(b))\int_a^x \mathbf{K}(\psi(s))^{-1}\cdot\mathbf{e}_n\left[\frac{-f(s)}{p_0(s)}\right]ds \\ &- \psi(x)^T\mathbf{D}^{-1}\mathbf{W}_b\mathbf{K}(\psi(b))\int_x^b \mathbf{K}(\psi(s))^{-1}\cdot\mathbf{e}_n\left[\frac{-f(s)}{p_0(s)}\right]ds \\ &+ \psi(x)^T\mathbf{D}^{-1}\mathbf{W}_a\mathbf{K}(\psi(a))\int_a^x \mathbf{K}(\psi(s))^{-1}\cdot\mathbf{e}_n\left[\frac{-f(s)}{p_0(s)}\right]ds \\ &+ \psi(x)^T\mathbf{D}^{-1}\mathbf{W}_b\mathbf{K}(\psi(b))\int_a^x \mathbf{K}(\psi(s))^{-1}\cdot\mathbf{e}_n\left[\frac{-f(s)}{p_0(s)}\right]ds \end{aligned}$$

The first and last terms cancel to give

$$u(x) = \int_a^b G(x,s)f(s)\,ds$$

where

$$G(x,s) = \begin{cases} -\mathbf{e}_1^T\mathbf{K}(\psi(x))\mathbf{D}^{-1}\mathbf{W}_a\mathbf{K}(\psi(a))\mathbf{K}^{-1}(\psi(s))\frac{\mathbf{e}_n}{p_0(s)}, & a < s < x \\ \mathbf{e}_1^T\mathbf{K}(\psi(x))\mathbf{D}^{-1}\mathbf{W}_b\mathbf{K}(\psi(b))\mathbf{K}^{-1}(\psi(s))\frac{\mathbf{e}_n}{p_0(s)}, & x < s < b \end{cases} \tag{19.41}$$

Example 19.5. Consider the second-order operator with Dirichlet boundary conditions

$$\frac{d^2}{dx^2}; \quad u(0) = 0, \quad u(1) = 0$$

$\Longrightarrow$

$$\mathbf{W} = \begin{pmatrix} 1 & 0 & 0 & 0 \\ 0 & 0 & 1 & 0 \end{pmatrix}$$

$$\psi_1(x) = x \quad \text{and} \quad \psi_2(x) = 1 - x$$

$$\mathbf{K}(\psi(x)) = \begin{pmatrix} x & 1-x \\ 1 & -1 \end{pmatrix}$$

$\Longrightarrow$

$$\mathbf{K}^{-1}(\psi(x)) = \begin{pmatrix} 1 & 1-x \\ 1 & -x \end{pmatrix}$$

$$\mathbf{K}(\psi(0)) = \begin{pmatrix} 0 & 1 \\ 1 & -1 \end{pmatrix} \quad \text{and} \quad \mathbf{K}(\psi(1)) = \begin{pmatrix} 1 & 0 \\ 1 & -1 \end{pmatrix}$$

$$\begin{aligned} \mathbf{D} &= \begin{pmatrix} 1 & 0 \\ 0 & 0 \end{pmatrix}\begin{pmatrix} 0 & 1 \\ 1 & -1 \end{pmatrix} + \begin{pmatrix} 0 & 0 \\ 1 & 0 \end{pmatrix}\begin{pmatrix} 1 & 0 \\ 1 & -1 \end{pmatrix} \\ &= \begin{pmatrix} 0 & 1 \\ 0 & 0 \end{pmatrix} + \begin{pmatrix} 0 & 0 \\ 1 & 0 \end{pmatrix} \\ &= \begin{pmatrix} 0 & 1 \\ 1 & 0 \end{pmatrix}, \end{aligned}$$

$\Longrightarrow$

$$\mathbf{D}^{-1} = \begin{pmatrix} 0 & 1 \\ 1 & 0 \end{pmatrix}$$

$$G = \begin{cases} -[x \quad (1-x)]\begin{pmatrix} 0 & 1 \\ 1 & 0 \end{pmatrix}\begin{pmatrix} 1 & 0 \\ 0 & 0 \end{pmatrix}\begin{pmatrix} 0 & 1 \\ 1 & -1 \end{pmatrix}\begin{pmatrix} 1 & 1-s \\ 1 & -s \end{pmatrix}\begin{pmatrix} 0 \\ 1 \end{pmatrix} \\ 0 < s < x \\ [x \quad (1-x)]\begin{pmatrix} 0 & 1 \\ 1 & 0 \end{pmatrix}\begin{pmatrix} 0 & 0 \\ 1 & 0 \end{pmatrix}\begin{pmatrix} 1 & 0 \\ 1 & -1 \end{pmatrix}\begin{pmatrix} 1 & 1-s \\ 1 & -s \end{pmatrix}\begin{pmatrix} 0 \\ 1 \end{pmatrix}, \\ x < s < 1 \end{cases}$$

$\Longrightarrow$

$$G(x,s) = \begin{cases} s(1-x), & 0 < s < x \\ x(1-s), & x < s < 1 \end{cases}$$

Example 19.6. Green's function for the second-order operator with mixed BCs

$$u'' = -f(x)$$

$$u(0) + u(1) = 0, \quad u'(0) + u'(1) = 0$$

$\Longrightarrow$

$$\mathbf{W}_a = \begin{pmatrix} 1 & 0 \\ 0 & 1 \end{pmatrix}, \quad \mathbf{W}_b = \begin{pmatrix} 1 & 0 \\ 0 & 1 \end{pmatrix}$$

$$\psi^T(x) = [1\ \ x]$$

$$\mathbf{K}(\psi(x)) = \begin{pmatrix} 1 & x \\ 0 & 1 \end{pmatrix}$$

$\Longrightarrow$

$$\mathbf{K}^{-1}(\psi(s)) = \begin{pmatrix} 1 & -s \\ 0 & 1 \end{pmatrix}$$

$$\begin{aligned}
\mathbf{D} &= \mathbf{W}_a\mathbf{K}(\psi(0)) + \mathbf{W}_b\mathbf{K}(\psi(1)) \\
&= \begin{pmatrix} 1 & 0 \\ 0 & 1 \end{pmatrix}\begin{pmatrix} 1 & 0 \\ 0 & 1 \end{pmatrix} + \begin{pmatrix} 1 & 0 \\ 0 & 1 \end{pmatrix}\begin{pmatrix} 1 & 1 \\ 0 & 1 \end{pmatrix} \\
&= \begin{pmatrix} 2 & 1 \\ 0 & 2 \end{pmatrix}
\end{aligned}$$

$$\mathbf{D}^{-1} = \begin{pmatrix} \frac{1}{2} & \frac{-1}{4} \\ 0 & \frac{1}{2} \end{pmatrix}$$

$\therefore$

$$G(x,s) = \begin{cases} -[1\ \ x]\begin{pmatrix} \frac{1}{2} & \frac{-1}{4} \\ 0 & \frac{1}{2} \end{pmatrix}\begin{pmatrix} 1 & 0 \\ 0 & 1 \end{pmatrix}\begin{pmatrix} 1 & 0 \\ 0 & 1 \end{pmatrix}\begin{pmatrix} 1 & -s \\ 0 & 1 \end{pmatrix}\begin{pmatrix} 0 \\ 1 \end{pmatrix}, & 0 < s < x \\ [1\ \ x]\begin{pmatrix} \frac{1}{2} & \frac{-1}{4} \\ 0 & \frac{1}{2} \end{pmatrix}\begin{pmatrix} 1 & 0 \\ 0 & 1 \end{pmatrix}\begin{pmatrix} 1 & 1 \\ 0 & 1 \end{pmatrix}\begin{pmatrix} 1 & -s \\ 0 & 1 \end{pmatrix}\begin{pmatrix} 0 \\ 1 \end{pmatrix}, & x < s < 1 \end{cases}$$

$$= \begin{cases} \frac{1}{4} - \frac{1}{2}(x-s), & 0 < s < x \\ \frac{1}{4} - \frac{1}{2}(s-x), & x < s < 1 \end{cases}$$

The Green's function is symmetric. It is easily verified that the given problem is self-adjoint even though the BCs are mixed.

Example 19.7. Green's function for the third-order operator

$$u''' = -f(x)$$

with boundary conditions

$$u(0) = 0, \quad u(1) = 0, \quad u'(0) - u'(1) = 0$$

$\Longrightarrow$

$$\mathbf{W}_a = \begin{pmatrix} 1 & 0 & 0 \\ 0 & 0 & 0 \\ 0 & 1 & 0 \end{pmatrix}, \quad \mathbf{W}_b = \begin{pmatrix} 0 & 0 & 0 \\ 1 & 0 & 0 \\ 0 & -1 & 0 \end{pmatrix}$$

$$\psi^T(x) = (1 \quad x \quad x^2)$$

$$\mathbf{K}(\psi(x)) = \begin{pmatrix} 1 & x & x^2 \\ 0 & 1 & 2x \\ 0 & 0 & 2 \end{pmatrix}$$

$\Longrightarrow$

$$\mathbf{K}^{-1}(\psi(s)) = \begin{pmatrix} 1 & -s & \frac{s^2}{2} \\ 0 & 1 & -s \\ 0 & 0 & \frac{1}{2} \end{pmatrix}$$

$$\mathbf{D} = \begin{pmatrix} 1 & 0 & 0 \\ 0 & 0 & 0 \\ 0 & 1 & 0 \end{pmatrix}\begin{pmatrix} 1 & 0 & 0 \\ 0 & 1 & 0 \\ 0 & 0 & 2 \end{pmatrix} + \begin{pmatrix} 0 & 0 & 0 \\ 1 & 0 & 0 \\ 0 & -1 & 0 \end{pmatrix}\begin{pmatrix} 1 & 1 & 1 \\ 0 & 1 & 2 \\ 0 & 0 & 2 \end{pmatrix}$$

$$= \begin{pmatrix} 1 & 0 & 0 \\ 0 & 0 & 0 \\ 0 & 1 & 0 \end{pmatrix} + \begin{pmatrix} 0 & 0 & 0 \\ 1 & 1 & 1 \\ 0 & -1 & -2 \end{pmatrix}$$

$$= \begin{pmatrix} 1 & 0 & 0 \\ 1 & 1 & 1 \\ 0 & 0 & -2 \end{pmatrix}$$

$$\mathbf{D}^{-1} = \begin{pmatrix} 1 & 0 & 0 \\ -1 & 1 & 1/2 \\ 0 & 0 & -\frac{1}{2} \end{pmatrix}$$

$\therefore$

$$G(x,s) = \begin{cases} -(1 \quad x \quad x^2)\cdot\begin{pmatrix} 1 & 0 & 0 \\ -1 & 1 & 1/2 \\ 0 & 0 & -\frac{1}{2} \end{pmatrix}\cdot\begin{pmatrix} 1 & 0 & 0 \\ 0 & 0 & 0 \\ 0 & 1 & 0 \end{pmatrix} \\ \quad \cdot\begin{pmatrix} 1 & 0 & 0 \\ 0 & 1 & 0 \\ 0 & 0 & 2 \end{pmatrix}\cdot\begin{pmatrix} 1 & -s & \frac{s^2}{2} \\ 0 & 1 & -s \\ 0 & 0 & \frac{1}{2} \end{pmatrix}\cdot\begin{pmatrix} 0 \\ 0 \\ 1 \end{pmatrix}, & 0 < s < x \\ (1 \quad x \quad x^2)\cdot\begin{pmatrix} 1 & 0 & 0 \\ -1 & 1 & 1/2 \\ 0 & 0 & -\frac{1}{2} \end{pmatrix}\cdot\begin{pmatrix} 0 & 0 & 0 \\ 1 & 0 & 0 \\ 0 & -1 & 0 \end{pmatrix} \\ \quad \cdot\begin{pmatrix} 1 & 1 & 1 \\ 0 & 1 & 2 \\ 0 & 0 & 2 \end{pmatrix}\cdot\begin{pmatrix} 1 & -s & \frac{s^2}{2} \\ 0 & 1 & -s \\ 0 & 0 & \frac{1}{2} \end{pmatrix}\cdot\begin{pmatrix} 0 \\ 0 \\ 1 \end{pmatrix}, & x < s < 1 \end{cases}$$

$$= \begin{cases} \frac{s}{2}(x-s)(1-x), & 0 < s < x \\ \frac{x}{2}(s-x)(s-1), & x < s < 1 \end{cases}$$

This Green's function is not symmetric.

The formula for Green's function given in equation (19.41) can be made less cumbersome by representing it in terms of the solutions of the adjoint equation. Let $\mathbf{v}(x)$ be a fundamental vector for the adjoint equation:

$$L^* v = 0$$

and suppose that we choose $\mathbf{u}(x)$ and $\mathbf{v}(x)$ such that

$$\mathbf{K}^T(\mathbf{v}(x))\mathbf{P}(x)\mathbf{K}(\mathbf{u}(x)) = \mathbf{I}$$

$\Longrightarrow$

$$\mathbf{K}^{-1}(\mathbf{u}(s)) = \mathbf{K}^T(\mathbf{v}(s))\mathbf{P}(s)$$

$\Longrightarrow$

$$\mathbf{K}^{-1}(\mathbf{u}(s))\mathbf{e}_n = \mathbf{K}^T(\mathbf{v}(s))\mathbf{P}(s)\mathbf{e}_n$$

Recalling the form of $\mathbf{P} = \begin{pmatrix} \cdot & \cdot & p_0 \\ \cdot & -p_0 & 0 \\ p_0 & 0 & 0 \end{pmatrix}$, we note that

$$\begin{aligned} \mathbf{P}\mathbf{e}_n &= \text{last column of } \mathbf{P} \\ &= p_0(s)\mathbf{e}_1 \end{aligned}$$

$\therefore$

$$\begin{aligned} \mathbf{K}^{-1}(\mathbf{u}(s))\mathbf{e}_n &= \mathbf{K}^T(\mathbf{v}(s))\mathbf{e}_1 p_0(s) \\ &= \mathbf{v}(s)p_0(s) \end{aligned}$$

Thus,

$$G(x,s) = \begin{cases} -\mathbf{u}^T(x)\mathbf{D}^{-1}\mathbf{W}_a\mathbf{K}(\mathbf{u}(a))\mathbf{v}(s), & a < s < x \\ \mathbf{u}^T(x)\mathbf{D}^{-1}\mathbf{W}_b\mathbf{K}(\mathbf{u}(b))\mathbf{v}(s), & x < s < b \end{cases} \tag{19.42}$$

This formula reveals the symmetric nature of the Green's function in terms of $\mathbf{u}(x)$ and $\mathbf{v}(s)$.

Theorem 19.1. *Regarded as a function of x with s fixed, the Green's function has the following properties:*

1. *Together with its first $(n-1)$ derivatives it is continuous in $[a, s)$ and $(s, b]$. At the point $x = s$, G and its first $n - 2$ derivatives have removable discontinuities while $(n - 1)$st derivative has an upward jump of $-\frac{1}{p_0(s)}$.*
2. *G satisfies the differential equation except at $x = s$. It satisfies the boundary conditions.*
3. *G is the only function with properties (1) and (2).*

Theorem 19.2. *Regarded as a function of s with x fixed, the Green's function has the following properties:*

1. *Together with its first $(n-1)$ derivatives, it is continuous on $[a, x)$ and $(x, b]$. At the point $s = x$, G and its first $(n-2)$ derivatives have removable discontinuities while the $(n-1)$st derivative has a jump of $\frac{(-1)^{n-1}}{p_0(x)}$*
2. *G satisfies the adjoint differential equation $(L^* v = 0)$ except at $s = x$. It satisfies the adjoint BCs.*

Theorem 19.3. *The solution of the adjoint BVP*

$$L^* v(s) = -f(s)$$
$$-\mathbf{k}^T(v(a))\mathbf{P}(a) = \mathbf{a}^T \mathbf{W}_a, \quad \mathbf{k}^T(v(b))\mathbf{P}(b) = \mathbf{a}^T \mathbf{W}_b$$

is given by

$$v(s) = \int_a^b G(x, s) f(x)\, dx. \tag{19.43}$$

For proofs of these theorems, we refer to the book by R. H. Cole [14].

19.4.1 Physical interpretation of the Green's function

Consider the BVP:

$$Lu = -f \tag{19.44}$$

$$\mathbf{W}\begin{pmatrix} \mathbf{k}(u(a)) \\ \mathbf{k}(u(b)) \end{pmatrix} = \mathbf{0} \tag{19.45}$$

and let the symbol $\mathcal{L}$ stand for the linear differential operator $-L$ and the boundary conditions (19.45). Then we may write (19.44)–(19.45) as

$$\mathcal{L}u = f \tag{19.46}$$

We represented the solution to (19.44) and (19.45) as

$$u(x) = \int_a^b G(x, s) f(s)\, ds \tag{19.47}$$

Let $\mathbb{G}$ stand for the operator defined by

$$\mathbb{G}f = \int_a^b G(x,s)f(s)\,ds \tag{19.48}$$

Then equation (19.47) becomes

$$u = \mathbb{G}f \tag{19.49}$$

Comparing equations (19.46) and (19.49), it is clear that

$$\mathbb{G} = \mathcal{L}^{-1},$$

which is also shown schematically in Figure 19.3.

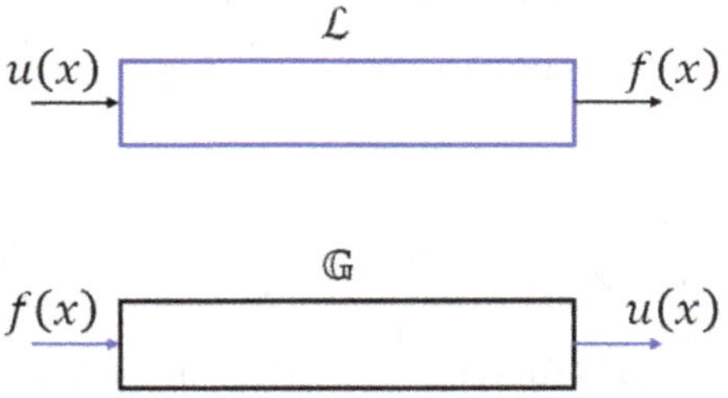

Figure 19.3: Schematic demonstration of the operator relationship $\mathbb{G} = \mathcal{L}^{-1}$.

[Remark: The analogy between discrete (matrix) and continuous (differential) operators can be seen from equations (19.46) and (19.49). When the BVP is discretized (e. g., using the second-order finite difference or finite volume methods), $\mathcal{L}$ becomes a matrix operator while $\mathbb{G}$ is the inverse matrix.]

It seems apparent from the formula

$$u(x) = \int_a^b G(x,s)f(s)\,ds \tag{19.50}$$

that $G(x,s)$ must be response at position x caused by a unit input at position s. Suppose that this is the case and there is a distribution of inputs $f(s)$. Then the response at x caused by an input at s of $f(s)\,ds$ must be $G(x,s)f(s)\,ds$ as schematically shown in Figure 19.4. Then $\int_a^b G(x,s)f(s)\,ds$ is the total response at x.

Moreover, $G(x,s)$ is the solution of

$$LG = -\delta(x-s); \quad \mathbf{W}\begin{pmatrix} \mathbf{k}(G(a,s)) \\ \mathbf{k}(G(b,s)) \end{pmatrix} = \mathbf{0} \tag{19.51}$$

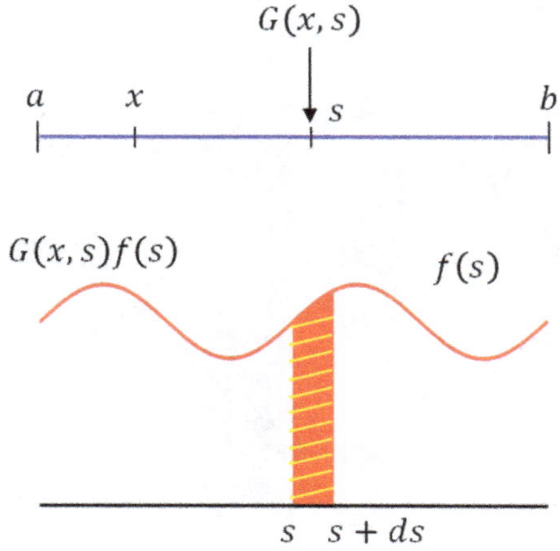

Figure 19.4: Schematic of the distributed function $f(s)$ and interpretation of Green's function.

or

$$LG = -\delta(s-x); \quad \mathbf{W}\begin{pmatrix} \mathbf{k}(G(x,a)) \\ \mathbf{k}(G(x,b)) \end{pmatrix} = \mathbf{0} \tag{19.52}$$

[Remark: In equation (19.51), G is considered to be a function of x, while in equation (19.52), it is considered to be a function of s].

The two examples given below illustrate this point clearly.

Example 19.8 (The deflection of a tightly stretched elastic string). Consider the deflection of a tightly stretched elastic string that is fixed at the end points as shown in Figure 19.5.

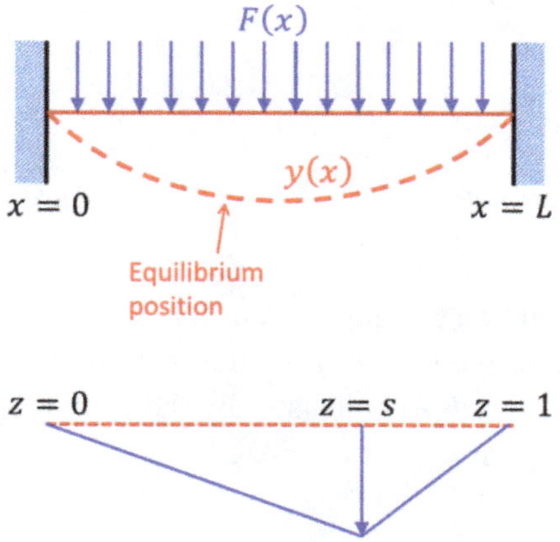

Figure 19.5: Deflection of tightly stretched elastic string under a load distribution (top) and a point load (bottom).

The amplitude of deflection $y(x)$ can be described by

$$T\frac{d^2y}{dx^2} = -F(x); \quad y(0) = 0, \quad y(L) = 0 \tag{19.53}$$

where T is the tension in the spring and $F(x)$ is the force distribution. In dimensionless form,

$$\frac{d^2y}{dz^2} = -f(z); \quad 0 < z < 1 \tag{19.54}$$

$$y(0) = y(1) = 0 \tag{19.55}$$

In this case, the Green's function

$$G(z,s) = \begin{cases} z(1-s), & 0 < s < z \\ s(1-z), & z < s < 1 \end{cases} \tag{19.56}$$

represents the deflection of the string at position z due to a unit force acting at position s and deflection for a distributed source can be expressed (same as in equation (19.50)) as

$$y(z) = \int_0^1 G(z,s)f(s)\,ds \tag{19.57}$$

Example 19.9 (Green's function for the one-dimensional diffusion-convection operator). Consider the one-dimensional diffusion-convection operator:

$$\frac{1}{\text{Pe}}\frac{d^2u}{dx^2} - \frac{du}{dx} = -\delta(x-s), \quad 0 < x < 1 \tag{19.58}$$

$$\frac{1}{\text{Pe}}\frac{du}{dx} - u = 0 \ @\ x = 0 \tag{19.59}$$

$$\frac{du}{dx} = 0 \ @\ x = 1 \tag{19.60}$$

Note that this is not a symmetric BVP. To determine the solution, we have for $0 < x < s$:

$$\frac{1}{\text{Pe}}\frac{d^2u}{dx^2} - \frac{du}{dx} = 0$$

$\Longrightarrow$

$$\frac{1}{\text{Pe}}\frac{du}{dx} - u = \text{Constant}$$

BC $\Longrightarrow$ Constant $= 0$

$$\frac{du}{dx} = \text{Pe} \cdot u$$

$\Longrightarrow$

$$u = c_1 e^{\mathrm{Pe}\,x} \tag{19.61}$$

$s < x < 1$:

$$\frac{1}{\mathrm{Pe}}\frac{d^2u}{dx^2} - \frac{du}{dx} = 0$$

$\Longrightarrow$

$$\frac{1}{\mathrm{Pe}}\frac{du}{dx} - u = \text{Constant} = c_2$$

$\Longrightarrow$

$$\begin{aligned} u' - \mathrm{Pe}\,u &= c_2.\,\mathrm{Pe} \\ ue^{-\mathrm{Pe}\,x} &= c_2.\,\mathrm{Pe}\int e^{-\mathrm{Pe}\,x}\,dx \\ &= -c_2.e^{-\mathrm{Pe}\,x} + c_3 \end{aligned}$$

$\Longrightarrow$

$$\begin{aligned} u &= c_3 e^{\mathrm{Pe}\,x} - c_2 \\ \frac{du}{dx} &= c_3\,\mathrm{Pe}\,e^{\mathrm{Pe}\,x}, \quad u'(1) = 0 \Longrightarrow c_3 = 0 \end{aligned}$$

$\Longrightarrow$

$$u(x) = -c_2 \quad \text{for } s < x < 1$$

$\therefore$

$$u(x) = \begin{cases} c_1 e^{\mathrm{Pe}\,x}, & 0 < x < s \\ -c_2, & s < x < 1 \end{cases} \tag{19.62}$$

$$u'(x) = \begin{cases} c_1\,\mathrm{Pe}\,e^{\mathrm{Pe}\,x}, & 0 < x < s \\ 0, & s < x < 1 \end{cases} \tag{19.63}$$

It follows from equation (19.58)

$$\frac{1}{\mathrm{Pe}}\frac{du}{dx} - u = -H(x - s)$$

$\Longrightarrow$

$$c_2 = -1$$

Continuity of $u(x)$ at $x = s$ gives

$$c_1 e^{\mathrm{Pe}\, s} = 1 \Longrightarrow c_1 = e^{-\mathrm{Pe}\, s}$$

$\therefore$

$$u(x,s) = G(x,s) = \begin{cases} e^{-\mathrm{Pe}\, s} e^{\mathrm{Pe}\, x}, & 0 < x < s \\ 1, & s < x < 1 \end{cases} \tag{19.64}$$

$\therefore$ The Green's function for the 1D diffusion-convection operator is given by

$$G(x,s) = \begin{cases} e^{-\mathrm{Pe}(s-x)}, & 0 < x < s \\ 1, & s < x < 1 \end{cases} \tag{19.65}$$

and is shown in Figure 19.6a for Pe = 4 and various values of s, while Figure 19.6b shows it for a fixed value of $s = 0.8$ and various values of Pe. We note that as Pe increases, $G(x,s)$ has a boundary/internal layer near $x = s$.

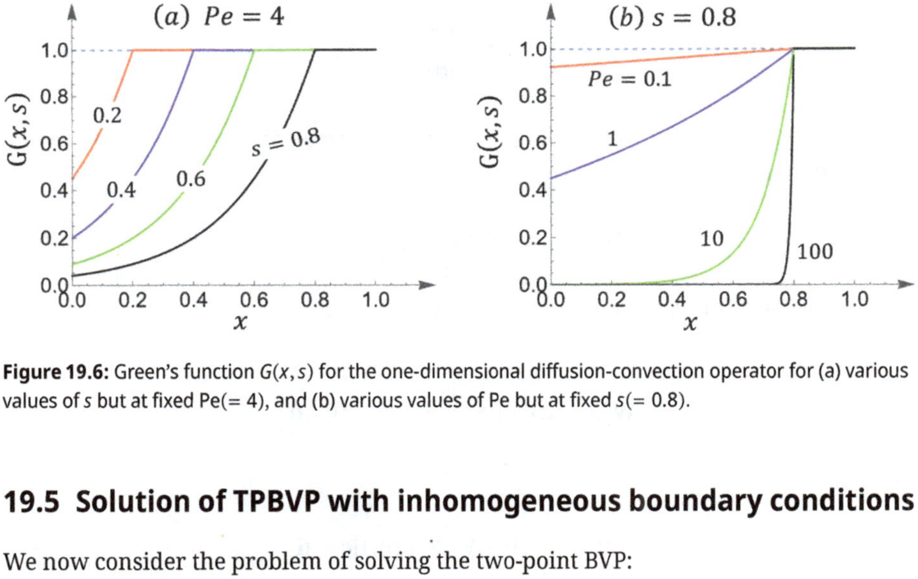

Figure 19.6: Green's function $G(x,s)$ for the one-dimensional diffusion-convection operator for (a) various values of s but at fixed Pe(= 4), and (b) various values of Pe but at fixed s(= 0.8).

19.5 Solution of TPBVP with inhomogeneous boundary conditions

We now consider the problem of solving the two-point BVP:

$$Lu = -f(x); \quad \mathbf{W}\begin{pmatrix} \mathbf{k}(u(a)) \\ \mathbf{k}(u(b)) \end{pmatrix} = \mathbf{d} \tag{19.66}$$

As stated in the previous chapter, using the principle of superposition, the solution of (19.66) may be expressed as

$$u = u_1 + u_2 \tag{19.67}$$

where $u_1(x)$ and $u_2(x)$ are the solutions of

$$Lu_1 = -f(x); \quad \mathbf{W}\begin{pmatrix} \mathbf{k}(u_1(a)) \\ \mathbf{k}(u_1(b)) \end{pmatrix} = \mathbf{0} \tag{19.68}$$

and

$$Lu_2 = 0; \quad \mathbf{W}\begin{pmatrix} \mathbf{k}(u_2(a)) \\ \mathbf{k}(u_2(b)) \end{pmatrix} = \mathbf{d} \tag{19.69}$$

We have already seen that (19.68) has a unique solution if the homogeneous problem is incompatible, i. e., $G(x,s)$ exists and is unique. Thus, we only need to solve (19.69) to get the complete solution to equation (19.66). We now consider equation (19.69) and present two methods for solving this BVP.

Method 1

Consider the BVP

$$Lu = 0; \quad \mathbf{W}\begin{pmatrix} \mathbf{k}(u(a)) \\ \mathbf{k}(u(b)) \end{pmatrix} = \mathbf{d} \tag{19.70}$$

Let $\mathbf{u}(x)^T = [u_1(x) \;\; u_2(x) \;\; . \; . \;\; u_n(x)]$ be a fundamental vector. Then any solution to equation (19.70) must be of the form

$$u(x) = \mathbf{u}(x)^T\mathbf{c} = c_1u_1(x) + c_2u_2(x) + \cdots + c_nu_n(x) \tag{19.71}$$

In order to determine $\mathbf{c}$, substitute equation (19.71) into the BCs:

$\Longrightarrow$

$$\mathbf{W}_a\mathbf{k}(\mathbf{u}(a)^T\mathbf{c}) + \mathbf{W}_b\mathbf{k}(\mathbf{u}(b)^T\mathbf{c}) = \mathbf{d}$$

$\Longrightarrow$

$$[\mathbf{W}_a\mathbf{K}(\mathbf{u}(a)) + \mathbf{W}_b\mathbf{K}(\mathbf{u}(b))]\mathbf{c} = \mathbf{d}$$

$\Longrightarrow$

$$\mathbf{D}\mathbf{c} = \mathbf{d}$$

Since the homogeneous problem is incompatible, $\mathbf{D}$ is invertible and

$$\mathbf{c} = \mathbf{D}^{-1}\mathbf{d}. \tag{19.72}$$

Example 19.10.

$$\frac{d^2u}{dx^2} = 0$$
$$u(0) = d_1, \quad u(1) = d_2$$
$$u_1(x) = x, \quad u_2(x) = (1-x)$$
$$\mathbf{K}(\mathbf{u}(x)) = \begin{pmatrix} x & 1-x \\ 1 & -1 \end{pmatrix}$$
$$\mathbf{D} = \begin{pmatrix} 1 & 0 \\ 0 & 0 \end{pmatrix}\begin{pmatrix} 0 & 1 \\ 1 & -1 \end{pmatrix} + \begin{pmatrix} 0 & 0 \\ 1 & 0 \end{pmatrix}\begin{pmatrix} 1 & 0 \\ 1 & -1 \end{pmatrix}$$
$$= \begin{pmatrix} 0 & 1 \\ 0 & 0 \end{pmatrix} + \begin{pmatrix} 0 & 0 \\ 1 & 0 \end{pmatrix} = \begin{pmatrix} 0 & 1 \\ 1 & 0 \end{pmatrix}$$
$$\mathbf{D}^{-1} = \begin{pmatrix} 0 & 1 \\ 1 & 0 \end{pmatrix}$$

$\therefore$

$$u(x) = [\ u_1 \quad u_2\ ]\begin{pmatrix} 0 & 1 \\ 1 & 0 \end{pmatrix}\begin{pmatrix} d_1 \\ d_2 \end{pmatrix}$$
$$= [\ u_1(x) \quad u_2(x)\ ]\begin{pmatrix} d_2 \\ d_1 \end{pmatrix}$$
$$= d_2 x + d_1(1-x)$$

Example 19.11.

$$u'' = 0$$
$$u(0) + u(1) = d_1, \quad u'(0) + u'(1) = d_2$$
$$\mathbf{u}(x)^T = [\ 1 \quad x\ ]$$
$$\mathbf{D} = \begin{pmatrix} 2 & 1 \\ 0 & 2 \end{pmatrix}$$

$\Longrightarrow$

$$\mathbf{D}^{-1} = \begin{pmatrix} \frac{1}{2} & -\frac{1}{4} \\ 0 & \frac{1}{2} \end{pmatrix}$$

$\therefore$

$$\mathbf{D}^{-1}\mathbf{d} = \begin{pmatrix} \frac{d_1}{2} - \frac{d_2}{4} \\ \frac{d_2}{2} \end{pmatrix}$$
$$u(x) = [1 \quad x]\begin{bmatrix} \frac{d_1}{2} - \frac{d_2}{4} \\ \frac{d_2}{2} \end{bmatrix} = \frac{2d_1 - d_2}{4} + \frac{d_2}{2}x$$

Method 2: (in terms of Green's function)
Consider the Lagrange identity

$$v(s)Lu(s) - u(s)L^*v(s) = \frac{d}{ds}[\mathbf{k}^T(v(s))\mathbf{P}(s)\mathbf{k}(u(s)))] \tag{19.73}$$

Let

$$u(s) = \mathbf{u}(s)^T\mathbf{c} \tag{19.74}$$

where $\mathbf{u}(s)$ is a fundamental vector of $(-Lu) = 0$. Now, let $v(s) = G(x,s)$ and substitute in equation (19.73) $\Longrightarrow$

$$-u(s)L^*G(x,s) = \frac{d}{ds}[\mathbf{k}^T(G(x,s))\mathbf{P}(s)\mathbf{k}(u(s)))] \tag{19.75}$$

We know that

$$L^*G(x,s) = 0, \quad a \leq s < x$$

and

$$L^*G(x,s) = 0, \quad x < s \leq b$$

Thus, integrate (19.75) from $s = a$ to $s = x^-$ and $s = x^+$ to $s = b$. The LHS is identically zero. $\Longrightarrow$

$$\mathbf{k}^T(G(x,x^-))\mathbf{P}(x^-)\mathbf{k}(u(x^-)) - \mathbf{k}^T(G(x,a))\mathbf{P}(a)\mathbf{k}(u(a)) = 0 \tag{19.76}$$
$$\mathbf{k}^T(G(x,b))\mathbf{P}(b)\mathbf{k}(u(b)) - \mathbf{k}^T(G(x,x^+))\mathbf{P}(x^+)\mathbf{k}(u(x^+)) = 0 \tag{19.77}$$

$\mathbf{P}(s)$ is continuous $\Longrightarrow$

$$\mathbf{P}(x^-) = \mathbf{P}(x^+) = \mathbf{P}(x)$$

Since the $(n-1)$ derivatives of $u(x)$ are continuous $\Longrightarrow$

$$\mathbf{k}(u(x^+)) = \mathbf{k}(u(x^-)) = \mathbf{k}(u(x)) = \mathbf{K}(\mathbf{u}(x))\mathbf{c}$$

Thus, adding equations (19.76) and (19.77) give

$$\begin{aligned}[\mathbf{k}^T(G(x,x^-)) - \mathbf{k}^T(G(x,x^+))]\mathbf{P}(x)\mathbf{k}(u(x)) &= \mathbf{k}^T(G(x,a))\mathbf{P}(a)\mathbf{k}(u(a)) \\ &\quad - \mathbf{k}^T(G(x,b))\mathbf{P}(b)\mathbf{k}(u(a))\end{aligned} \tag{19.78}$$

Since $G(x,s)$ and its first $(n-2)$ derivatives are continuous at $s = x$ and the $(n-1)$ derivative has a jump of $\frac{(-1)^{n-1}}{p_0(x)}$, the LHS of equation (19.78) may be simplified to

$$
\begin{aligned}
\text{LHS} &= \frac{(-1)^{n-1}}{p_0(x)} \mathbf{e}_n^T \mathbf{P}(x)\mathbf{k}(u(x)) \\
&= \frac{(-1)^{n-1}}{p_0(x)} \begin{bmatrix} 0 & . & . & 0 & 1 \end{bmatrix} \begin{bmatrix} . & . & . & p_0(x) \\ . & & -p_0(x) & 0 \\ . & . & 0 & 0 \\ (-1)^{n-1}p_0(x) & 0 & 0 & 0 \end{bmatrix} \begin{bmatrix} u(x) \\ u'(x) \\ . \\ u^{n-1}(x) \end{bmatrix} \\
&= \frac{(-1)^{n-1}}{p_0(x)} \begin{bmatrix} (-1)^{n-1}p_0(x) & 0 & . & . & 0 \end{bmatrix} \begin{bmatrix} u(x) \\ u'(x) \\ . \\ u^{n-1}(x) \end{bmatrix} \\
&= u(x)
\end{aligned}
$$

Similarly, the RHS of equation (19.78) simplifies to

$$
\begin{aligned}
\text{RHS} &= \begin{bmatrix} \mathbf{k}^T(G(x,a))\mathbf{P}(a) & -\mathbf{k}^T(G(x,b))\mathbf{P}(b) \end{bmatrix} \begin{bmatrix} \mathbf{k}(u(a)) \\ \mathbf{k}(u(b)) \end{bmatrix} \\
&= \begin{bmatrix} \mathbf{k}^T(G(x,a)) & \mathbf{k}^T(G(x,b)) \end{bmatrix} \begin{bmatrix} \mathbf{P}(a) & \mathbf{0} \\ \mathbf{0} & -\mathbf{P}(b) \end{bmatrix} \begin{bmatrix} \mathbf{k}(u(a)) \\ \mathbf{k}(u(b)) \end{bmatrix}
\end{aligned}
$$

$\therefore$

$$
u(x) = \begin{bmatrix} \mathbf{k}^T(G(x,a)) & \mathbf{k}^T G(x,b) \end{bmatrix} \begin{bmatrix} \mathbf{P}(a) & \mathbf{0} \\ \mathbf{0} & -\mathbf{P}(b) \end{bmatrix} \begin{bmatrix} \mathbf{k}(u(a)) \\ \mathbf{k}(u(b)) \end{bmatrix} \tag{19.79}
$$

But we have shown that $G(x,s)$ satisfies the homogeneous BCs at $s = a$ and $s = b$. Thus, the row vector

$$
\begin{bmatrix} \mathbf{k}^T(G(x,a)) & \mathbf{k}^T(G(x,b)) \end{bmatrix} \begin{bmatrix} \mathbf{P}(a) & \mathbf{0} \\ \mathbf{0} & -\mathbf{P}(b) \end{bmatrix} = \mathbf{h}(x)^T \mathbf{W} \tag{19.80}
$$

must belong to the row space of $\mathbf{W}$, and hence may be written as shown in equation (19.80). Substituting equation (19.80) in equation (19.79), we get

$$
\begin{aligned}
u(x) &= \mathbf{h}(x)^T \mathbf{W} \begin{bmatrix} \mathbf{k}(u(a)) \\ \mathbf{k}(u(b)) \end{bmatrix} \\
&= \mathbf{h}(x)^T \mathbf{d}
\end{aligned} \tag{19.81}
$$

which is the solution of equation (19.70). The vector function $\mathbf{h}(x)$ is determined from equation (19.80), or equivalently from

$$
\begin{aligned}
\mathbf{k}^T(G(x,a))\mathbf{P}(a) &= \mathbf{h}(x)^T \mathbf{W}_a \\
\mathbf{k}^T(G(x,b))\mathbf{P}(b) &= -\mathbf{h}(x)^T \mathbf{W}_b
\end{aligned} \tag{19.82}
$$

where n of these equations determine $\mathbf{h}(x)$ and the remaining n determine the adjoint BCs in terms of the Green's function.

Example 19.12.

$$\frac{d^2u}{dx^2} = -f(x)$$
$$u(0) = d_1, \quad u(1) = d_2.$$

For the operator $-\frac{d^2}{dx^2}$, $u(0) = 0$, $u(1) = 0$, we have

$$G(x,s) = \begin{cases} s(1-x), & 0 \le s < x \\ x(1-s), & x < s \le 1 \end{cases}$$

and

$$\mathbf{P}(x) = \begin{pmatrix} 0 & -1 \\ 1 & 0 \end{pmatrix}$$
$$p_0(x) = -1$$

∴ Equation (19.82) becomes

$$\begin{bmatrix} G(x,0) & \frac{\partial G}{\partial s}(x,0) \end{bmatrix} \begin{bmatrix} 0 & -1 \\ 1 & 0 \end{bmatrix} = \begin{bmatrix} h_1(x) & h_2(x) \end{bmatrix} \begin{bmatrix} 1 & 0 \\ 0 & 0 \end{bmatrix}$$
$$\begin{bmatrix} G(x,1) & \frac{\partial G}{\partial s}(x,1) \end{bmatrix} \begin{bmatrix} 0 & -1 \\ 1 & 0 \end{bmatrix} = -\begin{bmatrix} h_1(x) & h_2(x) \end{bmatrix} \begin{bmatrix} 0 & 0 \\ 1 & 0 \end{bmatrix}$$

$\Longrightarrow$

$$G(x,0) = 0, \quad G(x,1) = 0$$
$$h_1(x) = 1 - x, \quad h_2(x) = -\frac{\partial G}{\partial s}(x,1) = x$$

∴

$$u(x) = \int_0^1 G(x,s)f(s)\,ds + d_1(1-x) + d_2 x.$$

Problems

1. The deflection of a simply supported beam is described by the fourth-order boundary value problem

$$EI\frac{d^4u}{dx^4} = F(x), \quad 0 < x < L; \quad u(0) = u(L) = u''(0) = u''(L) = 0$$

where EI = flexural rigidity of the beam and $F(x)$ is the intensity of the distributed load (force per unit length). (a) Determine the Green's function of this system by evaluating the deflection curve for a unit load at $x = s$. (b) Evaluate the deflection curve for a triangularly distributed load, i. e.,

$$F(x) = \frac{wx}{L}$$

(c) Determine the maximum deflection for the load in (b)

2. Show that the Green's function for the operator

$$Lu = \frac{d}{dr}\left(r\frac{du}{dr}\right), \quad a < r < b; \quad u'(a) = 0, \quad u(b) = 0$$

is given by

$$G(r,s) = \begin{cases} \ln(b/r) & a < s < r \\ \ln(b/s) & r < s < b \end{cases}$$

Heat is generated in a thin annular disk with insulated faces. The conductivity is k in the radial direction. The internal circumference of the disk is insulated while the external circumferential area is held at temperature zero. The heat generation rate per unit volume is given by $\frac{a(b-r)}{(b-a)}$. Determine the temperature distribution in the disk using the Green's function.

3. (a) Derive a formula for the solution of the n-th order boundary value problem

$$Lu = -f(x), \quad a < x < b$$
$$\mathbf{W}_a\mathbf{k}(u(a)) + \mathbf{W}_b\mathbf{k}(u(b)) = \mathbf{d}$$

where the corresponding homogeneous problem is incompatible. (b) Determine a formula for the Green's matrix of the vector boundary value problem

$$\frac{d\mathbf{u}}{dx} - \mathbf{A}(x)\mathbf{u} = \mathbf{f}(x), \quad a < x < b$$
$$\mathbf{W}_a\mathbf{u}(a) + \mathbf{W}_b\mathbf{u}(b) = \mathbf{0}$$

4. For each of the following problems, find an equivalent integral equation:
(a) $(x^2u')' = -\lambda x^2 u$, $a < x < b$; $u'(a) = 0$, $u(b) = 0$
(b) $u'' = -f(u,x)$; $u'(0) = 0$, $u(1) = 0$ (Note: this is a nonlinear equation)
(c) $\frac{d^4u}{dx^4} = \lambda u$, $u(0) = u(1) = u''(1) = u''(1) = 0$.

20 Eigenvalue problems for differential operators

20.1 Definition of eigenvalue problems

Let

$$Ly = p_0(x)y^{[n]} + p_1(x)y^{[n-1]} + \cdots + p_n(x)y \tag{20.1}$$

be a regular linear differential operator, i. e., $p_0(x) \neq 0$ in $[a, b]$ and $p_j(x) \in C^{n-j}[a, b]$. Consider the homogeneous boundary value problem (BVP)

$$-Ly = \lambda y; \quad \mathbf{W}\begin{pmatrix} \mathbf{k}(y(a)) \\ \mathbf{k}(y(b)) \end{pmatrix} = \mathbf{0} \tag{20.2}$$

Definition. A real or complex number λ for which the BVP defined by equation (20.2) is compatible is called an eigenvalue and any corresponding nontrivial solution is called an eigenfunction. The set of all eigenvalues is called the *spectrum* of the BVP.

Remarks.

(1) For consistency with the finite-dimensional case, we should have defined the eigenvalue in equation (20.2) with a positive sign. However, the eigenvalues of most of the differential operators we encounter in our applications (e. g., the diffusion operator) are negative and we are following the literature notation here.

(2) Since the BVP given by equation (20.2) is homogeneous, the eigenfunctions associated with an eigenvalue form a subspace of $C^n[a, b]$.

The adjoint BVP is defined by

$$-L^*v = \bar{\lambda}v; \quad \mathbf{Q}\begin{pmatrix} \mathbf{k}(v(a)) \\ \mathbf{k}(v(b)) \end{pmatrix} = \mathbf{0} \tag{20.3}$$

Theorem. *If λ is an eigenvalue of equation* (20.2), *then $\bar{\lambda}$ is an eigenvalue of the adjoint BVP defined by equation* (20.3).

Proof. Suppose that $y(x)$ is the eigenfunction corresponding to the eigenvalue λ. Let $v(x)$ be any nonzero function in the domain of L^*. Then, by definition of the adjoint operator,

$$\langle Ly, v\rangle = \langle y, L^*v\rangle,$$

where the inner product is defined by

$$\langle y, v\rangle = \int_a^b y(x)\overline{v(x)}\,dx,$$

https://doi.org/10.1515/9783111598055-024

we obtain

$$\langle -\lambda y, v\rangle = \langle y, L^* v\rangle.$$

Using the properties of the inner product,

$$\begin{aligned} -\lambda\langle y, v\rangle &= \langle y, L^* v\rangle \\ \Longrightarrow \langle y, -\bar{\lambda} v\rangle &= \langle y, L^* v\rangle \\ \Longrightarrow \langle y, L^* v + \bar{\lambda} v\rangle &= 0 \end{aligned}$$

Thus, we have two possibilities: (i) $L^* v + \bar{\lambda} v = 0$, which implies that $\bar{\lambda}$ is an eigenvalue of L^* with $v(x)$ being the eigenfunction or (ii) the function $L^* v + \bar{\lambda} v$ is orthogonal to $y(x)$, i. e.,

$$\int_a^b (L^* v + \bar{\lambda} v)\overline{y(x)}\, dx = 0 \tag{20.4}$$

In case (i), we have already proved the result. Now, consider case (ii) and suppose that

$$L^* v + \bar{\lambda} v = -f(x) \tag{20.5}$$

Then equation (20.4) $\Longrightarrow$

$$\langle f, y\rangle = 0 \tag{20.6}$$

If $-(L^* + \bar{\lambda})v$ represented every continuous function $f(x)$, we have a contradiction since equation (20.6) $\Longrightarrow y(x) = 0$. Therefore, the other alternative must hold, i. e., $L^* v + \bar{\lambda} v = 0$ for some nonzero $v(x)$.

The above conclusion may also be reached using the following reasoning. Suppose that the operator $(L^* + \bar{\lambda})$ is invertible, i. e.,

$$(L^* + \bar{\lambda})v = 0 \Longrightarrow v = 0,$$

then the range of $(L^* + \bar{\lambda})$ consists of every continuous function. Hence, by choosing different $v(x)$ we can obtain different f for which $\langle f, y\rangle = 0$. Again, we have $y(x) = 0$, which is a contradiction. $\therefore$ $(L^* + \bar{\lambda})$ must be singular and there exists a nonzero $v(x)$ such that

$$L^* v = -\bar{\lambda} v$$

$\therefore$ $\bar{\lambda}$ is an eigenvalue of the adjoint system. □

Theorem. *The eigenfunctions of the BVP defined by equation* (20.2) *and the adjoint BVP, (equation* (20.3)*) corresponding to distinct eigenvalues are orthogonal, i. e.,*

$$\langle y_m(x), v_n(x)\rangle = 0, \quad m \neq n$$

Proof. To prove the biorthogonality property, let

$$
\begin{aligned}
Ly_m &= -\lambda_m y_m \quad \text{and} \quad L^* v_n = -\bar{\lambda}_n v_n \\
\Longrightarrow \langle Ly_m, v_n \rangle &= \langle y_m, L^* v_n \rangle \\
\Longrightarrow \langle -\lambda_m y_m, v_n \rangle &= \langle y_m, -\bar{\lambda}_n v_n \rangle \\
\Longrightarrow -\lambda_m \langle y_m, v_n \rangle &= -\lambda_n \langle y_m, v_n \rangle \\
\Longrightarrow -(\lambda_m - \lambda_n) \langle y_m, v_n \rangle &= 0
\end{aligned}
$$

Since $\lambda_m \neq \lambda_n \Longrightarrow$

$$\langle y_m, v_n \rangle = 0$$

□

Corollary. *If the BVP is self-adjoint, the eigenfunctions corresponding to distinct eigenvalues are orthogonal. Moreover, we can choose the eigenfunctions such that*

$$\langle y_m, y_n \rangle = \delta_{mn} = \begin{cases} 1, & m = n \\ 0, & m \neq n \end{cases}$$

As in the finite-dimensional case, we call such set of eigenfunctions an orthonormal set.

20.2 Determination of the eigenvalues

Consider the BVP

$$Ly = -\lambda y; \quad \mathbf{W} \begin{pmatrix} \mathbf{k}(u(a)) \\ \mathbf{k}(u(b)) \end{pmatrix} = \mathbf{0} \tag{20.7}$$

Let $y_1(x, \lambda), y_2(x, \lambda), \ldots, y_n(x, \lambda)$ be a set of linearly independent solutions of equation (20.7). Then any solution is of the form

$$
\begin{aligned}
y(x) &= \mathbf{y}(x, \lambda)^T \mathbf{c} \\
&= y_1(x, \lambda) c_1 + y_2(x, \lambda) c_2 + \cdots + y_n(x, \lambda) c_n
\end{aligned} \tag{20.8}
$$

where $\mathbf{c}$ is determined from

$$\mathbf{D}(\lambda)\mathbf{c} = \mathbf{0} \tag{20.9}$$

where

$$\mathbf{D}(\lambda) = \mathbf{W}_a \mathbf{K}(\mathbf{y}(a, \lambda)) + \mathbf{W}_b \mathbf{K}(\mathbf{y}(b, \lambda)) \tag{20.10}$$

is the characteristic matrix. If rank $\mathbf{D}(\lambda) = n$, then $\mathbf{c} = \mathbf{0}$ and the only solution to the BVP is the trivial one. Thus, to get a nontrivial solution, we require that

$$h(\lambda) \equiv \det \mathbf{D}(\lambda) = |\mathbf{D}(\lambda)| = 0 \tag{20.11}$$

Equation (20.11), which determines the eigenvalues of the BVP, is called the *characteristic equation*. The zeros of the scalar function $h(\lambda)$ give the eigenvalues. The corresponding eigenfunctions can be obtained by solving equation (20.9) for $\mathbf{c}$ and substituting in equation (20.8).

20.2.1 Relationship between the *n*-th order eigenvalue problem and the vector eigenvalue problem

Consider the scalar n-th order eigenvalue problem

$$Ly = -\lambda y; \quad \mathbf{W}\begin{pmatrix} \mathbf{k}(u(a)) \\ \mathbf{k}(u(b)) \end{pmatrix} = \mathbf{0} \tag{20.12}$$

where linear operator L is given by

$$Ly = p_0 y^{[n]} + p_1 y^{[n-1]} + \cdots + p_n y \tag{20.13}$$

Define

$$\begin{aligned} y_1(x) &= y(x) \\ y_2(x) &= y'(x) = y_1' \\ y_3(x) &= y''(x) = y_2' \\ &\;\;\vdots \\ y_n(x) &= y^{[n-1]}x = y_{n-1}' \end{aligned} \tag{20.14}$$

Then equation (20.12) may be written as

$$\begin{aligned} \frac{dy_1}{dx} &= y_2 \\ \frac{dy_2}{dx} &= y_3 \\ &\;\;\vdots \\ \frac{dy_{n-1}}{dx} &= y_n \\ \frac{dy_n}{dx} &= -\frac{p_1}{p_0}y_n - \frac{p_2}{p_0}y_{n-1} - \cdots - \frac{p_n}{p_0}y_1 - \frac{\lambda}{p_0}y_1 \end{aligned}$$

or defining

$$\mathbf{y} = \begin{pmatrix} y_1 \\ y_2 \\ . \\ . \\ y_n \end{pmatrix} \tag{20.15}$$

$$\frac{d\mathbf{y}}{dx} = \begin{pmatrix} 0 & 1 & 0 & . & 0 \\ 0 & 0 & 1 & . & 0 \\ . & & & & . \\ 0 & 0 & 0 & 1 & 0 \\ -\frac{p_n}{p_0} & -\frac{p_{n-1}}{p_0} & . & . & -\frac{p_1}{p_0} \end{pmatrix} \mathbf{y} + \lambda \begin{pmatrix} 0 & . & . & . & 0 \\ 0 & . & . & . & 0 \\ . & & & & \\ . & & & & \\ -\frac{1}{p_0} & 0 & . & . & 0 \end{pmatrix} \mathbf{y} \tag{20.16}$$

$\Longrightarrow$

$$\frac{d\mathbf{y}}{dx} = \mathbf{A}(x)\mathbf{y} + \lambda \mathbf{B}(x)\mathbf{y}$$

$\Longrightarrow$

$$\frac{d\mathbf{y}}{dx} = [\mathbf{A}(x) + \lambda \mathbf{B}(x)]\mathbf{y} \tag{20.17}$$

$$\mathbf{W}_a \mathbf{y}(a) + \mathbf{W}_b \mathbf{y}(b) = \mathbf{0}, \tag{20.18}$$

where the elements of the $n \times n$ matrices $\mathbf{A}(x)$ and $\mathbf{B}(x)$ are continuous functions. Obviously, this is a more general eigenvalue problem than that defined by equation (20.12).

Theorem. *Let* $\mathbf{A}$ *and* $\mathbf{B}$ *be continuous complex valued* $n \times n$ *matrices defined on the real* x*-interval* $[a, b]$. *Let* ξ *be any point in* (a, b) *and* $\mathbf{w}$ *be any constant vector in* C^n. *Consider the solution of equations* (20.17) *and* (20.18) *that passes through the point* $(\xi, \mathbf{w})$ *and denote it by* $\mathbf{y}(x, \lambda, \xi, \mathbf{w})$. *Then:*
1. *The solution* $\mathbf{y}(x, \lambda, \xi, \mathbf{w})$ *exists for all* x *in* (a, b) *and is continuous in* x, λ *and* $\mathbf{w}$ *for* x *in* (a, b) *and* $|\mathbf{w}| + |\lambda| < \infty$, *for each fixed* x *in* (a, b).
2. *It is analytic in* $\mathbf{w}$ *and* λ *for* $|\mathbf{w}| + |\lambda| < \infty$.

A proof of this theorem may be found in the book by Coddington and Levinson [13]. Another way of expressing this result is that if we generate a fundamental matrix of equation (20.17) by using the n conditions:

$$\mathbf{y} = w_j \mathbf{e}_j, \quad \xi \in [a, b] \quad j = 1, 2, 3, \ldots, n \tag{20.19}$$

Then, the fundamental matrix $\mathbf{Y}(x, \lambda, \xi, \mathbf{w})$ is an entire function of λ. Thus, the characteristic matrix

$$\mathbf{D}(\lambda) = \mathbf{W}_a \mathbf{Y}(a, \lambda, \xi, \mathbf{w}) + \mathbf{W}_b \mathbf{Y}(b, \lambda, \xi, \mathbf{w}) \tag{20.20}$$

has elements that are all entire functions

$\Longrightarrow h(\lambda) = \det \mathbf{D}(\lambda)$ is an entire function of λ. Thus, to study the nature (real or complex and distribution) of the eigenvalues, we need to study the zeros of entire functions. In the special case in which $\mathbf{A}(x)$ and $\mathbf{B}(x)$ are constant matrices, we get

$$\mathbf{Y}(x) = e^{(\mathbf{A}+\lambda\mathbf{B})\mathbf{x}}, \tag{20.21}$$

which is an entire function of λ. In this case,

$$\mathbf{D}(\lambda) = \mathbf{W}_a e^{(\mathbf{A}+\lambda\mathbf{B})a} + \mathbf{W}_b e^{(\mathbf{A}+\lambda\mathbf{B})b} \tag{20.22}$$

and

$$h(\lambda) = \det \mathbf{D}(\lambda). \tag{20.23}$$

It is clear that $h(\lambda)$ is an entire function of λ.

Remark. We can generate $h(\lambda)$ for the n-th order equation

$$Ly = -\lambda y,$$

which is an entire function by choosing the linearly independent solutions according to

$$\mathbf{k}(y(\xi, \lambda)) = \mathbf{e}_j, \quad a \leq \xi \leq b, j = 1, 2, \ldots, n$$

Other ways of generation of $h(\lambda)$ may not make it an entire function of λ.

20.3 Properties of the characteristic equation

Theorem. *Let $h(\lambda)$ be an entire function. Then:*
1. *The zeros of $h(\lambda)$ are discrete (isolated).*
2. *There cannot be an infinite number of zeros of $h(\lambda)$ in any closed region of the complex $\{\lambda\}$ plane, i. e., there is no cluster point of the zeros except at infinity.*
3. *If $h(\lambda)$ is real for λ real, the zeros must occur in complex conjugate pairs.*

Proof. Let us write z for λ.
1. Let $\mathcal{R}$ be any closed region in the complex plane and $z = a$ be a zero of $h(z)$. Since $h(z)$ is analytic in $\mathcal{R}$, we may write

   $$h(z) = \sum_{n=0}^{\infty} a_n (z-a)^n, \quad a_0 = 0$$

   This Taylor series expansion converges everywhere in $\mathcal{R}$. If $a_0 = a_1 = \cdots = a_{k-1} = 0$, then $z = a$ is a zero of multiplicity k.

$\Longrightarrow$

$$h(z) = (z - a)^k g(z),$$

where $g(z)$ is an analytical function with $g(a) \neq 0$. If k is not finite, then $h(z) \equiv 0$. If k is finite, then $\exists$ a region $|z - a| < \delta$ such that $g(z)$ has no zero, or equivalently $h(z)$ does not vanish. $\therefore$ The zeros are isolated

2. Suppose that $z_1, z_2, \ldots, z_n$ are the zeros in $\mathcal{R}$ such that

$$\lim_{n\to\infty} z_n = c, \quad \text{i. e.,}$$

c is a cluster point. Then

$$\lim_{n\to\infty} h(z_n) = h\Big(\lim_{n\to\infty} z_n\Big) = h(c) = 0$$

Hence, c is a zero of $h(z)$, which is not isolated contradicting $(i) \Longrightarrow$ No clustering of the zeros. Now if there are an infinite number of zeros in $\mathcal{R}$ then we can extract a convergent sequence $\{z_n\} \to a$, whose limit is in the domain. This contradicts the hypothesis. Hence, either $h(z) \equiv 0$ or $h(z)$ has only a finite number of zeros in $\mathcal{R}$.

3. Let

$$h(z) = a_0 + a_1 z + a_2 z^2 + a_3 z^3 + \cdots,$$

which converges for all z and is real for z real. Suppose that $z = b$ is a real number $\Rightarrow$

$$h(b) = a_0 + a_1 b + a_2 b^2 + a_3 b^3 + \cdots$$

If $b = 0$, h is real $\Longrightarrow a_0$ is real,

$$h'(z) = a_1 + 2a_2 z + \cdots$$

$\Rightarrow$

$$h'(0) \text{ is real } \Longrightarrow a_1 \text{ is real}$$

$\therefore$ $\{a_i\}$ are real. Now suppose

$$h(\alpha) = 0$$

$\Longrightarrow$

$$0 = a_0 + a_1\alpha + a_2\alpha^2 + \cdots = h(\alpha)$$
$$0 = a_0 + a_1\bar{\alpha} + a_2\bar{\alpha}^2 + \cdots = h(\bar{\alpha})$$

$\therefore \bar{a}$ is also a zero of $h(z)$. If $h(z)$ is real for z real and $h(z)$ is an entire function, then its zeros must occur in conjugate pairs. □

Further properties of the characteristic equation

The characteristic equation is given by

$$h(\lambda) = \det \mathbf{D}(\lambda) = \begin{vmatrix} d_{11}(\lambda) & d_{12}(\lambda) & . & d_{1n}(\lambda) \\ . & & & \\ . & & & \\ d_{n1}(\lambda) & d_{n2}(\lambda) & . & d_{nn}(\lambda) \end{vmatrix} \qquad (20.24)$$

$\Longrightarrow$

$$h'(\lambda) = \begin{vmatrix} d'_{11} & d'_{12} & . & d'_{1n} \\ . & & & . \\ . & & & . \\ d_{n1} & d_{n2} & . & d_{nn} \end{vmatrix} + \cdots + \begin{vmatrix} d_{11} & d_{12} & . & d_{1n} \\ . & & & . \\ . & & & . \\ d'_{n1} & d'_{n2} & . & d'_{nn} \end{vmatrix} \qquad (20.25)$$

If we expand each determinant by elements of the differentiated rows, we obtain a linear combination of determinants of order $(n-1)$, which are minors of $\mathbf{D}$. If the rank of $\mathbf{D}$ is $(n-2)$, then all these determinants vanish

$\Longrightarrow$

$$h'(\lambda) = 0.$$

Thus, if the eigenvalues are to be simple, the rank of $\mathbf{D}(\lambda)$ must be $(n-1)$. However, the converse is not true, i. e., $h(\lambda) = h'(\lambda) = 0$ does not imply rank $\mathbf{D} = n-2$. To illustrate, consider

$$\mathbf{D} = \begin{pmatrix} \lambda & 1 \\ 0 & \lambda \end{pmatrix}$$

which gives

$$h = \lambda^2, \quad h' = 2\lambda$$

Thus,

$$h = h' = 0 \,@\, \lambda = 0$$

but $\mathbf{D}(0)$ has rank one. We now return to equations (20.24) and (20.25). By Laplace's expansion, $h''(\lambda)$ is a linear combination of determinants of order $(n-2)$ and $(n-1)$ which are minors of $\mathbf{D}$. Thus, if rank $\mathbf{D} = (n-3)$,

$\Longrightarrow$

$$h(\lambda) = h'(\lambda) = h''(\lambda) = 0$$

Thus, the multiplicity of the eigenvalue = 3. Again, the converse is not true, i. e., $h(\lambda), h'(\lambda)$ and $h''(\lambda)$ can be zero even if rank $\mathbf{D} = (n-1)$. Thus, the multiplicity of the eigenvalue is at least k if rank $\mathbf{D} = n - k$.

Definition. The eigenvalue λ_i is said to have *algebraic multiplicity* k if

$$\begin{aligned} h(\lambda_i) &= 0 \\ \frac{dh}{d\lambda}(\lambda_i) &= 0 \\ &\cdot \\ &\cdot \\ \frac{d^{k-1}h}{d\lambda^{k-1}}(\lambda_i) &= 0 \quad \text{but } \frac{d^k h}{d\lambda^k}(\lambda_i) \neq 0 \end{aligned} \tag{20.26}$$

The eigenvalue λ_i is said to have *geometric multiplicity* k if rank $\mathbf{D}(\lambda_i) = n - k$. If both these multiplicities are equal and $k = 1$, the eigenvalue is said to be simple, $k = 2$, and the eigenvalue is said to be double, etc.

Example 20.1. Consider the second-order eigenvalue problem

$$\frac{d^2y}{dx^2} = -\lambda y; \quad y(0) = 0; \quad y(1) = 0. \tag{20.27}$$

We take

$$y_1 = \cos\sqrt{\lambda}x, \quad y_2 = \frac{\sin\sqrt{\lambda}x}{\sqrt{\lambda}},$$

which satisfy

$$\begin{aligned} \mathbf{k}(y_1(0)) &= \mathbf{e}_1 \\ \mathbf{k}(y_2(0)) &= \mathbf{e}_2 \end{aligned}$$

$\Longrightarrow$

$$\begin{aligned} \mathbf{K}(\mathbf{y}(x,\lambda)) &= \begin{pmatrix} \cos\sqrt{\lambda}x & \sin\sqrt{\lambda}x/\sqrt{\lambda} \\ -\sqrt{\lambda}\sin\sqrt{\lambda}x & \cos\sqrt{\lambda}x \end{pmatrix} \\ \mathbf{D}(\lambda) &= \mathbf{W}_a\mathbf{K}(\mathbf{y}(0)) + \mathbf{W}_b\mathbf{K}(\mathbf{y}(1)) \\ &= \begin{pmatrix} 1 & 0 \\ 0 & 0 \end{pmatrix}\begin{pmatrix} 1 & 0 \\ 0 & 1 \end{pmatrix} + \begin{pmatrix} 0 & 0 \\ 1 & 0 \end{pmatrix}\begin{pmatrix} \cos\sqrt{\lambda} & \sin\sqrt{\lambda}/\sqrt{\lambda} \\ -\sqrt{\lambda}\sin\sqrt{\lambda} & \cos\sqrt{\lambda} \end{pmatrix} \end{aligned}$$

$$= \begin{pmatrix} 1 & 0 \\ 0 & 0 \end{pmatrix} + \begin{pmatrix} 0 & 0 \\ \cos\sqrt{\lambda} & \sin\sqrt{\lambda}/\sqrt{\lambda} \end{pmatrix}$$

$$= \begin{pmatrix} 1 & 0 \\ \cos\sqrt{\lambda} & \sin\sqrt{\lambda}/\sqrt{\lambda} \end{pmatrix} \tag{20.28}$$

$\Rightarrow$

$$h(\lambda) = \frac{\sin\sqrt{\lambda}}{\sqrt{\lambda}} = 0 \Longrightarrow \sqrt{\lambda} = n\pi, \quad n = \pm 1, \pm 2, \ldots$$

The eigenvalues are given by

$$\lambda_n = n^2\pi^2, \quad n = 1, 2, \ldots$$

$$\mathbf{D}(\lambda) = \begin{pmatrix} 1 & 0 \\ \cos\sqrt{\lambda} & \frac{\sin\sqrt{\lambda}}{\sqrt{\lambda}} \end{pmatrix}, \quad \mathbf{D}(\lambda_n) = \begin{pmatrix} 1 & 0 \\ (-1)^n & 0 \end{pmatrix},$$

$$\mathbf{D}(\lambda_n)\mathbf{c} = \mathbf{0} \Longrightarrow c_1 = 0$$

$$\therefore y_n(x) = c_1 y_1 + c_2 y_2 = c_2 \sin n\pi x$$

Note that $n = \pm k$ gives the same eigenfunction and eigenvalue. Choose c_2 such that

$$\langle y_n, y_n \rangle = \int_0^1 y_n^2 \, dx = 1 \Longrightarrow c_2 = \sqrt{2}$$

$\therefore$ Normalized eigenfunctions are given by

$$y_n(x) = \sqrt{2} \sin n\pi x \tag{20.29}$$

Thus, there are infinite number of eigenvalues and the corresponding eigenfunctions. In addition,

$$h(\lambda) = \det \mathbf{D}(\lambda) = \frac{\sin\sqrt{\lambda}}{\sqrt{\lambda}}$$

$$\frac{dh}{d\lambda}(\lambda_n) = -\frac{\sin n\pi}{2n^3\pi^3} + \frac{\cos n\pi}{2n\pi} = \frac{(-1)^n}{2n\pi} \neq 0 \tag{20.30}$$

Thus, all the eigenvalues are simple. The eigenspace corresponding to each eigenvalue has dimension one.

Example 20.2. Consider the eigenvalue problem

$$y'' = -\lambda y; \quad y(0) = 0, \quad y(1) - 2y'(1) = 0 \tag{20.31}$$

Since the operator is the same, we can use the same Wronskian matrix as that in the previous example (but the boundary conditions are different). We have

$$\mathbf{W}_a = \begin{pmatrix} 1 & 0 \\ 0 & 0 \end{pmatrix}, \quad \mathbf{W}_b = \begin{pmatrix} 0 & 0 \\ 1 & -2 \end{pmatrix}$$

∴

$$\begin{aligned} D(\lambda) &= \begin{pmatrix} 1 & 0 \\ 0 & 0 \end{pmatrix}\begin{pmatrix} 1 & 0 \\ 0 & 1 \end{pmatrix} + \begin{pmatrix} 0 & 0 \\ 1 & -2 \end{pmatrix}\begin{pmatrix} \cos\sqrt{\lambda} & \sin\sqrt{\lambda}/\sqrt{\lambda} \\ -\sqrt{\lambda}\sin\sqrt{\lambda} & \cos\sqrt{\lambda} \end{pmatrix} \\ &= \begin{pmatrix} 1 & 0 \\ 0 & 0 \end{pmatrix} + \begin{pmatrix} 0 & 0 \\ \cos\sqrt{\lambda} + 2\sqrt{\lambda}\sin\sqrt{\lambda} & \frac{\sin\sqrt{\lambda}}{\sqrt{\lambda}} - 2\cos\sqrt{\lambda} \end{pmatrix} \\ &= \begin{pmatrix} 1 & 0 \\ \cos\sqrt{\lambda} + 2\sqrt{\lambda}\sin\sqrt{\lambda} & \frac{\sin\sqrt{\lambda}}{\sqrt{\lambda}} - 2\cos\sqrt{\lambda} \end{pmatrix} \end{aligned} \tag{20.32}$$

∴

$$h(\lambda) = \frac{\sin\sqrt{\lambda}}{\sqrt{\lambda}} - 2\cos\sqrt{\lambda} = 0 \Longrightarrow \tan\sqrt{\lambda} = 2\sqrt{\lambda} \tag{20.33}$$

This is the characteristic equation for determining the eigenvalues. This is a transcendental equation and has an infinite number of roots as shown in Figure 20.1, where the two curves (LHS and RHS of equation (20.33)) intersect at infinite number of points. We note that for large values of $\sqrt{\lambda}$, the points of intersection are close to odd multiples of $\frac{\pi}{2}$.

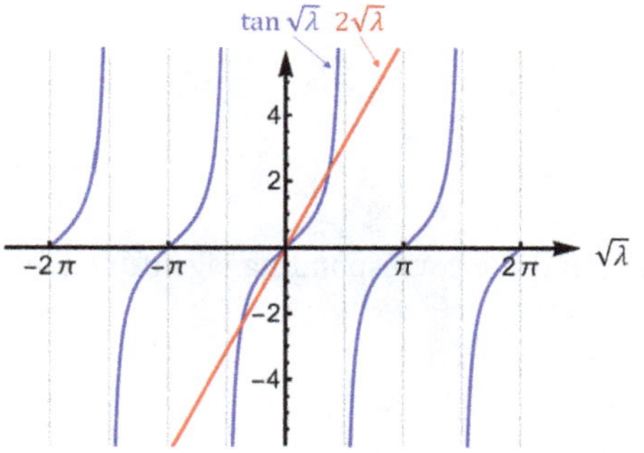

Figure 20.1: The roots of characteristic equation (intersection of the two curves).

This leads to infinite number of eigenvalues

$$\pm\sqrt{\lambda_1}, \pm\sqrt{\lambda_2}, \ldots$$

or

$$\begin{aligned} \lambda &= \lambda_1, \lambda_2, \lambda_3, \ldots \\ \mathbf{D}(\lambda_n)\mathbf{c} &= \mathbf{0} \Longrightarrow c_1 = 0 \end{aligned}$$

$\therefore$

$$y_n(x) = c_2 \sin\sqrt{\lambda_n}x$$

$$\langle y_n, y_n\rangle = 1 \Longrightarrow c_2^2 \int_0^1 \sin^2\sqrt{\lambda_n}x\,dx = 1 \Longrightarrow c_2^2 \int_0^1 \frac{1-\cos 2\sqrt{\lambda_n}x}{2}\,dx = 1$$

$\Longrightarrow$

$$\begin{aligned}\frac{2}{c_2^2} &= 1 - \left.\frac{\sin 2\sqrt{\lambda_n}x}{2\sqrt{\lambda_n}}\right|_0^1 = 1 - \frac{\sin 2\sqrt{\lambda_n}}{2\sqrt{\lambda_n}}\\ &= 1 - \frac{2\sin\sqrt{\lambda_n}\cos\sqrt{\lambda_n}}{2\sqrt{\lambda_n}}\\ &= 1 - 2\cos^2\sqrt{\lambda_n} \quad \text{(using Eq. (20.33))}\\ &= -\cos 2\sqrt{\lambda_n}\end{aligned}$$

$\Longrightarrow$

$$c_2^2 = \frac{2}{-\cos 2\sqrt{\lambda_n}} \quad \text{or} \quad c_2^2 = \frac{2}{1-2\cos^2\sqrt{\lambda_n}}$$

$\Longrightarrow$

$$c_2 = \sqrt{\frac{2}{1-2\cos^2\sqrt{\lambda_n}}}$$

$\therefore$ The normalized eigenfunctions are given by

$$y_n(x) = \frac{\sqrt{2}\sin\sqrt{\lambda_n}x}{\sqrt{1-2\cos^2\sqrt{\lambda_n}}}, \qquad n = 1, 2, \ldots \tag{20.34}$$

In addition, it can be shown that $h'(\lambda_n) \neq 0$, i. e., the eigenvalues are simple.

Example 20.3. Consider the eigenvalue problem

$$y'' = -\lambda y; \quad y(0) = y(1); \quad y'(0) = y'(1) \tag{20.35}$$

Again, since the operator is the same, we can use the same Wronskian matrix as that in Example 20.1. We have

$$\mathbf{W}_a = \begin{pmatrix} 1 & 0 \\ 0 & 1 \end{pmatrix}, \quad \mathbf{W}_b = \begin{pmatrix} -1 & 0 \\ 0 & -1 \end{pmatrix}$$

$$\mathbf{D}(\lambda) = \begin{pmatrix} 1 & 0 \\ 0 & 1 \end{pmatrix}\begin{pmatrix} 1 & 0 \\ 0 & 1 \end{pmatrix} + \begin{pmatrix} -1 & 0 \\ 0 & -1 \end{pmatrix}\begin{pmatrix} \cos\sqrt{\lambda} & \sin\sqrt{\lambda}/\sqrt{\lambda} \\ -\sqrt{\lambda}\sin\sqrt{\lambda} & \cos\sqrt{\lambda} \end{pmatrix}$$

$$= \begin{pmatrix} 1 & 0 \\ 0 & 1 \end{pmatrix} + \begin{pmatrix} -\cos\sqrt{\lambda} & -\sin\sqrt{\lambda}/\sqrt{\lambda} \\ \sqrt{\lambda}\sin\sqrt{\lambda} & -\cos\sqrt{\lambda} \end{pmatrix}$$

$$= \begin{pmatrix} 1-\cos\sqrt{\lambda} & -\sin\sqrt{\lambda}/\sqrt{\lambda} \\ \sqrt{\lambda}\sin\sqrt{\lambda} & 1-\cos\sqrt{\lambda} \end{pmatrix} \tag{20.36}$$

$\Rightarrow$

$$\begin{aligned} h(\lambda) &= \sin^2\sqrt{\lambda} + (1-\cos\sqrt{\lambda})^2 \\ &= 2[1-\cos\sqrt{\lambda}] = 0 \end{aligned} \tag{20.37}$$

The eigenvalues are given by

$$\begin{aligned} \sqrt{\lambda} &= 2n\pi \\ \lambda_n &= 4n^2\pi^2, \quad n = 0, 1, 2, \dots \end{aligned}$$

$\lambda = 0$ gives a nontrivial solution, $y_0(x) = 1$. For $n \geq 1$, we have

$$\mathbf{D}(\lambda_n) = \begin{pmatrix} 0 & 0 \\ 0 & 0 \end{pmatrix}, \quad \text{rank}\,\mathbf{D}(\lambda_n) = 0$$

Thus, we have two eigenfunctions

$$y_{ns}(x) = \sin 2n\pi x \quad \text{and} \quad y_{nc}(x) = \cos 2n\pi x, \quad n = 1, 2, 3, \dots$$

We note that each eigenvalue except $\lambda_0 = 0$ is double:

$$\frac{dh}{d\lambda} = \frac{\sin\sqrt{\lambda}}{\sqrt{\lambda}}$$

$$\left.\frac{dh}{d\lambda}\right|_{\lambda=\lambda_n} = 0, \quad \left.\frac{d^2h}{d\lambda^2}\right|_{\lambda=\lambda_n} = \frac{1}{2\lambda_n} \neq 0$$

Thus, the eigenspace corresponding to each eigenvalue λ_n is two-dimensional. To summarize, we have

Eigenvalues:

$$\lambda_0 = 0 \tag{20.38}$$

$$\lambda_n = 4n^2\pi^2, \quad n = 1, 2, 3, \dots \tag{20.39}$$

Eigenfunctions:

$$y_0(x) = 1 \tag{20.40}$$

$$y_{ns}(x) = \sin 2n\pi x \tag{20.41}$$

$$y_{nc}(x) = \cos 2n\pi x \tag{20.42}$$

Example 20.4. Consider the eigenvalue problem

$$y'' = -\lambda y; \quad y(0) = 0; \quad y'(0) = \alpha y(1); \quad \alpha \text{ real} \tag{20.43}$$

$\Longrightarrow$

$$\mathbf{W}_a = \begin{pmatrix} 1 & 0 \\ 0 & 1 \end{pmatrix}, \quad \mathbf{W}_b = \begin{pmatrix} 0 & 0 \\ -\alpha & 0 \end{pmatrix}$$

$$\begin{aligned} D(\lambda) &= \begin{pmatrix} 1 & 0 \\ 0 & 1 \end{pmatrix}\begin{pmatrix} 1 & 0 \\ 0 & 1 \end{pmatrix} + \begin{pmatrix} 0 & 0 \\ -\alpha & 0 \end{pmatrix}\begin{pmatrix} \cos\sqrt{\lambda} & \frac{\sin\sqrt{\lambda}}{\sqrt{\lambda}} \\ -\sqrt{\lambda}\sin\sqrt{\lambda} & \cos\sqrt{\lambda} \end{pmatrix} \\ &= \begin{pmatrix} 1 & 0 \\ 0 & 1 \end{pmatrix} + \begin{pmatrix} 0 & 0 \\ -\alpha\cos\sqrt{\lambda} & -\alpha\frac{\sin\sqrt{\lambda}}{\sqrt{\lambda}} \end{pmatrix} \\ &= \begin{pmatrix} 1 & 0 \\ -\alpha\cos\sqrt{\lambda} & 1-\alpha\frac{\sin\sqrt{\lambda}}{\sqrt{\lambda}} \end{pmatrix} \end{aligned} \tag{20.44}$$

$$h(\lambda) = |\mathbf{D}(\lambda)| = 1 - \frac{\alpha\sin\sqrt{\lambda}}{\sqrt{\lambda}} \tag{20.45}$$

Characteristic equation:

$$\frac{\sin\sqrt{\lambda}}{\sqrt{\lambda}} = \frac{1}{\alpha}$$

Note: For the special case of $\alpha = 1$, $\lambda = 0$ is an eigenvalue with eigenfunction

$$y_0(x) = x$$

We consider different cases

$$\begin{array}{llllll} \text{(A)} & \alpha = 1 & \text{(B)} & \alpha > 1 & \text{(C)} & 0 < \alpha < 1 \\ \text{(D)} & \alpha = \infty & \text{(E)} & \alpha < 0 & & \end{array} \tag{20.46}$$

Case A: $\alpha = 1$
Let $\sqrt{\lambda} = a + ib \Longrightarrow$

$$\sin(a+ib) = a + ib$$

$$\sin a\cosh b + i\cos a\sinh b = a + ib$$

$$a = \sin a\cosh b \tag{20.47}$$

$$b = \cos a\sinh b \tag{20.48}$$

$$\text{Equation (20.47)} \Longrightarrow b = \cosh^{-1}\left(\frac{a}{\sin a}\right) \tag{20.49}$$

$$\text{Equation (20.48)} \Longrightarrow \cosh^{-1}\left(\frac{a}{\sin a}\right) - \cos a \sinh\left\{\cosh^{-1}\left(\frac{a}{\sin a}\right)\right\} = 0 \tag{20.50}$$

Solve for a and get b from equation (20.49).

Asymptotic values

For $b \gg 1$, $\frac{b}{\sinh b} \to 0$. Therefore, equation (20.48) $\Longrightarrow \cos a = 0$:

$$a \approx (2n+1)\frac{\pi}{2}, \quad n = 0, 1, 2, \ldots \quad \Longrightarrow \sin a = (-1)^n$$

For n odd, the equation $\cosh b = \frac{a}{\sin a}$ cannot be satisfied. Thus,

$$\begin{aligned} a_m &= (2m+1)\frac{\pi}{2}, \quad m = 0, 2, 4, 6, \ldots \\ &= (4m+1)\frac{\pi}{2}, \quad m = 0, 1, 2, 3, \ldots \end{aligned} \tag{20.51}$$

$$\begin{aligned} b_m &= \cosh^{-1}\left\{(4m+1)\frac{\pi}{2}\right\} \\ &= \ln\left\{(4m+1)\frac{\pi}{2} + \sqrt{(4m+1)^2\frac{\pi^2}{4} - 1}\right\} \end{aligned} \tag{20.52}$$

∴

$$\sqrt{\lambda_m} = a_m + ib_m \Rightarrow \lambda_m = a_m^2 - b_m^2 \pm i2a_m b_m. \tag{20.53}$$

Exact eigenvalues:

$$\begin{aligned} \lambda_0 &= 0 \\ \lambda_1 &= 48.56 \pm i41.5 \\ \lambda_2 &= 181.97 \pm i93.2 \\ \lambda_m &\approx (4m+1)^2\frac{\pi^2}{4} - [\ln(4m+1)\pi]^2 \\ &\qquad \pm i(4m+1)\pi[\ln(4m+1)\pi], \quad m \geq 3 \end{aligned} \tag{20.54}$$

Case D: $\alpha = \infty$

In this case,

$$h(\lambda) = \frac{\sin\sqrt{\lambda}}{\sqrt{\lambda}} = 0 \Longrightarrow \sqrt{\lambda} = n\pi, \quad n = 1, 2, \ldots \tag{20.55}$$

$$\Longrightarrow \lambda_n = n^2\pi^2; \quad n = 1, 2, 3, \ldots, \infty \tag{20.56}$$

No complex eigenvalues for $\alpha = \infty$.

General α (Cases B, C and E)

$$\sqrt{\lambda} = a + ib \Longrightarrow \alpha \sin(a + ib) = a + ib$$

$\Longrightarrow$

$$a = \alpha \sin a \cosh b \tag{20.57}$$

$$b = \alpha \cos a \sinh b \tag{20.58}$$

$$\frac{a}{b} = \frac{\tan a}{\tanh b}$$

Equation (20.57) $\Longrightarrow$

$$b = \cosh^{-1}\left\{\frac{a}{\alpha \sin a}\right\} \tag{20.59}$$

Substituting equation (20.59) in equation (20.58) $\Longrightarrow$

$$\cosh^{-1}\left\{\frac{a}{\alpha \sin a}\right\} - \alpha \cos a \sinh \cosh^{-1}\left\{\frac{a}{\alpha \sin a}\right\} = 0 \tag{20.60}$$

Asymptotic values

For $b \gg 1$

$$\begin{aligned}\sqrt{\lambda_m} &= a_m + ib_m \\ &= (4m + 1)\frac{\pi}{2} + i \cosh^{-1}\left\{(4m + 1)\frac{\pi}{2\alpha}\right\}\end{aligned} \tag{20.61}$$

Summary

(i) For $\alpha = \infty$, only real eigenvalues

$$\lambda_n = n^2\pi^2$$

(ii) For $1 < \alpha < \infty$, a finite number of real eigenvalues, infinite number of complex eigenvalues

(iii) For $\alpha = 1$, one real eigenvalue all others are complex eigenvalues

(iv) For $0 < \alpha < 1$, all eigenvalues are complex and move to ∞ as $\alpha \to 0$.

This example illustrates the fact that for nonsymmetric or non-selfadjoint eigenvalue problems, the determination of the nature of the spectrum can be nontrivial.

Problems

1. Given the eigenvalue problem,

$$y'' = -\lambda y;$$
$$y(0) = y'(1)$$
$$y(1) = y'(0)$$

   (a) Find the eigenvalues and determine multiplicity.
   (b) Find the eigenfunctions.
   (c) Find the eigenvalues and eigenfunctions of the adjoint system.
   (d) Verify the biorthogonality property.

2. Given the eigenvalue problem,

$$\frac{d^4y}{dx^4} = \lambda y, \quad 0 < x < 1$$
$$y'''(0) + h_1 y(0) = 0; \quad y'''(1) - h_3 y(1) = 0$$
$$y''(0) - h_2 y'(0) = 0; \quad y''(1) + h_4 y'(1) = 0$$

   (a) Show that the eigenvalues are real.
   (b) If $h_j > 0$ for $j = 1, 2, 3, 4$, show that the eigenvalues are all nonnegative by integrating

$$\int_0^1 y \frac{d^4y}{dx^4} \, dx$$

   by parts twice.
   (c) Show that the eigenfunctions corresponding to different eigenvalues are orthogonal.

3. Consider the eigenvalue problem

$$\frac{d^2u}{dx^2} = -\lambda u; \quad 0 < x < 1$$
$$u(0) - u(1) = 0, \quad u'(0) - u'(1) = 0$$

   Determine the eigenvalues, eigenfunctions and adjoint eigenfunctions.

4. The buckling of a long column with pinned ends (see Figure 20.2) is described by the eigenvalue problem:

$$\frac{d^2y}{dx^2} = -\frac{P}{EI} y$$
$$y(0) = y(L) = 0$$
$$y = \text{displacement}$$

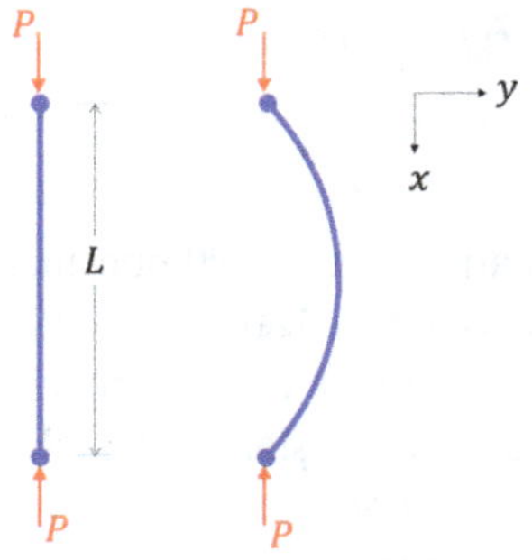

Figure 20.2: Buckling of a column with pinned endpoints.

$$P = \text{applied load}$$
$$EI = \text{flexural rigidity of the column}$$

What is the critical (smallest) load at which buckling occurs? [Note: the resulting formula is known as the Euler's column formula.]

5. The following sixth-order eigenvalue problem arises in the stability analysis of a fluid layer confined between two parallel plates kept at different temperatures (Rayleigh–Benard convection):

$$\phi^{[4]} - 2k^2\phi'' + k^4\phi - \mathrm{Ra}\,k^2\psi = \lambda(\phi'' - k^2\phi)$$
$$\psi'' - k^2\psi + \phi = \lambda\,\mathrm{Pr}\,\psi, \quad 0 < x < 1$$
$$\phi(0) = \phi'(0) = \psi(0) = 0$$
$$\phi(1) = \phi'(1) = \psi(1) = 0$$

Here, k is the wave number, Pr is the Prandtl number and Ra is the Rayleigh number. (a) Show that the eigenvalues are real. (b) Compute the smallest value of Ra for which an eigenvalue can be zero [this value of Ra defines the critical value beyond which the conduction state is not stable and leads to convective motions].

6. Consider the EVP

$$v'' = -\lambda v, \quad 0 < x < 1$$
$$v'(1) + \frac{\mathrm{Pe}}{2}v(1) = 0$$
$$v'(0) - \frac{\mathrm{Pe}}{2}v(0) + \mathrm{Pe}\,\mathrm{Re}^{\mathrm{Pe}/2}\,v(1) = 0,$$

where $\mathrm{Pe} > 0$ is the Peclet number and $R > 0$ is the recycle ratio. Determine the nature of the eigenvalues for different choices of Pe and R.

21 Sturm–Liouville theory and eigenfunction expansions

We have seen in Part I that the eigenvectors of a symmetric matrix form an orthonormal set, which can be used to expand any arbitrary vector in terms of the eigenvectors. We have also used such expansions to solve linear equations in which the symmetric matrix appeared. This chapter is a generalization of the same to infinite dimensions. While the general approach remains the same, as we have seen in Chapter 10, going from finite- to infinite-dimensional vector spaces can lead to some subtle differences in the manner in which the sequences converge, and hence the interpretation of the expansions. While the general terminology of "Fourier series" is used for these expansions, it should be pointed out that it took scientists and mathematicians nearly 200 years to clarify the various concepts related to eigenfunction expansions. While we are mainly concerned with the use of these expansions to solve linear differential equations, this topic has many other applications.

21.1 Sturm–Liouville (S-L) theory

Consider the self-adjoint second-order boundary value problem

$$-\frac{d}{dx}\left(p(x)\frac{dy}{dx}\right) + q(x)y = \lambda\rho(x)y, \quad a < x < b \tag{21.1}$$

with one of the following three types of boundary conditions:
unmixed BCs

$$\alpha_1 y(a) + \alpha_2 y'(a) = 0, \quad (\alpha_1^2 + \alpha_2^2 > 0) \tag{21.2}$$

$$\beta_1 y(b) + \beta_2 y'(b) = 0 \quad (\beta_1^2 + \beta_2^2 > 0), \tag{21.3}$$

or periodic BCs

$$p(a)y(a) = p(b)y(b) \tag{21.4}$$

$$y'(a) = y'(b), \tag{21.5}$$

or Dirichlet BCs

$$y(a) = 0 \tag{21.6}$$

$$y(b) = 0. \tag{21.7}$$

The BVP defined by equation (21.1) and either BCs given in equation (21.2)–(21.3), (21.4)–(21.5) or (21.6)–(21.7) is called a regular Sturm–Liouville (S-L) problem if $p(x) \neq 0$ in $[a, b]$, $-\infty < a < b < \infty$ (interval $[a, b]$ is finite). $\rho(x)$ is called the density or weight

https://doi.org/10.1515/9783111598055-025

function. It is assumed that $\rho(x) > 0$ for $a < x < b$. We discuss here the S-L theory for Dirichlet BCs. The extension of the theory to other cases is straightforward. (When $p(x)$ vanishes in $[a,b]$ or when the interval (a,b) is infinite the BVP is called irregular. These cases will be considered in Chapters 24 and 25.)

Remark. The second order equation

$$y'' + f(x)y' + g(x)y = -\lambda h(x)y \tag{21.8}$$

with either of the three sets of BCs may be written in a self-adjoint form using the transformation

$$p(x) = \exp\left\{\int_a^x f(x)\,dx\right\} \tag{21.9}$$

$\Longrightarrow$

$$(p(x)y')' = p(x)y'' + p(x)y'f(x)$$

Multiplying both sides of equation (21.8) by the positive function $p(x)$ gives

$$(p(x)y')' + p(x)g(x)y = -\lambda p(x)h(x)y, \tag{21.10}$$

which is same as equation (21.1) if we define

$$q(x) = -p(x)g(x) \tag{21.11}$$

$$\rho(x) = p(x)h(x). \tag{21.12}$$

Theorem. *Consider the self-adjoint BVP*

$$(p(x)y')' - q(x)y = -\lambda\rho(x)y, \quad a < x < b \tag{21.13}$$

$$y(a) = 0, \quad y(b) = 0 \tag{21.14}$$

(i) *the eigenvalues are real*
(ii) *the eigenvalues are positive if* $q(x) \geq 0$ *and* $p(x) > 0$ *in* $[a,b]$
(iii) *the eigenfunctions corresponding to distinct eigenvalues are orthogonal with respect to the weight function* $\rho(x)$
(iv) *the eigenvalues are isolated. There are an infinite number of them with no cluster point, i. e.,* $\lambda_n \longrightarrow \infty$ *for* $n \to \infty$.

Proof. (i) We have already proved this for an n-th order self-adjoint BVP.
(ii) Let λ_n be an eigenvalue with $y_n(x)$ as the corresponding eigenfunction:

$$(p(x)y_n'(x))' - q(x)y_n = -\lambda_n\rho(x)y_n \tag{21.15}$$

multiply by y_n both sides and integrate.

$\Longrightarrow$

$$\int_a^b y_n \frac{d}{dx}(p(x)y_n')\,dx - \int_a^b q(x)y_n^2\,dx = -\lambda_n \int_a^b \rho(x)y_n^2\,dx$$

$\Longrightarrow$

$$py_n'y_n|_a^b - \int_a^b p(x)(y_n')^2\,dx - \int_a^b qy_n^2\,dx = -\lambda_n \int_a^b \rho(x)y_n^2\,dx$$

The first term vanishes since

$$y_n(a) = y_n(b) = 0$$

$\Longrightarrow$

$$\lambda_n = \frac{\int_a^b p(x)(y_n')^2\,dx + \int_a^b q(x)y_n^2\,dx}{\int_a^b \rho(x)y_n^2\,dx} \tag{21.16}$$

[Remark: The RHS of equation (21.16) is the Rayleigh quotient encountered in Section 7.3. It is also a functional.] Thus, $\lambda_n > 0$, if $p(x) > 0$, $q(x) \geq 0$
(iii) To prove (iii), let λ_n and λ_m be eigenvalues ($\lambda_n \neq \lambda_m$) with eigenfunctions $y_n(x)$ and $y_m(x)$, respectively. Then,

$$(py_n')' - qy_n = -\lambda_n \rho y_n \tag{21.17}$$

$$(py_m')' - qy_m = -\lambda_m \rho y_m \tag{21.18}$$

Multiply equation (21.17) by y_m and equation (21.18) by y_n, subtract and integrate by parts
$\Longrightarrow$

$$\int_a^b [(py_n')'y_m - (py_m')'y_n]\,dx - \int_a^b q(y_ny_m - y_my_n)\,dx$$
$$= (\lambda_m - \lambda_n)\int_a^b \rho y_n y_m\,dx$$

$\Longrightarrow$

$$(\lambda_m - \lambda_n)\int_a^b \rho y_n y_m\,dx = [py_n'y_m - py_m'y_n]_a^b - \int_a^b py_n'y_m'\,dx + \int_a^b py_m'y_n'\,dx$$
$$= p(b)y_n'(b)y_m(b) - p(b)y_m'(b)y_n(b) - p(a)y_n'(a)y_m(a)$$

$$+ p(a)y'_m(a)y_n(a)$$
$$= 0$$

$\Longrightarrow$

$$(\lambda_m - \lambda_n) \int_a^b \rho y_n y_m \, dx = 0 \tag{21.19}$$

Since $\lambda_m \neq \lambda_n \Longrightarrow$

$$\int_a^b \rho(x) y_n(x) y_m(x) \, dx = 0$$

(iv) We have already shown that the eigenvalues are the zeros of an entire function. Thus, the zeros are isolated. To prove that there are an infinite number of them, we refer to the book by Coddington and Levinson [13].

We showed that the BVP (equations (21.13)–(21.14)) may be written as an integral equation:

$$y(x) = \int_a^b \lambda \rho(s) y(s) G(x, s) \, ds$$

or

$$\frac{1}{\lambda} y(x) = \int_a^b G(x, s) \rho(s) y(s) \, ds \tag{21.20}$$

where $G(x, s)$ is the Green's function. Thus, if we define an integral operator $\mathcal{G}$ by

$$\mathcal{G}y = \int_a^b G(x, s) \rho(s) y(s) \, ds \tag{21.21}$$

Then the EVP is equivalent to

$$\mathcal{G}y = \mu y, \quad \left(\mu = \frac{1}{\lambda}\right) \tag{21.22}$$

i. e., the eigenvalues of $\mathcal{G}$ are reciprocals of those of equations (21.13)–(21.14) and the eigenfunctions are the same.

$\mathcal{G}$: $C[a, b] \to C[a, b]$ is a linear operator that is bounded and continuous w. r. t. the norm induced by the inner product

$$\langle u, v\rangle = \int_a^b u(x)v(x)\rho(x)\,dx$$

Note also that

$$\langle \mathcal{G}y, u\rangle = \langle y, \mathcal{G}u\rangle$$

i. e., $\mathcal{G}$ is self-adjoint

$$\begin{aligned}\langle \mathcal{G}y, u\rangle &= \int_a^b \left[\int_a^b G(x,s)\rho(s)y(s)\,ds\right].u(x)\rho(x)\,dx\\ &= \int_a^b \left[\int_a^b G(x,s)\rho(x)u(x)\,dx\right].\rho(s)y(s)\,ds = \langle y, \mathcal{G}u\rangle,\end{aligned}$$

since $G(x,s) = G(s,x)$. These properties may be used to prove spectral theorem for the operator $\mathcal{G}$. Again, we refer to the book of Coddington and Levinson [13] for further details. □

Asymptotic distribution of eigenvalues for Sturm–Liouville systems

Consider again the S–L eigenvalue problem

$$-\frac{d}{dx}\left(p(x)\frac{dy}{dx}\right) + q(x)y = \lambda\rho(x)y(x) \quad a < x < b \tag{21.23}$$

$$\alpha_1 y(a) + \alpha_2 y'(a) = 0 \quad (\alpha_1^2 + \alpha_2^2 \neq 0) \tag{21.24}$$

$$\beta_1 y(b) + \beta_2 y'(b) = 0 \quad (\beta_1^2 + \beta_2^2 \neq 0) \tag{21.25}$$

Recall that this is a regular S–L problem if (i) $-\infty < a < b < \infty$, (ii) $p(x) \in C^1(a,b)$ and $q(x), \rho(x) \in C^0(a,b)$ (iii) $\exists \quad p_0 > 0$ and $\rho_0 > 0$ such that $p(x) \geq p_0$ and $\rho(x) \geq \rho_0$ in $[a,b]$. For this case, the following asymptotic result on the eigenvalues may be obtained (for details, we refer to the books by Courant and Hilbert [15] and Morse and Feshback [23]).

Asymptotic results

For $n \to \infty$

$$\lambda_n^{1/2} \approx \frac{(n\pi + \mu)}{\left\{\int_a^b \left|\frac{\rho(x)}{p(x)}\right|^{1/2} dx\right\}} + \frac{C}{n} + O(n^{-2}) \tag{21.26}$$

where the constants μ and C depend only on the BCs, and

$$y_n(x) \approx \frac{1}{[p(x)\rho(x)]^{\frac{1}{4}}} \cos\left\{\sqrt{\lambda_n}\int_a^x \sqrt{\frac{\rho(z)}{p(z)}}\,dz + \theta\right\}, \tag{21.27}$$

where the phase angle θ depends on the boundary conditions and

$$\int_a^b \rho(x) y_n^2(x)\, dx = 1 \tag{21.28}$$

In particular, $\exists$ a constant $c > 0 \ni$

$$|y_n(x)| \leq c \quad \text{for all } a \leq x \leq b \tag{21.29}$$

We provide below some examples of S–L eigenvalue problems, where $\rho(x)$ is not a constant.

Example 21.1. Consider the Graetz–Nusselt EVP that arises in heat/mass transfer analysis in laminar flow between parallel plates

$$\begin{aligned} y'' &= -\lambda \rho(x) y; \quad \rho(x) = \frac{3}{2}(1 - x^2), \quad 0 < x < 1 \\ y'(0) &= 0, \quad y(1) = 0 \end{aligned}$$

Here, $p(x) = 1.0$, $q(x) = 0$ and $\rho(x) = \frac{3}{2}(1 - x^2)$ is the velocity profile. The first few eigenvalues are listed in Table 21.1.

Table 21.1: First few eigenvalues of Gratz–Nusselt EVP.

n	λ_n
1	1.88517
2	21.4315
3	62.3166
4	124.537
5	208.091

We note that the Graetz-Nusselt eigenfunctions satisfy the orthogonality relation based on weighted inner product

$$\int_0^1 \rho(x) y_n(x) y_m(x)\, dx = 0, \quad m \neq n.$$

A plot of these functions with the normalization condition $y_n(0) = 1$ is shown in Figure 21.1.

Example 21.2. Consider the Airy EVP

$$\begin{aligned} y'' &= -\lambda x y, \quad 0 < x < 1 \\ y(0) &= y(1) = 0. \end{aligned}$$

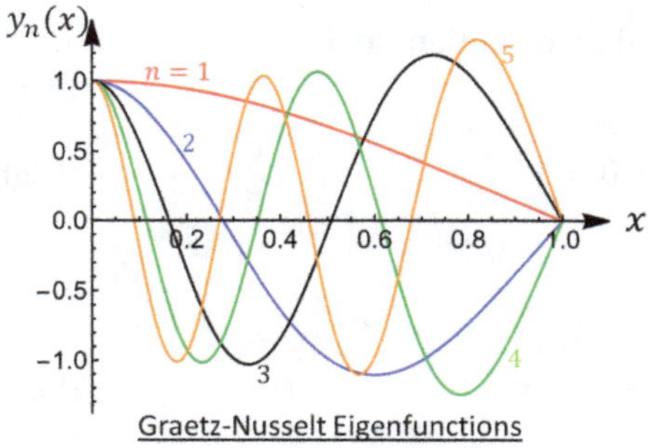

Figure 21.1: First five Graetz–Nusselt eigenfunctions $y_n(x)$ satisfying $y_n(0) = 1$.

Table 21.2: First few eigenvalues of Airy EVP.

n	λ_n
1	18.9563
2	81.8866
3	189.221
4	340.967
5	537.126

Here, $p(x) = 1$, $q(x) = 0$ and $\rho(x) = x$. The first few eigenvalues are listed in Table 21.2.

The eigenfunctions satisfy the orthogonality condition

$$\int_0^1 x y_n(x) y_m(x)\, dx = 0, \quad m \neq n.$$

Figure 21.2 shows first five of these eigenfunctions with constraint $y_n'(0) = 1$.

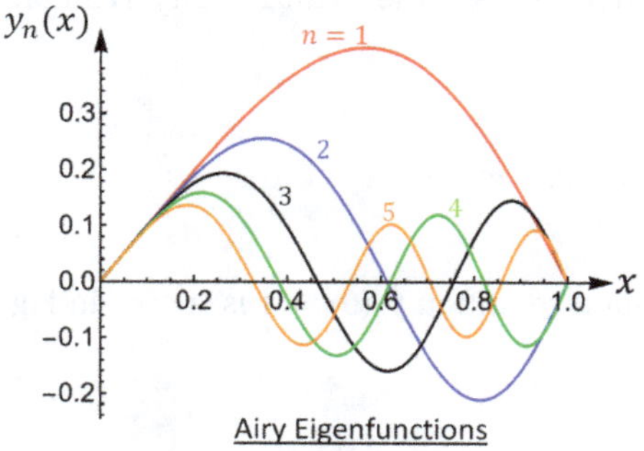

Figure 21.2: First five Airy eigenfunctions $y_n(x)$ satisfying $y_n'(0) = 1$.

21.2 Eigenfunction expansions

We have shown that the self-adjoint BVP

$$(p(x)y')' - q(x)y = -\lambda\rho(x)y, \quad a < x < b \tag{21.30}$$

$$y(a) = 0, \quad y(b) = 0 \tag{21.31}$$

has an infinite number of eigenvalues $\{\lambda_i\}$ and an orthonormal set of eigenfunctions $\{y_i(x)\}$

$$\begin{aligned}\langle y_i, y_j\rangle &= \int_a^b \rho(x)y_i(x)y_j(x)\,dx \\ &= \delta_{ij} = \begin{cases} 1 & \text{if } i = j \\ 0 & \text{otherwise}\end{cases}\end{aligned} \tag{21.32}$$

We now consider the expansion of any arbitrary function $f(x)$ in terms of the eigenfunctions $\{y_i(x)\}$. We write

$$f(x) = \sum_{i=1}^{\infty} c_i y_i(x) \tag{21.33}$$

In order to determine c_i, multiply both sides of equation (21.33) by $\rho(x)y_j(x)$ and integrate from a to b $\Longrightarrow$

$$\int_a^b f(x)y_j(x)\rho(x)\,dx = \int_a^b \rho(x)y_j(x)\left(\sum_{i=1}^{\infty} c_i y_i(x)\right)dx$$

Assuming that the summation and integration can be interchanged (this is so if the series in (21.33) converges uniformly), we get

$$\begin{aligned}\int_a^b f(x)y_j(x)\rho(x)\,dx &= \left(\sum_{i=1}^{\infty} c_i \int_a^b \rho(x)y_i(x)y_j(x)\,dx\right)dx \\ &= c_j \quad \text{[using equation (21.32)]}\end{aligned}$$

Thus,

$$c_j = \int_a^b f(x)y_j(x)\rho(x)\,dx = \langle f, y_j\rangle \tag{21.34}$$

and

$$f(x) = \sum_{i=1}^{\infty} \langle f, y_i\rangle y_i \tag{21.35}$$

Definition. Let $\{y_i(x)\}$ be an infinite system of orthonormal functions relative to the weight function $\rho(x)$ on an interval $[a, b]$. If $f(x)$ is any function for which the integrals in equation (21.34) exist, the infinite series in equation (21.33) is called the *eigenfunction expansion* or *Fourier series* of $f(x)$ relative to the system $\{y_i(x), \rho(x)\}$. The coefficients c_i are called *Fourier coefficients* of $f(x)$ relative to $\{y_i(x)\}$.

Remarks.

1. Historically, the term Fourier series is used when the orthonormal functions $\{y_i(x)\}$ are the sine functions, cosine functions, or sine and cosine functions. Each of these functions are generated from the following self-adjoint problems:

   (S)

   $$y'' = -\lambda y, \quad 0 < x < 1$$
   $$y(0) = y(1) = 0 \quad \text{(Sines)}$$

   with eigenvalues and orthonormal eigenfunctions,

   $$\lambda_n = n^2\pi^2, \quad y_n(x) = \sqrt{2}\sin[n\pi x], \quad n = 1, 2, \ldots$$

   (C)

   $$y'' = -\lambda y, \quad 0 < x < 1$$
   $$y'(0) = 0, \quad y'(1) = 0 \quad \text{(Cosines)}$$

   with eigenvalues and orthonormal eigenfunctions,

   $$\lambda_0 = 0, \quad y_1(x) = 1, \quad \lambda_n = n^2\pi^2, \quad y_n(x) = \sqrt{2}\cos[n\pi x], \quad n = 1, 2, \ldots$$

   (P)

   $$y'' = -\lambda y, \quad 0 < x < 1$$
   $$y(0) = y(1), \quad y'(0) = y'(1) \quad \text{(Sines and Cosines)}$$

   with eigenvalues and orthonormal eigenfunctions,

   $$\lambda_0 = 0, \quad y_0(x) = 1, \quad \lambda_n = n^2\pi^2, \quad y_{ns}(x) = \sqrt{2}\sin[n\pi x],$$
   $$y_{nc}(x) = \sqrt{2}\cos[n\pi x], \quad n = 1, 2, \ldots$$

2. If the eigenfunctions are not normalized, then the expansion (equation (21.35)) is of the form

   $$f(x) = \sum_{i=1}^{\infty} c_i y_i(x), \quad c_i = \frac{\langle f, y_i \rangle}{\langle y_i, y_i \rangle}. \tag{21.36}$$

3. When the functions $\{y_i(x)\}$ are not sines or cosines or when the weight function is not unity, the expansion (21.33) is called *eigenfunction expansion*, or *generalized Fourier series*.

21.3 Convergence in function spaces and introduction to Banach and Hilbert spaces

This section is a brief introduction to orthogonal expansions in infinite-dimensional vector spaces and convergence of such expansions.

21.3.1 Cauchy sequence

Definition. A sequence $\{S_n\}$ contained in a normed linear space is called a *Cauchy sequence* if given $\varepsilon > 0$, $\exists$ an N such that

$$\|S_n - S_m\| < \varepsilon, \quad \forall n, m > N.$$

From this definition, it is seen that any convergent sequence is necessarily a Cauchy sequence but the converse is not true.

Definition. A normed linear space V is said to be *complete* if every Cauchy sequence in the space converges to some element in V.

Example 21.3.

(i) Let V = set of rational numbers in $[0, 1]$ and for $x, y \in V$, define the metric/distance function $d(x, y) = |x - y|$. This space is not complete. The reason being sequences such as $S_n = (\frac{n}{n+1})^n$, converge but not to a point in V. In this case, $\lim_{n\to\infty} S_n$ is $\frac{1}{e}$, which is irrational.

(ii) Let $V = \mathbb{R}$ and $d(x, y) = |x - y|$. This space is complete.

(iii) Let $V = C[a, b]$, and

$$d_\infty(f, g) = \sup_{a \le x \le b} |f(x) - g(x)|.$$

This space is complete w. r. t. the supremum norm as the convergence is uniform.

(iv) let $V = C[a, b]$, and

$$d_2(f, g) = \int_a^b [f(x) - g(x)]^2 \, dx.$$

This space is not complete. For example, sequences such as $\{e^{-nx}\}$, $\{\tanh[nx]\} \in C[a, b]$ for finite n, but for $n \to \infty$, the limiting functions are not in $C[a, b]$.

Normed linear spaces may be completed by appending the missing element to the space. Thus, to the rational numbers, we add all the limits of convergent sequences, i. e., we add all the irrationals, to get the real number system, which is complete.

Similarly, to the space $C[a,b]$ with the metric d_2, we add the limits of convergent sequences, we get the space $\mathcal{L}_2[a,b]$, the space of all Lebesque square integrable functions on $[a,b]$. This space includes all square integrable continuous functions as well as those with finite or an infinite number of discontinuities.

21.3.2 Riemann and Lebesque integration

Recall the definition of the Riemann integral for a continuous or piecewise continuous function $f(x)$ over an interval $[a,b]$. We divide the interval $[a,b]$ into n subintervals and form the upper and lower Riemann sums. If these sums converge to the same value when $n \to \infty$ and the largest size of the subinterval goes to zero, then the Riemann integral exists. Now, consider the Dirichlet function of Section 10.1, for which the Riemann integral does not exist. However, in the Lebesque theory of integration, we ignore sets of measure zero (also referred to as null sets) in the integration process. Thus, the Lebesque integral exists for the Dirichlet function. The set of all Lebesque integrable functions is denoted by $\mathcal{L}[a,b]$, while the set of Riemann integrable functions is denoted by $R[a,b]$. It is clear that $C[a,b] \subset R[a,b] \subset \mathcal{L}[a,b]$.

21.3.3 Banach and Hilbert spaces

Definition. A normed linear space that is complete is called a *Banach space*.

Example 21.4.
(i) $\{C[a,b], d_\infty\}$ is a Banach space
(ii) $\{C[a,b], d_2\}$ is not a Banach space
(iii) $\{\mathbb{R}^n, d_p, 1 \le p < \infty\}$ is a Banach space
(iv) $\{\mathbb{C}^n, d_p, 1 \le p < \infty\}$ is a Banach space

Definition. An inner-product space that is complete is called a *Hilbert space*.

Every Hilbert space is a Banach space but the converse is not true.

Example 21.5.
(i) $\{\mathbb{R}^n, \langle \mathbf{u}, \mathbf{v}\rangle = \sum_{j=1}^n u_j v_j\}$ is a Hilbert space.
(ii) $\{\mathbb{R}^n, \langle \mathbf{u}, \mathbf{v}\rangle = \mathbf{v}^T\mathbf{G}\mathbf{u}, \mathbf{G}$ is positive definite$\}$ is a Hilbert space.
(iii) $\{\mathbb{C}^n, \langle \mathbf{u}, \mathbf{v}\rangle = \sum_{j=1}^n u_j \overline{v_j}\}$ is a Hilbert space.
(iv) $\{C[a,b], \langle f,g\rangle = \int_a^b f(x)g(x)\,dx\}$ is not a Hilbert space.
(v) $\{\mathcal{L}_2[a,b], \langle f,g\rangle = \int_a^b \rho(x)f(x)g(x)\,dx, \rho(x) > 0\}$ is a Hilbert space.

(vi) $\{\mathcal{L}_{2C}[a,b], \langle f,g\rangle = \int_a^b f(x)\overline{g(x)}\,dx\}$ is a Hilbert space of complex valued functions of a real variable.

Note that in the Hilbert space of example (v) above, the eigenfunction expansion

$$f(x) = \sum_{n=1}^{\infty} \langle f, y_n(x)\rangle y_n(x)$$

converges in the mean square sense w. r. t. the norm $\|\cdot\|_2$, i. e.,

$$\lim_{N\to\infty} \int_a^b \left\{ f(x) - \sum_{n=1}^{N} \langle f, y_n\rangle y_n \right\}^2 \rho(x)\,dx = 0$$

and when we write $f(x) \stackrel{\circ}{=} g(x)$, the two functions may disagree on a set of points having measure zero.

Definition. Let V be a normed linear space and U be a subset of V. Then we say that U is a *dense subset* of V if given $x \in V$, $\exists x_0 \in U$ such that

$$\|x - x_0\| < \varepsilon \quad \text{for every } \varepsilon > 0.$$

Example 21.6.

(i) The set of rational numbers is dense in the real line. Every irrational number can be approximated as closely as required by a rational number.

(ii) The set of polynomials is dense in $C[a,b]$ with the metric

$$d_\infty(f,g) = \sup_{a\le x\le b} |f(x) - g(x)|.$$

This is the well-known Weirstrass theorem.
Similarly, $C[a,b]$ is dense in $R[a,b]$ and $R[a,b]$ is dense in $\mathcal{L}[a,b]$. Thus, $C[a,b]$ is dense in $\mathcal{L}[a,b]$.

21.3.4 Convergence theorems for eigenfunction expansions

Theorem 21.1. *Let $f(x)$ be defined and continuous with two continuous derivatives on $[a,b]$ and $f(x)$ satisfies the same boundary conditions as the eigenfunctions $\{\rho(x); \phi_n(x), n = 1,2,\ldots\}$, then the eigenfunction expansion*

$$f(x) = \sum_{n=1}^{\infty} c_n \phi_n(x) \tag{21.37}$$

converges uniformly to $f(x)$ on $[a,b]$.

Theorem 21.2. *Let $f(x)$ be piecewise smooth on $[a,b]$. Then, for each x in $[a,b]$, the eigenfunction expansion converges, and*

$$\frac{f(x^+)+f(x^-)}{2} = \sum_{n=1}^{\infty} c_n\phi_n(x), \quad a < x < b \tag{21.38}$$

Mean square convergence
A sequence of functions

$$S_n(x) = \sum_{j=1}^{n} c_j\phi_j(x)$$

is said to converge to $f(x)$ in mean square relative to $\rho(x)$ on $[a,b]$ provided

$$\lim_{n\to\infty} \int_a^b [f(x) - S_n(x)]^2 \rho(x)\,dx = 0$$

21.3.5 Fourier series (eigenfunction expansions) and Parseval's theorem

Let $\{\rho(x); \phi_n(x), n = 1, 2, \ldots\}$ be the orthonormal set of eigenfunctions of a S–L problem. Let $f(x) \in \mathcal{L}_2[a,b]$ and

$$f(x) = \sum_{n=1}^{\infty} c_n\phi_n(x) \tag{21.39}$$

$$c_n = \langle f, \phi_n \rangle = \int_a^b f(x)\phi_n(x)\rho(x)\,dx. \tag{21.40}$$

Then

$$\|f\|^2 = \int_a^b \rho(x)f(x)^2\,dx = \sum_{n=1}^{\infty} c_n^2. \tag{21.41}$$

Equation (21.41) is a generalization of the orthogonal expansion from finite to infinite dimensions, and is known as Parseval's relation.

If the function $f(x)$ is such that $\|f\| = 1$, then Parseval's relation may be used to determine the number of terms needed in equation (21.41) so that

$$\left(\|f\|^2 - \sum_{j=1}^{n} c_j^2\right) < \varepsilon \tag{21.42}$$

where $\varepsilon > 0$ is the desired accuracy.

21.3.6 Example of Fourier series (eigenfunction expansions)

Example 21.7. Consider the function $f(x) = x(1-x)$ and EVP

$$
\begin{aligned}
u'' &= -\lambda u, \quad 0 < x < 1 \\
u(0) &= u(1) = 0
\end{aligned}
$$

⇒

$$
\lambda_n = n^2\pi^2, \quad u_n(x) = \sqrt{2}\sin n\pi x, \quad n = 1, 2, \ldots
$$

Thus, the function $f(x)$ can be expanded as

$$
f(x) = \sum_{j=1}^{\infty} c_j u_j(x)
$$

where

$$
\begin{aligned}
c_j = \langle f, u_j \rangle &= \int_0^1 f(\xi)\sqrt{2}\sin(j\pi\xi)\,d\xi \\
&= \int_0^1 x(1-x)\sqrt{2}\sin j\pi x\,dx \\
&= \frac{2\sqrt{2}}{j^3\pi^3}(1-\cos j\pi) = \begin{cases} 0 & \text{if } j \text{ is even} \\ \frac{4\sqrt{2}}{j^3\pi^3} & \text{if } j \text{ is odd} \end{cases}
\end{aligned}
$$

Thus, the Fourier series expansion of $x(1-x)$ can be expressed as

$$
\begin{aligned}
x(1-x) &= \frac{8}{\pi^3}\left[\sin \pi x + \frac{\sin 3\pi x}{27} + \frac{\sin 5\pi x}{125} + \cdots\right] \\
&= \frac{8}{\pi^3}\sum_{k=0}^{\infty}\frac{\sin(2k+1)\pi x}{(2k+1)^3} \\
&= \frac{8}{\pi^3}\sum_{k=0}^{N-1}\frac{\sin(2k+1)\pi x}{(2k+1)^3} + R_N(x) \qquad (21.43)
\end{aligned}
$$

where R_N is the remainder term

$$
R_N(x) = \frac{8}{\pi^3}\sum_{k=N}^{\infty}\frac{\sin(2k+1)\pi x}{(2k+1)^3}.
$$

Note that

$$|R_N(x)| = \left|\frac{8}{\pi^3}\sum_{k=N}^{\infty}\frac{\sin(2k+1)\pi x}{(2k+1)^3}\right|$$
$$\leq \frac{8}{\pi^3}\sum_{k=N}^{\infty}\frac{1}{(2k+1)^3}$$
$$< \frac{8}{\pi^3}\left|\int_{N-1}^{\infty}\frac{dN}{(2N+1)^3}\right|$$
$$\therefore \quad |R_N(x)| < \frac{8}{\pi^3}\frac{1}{4(2N-1)^2}$$

For $N = 3 \Longrightarrow$

$$|R_N(x)| < 0.00258$$

Thus,

$$S_3(x) = \frac{8}{\pi^3}\left[\sin \pi x + \frac{\sin 3\pi x}{27} + \frac{\sin 5\pi x}{125}\right]$$

approximates $f(x) = x(1-x)$ within an accuracy of 0.0026 for all x. Figure 21.3 shows plots of exact function $f(x)$ along with Fourier series expansion using first few terms.

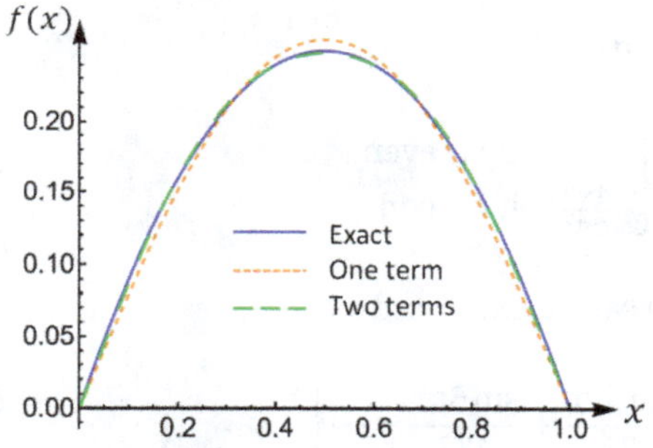

Figure 21.3: Plot of function $f(x) = x(1-x)$ and its representation with Fourier series expansion using only first term and first two terms.

In this case, the function $f(x)$ is twice differentiable and satisfies the same BCs as the eigenfunctions. Thus, the convergence is uniform.

Example 21.8. Consider the function

$$f(x) = \begin{cases} 1, & 0 \leq x \leq \frac{\pi}{2} \\ 0, & \frac{\pi}{2} < x \leq \pi. \end{cases}$$

Let us expand this function in terms of the eigenfunctions of the S–L problem

$$y'' = -\lambda y, \quad 0 \le x \le \pi$$
$$y'(0) = 0; \quad y'(\pi) = 0.$$

The eigenvalues are

$$\lambda_n = \begin{cases} 0, & n = 0 \\ n^2, & n = 1, 2, 3, \ldots \end{cases}$$

and the normalized eigenfunctions are

$$y_n(x) = \begin{cases} \frac{1}{\sqrt{\pi}}, & n = 0 \\ \sqrt{\frac{2}{\pi}} \cos nx, & n = 1, 2, 3, \ldots. \end{cases}$$

Thus, the Fourier series expansion of $f(x)$ can be expressed as

$$f(x) = \sum_{j=0}^{\infty} c_j y_j(x); \quad c_j = \langle f, y_j \rangle$$

$\Longrightarrow$

$$c_0 = \langle y_0, f \rangle = \frac{1}{\sqrt{\pi}} \int_0^{\pi} f(x)\, dx = \frac{1}{\sqrt{\pi}} \cdot \frac{\pi}{2} = \frac{\sqrt{\pi}}{2}$$

and

$$c_j = \langle y_j, f \rangle = \sqrt{\frac{2}{\pi}} \int_0^{\pi/2} \cos jx\, dx = \sqrt{\frac{2}{\pi}} \frac{\sin(\frac{j\pi}{2})}{j}$$

Thus,

$$c_0 = \frac{\sqrt{\pi}}{2}$$
$$c_{2k} = 0 \quad \text{and} \quad c_{2k+1} = (-1)^k \sqrt{\frac{2}{(2k+1)^2 \pi}}, \qquad k = 0, 1, 2, 3$$

$\Longrightarrow$

$$\begin{aligned} f(x) &= \frac{\sqrt{\pi}}{2} \cdot \frac{1}{\sqrt{\pi}} + \sum_{k=0}^{\infty} \sqrt{\frac{2}{\pi}} \frac{(-1)^k}{(2k+1)} \cdot \sqrt{\frac{2}{\pi}} \cos(2k+1)x \\ &= \frac{1}{2} + \frac{2}{\pi} \sum_{k=0}^{\infty} \frac{(-1)^k}{(2k+1)} \cos(2k+1)x. \end{aligned} \qquad (21.44)$$

Figure 21.4 shows the plot of this function and representation from Fourier series expansion with one, three, five and hundred terms. Note the Gibb's phenomena of over and

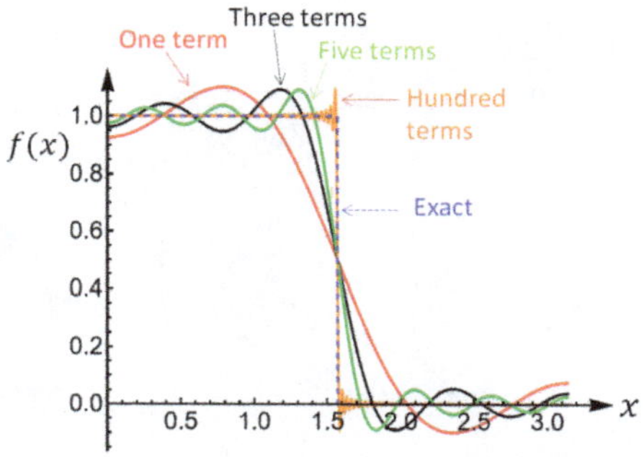

Figure 21.4: Representation with Fourier series expansion using only first 1, 3, 5 and 100 terms, demonstrating the Gibb's phenomena of over and undershoot at the point of discontinuities.

undershoot at the point of discontinuity. In this example, the function $f(x)$ is not differentiable at $x = \frac{\pi}{2}$. Hence, the eigenfunction expansion converges in the mean square norm.

Parseval's relation gives

$$\sum_{k=0}^{\infty} c_k^2 = \int_a^b f(x)^2 \rho(x)\, dx$$

$\Longrightarrow$

$$\frac{\pi}{4} + \sum_{k=0}^{\infty} \frac{2}{\pi} \cdot \frac{1}{(2k+1)^2} = \int_0^{\pi/2} dx = \frac{\pi}{2}$$

$\Longrightarrow$

$$\frac{2}{\pi} \sum_{k=0}^{\infty} \frac{1}{(2k+1)^2} = \frac{\pi}{4}$$

or

$$\sum_{k=0}^{\infty} \frac{1}{(2k+1)^2} = \frac{\pi^2}{8} \tag{21.45}$$

If we set $x = 0$ on both sides of equation (21.44), we get

$$1 = \frac{1}{2} + \frac{2}{\pi} \sum_{k=0}^{\infty} \frac{(-1)^k}{(2k+1)}$$

$\Longrightarrow$

$$\frac{\pi}{4} = \sum_{k=0}^{\infty} \frac{(-1)^k}{(2k+1)} = 1 - \frac{1}{3} + \frac{1}{5} - \frac{1}{7} + \frac{1}{9} - \cdots \tag{21.46}$$

Many such relations (equations (21.45) and (21.46)) can be derived using eigenfunction expansions.

Example 21.9 (Eigenfunction expansion in two variables). Consider the function $f(x) = x(1-x)y(1-y)$ and EVP

$$\frac{\partial^2 u}{\partial x^2} + \frac{\partial^2 u}{\partial y^2} = -\lambda u, \quad 0 < x < 1, \quad 0 < y < 1$$
$$u(0,y) = u(1,y) = u(x,0) = u(x,1) = 0$$

$\Rightarrow$

$$\lambda_n = (n^2 + m^2)\pi^2, \quad u_{nm}(x) = 2 \sin n\pi x \sin m\pi y, \quad n = 1, 2, \ldots, \quad m = 1, 2, \ldots$$

Thus, the function $f(x)$ can be expanded as

$$f(x,y) = \sum_{i=1}^{\infty}\sum_{j=1}^{\infty} c_{ij} u_{ij}(x,y)$$

where

$$\begin{aligned}
c_{ij} = \langle f, u_{ij}\rangle &= \int_0^1 f(x',y') 2 \sin(i\pi x') \sin(j\pi y')\, dx'\, dy' \\
&= \left(\int_0^1 x'(1-x')\sqrt{2} \sin i\pi x'\, dx'\right) \cdot \left(\int_0^1 y'(1-y')\sqrt{2} \sin j\pi y'\, dy'\right) \\
&= \left[\frac{2\sqrt{2}}{i^3\pi^3}(1 - \cos i\pi)\right] \cdot \left[\frac{2\sqrt{2}}{j^3\pi^3}(1 - \cos j\pi)\right] = \begin{cases} 0 & \text{if } i \text{ or } j \text{ is even} \\ \frac{32}{i^3 j^3 \pi^6} & \text{if } i \text{ and } j \text{ are odd} \end{cases}
\end{aligned}$$

Thus, the Fourier series expansion of $x(1-x)y(1-y)$ can be expressed as

$$\begin{aligned}
&x(1-x)y(1-y) \\
&= \frac{64}{\pi^6} \sum_{k=0}^{\infty}\sum_{l=0}^{\infty} \frac{\sin(2k+1)\pi x \sin(2l+1)\pi y}{(2k+1)^3(2l+1)^3} \\
&= \frac{64}{\pi^6} \left[\sum_{k=0}^{\infty} \frac{\sin(2k+1)\pi x}{(2k+1)^3}\right]\left[\sum_{l=0}^{\infty} \frac{\sin(2l+1)\pi y}{(2l+1)^3}\right] \\
&= \frac{64}{\pi^6}\left[\sin \pi x + \frac{\sin 3\pi x}{27} + \frac{\sin 5\pi x}{125} + \cdots\right]\left[\sin \pi y + \frac{\sin 3\pi y}{27} + \frac{\sin 5\pi y}{125} + \cdots\right]
\end{aligned} \tag{21.47}$$

Figure 21.5 shows the plots of exact function $f(x,y)$ along with 2D Fourier series expansion using first few terms. The convergence here is uniform.

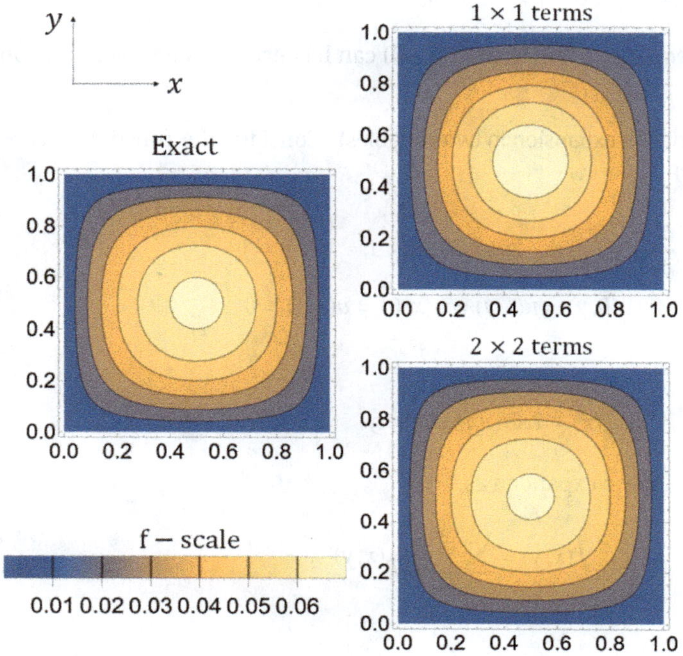

Figure 21.5: Plot of function $f(x,y) = x(1-x)y(1-y)$ and its representation with Fourier series expansion using 1×1 and 2×2 terms.

21.3.7 Fourier series (eigenfunction expansion) of the Green's function

Consider the S–L problem

$$(p(x)y')' - q(x)y = -\lambda\rho(x)y, \quad a < x < b \tag{21.48}$$

$$y(a) = 0, \quad y(b) = 0 \tag{21.49}$$

Let $\{\lambda_i\}$, $\{y_i(x)\}$, $i = 1, 2, 3, \ldots$ be the eigenvalues and normalized eigenfunctions, respectively. Let $G(x,s)$ be the Green's function. We have seen that equations (21.48)–(21.49) may be written as

$$y(x) = \lambda \int_a^b G(x,s)\rho(s)y(s)\,ds \tag{21.50}$$

Thus,

$$\frac{y_i(x)}{\lambda_i} = \int_a^b G(x,s)\rho(s)y_i(s)\,ds \tag{21.51}$$

Now, consider the expansion of $G(x,s)$ (considered as a function of s) in terms of the eigenfunctions $\{y_i(s)\}$

$$G(x,s) = \sum_{i=1}^{\infty} c_i y_i(s)$$

$\Rightarrow$

$$c_i = \langle G(x,s), y_i(s)\rangle = \int_a^b \rho(s)G(x,s)y_i(s)\,ds = \frac{y_i(x)}{\lambda_i}$$

Thus, we obtain Mercer's expansion

$$G(x,s) = \sum_{i=1}^{\infty} \frac{y_i(x)y_i(s)}{\lambda_i}. \tag{21.52}$$

Now consider parseval's equation

$$\sum_{i=1}^{\infty} c_i^2 = \int_a^b \rho(s)f(s)^2\,ds$$

Let $f(s) = G(x,s)$

$\Longrightarrow$

$$\sum_{i=1}^{\infty} \frac{y_i^2(x)}{\lambda_i^2} = \int_a^b G(x,s)^2\rho(s)\,ds$$

Multiply both sides by $\rho(x)$ and integrate from a to b and use the relation:

$$\int_a^b \rho(x)y_i^2(x)\,dx = 1$$

$\Longrightarrow$

$$\sum_{i=1}^{\infty} \frac{1}{\lambda_i^2} = \int_a^b \left[\int_a^b G(x,s)^2\rho(s)\,ds\right]\rho(x)\,dx \tag{21.53}$$

Since $G(x,s)$ is continuous, the integral on the RHS of equation (21.53) is finite. Therefore, $\sum_{i=1}^{\infty}\frac{1}{\lambda_i^2}$ converges:

$\Longrightarrow$

$$\frac{1}{\lambda_i^2} \to 0 \quad \text{for } i \to \infty \Longrightarrow \lambda_i^2 \to \infty \text{ as } i \to \infty$$

Example 21.10.

$$\begin{aligned} y'' &= -\lambda y, \quad 0 < x < 1 \\ y(0) &= y(1) = 0 \\ G(x,s) &= \begin{cases} s(1-x) & 0 < s < x \\ x(1-s) & x < s < 1 \end{cases} \end{aligned}$$

$\Rightarrow$

$$\begin{aligned} &\int_0^1 \left[\int_0^1 G(x,s)^2 \rho(s)\, ds \right] \rho(x)\, dx \\ &= \int_0^1 \left[\int_0^1 G(x,s)^2\, ds \right] dx \\ &= \int_0^1 \left[\int_0^x s^2(1-x)^2\, ds + \int_x^1 x^2(1-s)^2\, ds \right] dx \\ &= \int_0^1 \left[\frac{x^3(1-x)^2}{3} + \frac{x^2(1-x)^3}{3} \right] dx \\ &= \frac{1}{3} \int_0^1 x^2(1-x)^2\, dx \\ &= \frac{1}{90} \end{aligned}$$

Eigenvalues

$$\lambda_n = n^2\pi^2, \quad n = 1, 2, 3, \ldots$$

$\Longrightarrow$

$$\begin{aligned} \sum_{i=1}^{\infty} \frac{1}{\lambda_i^2} &= \int_0^1 \int_0^1 G(x,s)^2\, ds\, dx \\ \sum_{n=1}^{\infty} \frac{1}{n^4\pi^4} &= \frac{1}{90} \\ \Longrightarrow \sum_{n=1}^{\infty} \frac{1}{n^4} &= \frac{\pi^4}{90}. \end{aligned}$$

Problems

1. Consider the eigenvalue problem

$$y'' = -\lambda y, \quad 0 < x < 1$$
$$y'(0) = 0, \quad y'(1) + \mathrm{Bi}\, y(1) = 0 \quad (\mathrm{Bi} = \text{ Biot number}),$$

which arises in solving the unsteady-state heat (mass) diffusion problem in a flat plate.

(a) Show that the eigenvalue problem is self-adjoint.

(b) Determine the first eigenvalue as a function of the Biot number, i. e., compute λ_1 for Bi = 0.01, 0.02, 0.05, 0.1, 0.2, 0.5, 1.0, 2.0, 5.0, 10.0, 20.0, 50.0, 100.0. Identify the two asymptotes.

(c) Determine the eigenvalues and eigenfunctions for the two limiting cases of no external resistance ($\mathrm{Bi} \to \infty$) and no internal resistance ($\mathrm{Bi} \to 0$).

2. Consider the eigenvalue problem

$$\frac{1}{\mathrm{Pe}} y'' - y' = -\lambda y, \quad 0 < x < 1$$
$$\frac{1}{\mathrm{Pe}} y'(0) - y(0) = 0, \quad y'(1) = 0 \quad (\mathrm{Pe} = \text{Peclet number}),$$

which arises in solving unsteady-state diffusion–convection–reaction problems with Danckwert's boundary conditions

(a) Show that the substitutions

$$y = w \exp\left(\frac{x\,\mathrm{Pe}}{2}\right), \quad \Lambda = \lambda\,\mathrm{Pe} - \frac{\mathrm{Pe}^2}{4}$$

transform the eigenvalue problem to a self-adjoint form

$$w'' = -\Lambda w; \quad w'(0) - \frac{\mathrm{Pe}}{2} w(0) = 0, \quad w'(1) + \frac{\mathrm{Pe}}{2} w(1) = 0$$

(b) Determine the first eigenvalue as a function of the Peclet number, i. e., compute λ_1 for Pe = 0.01, 0.1, 1, 10, 100. Identify the two asymptotes.

(c) Determine and sketch the first eigenfunction for different values of the Peclet number.

3. Consider the Schrodinger equation in one spatial dimension

$$-\frac{h^2}{8\pi^2 m}\frac{d^2\psi}{dx^2} + V(x)\psi(x) = E\psi(x)$$

where $\psi(x)$ is the wave function. E is the total energy of the particle (electron) and $V(x)$ is the potential energy of the particle at position x. For a square well potential, $V(x)$ is zero for $0 < x < L$ and is infinity outside this region. The appropriate boundary conditions for the wave function are

$$\psi(0) = \psi(L) = 0$$

(a) Determine the permitted energy levels for the particle. What is the separation between neighboring quantum levels?

(b) Sketch the first five wave functions.

4. The following fourth-order eigenvalue problem arises in the stability analysis of a fluid filled porous medium confined between two parallel plates kept at different temperatures (Lapwood convection):

$$\begin{aligned} \phi'' - k^2\phi + \psi - \lambda\phi &= 0 \\ \psi'' - k^2\psi + \mathrm{Ra}\, k^2\phi &= 0 \\ \phi(0) = \phi(1) &= 0 \\ \psi(0) = \psi(1) &= 0 \end{aligned}$$

Here, k is the wave number and Ra is the Rayleigh number. Show that the eigenvalues are real.

5. Consider the eigenvalue problem

$$y'' + \lambda y = 0, \quad 0 < x < \pi; \quad y'(0) = 0, \quad y'(\pi) = 0$$

(a) Determine the eigenvalues and corresponding orthonormal set of eigenfunctions.

(b) Suppose that $f(x)$ has continuous first and second derivatives in $[0, \pi]$ and satisfies the boundary conditions. Show that the eigenfunction expansion converges to $f(x)$ uniformly on $[0, \pi]$.

(c) Determine the expansion of the function $f(x) = x^2(\pi - x)^2$.

6. Eigenfunctions satisfying four boundary conditions appear frequently in the solution of many transport and reaction problems.

(a) Find an orthonormal set of eigenfunctions, which vanish along with their first derivative at the end points of the interval $[-1, 1]$. Sketch the first three eigenfunctions.

(b) Determine formulas for the expansion of an arbitrary function $f(x)$ in terms of the above eigenfunctions.

7. Consider the eigenvalue problem

$$y'' = -\lambda xy, \quad y(0) = 0, \quad y(1) = 0$$

(a) Compute an approximate first eigenvalue using the Rayleigh quotient.

(b) Apply the Rayleigh method to the inverse operator G to show that

$$\lambda_1 = \mathrm{Min} \frac{\langle y, y \rangle}{\langle \mathrm{G}y, y \rangle},$$

and hence compute a better approximation to λ_1.

(c) This equation may be transformed into a Bessel's equation and eigenvalues obtained exactly. Show that the eigenvalues are the roots of the equation

$$J_{1/3}\left(\frac{2}{3}\sqrt{\lambda}\right) = 0$$

Find the eigenvalue from appropriate tables (Abromowitz and Stegun [2]).

8. (a) Let $C^1[a,b]$ be the space of continuously differentiable functions on $[a,b]$ with the norm

$$\|f(x)\| = \text{Sup}\{|f(x)| + |f'(x)|\}; \quad a \le x \le b$$

Is the normed space complete (Banach space)?

(b) Repeat (a) for the norm

$$\|f(x)\| = \sqrt{\int_a^b (|f(x)|^2 + |f'(x)|^2)\,dx}$$

(c) Consider the following inner product in $C^1[a,b]$:

$$\langle f, g\rangle = \int_a^b [f(x)g(x) + f'(x)g'(x)]\,dx$$

Identify the missing elements that need to be added to $C^1[a,b]$ to make it a Hilbert space (the resulting Hilbert space is called a Sobolev space).

22 Introduction to the solution of linear integral equations

22.1 Introduction

An integral equation (IE) is an equation in which the unknown function appears under one or more integrals. IEs appear in applications that deal with particulate processes or population balances. In addition, as we have seen in previous chapters, the solution of many initial and boundary value problems may be expressed in terms of integral equations. This chapter is a brief introduction to the theory of linear integral equations in one dependent and one independent variable.

The general form of a linear IE is

$$h(x)u(x) = f(x) + \lambda \int_a^x K(x,s)u(s)\,ds, \tag{22.1}$$

where $u(x)$ is the unknown function; $h(x), f(x)$ and $K(x,s)$ are known functions and $\lambda(\neq 0)$ is a real or complex parameter. If $f(x) = 0$, the IE is called homogeneous. If the upper limit of integration is fixed (e. g., $x = b$), then it is called a *Fredholm equation*. Otherwise, it is called a *Volterra equation*. If the unknown function $u(x)$ appears only under the integral sign (e. g., $h(x) = 0$), it is called the *integral equation of the first kind*. If it appears both inside and outside of the integral sign, it is called the *integral equation of the second kind*.

Examples.

1.

$$f(x) = \int_a^b K(x,s)u(s)\,ds \tag{22.2}$$

is a Fredholm equation of the first kind.

2.

$$u(x) = f(x) + \lambda \int_a^b K(x,s)u(s)\,ds \tag{22.3}$$

is a Fredholm equation of the second kind.

3.

$$f(x) = \int_a^x K(x,s)u(s)\,ds \tag{22.4}$$

https://doi.org/10.1515/9783111598055-026

is a Volterra equation of the first kind.

4.

$$u(x) = f(x) + \lambda \int_a^x K(x,s)u(s)\,ds \tag{22.5}$$

is a Volterra equation of the second kind.

In these equations, $K(x,s)$ is called the *kernel*. When one or both limits of integration become infinity or when the kernel becomes infinity within the range of integration, the IE is called a *singular integral equation*.

For example, the Laplace transform defined by

$$\hat{u}(s) = \int_0^\infty e^{-st}u(t)\,dt \tag{22.6}$$

is a singular IE of the first kind with kernel $K(t,s) = e^{-st}$. The Fourier transform defined by

$$\hat{f}(\alpha) = \int_{-\infty}^{\infty} e^{-i\alpha x}f(x)\,dx \tag{22.7}$$

is also a singular integral equation of the first kind with kernel $K(x,\alpha) = e^{-i\alpha x}$. The other types of kernels that are of interest are

(i) Separable kernel: $K(x,s) = \sum_{j=1}^{n} a_j(x)b_j(s)$.
(ii) Symmetric kernel: $K(x,s) = K(s,x)$.
(iii) Convolution kernel: $K(x,s) = h(x-s)$.

In the next two sections, we review and outline the procedure for converting initial and boundary value problems into integral equations.

22.2 Transformation of an IVP into an IE of Volterra type

Consider the linear second-order IVP

$$\frac{d^2u}{dt^2} + a_1(t)\frac{du}{dt} + a_2(t)u = g(t), \quad t > 0 \tag{22.8}$$

$$u(0) = \alpha_0, \quad u'(0) = \alpha_1 \tag{22.9}$$

and assume that the functions $a_i(t)$ are continuous. Define

$$\frac{d^2u}{dt^2} = h(t) \tag{22.10}$$

and integrate once to obtain

$$\frac{du}{dt} = a_1 + \int_0^t h(s)\,ds \tag{22.11}$$

Integrating again gives

$$u(t) = a_0 + a_1 t + \int_0^t \int_0^{t'} h(s)\,ds\,dt'. \tag{22.12}$$

Changing the order of integration in the double integral (in equation (22.12)) and simplifying gives

$$u(t) = a_0 + a_1 t + \int_0^t (t-s)h(s)\,ds \tag{22.13}$$

Now, multiplying $\frac{d^j u}{dt^j}$ by $a_{2-j}(t)$ and summing from $j = 0$ to 2 with $a_0(t) = 1$ gives

$$h(t) = f(t) + \int_0^t K(t,s)h(s)\,ds \tag{22.14}$$

where

$$\begin{aligned} f(t) &= g(t) - \alpha_1 a_1(t) - (\alpha_0 + \alpha_1 t)a_2(t) \\ K(t,s) &= -a_1(t) - (t-s)a_2(t). \end{aligned}$$

Once $h(t)$ is known, we can determine $u(t)$ from equation (22.13). Thus, the IVP is reduced to a Volterra integral equation of the second kind.

The above procedure can be extended to the n-th order IVP. In this case, equation (22.13) becomes

$$u(t) = \alpha_0 + \alpha_1 t + \cdots + \alpha_{n-1}\frac{t^{n-1}}{(n-1)!} + \int_0^t \frac{(t-s)^{n-1}}{(n-1)!}h(s)\,ds \tag{22.15}$$

and equation (22.14) remains the same with

$$f(t) = g(t) - \alpha_{n-1}a_1(t) - \cdots - \left[\alpha_0 + \alpha_1 t + \cdots + \alpha_{n-1}\frac{t^{n-1}}{(n-1)!}\right]a_n(t) \tag{22.16}$$

$$K(t,s) = -\sum_{j=1}^{n} \frac{(t-s)^{j-1}}{(j-1)!} a_j(t) \tag{22.17}$$

We note that for the special case of homogeneous initial conditions ($\alpha_j = 0$), $f(t) = g(t)$ and the kernel depends only on the coefficient functions $a_j(t)$. Further, for the special case in which $a_j(t)$ are constants, the kernel is of convolution type.

22.3 Transformation of TPBVP into an IE of Fredholm type

We have already shown (in Chapter 19) that the n-th order TPBVP

$$Lu = -f(x), \quad a < x < b \tag{22.18}$$

$$\mathbf{W}_a \mathbf{k}(u(a)) + \mathbf{W}_b \mathbf{k}(u(b)) = \mathbf{0} \tag{22.19}$$

may be transformed to an integral equation of the form

$$u(x) = \int_a^b G(x,s) f(s)\, ds \tag{22.20}$$

where $G(x,s)$ is the Green's function. We note that equation (22.20) is valid when $f(x)$ is replaced by a more general (and possibly nonlinear) source term of the form $h(x, u(x))$, in which case equation (22.20) becomes a nonlinear IE of the form

$$u(x) = \int_a^b G(x,s) h(s, u(s))\, ds \tag{22.21}$$

Thus, two-point BVPs can be transformed into Fredholm integral equations with the kernel being the Green's function. We have also seen that for the special case in which the homogeneous two-point BVP is self-adjoint, the Green's function (kernel) is symmetric. The kernel can also be made symmetric for the more general case in which the weight function in the inner product is not unity. For example, the Sturm–Liouville eigenvalue problem

$$\frac{d}{dx}\left(p(x)\frac{du}{dx}\right) - q(x)u(x) = -\lambda \rho(x) u(x), \quad a < x < b \tag{22.22}$$

$$u(a) = 0, \quad u(b) = 0 \tag{22.23}$$

can be converted to an integral equation

$$u(x) = \lambda \int_a^b G(x,s) \rho(s) u(s)\, ds. \tag{22.24}$$

However, the kernel $G(x,s)\rho(s)$ is not symmetric when $\rho(s)$ is not unity. By defining

$$\phi(x) = \sqrt{\rho(s)}u(x), \tag{22.25}$$

Equation (22.24) may be written as a homogeneous Fredholm equation with symmetric kernel:

$$\phi(x) = \lambda \int_a^b K(x,s)\phi(s)\,ds \tag{22.26}$$

where

$$K(x,s) = G(x,s)\sqrt{\rho(x)}\sqrt{\rho(s)}. \tag{22.27}$$

This is possible since the density function $\rho(x)$ is strictly positive in (a,b).

22.4 Solution of Fredholm integral equations with separable kernels

Consider the Fredholm IE of the second kind as given by

$$u(x) = f(x) + \lambda \int_a^b K(x,s)u(s)\,ds \tag{22.28}$$

We consider the case of separable kernel as the solution procedure for this case may be related to that of linear algebraic equations. Express the kernel as

$$K(x,s) = \sum_{i=1}^{N} a_i(x)b_i(s). \tag{22.29}$$

Without loss of generality, we assume that the functions $a_i(x)$ and $b_i(s)$ are linearly independent. If they are not, we can combine the terms and reduce the number of terms in the summation.

22.4.1 Homogeneous equation

We first consider the homogeneous Fredholm IE, i. e., equation (22.28) with $f(x) = 0$, with separable kernel (equation (22.29)), which can be expressed as

$$u(x) = \lambda \int_a^b \left(\sum_{i=1}^{N} a_i(x)b_i(s) \right) u(s)\,ds. \tag{22.30}$$

Interchanging the summation and integration gives

$$u(x) = \lambda \sum_{i=1}^{N} a_i(x) \int_a^b b_i(s)u(s)\,ds. \tag{22.31}$$

Let

$$c_i = \int_a^b b_i(s)u(s)\,ds \tag{22.32}$$

then equation (22.31) implies

$$u(x) = \lambda \sum_{j=1}^{N} c_j a_j(x). \tag{22.33}$$

To determine c_i, we substitute equation (22.33) in equation (22.32), which leads to

$$\begin{aligned} c_i &= \int_a^b b_i(s)\lambda \sum_{j=1}^{N} c_j a_j(s)\,ds \\ &= \lambda \sum_{j=1}^{N} \left(\int_a^b b_i(s)a_j(s)\,ds \right) c_j \\ \Rightarrow c_i &= \lambda \sum_{j=1}^{N} A_{ij} c_j \end{aligned} \tag{22.34}$$

where

$$A_{ij} = \int_a^b a_j(s)b_i(s)\,ds, \quad i = 1, 2, \ldots, N \text{ and } j = 1, 2, \ldots, N \tag{22.35}$$

Thus, equation (22.34) and equation (22.35) lead to

$$\mathbf{c} = \lambda \mathbf{A}\mathbf{c}$$

or the homogeneous linear algebraic equations

$$(\mathbf{I} - \lambda \mathbf{A})\mathbf{c} = \mathbf{0}. \tag{22.36}$$

Let

$$D(\lambda) = |\mathbf{I} - \lambda \mathbf{A}| = \det(\mathbf{I} - \lambda \mathbf{A}). \tag{22.37}$$

If $D(\lambda) \neq 0$, then the only solution to equation (22.36) is the trivial one, i. e., $\mathbf{c} = \mathbf{0}$, which implies $u(x) \equiv 0$ is the only solution to the homogeneous equation (22.30). The λ-values for which $D(\lambda) = 0$ are called the eigenvalues of the kernel. There are at most N of them. The nontrivial solution $\mathbf{c}$ corresponding to an eigenvalue gives a nontrivial $u(x) = \lambda\sum_{j=1}^{N} c_j a_j(x)$. These are the *eigenfunctions of the kernel.*

22.4.2 Inhomogeneous equation

Now, consider the inhomogeneous Fredholm IE (equation (22.28)) with separable kernel (equation (22.29)), which can be expressed as

$$u(x) = f(x) + \lambda \int_a^b \left[\sum_{i=1}^{N} a_i(x) b_i(s) \right] u(s)\, ds \tag{22.38}$$

$\Rightarrow$

$$u(x) = f(x) + \lambda \sum_{i=1}^{N} a_i(x) c_i \tag{22.39}$$

where

$$\begin{aligned} c_i &= \int_a^b b_i(s) u(s)\, ds \\ &= \int_a^b b_i(s) \left[f(s) + \lambda \sum_{j=1}^{N} a_j(s) c_j \right] ds \end{aligned}$$

$\Rightarrow$

$$\mathbf{c} = \mathbf{f} + \lambda \mathbf{A}\mathbf{c} \tag{22.40}$$

where

$$f_i = \int_a^b b_i(s) f(s)\, ds, \quad i = 1, 2, \dots, N$$

Equation (22.40) can be expressed as

$$(\mathbf{I} - \lambda \mathbf{A})\mathbf{c} = \mathbf{f}. \tag{22.41}$$

If $D(\lambda) \neq 0$, then equation (22.40) can be inverted as

$$\mathbf{c} = (\mathbf{I} - \lambda \mathbf{A})^{-1} \mathbf{f}$$

$\Rightarrow$

$$c_i = \frac{1}{D(\lambda)} \sum_{j=1}^{N} D_{ij}(\lambda) f_j \tag{22.42}$$

where $D_{ij}(\lambda) = (i,j)^{\text{th}}$ element of the classical adjoint of $(\mathbf{I} - \lambda\mathbf{A})$, i. e., matrix of cofactors. Substituting equation (22.42) into equation (22.39) gives

$$\begin{aligned} u(x) &= f(x) + \lambda \sum_{i=1}^{N} \frac{a_i(x)}{D(\lambda)} \left(\sum_{j=1}^{N} D_{ij}(\lambda) f_j \right) \\ &= f(x) + \lambda \sum_{i=1}^{N} \sum_{j=1}^{N} \frac{a_i(x)}{D(\lambda)} D_{ij}(\lambda) \int_a^b b_j(s) f(s)\, ds \end{aligned}$$

$\Rightarrow$

$$u(x) = f(x) + \lambda \int_a^b \Gamma(x,s,\lambda) f(s)\, ds \tag{22.43}$$

where

$$\Gamma(x,s,\lambda) = \sum_{i=1}^{N} \sum_{j=1}^{N} \frac{a_i(x) D_{ij}(\lambda) b_j(s)}{D(\lambda)} \tag{22.44}$$

is called the *resolvent kernel*. Thus, when the kernel is separable and $D(\lambda) \neq 0$, the solution of the Fredholm equation of the second kind is given by equations (22.43) and (22.44). Further, it can be shown that the solution in this case is unique.

Example 22.1. Consider the equation

$$u(x) = f(x) + \lambda \int_0^1 (x+s) u(s)\, ds.$$

Here,

$$\begin{aligned} K(x,s) &= (x+s) \\ &= x.1 + 1.s \\ &= a_1(x) b_1(s) + a_2(x) b_2(s) \end{aligned}$$

Thus,

$$A_{ij} = \int_a^b a_j(s) b_i(s)\, ds$$

leads to

$$A_{11} = \int_0^1 s.1\,ds = \frac{1}{2}, \quad A_{12} = \int_0^1 1.1\,ds = 1,$$

$$A_{21} = \int_0^1 s.s\,ds = \frac{1}{3}, \quad A_{22} = \int_0^1 1.s\,ds = \frac{1}{2},$$

$\Rightarrow$

$$\mathbf{A} = \begin{pmatrix} \frac{1}{2} & 1 \\ \frac{1}{3} & \frac{1}{2} \end{pmatrix} \Rightarrow (\mathbf{I} - \lambda\mathbf{A}) = \begin{pmatrix} 1 - \frac{\lambda}{2} & -\lambda \\ -\frac{\lambda}{3} & 1 - \frac{\lambda}{2} \end{pmatrix}$$

$$\Rightarrow D(\lambda) = 1 - \lambda - \frac{\lambda^2}{12}$$

$\Rightarrow$

$$\mathbf{c} = \frac{1}{1 - \lambda - \frac{\lambda^2}{12}} \begin{pmatrix} 1 - \frac{\lambda}{2} & \lambda \\ \frac{\lambda}{3} & 1 - \frac{\lambda}{2} \end{pmatrix} \begin{pmatrix} f_1 \\ f_2 \end{pmatrix}$$

$$\Rightarrow \begin{pmatrix} c_1 \\ c_2 \end{pmatrix} = \frac{1}{12 - 12\lambda - \lambda^2} \begin{pmatrix} 12\lambda f_2 + (12 - 6\lambda) f_1 \\ 4\lambda f_1 + (12 - 6\lambda) f_2 \end{pmatrix}$$

where

$$f_1 = \int_0^1 f(s)\,ds \quad \text{and} \quad f_2 = \int_0^1 s f(s)\,ds$$

Thus, from equation (22.39),

$$\begin{aligned} u(x) &= f(x) + \lambda(x c_1 + c_2) \\ &= f(x) + \lambda \int_0^1 \Gamma(x, s, \lambda) f(s)\,ds \end{aligned}$$

where

$$\Gamma(x, s, \lambda) = \frac{6(\lambda - 2)(x + s) - 12\lambda x s - 4\lambda}{\lambda^2 + 12\lambda - 12}.$$

We note that $D(\lambda) = 0 \Rightarrow \lambda_{1,2} = -6 \pm 4\sqrt{3}$. For these values of λ, the homogeneous equation has nontrivial solutions. Hence, the above solution is valid only if $\lambda \neq \lambda_1, \lambda_2$.

To determine the nontrivial solutions (or eigenfunctions) of the homogeneous equation for $\lambda = \lambda_1, \lambda_2$, we have

$$(\mathbf{I} - \lambda_1 \mathbf{A})\mathbf{c} = \mathbf{0} \Rightarrow \mathbf{c}_1 = 1, \quad c_2 = \sqrt{3}$$

$$(\mathbf{I} - \lambda_2 \mathbf{A})\mathbf{c} = \mathbf{0} \Rightarrow \mathbf{c}_1 = 1, \quad c_2 = -\sqrt{3}$$

$\Rightarrow$

$$\begin{aligned} u_1(x) &= \lambda_1 \sum_{j=1}^{2} c_j a_j(x) = \lambda_1(\sqrt{3}x + 1) \\ u_2(x) &= \lambda_2(-\sqrt{3}x + 1) \end{aligned}$$

are eigenfunctions (or any multiple of these).

Example 22.2. Consider the equation

$$u(x) = f(x) + \lambda \int_{-1}^{1} (xs + x^2 s^2) u(s)\, ds.$$

Here,

$$\begin{aligned} K(x,s) &= xs + x^2 s^2 \\ &= a_1(x)b_1(s) + a_2(x)b_2(s) \end{aligned}$$

$\Rightarrow$

$$\mathbf{A} = \begin{pmatrix} \frac{2}{3} & 0 \\ 0 & \frac{2}{5} \end{pmatrix}$$

$$\Rightarrow c_1 = \frac{f_1}{1 - \frac{2}{3}\lambda}, \quad c_2 = \frac{f_2}{1 - \frac{2}{5}\lambda}$$

where

$$f_1 = \int_{-1}^{1} s f(s)\, ds \quad \text{and} \quad f_2 = \int_{-1}^{1} s^2 f(s)\, ds$$

$\Rightarrow$

$$u(x) = f(x) + \lambda \int_{-1}^{1} \Gamma(x,s,\lambda) f(s)\, ds$$

where

$$\Gamma(x,s,\lambda) = \frac{sx}{1 - \frac{2\lambda}{3}} + \frac{s^2 x^2}{1 - \frac{2\lambda}{5}}$$

is the resolvent kernel. Thus, the solution is unique for any λ except for $\lambda = 3/2$ or $\lambda = 5/2$. For $\lambda = 3/2$, $u_1(x) = x$ or for $\lambda = 5/2$, $u_2(x) = x^2$ is a solution to the homogeneous equation.

22.5 Solution procedure for Volterra integral equations of the second kind

Recall that a Volterra integral equation (VIE) of the second kind is given by

$$u(t) = f(t) + \lambda \int_a^t K(t,s)u(s)\,ds, \tag{22.45}$$

where $K(t,s)$ is the kernel of the equation and $f(t)$ is a continuous function and λ is a parameter. There exist various methods for solving equation (22.45). We discuss here two of these.

22.5.1 Method of successive approximation

Let $u_0(t)$ be the initial guess, then VIE of the second kind (equation (22.45)) gives the sequence $u_0(t), u_1(t), \ldots, u_n(t)$ that are expressed by

$$u_1(t) = f(t) + \lambda \int_0^t K(t,s)u_0(s)\,ds \tag{22.46}$$

$$u_2(t) = f(t) + \lambda \int_0^t K(t,s)u_1(s)\,ds \tag{22.47}$$

and so on. Thus, they satisfy the following recurrence relation:

$$u_n(t) = f(t) + \lambda \int_0^t K(t,s)u_{n-1}(s)\,ds, \quad n = 1, 2, \ldots \tag{22.48}$$

Let

$$u(t) = \lim_{n\to\infty} u_n(t)$$

when it exists. In the so-called *Picard's method*, $u_0(t) = f(t)$, and the recurrence relation leads to

$$u_0(t) = f(t) \tag{22.49}$$

$$u_1(t) = f(t) + \lambda \int_0^t K(t,s)f(s)\,ds \tag{22.50}$$

$$u_2(t) = f(t) + \lambda \int_0^t K(t,s)\left[f(s) + \lambda \int_0^s K(s,s')f(s')\,ds'\right] ds$$
$$= f(t) + \lambda \int_0^t K(t,s)f(s)\,ds + \lambda^2 \int_0^t \int_0^s K(t,s)K(s,s')f(s')\,ds'\,ds \tag{22.51}$$

$\Rightarrow$

$$u_2 - u_1 = \lambda^2 \int_0^t \int_0^s K(t,s)K(s,s')f(s')\,ds'\,ds$$

which, after interchanging the order of integration and simplifying further, gives

$$u_2 - u_1 = \lambda^2 \int_0^t f(s')\left[\int_{s'}^t K(t,s)K(s,s')\,ds\right] ds'$$
$$= \lambda^2 \int_0^t K_2(t,s')f(s')\,ds'$$

where

$$K_2(t,s') = \int_{s'}^t K(t,s)K(s,s')\,ds$$

Similarly,

$$u_3 - u_2 = \lambda^3 \int_0^t K_3(t,s')f(s')\,ds'; \quad K_3(t,s') = \int_{s'}^t K(t,s)K_2(s,s')\,ds$$

and so on. Continuing sequentially, we get

$$u_n - u_{n-1} = \lambda^n \int_0^t K_n(t,s')f(s')\,ds'; \tag{22.52}$$

with

$$K_n(t,s') = \int_{s'}^t K(t,s)K_{n-1}(s,s')\,ds, \quad \text{and} \quad K_1(t,s) = K(t,s) \tag{22.53}$$

or

$$u_n(t) = f(t) + \lambda \sum_{i=1}^{n} \lambda^{i-1} \int_0^t K_i(t, s') f(s')\, ds' \tag{22.54}$$

that leads to the Neumann series solution as

$$u(t) = \lim_{n\to\infty} u_n = f(t) + \lambda \int_0^t \Gamma(t, s', \lambda) f(s')\, ds' \tag{22.55}$$

where $\Gamma(t, s', \lambda)$ is the resolvent kernel given by

$$\Gamma(t, s', \lambda) = \sum_{i=1}^{\infty} \lambda^{i-1} K_i(t, s') \tag{22.56}$$

with iterated kernels as defined in (22.53).

Example 22.3.

$$u(t) = f(t) + \lambda \int_0^t e^{t-s} u(s)\, ds.$$

Here,

$$\begin{aligned}
K(t, s') &= e^{t-s'} \\
K_2(t, s') &= \int_{s'}^{t} K(t, s) K(s, s')\, ds = \int_{s'}^{t} e^{t-s} e^{s-s'}\, ds \\
&= \int_{s'}^{t} e^{t-s'}\, ds = (t - s') e^{t-s'} \\
K_3(t, s') &= \int_{s'}^{t} K(t, s) K_2(s, s')\, ds = \int_{s'}^{t} e^{t-s} (s - s') e^{s-s'}\, ds \\
&= \int_{s'}^{t} e^{t-s'} (s - s')\, ds = e^{t-s'} \int_{s'}^{t} (s - s')\, ds = e^{t-s'} \frac{(t - s')^2}{2!}
\end{aligned}$$

Thus, the resolvent kernel is given by

$$\begin{aligned}
\Gamma(t, s', \lambda) &= e^{t-s'} + \lambda (t - s') e^{t-s'} + \lambda^2 e^{t-s'} \frac{(t - s')^2}{2!} + \cdots \\
&= e^{t-s'} \left[1 + \lambda (t - s') + \frac{\lambda^2 (t - s')^2}{2!} + \cdots \right] \\
&= e^{t-s'} e^{\lambda (t-s')} = \exp[(1 + \lambda)(t - s')]
\end{aligned}$$

The solution is given by

$$u(t) = f(t) + \lambda \int_0^t e^{(1+\lambda)(t-s')} f(s')\, ds'.$$

Theorem. *Consider the VIE of the second kind*

$$u(t) = f(t) + \lambda \int_a^t K(t,s)u(s)\, ds$$

where a is a constant and assume that:
1. *$K(t,s)$ is continuous in the rectangle R, for which $a \leq t \leq b$, $a \leq s \leq b$, $|K(t,s)| \leq M$ in R and $K(t,s) \neq 0$.*
2. *$f(t) \neq 0$ is real and continuous in the interval $I : a \leq t \leq b$.*
3. *λ is a constant, then the integral equation has one and only one continuous solution $u(t)$ in I, and this solution is given by the absolutely and uniquely convergent series*

$$u(t) = f(t) + \lambda \sum_{n=1}^{\infty} \lambda^{n-1} \int_0^t K_n(t,s') f(s')\, ds'.$$

Remarks. (a) The above series is called the Neumann series. (b) When the kernel is bounded, the Neumann series converges since repeated integration leads to terms of the form $\frac{(t-s')^i}{i!}$ for the iterated kernel.

22.5.2 Adomian decomposition method

For VIE (equation (22.45)), the solution can be expressed in the form of

$$u(t) = \sum_{n=0}^{\infty} u_n(t) \tag{22.57}$$

where

$$u_0(t) = f(t), \quad \text{and} \quad u_n(t) = \lambda \int_0^t K(t,s) u_{n-1}(s)\, ds \quad \forall n \geq 1. \tag{22.58}$$

In this Adomian decomposition method, the kernel remains the same but we add higher-order terms in λ and $f(t)$ to the solution.

Example 22.4. Solve

$$u(t) = t + \int_0^t (s - t)u(s)\, ds$$

Equation (22.58) with $\lambda = 1$ gives

$$\begin{aligned} u_0(t) &= t \\ u_1(t) &= \int_0^t (s - t)s\, ds = -\frac{t^3}{3!} \\ u_2(t) &= -\int_0^t (s - t)\frac{s^3}{3!}\, ds = \frac{t^5}{5!} \end{aligned}$$

and so on. The solution is given by

$$u(t) = \frac{t}{1!} - \frac{t^3}{3!} + \frac{t^5}{5!} - \cdots = \sin t$$

22.6 Solution procedure for Volterra integral equations of the first kind

Consider the nonhomogeneous VIE of the first kind as

$$\lambda \int_0^t K(t, s)u(s)\, ds = f(t) \tag{22.59}$$

There are two ways to convert equation (22.59) into a VIE of the second kind:
(i) by differentiation of equation (22.59)
(ii) by integrating equation (22.59) by parts.

22.6.1 Differentiation approach

Differentiating equation (22.59) gives

$$\lambda K(t, t)u(t) + \lambda \int_0^t \frac{\partial K}{\partial t}(t, s)u(s)\, ds = f'(t) \tag{22.60}$$

If $K(t, t) \neq 0$ and $\lambda \neq 0$, we can write

$$u(t) + \int_0^t \frac{\partial K}{\partial t}(t, s)\frac{1}{K(t, t)}u(s)\, ds = \frac{f'(t)}{\lambda K(t, t)}.$$

Thus, defining

$$K^*(t,s) = \frac{-1}{\lambda K(t,t)}\frac{\partial K(t,s)}{\partial t}; \quad \text{and} \quad f^*(t) = \frac{f'(t)}{\lambda K(t,t)} \tag{22.61}$$

we get

$$u(t) = f^*(t) + \lambda \int_0^t K^*(t,s)u(s)\,ds \tag{22.62}$$

which is a VIE of the second kind and can be solved by methods discussed previously.

22.6.2 Integration approach

Defining

$$\phi(t) = \int_0^t u(s)\,ds \tag{22.63}$$

and integrating equation (22.59) by parts, we get

$$f(t) = \lambda\left[K(t,t)\phi(t) - \int_0^t \frac{\partial K(t,s)}{\partial s}\phi(s)\,ds\right]$$

Assuming $K(t,t) \neq 0$, we can rearrange the above equation as

$$\phi(t) = \frac{f(t)}{\lambda K(t,t)} + \int_0^t \frac{\frac{\partial K(t,s)}{\partial s}}{K(t,t)}\phi(s)\,ds$$

Thus, defining

$$\widehat{f}(t) = \frac{f(t)}{\lambda K(t,t)}; \quad \text{and} \quad \widehat{K}(t,s) = \frac{\frac{\partial K(t,s)}{\partial s}}{K(t,t)} \tag{22.64}$$

we get

$$\phi(t) = \widehat{f}(t) + \int_0^t \widehat{K}(t,s)\phi(s)\,ds, \tag{22.65}$$

which is a VIE of the second kind. Once $\phi(t)$ is known, $u(t)$ can be obtained from the relation $u(t) = \phi'(t)$. Note that this method does not require the function $f(t)$ to be differentiable.

22.7 Volterra integral equations with convolution kernel

Consider a VIE of the second kind:

$$u(t) = f(t) + \lambda \int_0^t K(t,t')u(t')\,dt' \tag{22.66}$$

where

$$K(t,t') = g(t-t') \tag{22.67}$$

⇒

$$u(t) = f(t) + \lambda \int_0^t g(t-t')u(t')\,dt'. \tag{22.68}$$

This equation can be solved by the Laplace transform method using the convolution property of LT. Let

$$\mathcal{L}[u(t)] = \int_0^\infty e^{-st}u(t)\,dt = \hat{u}(s) = L.T.\text{ of } u(t). \tag{22.69}$$

⇒

$$\mathcal{L}\left[\int_0^t g(t-t')u(t')\,dt'\right] = \mathcal{L}[g(t) * u(t)] = \hat{g}(s)\hat{u}(s).$$

Then taking LT, equation (22.68) gives

$$\hat{u}(s) = \hat{f}(s) + \lambda\hat{g}(s)\hat{u}(s)$$
$$\Rightarrow \hat{u}(s) = \frac{\hat{f}(s)}{1-\lambda\hat{g}(s)} \tag{22.70}$$

Let

$$\mathcal{L}^{-1}\left[\frac{1}{1-\lambda\hat{g}(s)}\right] = G(t) \tag{22.71}$$

⇒

$$u(t) = \int_0^t G(t-t')f(t')\,dt' \tag{22.72}$$

If we can expand

$$\frac{1}{1-\lambda\hat{g}(s)} = 1 + \lambda\hat{g}(s) + \lambda^2\hat{g}(s)^2 + \cdots$$

then

$$\mathcal{L}^{-1}\left[\frac{1}{1-\lambda\hat{g}(s)}\right] = G(t) = \delta(t) + \lambda g_1(t) + \lambda^2 g_2(t) + \cdots; \tag{22.73}$$

$$g_i(t) = \mathcal{L}^{-1}[\hat{g}(s)^i], \quad i = 1, 2, \ldots \tag{22.74}$$

⇒

$$\begin{aligned}
u(t) &= \int_0^t G(t-t')f(t')\,dt' \\
&= \int_0^t [\delta(t-t') + \lambda g_1(t-t') + \lambda^2 g_2(t-t')]f(t')\,dt' \\
&= f(t) + \lambda \int_0^t \Gamma(t,t',\lambda)f(t')\,dt'
\end{aligned}$$

where

$$\Gamma(t,t',\lambda) = g_1(t-t') + \lambda g_2(t-t') + \cdots$$

is the resolvent kernel.

Example 22.5. Solve Abel's equation:

$$f(t) = \int_0^t \frac{u(s)}{\sqrt{t-s}}\,ds = \int_0^t \frac{u(t')}{\sqrt{t-t'}}\,dt'.$$

The Laplace transform gives

$$\hat{f}(s) = \hat{u}(s).\sqrt{\frac{\pi}{s}}, \quad (\because \mathcal{L}\left[\frac{1}{\sqrt{t}}\right] = \sqrt{\frac{\pi}{s}})$$

⇒

$$\begin{aligned}
\hat{u}(s) &= \frac{\sqrt{s}\hat{f}(s)}{\sqrt{\pi}} = \frac{s}{\pi}\sqrt{\frac{\pi}{s}}\hat{f}(s) \\
\Rightarrow u(t) &= \frac{1}{\pi}\frac{d}{dt}\mathcal{L}^{-1}\left[\sqrt{\frac{\pi}{s}}\hat{f}(s)\right] = \frac{1}{\pi}\frac{d}{dt}\left[\int_0^t \frac{f(t')}{\sqrt{t-t'}}\,dt'\right]
\end{aligned}$$

22.8 Fredholm integral equations of the second kind

The Neumann series and the Adomian decomposition method can also be used to solve Fredholm integral equations (FIE) as illustrated by the example below.

Diffusion–reaction problem

Consider a diffusion–reaction problem given by

$$\frac{d^2c}{dx^2} = \phi^2 R(c), \quad 0 < x < 1 \tag{22.75}$$

$$c'(0) = 0, \quad \text{and} \quad c(1) = 1 \tag{22.76}$$

where $c(x)$ is concentration, $R(c)$ is rate of reaction and ϕ is the Thiele modulus. We can express equations (22.75)–(22.76) as an integral equation by integrating twice and changing the order of integration, which leads to the integral equation as

$$c(x) = 1 - \phi^2 \int_0^1 K(x,s)R(c(s))\,ds \tag{22.77}$$

where

$$K(x,s) = \begin{cases} 1-x, & 0 < s < x \\ 1-s, & x < s < 1 \end{cases} \tag{22.78}$$

For linear kinetics, i. e., $R(c) = c$, we can rewrite equation (22.77) as

$$c(x) = 1 - \phi^2 \int_0^1 K(x,s)c(s)\,ds \tag{22.79}$$

which is a FIE of the second kind.

22.8.1 Solution by successive substitution

FIE (equation (22.79)) can be written with equation (22.78) as

$$c(x) = 1 - \phi^2 \int_0^x (1-x)c(s)\,ds - \phi^2 \int_x^1 (1-s)c(s)\,ds \tag{22.80}$$

Thus, writing the recurrence relation as

$$c_j(x) = 1 - \phi^2 \int_0^x (1-x)c_{j-1}(s)\,ds - \phi^2 \int_x^1 (1-s)c_{j-1}(s)\,ds \tag{22.81}$$

with initial guess of $c_0(x) = 1$, and defining the effectiveness factor as

$$\eta_j = \int_0^1 c_j(s)\,ds \tag{22.82}$$

we get

$$\begin{aligned}
&c_0(x) = 1 \Rightarrow \eta_0 = 1\\
&c_1(x) = 1 - \frac{\phi^2}{2}(1-x^2) \Rightarrow \eta_1 = 1 - \frac{\phi^2}{3}\\
&c_2(x) = 1 - \frac{\phi^2}{2}(1-x^2) + \frac{\phi^4}{24}(1-x^2)(5-x^2)\\
&\Rightarrow \eta_2 = 1 - \frac{\phi^2}{3} + \frac{2}{15}\phi^4
\end{aligned}$$

and so on.

The higher-order terms can be obtained following the sequence, where it can be shown that the solutions for concentration profile and effectiveness factor are the Taylor series expansion (in ϕ^2) of the functions

$$c = c_\infty = \frac{\cosh\phi x}{\cosh\phi}; \quad \text{and} \quad \eta = \eta_\infty = \frac{\tanh\phi}{\phi}. \tag{22.83}$$

In this case, the solution converges for all values of ϕ^2, though the convergence may be slow for $\phi^2 > 1$.

22.8.2 Solution by Adomian decomposition method

In this approach, we take

$$c_0(x) = 1 \quad \text{and} \quad c(x) = \sum_{i=0}^{\infty} c_i(x) \Rightarrow \eta = \sum_{i=0}^{\infty} \eta_i \tag{22.84}$$

Thus, substituting equation (22.84) in FIE (equation (22.79)), we get

$$(c_0 + c_1 + c_2 + \cdots) = 1 - \phi^2 \int_0^1 K(x,s)[c_0 + c_1 + c_2 + \cdots]\,ds. \tag{22.85}$$

$\Rightarrow$

$$c_1(t) = -\phi^2 \int_0^1 K(x,s)c_0(s)\,ds \tag{22.86}$$

$$c_2(t) = -\phi^2 \int_0^1 K(x,s)c_1(s)\,ds \tag{22.87}$$

$$c_3(t) = -\phi^2 \int_0^1 K(x,s)c_2(s)\,ds, \quad \text{and so on.} \tag{22.88}$$

Thus, using the symbolic manipulation program Mathematica®, we can obtain

$$c_0(x) = 1 \Rightarrow \eta_0 = 1 \tag{22.89}$$

$$c_1(x) = -\frac{\phi^2}{2}(1-x^2) \Rightarrow \eta_1 = -\frac{\phi^2}{3} \tag{22.90}$$

$$c_2(x) = \frac{\phi^4}{24}(5-6x^2+x^4) \Rightarrow \eta_2 = \frac{2\phi^4}{15} \tag{22.91}$$

$$c_3(x) = -\frac{\phi^6}{720}(61-75x^2+15x^4-x^6) \Rightarrow \eta_3 = -\frac{17\phi^6}{315}, \quad \text{and so on} \tag{22.92}$$

Solving above recurrence relation sequentially, we can obtain the solution to any order.

22.9 Fredholm integral equations with symmetric kernels

Consider a homogeneous Fredholm integral equation (i. e., $f(x) = 0$) given by

$$u(x) = \lambda \int_a^b K(x,s)u(s)\,ds \tag{22.93}$$

with symmetric kernel, i. e.,

$$K(x,s) = K(s,x) \tag{22.94}$$

The eigenvalues and eigenfunctions of the kernel are defined by

$$\phi_n(x) = \lambda_n \int_a^b K(x,s)\phi_n(s)\,ds, \quad \phi_n(x) \neq 0 \tag{22.95}$$

Theorem.

1. *The eigenvalues of a symmetric kernel are real.*
2. *The eigenfunctions of a symmetric kernel corresponding to distinct eigenvalues are orthogonal.*

3. *The multiplicity of an eigenvalue is finite if kernel is symmetric and square integrable.*
4. *If λ_j is repeated p times, there are p linearly independent eigenfunctions corresponding to λ_j and these can be made orthogonal to each other.*
5. *The sequence of eigenfunctions of a symmetric kernel $K(x,s)$ can be made orthonormal, i. e.,*

$$\int_a^b \phi_i(x)\phi_j(x)\,dx = \delta_{ij} = \begin{cases} 1, & i = j \\ 0, & i \neq j. \end{cases}$$

6. *The eigenvalues of a symmetric nonseparable integrable kernel form an infinite sequence with no finite limit point [if the kernel is separable, then eigenvalues form a finite sequence]. If we include each eigenvalue in the sequence a number of times equal to its multiplicity, then*

$$\sum_{n=1}^{\infty} \frac{1}{\lambda_n^2} \leq \int_a^b \int_a^b K(x,s)^2\,dx\,ds$$

Equivalently, $\frac{1}{\lambda_n} \to 0$ for $n \to \infty$. When the equality sign holds, the kernel is said to be closed.
7. *The set of eigenvalues of the second iterated kernel*

$$K_2(x,s) = \int_a^b K(x,s')K(s',s)\,ds'$$

are squares of eigenvalues of the given kernel $K(x,s)$.
8. *If λ_1 is the smallest eigenvalue of the kernel, then*

$$\frac{1}{|\lambda_1|} = \max\left|\langle K\phi, \phi\rangle\right|, \quad \|\phi\| = 1$$

and the maximum on the RHS is attained when $\phi(x)$ is an eigenfunction of the symmetric $\mathcal{L}_2$-kernel corresponding to the smallest eigenvalue.

Proofs of the various statements in the above theorem may be found in the book by Courant and Hilbert [15].

Mercer's theorem. *If the kernel $K(x,s)$ is symmetric and square integrable on the square $\{(x,s) : a \leq x \leq b, a \leq s \leq b\}$, continuous and has only positive eigenvalues or at most a finite number of negative eigenvalues, then the series*

$$\sum_{n=1}^{\infty} \frac{\phi_n(x)\phi_n(s)}{\lambda_n}$$

converges absolutely and uniformly, and

$$K(x,s) = \sum_{n=1}^{\infty} \frac{\phi_n(x)\phi_n(s)}{\lambda_n}.$$

22.10 Adjoint operator and Fredholm alternative

If we define an integral operator $T : C[a,b] \to C[a,b]$ by

$$Tu(x) = \int_a^b K(x,s)u(s)\,ds, \tag{22.96}$$

it is easily seen that the adjoint operator with respect to the usual inner product

$$\langle u(x), v(x) \rangle = \int_a^b u(x)\overline{v(x)}\,dx$$

is given by

$$T^* v(x) = \int_a^b \overline{K(s,x)}v(s)\,ds. \tag{22.97}$$

From this result, it follows that the adjoint homogeneous equation to

$$u(x) = \lambda \int_a^b K(x,s)u(s)\,ds \tag{22.98}$$

is given by

$$v(x) = \overline{\lambda} \int_a^b \overline{K(s,x)}v(s)\,ds \tag{22.99}$$

We state now an important theorem related to the existence and uniqueness of solutions to Fredholm integral equations. A proof of this theorem may be found in the book by Courant and Hilbert [15].

Theorem (Fredholm alternative). *Either the integral equation*

$$u(x) = f(x) + \lambda \int_a^b K(x,s)u(s)\,ds \tag{22.100}$$

with fixed λ has one and only one solution $u(x)$ for arbitrary $\mathcal{L}_2$-functions $f(x)$ and $K(x,s)$, in particular, the solution $u(x) \equiv 0$ for $f(x) = 0$, or the homogeneous equation

$$u(x) = \lambda \int_a^b K(x,s)u(s)\,ds$$

possesses a finite number of r linearly independent solutions $u_{hi}(x)$, $i = 1, 2, \ldots, r$. In the first case, the adjoint equation

$$v(x) = f(x) + \overline{\lambda} \int_a^b \overline{K(s,x)}v(s)\,ds$$

also has a unique solution. In the second case, the adjoint homogeneous equation

$$v(x) = \overline{\lambda} \int_a^b \overline{K(s,x)}v(s)\,ds$$

also has r linearly independent solutions $v_{hi}(x)$, $i = 1, 2, \ldots, r$. The inhomogeneous equation has a solution if and only if the function $f(x)$ satisfies

$$\langle f, v_{hi}\rangle = \int_a^b f(x)\overline{v_{hi}(x)}\,dx = 0, \quad i = 1, 2, \ldots, r.$$

In this case, the solution to equation (22.100) *is determined only up to an additive linear combination $\sum_{i=1}^{r} c_i u_{hi}(x)$; it may be determined uniquely by the additional requirements*

$$\langle u, u_{hi}\rangle = 0, \quad i = 1, 2, \ldots, r.$$

[Remark: Compare this theorem with the version for algebraic equations discussed in Section 4.5.2.]

22.11 Solution of FIE of the second kind with symmetric kernels

Consider the FIE of the second kind as

$$u(x) = f(x) + \lambda \int_a^b K(x,s)u(s)\,ds \tag{22.101}$$

with symmetric kernel, i. e.,

$$K(x,s) = K(s,x) \tag{22.102}$$

Here, the resolvent kernel is given by

$$\Gamma(x,s,\lambda) = \sum_{n=1}^{\infty} \frac{\phi_n(x)\phi_n(s)}{(\lambda_n - \lambda)} \tag{22.103}$$

and

$$u(x) = f(x) + \lambda \sum_{n=1}^{\infty} \frac{a_n \phi_n(x)}{(\lambda_n - \lambda)}; \quad a_n = \int_a^b \phi_n(s) f(s)\, ds \tag{22.104}$$

where λ_n and ϕ_n are eigenvalues and eigenfunctions of the kernel $K(x,s)$ as defined in equation (22.95).

A solution exists and is unique if $\lambda \neq \lambda_n (n = 1, 2, \ldots)$. If $\lambda = \lambda_n$ for some n, a solution may exist but it is not unique.

Let

$$f(x) = \sum_{n=1}^{\infty} a_n \phi_n(x) \Longleftrightarrow a_n = \int_a^b \phi_n(s) f(s)\, ds \tag{22.105}$$

Write

$$u(x) = \sum_{n=1}^{\infty} b_n \phi_n(x) \tag{22.106}$$

This can be done since $u(x) \in \mathcal{L}_2$ and the eigenfunctions form a basis for $\mathcal{L}_2$. To determine b_n, substitute equation (22.106) into equation (22.101) to get

$$\sum_{n=1}^{\infty} b_n \phi_n(x) = \sum_{n=1}^{\infty} a_n \phi_n(x) + \lambda \int_a^b K(x,s) \sum_{n=1}^{\infty} b_n \phi_n(s)\, ds$$

Multiply both sides by $\phi_j(x)$ and integrate and use the orthogonal property of the eigenfunctions, which leads to

$$b_j = \frac{a_j \lambda_j}{\lambda_j - \lambda} \tag{22.107}$$

$\Rightarrow$

$$u(x) = \sum_{j=1}^{\infty} \frac{a_j \lambda_j}{\lambda_j - \lambda} \phi_j(x) \tag{22.108}$$

This solution exists only if $\lambda \neq \lambda_j, j = 1, 2, \ldots$.

Equation (22.108) implies

$$
\begin{aligned}
u(x) &= \sum_{j=1}^{\infty} \frac{\lambda_j \phi_j(x)}{\lambda_j - \lambda} \int_a^b \phi_j(s) f(s)\, ds \\
&= \int_a^b \sum_{j=1}^{\infty} \frac{\lambda_j \phi_j(x)\phi_j(s)}{\lambda_j - \lambda} f(s)\, ds \\
&= \int_a^b \sum_{j=1}^{\infty} \left(1 + \frac{\lambda}{\lambda_j - \lambda}\right) \phi_j(x)\phi_j(s) f(s)\, ds \\
&= \sum_{j=1}^{\infty} \phi_j(x) \int_a^b \phi_j(s) f(s)\, ds + \lambda \int_a^b \sum_{j=1}^{\infty} \frac{\phi_j(x)\phi_j(s)}{\lambda_j - \lambda} f(s)\, ds
\end{aligned}
\tag{22.109}
$$

$\Rightarrow$

$$
u(x) = f(x) + \lambda \int_a^b \Gamma(x, s, \lambda) f(s)\, ds \tag{22.110}
$$

where

$$
\Gamma(x, s, \lambda) = \sum_{j=1}^{\infty} \frac{\phi_j(x)\phi_j(s)}{\lambda_j - \lambda} \tag{22.111}
$$

is the resolvent kernel.

When $\lambda = \lambda_j$ for some j, then b_j is indeterminate and equation (22.101) is consistent iff $a_j = 0$, i. e., $f(x)$ is orthogonal to $\phi_j(x)$. In this case, the solution is not unique and is of the form:

$$
u(x) = c_j \phi_j(x) + \sum_{\substack{n=1 \\ n \neq j}}^{\infty} \frac{a_n \lambda_n}{\lambda_n - \lambda} \phi_n(x) \tag{22.112}
$$

where c_j is any arbitrary constant.

Example 22.6. The Fredholm equation

$$
u(x) = x + 4\pi^2 \int_0^1 K(x, s) u(s)\, ds
$$

with

$$
K(x, s) = \begin{cases} s(1 - x), & 0 \le s \le x \\ x(1 - s), & x \le s \le 1 \end{cases}
$$

is not solvable since $\lambda = 4\pi^2$ is an eigenvalue of the kernel with eigenfunction $\sqrt{2}\sin(2\pi x)$ and

$$\int_0^1 x.\sqrt{2}\sin(2\pi x)\,dx = \frac{-1}{\sqrt{2}\pi} \neq 0.$$

Example 22.7. The Fredholm equation

$$u(x) = \sin(3\pi x) + 4\pi^2 \int_0^1 K(x,s)u(s)\,ds$$

with kernel $K(x,s)$ same as in Example 22.6, is solvable with but the solution is not unique. It may be shown that

$$u = \frac{9}{5}\sin(3\pi x) + c_2 \sin(2\pi x)$$

is a solution for any c_2.

Problems

1. Apply the IE method and the Adomian decomposition method to solve vector form of one-dimensional diffusion–reaction model with linear kinetics.
2. Consider the Fredholm integral equation of the first kind

   $$f(x) = \int_a^b K(x,s)u(s)\,ds \quad (1)$$

   with a degenerate kernel of the form

   $$K(x,s) = \sum_{i=1}^{N} a_i(x)b_i(s)$$

   where $\{a_i(x), i = 1,2\ldots,N\}$ and $\{b_i(s), i = 1,2,\ldots,N\}$ are linearly independent sets. (a) Reason that the equation does not have solution unless the function $f(x)$ can be expressed as a linear combination of $a_i(x)$, (b) Reason that when equation (1) has a solution, it is not unique, i. e., there could be infinitely many solutions, (c) Consider equation (1) with a continuous kernel but not separable and continuous $f(x)$. Is the solution also continuous? Comment on the possible types of solutions, (d) How do the results in (b) and (c) change if the kernel is also symmetric?
3. (a) Determine the eigenvalues, eigenfunctions and the resolvent kernel for the Fredholm equation

$$u(x) = f(x) + \lambda \int_{-1}^{1} [xs + x^2 s^2] u(s)\, ds$$

(b) Determine the solution.

4. Consider the axial dispersion model

$$\frac{1}{\text{Pe}} \frac{d^2 c}{dx^2} - \frac{dc}{dx} - \text{Da}\, R(c) = 0; \quad 0 < x < 1$$

$$\frac{1}{\text{Pe}} \frac{dc}{dx} = c - 1 @ x = 0$$

$$\frac{dc}{dx} = 0 @ x = 1$$

where $R(c)$ is the dimensionless reaction rate and Pe and Da are the Peclet and Damkohler numbers, respectively. (a) Convert the boundary value problem into a Fredholm integral equation and (b) Solve the equation in (a) for the case of linear kinetics using the Neumann series method and determine the exit concentration as a function of Da up to the second order term in Da.

5. Consider the Volterra integral equation for human population $N(t)$ at time t:

$$N(t) = N_0 f(t) + k \int_0^t f(t - \tau) N(\tau) d\tau$$

where $f(t)$ is the survival function and k is a constant describing the rate of population variation per capita [or the birth rate is k times $N(t)$] (a) Solve the equation assuming a survival function of the form

$$f(t) = \exp\left[-\frac{t}{T}\right]$$

where T is the average life span of a person (b) Use the result in (a) to show that the population increases exponentially if $kT > 1$ and decreases exponentially if $kT < 1$.

Part V: **Fourier transforms and solution of boundary and initial-boundary value problems**

23 Finite Fourier Transforms

Finite Fourier Transform (FFT) and its various extensions is the most important tool available for scientists and engineers to solve many practical problems. The concepts of FFT appear in the analysis of time series, spatial profiles, length and time scales, data analysis and compression, development of numerical algorithms and so forth. In this chapter, we discuss mainly one application of FFT, namely, the solution of linear boundary and initial–boundary value problems (partial differential equations).

23.1 Definition and general properties

Let $\{\lambda_j, w_j(x), \rho(x)\}$, $j = 1, 2, 3, \ldots$ be the eigenvalues and the corresponding eigenfunctions of a self-adjoint eigenvalue problem in the interval $[a, b]$. Assume that the eigenfunctions are normalized so that

$$\int_a^b \rho(x) w_i(x) w_j(x)\, dx = \delta_{ij} \tag{23.1}$$

The eigenfunctions $\{w_j(x)\}$ form a basis for $\mathcal{L}_2[a, b]$, the Hilbert space of Lebesque square integrable (real valued) functions defined on the interval $[a, b]$. If $f(x) \in \mathcal{L}_2[a, b]$, we have

$$\int_a^b \rho(x) f(x)^2\, dx < \infty \tag{23.2}$$

and we can write

$$f(x) = \sum_{i=1}^{\infty} c_i w_i(x) \tag{23.3}$$

where

$$c_j = \langle f, w_j \rangle = \int_a^b \rho(x) f(x) w_j(x)\, dx \tag{23.4}$$

and

$$\sum_{j=1}^{\infty} c_j^2 = \int_a^b \rho(x) f(x)^2\, dx = \|f\|^2 \quad \text{(Parseval's relation)} \tag{23.5}$$

and the integrals are all defined in the Lebesque sense. The expansion given by equation (23.3) converges in $\mathcal{L}_2[a, b]$, i. e., the LHS and RHS of equation (23.3) are equal almost everywhere (except perhaps on a set of measure zero in $[a, b]$). Equations (23.3) and (23.4) define the *Finite Fourier Transform*, i. e., given any $f(x) \in \mathcal{L}_2[a, b]$ we define the

https://doi.org/10.1515/9783111598055-028

Finite Fourier Transform (FFT) of $f(x)$ to be the infinite sequence of constants $\{c_i\}$, which give the coordinates of $f(x)$ in the Hilbert space $\mathcal{L}_2[a,b]$. We write

$$\begin{aligned}\mathcal{F}\{f(x)\} &= \text{FFT of } f(x)\\ &= \langle f, w_i\rangle\\ &= c_i\end{aligned} \tag{23.6}$$

The inverse transform uses the coordinates $\{c_i\}$ and the basis (eigen) functions to reconstruct the function (vector) $f(x)$. Thus,

$$\mathcal{F}^{-1}\{c_i\} = \sum_{i=1}^{\infty} c_i w_i(x) = f(x) \tag{23.7}$$

Thus,

$$\mathcal{F}\mathcal{F}^{-1} = \mathcal{F}^{-1}\mathcal{F} = \text{identity}. \tag{23.8}$$

The Finite Fourier Transform may be used to simplify and solve many linear differential equations in which the spatial self-adjoint operator $\mathbb{L}$ (whose eigenvalues and eigenfunctions are λ_i and $w_i(x)$, respectively) appears. We outline below the general procedure and illustrate with several examples.

23.1.1 Example 1 (solution of Poisson's equation)

Consider the general form of Poisson's equation:

$$\mathbb{L}u = f \tag{23.9}$$

where $\mathbb{L} = \mathbb{L}^*$ is a symmetric or self-adjoint operator. Let λ_n $(n = 1, 2, \dots)$ be eigenvalues of $\mathbb{L}$ with normalized eigenfunctions w_n, i. e., $\mathbb{L}w_n = \lambda_n w_n$, $\langle w_n, w_n\rangle = 1$. Then, FFT of equation (23.9) gives

$$\begin{aligned}\langle \mathbb{L}u, w_n\rangle &= \langle f, w_n\rangle \implies \langle u, \mathbb{L}w_n\rangle = \langle f, w_n\rangle\\ \implies \lambda_n\langle u, w_n\rangle &= \langle f, w_n\rangle\\ \implies \langle u, w_n\rangle &= \frac{1}{\lambda_n}\langle f, w_n\rangle \quad \text{if } \lambda_n \neq 0\end{aligned} \tag{23.10}$$

$\implies$

$$u = \sum_{n=1}^{\infty} \langle u, w_n\rangle w_n = \sum_{n=1}^{\infty} \frac{1}{\lambda_n}\langle f, w_n\rangle w_n \tag{23.11}$$

is the formal solution.

Remarks. (a) If $\mathbb{L}$ is an operator in two or three spatial dimensions, the sum may be a double or triple sum, (b) If the BCs are inhomogeneous, the solution will have additional terms and (c) In most of our applications, $\mathbb{L}$ is usually an elliptic differential operator such as the Laplacian operator, i. e., $\mathbb{L} = -\nabla^2$, which is self-adjoint and has a discrete spectrum on finite domains with Dirichlet, Neumann or Robin boundary conditions.

23.1.2 Example 2 (solution of heat/diffusion equation)

Consider the heat/diffusion equation:

$$\frac{\partial u}{\partial t} = -\mathbb{L}u, \quad t > 0; \quad u = f \text{ @ } t = 0 \tag{23.12}$$

with a self-adjoint operator $\mathbb{L}$. Taking FFT $\Longrightarrow$

$$\frac{d}{dt}\langle u, w_n\rangle = -\lambda_n\langle u, w_n\rangle, \quad t > 0; \tag{23.13}$$

$$\langle u, w_n\rangle = \langle f, w_n\rangle \text{ @ } t = 0 \tag{23.14}$$

$$u = \sum_{n=1}^{\infty}\langle u, w_n\rangle w_n = \sum_{n=1}^{\infty}\exp(-\lambda_n t)\langle f, w_n\rangle w_n \tag{23.15}$$

is the formal solution. In addition, when the operator $\mathbb{L}$ is a 2/3D spatial operator, the sum may be a double or triple sum.

23.1.3 Example 3 (solution of the wave equation)

Consider the wave equation:

$$\frac{\partial^2 u}{\partial t^2} = -\mathbb{L}u, \quad t > 0; \tag{23.16}$$

$$u = f \text{ @ } t = 0 \quad \text{(initial position)} \tag{23.17}$$

$$\frac{\partial u}{\partial t} = g \text{ @ } t = 0 \quad \text{(initial velocity)} \tag{23.18}$$

with a self-adjoint operator $\mathbb{L}$. Taking FFT, we get

$$\frac{d^2}{dt^2}\langle u, w_n\rangle = -\lambda_n\langle u, w_n\rangle, \quad t > 0;$$

$$\langle u, w_n\rangle = \langle f, w_n\rangle \text{ @ } t = 0$$

$$\frac{d}{dt}\langle u, w_n\rangle = \langle g, w_n\rangle \text{ @ } t = 0$$

$\Longrightarrow$

$$\langle u, w_n \rangle = \langle f, w_n \rangle \cos[\sqrt{\lambda_n} t] + \frac{1}{\sqrt{\lambda_n}} \langle g, w_n \rangle \sin[\sqrt{\lambda_n} t] \tag{23.19}$$

$\Longrightarrow$

$$u = \sum_{n=1}^{\infty} \langle f, w_n \rangle \cos[\sqrt{\lambda_n} t] w_n + \sum_{n=1}^{\infty} \frac{1}{\sqrt{\lambda_n}} \langle g, w_n \rangle \sin[\sqrt{\lambda_n} t] w_n \tag{23.20}$$

is the formal solution.

We now illustrate the application of FFT with specific examples in the rest of this chapter.

23.2 Application of FFT for BVPs in 1D

23.2.1 Example 1 (Poisson's equation in 1-D)

Consider the boundary value problem

$$\frac{d^2 u}{dx^2} = -f(x), \quad 0 < x < 1 \tag{23.21}$$

$$u(0) = 0, \quad u(1) = 0 \tag{23.22}$$

To solve, we consider the operator $\mathbb{L}$ defined by the self-adjoint eigenvalue problem

$$\mathbb{L} w = -\frac{d^2 w}{dx^2} = \lambda w, \quad 0 < x < 1 \tag{23.23}$$

$$w(0) = w(1) = 0 \tag{23.24}$$

We have the eigenvalues and normalized eigenfunctions

$$\lambda_n = n^2 \pi^2, \quad w_n(x) = \sqrt{2} \sin n\pi x; \quad n = 1, 2, \ldots \tag{23.25}$$

Taking FFT of equations (23.21)–(23.22), i. e., multiplying by $w_j(x)$ and integrating from 0 and 1, gives

$$\left\langle \frac{d^2 u}{dx^2}, w_j \right\rangle = \langle -f, w_j \rangle$$

$\Rightarrow$

$$\left(\frac{du}{dx} w_j - u \frac{dw_j}{dx} \right)_{x=0}^{x=1} + \left\langle u, \frac{d^2 w_j}{dx^2} \right\rangle = \langle -f, w_j \rangle$$

The first term (or the concomitant) vanishes as both $u(x)$ and $w_j(x)$ satisfy the homogeneous boundary conditions. Thus, we have

$$\left\langle u, \frac{d^2 w_j}{dx^2} \right\rangle = \langle -f, w_j \rangle$$

$\Rightarrow$

$$-\lambda_j \langle u, w_j \rangle = -\langle f, w_j \rangle$$

$\Longrightarrow$

$$\langle u, w_j \rangle = \frac{\langle f, w_j \rangle}{\lambda_j}. \tag{23.26}$$

Taking the inverse transform gives

$$\begin{aligned} u &= \sum_{j=1}^{\infty} \langle u, w_j \rangle w_j \\ &= \sum_{j=1}^{\infty} \frac{\langle f, w_j \rangle}{\lambda_j} w_j \\ &= \sum_{j=1}^{\infty} \frac{\sqrt{2} \sin j\pi x}{j^2 \pi^2} \int_0^1 f(\xi) \sqrt{2} \sin j\pi\xi \, d\xi \end{aligned} \tag{23.27}$$

We consider some special cases of this solution.

Special case 1: $f(x) = 1$

For $f(x) = 1$,

$$\int_0^1 f(\xi) \sqrt{2} \sin j\pi\xi \, d\xi = \begin{cases} \frac{2\sqrt{2}}{j\pi}, & j \text{ odd} \\ 0, & j \text{ even} \end{cases}$$

$\Longrightarrow$ Equation (23.27) can be simplified as

$$u(x) = \frac{4}{\pi^3} \sum_{k=1}^{\infty} \frac{\sin[(2k-1)\pi x]}{(2k-1)^3}. \tag{23.28}$$

For $f(x) = 1$, equations (23.21)–(23.22) can also be solved by integrating twice, which leads to another form of the solution

$$u(x) = \frac{1}{2} x(1-x). \tag{23.29}$$

In other words, equation (23.28) is the Fourier series expansion of the function given by equation (23.29). The exact solution (equation (23.29)) and Fourier series expansion (equation (23.28)) with only two terms are plotted in Figure 23.1.

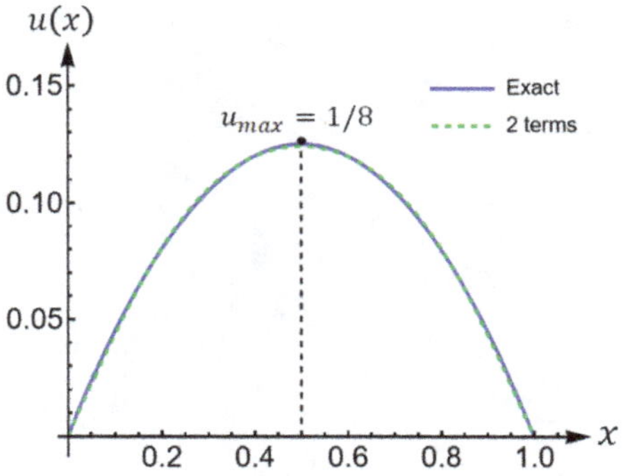

Figure 23.1: Solution of Poisson's equation with $f(x) = 1$: exact solution compared with the two terms of the Fourier series solution.

It can be seen from this figure that the Fourier series solution with only two terms represents the solution accurately in this example. The maximum value predicted by Fourier series solution with only two terms is 0.124 as compared to 0.125 predicted by the exact solution.

Consider again the solution given by equation (23.27)

$$u = \sum_{j=1}^{\infty} \frac{\sqrt{2}\sin j\pi x}{j^2\pi^2} \int_0^1 f(\xi)\sqrt{2}\sin j\pi\xi \, d\xi = \int_0^1 G(x,\xi)f(\xi)\, d\xi,$$

where

$$G(x,\xi) = \sum_{j=1}^{\infty} \frac{2\sin j\pi x \sin j\pi\xi}{j^2\pi^2} = \sum_{j=1}^{\infty} \frac{w_j(x)w_j(\xi)}{\lambda_j} \tag{23.30}$$

is the Green's function of $\mathbb{L}$. The Green's function may also be written as

$$G(x,\xi) = \begin{cases} (1-x)\xi, & 0 < \xi < x \\ x(1-\xi), & x < \xi < 1. \end{cases} \tag{23.31}$$

Special case 2: $f(x) = \delta(x - \frac{1}{2})$

For $f(x) = \delta(x - \frac{1}{2})$, a unit point source at the center of the domain, the direct integration method gives the solution of equations (23.21)–(23.22) as follows:

$$u(x) = \begin{cases} \frac{x}{2}, & 0 \le x \le \frac{1}{2} \\ \frac{(1-x)}{2}, & \frac{1}{2} \le x \le 1. \end{cases} \tag{23.32}$$

Alternatively, the solution (equation (23.27)) obtained from the eigenfunction expansion method simplifies as follows:

$$u = \sum_{j=1}^{\infty} \frac{\sqrt{2}\sin j\pi x}{j^2\pi^2}\sqrt{2}\sin\frac{j\pi}{2}$$
$$= \frac{2}{\pi^2}\sum_{k=1}^{\infty}(-1)^{k-1}\frac{\sin[(2k-1)\pi x]}{(2k-1)^2}. \tag{23.33}$$

Again, both expressions are equivalent. The exact solution and the two term FFT solution are plotted in Figure 23.2.

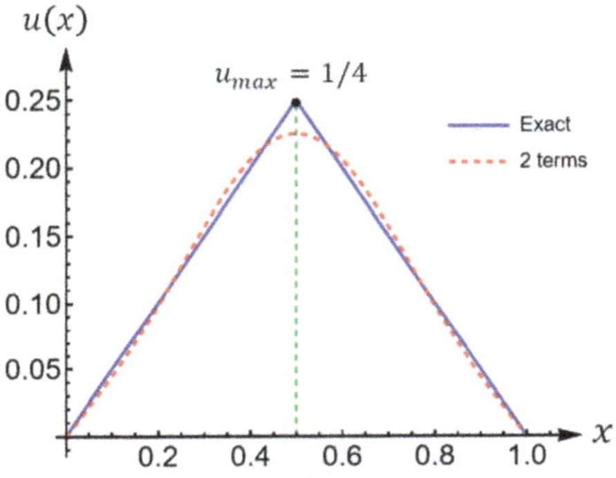

Figure 23.2: Solution of Poisson's equation with $f(x) = \delta(x - \frac{1}{2})$: exact solution compared with the two terms Fourier series solution.

It can be seen from Figure 23.2 again that only two terms can predict the solution with good accuracy. The largest error occurs in this case at maxima ($x = \frac{1}{2}$), where exact value is $u_{\text{exact}} = \frac{1}{4}$. This error can be reduced further by including more number of terms in the FFT solution as shown in Figure 23.3.

Remarks.

(1) If the interval is $(0, a)$, the eigenvalues are modified to

$$\lambda_n = \frac{n^2\pi^2}{a^2} \tag{23.34}$$

and the normalized eigenfunctions to

$$w_n(x) = \sqrt{\frac{2}{a}}\sin\left(\frac{n\pi x}{a}\right) \tag{23.35}$$

and the solution is given by

$$u(x) = \frac{2a}{\pi^2}\sum_{n=1}^{\infty}\frac{\sin(\frac{n\pi x}{a})}{n^2}\int_0^a f(\xi)\sin\left(\frac{n\pi\xi}{a}\right)d\xi \tag{23.36}$$

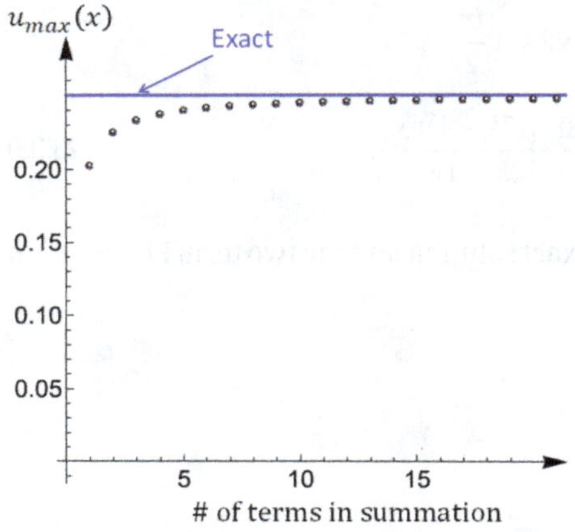

Figure 23.3: Demonstration of the convergence of Fourier series expansion with the solution of Poisson's equation with $f(x) = \delta(x - \frac{1}{2})$.

(2) The solution of the inhomogeneous problem

$$\frac{d^2u}{dx^2} = -f(x), \quad 0 < x < 1 \tag{23.37}$$

$$u(0) = \alpha_1, \quad u(1) = \alpha_2 \tag{23.38}$$

may be obtained by writing

$$u(x) = u_1(x) + u_2(x) \tag{23.39}$$

where

$$\frac{d^2u_1}{dx^2} = -f(x) \tag{23.40}$$

$$u_1(0) = 0, \quad u_1(1) = 0 \tag{23.41}$$

and u_2 satisfies the homogeneous equation with nonhomogeneous boundary conditions, i. e.,

$$\frac{d^2u_2}{dx^2} = 0 \tag{23.42}$$

$$u_1(0) = \alpha_1, \quad u_2(1) = \alpha_2 \tag{23.43}$$

Solving equations (23.42) and (23.43) gives

$$u_2 = (\alpha_2 - \alpha_1)x + \alpha_1 \tag{23.44}$$

while $u_1(x)$ is given by equation (23.27).

23.2.2 Example 2: higher-order boundary value problems (coupled equations) in 1D

Consider the vector equation

$$\frac{d^2\mathbf{u}}{dx^2} - \mathbf{A}\mathbf{u} = -\mathbf{f}(x); \quad \mathbf{u}(0) = \mathbf{0} = \mathbf{u}(1) \tag{23.45}$$

where $\mathbf{u}$ and $\mathbf{f}(x)$ are vectors with n-components and $\mathbf{A}$ is a constant $n \times n$ matrix. For $n = 2$, let $a_{ij}\{i,j = 1,2\}$ be real constants and consider the BVP in two dependent variables as defined by

$$\left.\begin{aligned} \frac{d^2u_1}{dx^2} - a_{11}u_1 - a_{12}u_2 = -f_1(x) \\ \frac{d^2u_2}{dx^2} - a_{21}u_1 - a_{22}u_2 = -f_2(x) \end{aligned}\right\}, \quad 0 < x < 1 \tag{23.46}$$

with homogeneous Dirichlet boundary conditions

$$u_j(0) = u_j(1) = 0; \quad j = 1,2. \tag{23.47}$$

Let

$$c_{im} = \langle u_i, w_m \rangle = \text{FFT of } u_i \tag{23.48}$$

Taking FFT, we get

$$\begin{aligned} -m^2\pi^2 c_{1m} - a_{11}c_{1m} - a_{12}c_{2m} &= -f_{1m} \\ -m^2\pi^2 c_{2m} - a_{21}c_{1m} - a_{22}c_{2m} &= -f_{2m}; \quad m = 1,2,\ldots \end{aligned}$$

$\Longrightarrow$

$$\begin{pmatrix} a_{11} + m^2\pi^2 & a_{12} \\ a_{21} & a_{22} + m^2\pi^2 \end{pmatrix} \begin{pmatrix} c_{1m} \\ c_{2m} \end{pmatrix} = \begin{pmatrix} f_{1m} \\ f_{2m} \end{pmatrix}$$

$$\mathbf{A}_m\mathbf{c}_m = \mathbf{f}_m, \quad \mathbf{A}_m = \mathbf{A} + m^2\pi^2\mathbf{I}. \tag{23.49}$$

We can solve these linear algebraic equations by the biorthogonal expansion

$$\mathbf{c}_m = \sum_{k=1}^{2} \frac{\mathbf{y}_{km}^*\mathbf{f}_m}{\mathbf{y}_{km}^*\mathbf{x}_{km}} \frac{1}{\lambda_{km}} \mathbf{x}_{km}, \tag{23.50}$$

where λ_{km}, $\mathbf{x}_{km}$ and $\mathbf{y}_{km}^*$ are eigenvalues, eigenvectors and eigenrows of $\mathbf{A}_m$. Taking the inverse Fourier transform of equation (23.50), we get

$$\mathbf{u}(x) = \sum_{m=1}^{\infty} \mathbf{c}_m w_m(x)$$

$$= \sum_{m=1}^{\infty} \sum_{k=1}^{2} \sqrt{2} \sin m\pi x \frac{\mathbf{y}_{km}^{*} \mathbf{f}_m}{\mathbf{y}_{km}^{*} \mathbf{x}_{km}} \frac{1}{\lambda_{km}} \mathbf{x}_{km} \tag{23.51}$$

where

$$\mathbf{f}_m = \int_0^1 \begin{bmatrix} f_1(\xi) \\ f_2(\xi) \end{bmatrix} \sqrt{2} \sin m\pi\xi \, d\xi. \tag{23.52}$$

Various special cases of this solution may be examined as in the previous example. The above procedure needs to be modified if λ_{km} can vanish for some k or m, since the homogeneous equation has nontrivial solution(s) when $\lambda_{km} = 0$.

23.3 FFT for parabolic, hyperbolic and elliptic PDEs (two independent variables)

23.3.1 Example 3: heat/diffusion equation in a finite domain

Consider the heat equation in dimensionless form

$$\frac{\partial u}{\partial t} = \frac{\partial^2 u}{\partial x^2}; \quad 0 < x < 1, \quad t > 0 \tag{23.53}$$

with homogeneous boundary conditions

$$u(0, t) = 0, \quad u(1, t) = 0 \tag{23.54}$$

and the initial condition

$$u(x, 0) = f(x). \tag{23.55}$$

Once again, the relevant spatial operator is

$$\mathbb{L}w : -\frac{d^2 w}{dx^2}, \quad w(0) = 0, \quad w(1) = 0, \tag{23.56}$$

which has eigenvalues

$$\lambda_n = n^2\pi^2; \; n = 1, 2, \ldots \tag{23.57}$$

and normalized eigenfunctions:

$$w_n(x) = \sqrt{2} \sin n\pi x \tag{23.58}$$

Taking inner product of equations (23.53)–(23.55) with $w_n(x)$ gives

$$\begin{aligned}\frac{d}{dt}\langle u, w_n\rangle &= \left\langle \frac{\partial^2 u}{\partial x^2}, w_n \right\rangle \\ &= \left\langle u, \frac{d^2 w_n}{dx^2} \right\rangle + \left(\frac{\partial u}{\partial x} w_n - u \frac{dw_n}{dx} \right)_{x=0}^{x=1} \\ &= \langle u, -\lambda_n w_n\rangle \\ &= -\lambda_n \langle u, w_n\rangle\end{aligned}$$

(Remark: The concomitant vanishes again as both $u(x,t)$ and $w_n(x)$ satisfy the homogeneous boundary conditions.) Integrating this equation using initial condition, we get

$$\langle u, w_n\rangle = \langle f, w_n\rangle e^{-\lambda_n t} \tag{23.59}$$

Thus,

$$u(x,t) = \sum_{n=1}^{\infty} \langle f, w_n\rangle e^{-\lambda_n t} w_n(x) \tag{23.60}$$

is the solution. Substituting for λ_n and w_n gives

$$u(x,t) = \sum_{n=1}^{\infty} (\sqrt{2}\sin n\pi x) e^{-n^2\pi^2 t} \int_0^1 f(\xi)\sqrt{2}\sin n\pi\xi \, d\xi$$

$\Longrightarrow$

$$u(x,t) = 2\sum_{n=1}^{\infty} e^{-n^2\pi^2 t} \sin n\pi x \int_0^1 f(\xi)\sin n\pi\xi \, d\xi \tag{23.61}$$

We now examine some special cases of solution given by equation (23.61).

Special case 1: $f(\xi) = 1 \Rightarrow$

$$\begin{aligned}\int_0^1 f(\xi)\sin n\pi\xi \, d\xi &= -\frac{\cos n\pi\xi}{n\pi}\bigg|_0^1 \\ &= \frac{1-\cos n\pi}{n\pi} \\ &= \begin{cases} 0 & \text{if } n = 2k \\ \frac{2}{(2k-1)\pi} & \text{if } n = 2k-1,\ k = 1,2,\dots \end{cases} \\ \Rightarrow u(x,t) &= \frac{4}{\pi}\sum_{k=1}^{\infty} \frac{e^{-(2k-1)^2\pi^2 t}\sin(2k-1)\pi x}{(2k-1)}\end{aligned} \tag{23.62}$$

The maximum value of u (i. e. $u_{\max}$) occurs at $x = \frac{1}{2}$ and is given by

$$u_{\max} = u\left(\frac{1}{2}, t\right) = \frac{4}{\pi}\sum_{k=1}^{\infty}\frac{(-1)^{k-1}e^{-(2k-1)^2\pi^2 t}}{(2k-1)}$$
$$= \frac{4}{\pi}\left[e^{-\pi^2 t} - \frac{1}{3}e^{-9\pi^2 t} + \frac{1}{5}e^{-25\pi^2 t} - \cdots\right] \approx \frac{4}{\pi}e^{-\pi^2 t} \quad \text{for } t \to \infty \tag{23.63}$$

The spatial average value of u ($\langle u\rangle$) is given by

$$\langle u\rangle = \int_0^1 u\,dx = \frac{4}{\pi^2}\sum_{k=1}^{\infty}\frac{e^{-(2k-1)^2\pi^2 t}}{(2k-1)^2}[-\cos(2k-1)\pi x]_0^1$$
$$= \frac{8}{\pi^2}\sum_{k=1}^{\infty}\frac{e^{-(2k-1)^2\pi^2 t}}{(2k-1)^2} = \frac{8}{\pi^2}\left[e^{-\pi^2 t} + \frac{1}{9}e^{-9\pi^2 t} + \frac{1}{25}e^{-25\pi^2 t} + \cdots\right] \tag{23.64}$$

The profiles for different times are shown in Figure 23.4. From equation (23.64), it may be observed that for $t > 0.0016$, the first term is sufficient to estimate the average value of $u(x, t)$ to within 5 %.

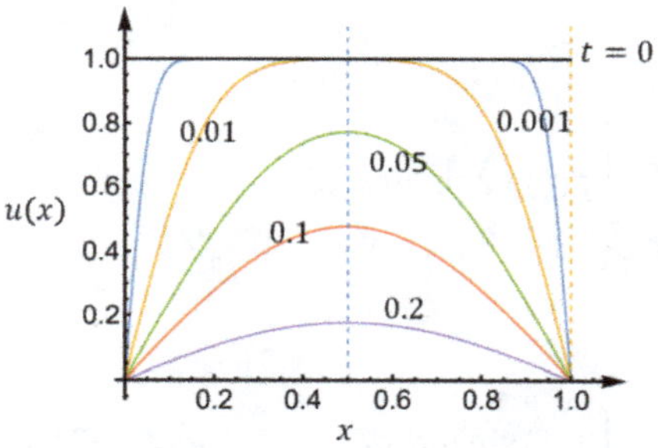

Figure 23.4: Dimensionless temperature distribution and its evolution with time with uniform initial source.

Special case 2: $f(x) = \sin[m\pi x]$, $m = 1, 2, \ldots \Longrightarrow$

$$2\int_0^1 f(\xi)\sin n\pi\xi\,d\xi = \delta_{nm} = \begin{cases} 1, & n = m \\ 0, & n \neq m \end{cases}$$

Equation (23.61) $\Longrightarrow$

$$u(x, t) = \sum_{n=1}^{\infty} e^{-n^2\pi^2 t}(\sin n\pi x)\,\delta_{nm} = e^{-m^2\pi^2 t}\sin[m\pi x] \tag{23.65}$$

Thus, when the initial condition corresponds to a mode represented by an eigenfunction, the solution remains in the same mode with amplitude decaying exponentially in time [with reciprocal of the eigenvalue as the decay constant]. The solutions corresponding to the first two modes $m = 1$ and $m = 2$ are shown in Figure 23.5.

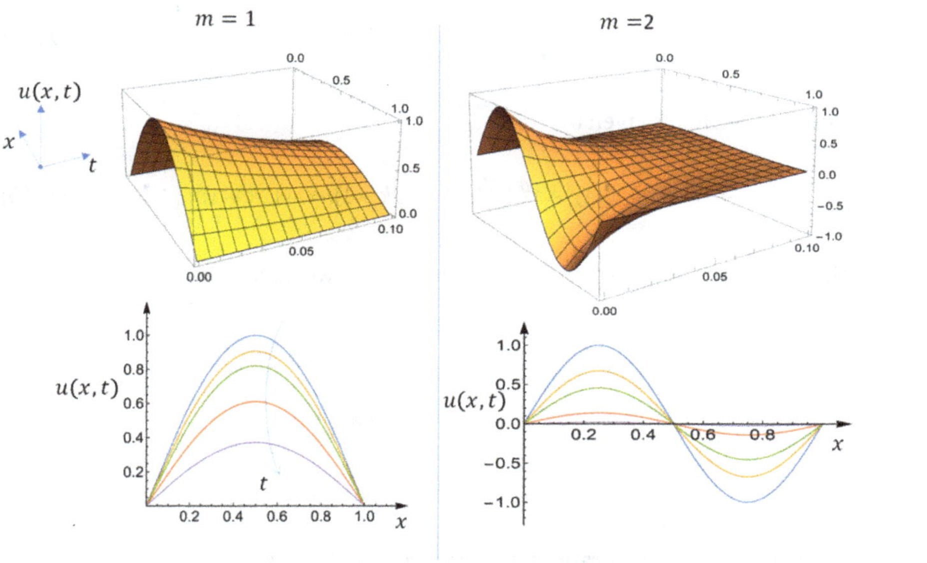

Figure 23.5: Solution of heat equation with initial conditions given by $f(x) = \sin[m\pi x]$, $m = 1, 2$.

Special case 3: $f(x) = \delta(x - s) \Longrightarrow$

$$u(x,t) = 2\sum_{n=1}^{\infty} e^{-n^2\pi^2 t} \sin n\pi x \sin n\pi s \tag{23.66}$$

Take $s = \frac{1}{2}$ (mid-point of the interval)

$\Longrightarrow$

$$\sin\frac{n\pi}{2} = \begin{cases} 0, & n \text{ even} \\ \pm 1, & n \text{ odd} \end{cases}$$

Take $n = 2k + 1$, $k = 0, 1, 2, \ldots$

$\Longrightarrow$

$$\begin{aligned} \sin\left\{\frac{(2k+1)\pi}{2}\right\} &= \sin\left(k\pi + \frac{\pi}{2}\right) \\ &= (-1)^k \end{aligned}$$

$\therefore$

$$u(x,t) = 2\sum_{k=0}^{\infty} (-1)^k e^{-(2k+1)^2\pi^2 t} \sin[(2k+1)\pi x]. \tag{23.67}$$

The spatial average value of $u(x,t)$ is given by

$$\langle u\rangle = \frac{4}{\pi}\sum_{k=0}^{\infty}\frac{(-1)^k}{(2k+1)}e^{-(2k+1)^2\pi^2 t}. \tag{23.68}$$

The dimensionless flux is given by

$$\frac{\partial u}{\partial x} = 2\pi\sum_{k=0}^{\infty}(-1)^k e^{-(2k+1)^2\pi^2 t}[\cos(2k+1)\pi x]\cdot(2k+1)\pi. \tag{23.69}$$

Evaluating the dimensionless flux at the left boundary, we have

$$\begin{aligned}\left.\frac{\partial u}{\partial x}\right|_{x=0} &= 2\pi\sum_{k=0}^{\infty}(-1)^k e^{-(2k+1)^2\pi^2 t}\cdot(2k+1)\pi\\ &= 2\sum_{k=0}^{\infty}(-1)^k(2k+1)e^{-(2k+1)^2\pi^2 t}\\ &= 2[e^{-\pi^2 t} - 3e^{-9\pi^2 t} + 5e^{-25\pi^2 t} - \cdots]\end{aligned} \tag{23.70}$$

The peak value of the concentration/temperature is at the center and is given by

$$\begin{aligned}u\left(\frac{1}{2},t\right) &= 2\sum_{k=0}^{\infty}e^{-(2k+1)^2\pi^2 t}\\ &= 2[e^{-\pi^2 t} + e^{-9\pi^2 t} + e^{-25\pi^2 t} + \cdots]\end{aligned} \tag{23.71}$$

This series converges for all $t > 0$. However, the convergence may be very slow for t values close to zero. Some schematic profiles are shown in Figure 23.6.

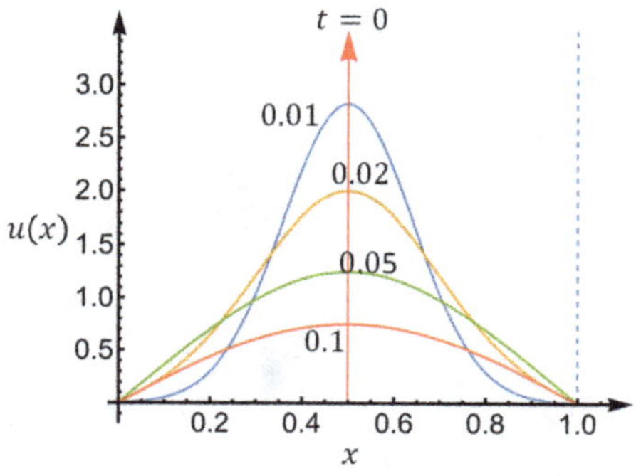

Figure 23.6: Dimensionless temperature distribution and its evolution with time with point initial source at the center of the domain.

23.3.2 Example 4: Green's function for the heat/diffusion equation in a finite domain

Consider the heat/diffusion equation with a source term

$$\frac{\partial u}{\partial t} - \frac{\partial^2 u}{\partial x^2} = f(x,t), \quad 0 < x < 1, \quad t > 0 \tag{23.72}$$

IC:

$$u(x,0) = 0 \tag{23.73}$$

BCs:

$$u(0,t) = u(1,t) = 0 \tag{23.74}$$

Let $\lambda_n = n^2\pi^2$ and $w_n(x) = \sqrt{2}\sin n\pi x$ be the eigenvalues and normalized eigenfunctions of the operator $-\frac{d^2w}{dx^2}$ with $w(0) = w(1) = 0$. Let

$$u_n = \langle u, w_n \rangle$$
$$f_n = \langle f, w_n \rangle$$

Taking FFT of equations (23.72)–(23.74) gives

$$\frac{du_n}{dt} + n^2\pi^2 u_n = f_n(t) \tag{23.75}$$

$$u_n = 0 \; @ \; t = 0 \tag{23.76}$$

Integrating equation (23.75), we get

$$u_n e^{n^2\pi^2 t} = \int_0^t e^{n^2\pi^2 t'} f_n(t')\, dt' + c_n$$

Equation (23.76) gives $c_n = 0$. Thus,

$$u_n = \int_0^t e^{n^2\pi^2 (t'-t)} f_n(t')\, dt'$$

$\Longrightarrow$

$$u = \sum_{n=1}^{\infty} \sqrt{2}\sin n\pi x \int_0^t e^{n^2\pi^2(t'-t)} \left[\int_0^1 \sqrt{2} f(s,t') \sin n\pi s \, ds \right] dt'$$

$\Longrightarrow$

$$u(x,t) = 2\sum_{n=1}^{\infty} \sin n\pi x \int_0^t \int_0^1 e^{n^2\pi^2(t'-t)} f(s,t') \sin n\pi s \, ds \, dt' \tag{23.77}$$

$$= \int_0^t \int_0^1 G(x,s,t,t') f(s,t') \, ds \, dt' \tag{23.78}$$

where

$$G(x,s,t,t') = 2\sum_{n=1}^{\infty} e^{n^2\pi^2(t'-t)} \cdot \sin n\pi x \sin n\pi s \tag{23.79}$$

is the Green's function. We note that if

$$f(x,t) = \delta(x-\xi)\delta(t-\tau)$$

then

$$u(x,t) = G(x,\xi,t,\tau).$$

Thus, $G(x,\xi,t,\tau)$ is the temperature (or concentration) at position x and time t due to a unit source at position ξ at time $\tau(t > \tau)$. We now consider some special cases of the solution given above.

(i) $f(x,t) = g_1(x)\delta(t)$

For this case, the solution is given by

$$u(x,t) = \sum_{n=1}^{\infty} e^{-n^2\pi^2 t}(2\sin n\pi x) \int_0^1 g_1(s) \sin n\pi s \, ds \tag{23.80}$$

This solution is identical to the solution obtained when we take the initial condition to be $u(x,0) = g_1(x)$

(ii) $f(x,t) = g_2(t)\delta(x-x_0)$

This corresponds to a point source at $x = x_0$ whose magnitude $g_2(t)$ varies with time. For this case, the solution simplifies to

$$u(x,t) = 2\sum_{n=1}^{\infty} (\sin n\pi x)(\sin n\pi x_0) \int_0^t e^{n^2\pi^2(t'-t)} g_2(t') \, dt'. \tag{23.81}$$

23.3.3 Example 5: heat/diffusion equation in a finite domain with time dependent boundary condition

Consider

$$\frac{\partial u}{\partial t} = \frac{\partial^2 u}{\partial x^2}; \quad 0 < x < 1, \quad t > 0 \tag{23.82}$$

IC:

$$u(x,0) = 0 \tag{23.83}$$

BCs:

$$u(0,t) = f(t); \quad u(1,t) = 0 \tag{23.84}$$

Taking FFT of equations (23.82)–(23.83), we get

$$\begin{aligned}
\frac{du_n}{dt} &= \int_0^1 \frac{\partial^2 u}{\partial x^2} \cdot w_n(x)\,dx \\
&= \int_0^1 u \frac{d^2 w_n}{dx^2}\,dx + \left[\frac{\partial u}{\partial x} \cdot w_n - u w_n'(x)\right]_{x=0}^{x=1} \\
&= -n^2\pi^2 u_n + \frac{\partial u}{\partial x}(1,t) w_n(1) - u(1,t) w_n'(1) \\
&\quad - \frac{\partial u}{\partial x}(0,t) w_n(0) + u(0,t) w_n'(0)
\end{aligned}$$

[Remark: Now, the concomitant does not vanish due to the inhomogeneous boundary condition at $x = 0$.] $\Longrightarrow$

$$\frac{du_n}{dt} = -n^2\pi^2 u_n + f(t) w_n'(0)$$

with

$$w_n = \sqrt{2}\sin n\pi x, \quad w_n' = \sqrt{2}n\pi \cos n\pi x \Longrightarrow w_n'(0) = \sqrt{2}n\pi \tag{23.85}$$

$\therefore$

$$\frac{du_n}{dt} = -n^2\pi^2 u_n + \sqrt{2}n\pi f(t) \tag{23.86}$$

$$u_n = 0 \ @\ t = 0 \tag{23.87}$$

$\Longrightarrow$

$$u_n = \int_0^t e^{-n^2\pi^2(t-t')} \sqrt{2}n\pi f(t')\,dt' \tag{23.88}$$

$\Longrightarrow$

$$u(x,t) = \sum_{n=1}^{\infty} \sqrt{2}\sin n\pi x \int_0^t e^{-n^2\pi^2(t-t')} \sqrt{2}n\pi f(t')\,dt'$$

$$= \int_0^t \sum_{n=1}^{\infty} (2n\pi \sin n\pi x) e^{-n^2\pi^2(t-t')} f(t')\, dt' \tag{23.89}$$

Special case:

$$f(t) = H(t) = \begin{cases} 1, & t > 0 \\ 0, & t < 0 \end{cases}$$

$$\int_0^t e^{-n^2\pi^2(t-t')}\, dt' = \left. \frac{e^{-n^2\pi^2 t} \cdot e^{n^2\pi^2 t'}}{n^2\pi^2} \right|_0^t = \frac{1}{n^2\pi^2} [1 - e^{-n^2\pi^2 t}] \tag{23.90}$$

$\Longrightarrow$

$$u(x,t) = \sum_{n=1}^{\infty} \frac{2 \sin n\pi x}{n\pi} [1 - e^{-n^2\pi^2 t}]$$

$\Longrightarrow$

$$u(x,\infty) = \sum_{n=1}^{\infty} \frac{2 \sin n\pi x}{n\pi} = 1 - x$$

$\Longrightarrow$

$$u(x,t) = 1 - x - \sum_{n=1}^{\infty} e^{-n^2\pi^2 t} \frac{2 \sin n\pi x}{n\pi} \tag{23.91}$$

$$= u_s(x) - \frac{2}{\pi} \sum_{n=1}^{\infty} \frac{e^{-n^2\pi^2 t} \sin n\pi x}{n} \tag{23.92}$$

where $u_s(x)$ is the steady-state profile. The profiles at various times are shown in Figure 23.7.

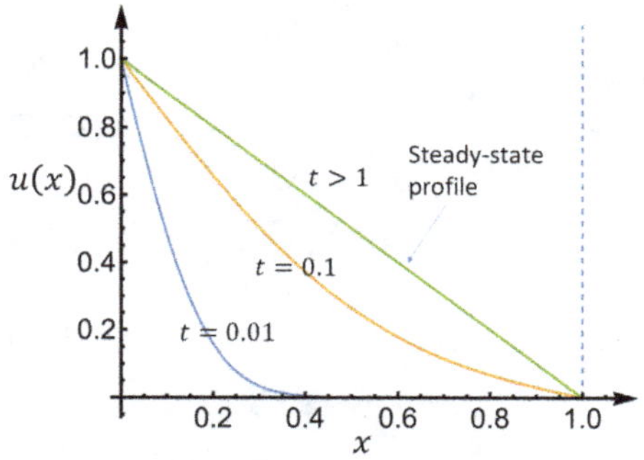

Figure 23.7: Dimensionless temperature distribution and steady-state profile for sudden rise of temperature to unity at the left boundary.

23.3.4 Example 6: heat/diffusion equation in a finite domain with general initial and boundary conditions

Consider

$$\frac{\partial u}{\partial t} = \frac{\partial^2 u}{\partial x^2}, \quad 0 < x < 1 \tag{23.93}$$

BCs:

$$u(0,t) = f_1(t), \quad u(1,t) = f_2(t) \tag{23.94}$$

IC:

$$u(x,0) = g(x) \tag{23.95}$$

We use the principle of superposition and write

$$u = u_1 + u_2 + u_3 \tag{23.96}$$

where

$$\frac{\partial u_1}{\partial t} = \frac{\partial^2 u_1}{\partial x^2}, \quad \begin{cases} \text{BCs:} & u_1(0,t) = f_1(t), \quad u_1(1,t) = 0 \\ \text{I. C.:} & u_1(x,0) = 0 \end{cases} \tag{23.97}$$

$$\frac{\partial u_2}{\partial t} = \frac{\partial^2 u_2}{\partial x^2}, \quad \begin{cases} \text{BCs:} & u_2(0,t) = 0, \quad u_2(1,t) = f_2(t) \\ \text{I. C.:} & u_2(x,0) = 0 \end{cases} \tag{23.98}$$

$$\frac{\partial u_3}{\partial t} = \frac{\partial^2 u_3}{\partial x^2}, \quad \begin{cases} \text{BCs:} & u_3(0,t) = 0, \quad u_3(1,t) = 0 \\ \text{I. C.:} & u_3(x,0) = g(x) \end{cases} \tag{23.99}$$

Each of these problems has been solved before.

23.3.5 Example 7 (wave equation)

Consider the wave equation

$$\frac{\partial^2 u}{\partial t^2} = c^2 \frac{\partial^2 u}{\partial x^2}, \quad 0 < x < 1, \quad t > 0 \tag{23.100}$$

BCs:

$$u(0,t) = 0, \quad u(1,t) = 0 \quad \text{(fixed ends)} \tag{23.101}$$

ICs:

$$u(x,0) = f(x) \quad \text{(initial displacement)} \tag{23.102}$$

$$\frac{\partial u}{\partial t}(x,0) = g(x) \quad \text{(initial velocity)} \tag{23.103}$$

For the operator,

$$\frac{d^2w}{dx^2}, \quad 0 < x < 1, \quad w(0) = 0, \quad w(1) = 0 \tag{23.104}$$

the eigenvalues are

$$\lambda_n = -n^2\pi^2, \quad n = 1, 2, \ldots \tag{23.105}$$

while the normalized eigenfunctions are

$$w_n(x) = \sqrt{2} \sin n\pi x \tag{23.106}$$

Taking inner product of equations (23.100)–(23.103) with $w_n(x)$ gives

$$\frac{d^2}{dt^2}\langle u, w_n\rangle = -c^2n^2\pi^2\langle u, w_n\rangle \Rightarrow \langle u, w_n\rangle = c_{1n}\cos n\pi ct + c_{2n}\sin n\pi ct \tag{23.107}$$

IC1:

$$c_{1n} = \langle f, w_n\rangle \tag{23.108}$$

IC2:

$$c_{2n} \cdot n\pi c = \langle g, w_n\rangle \tag{23.109}$$

$\therefore$

$$\langle u, w_n\rangle = \langle f, w_n\rangle \cos n\pi ct + \frac{\langle g, w_n\rangle}{n\pi c}\sin n\pi ct \tag{23.110}$$

Taking inverse FFT gives

$$u(x,t) = \sum_{n=1}^{\infty} \sqrt{2}\sin n\pi x \left[\cos n\pi ct \int_0^1 f(\xi)\sqrt{2}\sin n\pi\xi \, d\xi + \frac{\sin n\pi ct}{n\pi c}\int_0^1 g(\xi)\sqrt{2}\sin n\pi\xi \, d\xi\right]$$

$\Rightarrow$

$$u(x,t) = 2\sum_{n=1}^{\infty} \sin n\pi x \left[\cos n\pi ct \int_0^1 f(\xi)\sin n\pi\xi \, d\xi + \frac{\sin n\pi ct}{n\pi c}\int_0^1 g(\xi)\sin n\pi\xi \, d\xi\right] \tag{23.111}$$

Consider the special case in which

$$g(\xi) = 0 \quad \text{(zero initial velocity)} \tag{23.112}$$

$$f(\xi) = \alpha \sin j\pi\xi \quad \text{(initial displacement is in the shape of } j\text{-th eigenfunction).} \tag{23.113}$$

The solution simplifies to

$$\begin{aligned} u(x,t) &= 2 \cdot \sin j\pi x \cos j\pi ct \cdot \frac{1}{2}\alpha \\ &= (\cos j\pi ct)(\alpha \sin j\pi x) \end{aligned} \tag{23.114}$$

This is a pure mode of vibration, which is periodic in time with a period

$$T = \frac{2\pi}{j\pi c} = \frac{2}{jc} \tag{23.115}$$

and

$$\text{cyclic frequency} = \frac{jc}{2} \tag{23.116}$$

The solution profile for specific case of $c = 1$ and $j = 10$ is shown in Figure 23.8.

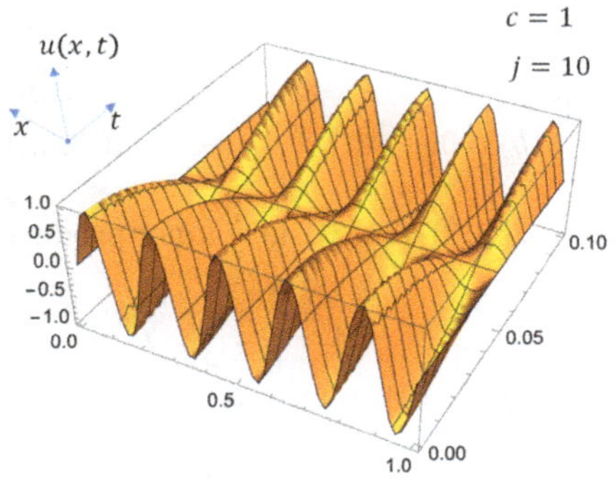

Figure 23.8: Solution of wave equation for $c = 1$ and initial displacement $f(x) = \sin[10\pi x]$.

[Remark: The cyclic frequency of mode $j = 1$ is called the *fundamental frequency* while the modes with $j \geq 2$ are referred to as *overtones* or *harmonics*.]

23.3.6 Example 8 (Poisson's equation in 2-D)

Consider the equation

$$\frac{\partial^2 u}{\partial x^2} + \frac{\partial^2 u}{\partial y^2} = -f(x,y); \quad 0 < x < a; \quad 0 < y < b \tag{23.117}$$

$$u = 0 \quad \text{on } \partial\Omega \tag{23.118}$$

This is the Poisson equation in a rectangle ($\Omega = (x, y) : 0 < x < a, 0 < y < b$) and in applications, $u(x, y)$ represents the temperature, concentration, pressure or velocity (at steady-state) due to a source $f(x, y)$. Equations (23.117)–(23.118) contain two operators:

$$\mathbb{L}_1 : -\frac{d^2}{dx^2}, \quad w(0) = 0, \quad w(a) = 0$$

eigenvalues $\lambda_n = \frac{n^2\pi^2}{a^2}$
eigenfunctions $w_n(x) = \sqrt{\frac{2}{a}} \sin(\frac{n\pi x}{a})$; $n = 1, 2, \ldots$

$$\mathbb{L}_2 : -\frac{d^2}{dy^2}, \quad w(0) = 0, \quad w(b) = 0$$

eigenvalues $\lambda_m = \frac{m^2\pi^2}{b^2}$
eigenfunctions $w_m(y) = \sqrt{\frac{2}{b}} \sin(\frac{m\pi y}{b})$; $m = 1, 2, \ldots$
Consider the eigenvalue problem

$$\frac{\partial^2 w}{\partial x^2} + \frac{\partial^2 w}{\partial y^2} = -\lambda w \tag{23.119}$$

$$w(0, y) = 0, \quad w(a, y) = 0 \tag{23.120}$$

$$w(x, 0) = 0, \quad w(x, b) = 0 \tag{23.121}$$

To determine the eigenvalues and eigenfunction of (23.119)–(23.121), take inner product with $w_n(x) \Longrightarrow$

$$-\lambda_n\langle w(x, y), w_n(x)\rangle + \frac{d^2}{dy^2}\langle w(x, y), w_n(x)\rangle = -\lambda\langle w(x, y), w_n(x)\rangle$$

Take the inner product again w. r. t. $w_m(y)$
$\Longrightarrow$

$$-\lambda_n\langle\langle w, w_n\rangle, w_m\rangle - \lambda_m\langle\langle w, w_n\rangle, w_m\rangle = -\lambda\langle\langle w, w_n\rangle, w_m\rangle \tag{23.122}$$

If w_{nm} is an eigenfunction and λ_{nm} is an eigenvalue of (23.119)–(23.121), then equation (23.122) $\Longrightarrow$

$$\lambda_{nm} = \lambda_n + \lambda_m = \pi^2\left(\frac{n^2}{a^2} + \frac{m^2}{b^2}\right) \tag{23.123}$$

$$\begin{aligned} w_{nm}(x, y) &= w_n(x)w_m(y) \\ &= \sqrt{\frac{4}{ab}} \sin\left(\frac{n\pi x}{a}\right) \sin\left(\frac{n\pi y}{b}\right) \end{aligned} \tag{23.124}$$

These are the eigenvalues and eigenfunctions of the Laplacian operator

$$\mathbb{L}w : -\left(\frac{\partial^2 w}{\partial x^2} + \frac{\partial^2 w}{\partial y^2}\right), \quad \begin{cases} w(0,y) = 0, & w(a,y) = 0 \\ w(x,0) = 0, & w(x,b) = 0 \end{cases} \tag{23.125}$$

in the domain Ω. Now, to solve equations (23.117)–(23.118) take inner product with w_{nm}, $\Longrightarrow$

$$\begin{aligned} \langle \mathbb{L}u, w_{nm}\rangle &= \langle f, w_{nm}\rangle \\ \langle u, \mathbb{L}^* w_{nm}\rangle &= \langle f, w_{nm}\rangle; \quad \mathbb{L}^* = \mathbb{L} \\ \lambda_{nm}\langle u, w_{nm}\rangle &= \langle f, w_{nm}\rangle \end{aligned}$$

$\Longrightarrow$

$$\langle u, w_{nm}\rangle = \frac{\langle f, w_{nm}\rangle}{\lambda_{nm}}$$

$\Longrightarrow$

$$\begin{aligned} u &= \sum_{n=1}^{\infty}\sum_{m=1}^{\infty} \langle u, w_{nm}\rangle w_{nm} \\ &= \sum_{n=1}^{\infty}\sum_{m=1}^{\infty} \frac{1}{\pi^2(\frac{n^2}{a^2} + \frac{m^2}{b^2})} \frac{2}{\sqrt{ab}} \sin\left(\frac{n\pi x}{a}\right)\sin\left(\frac{m\pi y}{b}\right)\langle f, w_{nm}\rangle \\ \langle f, w_{nm}\rangle &= \int_0^a\int_0^b f(\xi,\eta)\frac{2}{\sqrt{ab}} \sin\left(\frac{n\pi\xi}{a}\right)\sin\left(\frac{m\pi\eta}{b}\right) d\xi d\eta \end{aligned}$$

$\Longrightarrow$

$$u(x,y) = \frac{4}{ab\pi^2}\sum_{n=1}^{\infty}\sum_{m=1}^{\infty} \frac{\sin(\frac{n\pi x}{a})\sin(\frac{m\pi y}{b})}{(\frac{n^2}{a^2} + \frac{m^2}{b^2})} \int_0^a\int_0^b f(\xi,\eta)\sin\left(\frac{n\pi\xi}{a}\right)\sin\left(\frac{m\pi\eta}{b}\right) d\xi d\eta. \tag{23.126}$$

We now consider some special cases of this solution.

Special case 1: $f(x,y) = 1$

For special case of $f(x,y) = 1$, the integration term in equation (23.126) simplifies to

$$\begin{aligned} I &= \int_0^a\int_0^b f(\xi,\eta)\sin\left(\frac{n\pi\xi}{a}\right)\sin\left(\frac{m\pi\eta}{b}\right) d\xi d\eta \\ &= \int_0^a \sin\left(\frac{n\pi\xi}{a}\right) d\xi \int_0^b \sin\left(\frac{m\pi\eta}{b}\right) d\eta \\ &= \begin{cases} \frac{4ab}{nm\pi^2}, & n \text{ and } m \text{ odd} \\ 0, & \text{if } n \text{ or } m \text{ even} \end{cases} \end{aligned}$$

$\Longrightarrow$ from equation (23.126)

$$u(x,y) = \frac{16}{\pi^4}\sum_{i=1}^{\infty}\sum_{j=1}^{\infty}\frac{\sin[\frac{(2i-1)\pi x}{a}]\sin[\frac{(2j-1)\pi y}{b}]}{(\frac{(2i-1)^2}{a^2}+\frac{(2j-1)^2}{b^2})(2i-1)(2j-1)} \tag{23.127}$$

which further implies the solution for $a = 1$ and $b = 1$ (unit square) as

$$u(x,y) = \frac{16}{\pi^4}\sum_{i=1}^{\infty}\sum_{j=1}^{\infty}\frac{\sin[(2i-1)\pi x]\sin[(2j-1)\pi y]}{((2i-1)^2+(2j-1)^2)(2i-1)(2j-1)} \tag{23.128}$$

The solution (equation (23.128)) is plotted in Figure 23.9.

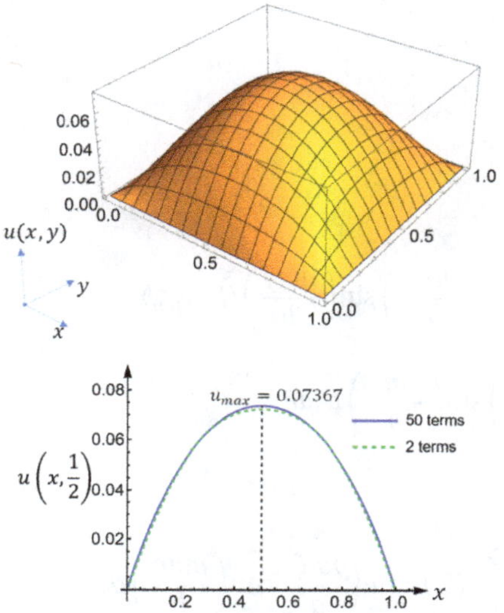

Figure 23.9: Solution of 2D Poisson's equation in a square domain with $f(x,y) = 1$.

The top diagram shows a 3D plot of the solution in xy domain with $50 \times 50 = 2500$ terms in the summation, while the bottom diagram corresponds to $u(x, y = \frac{1}{2})$ versus x using the Fourier series expansion with $2 \times 2 = 4$ terms (green dashed line) and 50×50 terms (blue solid lines). It can be seen from the bottom plot that the Fourier series solution with only 2×2 terms is sufficient to predict the solution with good accuracy in this case. The maximum value $u_{\max} = u(\frac{1}{2}, \frac{1}{2})$ is 0.07219 using the 2×2 terms as compared to 0.07367 using 50×50 terms in summation. [Remark: Extending the result to 3D, it may be shown that $u_{\max} = 0.05621$. Also, recall from example 1, for the 1D case, $u_{\max} = 0.125$.]

Special case 2: $f(x,y) = \delta(x - \frac{a}{2})\delta(y - \frac{b}{2})$

Another special case of interest is a point source at the center of the domain, i. e.,

$$f(x,y) = \delta\left(x - \frac{a}{2}\right)\delta\left(y - \frac{b}{2}\right).$$

The solution (equation (23.126)) can be simplified for this case as follows:

$$u(x,y) = \frac{4}{ab\pi^2}\sum_{j=1}^{\infty}\sum_{k=1}^{\infty}(-1)^{k+j}\frac{\sin[\frac{(2k-1)\pi x}{a}]\sin[\frac{(2j-1)\pi y}{b}]}{[(\frac{2k-1}{a})^2+(\frac{2j-1}{b})^2]}, \tag{23.129}$$

which can further be simplified for the case of $a = 1$ and $b = 1$ (unit square) as

$$u(x,y) = \frac{4}{\pi^2}\sum_{j=1}^{\infty}\sum_{k=1}^{\infty}(-1)^{k+j}\frac{\sin[(2k-1)\pi x]\sin[(2j-1)\pi y]}{[(2k-1)^2+(2j-1)^2]}. \tag{23.130}$$

A plot of this solution is shown in Figure 23.10.

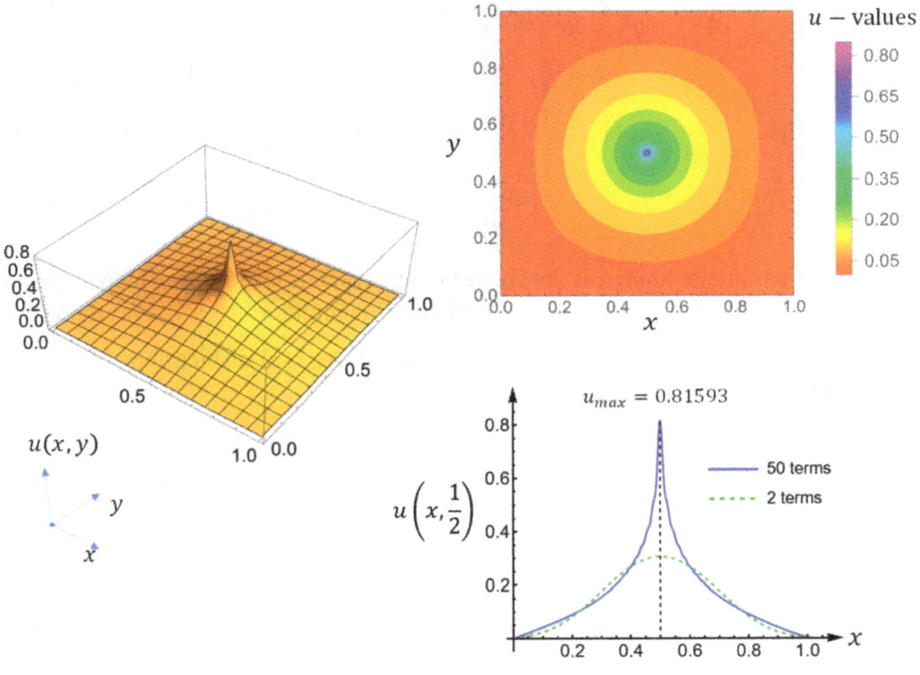

Figure 23.10: Solution of 2D Poisson's equation in a square domain with $f(x,y) = \delta(x-\frac{1}{2})\delta(y-\frac{1}{2})$.

The left diagram shows a 3D plot of the solution and the top-right diagram shows a contour plot of solution in xy domain with $50 \times 50 = 2500$ terms in the summation, while the bottom-right diagram corresponds to $u(x, y = \frac{1}{2})$ versus x using the Fourier series expansion with $2 \times 2 = 4$ terms (green dashed line) and 50×50 terms (blue solid lines). It can be seen from the bottom plot that the Fourier series solution with only 2×2 terms is sufficient to predict the solution in most of the domain. However, the maximum error occurs at the center (i. e., $x = \frac{1}{2}$), where the maximum value $u_{\max} = u(\frac{1}{2}, \frac{1}{2})$ is 0.3062 using the 2 × 2 terms as compared to 0.81593 using the 50 × 50 terms in summation. In this case, the convergence to the value at the center is very slow as it is the location of the source.

23.4 Additional applications of FFT in rectangular coordinates

23.4.1 Example 9 (diffusion and reaction in a catalyst cube)

The dimensionless model describing the problem of diffusion–reaction in a catalyst particle that is in the form of a cube may be expressed as

$$\frac{\partial^2 Y}{\partial \xi^2} + \frac{\partial^2 Y}{\partial \eta^2} + \frac{\partial^2 Y}{\partial \rho^2} - \Lambda^2 Y = 0 \tag{23.131}$$

$$\frac{\partial Y}{\partial \xi} = 0 \,@\, \xi = 0; \quad Y = 1 \,@\, \xi = 1 \tag{23.132}$$

$$\frac{\partial Y}{\partial \eta} = 0 \,@\, \eta = 0; \quad Y = 1 \,@\, \eta = 1 \tag{23.133}$$

$$\frac{\partial Y}{\partial \rho} = 0 \,@\, \rho = 0; \quad Y = 1 \,@\, \rho = 1 \tag{23.134}$$

where Λ is a parameter known as the Thiele modulus [Remark: The above model applies to $\frac{1}{8}$th of a cube. If we define the normalized Thiele modulus as $\Phi^2 = \frac{k}{D}(\frac{V_p}{S_p})^2$, then $\Phi^2 = \frac{\Lambda^2}{9}$. Here, V_p is the volume of the particle and S_p is the external surface area.]

Consider the operator

$$\mathbb{L}_1 : -\frac{\partial^2}{\partial \xi^2}, \quad Y'(0) = 0, \quad Y(1) = 0 \tag{23.135}$$

Eigenvalues

$$\lambda_n = (2n-1)^2 \frac{\pi^2}{4} \tag{23.136}$$

Eigenfunctions (normalized)

$$\sqrt{2}\cos\left\{\frac{(2n-1)\pi\xi}{2}\right\} \tag{23.137}$$

Define $U = 1 - Y$. Then equations (23.131)–(23.134) $\Longrightarrow$

$$\frac{\partial^2 U}{\partial \xi^2} + \frac{\partial^2 U}{\partial \eta^2} + \frac{\partial^2 U}{\partial \rho^2} + \Lambda^2 - \Lambda^2 U = 0 \tag{23.138}$$

$$\frac{\partial U}{\partial \xi} = 0 \,@\, \xi = 0; \quad U = 0 \,@\, \xi = 1 \tag{23.139}$$

$$\frac{\partial U}{\partial \eta} = 0 \ @ \ \eta = 0; \quad U = 0 \ @ \ \eta = 1 \tag{23.140}$$

$$\frac{\partial U}{\partial \rho} = 0 \ @ \ \rho = 0; \quad U = 0 \ @ \ \rho = 1 \tag{23.141}$$

The relevant eigenvalue problem is defined by

$$\frac{\partial^2 w}{\partial \xi^2} + \frac{\partial^2 w}{\partial \eta^2} + \frac{\partial^2 w}{\partial \rho^2} = -\lambda w \tag{23.142}$$

with BCs given by equation (23.139), (23.140), (23.141).
Eigenvalues

$$\lambda_{nml} = \frac{\pi^2}{4}[(2n-1)^2 + (2m-1)^2 + (2l-1)^2] \tag{23.143}$$

Eigenfunctions (normalized)

$$w_{nml} = 2\sqrt{2}\cos\left\{\frac{(2n-1)\pi\xi}{2}\right\}\cos\left\{\frac{(2m-1)\pi\eta}{2}\right\}\cos\left\{\frac{(2l-1)\pi\rho}{2}\right\} \tag{23.144}$$

Inner product of equations (23.138)–(23.141) with $w_{nml} \Longrightarrow$

$$-\lambda_{nml}\langle U, w_{nml}\rangle - \Lambda^2\langle U, w_{nml}\rangle + \Lambda^2\langle 1, w_{nml}\rangle = 0$$

$\Longrightarrow$

$$\begin{aligned}
\langle U, w_{nml}\rangle &= \frac{\Lambda^2\langle 1, w_{nml}\rangle}{\langle \Lambda^2 + \lambda_{nml}\rangle} \\
\langle 1, w_{nml}\rangle &= \left(\int_0^1 \sqrt{2}\cos\frac{(2n-1)\pi\xi}{2}\,d\xi\right) \\
&\quad \times \left(\int_0^1 \sqrt{2}\cos\frac{(2m-1)\pi\eta}{2}\,d\eta\right)\left(\int_0^1 \sqrt{2}\cos\frac{(2l-1)\pi\rho}{2}\,d\rho\right) \\
&= \frac{2\sqrt{2}(-1)^{n-1}}{(2n-1)\pi}\cdot\frac{2\sqrt{2}(-1)^{m-1}}{(2m-1)\pi}\cdot\frac{2\sqrt{2}(-1)^{l-1}}{(2l-1)\pi} \\
&= \frac{(2\sqrt{2})^3(-1)^{n+m+l-3}}{(2n-1)(2m-1)(2l-1)\pi^3}
\end{aligned}$$

$\therefore$

$$U(\xi,\eta,\rho) = \frac{64}{\pi^3}\sum_{n=1}^{\infty}\sum_{m=1}^{\infty}\sum_{l=1}^{\infty}\frac{\Lambda^2(-1)^{n+m+l-3}\cos[(2n-1)\frac{\pi\xi}{2}]\cos[(2m-1)\frac{\pi\eta}{2}]\cos[(2l-1)\frac{\pi\rho}{2}]}{(2n-1)(2m-1)(2l-1)[\Lambda^2+\lambda_{nml}]} \tag{23.145}$$

The effectiveness factor (or the average value of Y) is given by

$$\hat{\eta} = \int_0^1\int_0^1\int_0^1 Y\,d\xi\,d\eta\,d\rho = 1 - \int_0^1\int_0^1\int_0^1 U\,d\xi\,d\eta\,d\rho \tag{23.146}$$

$\Longrightarrow$

$$\hat{\eta} = 1 - \frac{512\Lambda^2}{\pi^6}\sum_{n=1}^{\infty}\sum_{m=1}^{\infty}\sum_{l=1}^{\infty}\frac{1}{(2n-1)^2(2m-1)^2(2l-1)^2[\Lambda^2+\lambda_{nml}]} \tag{23.147}$$

Remarks.

1. Solution of the same problem in the square geometry (two-dimensional case for which $\Phi^2 = \frac{\Lambda^2}{4}$) gives

$$\hat{\eta} = 1 - \frac{64\Lambda^2}{\pi^4}\sum_{n=1}^{\infty}\sum_{m=1}^{\infty}\frac{1}{(2n-1)^2(2m-1)^2[\Lambda^2+\frac{\pi^2}{4}(2n-1)^2+(2m-1)^2]} \tag{23.148}$$

   while for the slab geometry (one-dimensional case for which $\Phi^2 = \Lambda^2$),

$$\hat{\eta} = 1 - \frac{8\Lambda^2}{\pi^2}\sum_{n=1}^{\infty}\frac{1}{(2n-1)^2[\Lambda^2+\frac{\pi^2}{4}(2n-1)^2]} = \frac{\tanh\Lambda}{\Lambda} \tag{23.149}$$

   A plot of the effectiveness factor for 1D, 2D and 3D solutions is shown in Figure 23.11.
2. The above formulae may be used to obtain the small and large Φ asymptotes for all three cases. These asymptotes can also be visualized from Figure 23.11.

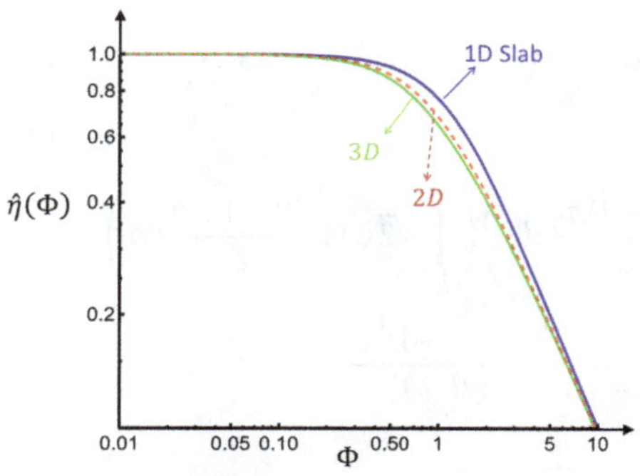

Figure 23.11: Effectiveness factor for 1D, 2D and 3D diffusion–reaction problems in rectangular coordinates.

23.4.2 Example 10 (axial dispersion model)

Consider the 1D transient diffusion–convection model with a position and time dependent source term:

$$\frac{\partial c}{\partial t} + \langle u\rangle \frac{\partial c}{\partial x} = D\frac{\partial^2 c}{\partial x^2} + S(x,t), \quad 0 < x < L, \quad t > 0 \tag{23.150}$$

with inlet (Danckwerts) BCs

$$-D\frac{\partial c}{\partial x} = \langle u\rangle [c_0(t) - c], \quad @\, x = 0, \tag{23.151}$$

exit condition

$$\frac{\partial c}{\partial x} = 0, \quad @\, x = L$$

and initial condition (IC)

$$c = c_i(x), \quad 0 < x < L, \quad t = 0. \tag{23.152}$$

The model defined by the above equations is known as the *axial dispersion model.*

Let c^* = a reference concentration. Define dimensionless variables

$$z = \frac{x}{L}, \quad \tau = \frac{\langle u\rangle t}{L} \tag{23.153}$$

$$C = \frac{c(x,t)}{c^*}, \quad \mathrm{Pe} = \frac{\langle u\rangle L}{D} \tag{23.154}$$

$\Longrightarrow$

$$c^* \frac{\partial C}{\partial \tau} \cdot \frac{\langle u\rangle}{L} + \frac{\langle u\rangle c^*}{L}\frac{\partial C}{\partial z} = \frac{Dc^*}{L^2}\frac{\partial^2 C}{\partial z^2} + S\left(Lz, \frac{L}{\langle u\rangle}\tau\right)$$

$\Longrightarrow$

$$\frac{\partial C}{\partial \tau} + \frac{\partial C}{\partial z} = \frac{1}{\mathrm{Pe}}\frac{\partial^2 C}{\partial z^2} + \frac{1}{c^*}\frac{L}{\langle u\rangle} S\left(Lz, \frac{L}{\langle u\rangle}\tau\right)$$

Let

$$s(z,\tau) = \frac{L}{\langle u\rangle c^*} S\left(Lz, \frac{L\tau}{\langle u\rangle}\right) = \text{dimensionless source term} \tag{23.155}$$

$\Rightarrow$

$$\frac{\partial C}{\partial \tau} + \frac{\partial C}{\partial z} = \frac{1}{\mathrm{Pe}}\frac{\partial^2 C}{\partial z^2} + s(z,\tau) \tag{23.156}$$

BC1 $\Rightarrow$

$$-\frac{D}{L}\frac{\partial C}{\partial z} = \langle u\rangle\left[\frac{c_0(\frac{L\tau}{\langle u\rangle})}{c^*} - C\right]$$

⇒

$$-\frac{1}{\text{Pe}}\frac{\partial C}{\partial z} = \frac{1}{c^*}c_0\left(\frac{L\tau}{\langle u\rangle}\right) - C$$

⇒

$$-\frac{1}{\text{Pe}}\frac{\partial C}{\partial z} + C = \hat{c}_0(\tau) \,@\, z = 0, \tag{23.157}$$

where

$$\hat{c}_0(\tau) = \frac{c_0(\frac{L\tau}{\langle u\rangle})}{c^*}. \tag{23.158}$$

BC2⇒

$$\frac{\partial C}{\partial z} = 0 \,@\, z = 1 \tag{23.159}$$

IC⇒

$$C = \frac{c_i(Lz)}{c^*} = \hat{c}_i(z) \,@\, \tau = 0 \tag{23.160}$$

Thus, the dimensionless form of the model is

$$\frac{\partial C}{\partial \tau} + \frac{\partial C}{\partial z} = \frac{1}{\text{Pe}}\frac{\partial^2 C}{\partial z^2} + s(z,\tau) \tag{23.161}$$

$$\frac{1}{\text{Pe}}\frac{\partial C}{\partial z} - C = -\hat{c}_0(\tau) \,@\, z = 0 \tag{23.162}$$

$$\frac{\partial C}{\partial z} = 0 \,@\, z = 1 \tag{23.163}$$

$$C = \hat{c}_i(z) \,@\, \tau = 0 \tag{23.164}$$

The dimensionless group Pe is the Peclet number, which is the ratio of diffusion to convection time scales. As noted in Chapter 17, it is difficult to solve the problem defined by equations (23.161)–(23.164) by the Laplace transform method except for some special cases of the source function $s(z,t)$ and initial conditions $\hat{c}_i(z)$. We obtain a formal solution for the general case using FFT, and examine various special cases.

The spatial operator appearing in equation (23.161) is not formally self-adjoint. To put in a self-adjoint form, we define

$$C = w(z,\tau).\exp\left[\frac{\text{Pe}\, z}{2}\right] \tag{23.165}$$

⇒

$$\frac{\partial C}{\partial \tau} = \frac{\partial w}{\partial \tau}\exp\left[\frac{\text{Pe}\, z}{2}\right]$$
$$\frac{\partial C}{\partial z} = \frac{\partial w}{\partial z}\exp\left[\frac{\text{Pe}\, z}{2}\right] + \frac{\text{Pe}}{2} w \exp\left[\frac{\text{Pe}\, z}{2}\right]$$
$$\frac{\partial^2 C}{\partial z^2} = \frac{\partial^2 w}{\partial z^2}\exp\left[\frac{\text{Pe}\, z}{2}\right] + \text{Pe}\frac{\partial w}{\partial z}\exp\left[\frac{\text{Pe}\, z}{2}\right] + \frac{\text{Pe}^2}{4} w \exp\left[\frac{\text{Pe}\, z}{2}\right]$$

⇒ Substituting in equation (23.161)

$$\frac{\partial w}{\partial \tau} + \frac{\partial w}{\partial z} + \frac{\text{Pe}}{2} w = \frac{1}{\text{Pe}}\left[\frac{\partial^2 w}{\partial z^2} + \text{Pe}\frac{\partial w}{\partial z} + \frac{\text{Pe}^2}{4} w\right] + s(z,\tau) e^{-\frac{\text{Pe}\, z}{2}}.$$

Denoting the last term by $s^*(z,\tau)$, the model becomes

$$\frac{\partial w}{\partial \tau} = \frac{1}{\text{Pe}}\frac{\partial^2 w}{\partial z^2} - \frac{\text{Pe}}{4} w + s^*(z,\tau) \tag{23.166}$$
$$\frac{1}{\text{Pe}}\frac{\partial w}{\partial z} - \frac{w}{2} = -\hat{c}_0(\tau) \;@\; z = 0 \tag{23.167}$$
$$\frac{1}{\text{Pe}}\frac{\partial w}{\partial z} + \frac{w}{2} = 0 \;@\; z = 1 \tag{23.168}$$
$$w = c_i(z)\exp\left(-\frac{\text{Pe}\, z}{2}\right) \;@\; \tau = 0$$
$$\triangleq \hat{w}_i(z) \tag{23.169}$$

Consider the self-adjoint EVP

$$\frac{d^2\psi}{dz^2} = -\Lambda^2\psi, \quad 0 < z < 1 \tag{23.170}$$
$$\psi'(0) - \frac{\text{Pe}}{2}\psi(0) = 0, \quad \psi'(1) + \frac{\text{Pe}}{2}\psi(1) = 0 \tag{23.171}$$

Let Λ_n^2 be the eigenvalues and $\psi_n(z)$ be the eigenfunctions (normalized). Taking inner product of equation (23.166) with $\psi_n(z)$ gives

⇒

$$\frac{d}{d\tau}\langle w, \psi_n\rangle = \frac{1}{\text{Pe}}\left\langle \frac{\partial^2 w}{\partial z^2}, \psi_n \right\rangle - \frac{\text{Pe}}{4}\langle w, \psi_n\rangle + \langle s^*, \psi_n\rangle \tag{23.172}$$

Now,

$$\left\langle \frac{\partial^2 w}{\partial z^2}, \psi_n \right\rangle = \left|\frac{\partial w}{\partial z}\psi_n - w\psi_n'\right|_0^1 + \langle w, -\Lambda_n^2\psi_n\rangle$$
$$= \frac{\partial w}{\partial z}(1,\tau)\psi_n(1) - w(1,\tau)\psi_n'(1) - \frac{\partial w}{\partial z}(0,\tau)\psi_n(0) + w(0,\tau)\psi_n'(0) - \Lambda_n^2\langle w, \psi_n\rangle$$

$$= -\frac{\mathrm{Pe}\,w(1,\tau)}{2}\psi_n(1) - w(1,\tau)\left[\frac{-\mathrm{Pe}\,\psi_n(1)}{2}\right] - \psi_n(0)\left[\frac{\mathrm{Pe}\,w(0,\tau)}{2} - \mathrm{Pe}\,\hat{c}_0(\tau)\right]$$
$$+ w(0,\tau)\cdot\frac{\mathrm{Pe}}{2}\psi_n(0) - \Lambda_n^2\langle w,\psi_n\rangle$$
$$= \mathrm{Pe}\,\psi_n(0)\hat{c}_0(\tau) - \Lambda_n^2\langle w,\psi_n\rangle$$

Thus,

$$\frac{d}{d\tau}\langle w,\psi_n\rangle = \psi_n(0)\hat{c}_0(\tau) - \frac{\Lambda_n^2}{\mathrm{Pe}}\langle w,\psi_n\rangle - \frac{\mathrm{Pe}}{4}\langle w,\psi_n\rangle + \langle s^*,\psi_n\rangle$$

$\Longrightarrow$

$$\frac{d}{d\tau}\langle w,\psi_n\rangle + \left(\frac{\Lambda_n^2}{\mathrm{Pe}} + \frac{\mathrm{Pe}}{4}\right)\langle w,\psi_n\rangle = \psi_n(0)\hat{c}_0(\tau) + \langle s^*,\psi_n\rangle \tag{23.173}$$

with initial condition

$$\langle w,\psi_n\rangle = \langle \hat{w}_i(z),\psi_n\rangle \ @\ \tau = 0\,. \tag{23.174}$$

Let

$$\mu_n = \frac{\Lambda_n^2}{\mathrm{Pe}} + \frac{\mathrm{Pe}}{4} \tag{23.175}$$

$\Rightarrow$

$$\langle w,\psi_n\rangle e^{\mu_n\tau} = \int_0^\tau e^{\mu_n\tau'}\psi_n(0)\hat{c}(\tau')\,d\tau' + \int_0^\tau e^{\mu_n\tau'}\langle s^*(z,\tau'),\psi_n\rangle\,d\tau' + \text{constant}$$
$$\tau = 0 \Rightarrow \text{constant} = \langle \hat{w}_i(z),\psi_n\rangle$$

$\therefore$

$$\langle w,\psi_n\rangle = \langle \hat{w}_i(z),\psi_n\rangle e^{-\mu_n\tau} + \int_0^\tau e^{-\mu_n(\tau-\tau')}\psi_n(0)\hat{c}_0(\tau')\,d\tau'$$
$$+ \int_0^\tau e^{-\mu_n(\tau-\tau')}\langle s^*(z,\tau'),\psi_n\rangle\,d\tau' \tag{23.176}$$

$\Rightarrow$

$$w(z,\tau) = \sum_{n=1}^{\infty}\langle w,\psi_n\rangle\psi_n(z) \tag{23.177}$$

$$C(z,\tau) = e^{\frac{\mathrm{Pe}\,z}{2}} \sum_{n=1}^{\infty} \psi_n(z) \begin{bmatrix} e^{-\mu_n \tau} \int_0^1 \hat{w}_i(z')\psi_n(z')\,dz' \\ + \int_0^{\tau} e^{-\mu_n(\tau-\tau')}\psi_n(0)\hat{c}_0(\tau')\,d\tau' \\ + \int_0^{\tau} e^{-\mu_n(\tau-\tau')}\left(\int_0^1 s^*(z',\tau')\psi_n(z')\,dz'\right)d\tau' \end{bmatrix} \tag{23.178}$$

Equation (23.178) gives the general solution to the axial dispersion model. We note that the first term is due to the initial condition, the second term is due to the inlet condition and the third term is due to the source term. To evaluate this solution, we need to determine the eigenvalues and normalized eigenfunctions and their dependence on the Peclet number. This is considered below.

Eigenvalue problem

$$\frac{d^2\psi}{dz^2} = -\Lambda^2\psi, \quad 0 < z < 1 \tag{23.179}$$

$$\psi'(0) = \frac{\mathrm{Pe}}{2}\psi(0), \quad \psi'(1) = -\frac{\mathrm{Pe}}{2}\psi(1) \tag{23.180}$$

$$\psi = c_1 \sin \Lambda z + c_2 \cos \Lambda z$$

$$\psi' = c_1\Lambda \cos \Lambda z - c_2 \Lambda \sin \Lambda z$$

$\Rightarrow$

$$\psi'(0) = c_1\Lambda, \quad \psi(0) = c_2, \quad \mathrm{BC1} \Longrightarrow c_1\Lambda = \frac{\mathrm{Pe}}{2}c_2$$

BC2 $\Rightarrow$

$$c_1\Lambda\cos\Lambda - c_2\Lambda\sin\Lambda + \frac{\mathrm{Pe}}{2}[c_1\sin\Lambda + c_2\cos\Lambda] = 0$$

$\Rightarrow$

$$c_2\left[\frac{\mathrm{Pe}}{2}\cos\Lambda - \Lambda\sin\Lambda + \frac{\mathrm{Pe}}{2}\cdot\frac{\mathrm{Pe}}{2\Lambda}\sin\Lambda + \frac{\mathrm{Pe}}{2}\cos\Lambda\right] = 0$$

$c_2 \neq 0 \Rightarrow$

$$\mathrm{Pe}\cos\Lambda + \left(\frac{\mathrm{Pe}^2}{4\Lambda} - \Lambda\right)\sin\Lambda = 0$$

$\Longrightarrow$

$$\cot\Lambda = \frac{\Lambda}{\mathrm{Pe}} - \frac{\mathrm{Pe}}{4\Lambda} \quad \text{(Characteristic equation)}, \tag{23.181}$$

and the eigenfunctions:

$$\psi_n(z) = c_2\left[\frac{\mathrm{Pe}}{2}\frac{\sin\Lambda_n z}{\Lambda_n} + \cos\Lambda_n z\right]. \tag{23.182}$$

[Note that equation (23.181) is identical to equation (17.172).]

Normalized eigenfunctions

The eigenfunction can be normalized and the constant c_2 can be obtained by setting

$$\int_0^1 \psi_n(z)^2\,dz = 1$$

$\Rightarrow$

$$\begin{aligned}
\frac{1}{c_2^2} &= \int_0^1 \cos^2\Lambda_n z\,dz + \frac{\mathrm{Pe}^2}{4}\int_0^1 \frac{\sin^2\Lambda_n z}{\Lambda_n^2}\,dz + \frac{\mathrm{Pe}}{\Lambda_n}\int_0^1 \cos\Lambda_n z\sin\Lambda_n z\,dz \\
&= \left(\frac{1}{2} + \frac{\sin 2\Lambda_n}{4\Lambda_n}\right) + \frac{\mathrm{Pe}^2}{4\Lambda_n^2}\left(\frac{1}{2} - \frac{\sin 2\Lambda_n}{4\Lambda_n}\right) - \frac{\mathrm{Pe}}{\Lambda_n}\cdot\frac{\cos 2\Lambda_n - 1}{4\Lambda_n} \\
&= \frac{\sin\Lambda_n\cos\Lambda_n}{2\Lambda_n}\left(1 - \frac{\mathrm{Pe}^2}{4\Lambda_n^2}\right) + \frac{\mathrm{Pe}}{\Lambda_n}\cdot\frac{\sin^2\Lambda_n}{2\Lambda_n} + \frac{\mathrm{Pe}^2}{8\Lambda_n^2} + \frac{1}{2}.
\end{aligned}$$

Using equation (23.181), the above expression may be simplified as

$$\begin{aligned}
\frac{1}{c_2^2} &= \frac{1}{2} + \frac{\mathrm{Pe}^2}{8\Lambda_n^2} + \frac{\mathrm{Pe}}{2\Lambda_n^2}(\sin^2\Lambda_n + \cos^2\Lambda_n) \\
&= \frac{1}{2} + \frac{\mathrm{Pe}^2}{8\Lambda_n^2} + \frac{\mathrm{Pe}}{2\Lambda_n^2}
\end{aligned}$$

$$\Longrightarrow c_2 = \sqrt{\frac{8\Lambda_n^2}{\mathrm{Pe}^2 + 4\,\mathrm{Pe} + 4\Lambda_n^2}}. \tag{23.183}$$

Determination of the roots of the characteristic equation

The characteristic equation (23.181), which is the same as equation (17.172), can further be simplified (by solving for Pe in terms of Λ) into a more convenient form as

$$\frac{\mathrm{Pe}}{4} = \begin{cases} \frac{\Lambda}{2}\tan[\frac{\Lambda}{2}] \\ -\frac{\Lambda}{2}\cot[\frac{\Lambda}{2}], \end{cases} \tag{23.184}$$

which can be used to determine the eigenvalue Λ_n for a given Peclet number Pe as shown in Figure 23.12.

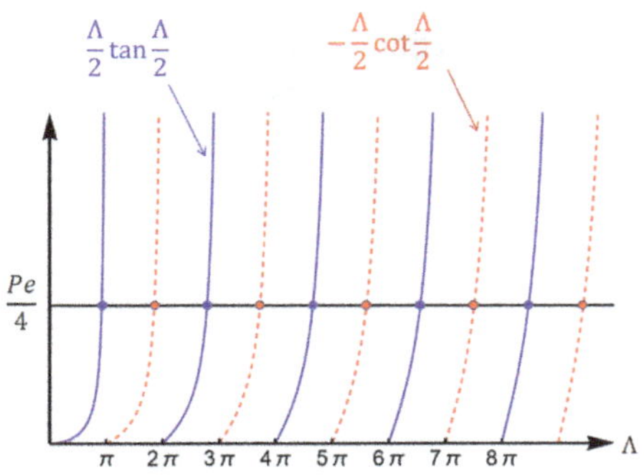

Figure 23.12: Roots of characteristic equation and determination of eigenvalues for the axial dispersion model.

It can be seen from this figure that for any Pe, the n-th root Λ_n lies in the interval $[(n-1)\pi, n\pi]$, i. e.,

$$(n-1)\pi \le \Lambda_n \le n\pi, \quad n = 1, 2, 3, \dots \text{ for any Pe}.$$

For large Pe (i. e., Pe $\gg$ 1), it can be shown that

$$\Lambda_n \approx \frac{n\pi}{(1 + \frac{4}{\text{Pe}})}, \quad n = 1, 2, \dots$$

while for small Pe (i. e., Pe $\ll$ 1), we have

$$\Lambda_1 \approx \sqrt{\text{Pe}} \quad \text{and}$$
$$\Lambda_n \approx (n-1)\pi + \frac{\text{Pe}}{(n-1)\pi} + \cdots, \quad n = 2, 3, \dots$$

The numerical values of the first six roots of the characteristic equation (23.181), Λ_n are tabulated for some Pe-values in Table 23.1.

Table 23.1: First six roots Λ_n of characteristic equations for some Pe values.

Pe	Λ_1	Λ_2	Λ_3	Λ_4	Λ_5	Λ_6
0.01	0.09996	3.14477	6.28478	9.42584	12.5672	15.7086
0.1	0.31492	3.1731	6.29906	9.43538	12.5743	15.7143
1.0	0.96019	3.43101	6.4382	9.52962	12.6454	15.7714
2.0	1.30654	3.67319	6.58462	9.63168	12.7232	15.8341
5.0	1.86151	4.21275	6.97179	9.9186	12.9478	16.0176
10.0	2.28445	4.76129	7.46368	10.3266	13.2862	16.3031
20.0	2.62768	5.30732	8.06714	10.9087	13.8192	16.7827
100.0	3.0209	6.04265	9.06603	12.0918	15.1206	18.153
1000.0	3.12908	6.25815	9.38723	12.5163	15.6454	18.7745

Similarly, for any value of Pe, the normalized eigenfunctions corresponding to these eigenvalues can be determined easily. As an example, Figure 23.13 shows the plots of normalized eigenfunctions for Pe = 10 corresponding to first six eigenvalues.

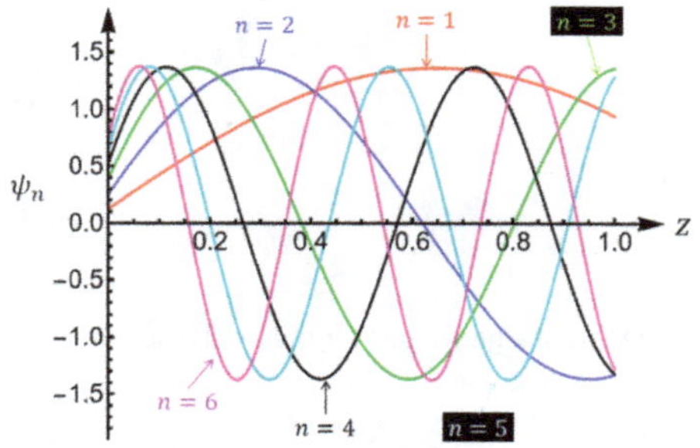

Figure 23.13: Normalized eignfunctions corresponding to the first six eigenvalues of the self-adjoint form of axial dispersion operator for Pe = 10.

Specific solutions

We consider some specific cases of the general solution of the axial dispersion model:

$$C(z,\tau) = e^{\frac{\mathrm{Pe}\,z}{2}} \sum_{n=1}^{\infty} e^{-\mu_n \tau}\psi_n(z)\left[\int_0^1 \hat{w}_i(z')\psi_n(z')\,dz' + \int_0^{\tau} e^{\mu_n \tau'}\psi_n(0)\hat{c}_0(\tau')\,d\tau' \right.$$
$$\left. + \int_0^{\tau} e^{\mu_n \tau'}\left(\int_0^1 s^*(z',\tau')\psi_n(z')\,dz' \right) d\tau' \right] \tag{23.185}$$

(1) Special case 1

$$\hat{c}_0(\tau') = 0 \quad [c_0(t) = 0] \tag{23.186}$$
$$\hat{w}_i(z') = 0 \quad [c_i(t) = 0] \tag{23.187}$$
$$S(x,t) = c^* L\delta(x)\delta(t) \tag{23.188}$$

i. e., unit point source at inlet at time zero

$\Rightarrow$

$$\begin{aligned} s(z,\tau) &= \frac{L}{\langle u\rangle c^*} c^* L\delta(Lz)\delta\left(\frac{L\tau}{\langle u\rangle}\right) \\ &= \frac{L^2}{\langle u\rangle}\delta(Lz)\delta\left(\frac{L\tau}{\langle u\rangle}\right) \end{aligned}$$

Using the result,

$$\delta(az) = \frac{1}{|a|}\delta(z)$$

$\Rightarrow$

$$s(z,\tau) = \delta(z)\delta(\tau) \tag{23.189}$$

$$s^*(z,\tau) = e^{-\frac{\mathrm{Pe}\,z}{2}}\delta(z)\delta(\tau), \quad \hat{w}_i(z') = 0, \quad \hat{c}(\tau') = 0 \tag{23.190}$$

Substituting equation (23.190) in equation (23.178) gives

$$\begin{aligned} C(z,\tau) &= \text{response to a unit point source at inlet at time zero} \\ &= e^{\frac{\mathrm{Pe}\,z}{2}} \sum_{n=1}^{\infty} \psi_n(z) a_n, \end{aligned} \tag{23.191}$$

where

$$\begin{aligned} a_n &= \int_0^{\tau} e^{-\mu_n(\tau-\tau')} \left[\int_0^1 e^{-\frac{\mathrm{Pe}\,z'}{2}} \psi_n(z')\delta(z')\,dz' \right] \delta(\tau')\,d\tau' \\ &= e^{-\mu_n \tau}\psi_n(0). \end{aligned} \tag{23.192}$$

The exit response to a unit impulse input is known as the resident time distribution (RTD) curve. Assuming $E(\tau)$ is the RTD curve, i. e., the exit response to a unit-pulse input, then

$$\begin{aligned} E(\tau) &= C(1,\tau) \\ &= e^{\frac{\mathrm{Pe}}{2}} \sum_{n=1}^{\infty} e^{-\mu_n \tau} \psi_n(1)\psi_n(0) \end{aligned} \tag{23.193}$$

$$\psi_n(0) = c_2 \quad \text{and} \quad \psi_n(1) = c_2\left[\frac{\mathrm{Pe}}{2} \cdot \frac{\sin\Lambda_n}{\Lambda_n} + \cos\Lambda_n\right]$$

$\Rightarrow$

$$\begin{aligned} \psi_n(0)\psi_n(1) &= c_2^2\left[\frac{\mathrm{Pe}}{2} \cdot \frac{\sin\Lambda_n}{\Lambda_n} + \cos\Lambda_n\right] \\ &= \frac{4\,\mathrm{Pe}\,\Lambda_n^2}{\mathrm{Pe}^2 + 4\,\mathrm{Pe} + 4\Lambda_n^2} \frac{\sin\Lambda_n}{\Lambda_n}\left[1 + \frac{2\Lambda_n}{\mathrm{Pe}}\left(\frac{\Lambda_n}{\mathrm{Pe}} - \frac{\mathrm{Pe}}{4\Lambda_n}\right)\right] \\ &= \frac{\Lambda_n \sin\Lambda_n}{\mathrm{Pe}^2 + 4\,\mathrm{Pe} + 4\Lambda_n^2} \cdot \left[2\,\mathrm{Pe} + \frac{8\Lambda_n^2}{\mathrm{Pe}}\right] \end{aligned}$$

$\Longrightarrow$

$$\psi_n(1)\psi_n(0) = \frac{2\Lambda_n \sin\Lambda_n}{\mathrm{Pe}^2 + 4\,\mathrm{Pe} + 4\Lambda_n^2}\left[\frac{\mathrm{Pe}^2 + 4\Lambda_n}{\mathrm{Pe}}\right]$$

$\Rightarrow$

$$E(\tau) = e^{\frac{\mathrm{Pe}}{2}} \sum_{n=1}^{\infty} \left(\frac{2}{\mathrm{Pe}}\right) \frac{\Lambda_n \sin\Lambda_n(\mathrm{Pe}^2 + 4\Lambda_n^2)}{(\mathrm{Pe}^2 + 4\,\mathrm{Pe} + 4\Lambda_n^2)} e^{-(\frac{\mathrm{Pe}^2 + 4\Lambda_n^2}{4\,\mathrm{Pe}})\tau}$$

$\Rightarrow$

$$\begin{aligned} E(\tau) &= \frac{2}{\mathrm{Pe}} e^{\frac{\mathrm{Pe}}{2}} \sum_{n=1}^{\infty} \frac{\Lambda_n \sin\Lambda_n(\mathrm{Pe}^2 + 4\Lambda_n^2)}{(\mathrm{Pe}^2 + 4\,\mathrm{Pe} + 4\Lambda_n^2)} e^{-(\frac{\mathrm{Pe}^2 + 4\Lambda_n^2}{4\,\mathrm{Pe}})\tau} \\ &= 8 e^{\frac{\mathrm{Pe}}{2}} \sum_{n=1}^{\infty} \frac{(-1)^{n-1}\Lambda_n^2}{(\mathrm{Pe}^2 + 4\,\mathrm{Pe} + 4\Lambda_n^2)} e^{-(\frac{\mathrm{Pe}^2 + 4\Lambda_n^2}{4\,\mathrm{Pe}})\tau} \end{aligned} \tag{23.194}$$

The last simplification follows by expressing $\sin\Lambda_n$ as

$$\sin\Lambda_n = \frac{(-1)^{n-1}\,\mathrm{Pe}\,\Lambda_n}{\Lambda_n^2 + \frac{\mathrm{Pe}^2}{4}}$$

and noting that the sign of $\sin\Lambda_n$ depends on n and such that $(n-1)\pi \le \Lambda_n \le n\pi$. [For n odd, the sign is positive and n even, sign is negative.] A plot of the solution given by equation (23.194) is shown in Figure 17.11 for $\mathrm{Pe} = 0.5,\ 2.0$ and 5.0.

(2) Special case 2

$$s^*(z', \tau') = 0, \quad \hat{c}_0(\tau') = 0 \tag{23.195}$$

with initial condition

$$\hat{w}_i(z') = \delta(z') = \text{ unit pulse at inlet at } t = 0 \tag{23.196}$$

$\Rightarrow$

$$C(z,\tau) \quad \text{is given by equations (23.191)–(23.192)}$$

(3) Special case 3

$$s^*(z', \tau') = 0, \quad \hat{w}_i(z') = 0 \tag{23.197}$$

$$\hat{c}_0(\tau') = \delta(\tau') \tag{23.198}$$

$\Rightarrow$ same solution as that given by equations (23.191)–(23.192).

Thus, we get the same solution (as can be expected intuitively) for the unit pulse appearing as a source term, in the initial condition or in the inlet boundary condition.

23.4.3 Example 11 (Fourier's ring problem)

As our next example, we consider the historical problem solved by Fourier, i. e., determining the temperature distribution in a circular ring (also known as Fourier's ring problem). The problem in dimensionless form is described by

$$\frac{\partial u}{\partial t} = \frac{\partial^2 u}{\partial \theta^2}, \quad 0 < \theta \leq 2\pi, \quad t > 0 \tag{23.199}$$

with BCs:

$$u(\theta, t) = u(\theta + 2\pi, t), \quad \frac{\partial u}{\partial \theta}(\theta, t) = \frac{\partial u}{\partial \theta}(\theta + 2\pi, t) \tag{23.200}$$

and IC:

$$u(\theta, 0) = f(\theta). \tag{23.201}$$

We note that the operator

$$\frac{\partial^2 w}{\partial \theta^2}, \quad 0 < \theta \leq 2\pi \tag{23.202}$$

with periodic boundary conditions

$$w(\theta) = w(\theta + 2\pi), \quad w'(\theta) = w'(\theta + 2\pi) \tag{23.203}$$

has eigenvalues:

$$\lambda_n = n^2, \quad n = 0, 1, 2, \ldots \tag{23.204}$$

and normalized eigenfunctions:

$$w_0 = \frac{1}{\sqrt{2\pi}}; \quad w_n = \begin{cases} \equiv w_{ns} = \frac{\sin n\theta}{\sqrt{\pi}} \\ \equiv w_{nc} = \frac{\cos n\theta}{\sqrt{\pi}}, \end{cases} \quad n = 1, 2, \ldots \tag{23.205}$$

Thus, the solution may be expressed as

$$\begin{aligned} u(\theta, t) &= \sum_{n=0}^{\infty} e^{-\lambda_n t} \langle f, w_n \rangle w_n(\theta) \\ &= \langle f, w_0 \rangle w_0 + \sum_{n=0}^{\infty} e^{-n^2 t} \frac{\sin n\theta}{\sqrt{\pi}} \int_0^{2\pi} f(\theta') \frac{\sin n\theta'}{\sqrt{\pi}} \, d\theta' \\ &\quad + \sum_{n=1}^{\infty} e^{-n^2 t} \frac{\cos n\theta}{\sqrt{\pi}} \int_0^{2\pi} f(\theta') \frac{\cos n\theta'}{\sqrt{\pi}} \, d\theta' \end{aligned}$$

$\Rightarrow$

$$u(\theta, t) = \frac{1}{2\pi} \int_0^{2\pi} f(\theta') \, d\theta' + \sum_{n=1}^{\infty} \frac{e^{-n^2 t}}{\pi} \int_0^{2\pi} [\sin n\theta \cdot \sin n\theta' + \cos n\theta \cos n\theta'] f(\theta') \, d\theta'$$

$$= \frac{1}{2\pi}\int_0^{2\pi} f(\theta')\,d\theta' + \sum_{n=1}^{\infty}\frac{e^{-n^2 t}}{\pi}\int_0^{2\pi}\cos[n(\theta-\theta')]f(\theta')\,d\theta' \tag{23.206}$$

$\Rightarrow$

$$u(\theta,0) = f(\theta) = \frac{1}{2\pi}\int_0^{2\pi} f(\theta')\,d\theta' + \frac{1}{\pi}\sum_{n=1}^{\infty}\int_0^{2\pi}\cos[n(\theta-\theta')]f(\theta')\,d\theta' \tag{23.207}$$

$$u(\theta,\infty) = \frac{1}{2\pi}\int_0^{2\pi} f(\theta')\,d\theta'. \tag{23.208}$$

Note that $u(\theta,\infty)$ is the average (steady-state) temperature.

23.4.4 Example 12: Coupled reaction–diffusion equations

As our final example, we consider the coupled reaction–diffusion equations

$$\left.\begin{aligned}\frac{\partial u_1}{\partial t} &= \frac{\partial^2 u_1}{\partial x^2} + a_{11}u_1 + a_{12}u_2\\ \frac{\partial u_2}{\partial t} &= \frac{\partial^2 u_2}{\partial x^2} + a_{21}u_1 + a_{22}u_2\end{aligned}\right\};\quad 0 < x < 1,\quad t > 0 \tag{23.209}$$

with homogeneous Dirichlet boundary conditions:

$$u_1(0,t) = u_2(0,t) = 0,\quad u_1(1,t) = u_2(1,t) = 0 \tag{23.210}$$

and initial conditions

$$u_i(x,0) = f_i(x);\quad i = 1,2. \tag{23.211}$$

The spatial operator

$$-\frac{\partial^2 w}{\partial x^2} = \lambda w,\quad w(0) = 0,\quad w(1) = 0 \tag{23.212}$$

has eigenvalues

$$\lambda_n = n^2\pi^2;\quad n = 1,2,\ldots \tag{23.213}$$

and normalized eigenfunctions

$$w_n(x) = \sqrt{2}\sin n\pi x. \tag{23.214}$$

Let

$$V_{jn} = \langle u_j, w_n\rangle;\quad j = 1,2 \tag{23.215}$$

Equations (23.209)–(23.211) ⇒

$$\frac{d}{dt}\begin{bmatrix} V_{1n} \\ V_{2n} \end{bmatrix} = \begin{bmatrix} a_{11} - n^2\pi^2 & a_{12} \\ a_{21} & a_{22} - n^2\pi^2 \end{bmatrix}\begin{bmatrix} V_{1n} \\ V_{2n} \end{bmatrix} \tag{23.216}$$

$$\begin{bmatrix} V_{1n} \\ V_{2n} \end{bmatrix} = \begin{bmatrix} g_{1n}^0 \\ g_{2n}^0 \end{bmatrix} @\, t = 0, \quad g_{in}^0 = \langle f_i, w_n \rangle \quad i = 1, 2 \tag{23.217}$$

Let

$$\mathbf{B}_n = \begin{bmatrix} a_{11} - n^2\pi^2 & a_{12} \\ a_{12} & a_{22} - n^2\pi^2 \end{bmatrix} \tag{23.218}$$

$$\mu_{1n}, \quad \mu_{2n} = \text{eigenvalues of } \mathbf{B}_n \tag{23.219}$$

$$\mathbf{x}_{1n}, \mathbf{x}_{2n} = \text{eigenvectors of } \mathbf{B}_n \tag{23.220}$$

$$\mathbf{y}_{1n}^*, \mathbf{y}_{2n}^* = \text{eigenrows of } \mathbf{B}_n \tag{23.221}$$

Solution of equations (23.216)–(23.217) is given by

$$\mathbf{V}_n = \sum_{j=1}^{2} \frac{\mathbf{y}_{jn}^* \mathbf{g}_n^0}{\mathbf{y}_{jn}^* \mathbf{x}_{jn}} e^{\mu_{jn} t} \mathbf{x}_{jn} \tag{23.222}$$

⇒

$$u = \begin{pmatrix} u_1(x,t) \\ u_2(x,t) \end{pmatrix} = \sum_{n=1}^{\infty} \sqrt{2} \sin n\pi x \left(\sum_{j=1}^{2} \frac{\mathbf{y}_{jn}^* \mathbf{g}_n^0}{\mathbf{y}_{jn}^* \mathbf{x}_{jn}} e^{\mu_{jn} t} \right) \mathbf{x}_{jn} \tag{23.223}$$

The growth or decay of the solution is determined by the eigenvalues of the matrix $\mathbf{B}_n (n = 1, 2, \ldots)$. If all μ_{jn} have a negative real part, the initial perturbations decay to the trivial solution, while spatial or spatiotemporal patterns may be formed when μ_{jn} cross the imaginary axis.

Problems

1. Consider the problem of unsteady-state heat/mass transfer in a flat plate given by the equations

$$\frac{\partial^2 \theta}{\partial x^2} = \frac{\partial \theta}{\partial \tau}; \quad 0 < x < 1, \quad \tau > 0$$

$$\theta(x, 0) = f(x)$$

$$\frac{\partial \theta}{\partial x}(0, \tau) = 0$$

$$\frac{\partial \theta}{\partial x}(1, \tau) + \mathrm{Bi}\, \theta(1, \tau) = 0$$

(a) Determine the solution using finite Fourier transformation. Compare your result with that in Carslaw and Jaeger [10].
(b) Consider the case in which $f(x) = 1$. What is the limiting form of the solution for the case no external resistance (Bi $\to \infty$) and no internal resistance (Bi $\to 0$)?

2. (a) Obtain the solution of the Poisson's equation

$$\nabla^2 u = -f \quad \text{in } \Omega$$
$$u = 0 \quad \text{on } \partial\Omega$$

in two and three dimensions when Ω is a rectangular region. Identify Green's function and give a physical interpretation.
(b) The velocity profile for slow viscous flow of a fluid in a rectangular channel is given by

$$\frac{\partial^2 u}{\partial x^2} + \frac{\partial^2 u}{\partial y^2} = \frac{\Delta p}{\mu L}$$
$$u = 0 \quad @x = 0 \text{ and } x = a$$
$$u = 0 \quad @y = 0 \text{ and } y = b$$

Obtain a complete solution. Use the solution to derive a rectangular analogue of Poiseuille's law (relation between pressure drop and flow rate).

3. Heat transfer between two infinite parallel plates in one-dimensional laminar flow neglecting conduction in the flow direction may be described by

$$k_f \frac{\partial^2 T}{\partial y^2} = \frac{3}{2} \langle u \rangle \rho C_p \left(1 - \frac{y^2}{a^2}\right) \frac{\partial T}{\partial x}$$
$$T = T_w \, @y = \pm a, \quad T = T_{in} F(y), \quad @x = 0$$

(a) Cast into dimensionless form and find the formal solution without determining the eigenvalues and eigenfunctions (Graetz functions) explicitly. (b) Determine an expression for the cup-mixing (velocity weighted) temperature (T_m) for the case of uniform inlet temperature, i. e., $F(y) = 1$. (c) If the heat transfer coefficient (h) is defined by

$$h(x) = \frac{-k_f \frac{\partial T}{\partial y}(x, y = a)}{T_m - T_w},$$

where T_w is the wall temperature, obtain an expression for the dimensionless heat transfer coefficient (or the local Nusselt number)

$$Nu(x) = \frac{h(x) a}{k_f}$$

(d) Determine the two asymptotes (short and long distance) of the Nusselt number as a function of position

4. (a) Given the operator

$$Lw = -\frac{d^2w}{dx^2}, \quad 0 < x < 1; \quad w'(0) = 0, \quad w'(1) = 0$$

determine the eigenvalues and orthonormal set of eigenfunctions.
(b) Determine the expansion of the function $f(x) = \delta(x - \frac{1}{2})$ in terms of the eigenfunctions determined in (a) above.
(c) Use the above results to solve the diffusion equation

$$\frac{\partial^2 u}{\partial x^2} = \frac{\partial u}{\partial t}; \quad 0 < x < 1, \quad t > 0$$

$$u'(0,t) = 0, \quad u'(1,t) = 0, \quad u(x,0) = \delta\left(x - \frac{1}{2}\right)$$

Show schematic profiles of $u(x,t)$ for $0 \le x \le 1$ for $t = 0$, $t \to \infty$ and a finite value of t. Give a physical interpretation of the solution.

5. (a) Solve Laplace's equation on the unit square:

$$\frac{\partial^2 u}{\partial x^2} + \frac{\partial^2 u}{\partial y^2} = 0, \quad 0 < x < 1, \quad 0 < y < 1$$

$$u(x,0) = f(x), \quad u(x,1) = 0; \quad u(0,y) = 0, \quad u(1,y) = 0$$

(b) Simplify the solution for the special case of $f(x) = 1$ and plot a few isotherms (corresponding to constant values of u).

6. Solve the problem

$$D\frac{\partial^2 C}{\partial x^2} = \frac{\partial C}{\partial t}; \quad 0 < x < L, \quad t > 0$$

$$\frac{\partial C}{\partial x} = 0, \quad @\, x = 0, \quad -D\frac{\partial C}{\partial x} = k_g[C - C_0(t)], \quad @\, x = L$$

$$C = 0, \quad @\, t = 0$$

where $C_0(t)$ is a given function of time.

7. Given the following set of equations, which describe adsorption in a fixed bed of adsorbent,

$$\varepsilon\left[-D\frac{\partial^2 C}{\partial x^2} + u\frac{\partial C}{\partial x} + \frac{\partial C}{\partial t}\right] + (1-\varepsilon)\frac{\partial n}{\partial t} = 0, \quad 0 < x < L$$

$$(1-\varepsilon)\frac{\partial n}{\partial t} = k_g a(C - C^*), \quad n = kC^* \quad \text{(equilibrium relation)}$$

$$-D\frac{\partial C}{\partial x} = u(C_0 - C), \quad x = 0 \quad (C_0 \text{ is a constant}), \quad \frac{\partial C}{\partial x} = 0, \quad x = L$$

$$C = f(x), \quad n = g(x), \quad t = 0$$

Determine the solution. This is adsorption for the case in which adsorption is mass transfer limited. $C(x,t)$ is the concentration in the interstitial fluid and $n(x,t)$ is the concentration in the solid phase.

8. Consider the steady-state problem of diffusion and surface reaction in a rectangular pore. The relevant equations are given by

$$\frac{\partial^2 C}{\partial y^2} + \frac{\partial^2 C}{\partial z^2} = 0, \quad -H < y < H, \quad 0 < z < L$$

$$C = C_0, \quad @z = 0; \quad \frac{\partial C}{\partial z} = 0, \quad @z = L$$

$$\pm D\frac{\partial C}{\partial y} + k_s C = 0; \quad @y = \pm H$$

Cast into dimensionless form and determine the solution. Use the solution to determine the effectiveness factor (ratio of the actual reaction rate in pore to that if concentration at all points inside is equal to C_0).

9. Transient convection–reaction problems in one spatial dimension are described by hyperbolic system of equations of the form

$$\mathbf{C}_1(x)\frac{\partial \mathbf{y}}{\partial t} + \mathbf{C}_2(x)\frac{\partial \mathbf{y}}{\partial x} = \mathbf{C}_3(x)\mathbf{y}; \quad a < x < b, \quad t > 0$$

$$\text{B.C.:} \quad \mathbf{W}_a\mathbf{y}(a,t) + \mathbf{W}_b\mathbf{y}(b,t) = 0, \quad t > 0$$

$$\text{IC:} \quad \mathbf{y}(x,0) = \mathbf{f}(x)$$

where $\mathbf{C}_i$, $i = 1,2,3$ are nonsingular $N \times N$ matrices that are continuous in x and $\mathbf{W}_a, \mathbf{W}_b$ are constant $N \times N$ matrices.

(a) Use the separation of variable technique and identify the eigenvalue problem that results. What is the adjoint eigenvalue problem?

(b) Use the biorthogonal expansion to obtain formal solution to the above transient equations. Comment on the usefulness of the solution when the eigenvalues are complex.

(Specific examples of this type of systems include heat exchangers, distillation columns, autothermal reactors, chromatographs, etc. Try to formulate the simplest transient models of any of these systems and put them in the above form.)

10. Axial dispersion model:

The dispersion of a tracer in unidirectional flows in pipes and channels is described by the axial dispersion model given by the equations

$$\frac{\partial c}{\partial t} + u\frac{\partial c}{\partial x} = D\frac{\partial^2 c}{\partial x^2} + S(x,t); \quad 0 < x < L, \quad t > 0$$

$$\text{BC:} \quad \left\{ -D\frac{\partial c}{\partial x} = u[c_0(t) - c] @ x = 0, \quad \frac{\partial c}{\partial x} = 0, \quad @ x = L, \quad t > 0 \right.$$

$$\text{I.C :} \quad c = c_i(x), \quad @ t = 0, \quad 0 < x < L.$$

Here, c is the concentration of the tracer, u is the average velocity of the stream, D is the effective axial dispersion coefficient, S is the sources/sinks of tracer, c_0 is the inlet concentration of tracer and c_i is the initial distribution of tracer.

(a) Cast the equations into dimensionless form.
(b) Obtain a formal solution to the model in (a).
(c) Simplify the solution for the special case of $c_0(t) = 0$, $c_i(x) = 0$ and unit pulse at the inlet at time zero.

(Note: The solution in case (c) gives the residence time distribution function for the axial dispersion model.)

11. Transient diffusion–convection–reaction problems for N chemical species are described by coupled parabolic equations of the form

$$\mathbf{D}\frac{\partial^2 \mathbf{c}}{\partial x^2} - u\frac{\partial \mathbf{c}}{\partial x} - \mathbf{K}\mathbf{c} = \frac{\partial \mathbf{c}}{\partial t}, \quad 0 < x < L, \quad t > 0$$

$$\text{B.C:} \quad \left\{ -\mathbf{D}\frac{\partial \mathbf{c}}{\partial x} = u(\mathbf{c}_0 - \mathbf{c}), \quad @ x = 0, \quad \frac{\partial \mathbf{c}}{\partial x} = 0, \quad @ x = L, \quad t > 0 \right.$$

$$\text{I.C:} \quad \mathbf{c} = \mathbf{f}(x), \quad t = 0$$

where $\mathbf{D}$ is the matrix of dispersion coefficients, u is the velocity, $\mathbf{K}$ is the matrix of rate constants and $\mathbf{c}$ is the concentration vector.

(a) Cast the equations into dimensionless form and identify the linear operators of interest.
(b) Indicate how the equations may be decoupled into N scalar equations.
(c) Obtain a formal solution to each scalar equation. Write down the form of the solution to the complete system of equations.

12. Obtain a formal solution to the system of partial differential equations

$$\left. \begin{aligned} \frac{\partial^2 u_1}{\partial x^2} + \frac{\partial^2 u_1}{\partial z^2} + a_1\frac{\partial u_2}{\partial x} &= f_1(x,z) \\ \frac{\partial^2 u_2}{\partial x^2} + \frac{\partial^2 u_2}{\partial z^2} + a_1\frac{\partial u_1}{\partial x} &= f_2(x,z) \end{aligned} \right\}; \quad 0 < x < 1, \quad 0 < z < 1$$

$$u_1(0,z) = u_1(1,z) = 0, \quad u_2(x,0) = u_2(x,1) = 0$$

$$\frac{\partial u_2}{\partial x}(0,z) = \frac{\partial u_2}{\partial x}(1,z) = 0, \quad u_1(x,0) = u_1(x,1) = 0.$$

Does a solution exist for every choice of f_1 and f_2? Explain.

13. (a) Obtain a formal solution to the following system of linear PDEs describing diffusion and reaction:

$$\frac{\partial \mathbf{u}}{\partial t} = \mathbf{D}\nabla^2\mathbf{u} + \mathbf{A}\mathbf{u}; \quad \text{in } \Omega$$

$$\mathbf{u} = \mathbf{0} \quad \text{on } \partial\Omega \quad \text{(Dirichlet boundary conditions)}$$

where Ω is a rectangle. Here, **u** is the vector of concentrations, **D** is the matrix of diffusion coefficients and **A** is a constant $n \times n$ matrix.

(b) Obtain a formal solution to (1) with Neumann boundary conditions,

$$\nabla \mathbf{u}.\mathbf{n} = \mathbf{0} \quad \text{on } \partial\Omega$$

Here, **n** is the unit outward normal to $\partial\Omega$.

(c) Show schematic diagrams of contour plots of the spatial eigenfunctions for cases (a) and (b).

14. Let **V** be the vector space of complex-values functions $u(x)$ defined on the interval $(0, a)$ satisfying the periodicity condition $u(0) = u(a)$. Let L be the linear operator on **V** defined by

$$Lu(x) = -i\frac{du}{dx}; \quad i = \sqrt{-1}$$

(a) Show that this operator is self-adjoint w. r. t. the usual inner product on **V**, i. e.,

$$\langle u, v\rangle = \int_0^a u(x)\overline{v(x)}\, dx$$

(b) Determine the eigenvalues and normalized eigenfunctions of L.

(c) Determine the coefficient in the expansion of an arbitrary complex values periodic function $f(x)$ with $f(x) = f(x + a)$ in terms of the eigenfunctions in (b).

(d) State the Parseval's relation for the expansion in (c).
[Remark: The expansion in (c) is the complex form of the Fourier series for a periodic function for a periodic function, which is often used in signal processing and numerical calculations.]

15. Consider the solution of the diffusion equation in a finite domain

$$\frac{\partial^2 u}{\partial x^2} = \frac{\partial u}{\partial t}; \quad 0 < x < 1, \quad t > 0$$
$$u(0, t) = 0, \quad u(1, t) = 0$$
$$u(x, 0) = f(x)$$

Simplify the solution for the special case of $f(x) = 1$ and show that for short times it reduces to the error function solution.

16. Consider the solution of Laplace's equation in the rectangle

$$\frac{\partial^2 u}{\partial x^2} + \frac{\partial^2 u}{\partial y^2} = 0; \quad -a < x < a, \quad 0 < y < b \quad (a, b > 0)$$

$$u(-a,y) = 0, \quad u(a,y) = 0, \quad u(x,b) = 0, \quad u(x,0) = f(x)$$

(a) Determine the solution using finite Fourier transform.

(b) Consider the case of a and b going to infinity (or the case of rectangle being extended to the upper half-plane). Show that for this limiting case, the solution may be simplified to the Poisson's formula for the upper half plane:

$$u(x,y) = \frac{1}{\pi} \int_{-\infty}^{\infty} \frac{yf(\xi)}{[(x-\xi)^2 + y^2]} \, d\xi$$

24 Fourier transforms on infinite intervals

Recall that a regular differential operator had two characteristics (i) it was defined on a finite interval, and (ii) the leading coefficient $p_0(x)$ did not vanish inside or at the end points of the interval. We now consider problems in which condition (i) is violated. This leads to the Fourier transform on infinite and semi-infinite domains.

24.1 Fourier transform on $(-\infty, \infty)$

Consider the eigenvalue problem

$$\frac{d^2u}{dx^2} = -\lambda u, \quad -a < x < a \tag{24.1}$$

$$u(-a) = u(a), \quad u'(-a) = u'(a) \quad \text{Periodic BCs.} \tag{24.2}$$

It is easily verified that this is a self-adjoint eigenvalue problem.
Eigenvalues:

$$\lambda_n = \frac{n^2\pi^2}{a^2}, \quad n = 0, 1, 2, \ldots \tag{24.3}$$

Normalized eigenfunctions:

$$y_0(x) = \frac{1}{\sqrt{2a}} \tag{24.4}$$

$$u_n(x) = \frac{1}{\sqrt{a}} \sin\left(\frac{n\pi x}{a}\right) \tag{24.5}$$

$$y_n(x) = \frac{1}{\sqrt{a}} \cos\left(\frac{n\pi x}{a}\right); \quad \text{each } \lambda_n (n > 0) \text{ is double} \tag{24.6}$$

Note that

$$\left.\begin{array}{ll} \int_{-a}^{a} y_0 u_n(x)\, dx = 0 & n \neq 0 \\ \int_{-a}^{a} y_0 y_n(x)\, dx = 0 & n \neq 0 \end{array}\right\} \quad \text{(orthogonality relations)} \tag{24.7}$$

and

$$\int_{-a}^{a} y_m(x) y_n(x)\, dx = \begin{cases} 0 & m \neq n \\ 1 & \text{if } m = n \end{cases} \quad \Rightarrow \int_{-a}^{a} y_n(x) y_m(x)\, dx = \delta_{mn} \tag{24.8}$$

$$\int_{-a}^{a} u_m(x) u_n(x)\, dx = \begin{cases} 0 & m \neq n \\ 1 & \text{if } m = n \end{cases} \quad \Rightarrow \int_{-a}^{a} u_n(x) u_m(x)\, dx = \delta_{mn} \tag{24.9}$$

https://doi.org/10.1515/9783111598055-029

$$\int_{-a}^{a} y_n(x) u_m(x)\, dx = 0 \quad \forall m, n\ (m = n \text{ included}) \tag{24.10}$$

$\{\lambda_n, u_n(x), y_n(x)\}$ form a basis for $\mathcal{L}_2[-a, a]$, Hilbert space of periodic functions with the standard inner product

$$\langle u, y \rangle = \int_{-a}^{a} u(x) y(x)\, dx \tag{24.11}$$

If $f(x) \in \mathcal{L}_2[-a, a]$, then

$$f(x) = a_0 y_0(x) + \sum_{n=1}^{\infty} [a_n y_n(x) + b_n u_n(x)] \tag{24.12}$$

$$a_0 = \langle f, y_0 \rangle = \int_{-a}^{a} \frac{1}{\sqrt{2a}} f(\xi)\, d\xi \tag{24.13}$$

$$a_n = \langle f, y_n \rangle = \int_{-a}^{a} \frac{1}{\sqrt{a}} \sin\left(\frac{n\pi\xi}{a}\right) f(\xi)\, d\xi \tag{24.14}$$

$$b_n = \langle f, u_n \rangle = \int_{-a}^{a} \frac{1}{\sqrt{a}} \cos\left(\frac{n\pi\xi}{a}\right) f(\xi)\, d\xi \tag{24.15}$$

$\Rightarrow$

$$\begin{aligned} f(x) = \frac{1}{2a} \int_{-a}^{a} f(\xi)\, d\xi &+ \sum_{n=1}^{\infty} \frac{1}{a} \sin\left(\frac{n\pi x}{a}\right) \int_{-a}^{a} \sin\left(\frac{n\pi\xi}{a}\right) f(\xi)\, d\xi \\ &+ \sum_{n=1}^{\infty} \frac{1}{a} \cos\left(\frac{n\pi x}{a}\right) \int_{-a}^{a} \cos\left(\frac{n\pi\xi}{a}\right) f(\xi)\, d\xi. \end{aligned}$$

Equation (24.12) is the classical Fourier series expansion of a periodic function in terms of the eigenfunctions (which in this case are sines and cosines). Now,

$$\begin{aligned} f(x) &= \frac{1}{2a} \int_{-a}^{a} f(\xi)\, d\xi + \frac{1}{a} \sum_{n=1}^{\infty} \int_{-a}^{a} \left[\sin\left(\frac{n\pi x}{a}\right) \sin\left(\frac{n\pi\xi}{a}\right) + \cos\left(\frac{n\pi x}{a}\right) \cos\left(\frac{n\pi\xi}{a}\right)\right] f(\xi)\, d\xi \\ &= \frac{1}{2a} \int_{-a}^{a} f(\xi)\, d\xi + \frac{1}{a} \sum_{n=1}^{\infty} \int_{-a}^{a} \cos\left[\frac{n\pi(x - \xi)}{a}\right] f(\xi)\, d\xi \end{aligned}$$

Thus,

$$f(x) = \frac{1}{2a}\int_{-a}^{a} f(\xi)\,d\xi + \frac{1}{a}\sum_{n=1}^{\infty}\int_{-a}^{a}\cos\left[\frac{n\pi(x-\xi)}{a}\right]f(\xi)\,d\xi \tag{24.16}$$

This is an identity for any $f \in \mathcal{L}_2[-a,a]$. We now obtain the Fourier integral formula from this equation.

24.1.1 Fourier integral formula

Assume that $f(x)$ is absolutely integrable, i. e.,

$$\int_{-\infty}^{\infty} |f(x)|\,dx < \infty \tag{24.17}$$

Then, as we let $a \longrightarrow \infty$, the first term on the RHS of equation (24.16) goes to zero. Define

$$\alpha_n = \frac{n\pi}{a} \Rightarrow \Delta\alpha_n = \alpha_{n+1} - \alpha_n = \frac{\pi}{a} \tag{24.18}$$

Then the second term on RHS of equation (24.16) may be written

$$T_2 = \frac{1}{\pi}\sum_{n=1}^{\infty}\int_{-a}^{a}\cos[\alpha_n(x-\xi)]f(\xi)\,d\xi\Delta\alpha_n \tag{24.19}$$

The sum

$$\sum_{n=1}^{\infty} F(\alpha_n)\Delta\alpha_n; \quad F(\alpha_n) = \frac{1}{\pi}\int_{-a}^{a}\cos[\alpha_n(x-\xi)]f(\xi)\,d\xi$$

is a Riemann sum for the integral

$$\int_{0}^{\infty} F(\alpha)\,d\alpha$$

Thus, taking limit $a \rightarrow \infty$, we get

$$f(x) = \frac{1}{\pi}\int_{0}^{\infty}\int_{-\infty}^{\infty} f(\xi)\cos[\alpha(x-\xi)]\,d\xi\,d\alpha \tag{24.20}$$

This formula is also an identity for any $f \in \mathcal{L}_2(-\infty,\infty)$. This is known as the *Fourier integral formula*. Since cosine is an even function and sine is an odd function, equation (24.20) may be written as

$$f(x) = \frac{1}{2\pi}\int_{-\infty}^{\infty}\int_{-\infty}^{\infty} f(\xi)\cos[\alpha(x-\xi)]\,d\xi\,d\alpha + \frac{i}{2\pi}\int_{-\infty}^{\infty}\int_{-\infty}^{\infty} f(\xi)\sin[\alpha(x-\xi)]\,d\xi\,d\alpha$$

(the second term is zero since $\sin\alpha$ is odd in α) (24.21)

$\Rightarrow$

$$f(x) = \frac{1}{2\pi}\int_{-\infty}^{\infty}\int_{-\infty}^{\infty} f(\xi)e^{i\alpha(x-\xi)}\,d\xi\,d\alpha$$
$$= \frac{1}{2\pi}\int_{-\infty}^{\infty}\left(\int_{-\infty}^{\infty} f(\xi)e^{-i\alpha\xi}\,d\xi\right)e^{i\alpha x}\,d\alpha. \tag{24.22}$$

If we define

$$F(\alpha) = \int_{-\infty}^{\infty} e^{-i\alpha\xi} f(\xi)\,d\xi = \text{the Fourier transform of } f(x), \tag{24.23}$$

then equation (24.22) gives the inversion formula:

$$f(x) = \frac{1}{2\pi}\int_{-\infty}^{\infty} e^{i\alpha x}F(\alpha)\,d\alpha. \tag{24.24}$$

This transform is useful in solving equations with a second derivative operator (as well as derivatives of other orders) on an infinite domain:

$$-\frac{d^2}{dx^2}, \quad -\infty < x < \infty.$$

For example, (i) in the solution of the heat equation,

$$\frac{\partial^2 u}{\partial x^2} = \frac{\partial u}{\partial t}, \quad -\infty < x < \infty$$
$$u(x,0) = f(x)$$

or (ii) in the solution of the wave equation

$$c^2\frac{\partial^2 u}{\partial x^2} = \frac{\partial^2 u}{\partial t^2}, \quad -\infty < x < \infty$$
$$u(x,0) = f(x), \quad \frac{\partial u}{\partial t}(x,0) = g(x)$$

or (iii) in the solution of Laplace's equation

$$\frac{\partial^2 u}{\partial x^2} + \frac{\partial^2 u}{\partial y^2} = 0, \quad -a < y < a, \quad -\infty < x < \infty$$

in an infinite strip $(-\infty < x < \infty, -a < y < a)$ with boundary conditions

$$u(x, -a) = f(x), \quad u(x, a) = g(x).$$

24.2 Finite Fourier transform and the Fourier transform

Consider the eigenvalue problem

$$\frac{d^2u}{dx^2} = -\lambda u, \quad -\infty < x < \infty \tag{24.25}$$

We note that $u = e^{i\alpha x}$ (with $\alpha = \pm\sqrt{\lambda}$) satisfies the equation and is bounded if α is real. Now,

$$u' = i\alpha u, \quad u'' = (i\alpha)^2 u = -\lambda u \tag{24.26}$$

Thus, every α^2 is an eigenvalue of equation (24.25) with eigenfunction $u = e^{i\alpha x}$. (Note that $\sin \alpha x$ and $\cos \alpha x$ are also eigenfunctions.) Thus, in this case we have a *continuous spectrum*. When the interval is of finite length, the spectrum is discrete. In the case of discrete spectrum, the finite Fourier transform

$$f(x) \to \langle f, u_n(x) \rangle \tag{24.27}$$

generates a countable infinite sequence of coefficients. As the size of the interval is increased to infinity the spectrum becomes continuous and the finite Fourier transform, becomes a continuous function or *"the Fourier transform,"*

$$F(f) = \int_{-a}^{a} f(\xi) u_n(\xi)\, d\xi, \quad n = 1, 2, \ldots; \quad \text{Finite FT, (discrete spectrum)}$$

$$a \to \infty, \quad \text{continuous spectrum} \Rightarrow F(\alpha) = \int_{-\infty}^{\infty} e^{-i\alpha\xi} f(\xi)\, d\xi \tag{24.28}$$

Thus, $F(\alpha)$ plays the role of the coefficients in the finite Fourier transform.

Remark. Since the cosine is an even function, the Fourier integral formula, given in equation (24.20), may also be written as

$$f(x) = \frac{1}{\pi} \int_{0}^{\infty} \int_{-\infty}^{\infty} f(\xi) \cos \alpha(x - \xi)\, d\xi\, d\alpha$$

$$= \frac{1}{\pi} \int_{0}^{\infty} \int_{-\infty}^{\infty} f(\xi) \cos \alpha(\xi - x)\, d\xi\, d\alpha \tag{24.29}$$

$\Longrightarrow$

$$f(x) = \frac{1}{2\pi} \int_{-\infty}^{\infty} \int_{-\infty}^{\infty} f(\xi) \cos[\alpha(\xi - x)]\, d\xi\, d\alpha + \frac{i}{2\pi} \int_{-\infty}^{\infty} \int_{-\infty}^{\infty} f(\xi) \sin[\alpha(\xi - x)]\, d\xi\, d\alpha$$

$\Rightarrow$

$$\begin{aligned} f(x) &= \frac{1}{2\pi} \int_{-\infty}^{\infty} \int_{-\infty}^{\infty} f(\xi) e^{i\alpha(\xi - x)}\, d\xi\, d\alpha \\ &= \frac{1}{2\pi} \int_{-\infty}^{\infty} \left(\int_{-\infty}^{\infty} f(\xi) e^{i\alpha\xi}\, d\xi \right) e^{-i\alpha x}\, d\alpha \end{aligned} \tag{24.30}$$

Thus, we can also define

$$F(\alpha) = \int_{-\infty}^{\infty} e^{i\alpha\xi} f(\xi)\, d\xi \Longrightarrow f(x) = \frac{1}{2\pi} \int_{-\infty}^{\infty} e^{-i\alpha x} F(\alpha)\, d\alpha. \tag{24.31}$$

Many authors (Carslaw and Jaeger [10]; Sneddon [28]) use these (equation (24.31)) as the transform pair. However, we shall follow the notation used by Churchill [12] and use the pair defined in equations (24.23) and (24.24) as

$$F(\alpha) = \int_{-\infty}^{\infty} f(\xi) e^{-i\alpha\xi}\, d\xi \tag{24.32}$$

$$f(x) = \frac{1}{2\pi} \int_{-\infty}^{\infty} F(\alpha) e^{i\alpha x}\, d\alpha \tag{24.33}$$

The transforms given in equations (24.31) and (24.32)–(24.33) differ only by a change of sign in α. Since α and ξ vary from $-\infty$ to ∞, they are equivalent.

Example 24.1. Consider the function (also known as the decaying pulse)

$$f(x) = \begin{cases} e^{-cx}, & x \geq 0 \\ 0, & x < 0 \end{cases} \tag{24.34}$$

where c is a real positive constant. The FT of this function is given by

$$\begin{aligned} F(\alpha) &= \int_{-\infty}^{\infty} e^{-i\alpha\xi} f(\xi)\, d\xi = \int_{0}^{\infty} e^{-i\alpha\xi} \cdot e^{-c\xi}\, d\xi \\ &= \int_{0}^{\infty} e^{-(i\alpha + c)\xi}\, d\xi = \left. \frac{e^{-(i\alpha + c)\xi}}{-(i\alpha + c)} \right|_0^{\infty} \\ &= \frac{1}{(c + i\alpha)} = \frac{c - i\alpha}{(c^2 + \alpha^2)} \end{aligned} \tag{24.35}$$

$\Rightarrow$

$$
\begin{aligned}
f(x) &= \frac{1}{2\pi}\int_{-\infty}^{\infty} \frac{1}{(c+i\alpha)} e^{i\alpha x}\, d\alpha \\
&= \frac{1}{2\pi}\int_{-\infty}^{\infty} \frac{c-i\alpha}{c^2+\alpha^2} e^{i\alpha x}\, d\alpha; \quad (c+i\alpha = s) \\
&= \frac{1}{2\pi i}\int_{c-i\infty}^{c+i\infty} \frac{1}{s} e^{(s-c)x}\, ds \\
&= e^{-cx}\cdot \mathcal{L}^{-1}\left\{\frac{1}{s}\right\} \quad \left(\mathcal{L}^{-1} \text{ is inverse Laplace Transform}\right) \\
&= \begin{cases} e^{-cx}, & x \geq 0 \\ 0, & x < 0 \end{cases}
\end{aligned}
$$

For an extensive table of Fourier transforms, see the book by Churchill [11].

We note that

$$
\|f\|^2 = \int_0^{\infty} e^{-2cx}\, dx = \frac{1}{2c} = \frac{1}{2\pi}\int_{-\infty}^{\infty} |F(\alpha)|^2\, d\alpha.
$$

This relation is similar to Parseval's theorem and is known as the Plancherel's theorem. We discuss it in more general form below.

24.2.1 Physical interpretation

Consider again the eigenvalue problem:

$$
\frac{d^2u}{dx^2} = -\alpha^2 u, \quad -\infty < x < \infty \tag{24.36}
$$

As stated earlier, every $\alpha^2 > 0$ is an eigenvalue with eigenfunction $u_\alpha(x) = e^{-i\alpha x}$. Note that $e^{i\alpha x}$ is also an eigenfunction corresponding to eigenvalue α^2. However if we let α vary from $-\infty$ to ∞ we need to consider only one of these eigenfunctions. Thus, we have a continuous spectrum. To show that the eigenvalue problem is self-adjoint, we consider two functions $u, v \in \mathcal{L}_2(-\infty, \infty)$, a Hilbert space with the usual inner product. Now,

$$
\begin{aligned}
\left\langle \frac{\partial^2 u(x)}{\partial x^2}, v \right\rangle &= \int_{-\infty}^{\infty} \frac{\partial^2 u(x)}{\partial x^2}\overline{v(x)}\, dx \\
&= \left(\frac{\partial u}{\partial x}\overline{v} - u\frac{\partial \overline{v}}{\partial x}\right)\Bigg|_{-\infty}^{\infty} + \int_{-\infty}^{\infty} u\frac{\partial^2 \overline{v}}{\partial x^2}\, dx \\
&= \langle u, Lv\rangle + \left(\frac{\partial u}{\partial x}\overline{v} - u\frac{\partial \overline{v}}{\partial x}\right)\Bigg|_{-\infty}^{\infty}
\end{aligned} \tag{24.37}
$$

Assuming that u and $\frac{\partial u}{\partial x}$ vanish at infinity (this assumption is reasonable since if this is not the case then u is not absolutely integrable),
$\Rightarrow$

$$\langle Lu, v\rangle = \langle u, Lv\rangle \tag{24.38}$$

Thus, we may use the formalism we had before. Consider again the Hilbert space $\mathcal{L}_2(-\infty,\infty)$ with the usual inner product

$$\langle u, v\rangle = \int_{-\infty}^{\infty} u(x)\overline{v(x)}\,dx.$$

If $f(x) \in \mathcal{L}_2(-\infty,\infty)$, and

$$F(\alpha) = \mathcal{F}(f(x)) = \text{Fourier transform of } f(x)$$

then

$$F(\alpha) = \langle f(x), e^{i\alpha x}\rangle = \int_{-\infty}^{\infty} e^{-i\alpha\xi} f(\xi)\,d\xi \tag{24.39}$$

and the inverse is given by

$$f(x) = \frac{1}{2\pi}\int_{-\infty}^{\infty} e^{i\alpha x} F(\alpha)\,d\alpha \tag{24.40}$$

Thus, $F(\alpha)$ plays the role of the coefficients in the finite Fourier transform, as already mentioned earlier.

24.2.2 Properties of the Fourier transform

The Fourier transform defined by

$$F(\alpha) = \mathcal{F}\{f(x)\} = \int_{-\infty}^{\infty} f(\xi)e^{-i\alpha\xi}\,d\xi \tag{24.41}$$

can be used to obtain the following properties:

1. Transform of derivatives:

$$\mathcal{F}\left\{\frac{d^m f}{dx^m}\right\} = (i\alpha)^m F(\alpha), \quad m = 1, 2, 3, \ldots \tag{24.42}$$

   if we assume that f and its derivatives vanish for $x \to \pm\infty$.

2. Multiplication by x:

$$\mathcal{F}\{xf(x)\} = i\,\frac{dF}{d\alpha} \tag{24.43}$$

3. Shift in x:

$$\mathcal{F}\{f(x+c)\} = e^{i\alpha c}F(\alpha), \quad c \text{ real} \tag{24.44}$$

4. Shift in α:

$$\mathcal{F}\{f(x)e^{icx}\} = F(\alpha - c), \quad c \text{ real} \tag{24.45}$$

5. Scaling in x:

$$\mathcal{F}\left\{\frac{1}{|c|}f\left(\frac{x}{c}\right)\right\} = F(\alpha c), \quad c \text{ real}, \quad c \neq 0 \tag{24.46}$$

6. Reflection in x:

$$\mathcal{F}\{f(-x)\} = F(-\alpha) \tag{24.47}$$

7. Transform of complex conjugate:

$$\mathcal{F}\{\overline{f(x)}\} = \overline{F(-\alpha)} \tag{24.48}$$

8. Convolution:

$$\mathcal{F}\left\{\int_{-\infty}^{\infty} f(x')g(x-x')\,dx'\right\} = F(\alpha)G(\alpha) \tag{24.49}$$

9. Transform and representation of Dirac delta function:

$$\mathcal{F}\{\delta(x-s)\} = e^{-i\alpha s}, \quad \delta(x-s) = \frac{1}{2\pi}\int_{-\infty}^{\infty} e^{i\alpha(x-s)}\,d\alpha \tag{24.50}$$

10. Moments theorem: We can expand $F(\alpha)$ in powers of α,

$$F(\alpha) = \sum_{k=0}^{\infty} \frac{(-i\alpha)^k}{k!} M_k \tag{24.51}$$

where M_k is k-th spatial moment of $f(x)$ [see further explanation in the next section].

11. Plancherel's theorem:

$$\int_{-\infty}^{\infty} f(x)\overline{g(x)}\,dx = \frac{1}{2\pi}\int_{-\infty}^{\infty} F(\alpha)\overline{G(\alpha)}\,d\alpha. \tag{24.52}$$

For the special case of $f(x) = g(x)$, equation (24.52) gives

$$\|f(x)\|^2 = \int_{-\infty}^{\infty} |f(x)|^2 \, dx = \frac{1}{2\pi} \|F(\alpha)\|^2. \tag{24.53}$$

The LHS of equation (24.53) may be interpreted as the total energy content of $f(x)$ while the RHS represents the same in the frequency domain.

24.2.3 Moments theorem for Fourier transform

Let $F(\alpha) = \mathcal{F}\{f(x)\}$, i. e.,

$$\begin{aligned} F(\alpha) &= \int_{-\infty}^{\infty} f(\xi) e^{-i\alpha\xi} \, d\xi \\ &= \int_{-\infty}^{\infty} f(\xi) \sum_{k=0}^{\infty} \frac{(-i\alpha\xi)^k}{k!} \, d\xi \end{aligned}$$

Interchanging the sum and integral gives

$$\begin{aligned} F(\alpha) &= \sum_{k=0}^{\infty} \frac{(-i\alpha)^k}{k!} \int_{-\infty}^{\infty} \xi^k f(\xi) \, d\xi \\ &= \sum_{k=0}^{\infty} \frac{(-i\alpha)^k}{k!} M_k, \end{aligned} \tag{24.54}$$

where M_k is the k-th spatial moment of $f(\xi)$, defined by

$$M_k = \int_{-\infty}^{\infty} \xi^k f(\xi) \, d\xi, \quad k = 0, 1, 2, \ldots \tag{24.55}$$

while the central moments are defined by

$$m_k = \int_{-\infty}^{\infty} (\xi - M_1)^k f(\xi) \, d\xi, \quad k = 1, 2, 3, \ldots \tag{24.56}$$

Thus, if the Fourier transform of a function $f(x)$ is known, we can determine the kth spatial moment [or temporal moment for a function of time] without inverting the Fourier transform using equation (24.51):

$$M_0 = F|_{\alpha=0}$$

$$M_1 = -i\frac{dF}{d\alpha}\bigg|_{\alpha=0}$$

$$M_2 = (-i)^2\frac{d^2F}{d\alpha^2}\bigg|_{\alpha=0}$$

$$M_k = (-i)^k\frac{d^kF}{d\alpha^k}\bigg|_{\alpha=0}, \quad k = 0, 1, 2, \ldots \tag{24.57}$$

The use of the moment theorem is illustrated in the example below.

Example 24.2. Consider the convective-diffusion equation in an infinite domain

$$\frac{\partial c}{\partial t} + u\frac{\partial c}{\partial x} = D_m\frac{\partial^2 c}{\partial x^2}, \quad -\infty < x < \infty, \quad t > 0$$

with a point source initial condition

$$c(x, 0) = \delta(x);$$

and the conditions

$$c, \frac{\partial c}{\partial x} \to 0 \quad \text{for } x \to \pm\infty.$$

Here, u is the mean (convective) velocity, D_m is the diffusivity and $c(x, t)$ is the solute concentration. Let

$$\hat{c}(\alpha, t) = \mathcal{F}\{c(x, t)\}.$$

Taking the Fourier transform of the evolution equation and initial condition gives

$$\frac{d\hat{c}}{dt} + u(i\alpha)\hat{c} = D_m(i\alpha)^2\hat{c}; \quad \hat{c}(t = 0) = 1$$

$\Longrightarrow$

$$\begin{aligned}\hat{c} &= \exp[-(i\alpha u + \alpha^2 D_m)t] \\ &= 1 - (i\alpha u + \alpha^2 D_m)t + \frac{(i\alpha u + \alpha^2 D_m)^2 t^2}{2!} - \cdots \\ &= 1 - i\alpha ut - \alpha^2\left(D_m t + \frac{u^2t^2}{2}\right) + \cdots\end{aligned}$$

$\Longrightarrow$

$$\begin{aligned}M_0 &= 1, \\ M_1 &= ut, \\ M_2 &= 2D_m t + u^2t^2,\end{aligned}$$

$$M_3 = 6D_m t^2 u + u^3 t^3,$$
$$M_4 = 12D_m^2 t^2 + 12D_m t^3 u^2 + u^4 t^4, \quad \text{etc.}$$

$\Longrightarrow$ The central moments are given by

$$m_2 = \sigma^2 = M_2 - M_1^2 = 2D_m t$$
$$m_3 = 0$$
$$m_4 = 12D_m^2 t^2, \quad \text{and so on.}$$

It can be shown that all the odd central moments are zero. Thus, the dispersion is symmetric around the centroid located at ut. [Remark: This result or the symmetric spreading of the solute around the centroid does not hold in a finite domain due to inlet and exit boundary conditions.]

24.2.4 Fourier transform in spatial and cyclic frequencies

We have defined the Fourier transform pair in spatial frequency α (rad/cm) as

$$F(\alpha) = \mathcal{F}\{f(x)\} = \int_{-\infty}^{\infty} e^{-i\alpha x} f(x)\,dx$$
$$f(x) = \mathcal{F}^{-1}\{F(\alpha)\} = \frac{1}{2\pi}\int_{-\infty}^{\infty} e^{i\alpha x} F(\alpha)\,d\alpha \tag{24.58}$$

If we use cyclic frequency $\omega = \frac{\alpha}{2\pi}$ (cycle/cm), the transform pair is defined by

$$F(\omega) = \mathcal{F}\{f(x)\} = \int_{-\infty}^{\infty} e^{-2\pi i\omega x} f(x)\,dx$$
$$f(x) = \mathcal{F}^{-1}\{F(\omega)\} = \int_{-\infty}^{\infty} e^{2\pi i\omega x} F(\alpha)\,d\omega \tag{24.59}$$

Note that the constant multiplier 2π in inverse FT in spatial frequency has disappeared when cyclic frequency is used. The cyclic frequency definition is used in analyzing signals in time or spatially periodic structures.

Fourier transforms in 2D and 3D

Let $f(x,y)$ denote a 2D intensity image where x and y are spatial variables (having length units). The FT pair of $f(x,y)$ in spatial frequency is defined by

$$F(\alpha_1,\alpha_2) = \mathcal{F}\{f(x,y)\} = \int_{-\infty}^{\infty}\int_{-\infty}^{\infty} e^{-i\alpha_1 x - i\alpha_2 y} f(x,y)\,dx\,dy$$
$$f(x,y) = \mathcal{F}^{-1}\{F(\alpha_1,\alpha_2)\} = \frac{1}{(2\pi)^2}\int_{-\infty}^{\infty}\int_{-\infty}^{\infty} e^{i\alpha_1 x + i\alpha_2 y} F(\alpha_1,\alpha_2)\,d\alpha_1\,d\alpha_2 \tag{24.60}$$

where α_1 and α_2 are in rad/cm. If we use cyclic frequencies $\omega_1 = \frac{\alpha_1}{2\pi}$ and $\omega_2 = \frac{\alpha_2}{2\pi}$ in cycle/cm, the transform pair can be defined by

$$F(\omega_1,\omega_2) = \mathcal{F}\{f(x,y)\} = \int_{-\infty}^{\infty}\int_{-\infty}^{\infty} e^{-2\pi i(\omega_1 x + \omega_2 y)} f(x,y)\,dx\,dy$$
$$f(x,y) = \mathcal{F}^{-1}\{F(\omega_1,\omega_2)\} = \int_{-\infty}^{\infty}\int_{-\infty}^{\infty} e^{2\pi i(\omega_1 x + \omega_2 y)} F(\omega_1,\omega_2)\,d\omega_1\,d\omega_2 \tag{24.61}$$

Similarly, in 3D, we can define the pair using the spatial frequency vector $\boldsymbol{\alpha} = \begin{pmatrix}\alpha_1\\ \alpha_2\\ \alpha_3\end{pmatrix}$ as

$$F(\boldsymbol{\alpha}) = \mathcal{F}\{f(\mathbf{x})\} = \int_{-\infty}^{\infty}\int_{-\infty}^{\infty}\int_{-\infty}^{\infty} e^{-i\boldsymbol{\alpha}.\mathbf{x}} f(\mathbf{x})\,d\mathbf{x}$$
$$f(\mathbf{x}) = \mathcal{F}^{-1}\{F(\boldsymbol{\alpha})\} = \frac{1}{(2\pi)^3}\int_{-\infty}^{\infty}\int_{-\infty}^{\infty}\int_{-\infty}^{\infty} e^{i\boldsymbol{\alpha}.\mathbf{x}} F(\boldsymbol{\alpha})\,d\boldsymbol{\alpha} \tag{24.62}$$

or in the cyclic frequency vector $\boldsymbol{\omega} = \begin{pmatrix}\omega_1\\ \omega_2\\ \omega_3\end{pmatrix}$ as

$$F(\boldsymbol{\omega}) = \mathcal{F}\{f(\mathbf{x})\} = \int_{-\infty}^{\infty}\int_{-\infty}^{\infty}\int_{-\infty}^{\infty} e^{-2\pi i\boldsymbol{\omega}.\mathbf{x}} f(\mathbf{x})\,d\mathbf{x}$$
$$f(\mathbf{x}) = \mathcal{F}^{-1}\{F(\boldsymbol{\omega})\} = \int_{-\infty}^{\infty}\int_{-\infty}^{\infty}\int_{-\infty}^{\infty} e^{2\pi i\boldsymbol{\omega}.\mathbf{x}} F(\boldsymbol{\omega})\,d\boldsymbol{\omega} \tag{24.63}$$

where $\mathbf{x} = \begin{pmatrix}x\\ y\\ z\end{pmatrix}$ is the vector of spatial coordinates, and $\boldsymbol{\alpha}.\mathbf{x} = \alpha_1 x + \alpha_2 y + \alpha_3 z$ and $\boldsymbol{\omega}.\mathbf{x} = \omega_1 x + \omega_2 y + \omega_3 z$ represent the usual dot product in $\mathbb{R}^3$.

24.2.5 Fourier transform and Plancherel's theorem

Let

$$F(\alpha) = \mathcal{F}\{f(x)\} \quad \text{and} \quad G(\alpha) = \mathcal{F}\{g(x)\} \tag{24.64}$$

then

$$\int_{-\infty}^{\infty} f(x)\overline{g(x)}\,dx = \frac{1}{2\pi}\int_{-\infty}^{\infty} F(\alpha)\overline{G(\alpha)}\,d\alpha \tag{24.65}$$

[As shown below, the 2π factor disappears if we use cyclic frequencies.] As stated earlier, this is known as *Plancherel's theorem*. A proof of this theorem uses integral representation of the Dirac delta function:

$$\delta(\alpha-\alpha') = \frac{1}{2\pi}\int_{-\infty}^{\infty} e^{ix(\alpha-\alpha')}\,dx \tag{24.66}$$

Since we have

$$f(x) = \frac{1}{2\pi}\int_{-\infty}^{\infty} e^{i\alpha x}F(\alpha)\,d\alpha$$

and property (24.48) $\Longrightarrow$

$$\overline{g(x)} = \frac{1}{2\pi}\int_{-\infty}^{\infty} e^{-i\alpha' x}\overline{G(\alpha')}\,d\alpha',$$

we can write

$$\begin{aligned}\int_{-\infty}^{\infty} f(x)\overline{g(x)}\,dx &= \frac{1}{(2\pi)^2}\int_{-\infty}^{\infty}\int_{-\infty}^{\infty}\int_{-\infty}^{\infty} e^{i\alpha x}F(\alpha)e^{-i\alpha' x}\overline{G(\alpha')}\,d\alpha'\,d\alpha\,dx\\ &= \frac{1}{2\pi}\int_{-\infty}^{\infty}\int_{-\infty}^{\infty} F(\alpha)\overline{G(\alpha')}\left(\frac{1}{2\pi}\int_{-\infty}^{\infty} e^{i\alpha x}e^{-i\alpha' x}\,dx\right)d\alpha'\,d\alpha\\ &= \frac{1}{2\pi}\int_{-\infty}^{\infty}\int_{-\infty}^{\infty} F(\alpha)\overline{G(\alpha')}\delta(\alpha-\alpha')\,d\alpha'\,d\alpha\\ &= \frac{1}{2\pi}\int_{-\infty}^{\infty} F(\alpha)\overline{G(\alpha)}\,d\alpha. \quad \square\end{aligned}$$

Taking $g(x) = f(x)$, we get from equation (24.65)

$$\int_{-\infty}^{\infty} |f(x)|^2\,dx = \frac{1}{2\pi}\int_{-\infty}^{\infty} |F(\alpha)|^2\,d\alpha \tag{24.67}$$

$$= \int_{-\infty}^{\infty} |F(2\pi\omega)|^2\,d\omega, \quad \omega = \frac{\alpha}{2\pi}. \tag{24.68}$$

24.3 Solution of BVPs and IBVPs in infinite intervals using the FT

In this section, we illustrate the application of the Fourier transform to the solution of linear differential equations on infinite and semi-infinite domains.

24.3.1 Heat equation in an infinite rod

Consider solving the heat equation (in dimensionless form)

$$\frac{\partial^2 u}{\partial x^2} = \frac{\partial u}{\partial t}, \quad -\infty < x < \infty \tag{24.69}$$

with initial condition

$$u(x,0) = f(x). \tag{24.70}$$

Assume that u and its derivatives w. r. t. x are bounded at $x = \pm\infty$. Let

$$\hat{u}(\alpha,t) = \mathcal{F}\{u(x,t)\} = \int_{-\infty}^{\infty} e^{-i\alpha\xi} u(\xi,t)\, d\xi \tag{24.71}$$

$$F(\alpha) = \mathcal{F}\{f(x)\} = \hat{u}(\alpha, t=0) \tag{24.72}$$

Take the inner product of equations (24.69)–(24.70) with eigenfunctions

$$\left\langle \frac{\partial^2 u}{\partial x^2}, e^{i\alpha x} \right\rangle = \left\langle \frac{\partial u}{\partial t}, \quad e^{i\alpha x} \right\rangle$$

$\Rightarrow$

$$-\alpha^2 \langle u, e^{i\alpha x} \rangle = \frac{d}{dt}\langle u, e^{+i\alpha x} \rangle = \frac{d\hat{u}}{dt}$$

$\Rightarrow$

$$\hat{u} = \langle u, e^{+i\alpha x} \rangle = Ke^{-\alpha^2 t}$$

$\Rightarrow$

$$\lim_{t\to 0}\langle u, e^{+i\alpha x} \rangle = K \lim_{t\to 0} e^{-\alpha^2 t} \quad \text{or} \quad \left\langle \lim_{t\to 0} u, e^{+i\alpha x} \right\rangle = K$$

$\Rightarrow$

$$K = \langle f(x), e^{+i\alpha x} \rangle = F(\alpha)$$

∴

$$\hat{u} = \langle u, e^{iax} \rangle = F(a)e^{-a^2 t} \tag{24.73}$$

⇒

$$\begin{aligned} u(x,t) &= \frac{1}{2\pi} \int_{-\infty}^{\infty} e^{iax} F(a) e^{-a^2 t}\, da \\ &= \frac{1}{2\pi} \int_{-\infty}^{\infty} e^{iax} e^{-a^2 t} \left(\int_{-\infty}^{\infty} e^{-ia\xi} f(\xi)\, d\xi \right) da \\ &= \frac{1}{2\pi} \int_{-\infty}^{\infty} \int_{-\infty}^{\infty} e^{ia(x-\xi)-a^2 t} f(\xi)\, d\xi\, da \end{aligned} \tag{24.74}$$

This is the formal solution of the heat/diffusion equation in an infinite domain.

The integral in equation (24.74) w.r.t the wave number a can be evaluated either directly or by using Cauchy's theorem as shown below.

Direct method

Let

$$\begin{aligned} I &= \int_{-\infty}^{\infty} e^{-a^2 t + ia(x-\xi)}\, da \\ &= \int_{-\infty}^{\infty} e^{-a^2 t} \cdot [\cos a(x-\xi) + i \sin a(x-\xi)]\, da \\ &= \int_{-\infty}^{\infty} e^{-a^2 t} \cos a(x-\xi)\, da + i \int_{-\infty}^{\infty} e^{-a^2 t} \sin a(x-\xi)\, da \\ &= 2 \int_{0}^{\infty} e^{-a^2 t} \cos a(x-\xi)\, da \\ &= 2 \cdot \sqrt{\frac{\pi}{4t}} e^{-\frac{(x-\xi)^2}{4t}} \end{aligned}$$

∴ Equation (24.74) ⟹

$$\begin{aligned} u(x,t) &= \frac{1}{2\pi} \int_{-\infty}^{\infty} 2 \cdot \sqrt{\frac{\pi}{4t}} e^{-\frac{(x-\xi)^2}{4t}} f(\xi)\, d\xi \\ &= \frac{1}{\sqrt{4\pi t}} \int_{-\infty}^{\infty} e^{-\frac{(x-\xi)^2}{4t}} f(\xi)\, d\xi \end{aligned} \tag{24.75}$$

Method based on Cauchy's theorem

Let

$$I = \int_{-\infty}^{\infty} e^{-a^2 t + ia(x-\xi)}\, da$$

$$= \int_{-\infty}^{\infty} e^{-\frac{1}{t}[at - \frac{i}{2}(x-\xi)]^2} \cdot e^{-\frac{(x-\xi)^2}{4t}}\, da$$

$$= e^{-\frac{(x-\xi)^2}{4t}} \int_{-\infty}^{\infty} e^{-t[a-i\delta]^2}\, da \tag{24.76}$$

where

$$\delta = \frac{(x-\xi)}{2t} \tag{24.77}$$

∴

$$I = e^{-\frac{(x-\xi)^2}{4t}} J, \tag{24.78}$$

$$J = \int_{-\infty}^{\infty} e^{-t(a-i\delta)^2}\, da\,. \tag{24.79}$$

To evaluate J, we use Cauchy's theorem around the contour shown in Figure 24.1.

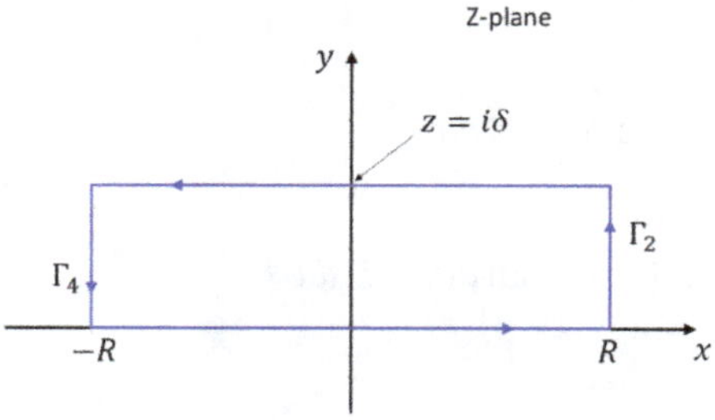

Figure 24.1: Schematic of the contour for Cauchy theorem to evaluate integral.

Since $g(z) = e^{-t(z-i\delta)^2}$ is analytic inside and on the boundary of C,

$$\int_C e^{-t(z-i\delta)^2}\, dz = 0,$$

⇒

$$\int_{-R}^{R} e^{-t(x-i\delta)^2}\, dx + \int_{\Gamma_2} e^{-t(z-i\delta)^2}\, dz + \int_{R}^{-R} e^{-tx^2}\, dx + \int_{\Gamma_4} e^{-t(z-i\delta)^2}\, dz = 0 \tag{24.80}$$

Take limit $R \to \infty$, then

$$\int_{\Gamma_2} e^{-t(z-i\delta)^2}\, dz = i \int_0^{\delta} e^{-t[R+iu-i\delta]^2}\, du$$

$$= ie^{-tR^2} \int_0^{\delta} e^{t[u-\delta]^2} \cdot e^{-2iRt(u-\delta)}\, du = 0 \quad \text{for } R \to \infty$$

Similarly,

$$\lim_{R\to\infty} \int_{\Gamma_4} e^{-t(z-i\delta)^2}\, dz = 0$$

Equation (24.80) $\Rightarrow$

$$J = \int_{-\infty}^{\infty} e^{-t(x-i\delta)^2}\, dx = \int_{-\infty}^{\infty} e^{-tx^2}\, dx = \sqrt{\frac{\pi}{t}} \tag{24.81}$$

$\therefore$

$$u(x,t) = \frac{1}{\sqrt{4\pi t}} \int_{-\infty}^{\infty} e^{-\frac{(x-\xi)^2}{4t}} f(\xi)\, d\xi \tag{24.82}$$

is the solution to equations (24.69)–(24.70).

If we assume that $\int_{-\infty}^{\infty} |f(\xi)|\, d\xi$ exists, then so long as $t > 0$, it can be shown that the integral on the RHS of equation (24.82) converges absolutely and uniformly in both x and t for $t > 0$, as well as all of its derivatives w. r. t. x and t. Thus, differentiation under the integral sign is valid. To verify the initial condition, let

$$\frac{x-\xi}{\sqrt{4t}} = y \Rightarrow \xi = x - y\sqrt{4t} \Rightarrow d\xi = -\sqrt{4t}\, dy \tag{24.83}$$

$\Longrightarrow$

$$u(x,t) = \frac{1}{\sqrt{4\pi t}} \int_{\infty}^{-\infty} e^{-y^2} f(x - 2y\sqrt{t}) \cdot (-\sqrt{4t})\, dy$$

$$= \frac{1}{\sqrt{\pi}} \int_{-\infty}^{\infty} e^{-y^2} f(x - 2y\sqrt{t})\, dy \tag{24.84}$$

If $f(x)$ is sectionally smooth or piecewise continuous, then the integral converges uniformly and absolutely in x and t. Hence, we can take the limit under the integral sign,

$$\lim_{t\to 0} u(x,t) = \frac{1}{\sqrt{\pi}} \int_{-\infty}^{\infty} e^{-y^2} \lim_{t\to 0} f(x - 2y\sqrt{t})\, dy$$
$$= f(x) \tag{24.85}$$

∴ For all $f(x)$ for which $\int_{-\infty}^{\infty} |f(x)|\, dx$ exists, the solution of equations (24.69)–(24.70) is given by equation (24.82).

Physical interpretation

Let

$$f(x) = \delta(x - s)$$
$$= \text{unit source of heat at position } x = s \text{ at time } t = 0 \tag{24.86}$$

⇒

$$u(x,t) = \frac{1}{\sqrt{4\pi t}} e^{-\frac{(x-s)^2}{4t}}$$
$$= \text{temperature at position } x \text{ and time } t \text{ due to a unit source at position } s \text{ at } t = 0 \tag{24.87}$$

More generally,

$$W(x,t,s,\tau) = \frac{1}{\sqrt{4\pi(t-\tau)}} e^{-\frac{(x-s)^2}{4(t-\tau)}}$$
$$= \text{temperature at position } x \text{ at time } t \text{ due to a unit source at position } s \text{ at time } \tau (t > \tau) \tag{24.88}$$

W is called a *fundamental solution* (or the Green's function) of the heat equation. It satisfies the equation,

$$\frac{\partial W}{\partial t} - \frac{\partial^2 W}{\partial x^2} = \delta(x - s)\delta(t - \tau), \tag{24.89}$$

as well as the adjoint heat equation

$$-\frac{\partial W}{\partial \tau} - \frac{\partial^2 W}{\partial s^2} = \delta(x - s)\delta(t - \tau). \tag{24.90}$$

We note that $f(\xi)\, d\xi$ = amount of heat between ξ and $\xi + d\xi$ at time $t = 0$. Thus, considering the distributed source as the sum(integral) of point sources and using the principle of superposition, the temperature at x and time t is

$$u(x,t) = \frac{1}{\sqrt{4\pi t}} \int_{-\infty}^{\infty} e^{-\frac{(x-\xi)^2}{4t}} f(\xi)\, d\xi \tag{24.91}$$

This solution is valid for any piecewise continuous $f(\xi)$ for which the integral in equation (24.91) exists. We now consider a special case:

Special case

Consider the special case of the IC being a unit step function:

$$f(\xi) = \begin{cases} 1, & \xi \leq 0 \\ 0, & \xi > 0 \end{cases} \tag{24.92}$$

$\Rightarrow$

$$u = \frac{1}{\sqrt{4\pi t}} \int_{-\infty}^{0} e^{-\frac{(x-\xi)^2}{4t}}\, d\xi \tag{24.93}$$

Let

$$\frac{x-\xi}{\sqrt{4t}} = \eta$$

$$\begin{aligned} u &= \frac{1}{\sqrt{4\pi t}} \int_{\infty}^{\frac{x}{\sqrt{4t}}} e^{-\eta^2}(-\sqrt{4t})\, d\eta = \frac{1}{\sqrt{\pi}} \int_{\frac{x}{\sqrt{4t}}}^{\infty} e^{-\eta^2}\, d\eta \\ &= \frac{1}{2} \cdot \frac{2}{\sqrt{\pi}} \left[\int_{0}^{\frac{x}{\sqrt{4t}}} e^{-\eta^2}\, d\eta + \int_{\frac{x}{\sqrt{4t}}}^{\infty} e^{-\eta^2}\, d\eta - \int_{0}^{\frac{x}{\sqrt{4t}}} e^{-\eta^2}\, d\eta \right] \\ &= \frac{1}{2}\left[1 - \frac{2}{\sqrt{\pi}} \int_{0}^{\frac{x}{\sqrt{4t}}} e^{-\eta^2}\, d\eta \right] \end{aligned}$$

$\Rightarrow$

$$u(x,t) = \frac{1}{2}\left[1 - \operatorname{erf}\left(\frac{x}{\sqrt{4t}}\right)\right] \tag{24.94}$$

Figure 24.2 shows the solution profile at various times. [Here, $\operatorname{erf}\eta$ denotes the *error function.*]

Remark. If the equation is

$$D\frac{\partial^2 u}{\partial x^2} = \frac{\partial u}{\partial t} \tag{24.95}$$

then the solution for the same initial condition is

$$u = \frac{1}{2}\left[1 - \operatorname{erf}\left(\frac{x}{\sqrt{4Dt}}\right)\right]. \tag{24.96}$$

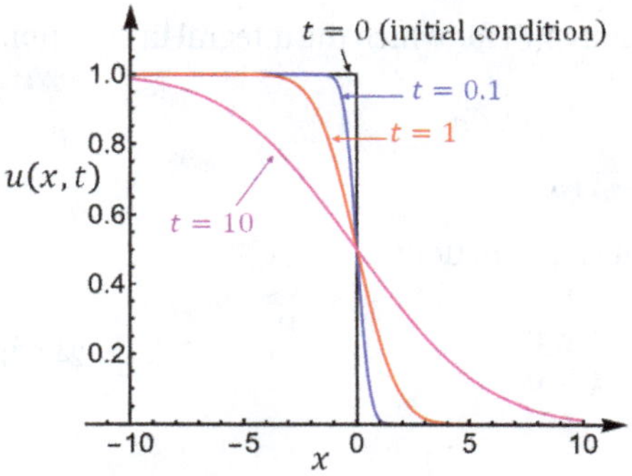

Figure 24.2: Temporal evolution of dimensionless temperature distribution for 1D transient diffusion in infinite domain.

Note that $u(x,t) > 0$ for all $x > 0$ and $t > 0$ even though $u(x,t=0) = 0$ for $x > 0$. This implies an infinite speed of propagation, which is a property of parabolic equations.

24.3.2 Solution of the heat equation in semi-infinite domain

Consider the equation

$$\frac{\partial^2 u}{\partial x^2} = \frac{\partial u}{\partial t}, \quad 0 < x < \infty, \quad t > 0 \tag{24.97}$$

with boundary condition

$$u = 0 \text{ @ } x = 0 \tag{24.98}$$

and initial condition

$$u(x,0) = f(x) \tag{24.99}$$

In the solution (equation (24.91)), suppose that we assume that $f(\xi)$ is odd, i. e.,

$$f(-\xi) = -f(\xi) \tag{24.100}$$

Then

$$u(x,t) = \frac{1}{\sqrt{4\pi t}} \left[\int_{-\infty}^{0} e^{-\frac{(x-\xi)^2}{4t}} f(\xi)\, d\xi + \int_{0}^{\infty} e^{-\frac{(x-\xi)^2}{4t}} f(\xi)\, d\xi \right] \tag{24.101}$$

Setting $\xi = -s$, and $f(-s) = -f(s)$ in the first integral, gives

$$\begin{aligned} I_1 &= -\int_{\infty}^{0} e^{-\frac{(x+s)^2}{4t}} f(-s)\,ds \\ &= \int_{\infty}^{0} f(s) e^{-\frac{(x+s)^2}{4t}}\,ds \\ &= -\int_{0}^{\infty} f(\xi) e^{-\frac{(x+\xi)^2}{4t}}\,d\xi \end{aligned}$$

Thus,

$$u(x,t) = \frac{1}{\sqrt{4\pi t}}\left[\int_0^{\infty} \left(e^{-\frac{(x-\xi)^2}{4t}} - e^{-\frac{(x+\xi)^2}{4t}}\right) f(\xi)\,d\xi\right]. \tag{24.102}$$

Since the integral (in equation (24.102)) converges uniformly and absolutely, we can take the limit under the integral sign:

$\Rightarrow$

$$u(0,t) = 0$$

Thus, the solution to equations (24.97)–(24.99) is given by equation (24.102).

Example: $f(\xi) = 1$

$\Rightarrow$ The solution given by equation (24.102) simplifies to

$$\begin{aligned} u(x,t) &= \frac{1}{\sqrt{4\pi t}}\int_0^{\infty} e^{-\frac{(x-\xi)^2}{4t}}\,d\xi - \frac{1}{\sqrt{4\pi t}}\int_0^{\infty} e^{-\frac{(x+\xi)^2}{4t}}\,d\xi \\ &= \frac{1}{2}\left[1 + \operatorname{erf}\left(\frac{x}{\sqrt{4t}}\right)\right] - \frac{1}{2}\left[1 - \operatorname{erf}\left(\frac{x}{\sqrt{4t}}\right)\right] \\ &= \operatorname{erf}\left(\frac{x}{\sqrt{4t}}\right). \end{aligned} \tag{24.103}$$

This is the solution to the heat equation in a semi-infinite domain with initial temperature of unity and boundary ($x = 0$) temperature of zero (for $t > 0$). Figure 24.3 shows the spatial profiles of the solution at various times.

Nonhomogeneous problem

Consider the heat/diffusion equation in a semi-infinite domain

$$\frac{\partial^2 u}{\partial x^2} = \frac{\partial u}{\partial t}, \quad 0 < x < \infty \tag{24.104}$$

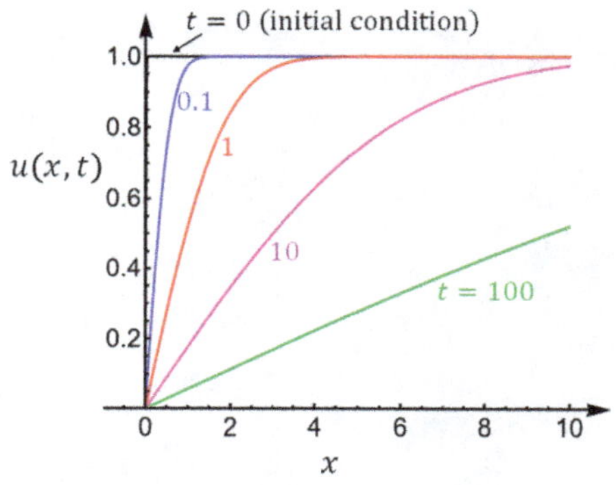

Figure 24.3: Spatial profiles of solution $u(x, t)$ at various times for heat equation in semi-infinite domain.

with boundary and initial conditions

$$u = 1 @ x = 0, \quad t > 0 \quad \text{and} \quad u(x, 0) = f(x). \tag{24.105}$$

Define

$$w = u - 1 \tag{24.106}$$

⇒

$$\frac{\partial^2 w}{\partial x^2} = \frac{\partial w}{\partial t}, \tag{24.107}$$

$$w(0, t) = 0, \quad w(x, 0) = f(x) - 1 = F(x) \tag{24.108}$$

⇒

$$u = 1 + \frac{1}{\sqrt{4\pi t}}\left[\int_0^\infty [e^{-\frac{(x-\xi)^2}{4t}} - e^{-\frac{(x+\xi)^2}{4t}}][f(\xi) - 1]\, d\xi\right] \tag{24.109}$$

For special case of $f(x) = 0$, the solution given in equation (24.109) reduces to

$$u(x, t)|_{f=0} = u_0(x, t) = 1 - \operatorname{erf}\left(\frac{x}{\sqrt{4t}}\right) = \operatorname{erfc}\left(\frac{x}{\sqrt{4t}}\right). \tag{24.110}$$

Thus, equation (24.110) is the solution to

$$\frac{\partial^2 u}{\partial x^2} = \frac{\partial u}{\partial t}, \quad u(0, t > 0) = 1, \quad u(x, 0) = 0 \tag{24.111}$$

Consider the more general case of a nonhomogeneous problem given by

$$\frac{\partial^2 u}{\partial x^2} = \frac{\partial u}{\partial t}, \quad 0 < x < \infty \tag{24.112}$$

$$u = g(t) \ @ \ x = 0, \quad u = f(x) \ @ \ t = 0 \tag{24.113}$$

We have seen how to solve this problem for $g(t) = 1$. Call this solution $U(x,t)$, i. e.,

$$U(x,t) = 1 + \frac{1}{\sqrt{4\pi t}}\left[\int_0^\infty [e^{-\frac{(x-\xi)^2}{4t}} - e^{-\frac{(x+\xi)^2}{4t}}][f(\xi) - 1]\, d\xi\right] \tag{24.114}$$

Differentiate this solution w. r. t. time to get the unit impulse response, i. e., let

$$W(x,t) = \frac{\partial U}{\partial t}(x,t) \tag{24.115}$$

Then

$$u(x,t) = \int_0^t W(x,t-\tau)g(\tau)\, d\tau \Rightarrow u(x,t) = \int_0^t \frac{\partial U}{\partial t}(x,t-\tau)g(\tau)\, d\tau \tag{24.116}$$

is the solution to equations (24.112)–(24.113). This formula is often called *Duhammel's formula* in the engineering literature. If $f(x) = 0$, the solution to equations (24.112)–(24.113) is given by

$$u(x,t) = \frac{2}{\sqrt{\pi}} \int_{\frac{x}{\sqrt{4t}}}^\infty g\left(t - \frac{x^2}{4\lambda^2}\right)e^{-\lambda^2}\, d\lambda \tag{24.117}$$

and the complete solution to equations (24.112)–(24.113) is

$$u(x,t) = \frac{2}{\sqrt{\pi}} \int_{\frac{x}{\sqrt{4t}}}^\infty g\left(t - \frac{x^2}{4\lambda^2}\right)e^{-\lambda^2}\, d\lambda + \frac{1}{\sqrt{4\pi t}}\left[\int_0^\infty [e^{-\frac{(x-\xi)^2}{4t}} - e^{-\frac{(x+\xi)^2}{4t}}]f(\xi)\, d\xi\right]. \tag{24.118}$$

Heat equation in a semi-infinite domain with zero flux at $x = 0$

Consider again the solution

$$u(x,t) = \frac{1}{\sqrt{4\pi t}} \int_{-\infty}^\infty e^{-\frac{(x-\xi)^2}{4t}} f(\xi)\, d\xi$$

of the equation

$$u_{xx} = u_t, \quad -\infty < x < \infty \quad \text{with } u(x,0) = f(x)$$

Let $f(x)$ be an even function, i. e.,

$$f(-x) = f(x)$$

⇒

$$\begin{aligned}
u(x,t) &= \frac{1}{\sqrt{4\pi t}}\left[\int_{-\infty}^{0} e^{-\frac{(x-\xi)^2}{4t}} f(\xi)\,d\xi + \int_{0}^{\infty} e^{-\frac{(x-\xi)^2}{4t}} f(\xi)\,d\xi\right] \\
&= \frac{1}{\sqrt{4\pi t}}\left[-\int_{\infty}^{0} e^{-\frac{(x+s)^2}{4t}} f(-s)\,ds + \int_{0}^{\infty} e^{-\frac{(x-\xi)^2}{4t}} f(\xi)\,d\xi\right] \\
&= \frac{1}{\sqrt{4\pi t}}\int_{0}^{\infty} [e^{-\frac{(x+\xi)^2}{4t}} + e^{-\frac{(x-\xi)^2}{4t}}] f(\xi)\,d\xi
\end{aligned}$$

⇒

$$\frac{\partial u}{\partial x} = \frac{1}{\sqrt{4\pi t}}\int_{0}^{\infty}\left[-\frac{2(x+\xi)}{4t}e^{-\frac{(x+\xi)^2}{4t}} - \frac{2(x-\xi)}{4t}e^{-\frac{(x-\xi)^2}{4t}}\right] f(\xi)\,d\xi$$

$$\left.\frac{\partial u}{\partial x}\right|_{x=0} = 0$$

∴ The solution of

$$\begin{aligned}
\frac{\partial^2 u}{\partial x^2} &= \frac{\partial u}{\partial t}, \quad 0 < x < \infty \\
\frac{\partial u}{\partial x}(0,t) &= 0 \\
u(x,0) &= f(x)
\end{aligned} \tag{24.119}$$

is given by

$$u(x,t) = \frac{1}{\sqrt{4\pi t}}\int_{0}^{\infty} [e^{-\frac{(x+\xi)^2}{4t}} + e^{-\frac{(x-\xi)^2}{4t}}] f(\xi)\,d\xi. \tag{24.120}$$

Extension of the solutions from finite to infinite domains

In order to solve the heat/diffusion equation in a semi-infinite domain as given by

$$\begin{aligned}
\frac{\partial u}{\partial t} &= \frac{\partial^2 u}{\partial x^2}, \quad 0 < x < \infty \\
u(x,0) &= f(x), \\
u(0,t) &= 0,
\end{aligned}$$

we first consider the same problem in a finite domain of length a (with $u(a,t) = 0$). The solution for this case is given by

$$u(x,t) = \frac{2}{a}\sum_{n=1}^{\infty} e^{-\frac{n^2\pi^2}{a^2}t}\sin\left(\frac{n\pi x}{a}\right)\int_0^a f(s)\sin\left(\frac{n\pi s}{a}\right)ds$$

To extend this solution to semi-infinite domain, let

$$\alpha_n = \frac{n\pi}{a} \Rightarrow \Delta\alpha_n = \frac{\pi}{a} \quad \text{and} \quad a \to \infty$$

$\Rightarrow$

$$\begin{aligned} u(x,t) &= \frac{2}{\pi}\int_0^{\infty} e^{-\alpha^2 t}\sin\alpha x\left(\int_0^{\infty} f(s)\sin\alpha s\,ds\right)d\alpha \\ &= \frac{2}{\pi}\int_0^{\infty}\int_0^{\infty} e^{-\alpha^2 t}f(s)\left[\frac{e^{i\alpha x}-e^{-i\alpha x}}{2i}\right]\left[\frac{e^{i\alpha s}-e^{-i\alpha s}}{2i}\right]ds \\ &= \frac{1}{\pi}\int_0^{\infty}\int_0^{\infty} e^{-\alpha^2 t}[\cos\alpha(x-s)-\cos\alpha(x+s)]f(s)\,d\alpha\,ds \end{aligned}$$

Using the result,

$$\int_0^{\infty} e^{-\alpha^2 t}\cos\alpha y\,d\alpha = \sqrt{\frac{\pi}{4t}}e^{-\frac{y^2}{4t}}$$

$\Rightarrow$

$$u(x,t) = \frac{1}{\sqrt{4\pi t}}\int_0^{\infty}[e^{-\frac{(x-s)^2}{4t}} - e^{-\frac{(x+s)^2}{4t}}]f(s)\,ds.$$

Thus, the solution in a semi-infinite domain may be obtained from that of finite domain by taking the limiting process. Other problems of infinite and semi-infinite domains may also be solved in a similar way.

24.3.3 Transforms on the half-line

Consider the Fourier integral identity

$$\begin{aligned} f(x) &= \frac{1}{\pi}\int_0^{\infty}\int_{-\infty}^{\infty} f(\xi)\cos[\alpha(x-\xi)]\,d\xi\,d\alpha \\ &= \frac{1}{\pi}\int_0^{\infty}\int_{-\infty}^{\infty} f(\xi)[\cos\alpha x\cos\alpha\xi + \sin\alpha x\sin\alpha\xi]\,d\xi\,d\alpha \end{aligned} \tag{24.121}$$

If $f(\xi)$ is odd, the first term vanishes and the identity becomes

$$f(x) = \frac{2}{\pi}\int_0^\infty\int_0^\infty f(\xi)\sin\alpha x\sin\alpha\xi\cdot d\xi\,d\alpha$$
$$= \int_0^\infty\left(\sqrt{\frac{2}{\pi}}\sin\alpha x\right)\left(\int_0^\infty\sqrt{\frac{2}{\pi}}\sin\alpha\xi\cdot f(\xi)\,d\xi\right)d\alpha\,. \tag{24.122}$$

We define the transform pair

$$\left.\begin{aligned}F_s(\alpha) &= \sqrt{\tfrac{2}{\pi}}\int_0^\infty f(\xi)\sin\alpha\xi\cdot d\xi\\ f(x) &= \sqrt{\tfrac{2}{\pi}}\int_0^\infty F_s(\alpha)\sin\alpha x\cdot d\alpha\end{aligned}\right\}\quad\text{Fourier sine transform pair.} \tag{24.123}$$

Note the similarity with the finite Fourier transform pair

$$F(n) = \sqrt{\frac{2}{l}}\int_0^l f(\xi)\sin\left(\frac{n\pi\xi}{l}\right)d\xi \tag{24.124}$$

$$f(x) = \sqrt{\frac{2}{l}}\sum_{n=1}^\infty F(n)\sin\left(\frac{n\pi x}{l}\right) \tag{24.125}$$

The sine transform is useful when we have the operator

$$\frac{d^2}{dx^2},\ 0 < x < \infty \tag{24.126}$$
$$u(0) = 0. \tag{24.127}$$

If $f(x)$ is an even function, then we use the identity

$$f(x) = \frac{2}{\pi}\int_0^\infty\int_0^\infty f(\xi)\cos\alpha x\cos\alpha\xi\cdot d\xi\,d\alpha \tag{24.128}$$

Thus, we get the Fourier cosine transform

$$\left.\begin{aligned}F_c(\alpha) &= \sqrt{\tfrac{2}{\pi}}\int_0^\infty f(\xi)\cos\alpha\xi\,d\xi\\ f(x) &= \sqrt{\tfrac{2}{\pi}}\int_0^\infty F_c(\alpha)\cos\alpha x\,d\alpha\end{aligned}\right\}\quad\text{Fourier cosine transform pair.} \tag{24.129}$$

This transform is useful for the operator

$$\frac{d^2}{dx^2},\ 0 < x < \infty \tag{24.130}$$
$$u'(0) = 0 \tag{24.131}$$

24.3.4 Solution of heat/diffusion equation with radiation BC

Consider the solution of the heat/diffusion equation in a semi-infinite domain

$$\frac{\partial^2 u}{\partial x^2} = \frac{\partial u}{\partial t}, \quad 0 < x < \infty, \quad t > 0 \tag{24.132}$$

with boundary condition of type 3 (also known as radiation BC)

$$\frac{\partial u}{\partial x}(0,t) - \mathrm{Bi}\, u(0,t) = 0, \tag{24.133}$$

and initial condition

$$u(x,0) = f(x). \tag{24.134}$$

We consider the eigenvalue problem

$$\frac{d^2 w}{dx^2} = -\lambda w \tag{24.135}$$

$$w' - \mathrm{Bi}\, w = 0, \quad x = 0 \tag{24.136}$$

$$w' + \mathrm{Bi}\, w = 0, \quad x = l \tag{24.137}$$

which is self-adjoint. We solve the problem on the finite domain to obtain

$$f(x) = 2\sum_{n=1}^{\infty} \frac{\alpha_n \cos \alpha_n x + \mathrm{Bi} \sin \alpha_n x}{(\alpha_n^2 + \mathrm{Bi}^2)l + 2\,\mathrm{Bi}} \int_0^l f(\xi)[\alpha_n \cos \alpha_n \xi + \mathrm{Bi} \sin \alpha_n \xi]\, d\xi \tag{24.138}$$

where

$$\cot \alpha_n l = \frac{\alpha_n}{2\,\mathrm{Bi}} - \frac{\mathrm{Bi}}{2\alpha_n} \tag{24.139}$$

For large n, $\alpha_n \approx \frac{(n-1)\pi}{l}$

$$\Delta\alpha_n = \alpha_{n+1} - \alpha_n = \frac{\pi}{l}$$

Let $l \to \infty$ and replace the Riemann sum by an integral
$\Rightarrow$

$$f(x) = \frac{2}{\pi}\int_0^{\infty} \frac{\alpha \cos \alpha x + \mathrm{Bi} \sin \alpha x}{\alpha^2 + \mathrm{Bi}^2} \left(\int_0^{\infty} [\alpha \cos \alpha\xi + \mathrm{Bi} \sin \alpha\xi] f(\xi)\, d\xi \right) d\alpha \tag{24.140}$$

For some class of functions, this is an identity and is another type of Fourier transform. We define the transform pair by

$$F(\alpha) = \sqrt{\frac{2}{\pi}} \int_0^\infty f(\xi) \frac{(\alpha \cos \alpha\xi + \text{Bi} \sin \alpha\xi)}{\sqrt{\alpha^2 + \text{Bi}^2}} \, d\xi \tag{24.141}$$

$$f(x) = \sqrt{\frac{2}{\pi}} \int_0^\infty F(\alpha) \frac{(\alpha \cos \alpha x + \text{Bi} \sin \alpha x)}{\sqrt{\alpha^2 + \text{Bi}^2}} \, d\alpha \tag{24.142}$$

This transform pair is useful for solving the heat equation with radiation BC, i. e.,

$$\frac{d^2 u}{dx^2}, \quad 0 < x < \infty \tag{24.143}$$

$$u' - \text{Bi} \cdot u = 0 \; @ \; x = 0 \tag{24.144}$$

Let

$$\begin{aligned} \hat{u}(\alpha, t) &= \mathcal{F}(u(x,t)) \\ &= \sqrt{\frac{2}{\pi}} \int_0^\infty u(x,t) \frac{\alpha \cos \alpha x + \text{Bi} \sin \alpha x}{\sqrt{\alpha^2 + \text{Bi}^2}} \, dx \end{aligned} \tag{24.145}$$

then it is easily shown that

$$\mathcal{F}\left\{ \frac{\partial^2 u}{\partial x^2}(x,t) \right\} = -\alpha^2 \hat{u}(\alpha, t)$$

$$\hat{u}(\alpha, 0) = \mathcal{F}(f(x)) = F(\alpha) \, .$$

Thus, taking FT of the heat equation gives

$$\hat{u} = e^{-\alpha^2 t} \cdot F(\alpha) \tag{24.146}$$

$\Rightarrow$

$$\begin{aligned} u(x,t) &= \sqrt{\frac{2}{\pi}} \int_0^\infty e^{-\alpha^2 t} \frac{\alpha \cos \alpha x + \text{Bi} \sin \alpha x}{\sqrt{\alpha^2 + \text{Bi}^2}} \left(\int_0^\infty \sqrt{\frac{2}{\pi}} f(\xi) \frac{\alpha \cos \alpha\xi + \text{Bi} \sin \alpha\xi}{\sqrt{\alpha^2 + \text{Bi}^2}} \, d\xi \right) d\alpha \\ &= \frac{2}{\pi} \int_0^\infty \int_0^\infty \frac{e^{-\alpha^2 t} . f(\xi) [\alpha \cos \alpha x + \text{Bi} \sin \alpha x][\alpha \cos \alpha\xi + \text{Bi} \sin \alpha\xi]}{(\alpha^2 + \text{Bi}^2)} \, d\xi \, d\alpha \end{aligned} \tag{24.147}$$

Changing the order of integration,

$\Rightarrow$

$$u(x,t) = \frac{2}{\pi} \int_0^\infty f(\xi) \left(\int_0^\infty \frac{e^{-\alpha^2 t} . [\alpha \cos \alpha x + \text{Bi} \sin \alpha x][\alpha \cos \alpha\xi + \text{Bi} \sin \alpha\xi]}{(\alpha^2 + \text{Bi}^2)} \, d\alpha \right) d\xi \tag{24.148}$$

Let $\cos\theta = \frac{\alpha}{\sqrt{\alpha^2+\mathrm{Bi}^2}} \Rightarrow \sin\theta = \frac{\mathrm{Bi}}{\sqrt{\alpha^2+\mathrm{Bi}^2}}$. Then the second integral may be written as

$$I = \int_0^\infty e^{-\alpha^2 t}\cos(\alpha x - \theta)\cos(\alpha\xi - \theta)\,d\alpha \tag{24.149}$$

$$= \frac{1}{2}\int_0^\infty e^{-\alpha^2 t}[\cos(\alpha(x+\xi) - 2\theta) + \cos\alpha(x-\xi)]\,d\alpha \tag{24.150}$$

$$= \frac{1}{2}\int_0^\infty e^{-\alpha^2 t}\cos(\alpha x + \alpha\xi - 2\theta)\,d\alpha + \frac{1}{2}\int_0^\infty e^{-\alpha^2 t}\cos\alpha(x-\xi)\,d\alpha. \tag{24.151}$$

Thus, it may be shown after algebraic simplifications and evaluation of the integrals that

$$\begin{aligned} u(x,t) = {} & \frac{1}{\sqrt{4\pi t}}\int_0^\infty [e^{-\frac{(x+\xi)^2}{4t}} + e^{-\frac{(x-\xi)^2}{4t}}]f(\xi)\,d\xi \\ & - \mathrm{Bi}\int_0^\infty e^{\mathrm{Bi}^2 t + \mathrm{Bi}(x+\xi)}\operatorname{erf}c\left[\frac{x+\xi}{\sqrt{4t}} + \mathrm{Bi}\sqrt{t}\right]f(\xi)\,d\xi. \end{aligned} \tag{24.152}$$

Remarks.

(1) The solution of

$$u_t = u_{xx}, \quad x > 0,\ t > 0 \tag{24.153}$$

$$-\frac{\partial u}{\partial x}(0,t) + \mathrm{Bi}\,u(0,t) = \phi(t)\,\mathrm{Bi} \ @\ x = 0 \tag{24.154}$$

$$u(x,0) = f(x) \tag{24.155}$$

is given by

$$\begin{aligned} u = {} & \int_0^\infty \sqrt{\frac{1}{4\pi t}}[e^{-\frac{(x+\xi)^2}{4t}} + e^{-\frac{(x-\xi)^2}{4t}}]f(\xi)\,d\xi \\ & - \mathrm{Bi}\int_0^\infty e^{\mathrm{Bi}^2 t + \mathrm{Bi}(x+\xi)}\operatorname{erf}c\left[\frac{x+\xi}{\sqrt{4t}} + \mathrm{Bi}\sqrt{t}\right]f(\xi)\,d\xi \\ & + \mathrm{Bi}\int_0^\infty \left\{\frac{e^{-x^2/4(t-\tau)}}{\sqrt{\pi(t-\tau)}} - \mathrm{Bi}\,e^{\mathrm{Bi}^2 t + \mathrm{Bi}(x+\xi)}\operatorname{erf}c\left[\frac{x}{\sqrt{4(t-\tau)}} + \mathrm{Bi}\sqrt{t-\tau}\right]\right\}\phi(\tau)\,d\tau \end{aligned} \tag{24.156}$$

(2) Many problems in the infinite and semi-infinite regions can also be solved by using the Laplace transformation. We refer to the books by Carslaw and Jaeger [10] and Crank [16] for further examples.

24.3.5 Fourier transforms on an infinite domain: solution of the wave equation

Consider again the operator

$$\frac{d^2}{dx^2}, \quad -\infty < x < \infty, \tag{24.157}$$

which has continuous spectrum $\{\alpha^2 > 0\}$ with eigenfunction $\{e^{\pm i\alpha x}\}$. Let

$$\mathcal{F}\{f(x)\} = F(\alpha) = \int_{-\infty}^{\infty} f(x)e^{-i\alpha x}\,dx. \tag{24.158}$$

Inversion formula

$$f(x) = \frac{1}{2\pi}\int_{-\infty}^{\infty} F(\alpha)e^{i\alpha x}\,d\alpha \tag{24.159}$$

Wave equation

Consider the wave equation on an infinite domain

$$\frac{\partial^2 u}{\partial t^2} = c^2\frac{\partial^2 u}{\partial x^2}, \quad -\infty < x < \infty, \quad t > 0 \tag{24.160}$$

with initial conditions

$$u(x,0) = f(x), \quad \frac{\partial u}{\partial t}(x,0) = g(x), \quad -\infty < x < \infty\,. \tag{24.161}$$

Let

$$\hat{u}(\alpha,t) = \mathcal{F}[u(x,t)] = \int_{-\infty}^{\infty} u(x,t)e^{-i\alpha x}\,dx\,. \tag{24.162}$$

Now

$$\mathcal{F}\left\{\frac{\partial^n u}{\partial x^n}\right\} = (i\alpha)^n \cdot \mathcal{F}\{u(x,t)\}\,.$$

Let

$$F(\alpha) = \mathcal{F}[f(x)] \quad \text{and} \quad G(\alpha) = \mathcal{F}[g(x)]\,. \tag{24.163}$$

Then, taking FT on equations (24.160)–(24.161) ⇒

$$\frac{d^2\hat{u}}{dt^2} = -c^2\alpha^2\hat{u}$$

$$\hat{u}(0) = F(\alpha), \quad \frac{d\hat{u}}{dt}(0) = G(\alpha)$$

$\Rightarrow$

$$\hat{u}(\alpha, t) = F(\alpha)\cos[\alpha ct] + \frac{G(\alpha)}{\alpha c}\sin[\alpha ct] \tag{24.164}$$

$\therefore$

$$\begin{aligned} u(x,t) &= \frac{1}{2\pi}\int_{-\infty}^{\infty} \hat{u}(\alpha,t)e^{+i\alpha x}\,d\alpha \\ &= \frac{1}{2\pi}\int_{-\infty}^{\infty} F(\alpha)\cos\alpha ct e^{+i\alpha x}\,d\alpha + \frac{1}{2\pi}\int_{-\infty}^{\infty}\frac{G(\alpha)}{\alpha c}\sin\alpha ct e^{+i\alpha x}\,d\alpha\,. \end{aligned} \tag{24.165}$$

Consider the first integral

$$\begin{aligned} u_1(x,t) &= \frac{1}{2\pi}\int_{-\infty}^{\infty}\cos[\alpha ct]e^{+i\alpha x}\left(\int_{-\infty}^{\infty} f(\xi)e^{-i\alpha\xi}\,d\xi\right)d\alpha \\ &= \frac{1}{2\pi}\int_{-\infty}^{\infty}\int_{-\infty}^{\infty} e^{+i\alpha(x-\xi)}f(\xi)\cos[\alpha ct]\,d\xi\,d\alpha \\ \cos[\alpha ct] &= \frac{e^{i\alpha ct}+e^{-i\alpha ct}}{2} \end{aligned}$$

$\therefore$

$$u_1(x,t) = \frac{1}{4\pi}\int_{-\infty}^{\infty}\int_{-\infty}^{\infty} e^{i\alpha x - i\alpha\xi + i\alpha ct}f(\xi)\,d\xi\,d\alpha + \frac{1}{4\pi}\int_{-\infty}^{\infty}\int_{-\infty}^{\infty} e^{i\alpha x - i\alpha\xi - i\alpha ct}f(\xi)\,d\xi\,d\alpha. \tag{24.166}$$

In the Fourier integral formula,

$$f(z) = \frac{1}{2\pi}\int_{-\infty}^{\infty}\int_{-\infty}^{\infty} e^{-i\alpha\xi + i\alpha z}f(\xi)\,d\xi\,d\alpha,$$

let $z = x - ct$, $\quad x + ct$. Then equation (24.166) simplifies to

$$u_1(x,t) = \frac{1}{2}[f(x+ct) + f(x-ct)]\,. \tag{24.167}$$

Now consider the second integral in equation (24.165):

$$u_2(x,t) = \frac{1}{2\pi}\int_{-\infty}^{\infty} e^{i\alpha x}\frac{\sin[\alpha ct]}{\alpha c}\left(\int_{-\infty}^{\infty} e^{-i\alpha\xi}g(\xi)\,d\xi\right)d\alpha$$

$$= \frac{1}{2\pi c} \int_{-\infty}^{\infty} \int_{-\infty}^{\infty} e^{iax - ia\xi} g(\xi) \frac{\sin act}{a} \, d\xi \, da$$

$$= \frac{1}{2\pi c} \int_{-\infty}^{\infty} \int_{-\infty}^{\infty} e^{iax - ia\xi} g(\xi) \left(\frac{e^{iact} - e^{-iact}}{2ia} \right) d\xi \, da$$

$$= \frac{1}{2\pi c} \int_{-\infty}^{\infty} \int_{-\infty}^{\infty} \frac{e^{ia(x+ct) - ia\xi} g(\xi)}{2ia} \, d\xi \, da - \frac{1}{2\pi c} \int_{-\infty}^{\infty} \int_{-\infty}^{\infty} \frac{e^{ia(x-ct) - ia\xi} g(\xi)}{2ia} \, d\xi \, da$$

If $h(x) = \int_0^x f(x)\, dx$, then $\mathcal{F}\{h(x)\} = \frac{1}{ia} \mathcal{F}\{f(x)\} \Rightarrow \mathcal{F}^{-1}\{\frac{F(a)}{ia}\} = \int_0^x f(x)\, dx$,
∴

$$u_2(x,t) = \frac{-1}{2c} \int_0^{x-ct} g(\lambda)\, d\lambda + \frac{1}{2c} \int_0^{x+ct} g(\lambda)\, d\lambda$$

$$= \frac{1}{2c} \int_{x-ct}^{x+ct} g(\lambda)\, d\lambda \tag{24.168}$$

∴

$$u(x,t) = \frac{f(x+ct) + f(x-ct)}{2} + \frac{1}{2c} \int_{x-ct}^{x+ct} g(\lambda)\, d\lambda \tag{24.169}$$

is the solution to the initial value problem defined by equations (24.160)–(24.161). The relation $x + ct$ = constant indicates a wave moving to the left with velocity c, while $x - ct$ = constant is a wave moving to the right with velocity c. Therefore, the solution is a superposition of two waves, one moving to the left and one to the right with a velocity of c.

Related problems

(i)

$$\frac{\partial^2 u}{\partial t^2} = c^2 \frac{\partial^2 u}{\partial x^2}, \quad 0 < x < \infty, \quad t > 0 \tag{24.170}$$

BC:

$$u(0,t) = \phi(t) \tag{24.171}$$

IC:

$$u(x,0) = 0, \quad \frac{\partial u}{\partial t}(x,0) = 0 \tag{24.172}$$

Using the Fourier sine transform, it can be shown that

$$u(x,t) = \int_0^{\infty}\int_0^{ct} \phi(\tau)\sin\alpha(ct-\tau)\sin\alpha x\,d\alpha\,d\tau \tag{24.173}$$

$$= \begin{cases} \phi(ct-x), & 0 < x < ct \\ 0, & x > ct \end{cases} \tag{24.174}$$

(ii)

$$\frac{\partial^2 u}{\partial t^2} = \frac{\partial^2 u}{\partial x^2}, \quad 0 < x < \infty, \quad t > 0 \tag{24.175}$$

BC:

$$u_x(0,t) = 0 \tag{24.176}$$

IC:

$$u(x,0) = f(x), \quad \frac{\partial u}{\partial t}(x,0) = 0 \tag{24.177}$$

Using the cosine transform, one gets

$$u(x,t) = \begin{cases} \frac{f(x+t)+f(t-x)}{2}, & 0 < x < t \\ \frac{f(x+t)+f(x-t)}{2}, & x > t \end{cases} \tag{24.178}$$

(iii) The solution of the inhomogeneous wave equation

$$\frac{\partial^2 u}{\partial t^2} = c^2\frac{\partial^2 u}{\partial x^2} + q(x,t), \quad -\infty < x < \infty, \quad t > 0 \tag{24.179}$$

with IC

$$u(x,0) = 0, \quad \frac{\partial u}{\partial t}(x,0) = 0 \tag{24.180}$$

may be written in the form

$$u(x,t) = \frac{1}{2c}\int_0^{t}\int_{x-c(t-\tau)}^{x+c(t-\tau)} q(\xi,\tau)\,d\xi\,d\tau \tag{24.181}$$

24.3.6 Laplace's equation in infinite and semi-infinite domains

In this section, we consider the solution of Laplace's equation on unbounded domains using the Fourier transform method.

Laplace's equation in a strip

Consider Laplace's equation on an infinite strip

$$\frac{\partial^2 u}{\partial x^2} + \frac{\partial^2 u}{\partial y^2} = 0; \quad -\infty < x < \infty, \quad 0 < y < a \tag{24.182}$$

with boundary conditions

$$u(x,0) = f(x), \quad u(x,a) = 0\,. \tag{24.183}$$

Let

$$\hat{u} = \mathcal{F}(u(x,y)), \quad F(\alpha) = \mathcal{F}\{f(x)\} \tag{24.184}$$

⇒

$$\frac{d^2\hat{u}}{dy^2} - \alpha^2\hat{u} = 0$$
$$\hat{u}(a) = 0, \quad \hat{u}(0) = F(\alpha)$$

⇒

$$\hat{u} = c_1 \sinh \alpha y + c_2 \sinh \alpha(a-y)$$

$\hat{u}(a) = 0 \Rightarrow c_1 = 0$ and $\hat{u}(0) = F(\alpha) \Rightarrow c_2 = \frac{F(\alpha)}{\sinh \alpha a}$. Therefore,

$$\hat{u}(\alpha,y) = \frac{F(\alpha)\sinh \alpha(a-y)}{\sinh \alpha a} \tag{24.185}$$

$$u(x,y) = \frac{1}{2\pi}\int_{-\infty}^{\infty} e^{i\alpha x}\frac{\sinh \alpha(a-y)}{\sinh \alpha a}\left(\int_{-\infty}^{\infty} e^{-i\alpha\xi} f(\xi)\,d\xi\right) d\alpha \tag{24.186}$$

$$= \frac{1}{2\pi}\int_{-\infty}^{\infty}\int_{-\infty}^{\infty} e^{i\alpha x - i\alpha\xi}\frac{\sinh \alpha(a-y)}{\sinh \alpha a} f(\xi)\,d\xi\,d\alpha \tag{24.187}$$

Changing the order of integration,

⇒

$$u(x,y) = \frac{1}{2\pi}\int_{-\infty}^{\infty} f(\xi)\left(\int_{-\infty}^{\infty} e^{i\alpha(x-\xi)}\frac{\sinh \alpha(a-y)}{\sinh \alpha a}\,d\alpha\right) d\xi$$

This may be simplified to

$$u(x,y) = \frac{1}{a}\sinh\frac{\pi y}{a}\int_{-\infty}^{\infty}\frac{f(\xi)}{\cos\frac{\pi(a-y)}{a} + \cosh\frac{\pi(x-\xi)}{a}}\,d\xi. \tag{24.188}$$

Laplace's equation in a half-plane

Consider Laplace's equation

$$\frac{\partial^2 u}{\partial x^2} + \frac{\partial^2 u}{\partial y^2} = 0, \quad -\infty < x < \infty, \quad y > 0 \tag{24.189}$$

$$u(x,0) = f(x) \tag{24.190}$$

$u(x,y)$ is bounded. Let

$$\hat{u}_s = \sqrt{\frac{2}{\pi}} \int_0^\infty u(x,y) \sin \alpha y \, dy = \mathcal{F}_s\{u(x,y)\} \tag{24.191}$$

⇒

$$\mathcal{F}_s\left\{\frac{\partial^2 u}{\partial y^2}\right\} = -\alpha^2 \hat{u}_s + \sqrt{\frac{2}{\pi}}\, \alpha\, u(x,0)\,. \tag{24.192}$$

Thus, taking sine transform, equations (24.189)–(24.190) reduce to

$$\frac{d^2 \hat{u}_s}{dx^2} - \alpha^2 \hat{u}_s = -\sqrt{\frac{2}{\pi}}\, \alpha f(x), \quad -\infty < x < \infty \tag{24.193}$$

This equation is easily solved using the Fourier transform

$$\hat{u}_s = \frac{1}{2\pi}\sqrt{\frac{2}{\pi}} \int_{-\infty}^{\infty} \frac{\alpha e^{i\beta x}}{\alpha^2 + \beta^2}\left(\int_{-\infty}^{\infty} f(\xi) e^{-i\beta\xi}\, d\xi\right) d\beta \tag{24.194}$$

and

$$\begin{aligned} u(x,y) &= \sqrt{\frac{2}{\pi}} \int_0^\infty \hat{u}_s(x,\alpha) \sin \alpha y \, dy \\ &= \frac{1}{\pi^2} \int_0^\infty \left[\int_{-\infty}^{\infty} \frac{\alpha e^{i\beta x}}{\alpha^2 + \beta^2}\left(\int_{-\infty}^{\infty} f(\xi) e^{-i\beta\xi}\, d\xi\right) d\beta\right] \sin \alpha y \, d\alpha \end{aligned}$$

but

$$\int_{-\infty}^{\infty} \frac{\alpha e^{i\beta(x-\xi)}}{\alpha^2 + \beta^2}\, d\beta = 2\pi i \frac{\alpha e^{-\alpha(x-\xi)}}{2\alpha i} = \pi e^{-\alpha(x-\xi)}$$

⇒

$$u(x,y) = \frac{1}{\pi} \int_0^\infty \int_{-\infty}^{\infty} f(\xi) e^{-\alpha(x-\xi)} \sin \alpha y \, d\xi \, d\alpha$$

$$= \frac{y}{\pi} \int_{-\infty}^{\infty} \frac{f(\xi)}{(x-\xi)^2 + y^2} \, d\xi \tag{24.195}$$

This is the Poisson's integral formula. This problem is more easily solved by first taking FT w. r. t. x, solving the boundary value problem in y and taking the inverse transform.

Special case: $f(x)$ = boxcar function

For the special case of $f(x)$ being boxcar function:

$$f(x) = \begin{cases} 1, & |x| \leq 1 \\ 0, & |x| > 1, \end{cases}$$

the solution (24.195) is reduced to

$$u(x,y) = \frac{1}{\pi} \left\{ \tan^{-1}\left[\frac{1-x}{y}\right] + \tan^{-1}\left[\frac{1+x}{y}\right] \right\}$$

and a 3D plot of the solution is shown in Figure 24.4.

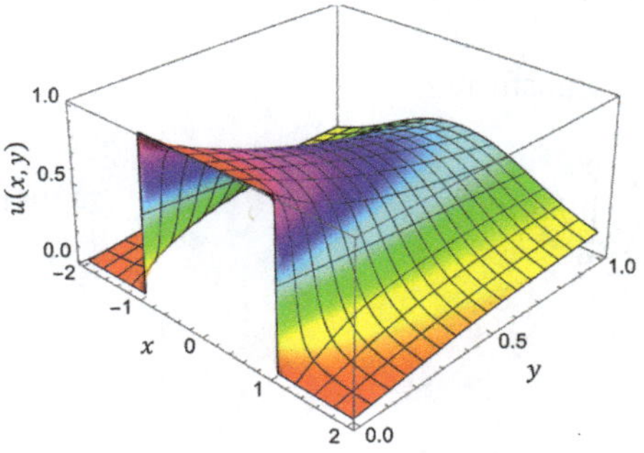

Figure 24.4: Solution profile for 2D Laplace equation in a half-plane with $u(x, y = 0) = \begin{cases} 1, & |x| \leq 1 \\ 0, & |x| > 1 \end{cases}$.

Laplace's equation in a half-strip

Consider Laplace's equation in a half-strip:

$$\frac{\partial^2 u}{\partial x^2} + \frac{\partial^2 u}{\partial y^2} = 0, \quad 0 < x < \infty, \quad 0 < y < 1 \tag{24.196}$$

with boundary conditions

$$u(x, 0) = f(x), \quad u(x, 1) = 0, \quad 0 < x < \infty \tag{24.197}$$

$$u(0, y) = 0, \quad 0 < y < 1. \tag{24.198}$$

We use the sine transform by defining

$$\hat{u} = \sqrt{\frac{2}{\pi}} \int_0^\infty u(\xi, y) \sin \alpha\xi \, d\xi \tag{24.199}$$

$$\hat{f} = \sqrt{\frac{2}{\pi}} \int_0^\infty f(\xi) \sin \alpha\xi \, d\xi. \tag{24.200}$$

Taking the sine transform gives

$$\frac{d^2\hat{u}}{dy^2} - \alpha^2 \hat{u} = 0$$

$$\hat{u} = \hat{f} \; @ \, y = 0, \quad \hat{u} = 0 \; @ \, y = 1$$

$\Rightarrow$

$$\hat{u} = c_1 \sinh \alpha y + c_2 \sinh \alpha(1 - y)$$

BCs $\Rightarrow$

$$c_1 = 0$$

$$c_2 = \frac{\hat{f}}{\sinh \alpha}$$

$\therefore$

$$\hat{u} = \frac{\hat{f} \sinh \alpha(1 - y)}{\sinh \alpha} \tag{24.201}$$

$$u(x, y) = \frac{2}{\pi} \int_0^\infty \frac{\sinh \alpha(1 - y)}{\sinh \alpha} \sin \alpha x \left(\int_0^\infty f(\xi) \sin \alpha\xi \, d\xi \right) d\alpha \tag{24.202}$$

is the formal solution.

Remark. If the BC at $x = 0$ is $\frac{\partial u}{\partial x} = 0$ then we can use the cosine transform.

24.3.7 Multiple Fourier transforms

Consider the solution of the heat equation in three dimensions

$$\frac{\partial u}{\partial t} = \frac{\partial^2 u}{\partial x^2} + \frac{\partial^2 u}{\partial y^2} + \frac{\partial^2 u}{\partial x^2}, \quad -\infty < x < \infty, \quad -\infty < y < \infty, \quad -\infty < z < \infty, \quad t > 0 \tag{24.203}$$

with initial condition

$$u(x,y,z,0) = f(x,y,z) \tag{24.204}$$

Eigenvalues

$$\{\alpha_1^2 + \alpha_2^2 + \alpha_3^2\} = |\boldsymbol{\alpha}|^2, \quad \boldsymbol{\alpha} = [\alpha_1 \quad \alpha_2 \quad \alpha_3]^T \tag{24.205}$$

eigenfunctions

$$e^{-i(\alpha_1 x + \alpha_2 y + \alpha_3 z)} = e^{-i\boldsymbol{\alpha}\cdot\mathbf{x}}, \quad \mathbf{x} = [x \quad y \quad z]^T \tag{24.206}$$

Define

$$\mathcal{F}\{f(x,y,z)\} = \int_{-\infty}^{\infty}\int_{-\infty}^{\infty}\int_{-\infty}^{\infty} e^{-i(\alpha_1 x + \alpha_2 y + \alpha_3 z)} f(x,y,z)\,dx\,dy\,dz = F(\alpha_1,\alpha_2,\alpha_3) \tag{24.207}$$

$$\mathcal{F}\{u(x,y,z,t)\} = \hat{u}(\alpha_1,\alpha_2,\alpha_3,t) \tag{24.208}$$

Then we get

$$\frac{d\hat{u}}{dt} = -(\alpha_1^2 + \alpha_2^2 + \alpha_3^2)\hat{u}$$
$$\hat{u}(0) = F(\alpha_1,\alpha_2,\alpha_3)$$

⇒

$$\begin{aligned}\hat{u} &= F(\alpha_1,\alpha_2,\alpha_3)e^{-(\alpha_1^2+\alpha_2^2+\alpha_3^2)t} \\ &= \int_{-\infty}^{\infty}\int_{-\infty}^{\infty}\int_{-\infty}^{\infty} f(\xi,\eta,\lambda)e^{-(\alpha_1^2+\alpha_2^2+\alpha_3^2)t - i(\alpha_1\xi+\alpha_2\eta+\alpha_3\lambda)}\,d\xi\,d\eta\,d\lambda\end{aligned} \tag{24.209}$$

⇒

$$\begin{aligned}u(x,y,z,t) = \frac{1}{(2\pi)^3}&\int_{-\infty}^{\infty}\int_{-\infty}^{\infty}\int_{-\infty}^{\infty} e^{-(\alpha_1^2+\alpha_2^2+\alpha_3^2)t + i(\alpha_1 x+\alpha_2 y+\alpha_3 z)}\,d\alpha_1\,d\alpha_2\,d\alpha_3 \\ &\times\left(\int_{-\infty}^{\infty}\int_{-\infty}^{\infty}\int_{-\infty}^{\infty} f(\xi,\eta,\zeta)e^{-i(\alpha_1\xi+\alpha_2\eta+\alpha_3\lambda)}\,d\xi\,d\eta\,d\zeta\right)\end{aligned}$$

This can be simplified in the same way as the one-dimensional problem. The final result is

$$u(x,y,z,t) = \frac{1}{(\sqrt{4\pi t})^3}\int_{-\infty}^{\infty}\int_{-\infty}^{\infty}\int_{-\infty}^{\infty} e^{-((x-\xi)^2+(y-\eta)^2+(z-\zeta)^2)/4t} f(\xi,\eta,\zeta)\,d\xi\,d\eta\,d\zeta \tag{24.210}$$

$$= \int_{-\infty}^{\infty}\int_{-\infty}^{\infty}\int_{-\infty}^{\infty} G(x,y,z,\xi,\eta,\zeta,t,0) f(\xi,\eta,\zeta)\, d\xi\, d\eta\, d\zeta \tag{24.211}$$

where

$$G = \frac{1}{[4\pi(t-\tau)]^{3/2}} \exp\{-[(x-\xi)^2 + (y-\eta)^2 + (z-\zeta)^2]/4(t-\tau)\} \tag{24.212}$$

is the temperature at position (x, y, z) and time t due to a unit point source at (ξ, η, ζ) at time $\tau (t > \tau)$.

G satisfies the heat equation

$$\frac{\partial G}{\partial t} - \frac{\partial^2 G}{\partial x^2} - \frac{\partial^2 G}{\partial y^2} - \frac{\partial^2 G}{\partial z^2} = \delta(x-\xi)\delta(y-\eta)\delta(z-\zeta)\delta(t-\tau) \tag{24.213}$$

as well as the adjoint heat equation

$$-\frac{\partial G}{\partial \tau} - \frac{\partial^2 G}{\partial \xi^2} - \frac{\partial^2 G}{\partial \eta^2} - \frac{\partial^2 G}{\partial \zeta^2} = \delta(x-\xi)\delta(y-\eta)\delta(z-\zeta)\delta(t-\tau) \tag{24.214}$$

In an analogous fashion, one can define double Fourier transforms, double sine transforms, double cosine transforms, triple cosine transforms, and so on.

24.4 Relationship between Fourier and Laplace transforms

We consider the Fourier integral formula

$$f(x) = \frac{1}{2\pi} \int_{-\infty}^{\infty} e^{i\alpha x} \left(\int_{-\infty}^{\infty} f(\xi) e^{-i\alpha\xi}\, d\xi \right) d\alpha \tag{24.215}$$

and assume that $f(x)$ is absolutely integrable. Then $f(x)$ is bounded, i. e., $\exists$ a constant M such that

$$|f(x)| < M. \tag{24.216}$$

Now, let

$$f(x) = \begin{cases} e^{-\gamma x}\phi(x), & x > 0 \\ 0, & x < 0 \end{cases} \tag{24.217}$$

Note that $\phi(x)$ is of exponential order, i. e.,

$$|\phi(x)| \le M e^{\gamma x}, \quad \gamma > 0 \tag{24.218}$$

Equation (24.215)$\Rightarrow$

$$e^{-\gamma x}\phi(x) = \frac{1}{2\pi}\int_{-\infty}^{\infty} e^{iax}\left(\int_{0}^{\infty} e^{-\gamma\xi}\cdot e^{-ia\xi}\phi(\xi)\,d\xi\right)da$$

$\Rightarrow$

$$\phi(x) = \frac{1}{2\pi}\int_{-\infty}^{\infty} e^{(\gamma+ia)x}\left(\int_{0}^{\infty} e^{-(\gamma+ia)\xi}\phi(\xi)\,d\xi\right)da \tag{24.219}$$

Let $s = \gamma + ia \Rightarrow da = \frac{1}{i}\,ds \Rightarrow$

$$\phi(x) = \frac{1}{2\pi i}\int_{\gamma-i\infty}^{\gamma+i\infty} e^{sx}\left(\int_{0}^{\infty} e^{-s\xi}\phi(\xi)\,d\xi\right)ds \tag{24.220}$$

This is an identity for some class of functions $\phi(x)$. Define the Laplace transform by

$$\ell\{\phi(x)\} = \Phi(s) = \int_{0}^{\infty} e^{-sx}\phi(x)\,dx \tag{24.221}$$

Then equation (24.220) gives the inversion formula

$$\phi(x) = \frac{1}{2\pi i}\int_{\gamma-i\infty}^{\gamma+i\infty} e^{sx}\Phi(s)\,ds. \tag{24.222}$$

We showed that $\Phi(s)$ is analytic in the right half-plane $\operatorname{Re} s > \gamma$. Thus, equations (24.221) and (24.222) define the Laplace transform and the inversion formula.

Consider again the Fourier transform pair

$$F(a) = \int_{-\infty}^{\infty} f(\xi)e^{-ia\xi}\,d\xi$$

$$f(x) = \frac{1}{2\pi}\int_{-\infty}^{\infty} F(a)e^{iax}\,da$$

and suppose that $f(\xi) = 0,\ \xi < 0 \Rightarrow$

$$F(a) = \int_{0}^{\infty} f(\xi)e^{-ia\xi}\,d\xi$$

$$f(x) = \frac{1}{2\pi}\int_{-\infty}^{\infty} F(a)e^{iax}\,da$$

or

$$\mathcal{F}\{f(x)\} = \int_0^\infty e^{-i\alpha x} f(x)\,dx$$

Compare this with the Laplace transform

$$\ell\{f(x)\} = \int_0^\infty e^{-sx} f(x)\,dx$$

Thus, in the Laplace transform, we replace s by $i\alpha$, we get the Fourier transform, provided both exist.

Example 24.3. Consider the decaying pulse

$$f(x) = \begin{cases} e^{-x}, & x > 0 \\ 0, & x < 0 \end{cases}$$

$$\ell\{f(x)\} = \frac{1}{1+s}$$

$$F(\alpha) = \mathcal{F}\{f(x)\} = \frac{1}{1+i\alpha} = \frac{1-i\alpha}{1+\alpha^2}.$$

$$\Longrightarrow |F(\alpha)|^2 = \frac{1}{1+\alpha^2}.$$

We note that

$$\int_{-\infty}^{\infty} |f(x)|^2\,dx = \frac{1}{2} = \frac{1}{2\pi}\int_{-\infty}^{\infty} |F(\alpha)|^2\,d\alpha,$$

which is Plancherel's theorem.

Problems

1. The evolution of small amplitude waves on a vertically falling film is described by the linear partial differential equation,

$$\frac{\partial h}{\partial t} + 3\frac{\partial h}{\partial x} + 3\frac{\partial^3 h}{\partial x^3} - \frac{5}{32}\,\mathrm{Re}\,\frac{\partial^2 h}{\partial x \partial t} - \frac{27}{160}\,\mathrm{Re}\,\frac{\partial^2 h}{\partial x^2} + \frac{\mathrm{Re}\,\mathrm{We}}{12}\frac{\partial^4 h}{\partial x^4} = 0, \quad -\infty < x < \infty,$$

where h is the film height, Re is the Reynolds number and We is the Weber number.
(a) Use the separation of variables method

$$h_\alpha(x,t) = \delta e^{i\alpha x} e^{\lambda t}$$

to determine the condition on the eigenvalue λ so that $h_\alpha(x,t)$ is a solution. (b) Use the result in (a) to determine the range of unstable wave numbers (for which the real part of the eigenvalue is positive), (c) Plot the neutral stability curve (that demarcates between the stable and unstable wave numbers) in the (α, We) plane.

2. The stability of the conduction state in a porous layer is determined by the homogeneous partial differential equations

$$\frac{\partial^2 \psi}{\partial x^2} + \frac{\partial^2 \psi}{\partial z^2} - \mathrm{Ra}\,\frac{\partial \theta}{\partial x} = 0$$

$$\frac{\partial^2 \theta}{\partial x^2} + \frac{\partial^2 \theta}{\partial z^2} + \frac{\partial \psi}{\partial x} = 0, \quad -\infty < x < \infty, \quad 0 < z < 1$$

with boundary conditions $\theta = 0, \psi = 0$ at $z = 0, 1$. Here, Ra is a physical parameter known as the Rayleigh number ($\psi(x,z)$ is the stream function while $\theta(x,z)$ is the temperature perturbation) (a) Use the separation of variables method or the appropriate Fourier transforms to determine the general form of the solution, i. e., how a typical term in the solution looks like, (b) Determine the condition(s) under which there exists a nontrivial solution, (c) What is the critical value of Ra below which only a trivial solution can exist? What is the wave number that leads to this destabilization? Give a physical interpretation of the eigenfunctions corresponding to this critical Ra and wave number.

3. (a) Determine the Fourier transform of the function

$$f(x) = e^{-a|x|}, \quad -\infty < x < \infty, \quad a > 0$$

(b) What is the relationship between the Fourier transforms of $f(x)$ and $f(x-\xi)$? (c) Use the above results to solve the boundary value problem

$$-\frac{d^2u}{dx^2} + a^2 u = \delta(x-\xi); \quad -\infty < x < \infty$$

(d) Use the result in (c) to solve the boundary value problem

$$-\frac{d^2u}{dx^2} + a^2 u = h(x); \quad -\infty < x < \infty$$

4. Consider the following evolution equation describing the dispersion of a tracer in laminar flow in a very long tube:

$$\frac{\partial c}{\partial t} + p\frac{\partial c}{\partial z} + \lambda p \frac{\partial^2 c}{\partial z \partial t} - \frac{\partial^2 c}{\partial z^2} = 0, \quad -\infty < z < \infty, \quad t > 0; \quad c(z,0) = \delta(z)$$

Here, p and λ are positive constants and $\delta(z)$ is the Dirac's delta function. (a) Use the Fourier transform (or any other method) to solve the equation and determine $c(z,t)$ for $\lambda = 0$ (b) If the k-th spatial moment ($k \geq 0$) of $c(z,t)$ is defined as

$$m_k(t) = \int_{-\infty}^{\infty} z^k c(z,t)\, dz$$

determine the first three spatial moments ($k = 0, 1, 2$) for any λ, and hence the variance (or second central moment) without solving for $c(z,t)$.

5. Solve the problem

$$D\frac{\partial^2 u}{\partial x^2} = \frac{\partial u}{\partial t}; \quad 0 < x < \infty, \quad t > 0$$
$$u = f(t); \quad @\, x = 0, \quad t > 0$$
$$u = g(x); \quad @\, t = 0, \quad x > 0$$

Apply your result to the special case $g = 0$ and $f = A\cos\omega t$ to determine how a periodic signal is attenuated.

6. (a) Consider the problem of a very long empty tubular reactor

$$D\frac{\partial^2 c}{\partial x^2} - u\frac{\partial c}{\partial x} - kc = \frac{\partial c}{\partial t}, \quad 0 < x < \infty, \quad t > 0$$
$$-D\frac{\partial c}{\partial x} = u(c_0 - c), \quad x = 0, \quad t > 0$$
$$c = f(x), \quad t = 0, \quad x > 0$$

Put it into self-adjoint form. Consider what transform on the semi-infinite interval might solve it.

(b) Consider the above problem with $c_0 = c_0(t)$. Cast the above equations into dimensionless form but leave $c(x,t)$ dimensional. Make successively, substitutions of the following form to put the equation in its simplest form:

$$c = \exp\left(\frac{x\,\mathrm{Pe}}{2}\right)v; \quad v = \exp\left\{-\left(\frac{\mathrm{Pe}^2}{4} + k\right)\tau\right\}w; \quad \phi = w - \frac{2}{\mathrm{Pe}}\frac{\partial w}{\partial x}$$

What is now the form for the problem and what is the solution? Having found ϕ, how does one find w?

7. A very long slab with two insulated opposite faces has arbitrary temperatures imposed on the other two faces so that the system is described by

$$\frac{\partial^2 u}{\partial x^2} + \frac{\partial^2 u}{\partial y^2} = 0, \quad -\infty < x < \infty, \quad 0 < y < L$$
$$u(x,0) = f(x), \quad -\infty < x < \infty$$
$$u(x,L) = g(x), \quad -\infty < x < \infty$$

Find a formal solution and show that it is identical to the Poisson's integral formula in the limit $L \to \infty$.

8. Use multiple Fourier transform to solve the problem of heat flow with heat production:

$$\frac{\partial u}{\partial t} - \operatorname{div}\operatorname{grad} u = q(x,y,z,t), \quad t > 0, \quad -\infty < x,y,z < \infty$$

$$u(x,y,z,0) = 0, \quad u(x,y,z,t) \ \text{bounded}.$$

25 Fourier transforms in cylindrical and spherical geometries

Recall that for a regular differential operator, the leading coefficient did not vanish inside or at the end points of the interval. We now consider problems in which this condition is violated. These problems arise mostly in cylindrical and spherical domains.

25.1 BVP and IBVP in cylindrical and spherical geometries

As in the case of rectangular coordinates, boundary and initial–boundary value problems usually involve the Laplacian operator ∇^2 with various types of boundary conditions. Examples of some of the well-known problems include but are not limited to:

1. Laplace's equation

$$\nabla^2 u = 0 \text{ in } \Omega; \quad u = g \text{ on } \partial\Omega \quad \text{(or other BCs)} \tag{25.1}$$

2. Poisson's equation

$$\nabla^2 u = f \text{ in } \Omega; \quad u = g \text{ on } \partial\Omega \quad \text{(or other BCs)} \tag{25.2}$$

3. Heat equation

$$\frac{\partial u}{\partial t} = \nabla^2 u \text{ in } \Omega, \quad t > 0; \tag{25.3}$$

$$u = u_0 \ @\ t = 0, \quad \text{in } \Omega \tag{25.4}$$

$$\alpha u + \beta \mathbf{n}.\nabla u = \gamma \text{ on } \partial\Omega, \quad t > 0 \tag{25.5}$$

4. Wave equation

$$\frac{\partial^2 u}{\partial t^2} = c^2 \nabla^2 u \text{ in } \Omega, \quad t > 0; \tag{25.6}$$

$$\text{BC:} \quad u = 0 \text{ on } \partial\Omega, \quad t > 0 \tag{25.7}$$

$$\text{IC:} \quad u = g_1 \text{ and } \frac{\partial u}{\partial t} = g_2 \text{ in } \Omega, \quad t = 0 \tag{25.8}$$

5. Helmholtz/diffusion–reaction equation

$$\nabla^2 u = \pm k^2 u \text{ in } \Omega; \quad u = g \text{ on } \partial\Omega \quad \text{(or other BCs)} \tag{25.9}$$

We examine the solution of some of these equations in spherical and cylindrical geometries using the Fourier transform and other methods. As discussed in Part IV, the case of hollow cylinder and sphere lead to regular Sturm–Liouville BVPs and

https://doi.org/10.1515/9783111598055-030

can be treated by the standard FFT method. Thus, the treatment below is mostly confined to a solid cylinder and sphere.

25.1.1 Cylindrical geometries

The domain of interest may include either the inside or outside of a cylindrical domain or the annular region or their combinations as shown schematically in Figure 25.1.

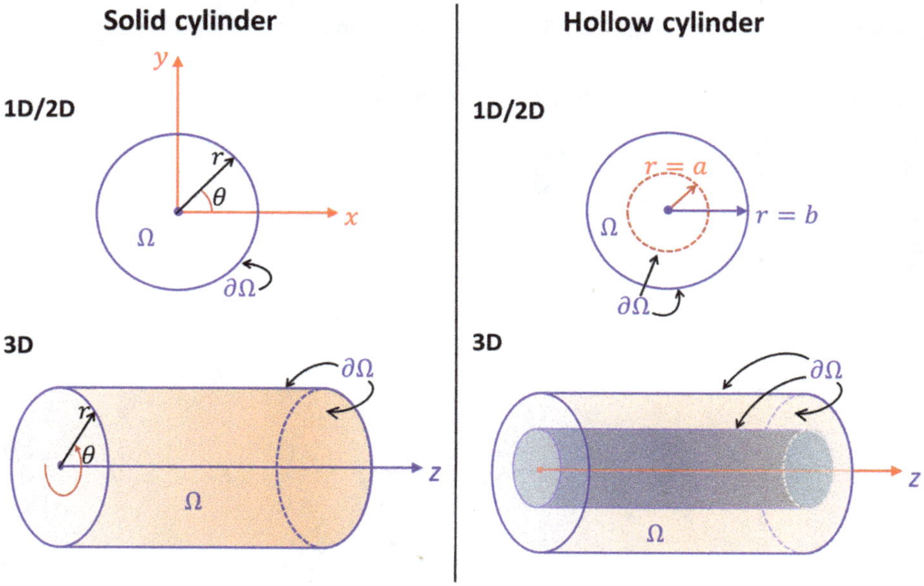

Figure 25.1: Schematic diagrams of cylindrical geometries: solid and hollow cylinders.

The Laplacian operator in cylindrical geometry can be obtained by the transformation from Cartesian coordinate to the cylindrical coordinate system as shown in Figure 25.2, and results in (r, θ, z) as

$$\nabla^2 u = \frac{1}{r}\frac{\partial}{\partial r}\left(r\frac{\partial u}{\partial r}\right) + \frac{1}{r^2}\frac{\partial^2 u}{\partial \theta^2} + \frac{\partial^2 u}{\partial z^2}, \tag{25.10}$$

where the finite cylindrical domain is represented as

$$\Omega \equiv \begin{cases} 0 < \theta < 2\pi \\ 0 < r < a \\ 0 < z < L, \end{cases} \tag{25.11}$$

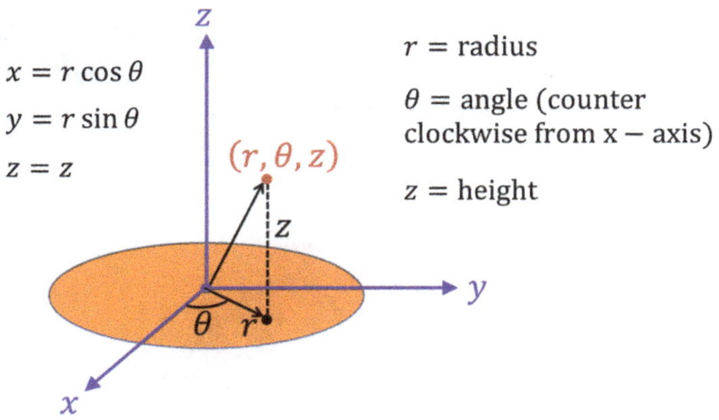

Figure 25.2: Cylindrical coordinate system.

with the boundary represented as

$$\partial\Omega \equiv \begin{cases} r = a, & 0 < z < L,\ 0 \leq \theta < 2\pi \\ z = 0, L, & 0 < r < a,\ 0 \leq \theta < 2\pi. \end{cases} \tag{25.12}$$

Similarly, the half-cylinder or sector can be represented by

$$\Omega \equiv \begin{cases} 0 < \theta < 2\pi \text{ or } \theta_1 < \theta < \theta_2 \\ 0 < r < a \\ 0 < z < L \end{cases} \tag{25.13}$$

and so forth. Similarly, the Laplacian operator in 1D (in r only) can be simplified to

$$\nabla^2 u = \frac{1}{r}\frac{\partial}{\partial r}\left(r\frac{\partial u}{\partial r}\right) \tag{25.14}$$

or in 2D as

$$(r, \theta) : \nabla^2 u = \frac{1}{r}\frac{\partial}{\partial r}\left(r\frac{\partial u}{\partial r}\right) + \frac{1}{r^2}\frac{\partial^2 u}{\partial \theta^2} \tag{25.15}$$

$$(r, z) : \nabla^2 u = \frac{1}{r}\frac{\partial}{\partial r}\left(r\frac{\partial u}{\partial r}\right) + \frac{\partial^2 u}{\partial z^2} \tag{25.16}$$

and so forth.

25.1.2 Spherical geometries

Similar to the cylindrical case, the BVPs and initial BVPs in spherical geometries usually involve solid or hollow spheres. The domain of interest may include either the inside or outside of the spherical domain or the annular region or their combinations. Figure 25.3 shows the spherical coordinate system, where the relationship between the Cartesian coordinate to the spherical coordinate system is as follows:

$x = r\sin\theta\cos\phi$

$y = r\sin\theta\sin\phi$

$z = r\cos\theta$

$r =$ radial distance

$\phi =$ longitude (azimuthal angle)
$\theta =$ latitude (polar angle)

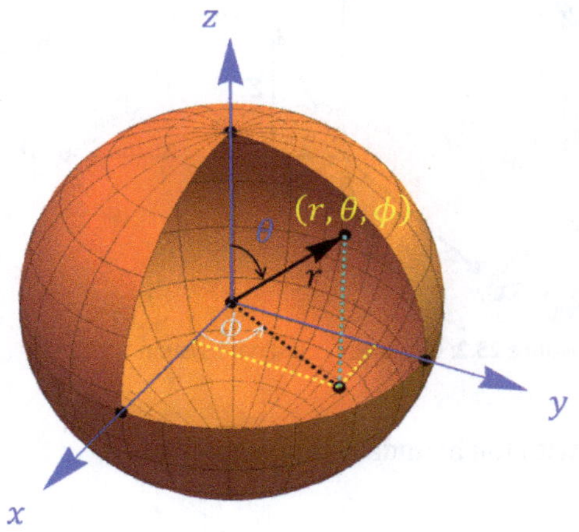

Figure 25.3: Spherical coordinate system.

which results in the Laplacian operator in 3D spherical coordinates in (r, θ, ϕ) as

$$\nabla^2 u = \frac{1}{r^2}\frac{\partial}{\partial r}\left(r^2\frac{\partial u}{\partial r}\right) + \frac{1}{r^2\sin\theta}\frac{\partial}{\partial\theta}\left(\sin\theta\frac{\partial u}{\partial\theta}\right) + \frac{1}{r^2\sin^2\theta}\frac{\partial^2 u}{\partial\phi^2}. \tag{25.17}$$

The interior domain of a sphere of radius a can be represented as

$$\Omega \equiv \begin{cases} 0 < r < a \\ 0 < \theta < \pi \\ 0 < \phi < 2\pi \end{cases} \tag{25.18}$$

with the boundary

$$\partial\Omega \equiv \{r = a, \quad 0 < \theta < \pi, \quad 0 < \phi < 2\pi. \tag{25.19}$$

Similarly, a hemisphere or cone can be represented by

$$\Omega = \begin{cases} 0 < \theta < \frac{\pi}{2} \text{ or } 0 < \theta < \theta_0 < \frac{\pi}{2} \\ 0 < r < a \\ 0 < \phi < 2\pi \end{cases} \tag{25.20}$$

and so forth.

[Note: The polar angle θ is also referred to as latitude, where $0 < \theta < \pi$, with $\theta = \frac{\pi}{2}$ denoting equator, $\theta = 0$ denoting north pole and $\theta = \pi$ denoting the south pole. Similarly, the azimuthal angle ϕ is also referred to as longitude where $0 < \phi < 2\pi$].

Similarly, the Laplacian operator in 1D (in r only) can be simplified to

$$\nabla^2 u = \frac{1}{r^2}\frac{\partial}{\partial r}\left(r^2\frac{\partial u}{\partial r}\right) \tag{25.21}$$

or in 2D as

$$(r,\theta) : \nabla^2 u = \frac{1}{r^2}\frac{\partial}{\partial r}\left(r^2\frac{\partial u}{\partial r}\right) + \frac{1}{r^2 \sin\theta}\frac{\partial}{\partial \theta}\left(\sin\theta\frac{\partial u}{\partial \theta}\right) \tag{25.22}$$

and so forth.

25.1.3 3D eigenvalue problems in cylindrical geometries

Consider the 3D EVP problem of the Laplacian operator in the interior of a finite cylinder:

$$\begin{aligned} \nabla^2 \psi = \frac{1}{r}\frac{\partial}{\partial r}\left(r\frac{\partial \psi}{\partial r}\right) + \frac{1}{r^2}\frac{\partial^2 \psi}{\partial \theta^2} + \frac{\partial^2 \psi}{\partial z^2} = -\lambda\psi, \\ 0 < r < a, \quad 0 < \theta < 2\pi, \quad 0 < z < L \end{aligned} \tag{25.23}$$

with homogeneous Dirichlet boundary condition on the surface

$$\psi = 0 \text{ on } r = a, \quad z = 0, L; \quad \psi = \text{finite}\left(\text{or, } \frac{\partial \psi}{\partial r} = 0\right) @ r = 0. \tag{25.24}$$

The above EVP can be solved using separation of variables. Let

$$\psi = R(r)\Theta(\theta)Z(z) \tag{25.25}$$

then substituting equation (25.25) into equations (25.23)–(25.24) and dividing by ψ, we get

$$\frac{1}{R}\frac{1}{r}\frac{d}{dr}\left(r\frac{dR}{dr}\right) + \frac{1}{\Theta}\frac{1}{r^2}\frac{d^2\Theta}{d\theta^2} + \lambda = -\frac{1}{Z}\frac{d^2 Z}{dz^2}, \tag{25.26}$$

$$R = 0 \text{ on } r = a; \quad R = \text{finite}\left(\text{or, } \frac{dR}{dr} = 0\right) @ r = 0 \tag{25.27}$$

$$Z = 0 \text{ at } z = 0, L; \tag{25.28}$$

Note that the LHS in equation (25.26) is function of (r, θ) while the RHS is a function of z only. Thus, requiring each to be a constant, we get

$$\frac{1}{Z}\frac{d^2 Z}{dz^2} = -\lambda_z; \quad Z = 0 @ z = 0, L \tag{25.29}$$

$\Longrightarrow$ the z-eigenvalues and eigenfunctions are given by

$$\lambda_z = \frac{n^2\pi^2}{L^2} = \lambda_n \quad \text{and} \quad Z_n = \sqrt{2}\sin\left[\frac{n\pi z}{L}\right], \quad n = 1, 2, 3, \ldots \tag{25.30}$$

which are orthonormal w. r. t. the standard inner product:

$$\langle Z_i, Z_j\rangle = \frac{1}{L}\int_0^L Z_i Z_j\, dz = \delta_{ij} \tag{25.31}$$

Thus, equation (25.26) can be rewritten by using equations (25.29) and (25.30) as follows:

$$\frac{1}{R} r \frac{d}{dr}\left(r\frac{dR}{dr}\right) + r^2(\lambda - \lambda_n) = -\frac{1}{\Theta}\frac{d^2\Theta}{d\theta^2} \tag{25.32}$$

which leads to the θ-EVP:

$$\frac{d^2\Theta}{d\theta^2} = -\lambda_\theta \Theta; \quad 0 < \theta < 2\pi \tag{25.33}$$

$$\Theta \text{ and } \Theta' \text{ periodic in } \theta \tag{25.34}$$

$\Longrightarrow$

$$\lambda_\theta = m^2 = \lambda_m \text{ and } \lambda_0 = 0 \text{ with } \Theta_0 = 1, \quad \Theta_{mc} = \sqrt{2}\cos m\theta, \quad \Theta_{ms} = \sqrt{2}\sin m\theta; \quad m = 1, 2, \ldots \tag{25.35}$$

which are orthonormal w. r. t. the standard inner product:

$$\langle \Theta_i, \Theta_j\rangle = \frac{1}{2\pi}\int_0^{2\pi} \Theta_i \Theta_j d\theta = \delta_{ij} \tag{25.36}$$

Thus, equation (25.32) can be rewritten by using equations (25.33) and (25.35) as follows:

$$\frac{1}{r}\frac{d}{dr}\left(r\frac{dR}{dr}\right) - \frac{m^2}{r^2}R = -(\lambda - \lambda_n)R \tag{25.37}$$

$$R = 0 \text{ on } r = a; \quad R = \text{finite}\left(\text{or, } \frac{dR}{dr} = 0\right) @\, r = 0 \tag{25.38}$$

Equations (25.37)–(25.38) suggest that the eigenfunction can be expressed in terms of Bessel function $J_m(\sqrt{\lambda - \lambda_n}r)$ for $\lambda > \lambda_n$. [Note: Since the domain of interest is the interior of the cylinder and we require the solution to be bounded, the second linearly independent solution of the Bessel equation $Y_m(\sqrt{\lambda - \lambda_n}r)$ does not appear here.] Thus, the normalized eigenfunctions and eigenvalues are given by

$$R_{mnk} = \frac{J_m(\sqrt{\lambda_{mnk} - \lambda_n}r)}{J_{m+1}(\sqrt{\lambda_{mnk} - \lambda_n}a)} \tag{25.39}$$

$$J_m(\sqrt{\lambda_{mnk} - \lambda_n}a) = 0; \quad \lambda_{mnk} > \lambda_n; \quad m = 0, 1, 2, \dots; \quad n = 1, 2, 3, \dots; \quad \text{and } k = 1, 2, 3, \dots \tag{25.40}$$

which are orthonormal w. r. t. the cylindrical inner product:

$$\langle R_{mnk}, R_{mnj} \rangle = \frac{1}{a^2} \int_0^a 2r R_{mnk} R_{mnj}\, dr = \delta_{jk} \tag{25.41}$$

Thus, the 3D Laplacian operator with Dirichlet boundary condition in cylindrical coordinate system has the eigenvalues and eigenfunctions

$$\begin{aligned} \lambda_{mnk} &= \frac{\mu_{mk}^2}{a^2} + \lambda_n, \\ \psi &= \psi_{mnk}(r, \theta, z) = R_{mnk}(r)\Theta_m(\theta)Z_n(z), \\ m &= 0, 1, 2, \dots; \quad n = 1, 2, 3, \dots; \quad \text{and } k = 1, 2, 3, \dots \end{aligned} \tag{25.42}$$

where R_{mnk}, Θ_m and Z_n are given in equations (21.26), (21.29) and (25.41), and μ_{mk} is the kth zero of the Bessel function J_m. Similarly, eigenvalues and eigenfunctions with other BCs (compatible with separation of variables) can also be obtained with a similar procedure. For obvious reasons, the integers n, m and k are known as the axial, azimuthal and radial mode numbers, respectively.

25.1.4 3D eigenvalue problems in spherical geometries

Consider the EVP problem for the Laplacian operator in 3D spherical coordinates:

$$\begin{aligned} \nabla^2\psi = \frac{1}{r^2}\frac{\partial}{\partial r}\left(r^2\frac{\partial\psi}{\partial r}\right) + \frac{1}{r^2 \sin\theta}\frac{\partial}{\partial\theta}\left(\sin\theta\frac{\partial\psi}{\partial\theta}\right) + \frac{1}{r^2\sin^2\theta}\frac{\partial^2\psi}{\partial\phi^2} = -\lambda\psi, \\ 0 < r < a, \quad 0 < \theta < \pi, \quad 0 < \phi < 2\pi \end{aligned} \tag{25.43}$$

with homogeneous Dirichlet boundary conditions

$$\psi = 0 \text{ on } r = a, \quad \psi = \text{finite}\left(\text{or, } \frac{\partial\psi}{\partial r} = 0\right) @\, r = 0. \tag{25.44}$$

Using the separation of variables approach, i. e., expressing $\psi(r, \theta, \phi)$ as

$$\psi = R(r)\Theta(\theta)\Phi(\phi) \tag{25.45}$$

and substituting equation (25.45) into equations (25.43)–(25.44) and dividing by $\frac{\psi}{r^2}$, we get

$$\frac{1}{R}\frac{d}{dr}\left(r^2\frac{dR}{dr}\right) + r^2\lambda = -\frac{1}{\Theta\sin\theta}\frac{d}{d\theta}\left(\sin\theta\frac{d\Theta}{d\theta}\right) - \frac{1}{\Phi\sin^2\theta}\frac{d^2\Phi}{d\phi^2}, \tag{25.46}$$

$$R = 0 \text{ on } r = a; \quad R = \text{finite} \left(\text{or, } \frac{\partial R}{\partial r} = 0 \right) @ r = 0 \tag{25.47}$$

Since the LHS in equation (25.46) is function of r while the RHS is a function of (θ, ϕ) only, both terms should be equal to a constant, and thus we can write

$$\frac{1}{\Theta \sin\theta} \frac{d}{d\theta}\left(\sin\theta \frac{d\Theta}{d\theta} \right) + \frac{1}{\Phi \sin^2\theta} \frac{d^2\Phi}{d\phi^2} = -\lambda_{\theta\phi}.$$

$\Longrightarrow$ by multiplying with $\sin^2\theta$ and rearranging,

$$\frac{\sin\theta}{\Theta} \frac{d}{d\theta}\left(\sin\theta \frac{d\Theta}{d\theta} \right) + \lambda_{\theta\phi} \sin^2\theta = -\frac{1}{\Phi} \frac{d^2\Phi}{d\phi^2} = \lambda_\phi \tag{25.48}$$

$$0 < \theta < \pi, \quad 0 < \phi < 2\pi$$

Thus, ϕ-eigenvalues and eigenfunctions are given by

$$\lambda_\phi = m^2 = \lambda_m, \quad \lambda_0 = 0, \quad \Phi_0 = 1, \quad \Phi_{mc} = \sqrt{2}\cos m\phi, \quad \Phi_{ms} = \sqrt{2}\sin m\phi; \quad m = 1, 2, \ldots \tag{25.49}$$

which are orthonormal w. r. t. the standard inner product:

$$\langle \Phi_i, \Phi_j \rangle = \frac{1}{2\pi} \int_0^{2\pi} \Phi_i \Phi_j \, d\phi = \delta_{ij}. \tag{25.50}$$

Similarly, θ-eigenfunctions can be expressed from equations (25.48) and (25.49) as

$$\frac{1}{\sin\theta} \frac{d}{d\theta}\left(\sin\theta \frac{d\Theta}{d\theta} \right) - \frac{m^2}{\sin^2\theta}\Theta = -\lambda_{\theta\phi}\Theta, \quad 0 < \theta < \pi \tag{25.51}$$

Since Θ is finite for $0 < \theta < \pi$ and $\sin\theta$ vanishes at $\theta = 0, \pi$, we define

$$z = \cos\theta \tag{25.52}$$

which coverts the EVP (equation (25.51)) to

$$(1 - z^2)\frac{d^2\Theta}{dz^2} - 2z\frac{d\Theta}{dz} + \left(\lambda_{\theta\phi} - \frac{m^2}{1 - z^2} \right)\Theta = 0, \quad -1 < z < 1 \tag{25.53}$$

with Θ being finite in the domain $(-1 < z < 1)$. Equation (25.53) is referred to as "Associated Legendre equation" [see Chapter 16, Sections 16.4.5 and 16.4.6 for a discussion of this equation]. It can be shown that the eigenvalues are

$$\lambda_{\theta\phi} = \lambda_{nm} = n(n + 1), \quad n = m, m + 1, \ldots, \quad \text{i. e. } 0 \le m \le n \tag{25.54}$$

with normalized eigenfunctions as

$$\Theta_{nm}(z) = \sqrt{\frac{(n-m)!(2n+1)}{(n+m)!}} P_n^m(z) \tag{25.55}$$

where P_n^m are the Associated Legendre polynomials [$m = 0$ corresponds to Legendre polynomials]:

$$P_n^m = \frac{(-1)^{n+m}}{2^n n!}(1-z^2)^{m/2}\frac{d^{n+m}}{dz^{n+m}}(1-z^2)^n. \tag{25.56}$$

Note that the constant multiplier in equations (25.55)–(25.56) makes the eigenfunctions $\Theta_{nm}(z)$ orthogonal with respect to the inner product:

$$\langle\Theta_{nm},\Theta_{km}\rangle = \frac{1}{2}\int_{-1}^{1}\Theta_{nm}\Theta_{km}\,dz = \delta_{kn}. \tag{25.57}$$

Thus, r-eigenfunction now can be expressed from equation (25.46) as

$$\frac{1}{R}\frac{d}{dr}\left(r^2\frac{dR}{dr}\right) + r^2\lambda = \lambda_{\theta\phi} = n(n+1), \quad n = m, m+1, \ldots \tag{25.58}$$

$$R = 0 \text{ on } r = a; \quad R = \text{finite}\left(\text{or, } \frac{\partial R}{\partial r} = 0\right) @ r = 0 \tag{25.59}$$

When $\lambda = 0$, equations (25.58)–(25.59) is referred to as Euler equation. When $\lambda \neq 0$, substituting

$$\xi = r\sqrt{\lambda} \tag{25.60}$$

in equations (25.58)–(25.59) gives

$$\xi^2\frac{d^2R}{d\xi^2} + 2\xi\frac{dR}{d\xi} + [\xi^2 - n(n+1)]R = 0, \quad n = m, m+1, \ldots \tag{25.61}$$

which is spherical Bessel equation (see the discussion in Chapter 16, Section 16.4.4). Thus, the eigenfunction can be expressed in terms of spherical Bessel functions $j_n(\xi) = j_n(r\sqrt{\lambda_{nk}})$ of first kind (due to the BC: R is finite at $r = 0$, the spherical Bessel function of second kind is omitted). Thus, using the boundary condition (equation (25.59)), the eigenvalues and normalized eigenfunctions are given by

$$\text{eigenvalues:} \quad j_n(a\sqrt{\lambda_{nk}}) = 0, \quad k = 1,2,3,\ldots \tag{25.62}$$

$$\text{eigenfunctions:} \quad R_{nk}(r) = \sqrt{\frac{2}{3}}\frac{j_n(r\sqrt{\lambda_{nk}})}{j_{n+1}(a\sqrt{\lambda_{nk}})}, \tag{25.63}$$

which are orthonormal with respect to the standard spherical inner product:

$$\langle R_{nk}, R_{nj} \rangle = \frac{1}{a^3} \int_0^a 3r^2 R_{nk} R_{nj}\, dr = \delta_{kj} \tag{25.64}$$

Thus, the 3D Laplacian operator with Dirichlet boundary condition on the surface and with domain being the interior of a sphere has the eigenfunctions

$$\begin{aligned} &\psi = \psi_{mnk}(r, \theta, z) = R_{nk}(r)\Theta_{nm}(\cos\theta)\Phi_m(\phi), \\ &m = 0, 1, 2, \ldots; \quad n = m, m+1, \ldots; \quad \text{and } k = 1, 2, 3, \ldots \end{aligned} \tag{25.65}$$

where R_{nk}, Θ_{mn} and Φ_m are given by equations (25.49)–(25.50), (25.54)–(25.57) and (25.62)–(25.64), respectively. The integers k, m and n are known as the radial, azimuthal and polar mode numbers, respectively.

The eigenfunctions in cylindrical and spherical geometries with other types of boundary conditions can be determined in a similar way. Similarly, when the domain is an annulus or an unbounded region with a cylindrical or spherical boundary, the eigenvalues and eigenfunctions can be determined using the separation of variables method.

25.2 FFT method for 1D problems in spherical and cylindrical geometries

In this section, we illustrate the use of FFT method for the solution of various 1D problems in cylindrical and spherical geometries and also compare the same with direct solution.

25.2.1 Steady-state diffusion and reaction in a cylindrical catalyst

Consider 1D diffusion–reaction problem in a cylindrical catalyst with Dirichlet boundary condition:

$$D_e \nabla^2 c = D_e \frac{1}{r}\frac{d}{dr}\left(r\frac{dc}{dr}\right) = kc, \quad \text{in } \Omega \equiv 0 < r < a \tag{25.66}$$

$$c = c_0 \text{ on } \partial\Omega \equiv r = a; \quad c = \text{finite} @ r = 0 \tag{25.67}$$

The quantity of interest is the effectiveness factor defined by

$$\eta = \frac{\text{actual reaction rate}}{\text{rate if } c = c_0 \quad \text{in } \Omega} = \frac{\int_0^a kc(r) 2\pi r\, dr}{\pi a^2 k c_0} \tag{25.68}$$

Nondimensionalization

Define

$$u = \frac{c}{c_0}; \quad \xi = \frac{r}{a}; \quad \phi^2 = \frac{a^2 k}{D_m} \tag{25.69}$$

$\Longrightarrow$

$$\frac{1}{\xi}\frac{d}{d\xi}\left(\xi\frac{du}{d\xi}\right) = \phi^2 u, \quad 0 < \xi < 1; \quad u(1) = 1; \quad u(0) \text{ finite} \tag{25.70}$$

and

$$\eta = \int_0^1 2\xi u(\xi)\, d\xi\,. \tag{25.71}$$

Direct solution

Equation (25.70) can be solved directly in terms of modified Bessel functions, which with given BCs can be expressed as

$$u(\xi) = \frac{I_0(\phi\xi)}{I_0(\phi)} \tag{25.72}$$

$\Longrightarrow \eta$(from equation (25.71)) as

$$\eta = \frac{2}{\phi^2}u'(1) = \frac{2}{\phi}\frac{I_1(\phi)}{I_0(\phi)} \tag{25.73}$$

Here, I_0 and I_1 are the modified Bessel functions of first kind of order *zero* and order *one*, respectively. The dimensionless group ϕ is the Thiele modulus.

FFT method

The model equation (25.70) can be rewritten by substituting

$$v = 1 - u \tag{25.74}$$

as

$$\frac{1}{\xi}\frac{d}{d\xi}\left(\xi\frac{dv}{d\xi}\right) - \phi^2 v = -\phi^2, \quad 0 < \xi < 1; \quad v(1) = 0; \quad v(0) \text{ finite} \tag{25.75}$$

In this case, the relevant EVP is

$$\frac{1}{\xi}\frac{d}{d\xi}\left(\xi\frac{dw}{d\xi}\right) = -\lambda w, \quad 0 < \xi < 1; \quad w(1) = 0; \quad w(0) \text{ finite} \tag{25.76}$$

As discussed in the earlier section, the eigenvalues and normalized eigenfunctions can be expressed as

$$J_0(\sqrt{\lambda_k}) = 0, \quad \text{and} \quad w_k = \frac{J_0(\sqrt{\lambda_k}\xi)}{J_1(\sqrt{\lambda_k})}, \quad k = 1, 2, 3, \ldots \tag{25.77}$$

which are orthonormal with respect to the standard cylindrical inner product:

$$\langle w_i, w_j \rangle = \int_0^1 2\xi w_i w_j \, d\xi = \delta_{ij}. \tag{25.78}$$

Take the inner product (FFT) of equation (25.75) with $w_j \Longrightarrow$

$$-\lambda_j \langle v, w_j \rangle - \phi^2 \langle v, w_j \rangle = -\phi^2 \langle 1, w_j \rangle \Longrightarrow \langle v, w_j \rangle = \frac{\phi^2 \langle 1, w_j \rangle}{\phi^2 + \lambda_j} \tag{25.79}$$

$\Longrightarrow$

$$v(\xi) = \sum_{j=1}^{\infty} \langle v, w_j \rangle w_j = \sum_{j=1}^{\infty} \frac{\phi^2 \langle 1, w_j \rangle}{\phi^2 + \lambda_j} w_j(\xi) \tag{25.80}$$

$\Longrightarrow$

$$\begin{aligned} \eta &= 1 - \langle 1, v \rangle \\ &= 1 - \sum_{j=1}^{\infty} \frac{\phi^2}{\phi^2 + \lambda_j} \langle 1, w_j \rangle^2 \\ &= 1 - \sum_{j=1}^{\infty} \frac{\phi^2}{\phi^2 + \lambda_j} \left\{ \int_0^1 2\xi \frac{J_0(\sqrt{\lambda_k}\xi)}{J_1(\sqrt{\lambda_k})} \, d\xi \right\}^2 \\ &= 1 - \sum_{j=1}^{\infty} \frac{4\phi^2}{\lambda_j(\phi^2 + \lambda_j)} \end{aligned} \tag{25.81}$$

where the eigenvalues λ_j are the roots of $J_0(\sqrt{\lambda_j}) = 0$. Figure 25.4 shows a comparison of effectiveness factor evaluated from direct solution (equation (25.73)) and FFT solution (equation (25.81)) with few terms included in the summation. Table 25.1 lists the eigenvalues of the 1D cylindrical Laplacian operator that are utilized in the summation [Remark: $\sqrt{\lambda_n}$ is the nth zero of $J_0(x)$.]

As expected, the high ϕ asymptote approaches $\frac{2}{\phi}$ and FFT solution becomes accurate as more terms are included in the summation. For large ϕ, the number of terms to be included to obtain η accurately is about equal to ϕ and is an indication of the boundary (reaction) layer thickness which is of the order $\frac{1}{\phi}$.

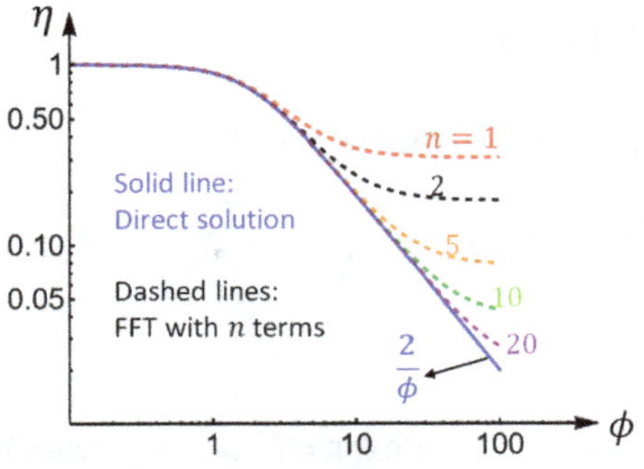

Figure 25.4: Effectiveness factor for a cylindrical catalyst from direct solution and FFT solution.

Table 25.1: First few eigenvalues λ_j of the Laplacian operator in 1D cylindrical coordinate: $J_0(\lambda_j) = 0$.

n	$\sqrt{\lambda_n}$	λ_n
1	2.40483	5.78319
2	5.52008	30.4713
3	8.65373	74.887
4	11.7915	139.04
5	14.9309	222.932
6	18.0711	326.563
7	21.2116	449.934
8	24.3525	593.043
9	27.4935	755.891
10	30.6346	938.479
11	33.7758	1140.81
12	36.9171	1362.87
13	40.0584	1604.68
14	43.1998	1866.22
15	46.3412	2147.51
16	49.4826	2448.53
17	52.6241	2769.29
18	55.7655	3109.79
19	58.907	3470.03
20	62.0485	3850.01

Remarks.
1. When the effective diffusion length $R_\Omega = \frac{V_\Omega}{A_\Omega} = \frac{a}{2}$ is used as the length scale to define the Thiele modulus as
$$\Phi^2 = \frac{R_\Omega^2 k}{D_e} = \frac{\phi^2}{4},$$
instead of the radius, then the effectiveness factor η approaches to $\frac{1}{\Phi}$ for $\Phi \gg 1$.
2. If the inner product in equation (25.78) is defined without the numerical factor 2, the normalized eigenfunction is multiplied by $\sqrt{2}$. The inner product as defined arises naturally and we note that $\eta = \langle u, 1\rangle$ or the average value of $u(\xi)$ w. r. t. this inner product.

25.2.2 Transient heat/mass transfer in an 1D infinite cylinder

Consider the unsteady-state heat/diffusion equation in 1D infinite cylinder:

$$\rho c_p \frac{\partial T}{\partial t} = k\nabla^2 T = \frac{k}{r}\frac{\partial}{\partial r}\left(r\frac{\partial T}{\partial r}\right), \quad 0 < r < a, \quad t > 0 \tag{25.82}$$

with BC and IC;

$$T = T_s \, @\, r = a; \quad T = T_0(r) \, @\, t = 0; \quad T \text{ finite } @\, r = 0. \tag{25.83}$$

Defining dimensionless variables

$$\xi = \frac{r}{a}; \quad u = \frac{T - T_s}{T_s}; \quad \tau = \frac{kt}{\rho c_p a^2}; \quad f(\xi) = \frac{T_0(a\xi) - T_s}{T_s}; \tag{25.84}$$

the dimensionless temperature u is given by the following initial-boundary value problem:

$$\frac{1}{\xi}\frac{\partial}{\partial \xi}\left(\xi\frac{\partial u}{\partial \xi}\right) = \frac{\partial u}{\partial \tau}, \quad 0 < \xi < 1, \tau > 0 \tag{25.85}$$

$$u(\xi = 1, \tau) = 0; \quad u(\xi = 0, \tau) \text{ finite}; \quad u(\xi, \tau = 0) = f(\xi) \tag{25.86}$$

The relevant EVP in this case is same as described previously in equations (25.76) and equations (25.77)–(25.78), i. e.,

$$J_0(\sqrt{\lambda_k}) = 0, \quad \text{and} \quad w_k = \frac{J_0(\sqrt{\lambda_k}\xi)}{J_1(\sqrt{\lambda_k})}, \quad k = 1, 2, 3, \ldots$$

which are orthonormal with respect to the standard cylindrical inner product (equation (25.78)). Thus, taking the inner product of model equations (25.85)–(25.86) with w_j, we get

$$-\lambda_j\langle u, w_j\rangle = \frac{d}{d\tau}\langle u, w_j\rangle, \quad \tau > 0; \quad \langle u, w_j\rangle = \langle f, w_j\rangle \, @\, \tau = 0$$

$\Rightarrow$

$$\langle u, w_j\rangle = \langle f, w_j\rangle e^{-\lambda_j \tau} \Longrightarrow u(\xi, \tau) = \sum_{j=1}^{\infty}\langle u, w_j\rangle w_j = \sum_{j=1}^{\infty} e^{-\lambda_j \tau}\langle f, w_j\rangle w_j \tag{25.87}$$

$\Longrightarrow$

$$u(\xi, \tau) = \sum_{j=1}^{\infty} e^{-\lambda_j \tau}\frac{J_0(\sqrt{\lambda_j}\xi)}{J_1(\sqrt{\lambda_j})}\int_0^1 2x f(x)\frac{J_0(\sqrt{\lambda_j}x)}{J_1(\sqrt{\lambda_j})}\, dx$$

$\Longrightarrow$

$$u(\xi,\tau) = 2\sum_{j=1}^{\infty} e^{-\lambda_j\tau}\frac{J_0(\sqrt{\lambda_j}\xi)}{J_1^2(\sqrt{\lambda_j})}\int_0^1 xf(x)J_0(\sqrt{\lambda_j}x)\,dx \tag{25.88}$$

Special case 1: $f(\xi) = 1$

For the special case of $f(\xi) = 1$, the integral in equation (25.88) simplifies to

$$\int_0^1 xJ_0(\sqrt{\lambda_j}x)\,dx = \frac{J_1(\sqrt{\lambda_j})}{\sqrt{\lambda_j}}$$

$\Rightarrow$

$$u(\xi,\tau) = 2\sum_{j=1}^{\infty}\frac{e^{-\lambda_j\tau}J_0(\sqrt{\lambda_j}\xi)}{\sqrt{\lambda_j}\cdot J_1(\sqrt{\lambda_j})}, \tag{25.89}$$

which leads to the average value as

$$\langle u\rangle(\tau) = \langle 1, u(\xi,\tau)\rangle = \int_0^1 2\xi u(\xi,\tau)\,d\xi = 4\sum_{j=1}^{\infty}\frac{e^{-\lambda_j\tau}}{\lambda_j}. \tag{25.90}$$

Figure 25.5 shows a 2D density plot of the dimensionless temperature $u(\xi,\tau)$ in (ξ,τ) space, as well as its average value $\langle u\rangle(\tau)$ and the value at the center $u(\xi = 0,\tau)$ as a function of time.

The numerical solution at $\tau = 0$ in the density plot shows oscillation in the solution instead of a constant value of unity. This is due to the Gibb's phenomena which arises due to the discontinuity in the initial condition at $\tau = 0$ (as discussed in Chapter 21).

Special case 2: $f(\xi) = \delta_2(\xi)$

Consider the special case $f(\xi)$ being the Dirac-delta function in cylindrical coordinate:

$$f(\xi) = \delta_2(\xi) = \lim_{\varepsilon\to 0}\frac{1}{\varepsilon}e^{-\frac{\xi^2}{\varepsilon}},$$

which satisfies

$$\int_0^{\infty} 2\xi\delta_2(\xi)\,d\xi = 1$$

and

$$\int_0^{\infty} 2\xi g(\xi)\delta_2(\xi)\,d\xi = g(0).$$

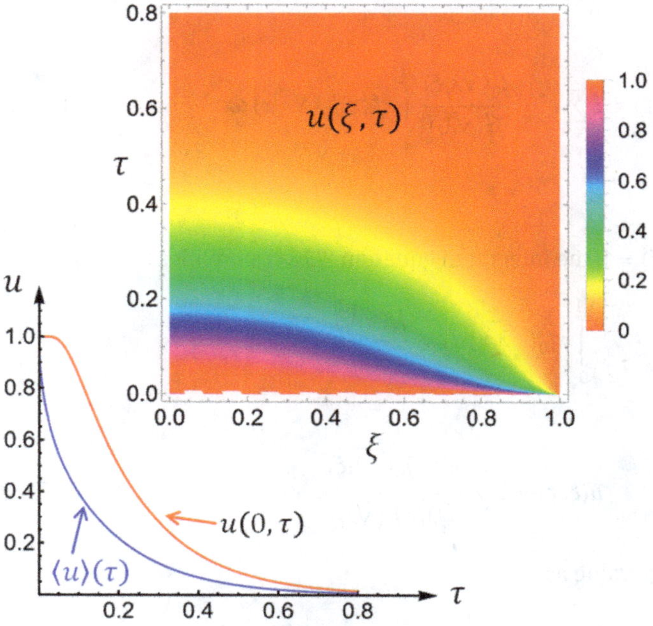

Figure 25.5: 2D density plot of $u(\xi,\tau)$ along with the temporal profile of average temperature $\langle u\rangle(\tau)$ and center temperature $u(\xi=0,\tau)$.

For this case, the solution given by equation (25.88) simplifies to

$$u(\xi,\tau)=\sum_{j=1}^{\infty}e^{-\lambda_j\tau}\frac{J_0(\sqrt{\lambda_j}\xi)}{J_1^2(\sqrt{\lambda_j})}.$$

A plot of the profile at different times is shown in Figure 25.6.

We note that more terms are needed in the summation to obtain $u(0,\tau)$ accurately for $\tau\to 0$.

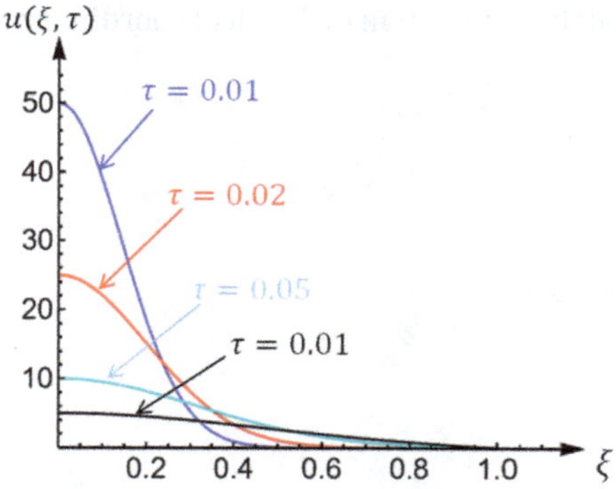

Figure 25.6: Spatial profile of the solution $u(\xi,\tau)$ at various times.

25.2.3 Steady-state 1D diffusion and reaction in a spherical catalyst particle

Consider again the 1D diffusion–reaction model with Dirichlet boundary condition but in a spherical catalyst particle:

$$D_e \nabla^2 c = D_e \frac{1}{r^2}\frac{d}{dr}\left(r^2 \frac{dc}{dr}\right) = kc, \quad \text{in } \Omega \equiv 0 < r < a \tag{25.91}$$

$$c = c_0 \text{ on } \partial\Omega \equiv r = a; \quad c = \text{finite} \ @\, r = 0 \tag{25.92}$$

In this case, the effectiveness factor is given by

$$\eta = \frac{\text{actual reaction rate}}{\text{rate if } c = c_0 \text{ in } \Omega} = \frac{\int_0^a kc(r) 4\pi r^2 \, dr}{\frac{4\pi}{3} a^3 k c_0} \tag{25.93}$$

Nondimensionalization

Define

$$u = \frac{c}{c_0}; \quad \xi = \frac{r}{a}; \quad \phi^2 = \frac{a^2 k}{D_m}; \quad R_\Omega = \frac{V_\Omega}{A_\Omega} = \frac{a}{3} \tag{25.94}$$

$\Longrightarrow$

$$\frac{1}{\xi^2}\frac{d}{d\xi}\left(\xi^2 \frac{du}{d\xi}\right) = \phi^2 u, \quad 0 < \xi < 1; \quad u(1) = 1; \quad u(0) \text{ finite} \tag{25.95}$$

and

$$\eta = \int_0^1 3\xi^2 u(\xi) \, d\xi \tag{25.96}$$

Direct solution

Equation (25.95) can be solved exactly in terms of modified spherical Bessel functions of order zero, which with given BCc can be expressed as

$$u(\xi) = \frac{\sinh(\phi\xi)}{\xi \sinh(\phi)} \tag{25.97}$$

$\Longrightarrow$ from equation (25.96) as

$$\eta = \frac{3}{\phi^2} u'(1) = \frac{3}{\phi^2}(\phi \coth \phi - 1). \tag{25.98}$$

FFT method

The model equation (25.95) can be rewritten by substituting

$$v = 1 - u \tag{25.99}$$

as

$$\frac{1}{\xi^2}\frac{d}{d\xi}\left(\xi^2\frac{dv}{d\xi}\right) - \phi^2 v = -\phi^2, \quad 0 < \xi < 1; \quad v(1) = 0; \quad v(0) \text{ finite} \tag{25.100}$$

In this case, the relevant EVP is

$$\frac{1}{\xi^2}\frac{d}{d\xi}\left(\xi^2\frac{dw}{d\xi}\right) = -\lambda w, \quad 0 < \xi < 1; \quad w(1) = 0; \quad w(0) \text{ finite} \tag{25.101}$$

As discussed in earlier section, the eigenvalues and normalized eigenfunctions can be expressed as spherical Bessel functions:

$$j_0(\sqrt{\lambda_k}) = \frac{\sin\sqrt{\lambda_k}}{\sqrt{\lambda_k}} = 0, \quad \Longrightarrow \lambda_k = k^2\pi^2, \quad k = 1, 2, 3, \dots \tag{25.102}$$

$$\text{and} \quad w_k = \sqrt{\frac{2}{3}}\frac{\sin k\pi\xi}{\xi}, \tag{25.103}$$

which are orthonormal with respect to the standard spherical inner product:

$$\langle w_i, w_j\rangle = \int_0^1 3\xi^2 w_i w_j \, d\xi = \delta_{ij}. \tag{25.104}$$

Take the inner product (FFT) of equation (25.100) with $w_j \Longrightarrow$

$$-\lambda_j\langle v, w_j\rangle - \phi^2\langle v, w_j\rangle = -\phi^2\langle 1, w_j\rangle \Longrightarrow \langle v, w_j\rangle = \frac{\phi^2\langle 1, w_j\rangle}{\phi^2 + \lambda_j} \tag{25.105}$$

$\Longrightarrow$

$$v(\xi) = \sum_{j=1}^{\infty}\langle v, w_j\rangle w_j = \sum_{j=1}^{\infty}\frac{\phi^2\langle 1, w_j\rangle}{\phi^2 + \lambda_j} w_j(\xi) \tag{25.106}$$

$\Longrightarrow$

$$\begin{aligned}\eta &= 1 - \langle 1, v\rangle \\ &= 1 - \sum_{j=1}^{\infty}\frac{\phi^2}{\phi^2 + \lambda_j}\langle 1, w_j\rangle^2\end{aligned}$$

$$
\begin{aligned}
&= 1 - \sum_{j=1}^{\infty} \frac{\phi^2}{\phi^2 + \lambda_j} \langle 1, w_j \rangle^2 \\
&= 1 - \sum_{j=1}^{\infty} \frac{\phi^2}{\phi^2 + j^2\pi^2} \left\{ \int_0^1 3\xi^2 \sqrt{\frac{2}{3}} \frac{\sin j\pi\xi}{\xi} \, d\xi \right\}^2 \\
&= 1 - \sum_{j=1}^{\infty} \frac{6\phi^2}{j^2\pi^2(\phi^2 + j^2\pi^2)}
\end{aligned} \tag{25.107}
$$

Figure 25.7 shows the comparison of effectiveness factors evaluated from the direct solution (equation (25.98)) and FFT solution (equation (25.107)) with few terms included in the summation.

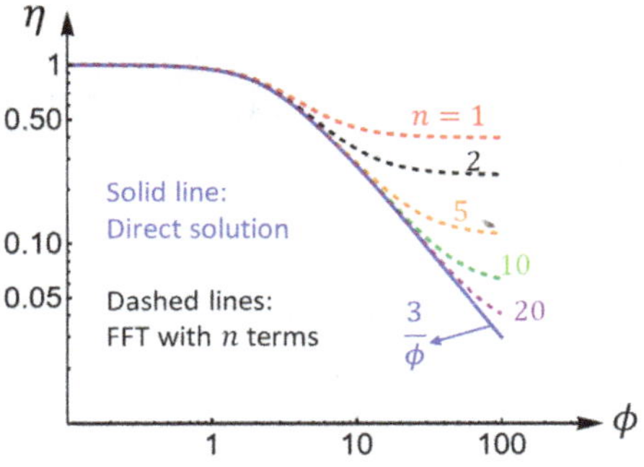

Figure 25.7: Effectiveness factor for a spherical catalyst from exact solution and FFT solution.

As expected, the high ϕ asymptote approaches $\frac{3}{\phi}$ and FFT solution becomes accurate as more terms are included in the summation.

Remark. If Thiele modulus is defined based on the effective diffusion length $R_\Omega = \frac{a}{3}$ instead of radius a, i. e.,

$$
\Phi^2 = \frac{R_\Omega^2 k}{D_3} = \frac{\phi^2}{9},
$$

then the effectiveness factor η approaches to $\frac{1}{\Phi}$ for $\Phi \gg 1$.

25.2.4 Transient 1D heat conduction in a spherical geometry

Consider the 1D transient heat equation (in dimensionless form)

$$
\frac{1}{\xi^2} \frac{\partial}{\partial \xi} \left(\xi^2 \frac{\partial u}{\partial \xi} \right) = \frac{\partial u}{\partial \tau}, \quad 0 < \xi < 1, \quad \tau > 0 \tag{25.108}
$$

$$\text{BC:}\quad u = 0 \;@\; \xi = 1, \quad \text{IC:}\quad u(\xi, 0) = f(\xi) \quad \text{and} \quad u(\xi, \tau) \text{ bounded} \tag{25.109}$$

The relevant EVP in this case is same as that described previously in equations (25.101) and equations (25.102)–(25.104), i. e.

$$\lambda_k = k^2\pi^2 \quad \text{and} \quad w_k = \sqrt{\frac{2}{3}}\frac{\sin k\pi\xi}{\xi}; \quad k = 1, 2, 3, \ldots$$

which are orthonormal w. r. t. the standard spherical inner product (equation (25.104)). Thus, taking the inner product of model equations (25.108)–(25.109) with w_j, we get

$$-\lambda_j\langle u, w_j\rangle = \frac{d}{d\tau}\langle u, w_j\rangle, \quad \tau > 0; \quad \langle u, w_j\rangle = \langle f, w_j\rangle \;@\; \tau = 0$$

$\Rightarrow$

$$\langle u, w_j\rangle = \langle f, w_j\rangle e^{-\lambda_j\tau} \Longrightarrow u(\xi, \tau) = \sum_{j=1}^{\infty}\langle u, w_j\rangle w_j = \sum_{j=1}^{\infty} e^{-\lambda_j\tau}\langle f, w_j\rangle w_j \tag{25.110}$$

$\Longrightarrow$

$$u(\xi, \tau) = \sum_{j=1}^{\infty} e^{-j^2\pi^2\tau}\frac{2}{3}\frac{\sin[j\pi\xi]}{\xi}\int_0^1 3x^2 f(x)\frac{\sin[j\pi x]}{x}\,dx$$

$\Longrightarrow$

$$u(\xi, \tau) = 2\sum_{j=1}^{\infty} e^{-j^2\pi^2\tau}\frac{\sin[j\pi\xi]}{\xi}\int_0^1 x f(x)\sin[j\pi x]\,dx \tag{25.111}$$

Special case: $f(\xi) = 1$

For the special case of $f(\xi) = 1$, the integral in equation (25.111) simplifies to

$$\int_0^1 x\sin[j\pi x]\,dx = \frac{(-1)^{j-1}}{j\pi}$$

$\Rightarrow$

$$u(\xi, \tau) = 2\sum_{j=1}^{\infty}(-1)^{j-1}e^{-j^2\pi^2\tau}\frac{\sin[j\pi\xi]}{j\pi\xi} \tag{25.112}$$

which leads to the average value as

$$\langle u\rangle(\tau) = \langle 1, u(\xi, \tau)\rangle = \int_0^1 3\xi^2 u(\xi, \tau)\,d\xi = 6\sum_{j=1}^{\infty}\frac{e^{-j^2\pi^2\tau}}{j^2\pi^2}. \tag{25.113}$$

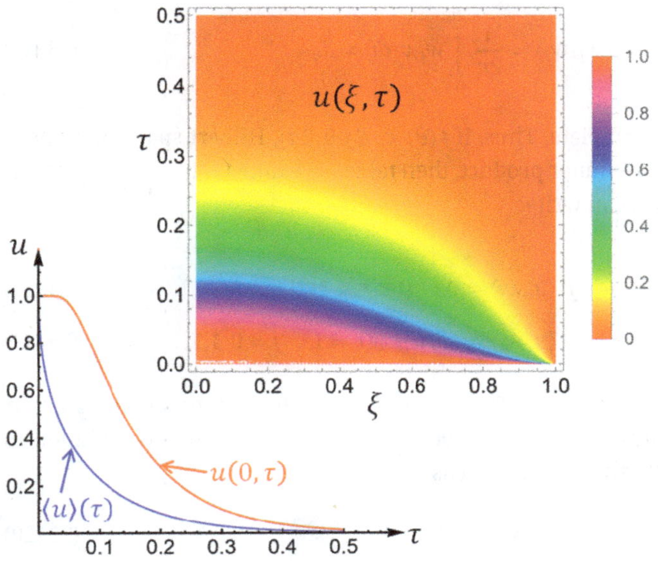

Figure 25.8: 2D density plot of $u(\xi,\tau)$ along with the temporal profile of average $\langle u\rangle(\tau)$ and center value $u(\xi=0,\tau)$.

Figure 25.8 shows the 2D density plot of the dimensionless profile $u(\xi,\tau)$ in $(\xi-\tau)$ space, as well as the temperature $\langle u\rangle(\tau)$ and center value $u(0,\tau)$ as a function of time τ.

It has similar characteristics as observed in cylindrical geometry.

25.3 2D and 3D problems in cylindrical geometry

25.3.1 Solution of Laplace's equation inside a unit circle

Consider the 2D Laplace equation inside a unit circle with Dirichlet boundary condition, i. e., the solution of

$$\nabla^2 u = \frac{1}{r}\frac{\partial}{\partial r}\left(r\frac{\partial u}{\partial r}\right) + \frac{1}{r^2}\frac{\partial^2 u}{\partial \theta^2} = 0, \qquad \begin{array}{c} 0 < r < 1 \\ 0 < \theta < 2\pi \end{array} \tag{25.114}$$

$$u(1,\theta) = f(\theta). \tag{25.115}$$

As shown earlier, the eigenvalues and eigenfunctions for the θ-operator are

$$\lambda_\theta = m^2 = \lambda_m \text{ and } w_0 = 1, \quad w_{mc} = \sqrt{2}\cos m\theta, \quad w_{ms} = \sqrt{2}\sin m\theta, \quad m = 1, 2, \ldots, \tag{25.116}$$

which are orthogonal with respect to the standard inner product:

$$\langle w_i, w_j \rangle = \frac{1}{2\pi} \int_0^{2\pi} w_i w_j \, d\theta = \delta_{ij} \tag{25.117}$$

These eigenfunctions are complete. Thus, if $f(\theta) \in \mathcal{L}_2[0, 2\pi]$, Hilbert space of 2π periodic functions with the above inner product, then taking the inner product of equations (25.114)–(25.115) with w_j, we can write

$$r^2 \frac{d^2 \langle u, w_j \rangle}{dr^2} + r \frac{d \langle u, w_j \rangle}{dr} - j^2 \langle u, w_j \rangle = 0, \quad 0 < r < 1 \tag{25.118}$$

$$\langle u, w_j \rangle = \langle f, w_j \rangle \; @ \, r = 1, \quad j = 0, 1, 2, \ldots \tag{25.119}$$

Equation (25.118) is Euler's equation. The two linearly independent solutions are r^j and r^{-j}. For bounded solution at $r \to 0$, we can only take r^j. Thus, the solution to equations (25.118)–(25.119) can be expressed as

$$\langle u, w_j \rangle = \langle f, w_j \rangle r^j \tag{25.120}$$

$\Longrightarrow$

$$u(r, \theta) = \sum_{j=0}^{\infty} \langle u, w_j \rangle w_j = \sum_{j=0}^{\infty} r^j \langle f, w_j \rangle w_j \tag{25.121}$$

$$\begin{aligned}
&= \frac{1}{2\pi} \int_0^{2\pi} f(\theta') \, d\theta' + \sum_{j=1}^{\infty} \frac{r^j}{\pi} \cos[j\theta] \int_0^{2\pi} f(\theta') \cos[j\theta'] \, d\theta' \\
&\quad + \sum_{j=1}^{\infty} \frac{r^j}{\pi} \sin[j\theta] \int_0^{2\pi} f(\theta') \sin[j\theta'] \, d\theta'
\end{aligned}$$

$$= \frac{1}{2\pi} \int_0^{2\pi} f(\theta') \left[1 + 2 \sum_{j=1}^{\infty} r^j \cos[j(\theta - \theta')] \right] d\theta' \tag{25.122}$$

Consider the summation term inside the integration:

$$\begin{aligned}
S &= \sum_{j=1}^{\infty} r^j \cos[j(\theta - \theta')] = \sum_{j=1}^{\infty} r^j \, \mathrm{Re}[e^{ij(\theta-\theta')}] \\
&= \mathrm{Re}\left[\sum_{j=1}^{\infty} \{ r e^{i(\theta-\theta')} \}^j \right] \\
&= \mathrm{Re}\left[\frac{r e^{i(\theta-\theta')}}{1 - r e^{i(\theta-\theta')}} \right] \quad \left(\because \sum_{j=1}^{\infty} z^j = z + z^2 + z^3 + \cdots = \frac{z}{1 - z} \right) \\
&= \frac{r \cos[\theta - \theta'] - r^2}{1 + r^2 - 2r \cos[\theta - \theta']}
\end{aligned}$$

$\Longrightarrow$ from equation (25.122)

$$u(r,\theta) = \frac{1}{2\pi}\int_0^{2\pi} f(\theta')\left[1 + 2\frac{r\cos[\theta-\theta'] - r^2}{1 + r^2 - 2r\cos[\theta-\theta']}\right] d\theta'$$

$$= \frac{1}{2\pi}\int_0^{2\pi} \frac{(1-r^2)f(\theta')}{1 + r^2 - 2r\cos[\theta-\theta']}\, d\theta' \tag{25.123}$$

which is also referred to as the Poisson's integral formula (for the interior of a circle).

Special cases

1. $f(\theta) = 1$: In this case,

$$\langle f, w_j\rangle = \langle 1, w_j\rangle = \langle w_0, w_j\rangle = \delta_{j0}$$

$\Longrightarrow$ from equation (25.121)

$$u(r,\theta) = \sum_{j=0}^{\infty} r^j \delta_{j0} w_j = 1$$

2. $f(\theta) = A\sin p\theta$ or $A\cos p\theta$, $p \in I$: In this case,

$$\langle f, w_j\rangle = \frac{A}{\sqrt{2}}\delta_{jp}$$

$\Longrightarrow$ from equation (25.121),

$$u(r,\theta) = \sum_{j=0}^{\infty} r^j \frac{A}{\sqrt{2}}\delta_{jps} w_j$$
$$= Ar^p \sin p\theta \quad \text{or} \quad Ar^p\cos p\theta.$$

As an example, taking cosine input with $p = 5$, i. e., $f(\theta) = \cos 5\theta$, the 3D plot and the density plot are shown in Figure 25.9.

25.3.2 Vibration of a circular membrane

Consider the vibrational motion of a circular membrane (such as a drum) described by the partial differential equation

$$\frac{\partial^2 u}{\partial t^2} = c^2\left(\frac{\partial^2 u}{\partial r^2} + \frac{1}{r}\frac{\partial u}{\partial r} + \frac{1}{r^2}\frac{\partial^2 u}{\partial \theta^2}\right), \quad 0 < r < 1, \quad 0 < \theta < 2\pi, \quad t > 0 \tag{25.124}$$

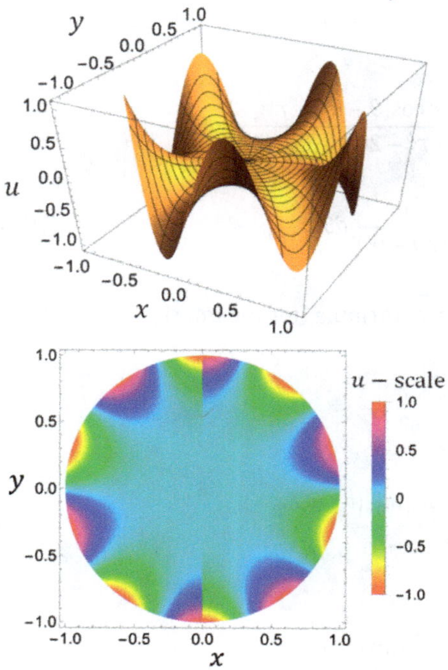

Figure 25.9: 3D plot (top) and density plot (bottom) for solution of Laplace equation with boundary value $f(\theta) = \cos 5\theta$.

BC

$$u = 0 \,@\, r = 1 \quad \text{(fixed end)} \tag{25.125}$$

ICs

$$u = f(r, \theta) \,@\, t = 0 \quad \text{(initial displacement)} \tag{25.126}$$

$$\frac{\partial u}{\partial t} = 0, \quad @\, t = 0 \quad \text{(zero initial velocity)} \tag{25.127}$$

Equation (25.124) contains two operators $\frac{\partial^2}{\partial \theta^2}$ and $\frac{1}{r}\frac{\partial}{\partial r}(r\frac{\partial}{\partial r})$. Due to the circular symmetry, the BCs for the first operator are periodicity in θ with period 2π. The eigenvalues and eigenfunctions for the θ-operator are given in equations (25.35)–(25.36) or (25.116)–(25.117):

$$\lambda_\theta = m^2 = \lambda_m \text{ and } w_0 = 1, \quad w_{mc} = \sqrt{2}\cos m\theta, \quad w_{ms} = \sqrt{2}\sin m\theta, \quad m = 1, 2, \ldots, \tag{25.128}$$

which are orthogonal w. r. t. standard inner product:

$$\langle w_i, w_j \rangle = \frac{1}{2\pi} \int_0^{2\pi} w_i w_j \, d\theta = \delta_{ij} \tag{25.129}$$

These eigenfunctions are complete. Thus, if $f(r,\theta) \in \mathcal{L}_2[0, 2\pi]$, then defining

$$u_m(r,t) = \langle u(r,\theta,t), w_m(\theta) \rangle, \quad m = 0, 1, 2, \tag{25.130}$$

and taking inner product of equation (25.124) with the eigenfunctions w_m, we get

$$\frac{\partial^2 u_m}{\partial t^2} = c^2 \left(\frac{\partial^2 u_m}{\partial r^2} + \frac{1}{r}\frac{\partial u_m}{\partial r} - \frac{m^2}{r^2} u_m \right), \quad 0 < r < 1, \quad t > 0 \tag{25.131}$$

BC

$$u_m = 0 \;@\; r = 1 \tag{25.132}$$

IC

$$u_m = f_m(r) = \langle f(r,\theta), w_m \rangle \;@\; t = 0 \tag{25.133}$$

$$\frac{\partial u_m}{\partial t} = 0 \;@\; t = 0 \tag{25.134}$$

The relevant eigenvalue problem (r-operator) in this case is

$$\frac{1}{r}\frac{d}{dr}\left(r \frac{d\psi}{dr} \right) - \frac{m^2}{r^2}\psi = -\lambda \psi$$

$$\psi(1) = 0, \quad \psi(0) \text{ finite}$$

which is a special case of equation (25.35), i. e., 3D to 2D. Thus, the eigenvalues and eigenfunctions are given by equations (25.39)–(25.41) as

$$\lambda = \{\lambda_{mk}\} : J_m(\sqrt{\lambda_{mk}}) = 0, \quad k = 1, 2, 3, \ldots \tag{25.135}$$

$$w_{mk}(r) = \frac{J_m(\sqrt{\lambda_{mk}}\, r)}{J_{m+1}(\sqrt{\lambda_{mk}})} \tag{25.136}$$

which are orthonormal with respect to the standard cylindrical inner product:

$$\langle w_{mi}, w_{mk} \rangle = \int_0^1 2r w_{mi} w_{mk} \, dr = \delta_{ik} \tag{25.137}$$

and form a complete set. Thus, taking inner product of equation (25.131) with $w_{mk}(r)$ gives

$$\frac{d^2U_{mk}}{dt^2} = -c^2\lambda_{mk}U_{mk}; \quad U_{mk} = \langle u_m, w_{mk}\rangle \tag{25.138}$$

$$U_{mk} = \langle f_m, w_{mk}\rangle = f_{mk} \text{ @ } t = 0, \quad U'_{mk} = 0 \text{ @ } t = 0 \tag{25.139}$$

$\Longrightarrow$

$$U_{mk} = f_{mk}\cos[c\sqrt{\lambda_{mk}}t] \tag{25.140}$$

$\therefore$

$$\begin{aligned}
u_m(r,t) &= \sum_{k=1}^{\infty}\langle u_m, w_{mk}\rangle w_{mk} = \sum_{k=1}^{\infty} U_{mk}(t)w_{mk}(r) \\
&= \sum_{k=1}^{\infty}\cos[c\sqrt{\lambda_{mk}}t]\cdot\frac{J_m(\sqrt{\lambda_{mk}}r)}{J_{m+1}(\sqrt{\lambda_{mk}})}\int_0^1 2\xi f_m(\xi)\frac{J_m(\sqrt{\lambda_{mk}}\xi)}{J_{m+1}(\sqrt{\lambda_{mk}})}\,d\xi \\
&= 2\sum_{k=1}^{\infty}\frac{\cos c\sqrt{\lambda_{mk}}tJ_m(\sqrt{\lambda_{mk}}r)}{J_{m+1}^2(\sqrt{\lambda_{mk}})}\cdot\int_0^1 \xi f_m(\xi)J_m(\sqrt{\lambda_{mk}}\xi)\,d\xi
\end{aligned} \tag{25.141}$$

$\Longrightarrow$

$$\begin{aligned}
u(r,\theta,t) &= \sum_{m=0}^{\infty} u_m(r,t)w_m(\theta) \\
&= u_0(r,t) + \sum_{m=1}^{\infty} u_{mc}\sqrt{2}\cos m\theta + \sum_{m=1}^{\infty} u_{ms}\sqrt{2}\sin m\theta \\
&= \sum_{k=1}^{\infty}\cos[c\sqrt{\lambda_{0k}}t]J_0(\sqrt{\lambda_{0k}}r)A_k^0 \\
&\quad + \sum_{k=1}^{\infty}\sum_{m=1}^{\infty}[A_{km}^c\cos m\theta + A_{km}^s\sin m\theta]J_m(\sqrt{\lambda_{mk}}r)\cos[c\sqrt{\lambda_{mk}}t]
\end{aligned} \tag{25.142}$$

where

$$A_k^0 = \frac{1}{\pi J_1^2(\sqrt{\lambda_{0k}})}\int_0^1\int_0^{2\pi} \xi J_0(\sqrt{\lambda_{0k}}\xi)f(\xi,\theta)\,d\theta\,d\xi \tag{25.143}$$

$$A_{km}^c = \frac{2}{\pi J_{m+1}^2(\sqrt{\lambda_{mk}})}\int_0^1\int_0^{2\pi} \xi J_m(\sqrt{\lambda_{mk}}\xi)f(\xi,\theta)\cos m\theta\,d\theta\,d\xi \tag{25.144}$$

$$A_{km}^s = \frac{2}{\pi J_{m+1}^2(\sqrt{\lambda_{mk}})}\int_0^1\int_0^{2\pi} \xi J_m(\sqrt{\lambda_{mk}}\xi)f(\xi,\theta)\sin m\theta\,d\theta\,d\xi \tag{25.145}$$

$$m = 1, 2, \ldots; \quad k = 1, 2, 3, \ldots$$

Special case: $f(r,\theta) = J_m(\sqrt{\lambda_{mk}}r)\cos m\theta$ or $J_m(\sqrt{\lambda_{mk}}r)\sin m\theta$ (i. e., pure eigenmodes)
For the special case when $f(r,\theta)$ is the pure mode of vibration, the solution remains in the same mode for all times. For example,

(i) when $f(r,\theta) = J_0(\sqrt{\lambda_{0k}}r)$, then equation (25.143), (25.144), (25.145) implies

$$A^c_{pq} = 0 = A^s_{pq}, \quad \forall p, q \text{ and } A^0_q = \delta_{0p}\delta_{kq}$$

and equation (25.142) simplifies to

$$u(r,\theta,t) = J_0(\sqrt{\lambda_{0k}}r)\cos[c\sqrt{\lambda_{0k}}t] = f(r,\theta)\cos[c\sqrt{\lambda_{0k}}t].$$

(ii) when $f(r,\theta) = J_m(\sqrt{\lambda_{mk}}r)\sin m\theta$ $(m > 0)$, then equations (25.143), (25.144), (25.145) implies

$$A^c_{pq} = 0 = A^0_q, \quad \forall p, q \text{ and } A^s_{pq} = \delta_{pm}\delta_{qk}$$

and equation (25.142) simplifies to

$$u(r,\theta,t) = \sin[m\theta]J_m(\sqrt{\lambda_{mk}}r)\cos[c\sqrt{\lambda_{mk}}t] = f(r,\theta)\cos[c\sqrt{\lambda_{mk}}t].$$

(iii) when $f(r,\theta) = J_m(\sqrt{\lambda_{mk}}r)\cos m\theta (m > 0)$, then equations (25.143)–(25.145) imply that all sine coefficients vanish, i. e.,

$$A^s_{pq} = 0 = A^0_q, \quad \forall p, q \text{ and } A^c_{pq} = \delta_{pm}\delta_{qk}$$

and equation (25.142) simplifies to

$$u(r,\theta,t) = \cos[m\theta]J_m(\sqrt{\lambda_{mk}}r)\cos[c\sqrt{\lambda_{mk}}t] = f(r,\theta)\cos[c\sqrt{\lambda_{mk}}t]$$

Thus, the solution remains in the same eigenmode with amplitude ratio $A.R.$ varying with time as

$$A.R. = \cos[c\sqrt{\lambda_{mk}}t].$$

The contour profile of first few modes of vibration (i. e., eigenmodes) are shown in Figure 25.10, while the eigenvalues corresponding to these modes are listed in Table 25.2.

Similarly, the 2D problems in cylindrical coordinate system in (r,z) or (z,θ) space can be solved. For additional chemical engineering applications, we refer to the articles by Balakotaiah and Gupta [6], Ratnakar and Balakotaiah [26], Balakotaiah [5] and Aris and Balakotaiah [4].

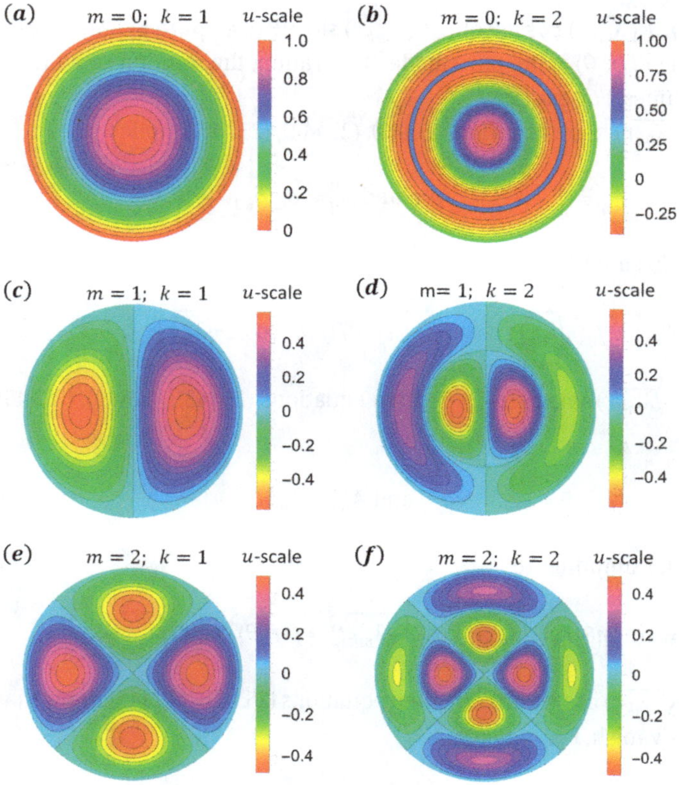

Figure 25.10: Contour profiles of first few eigenmodes: $\cos[m\theta]J_m[\sqrt{\lambda_{mk}}r]$.

Table 25.2: First few eigenvalues of cylindrical Laplace operator.

m	k	μ_{mk}	λ_{mk}
0	1	2.40483	5.78319
0	2	5.52008	30.4713
1	1	3.83171	14.682
1	2	7.01559	49.2185
2	1	5.13562	26.3746
2	2	8.41724	70.8499

25.3.3 Three-dimensional problems in cylindrical geometry

3D Poisson's equation

Consider the Poisson equation

$$\nabla^2 u = \frac{1}{r}\frac{\partial}{\partial r}\left(r\frac{\partial u}{\partial r}\right) + \frac{1}{r^2}\frac{\partial^2 u}{\partial \theta^2} + \frac{\partial^2 u}{\partial z^2} = -f(r,\theta,z) \quad \text{in } \Omega \tag{25.146}$$

$$\Omega \equiv 0 < r < a, \quad 0 < \theta < 2\pi, \quad 0 < z < L$$

with BC:

$$u = 0 \text{ on } \partial\Omega, \quad \text{i. e.,} \quad @\, r = R \text{ and } @\, z = 0, L \tag{25.147}$$

The eigenvalues and eigenfunctions are obtained in Section 25.1.3 for the Laplacian operator with Dirichlet condition in the cylindrical geometry (see equations (25.23)–(25.24) to (25.42)). Using these eigenfunctions and the corresponding inner product, the solution can be obtained by taking the inner product of equation (25.146) with ψ_{mnk} and using the boundary condition (25.147) as follows:

$$-\lambda_{mnk}\langle u, \psi_{mnk}\rangle = -\langle f, \psi_{nmk}\rangle \Longrightarrow \langle u, \psi_{mnk}\rangle = \frac{\langle f, \psi_{nmk}\rangle}{\lambda_{mnk}} \tag{25.148}$$

$\Rightarrow$

$$u = \sum_{m,n,k} \langle u, \psi_{mnk}\rangle \psi_{mnk} = \sum_{n=1}^{\infty}\sum_{m=0}^{\infty}\sum_{k=1}^{\infty} \frac{\langle f, \psi_{nmk}\rangle}{\lambda_{mnk}} \psi_{mnk} \tag{25.149}$$

where

$$\lambda_{mnk} = \frac{\mu_{mk}^2}{a^2} + \frac{n^2\pi^2}{L^2} : \quad J_m(\mu_{mk}) = 0 \tag{25.150}$$

$$\psi_{mnk} = \begin{cases} \psi_{0nk} = \sqrt{2}\frac{J_0(\sqrt{\mu_{0k}}r)}{J_1(\sqrt{\mu_{0k}}a)} \sin\frac{n\pi z}{L} \\ \psi_{mnk}^c = 2\frac{J_m(\sqrt{\mu_{mk}}r)}{J_{m+1}(\sqrt{\mu_{mk}}a)} \cos m\theta \sin\frac{n\pi z}{L} \\ \psi_{mnk}^s = 2\frac{J_m(\sqrt{\mu_{mk}}r)}{J_{m+1}(\sqrt{\mu_{mk}}a)} \sin m\theta \sin\frac{n\pi z}{L} \end{cases} \tag{25.151}$$

and

$$\langle f, \psi_{nmk}\rangle = \frac{1}{\pi a^2 L} \int_0^L \int_0^a \int_0^{2\pi} r f(r,\theta,z)\psi(r,\theta,z)\, d\theta\, dr\, dz \tag{25.152}$$

3D Heat/Diffusion equation

Consider the transient heat/diffusion equation

$$\frac{\partial u}{\partial t} = \nabla^2 u = \frac{1}{r}\frac{\partial}{\partial r}\left(r\frac{\partial u}{\partial r}\right) + \frac{1}{r^2}\frac{\partial^2 u}{\partial \theta^2} + \frac{\partial^2 u}{\partial z^2} \quad \text{in } \Omega \tag{25.153}$$

$$\Omega \equiv 0 < r < a, \quad 0 < \theta < 2\pi, \quad 0 < z < L$$

with BC:

$$u = 0 \text{ on } \partial\Omega, \quad \text{i. e. } @\, r = R \text{ and } @\, z = 0, L \tag{25.154}$$

and IC:

$$u(r,\theta,z,t=0)=f(r,\theta,z) \tag{25.155}$$

Again, using the same eigenvalues and eigenfunctions (obtained in Section 25.1.3 for the Laplacian operator with Dirichlet condition in the cylindrical geometries), and taking the inner product of equation (25.153) with ψ_{mnk}, we get:

$$\begin{aligned}\frac{d}{dt}\langle u,\psi_{mnk}\rangle &= -\lambda_{mnk}\langle u,\psi_{nmk}\rangle, \quad t>0;\\ \langle u,\psi_{mnk}\rangle &= \langle f,\psi_{nmk}\rangle \;@\, t=0\end{aligned}$$

$\Longrightarrow$

$$\langle u,\psi_{mnk}\rangle = \langle f,\psi_{nmk}\rangle e^{-\lambda_{mnk}t} \tag{25.156}$$

$\Rightarrow$

$$u=\sum_{m,n,k}\langle u,\psi_{mnk}\rangle\psi_{mnk}=\sum_{n=1}^{\infty}\sum_{m=0}^{\infty}\sum_{k=1}^{\infty}e^{-\lambda_{mnk}t}\langle f,\psi_{nmk}\rangle\psi_{mnk} \tag{25.157}$$

where eigenvalues and eigenfunctions are expressed in equations (25.150)–(25.152).

Similarly, other problems in cylindrical geometry with other types of boundary conditions can be solved once the relevant eigenvalue problems are identified.

25.4 2D and 3D problems in spherical geometry

Consider the spherical coordinate system shown in Figure 25.3. For a sphere of radius a, the interior domain is defined by $0<r<a$, $0<\theta<\pi$, $0<\phi<2\pi$.

25.4.1 Poisson's equation in a sphere

Consider the Poisson's equation in the interior of a sphere with Dirichlet boundary condition, i. e.,

$$\begin{aligned}&\nabla^2 u=\frac{1}{r^2}\frac{\partial}{\partial r}\left(r^2\frac{\partial u}{\partial r}\right)+\frac{1}{r^2\sin\theta}\frac{\partial}{\partial\theta}\left(\sin\theta\frac{\partial u}{\partial\theta}\right)+\frac{1}{r^2\sin^2\theta}\frac{\partial^2 u}{\partial\phi^2}=-f(r,\theta,\phi)\text{ in }\Omega\\ &\Omega\equiv 0<r<a,\quad 0<\theta<\pi,\quad 0<\phi<2\pi\end{aligned} \tag{25.158}$$

with

$$u=0\text{ on }\partial\Omega\equiv r=a,\quad u(0,\theta,\phi)=\text{finite} \tag{25.159}$$

FFT method

The eigenvalues and eigenfunctions of Laplacian operator in spherical coordinate are described in Section 25.1.4 (see equations (25.43)–(25.44) to (25.65)). Taking the inner product of equations (25.158)–(25.159) with eigenfunctions ψ_{mnk}, we get

$$-\lambda_{nk}\langle u,\psi_{mnk}\rangle = -\langle f,\psi_{mnk}\rangle \Longrightarrow \langle u,\psi_{mnk}\rangle = \frac{\langle f,\psi_{mnk}\rangle}{\lambda_{nk}} \tag{25.160}$$

where the eigenvalues λ_{nk} are given by

$$\lambda_{nk} = \mu_{nk}^2 : j_n(\mu_{nk}) = 0, \quad n = m, m+1, \ldots; \quad k = 1, 2, 3, \ldots \tag{25.161}$$

and eigenfunctions by

$$\psi_{mnk}(r,\theta,\phi) = \sqrt{\frac{2(n-m)!(2n+1)}{3(n+m)!}}\frac{j_n(\mu_{nk}r)}{j_{n+1}(\mu_{nk}a)}P_n^m(\cos\theta)\Phi_m(\phi) \tag{25.162}$$

$$\Phi_m(\phi) = \begin{cases} \Phi_0 = 1 & \\ \Phi_m^c = \sqrt{2}\cos m\phi & n = m, m+1, \ldots \\ \Phi_m^s = \sqrt{2}\sin m\phi, & \end{cases} \tag{25.163}$$

$\Longrightarrow$

$$u = \sum_{m,n,k}\langle u,\psi_{mnk}\rangle\psi_{mnk} = \sum_{m=0}^{\infty}\sum_{n=m}^{\infty}\sum_{k=1}^{\infty} = \frac{\langle f,\psi_{mnk}\rangle}{\lambda_{nk}}\psi_{mnk} \tag{25.164}$$

For the special case of azimuthal/longitudinal symmetry (i. e., symmetry w. r. t. ϕ), i. e., for the 2D case in (r,θ), we can disregard the m-operator, and only the $m = 0$ mode remains. In this case, the formal solution (equation (25.164)) simplifies to

$$u = \sum_{n,k}\langle u,\psi_{nk}\rangle\psi_{nk} = \sum_{n=0}^{\infty}\sum_{k=1}^{\infty} = \frac{\langle f,\psi_{nk}\rangle}{\lambda_{nk}}\psi_{nk} \tag{25.165}$$

where λ_{nk} is obtained by the same way (from equation (25.161)) and eigenfunctions simplify to

$$\psi_{nk}(r,\theta) = \sqrt{\frac{2(2n+1)}{3}}\frac{j_n(\mu_{nk}r)}{j_{n+1}(\mu_{nk}a)}P_n(\cos\theta) \tag{25.166}$$

The solution of equations (25.158)–(25.159) can also be obtained by using the FFT method in combination of direct or Green's function method, where FFT is used to reduce the problem to 1D (i. e., 3D to 1D or 2D to 1D), and then solving the 1D problem directly. We demonstrate this approach below.

Green's function-based method

Two-dimensional problem

We consider first a simplified case in which we assume longitudinal symmetry (i. e., symmetry with respect to ϕ). The relevant problem is

$$\frac{1}{r^2}\frac{\partial}{\partial r}\left(r^2\frac{\partial u}{\partial r}\right)+\frac{1}{r^2\sin\theta}\frac{\partial}{\partial\theta}\left(\sin\theta\frac{\partial u}{\partial\theta}\right)=-f(r,\theta), \tag{25.167}$$

$$0<r<a,\quad 0<\theta<\pi$$

$$u(a,\theta)=0,\quad u(r,\theta)\quad \text{finite} \tag{25.168}$$

Looking at the operator w. r. t. θ, we have the following eigenvalue problem:

$$\frac{1}{\sin\theta}\frac{d}{d\theta}\left(\sin\theta\frac{dy}{d\theta}\right)=-\lambda y,\quad 0<\theta<\pi \tag{25.169}$$

$$y(\theta)\text{ is finite for } 0<\theta<\pi \tag{25.170}$$

which is the special case of θ-operator in general case (see equation (25.51)). Thus, the eigenvalues and eigenfunction can be obtained from equations (25.54)–(25.56) with $m=0$, i. e., eigenvalues λ:

$$\lambda_n=n(n+1),\quad n=0,1,2,\ldots \tag{25.171}$$

eigenfunction y_n:

$$y_n(\theta)=\sqrt{2n+1}P_n(\cos\theta), \tag{25.172}$$

which are complete and orthonormal w. r. t. to the inner product

$$\langle y_n,y_m\rangle=\frac{1}{2}\int_0^\pi y_n(\theta)y_m(\theta)\sin\theta\,d\theta=\delta_{mn}. \tag{25.173}$$

It can be shown that they form a basis for the Hilbert space $\mathcal{L}_2[0,\pi]$ with the inner product defined above in equation (25.173), i. e., for any function $f(\theta)$, we can express

$$f(\theta)=\sum_{n=0}^{\infty}f_n y_n(\theta);\quad f_n=\langle f,y_n\rangle=\frac{1}{2}\int_0^\pi f(\theta)y_n(\theta)\sin\theta\,d\theta \tag{25.174}$$

Now consider the solution of the Poisson equation (25.167). Taking the inner product with eigenfunctions, we get

$$\frac{1}{r^2}\frac{d}{dr}\left(r^2\frac{du_n}{dr}\right)-\frac{n(n+1)}{r^2}u_n=-f_n(r);\quad n=0,1,2,\ldots \tag{25.175}$$

$$u_n(a)=0,\quad u_n(r)\text{ is finite} \tag{25.176}$$

where

$$u_n = \langle u, y_n \rangle = \frac{1}{2} \int_0^{\pi} u(r, \theta) y_n(\theta) \sin\theta \, d\theta \tag{25.177}$$

$$f_n = \langle f, y_n \rangle = \frac{1}{2} \int_0^{\pi} f(r, \theta) y_n(\theta) \sin\theta \, d\theta \tag{25.178}$$

Equations (25.175)–(25.176) may be solved using the Green's function method. We note that

$$u_1 = \left(\frac{r}{a}\right)^n \quad \text{and} \quad u_2 = \left(\frac{r}{a}\right)^{-n-1} - \left(\frac{r}{a}\right)^n \tag{25.179}$$

are two linearly independent solutions of the homogeneous equation satisfying the BCs at $r = 0$ and $r = a$, respectively. Thus,

$$u_n(r) = \int_0^a G_n(r, s) f_n(s) s^2 \, ds \tag{25.180}$$

where

$$G_n(r, s) = \frac{1}{(2n+1)a} \begin{cases} (\frac{s}{a})^n[(\frac{r}{a})^{-n-1} - (\frac{r}{a})^n], & s \leq r \\ (\frac{r}{a})^n[(\frac{s}{a})^{-n-1} - (\frac{s}{a})^n], & s \geqslant r \end{cases} \tag{25.181}$$

Therefore, the solution of equations (25.175)–(25.176) is given by

$$\begin{aligned} u(r, \theta) &= \sum_{n=0}^{\infty} u_n y_n(\theta) \\ &= \sum_{n=0}^{\infty} \frac{2n+1}{2} P_n(\cos\theta) \int_0^a G_n(r, s) s^2 \left(\int_0^{\pi} f(s, \alpha) P_n(\cos\alpha) \sin\alpha d\alpha \right) ds \\ &= \int_0^a \int_0^{\pi} \left[\sum_{n=0}^{\infty} \frac{2n+1}{2} G_n(r, s) P_n(\cos\theta) P_n(\cos\alpha) \right] f(s, \alpha) s^2 \sin\alpha d\alpha \, ds \\ &= \int_0^a \int_0^{\pi} G(r, s, \theta, \alpha) f(s, \alpha) s^2 \sin\alpha d\alpha \, ds \end{aligned} \tag{25.182}$$

where G is the Green's function for the operator on the LHS of equations (25.175)–(25.176), and can be expressed as

$$G(r, s, \theta, \alpha) = \sum_{n=0}^{\infty} G_n(r, s) K_n(\theta, \alpha) \tag{25.183}$$

where G_n is given in equation (25.181) and K_n is given by

$$K_n = \frac{2n+1}{2} P_n(\cos\theta) P_n(\cos\alpha) \tag{25.184}$$

Three-dimensional problem

Similar to the 2D problem, the solution to the 3D problem can be obtained by considering the eigenvalue problem in (θ, ϕ) and converting the 3D model to 1D model using FFT, and then using the direct solution. For example, equations (25.49)–(25.50) and (25.54)–(25.56) give the eigenvalues and eigenfunctions:

$$y_{nm} = \sqrt{\frac{(n-m)!(2n+1)}{(n+m)!}} P_n^m(\cos\theta)\Phi_m(\phi) \tag{25.185}$$

$$\Phi_m(\phi) = \begin{cases} \Phi_0 = 1 \\ \Phi_m^c = \sqrt{2}\cos m\phi & n = m, m+1, \dots \\ \Phi_m^s = \sqrt{2}\sin m\phi, \end{cases} \tag{25.186}$$

with eigenvalues $\lambda_{nm} = n(n+1)$. Thus, taking the inner product of equations (25.158)–(25.159) with y_{nm}, we get

$$\frac{1}{r^2}\frac{d}{dr}\left(r^2 \frac{du_{nm}}{dr}\right) - \frac{n(n+1)}{r^2} u_{nm} = -f_{nm}(r); \tag{25.187}$$

$$u_{nm}(a) = 0; \quad u_{nm}(r) \text{ is finite}; \quad m = 0, 1, 2, \dots; \quad n = m, m+1, \dots \tag{25.188}$$

where

$$u_{nm} = \langle u, y_{nm} \rangle = \frac{1}{4\pi} \int_0^{\pi}\int_0^{2\pi} u(r,\theta,\phi) y_{nm}(\theta,\phi) \sin\theta \, d\theta \, d\phi \tag{25.189}$$

$$f_{nm} = \langle f, y_{nm} \rangle = \frac{1}{4\pi} \int_0^{\pi}\int_0^{2\pi} f(r,\theta,\phi) y_{nm}(\theta,\phi) \sin\theta \, d\theta \, d\phi. \tag{25.190}$$

Note that equations (25.187)–(25.188) are the same as equations (25.175)–(25.176), and hence u_{nm} is given by (equation (25.180))

$$u_{nm}(r) = \int_0^a G_n(r,s) f_{nm}(s) s^2 \, ds, \tag{25.191}$$

where G_n is the same as given in equations (25.180)–(25.181). Thus, the solution can be expressed as

$$
\begin{aligned}
u(r,\theta,\phi) &= \sum_{m=0}^{\infty}\sum_{n=m}^{\infty} u_{nm}(r)y_{nm}(\theta,\phi) \\
&= \sum_{m=0}^{\infty}\sum_{n=m}^{\infty} \frac{1}{4\pi} y_{nm}(\theta,\phi)\int_0^a G_n(r,s)s^2 \int_0^{\pi}\int_0^{2\pi} f(r,\theta',\phi')y_{nm}(\theta',\phi')\sin\theta'\,d\theta'\,d\phi'\,ds \\
&= \int_0^a\int_0^{\pi}\int_0^{2\pi} G(r,s,\theta,\theta',\phi,\phi')f(s,\theta',\phi')s^2\sin\theta'\,d\theta' d\phi'\,ds.
\end{aligned}
\tag{25.192}
$$

Here, the Green's function $G(r,s,\theta,\theta',\phi,\phi')$ can be expressed as

$$
G(r,s,\theta,\theta',\phi,\phi') = \sum_{n=0}^{\infty} G_n(r,s)K_{nm}(\theta,\theta',\phi,\phi') \tag{25.193}
$$

where $G_n(r,s)$ is given in equations (25.180)–(25.181) and K_{nm} is given by

$$
K_{nm}(\theta,\theta',\phi,\phi') = \frac{1}{4\pi} y_{nm}(\theta,\phi)y_{nm}(\theta',\phi') \tag{25.194}
$$

Similarly, other problems in spherical geometry with other types of boundary conditions can also be solved using this approach. For further example of problems in cylindrical and spherical geometries, we refer to the books by Carslaw and Jaeger [10] and Crank [16].

Problems

1. Consider the problem of cooling a circular cylinder of length 10 cms and diameter 2 cms made of copper. Its initial temperature is 500 °C and it is suddenly plunged into an agitated bath at 100 °C temperature. Assume that the agitation is high enough so that the surface of the cylinder is at 100 °C.
   (a) Derive a mathematical model for the temperature history of the rod and cast into dimensionless form.
   (b) Find the solution of the model. Determine an expression for the maximum (center) temperature of the rod as a function of time. Compute this value after 0.05, 0.1, 1.0, 5.0 and 10 seconds.
   (c) Repeat part (b) by assuming the cylinder to be of infinite length and compare your result.
2. Consider the solution of the Laplace's equation in the domain outside of a circle of radius unity ($r > 1, 0 < \theta \leq 2\pi$) with Dirichlet condition $u = f(\theta)$ on the boundary ($r = 1, 0 < \theta \leq 2\pi$). Obtain a solution to this problem using Finite Fourier transform. Show that the solution obtained may be simplified to the Poisson's integral formula:

$$
u(r,\theta) = \frac{1}{2\pi}\int_0^{2\pi} \frac{(r^2-1)f(\phi)}{1-2r\cos(\theta-\phi)+r^2}\,d\phi
$$

3. The dispersion of a tracer in laminar flow in a pipe is described by the convective-diffusion equation

$$\frac{\partial C}{\partial t} + 2\hat{u}\left[1 - \frac{r^2}{R^2}\right]\frac{\partial C}{\partial z} = D_m \nabla^2 C + f(r,t); \quad 0 < r < R, \quad 0 < z < L, \quad t > 0$$

with no flux boundary conditions at the pipe wall and $C = C_0(r,t)$ at $z = 0$. Cast into dimensionless form and solve using finite Fourier transform. Determine an expression for the convected mean concentration at any axial position and time.

4. Consider the problem of tracer dispersion described in problem 3 above. Assume that the pipe is of infinite length. Cast into dimensionless form and solve using finite Fourier transform.

5. Consider the problem of unsteady-state diffusion and reaction in a porous spherical catalyst

$$\frac{\varepsilon D}{r^2}\frac{\partial}{\partial r}\left(r^2 \frac{\partial C}{\partial r}\right) - kC = \varepsilon \frac{\partial C}{\partial t}; \quad 0 < r < R, \quad t > 0$$

$$C = C_0(t), \quad @r = R, \quad t > 0; \quad C = F(r), \quad @t = 0, \quad 0 < r < R$$

Cast into dimensionless form and solve using finite Fourier transform.

6. Solve the problem of vibration of a sphere:

$$\frac{\partial^2 u}{\partial t^2} = \nabla^2 u \quad \text{for } r < 1, \quad t > 0$$

$$u(r,\theta,\phi,0) = f(r,\theta,\phi); \quad \frac{\partial u}{\partial t}(r,\theta,\phi,0) = 0; \quad u(1,\theta,\phi,t) = 0$$

7. Creeping flow around a sphere placed in an infinite stream of fluid moving at a constant velocity of U is described by the equations (assuming azimuthal symmetry)

$$\frac{1}{r^2}\frac{\partial}{\partial r}(r^2 v_r) + \frac{1}{r \sin\theta}\frac{\partial}{\partial \theta}(v_\theta \sin\theta) = 0 \quad \text{(continuity)}$$

$$-\frac{\partial p}{\partial r} + \mu\left[\frac{1}{r^2}\frac{\partial}{\partial r}\left(r^2 \frac{\partial v_r}{\partial r}\right) + \frac{1}{r^2 \sin\theta}\frac{\partial}{\partial \theta}\left(\sin\theta \frac{\partial v_r}{\partial \theta}\right) - 2\frac{v_r}{r^2} - \frac{2}{r^2}\frac{\partial v_\theta}{\partial \theta} - \frac{2}{r^2} v_\theta \cot\theta\right] = 0$$

(r-momentum)

$$-\frac{1}{r}\frac{\partial p}{\partial \theta} + \mu[\frac{1}{r^2}\frac{\partial}{\partial r}\left(r^2 \frac{\partial v_\theta}{\partial r}\right) + \frac{1}{r^2 \sin\theta}\frac{\partial}{\partial \theta}\left(\sin\theta \frac{\partial v_\theta}{\partial \theta}\right) + \frac{2}{r^2}\frac{\partial v_r}{\partial \theta} - \frac{v_\theta}{r^2 \sin^2\theta} = 0$$

(θ-momentum)

with boundary conditions

$$@r = R, \quad v_r = v_\theta = 0$$

$$r \to \infty, \quad v_r = U\cos\theta, \quad v_\theta = -U\sin\theta, \quad p = p_0$$

Obtain a solution using separation of variables.

8. (a) Obtain a formal solution to the following system of linear PDEs describing diffusion and reaction:

$$\frac{\partial \mathbf{u}}{\partial t} = \mathbf{D}\nabla^2\mathbf{u} + \mathbf{A}\mathbf{u}; \quad \text{in } \Omega \quad (1)$$

$$\mathbf{u} = 0 \quad \text{on } \partial\Omega$$

where Ω is a (i) circular disk (ii) sphere. Here, **u** is the vector of concentrations, **D** is the matrix of diffusion coefficients and **A** is a constant $n \times n$ matrix, (b) Obtain a formal solution to (1) with Neumann boundary conditions

$$\nabla\mathbf{u}.\mathbf{n} = 0 \quad \text{on } \partial\Omega$$

where **n** is the outward normal to $\partial\Omega$. (c) Show schematic diagrams of contour plots of the first few eigenfunctions for case a(i).

9. A first-order chemical reaction carried out in a tubular reactor that discharges into a stirred reactor. The entire assembly is held at a constant temperature. Reaction occurs in the tube as well as the stirred vessel.
   (a) Assuming that the axial dispersion model applies in the tube, develop a mathematical model for the system.
   (b) Devise a suitable self-adjoint formalism and determine a formal solution to the model.
   (c) Determine the residence time distribution function (response of the system for a pulse input when there is no reaction) of the model developed in (*a*).
   (d) How does your result change if the axial dispersion model is replaced by a two-dimensional model that accounts for axial as well as radial dispersion of the reacting species?

10. (a) Consider the problem of heat loss from a very long heated pipe buried vertically in the earth. Suppose that the pipe outside radius is R and its surface temperature is raised and kept constant at T_s. Assume that the initial earth temperature is T_a.
   (i) Set up an appropriate mathematical model and determine an expression for the heat flux at the surface of the pipe as a function of time.
   (ii) Use the result in (i) to determine the effective heat transfer coefficient as a function of time.

   (b) A circular cylinder of radius R and infinite length is immersed in a fluid at rest everywhere, and is suddenly made to move with steady velocity U parallel to its length. Determine the frictional force per unit length of the cylinder at time t after the motion has begun. With appropriate change of notation, show that this expression is identical to that determined in (a) above.

   (c) Determine the asymptotic form of the solutions in (a) and (b) for small and large times.

Part VI: **Formulation and solution of some classical chemical engineering problems**

Introduction

The aim of this last part is to demonstrate the use of mathematical tools, developed in earlier parts, in the solution of some classical problems encountered by chemical engineers. It is hoped that these representative problems combined with practical knowledge gained from experience can help the student in the formulation and mathematical analysis of other such problems.

https://doi.org/10.1515/9783111598055-031

26 The classical Graetz–Nusselt problem

The classical Graetz–Nusselt problem deals with the determination of the heat or mass transfer from a fluid in a duct to the wall in steady laminar flow. Here, we formulate the model for heat transfer, and show how the tools of linear analysis may be used to determine the local heat transfer coefficient.

26.1 Model formulations and formal solution

Consider the problem of describing the steady-state temperature of a fluid in laminar flow in a duct of an arbitrary but constant cross-section as shown schematically in Figure 26.1. Assume that the hydraulic diameter of the duct (d_h) is much smaller compared to the length L (i. e., $L/d_h \gg 1$) so that axial conduction may be neglected.

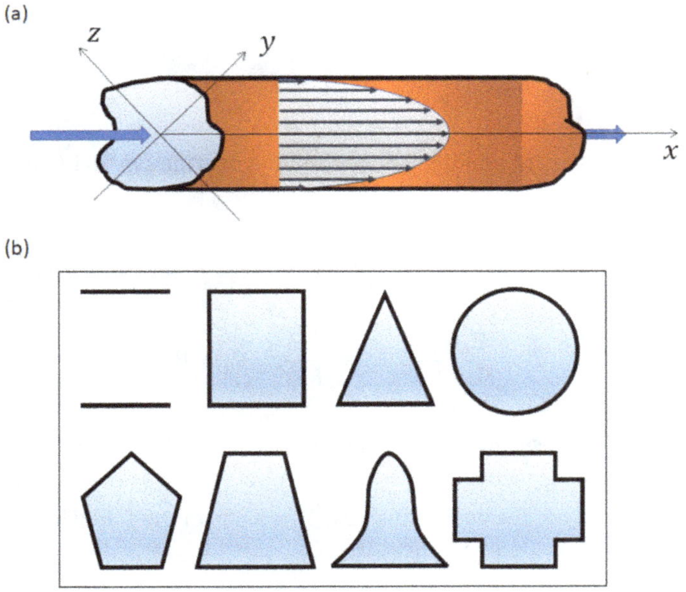

Figure 26.1: Schematics of (a) a laminar flow in a duct of an arbitrary cross-section, and (b) typical cross-sections used in various applications.

The mathematical model describing the fluid temperature in the duct with negligible axial conduction is given by

$$\rho_f C_{pf} u_f(y', z') \frac{\partial T}{\partial x'} = k_f \nabla_{\perp}'^2 T = k_f \left(\frac{\partial^2 T}{\partial y'^2} + \frac{\partial^2 T}{\partial z'^2} \right), \quad x' > 0, \quad (y', z') \in \Omega \tag{26.1}$$

https://doi.org/10.1515/9783111598055-032

with inlet and boundary conditions

$$T|_{x'=0} = T_{\text{in}}; \quad T|_{\partial\Omega} = T_w \quad \text{or} \quad (\mathbf{n}.k_f \nabla'_\perp T)|_{\partial\Omega_f} = q_w. \tag{26.2}$$

Here, T, ρ_f, C_{pf} and k_f are the temperature, density, specific heat capacity and thermal conductivity of the fluid; x' is the coordinate along flow direction, (y', z') are transverse coordinates; Ω_f and $\partial\Omega_f$ are transverse domain and its boundary, respectively; T_{in} and T_w are fluid temperatures at the inlet and at the transverse boundary (wall); q_w is the heat flux at the transverse boundary entering the domain Ω_f; $u_f(y', z')$ is the velocity profile; and $\mathbf{n}$ is unit outward normal to $\partial\Omega_f$. In the general case, there are two types of boundary conditions: (i) constant wall temperature and (ii) constant heat flux at the wall. Here, we discuss the first case and leave the second case for exercises.

26.1.1 Analysis of constant wall temperature boundary condition

We consider the boundary condition given by

$$T = T_w \, @ \, \partial\Omega_f \tag{26.3}$$

and define the following quantities to nondimensionalize the governing model (equations (26.1) and (26.3)):

$$R_{\Omega f} = \frac{A_{\Omega f}}{P_{\Omega f}}; \quad (y, z) = \frac{(y', z')}{R_{\Omega f}}; \quad u(y, z) = \frac{u_f(y', z')}{\overline{u_f}};$$
$$\alpha_f = \frac{k_f}{\rho_f C_{pf}}; \quad x = \frac{\alpha_f x'}{R^2_{\Omega f}\overline{u_f}}; \quad \theta = \frac{T_w - T}{T_w - T_{\text{in}}}; \quad \Omega = \frac{\Omega'_f}{R_{\Omega f}}; \quad \nabla_\perp = R_{\Omega f}\nabla'_\perp \tag{26.4}$$

Here, $A_{\Omega f}, P_{\Omega f}$ and $R_{\Omega f}$ are the cross-section area, wetted perimeter and hydraulic radius [hydraulic diameter $d_h = 4R_{\Omega f}$], respectively; $\overline{u_f} = \frac{1}{A_{\Omega f}} \int_{\Omega_f} u(y', z')\, dy'\, dz'$ is the average velocity in the fluid phase; α_f is the thermal diffusivity. This leads to the following dimensionless linear model:

$$L\theta = \frac{1}{u(y, z)}\left(\frac{\partial^2\theta}{\partial y^2} + \frac{\partial^2\theta}{\partial z^2}\right) = \frac{\partial\theta}{\partial x}, \quad x > 0, \quad (y, z) \in \Omega \tag{26.5}$$

with inlet and boundary conditions as

$$\theta|_{x=0} = 1; \quad \theta|_{\partial\Omega} = 0. \tag{26.6}$$

The same model is obtained for mass transfer from the duct interior to the wall with an infinitely fast wall reaction (so that the species concentration at the wall is zero).

The dimensionless model given by equations (26.5) and (26.6) can be solved by considering the eigenvalue problem (EVP) defined by

$$L\psi = \frac{1}{u(y,z)}\nabla_\perp^2\psi = \frac{1}{u(y,z)}\left(\frac{\partial^2\psi}{\partial y^2} + \frac{\partial^2\psi}{\partial z^2}\right) = -\lambda\psi \text{ in } \Omega; \quad \psi|_{\partial\Omega} = 0. \tag{26.7}$$

Note that the operator appearing in equation (26.7) is a self-adjoint operator (i. e., $L^* = L$) w. r. t. the inner product defined by

$$\langle\psi_i, \psi_j\rangle = \frac{1}{A_\Omega}\int_\Omega u(y,z)\psi_i\psi_j\,dy\,dz = \delta_{ij}; \quad A_\Omega = \int_\Omega dy\,dz \tag{26.8}$$

where

$$\delta_{ij} = \left\{\begin{array}{ll} 1, & i = j \\ 0, & i \neq j \end{array}\right\}$$

is the Kronecker delta. Hence, equation (26.7) defines a Sturm–Liouville EVP with eigenvalues λ_i and eigenfunctions ψ_i.

Thus, taking the inner product with ψ_i, equation (26.5) leads to

$$\frac{d}{dx}\langle\theta, \psi_i\rangle = \langle L\theta, \psi_i\rangle = \langle\theta, L^*\psi_i\rangle = \langle\theta, L\psi_i\rangle = -\lambda_i\langle\theta, \psi_i\rangle; \quad \langle\theta, \psi_i\rangle|_{x=0} = \langle 1, \psi_i\rangle$$

$\Longrightarrow$

$$\begin{aligned} \langle\theta, \psi_i\rangle &= \langle 1, \psi_i\rangle \exp(-\lambda_i x) \Longrightarrow \\ \theta(x,y,z) &= \sum_i \langle\theta, \psi_i\rangle\psi_i = \sum_i \exp(-\lambda_i x)\langle 1, \psi_i\rangle\psi_i(y,z) \end{aligned} \tag{26.9}$$

Heat transfer coefficient and Nusselt number

The heat transfer coefficient from wall to the fluid is defined by

$$h = \frac{q_w}{T_w - T_m} = \frac{1}{(T_w - T_m)P_{\Omega f}}\int_{\partial\Omega_f} (\mathbf{n}.k_f\nabla'_\perp T)\,dP_{\Omega f}, \tag{26.10}$$

which leads to the Nusselt number, Nu (or the dimensionless heat transfer coefficient) as

$$\mathrm{Nu}_\Omega = \frac{hR_{\Omega f}}{k_f} = \frac{-1}{\theta_m P_\Omega}\int_{\partial\Omega} (\mathbf{n}.\nabla_\perp\theta)\,dP_\Omega. \tag{26.11}$$

Here, T_m or θ_m is the cup-mixing (velocity weighted) temperature defined as

$$\theta_m(x) = \langle 1, \theta \rangle = \frac{1}{A_\Omega} \int_\Omega u(y,z)\theta(x,y,z)\,dy\,dz = \sum_i \exp(-\lambda_i x)\langle 1, \psi_i \rangle^2. \tag{26.12}$$

Thus, the Nusselt number can be expressed in terms of eigenvalues and eigenfunctions from equations (26.11)–(26.12) as follows:

$$\mathrm{Nu}_\Omega = \frac{\frac{1}{P_\Omega} \sum_i \exp(-\lambda_i x)\langle 1, \psi_i \rangle \int_{\partial\Omega} (-\mathbf{n}.\nabla_\perp \psi_i)\,dP_\Omega}{\sum_i \exp(-\lambda_i x)\langle 1, \psi_i \rangle^2}. \tag{26.13}$$

The above expression can be simplified further using the EVP (equation (26.7)) that suggests

$$\begin{aligned} -\lambda\langle 1, \psi \rangle = \langle 1, L\psi \rangle &= \frac{1}{A_\Omega} \int_\Omega \left(\frac{\partial^2 \psi}{\partial y^2} + \frac{\partial^2 \psi}{\partial z^2} \right) dA_\Omega \\ &= \frac{1}{A_\Omega} \int_{\partial\Omega} (\mathbf{n}.\nabla_\perp \psi)\,dP_\Omega = \frac{1}{P_\Omega} \int_{\partial\Omega} (\mathbf{n}.\nabla_\perp \psi)\,dP_\Omega \end{aligned} \tag{26.14}$$

Note that due to nondimensionlization of the spatial coordinate with hydraulic radius, the dimensionless domain Ω has the property: $A_\Omega = P_\Omega$. Thus, the general expression of the Nusselt number can be simplified from equations (26.13) and (26.14) as

$$\mathrm{Nu}_\Omega(x) = \frac{\sum_i \lambda_i \exp(-\lambda_i x)\langle 1, \psi_i \rangle^2}{\sum_i \exp(-\lambda_i x)\langle 1, \psi_i \rangle^2} = \frac{\sum_i \lambda_i \exp(-\lambda_i x)\beta_i}{\sum_i \exp(-\lambda_i x)\beta_i}, \tag{26.15}$$

where $\beta_i = \langle 1, \psi_i \rangle^2$ is the Fourier weight. Equation (26.15) shows that the local Nusselt number can be expressed in terms of the eigenvalues and Fourier weights.

[Remark: In the literature, the Nusselt number Nu is defined using hydraulic diameter d_h as the length scale. It is clear that $\mathrm{Nu} = 4\,\mathrm{Nu}_\Omega$.]

Long distance asymptote

In order to examine the limit of $\mathrm{Nu}_\Omega(x)$ for large x, we note that the linear operator L is self-adjoint operator implying that all eigenvalues λ_i are real. In addition, taking the inner product with ψ, the EVP defined in equation (26.7) leads to

$$\begin{aligned} -\lambda\langle \psi, \psi \rangle = \langle L\psi, \psi \rangle &= \frac{1}{A_\Omega} \int_\Omega \psi \nabla_\perp^2 \psi\,dy\,dz \\ &= \frac{1}{A_\Omega} \int_{\partial\Omega} \psi(\mathbf{n}.\nabla_\perp \psi)\,dP_\Omega - \frac{1}{A_\Omega} \int_\Omega (\nabla_\perp \psi.\nabla_\perp \psi)\,dy\,dz \quad \text{(Green's identity)} \\ &= 0 - \frac{1}{A_\Omega} \int_\Omega (\nabla_\perp \psi.\nabla_\perp \psi)\,dy\,dz \end{aligned}$$

$\Longrightarrow$

$$\lambda_i = \frac{1}{A_\Omega} \frac{\int_\Omega (\nabla_\perp \psi_i . \nabla_\perp \psi_i)\, dy\, dz}{\int_\Omega u(y,z) \psi_i^2\, dy\, dz} > 0 \quad \forall i \text{ (since } u \text{ does not change sign)}$$

Further, from expanding unity in terms of eigenfunctions ψ_i's, we get the Parseval's relation:

$$1 = \sum_i \langle 1, \psi_i \rangle \psi_i \Longrightarrow \sum_i \langle 1, \psi_i \rangle^2 = \sum_i \beta_i = \langle 1, 1 \rangle = 1$$

where the energy content β_i of the ith mode is less than unity. Thus, the long distance asymptote ($x \gg 1$) can be obtained from equation (26.15) as follows:

$$\mathrm{Nu}_{\Omega\infty} = \lim_{x \gg 1} \mathrm{Nu}_\Omega(x) = \lambda_1. \tag{26.16}$$

Thus, only the first eigenvalue determines the dimensionless heat transfer coefficient (Nu_Ω) at long distance from inlet.

Short distance asymptote (Leveque solution)

For short distance asymptote, the main gradient exists very close to the wall (i. e., the domain of interest is the boundary layer region near $\partial\Omega_f$). Assuming ξ is the dimensionless distance normal to the wall, the velocity profile near the wall can be approximated by

$$u(\xi) = u_0 \xi \tag{26.17}$$

and the governing model (equations (26.5) and (26.6)) can be simplified as follows:

$$\frac{1}{u_0 \xi} \frac{\partial^2 \theta}{\partial \xi^2} = \frac{\partial \theta}{\partial x}; \quad \theta|_{x=0} = 1; \quad \theta|_{\xi=0} = 0; \quad \theta|_{\xi \to \infty} \to 1 \tag{26.18}$$

[The velocity u_0 depends on the specific geometry and the definition of ξ as illustrated in the next two sections]. Thus, by defining a new variable (suggested by scaling considerations) as

$$z = \frac{\gamma \xi}{x^{1/3}}, \tag{26.19}$$

we get

$$\frac{\partial \theta}{\partial x} = \frac{\partial \theta}{\partial z} \frac{\partial z}{\partial x} = \frac{-\gamma \xi}{3x^{4/3}} \frac{\partial \theta}{\partial z}$$

and

$$\frac{\partial \theta}{\partial \xi} = \frac{\partial \theta}{\partial z} \frac{\partial z}{\partial \xi} = \frac{\gamma}{x^{1/3}} \frac{\partial \theta}{\partial z} \quad \text{or} \quad \frac{\partial^2 \theta}{\partial \xi^2} = \frac{\gamma^2}{x^{2/3}} \frac{\partial^2 \theta}{\partial z^2},$$

which expresses the governing model (equation (26.18)) as

$$\frac{1}{u_0\xi}\frac{\gamma^2}{x^{2/3}}\frac{\partial^2\theta}{\partial z^2} = \frac{-\gamma\xi}{3x^{4/3}}\frac{\partial\theta}{\partial z}$$

$\Longrightarrow$

$$\frac{\partial^2\theta}{\partial z^2} = -\frac{u_0}{3\gamma^3}z^2\frac{\partial\theta}{\partial z} = -3z^2\frac{\partial\theta}{\partial z} \quad \text{if } \gamma = \sqrt[3]{\frac{u_0}{9}} \tag{26.20}$$

with boundary conditions:

$$\theta = 1 \,@\, z \to \infty; \quad \text{and} \quad \theta = 0 \,@\, z = 0 \tag{26.21}$$

Integrating the above equation twice $\Longrightarrow$

$$\theta(x,\xi) = \theta(z) = \frac{1}{\Gamma(\frac{4}{3})}\int_0^{z=\frac{\gamma\xi}{x^{1/3}}} \exp(-t^3)\,dt \tag{26.22}$$

where Γ is the Gamma function.

Thus, the Nusselt number can be obtained from equations (26.10)–(26.11) as

$$\mathrm{Nu}_{\Omega 0} = \lim_{x\ll 1}\mathrm{Nu}_\Omega(x) = \lim_{x\ll 1}\frac{1}{\theta_m}\frac{\partial\theta}{\partial\xi}\bigg|_{\xi=0} = \frac{\gamma}{x^{1/3}}\frac{\partial\theta}{\partial z}\bigg|_{z=0} = \frac{\gamma}{\Gamma(\frac{4}{3})x^{1/3}} \tag{26.23}$$

$\Longrightarrow$

$$\mathrm{Nu}_{\Omega 0} = \lim_{x\ll 1}\mathrm{Nu}_\Omega(x) = \frac{\gamma}{\Gamma(\frac{4}{3})x^{1/3}} = \frac{1}{\Gamma(\frac{4}{3})}\left(\frac{u_0}{9x}\right)^{1/3}. \tag{26.24}$$

The long and short distance asymptotes can also be combined to obtain a simpler approximate expression for the Nusselt number $Nu_\Omega(x)$ as discussed in Gundlapally and Balakotaiah [19].

26.2 Parallel plate with fully-developed velocity profile

Consider the flow between parallel plates (Figure 26.1b) where hydraulic radius is the half-spacing between the plates. In this case, the fully-developed dimensionless velocity profile in laminar flow is given by

$$u(y) = \frac{3}{2}(1 - y^2) \tag{26.25}$$

and the governing model can be expressed in dimensionless form as follows:

$$\frac{3}{2}(1-y^2)\frac{\partial \theta}{\partial x} = \frac{\partial^2 \theta}{\partial y^2}, \quad x > 0 \text{ and } 0 < y < 1 \tag{26.26}$$

with inlet and boundary conditions as

$$\theta(x=0,y) = 1 \quad \text{and} \quad \theta(x,y=1) = 0, \quad \frac{\partial \theta}{\partial y}(x,y=0) = 0 \tag{26.27}$$

[In writing the above model, the symmetry of the solution in the y-coordinate around y=0 is utilized. Further, the velocity profile is scaled so that the cross-section average velocity is unity.] The EVP can be defined for parallel plate geometry from equation (26.7) as follows:

$$L\psi = \frac{2}{3(1-y^2)}\frac{\partial^2 \psi}{\partial y^2} = -\lambda\psi, \quad 0 < y < 1; \tag{26.28}$$

with boundary conditions

$$\psi|_{y=1} = 0; \quad \left.\frac{\partial \psi}{\partial y}\right|_{y=0} = 0, \tag{26.29}$$

which is self-adjoint w. r. t. the inner product:

$$\langle \psi_i, \psi_j \rangle = \int_0^1 \frac{3}{2}(1-y^2)\psi_i\psi_j \, dy = \delta_{ij} \tag{26.30}$$

The first few eigenvalues and Fourier coefficients $\beta_i = \langle 1, \psi_i \rangle^2$ are listed in Table 26.1. The corresponding Graetz eigenfunctions are shown in Figure 21.1.

Table 26.1: First few eigenvalues and Fourier coefficients of the Graetz problem for flow between parallel plates and constant wall temperature.

i	λ_i	$\beta_i = \langle 1, \psi_i \rangle^2$
1	1.8852	0.91035
2	21.431	0.05314
3	62.317	0.01528
4	124.54	0.00681
5	208.09	0.00374
6	312.98	0.00232

Using these values and equation (26.15), the Nusselt number can be plotted against the dimensionless axial coordinate x, which is shown in Figure 26.2.

It can be seen from this figure that it exhibits the two asymptotes in the limits of $x \gg 1$ and $x \ll 1$. The long distance asymptote ($x \gg 1$) can be obtained from equation (26.16), which simplifies to

$$\mathrm{Nu}_{\Omega\infty} = \lim_{x \gg 1} \mathrm{Nu}_\Omega(x) = \lambda_1 = 1.8852. \tag{26.31}$$

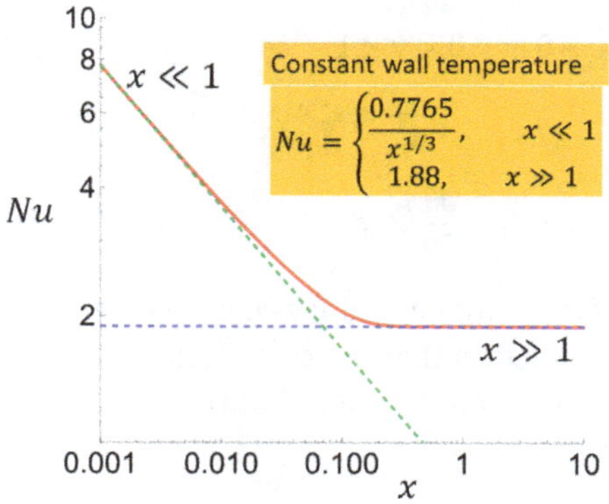

Figure 26.2: Local or position dependent Nusselt number (dimensionless heat transfer coefficient) for fully developed laminar flow through parallel plates for the case of constant wall temperature.

Similarly, the short distance asymptote can be obtained from equation (26.15), where velocity profile is simplified as

$$\lim_{y\to 1} u(y) = \lim_{y\longrightarrow 1} \frac{3}{2}(1-y^2) \to 3(1-y) = 3\xi \Longrightarrow u_0 = 3,$$

and hence the short distance asymptote of Nusselt number can be given from equation (26.24) as

$$\mathrm{Nu}_{\Omega 0} = \lim_{x\ll 1} \mathrm{Nu}_{\Omega}(x) = \frac{1}{\Gamma(\frac{4}{3})}\sqrt[3]{\frac{1}{3}}x^{-1/3} = 0.7765x^{-1/3}. \tag{26.32}$$

26.3 Circular channel with fully-developed velocity profile

Consider the fully developed laminar flow in a circular duct (Figure 26.1b). Here, hydraulic radius $R_{\Omega f}$ is half of the duct radius a, i. e., $R_{\Omega f} = \frac{a}{2}$. While the transverse coordinate can be nondimenionalized by hydraulic radius leading to $y \in [0, 2]$, it could also be nondimensionalized by the duct radius, which will lead $y \in [0, 1]$. In the general case, we have shown the use of hydraulic radius in earlier sections and applied it to parallel plate geometry. Now, we show the use of duct radius to nondimensionalize for the circular geometry, which leads to the fully-developed velocity profile in laminar flow as

$$u(y) = 2(1-y^2) \tag{26.33}$$

and the governing model can be expressed in dimensionless form as

$$2(1-y^2)\frac{\partial\theta}{\partial x} = \frac{1}{y}\frac{\partial}{\partial y}\left(y\frac{\partial\theta}{\partial y}\right), \quad x > 0 \text{ and } 0 < y < 1 \tag{26.34}$$

with inlet and boundary conditions as

$$\theta(x = 0, y) = 1 \quad \text{and} \quad \theta(x, y = 1) = 0, \quad \left.\frac{\partial \theta}{\partial y}\right|_{y=0} = 0. \tag{26.35}$$

The EVP can be defined as follows:

$$L\psi = \frac{1}{2(1-y^2)}\frac{1}{y}\frac{\partial}{\partial y}\left(y\frac{\partial \psi}{\partial y}\right) = -\lambda\psi, \quad 0 < y < 1; \tag{26.36}$$

with boundary conditions

$$\psi|_{y=1} = 0; \quad \left.\frac{\partial \psi}{\partial y}\right|_{y=0} = 0, \tag{26.37}$$

which is self-adjoint w. r. t. the inner product:

$$\langle \psi_i, \psi_j \rangle = \int_0^1 2yu(y)\psi_i\psi_j \, dy = \int_0^1 4y(1-y^2)\psi_i\psi_j \, dy = \delta_{ij}. \tag{26.38}$$

The first six eigenvalues and Fourier coefficients $\beta_i = \langle 1, \psi_i \rangle^2$ are listed in Table 26.2. The corresponding Graetz eigenfunctions are shown in Figure 26.3.

Table 26.2: First few eigenvalues and Fourier coefficients of the Graetz problem for flow through a circular duct and constant wall temperature.

i	λ_i	$\beta_i = \langle 1, \psi_i \rangle^2$
1	3.65679	0.819050
2	22.3047	0.097527
3	56.9605	0.032504
4	107.620	0.015440
5	174.282	0.008788
6	256.945	0.005584

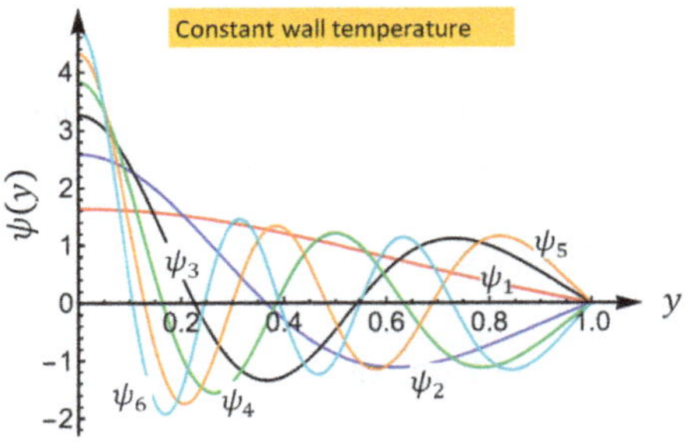

Figure 26.3: First six eigenfunctions of the Graetz–Nusselt problem for fully developed laminar flow through a circular duct with constant wall temperature.

Note that in this case, since duct radius is used to nondimensionlize the transverse coordinate, $A_\Omega \neq P_\Omega$ ($A_\Omega = 1$ and $P_\Omega = 2$), and hence we rewrite equation (26.14) as

$$-\lambda\langle 1, \psi\rangle = \langle 1, L\psi\rangle = \frac{1}{A_\Omega}\int_{\partial\Omega} (\mathbf{n}.\nabla_\perp \psi)\, dP_\Omega = \frac{2}{P_\Omega}\int_{\partial\Omega} (\mathbf{n}.\nabla_\perp \psi)\, dP_\Omega$$

$\Longrightarrow$

$$\frac{1}{P_\Omega}\int_{\partial\Omega} (\mathbf{n}.\nabla_\perp \psi_j)\, dP_\Omega = \frac{-\lambda_j}{2}\langle 1, \psi_j\rangle.$$

This simplifies the Nusselt number based on hydraulic diameter from equation (26.14) as

$$\mathrm{Nu_d} = \frac{2ha}{k_f} = 4\,\mathrm{Nu}_\Omega = \frac{\sum_i \lambda_i \exp(-\lambda_i x)\beta_i}{\sum_i \exp(-\lambda_i x)\beta_i}; \quad \beta_i = \langle 1, \psi_i\rangle^2, \tag{26.39}$$

which is plotted against the dimensionless axial coordinate x in Figure 26.4.

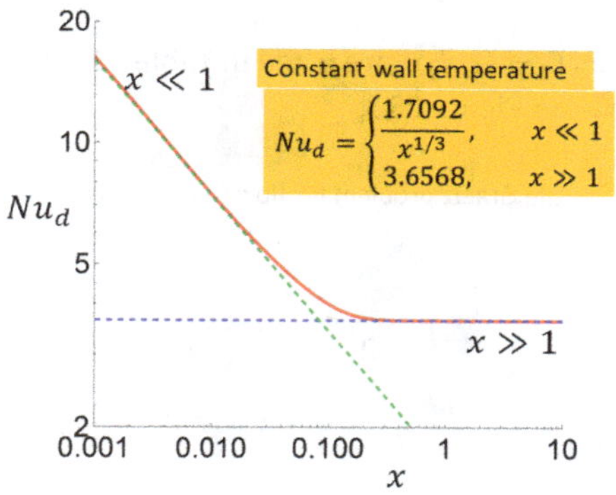

Figure 26.4: Local or position dependent Nusselt number (dimensionless heat transfer coefficient) based on hydraulic diameter for fully developed laminar flow through a circular duct for the case of constant wall temperature.

As expected, it exhibits the two asymptotes in the limits of $x \gg 1$ and $x \ll 1$. The long distance asymptote ($x \gg 1$) can be obtained from equation (26.39), which simplifies to

$$\mathrm{Nu}_{d\infty} = \lim_{x\gg 1} \mathrm{Nu}(x) = \lambda_1 = 3.6568. \tag{26.40}$$

Similarly, the short distance asymptote can be obtained from equation (26.15), where the velocity profile is simplified as

$$\lim_{y\to 1} u(y) = \lim_{y\longrightarrow 1} 2(1-y^2) \to 4(1-y) = 4\xi \Longrightarrow u_0 = 4,$$

and hence the short distance asymptote of the Nusselt number can be given from equation (26.24) as

$$\mathrm{Nu}_{d0} = \frac{2}{\Gamma(\frac{4}{3})}\sqrt[3]{\frac{4}{9}}x^{-1/3} = 1.7092x^{-1/3}. \tag{26.41}$$

Other duct geometries as well as flux and mixed boundary conditions at the wall can be analyzed in a similar way.

Problems

1. Consider the Graetz–Nusselt problem in a triangular duct with constant wall temperature boundary condition:
   (a) Identify the EVP and obtain the formal solution.
   (b) Determine the long and short distance asymptotes for the Nusselt number.
2. Consider the Graetz–Nusselt problem in a circular tube with constant wall flux boundary condition. Formulate the model equations and identify the EVP. Determine an expression for the Nusselt number and identify the short and long distance asymptotes.
3. Repeat problem 2 for parallel plate geometry.
4. The concentration of a reactant in a tubular catalytic reactor in which the flow is laminar and fully developed is given by (assuming centerline symmetry and neglecting axial diffusion)

$$2\langle u\rangle\left(1-\frac{r^2}{a^2}\right)\frac{\partial C}{\partial x} = D_m\frac{1}{r}\frac{\partial}{\partial r}\left(r\frac{\partial C}{\partial r}\right); \quad x>0,\ 0<r<a$$

   with boundary and initial conditions

$$\frac{\partial C}{\partial r} = 0 \text{ at } r=0; \quad -D_m\frac{\partial C}{\partial r} = k_s C \text{ at } r=a$$

$$C(r, x=0) = C_0$$

   (a) Cast the model equations in dimensionless form and solve using finite Fourier transform.
   (b) Use the solution in (*a*) to determine the local Sherwood number (or dimensionless mass transfer coefficient) defined by

$$Sh = \left(\frac{2a}{D_m}\right)k_c = \left(\frac{2a}{D_m}\right)\frac{[-D_m\frac{\partial C}{\partial r}(r=a)]}{[C_m - C(r=a)]},$$

   where C_m is the cupmixing (velocity weighted) concentration.
   (c) How does the result in (b) simplify for long ($L/a \gg 1$) channels?
   (d) Simplify the result in (b) for the two limiting cases of infinitely fast ($k_s \to \infty$) and slow reaction ($k_s \to 0$) and comment on your results.

5. The Graetz–Nusselt formulation assumes that the duct length is much larger than the hydraulic diameter and neglects the axial diffusion term (or takes axial Peclet number to be infinity). At the other extreme of the large hydraulic diameter compared to length for heat/mass transfer or reaction, for which the axial Peclet number goes to zero, the appropriate model is the so-called "short tube model." For the case of fixed temperature boundary condition, this model may be expressed as

$$k_f \nabla_{\perp}^2 T = \rho_f C_{pf} u_f(y', z') \left(\frac{T - T_{\text{in}}}{L} \right), \quad (y', z') \in \Omega$$

$$T|_{\partial\Omega} = T_w$$

   (a) Cast this model in dimensionless form.
   (b) Obtain the solution for the temperature profile and use it to determine the Nusselt number.
   (c) Identify the asymptotic behavior of Nu for small and large values of $P_h = \frac{R_\Omega^2 \langle u_f \rangle}{L a_f}$.
6. (a) Discuss the similarities and differences in the solution of the Graetz–Nusselt problem with fixed wall temperature for the different duct shapes shown in Figure 26.1.
   (b) Discuss how the short and long distance asymptotes may be combined to obtain an approximate expression for Nu for any arbitrary geometry.

27 Friction factors for steady-state laminar flow in ducts

27.1 Model formulations and formal solution

Consider steady-state laminar flow in a duct of a constant but arbitrarily shaped cross-section as shown schematically in Figure 27.1.

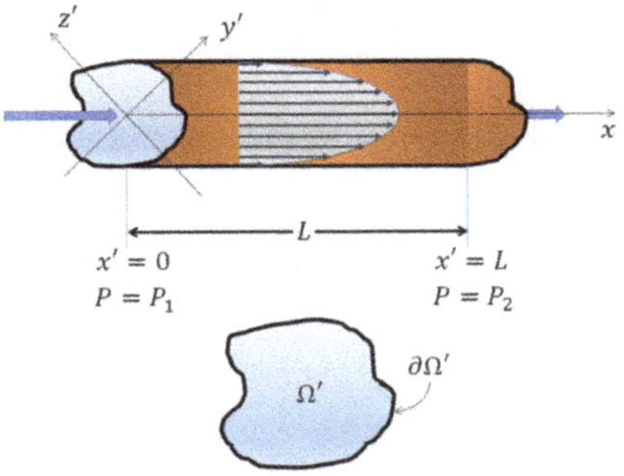

Figure 27.1: Schematic of laminar flow in a duct of constant but arbitrary cross-section.

For such unidirectional flow in a duct, the x-momentum balance reduces to

$$\mu\nabla'^2 u_x = \left(\frac{\Delta P}{L}\right) \text{ in } \Omega'; \quad u_x = 0 \text{ on } \partial\Omega' \quad \text{(no slip boundary condition)} \tag{27.1}$$

Here, μ is the fluid viscosity; u_x is the axial component of the velocity; $-\Delta P = P_1 - P_2 > 0$ is the pressure drop; P_1 and P_2 are pressures at the entrance ($x' = 0$) and the exit ($x' = L$), respectively; L is the length of the duct; Ω' and $\partial\Omega'$ are the transverse (cross-sectional) domain and its boundary.

Defining the effective transverse length (i. e., hydraulic radius) R_Ω and dimensionless transverse coordinates as

$$R_\Omega = \frac{\text{flow area}}{\text{wetted perimeter}} = \frac{A_{\Omega'}}{P_{\Omega'}}; \quad (y, z) = \frac{(y', z')}{R_\Omega}, \tag{27.2}$$

the characteristic velocity u^* and dimensionless velocity U as

$$u^* = \frac{R_\Omega^2}{\mu}\left(\frac{-\Delta P}{L}\right); \quad U(y, z) = \frac{u_x(y', z')}{u^*}, \tag{27.3}$$

the momentum balance (equation (27.1)) and the no-slip wall boundary condition can be expressed in dimensionless form as follows:

$$\nabla^2 U = -1 \text{ in } \Omega; \quad U = 0 \text{ on } \partial\Omega \tag{27.4}$$

https://doi.org/10.1515/9783111598055-033

Note that the hydraulic radius of scaled (dimensionless) domain Ω is unity.

An overall force balance on the duct gives

$$\tau_w P_{\Omega'} L = (-\Delta P) A_{\Omega'} \Longrightarrow \tau_w = \frac{(-\Delta P)}{L} R_{\Omega}. \tag{27.5}$$

Here, τ_w is the average wall shear stress. The friction factor f (or the dimensionless pressure drop) is defined by

$$f = \frac{\tau_w}{\frac{1}{2}\rho \langle u_x \rangle^2}, \tag{27.6}$$

where ρ is the fluid density and $\langle u_x \rangle$ is the average velocity. Defining the Reynolds number Re as

$$\mathrm{Re} = \frac{(4R_{\Omega})\langle u_x \rangle \rho}{\mu}, \tag{27.7}$$

the friction factor can be expressed as

$$f\,\mathrm{Re} = \frac{\frac{(-\Delta P)}{L} R_{\Omega}}{\frac{1}{2}\rho \langle u_x \rangle^2} \cdot \frac{(4R_{\Omega})\langle u_x \rangle \rho}{\mu} = \frac{8u^*}{\langle u_x \rangle} = \frac{8}{\langle U \rangle}. \tag{27.8}$$

Thus, to obtain the relationship between pressure drop (or f) and flow rate (or $\langle U \rangle$), we need to solve the Poisson equation (27.4) to determine $U(y, z)$ and then $\langle U \rangle$ can be determined by

$$\langle U \rangle = \frac{1}{A_{\Omega}} \int_{\Omega} U(y, z)\, dy\, dz. \tag{27.9}$$

Equation (27.8) gives the required relationship. Equation (27.4) can be solved either by direct method (e. g., 1D boundary value problem for symmetric geometries) or by finite Fourier transform.

FFT method

Let λ_i be the eigenvalue and ψ_i be the eigenfunction of the EVP:

$$L\psi = \nabla^2 \psi = -\lambda \psi \text{ in } \Omega; \quad \psi = 0 \text{ on } \partial\Omega \tag{27.10}$$

which is a self-adjoint operator (i. e., $L = L^*$) w. r. t. inner product defined by

$$\langle \psi_i, \psi_j \rangle = \frac{1}{A_{\Omega}} \int_{\Omega} \psi_i \psi_j\, dy\, dz = \delta_{ij} \tag{27.11}$$

Thus, taking the inner product with ψ_i, equation (27.4) leads to

$$-\langle 1, \psi_i \rangle = \langle \nabla^2 U, \psi_i \rangle = \langle U, \nabla^2 \psi_i \rangle = -\lambda_i \langle U, \psi_i \rangle$$

$\Longrightarrow$

$$\langle U, \psi_i \rangle = \frac{1}{\lambda_i} \langle 1, \psi_i \rangle, \tag{27.12}$$

which leads to the formal solution for velocity profile as

$$U(y, z) = \sum_i \langle U, \psi_i \rangle \psi_i(y, z) = \sum_i \frac{1}{\lambda_i} \langle 1, \psi_i \rangle \psi_i(y, z). \tag{27.13}$$

Thus, the average velocity can be expressed as

$$\langle U \rangle = \langle 1, U \rangle = \sum_i \frac{1}{\lambda_i} \langle 1, \psi_i \rangle^2 = \sum_i \frac{\beta_i}{\lambda_i}; \quad \beta_i = \langle 1, \psi_i \rangle^2 \tag{27.14}$$

where β_i are the Fourier weights that satisfy the Parseval's relation: $\sum_i \beta_i = 1$. Thus, the friction factor can be expressed in terms of eigenvalues of the Laplacian operator (with Dirichlet boundary condition) and Fourier weights as follows:

$$f\,\mathrm{Re} = \frac{8}{\langle U \rangle} = \frac{8}{\sum_i \frac{1}{\lambda_i} \langle 1, \psi_i \rangle^2} = \frac{8}{\sum_i \frac{\beta_i}{\lambda_i}}. \tag{27.15}$$

27.2 Specific example: parallel plates

Consider the parallel plate geometry with half-spacing of a. The hydraulic radius for this case is $R_\Omega = a$. The dimensionless model for velocity is given by

$$\frac{d^2 U}{dy^2} = -1; \quad U(y = \pm 1) = 0 \tag{27.16}$$

This model can either be solved directly or by FFT.

27.2.1 Direct solution

The direct solution of equation (27.16) can be obtained by integrating it twice, which with Dirichlet boundary condition leads to the velocity profile

$$U(y) = \frac{1}{2}(1 - y^2) \tag{27.17}$$

$\Longrightarrow$

$$\langle U \rangle = \int_0^1 U(y)\, dy = \int_0^1 \frac{1}{2}(1 - y^2)\, dy = \frac{1}{3} \tag{27.18}$$

$\Longrightarrow$

$$f\,\mathrm{Re} = \frac{8}{\langle U\rangle} = 24 \tag{27.19}$$

27.2.2 FFT approach

The EVP for parallel plates can be expressed as

$$L\psi = \frac{d^2\psi}{dy^2} = -\lambda\psi, \quad \psi[y = \pm 1] = 0, \tag{27.20}$$

which is self-adjoint w. r. t. the inner product defined by

$$\langle \psi_i, \psi_j\rangle = \frac{1}{2}\int_{-1}^{1} \psi_i\psi_j\, dy = \delta_{ij} \tag{27.21}$$

This gives the following eigenvalues and normalized eigenfunctions:

$$\lambda_i = \frac{(2i-1)^2\pi^2}{4}, \psi_i = \sqrt{2}\cos\left[(2i-1)\frac{\pi y}{2}\right], \quad i = 1, 2, 3, \dots \tag{27.22}$$

The average value of these eigenfunctions is

$$\langle 1, \psi_i\rangle = \frac{1}{2}\int_{-1}^{1} \sqrt{2}\cos\left[\left(i - \frac{1}{2}\right)\pi y\right] dy = \frac{(-1)^{i-1}2\sqrt{2}}{(2i-1)\pi}, \tag{27.23}$$

which gives the Fourier weight β as

$$\beta_i = \langle 1, \psi_i\rangle^2 = \frac{8}{(2i-1)^2\pi^2} \tag{27.24}$$

Thus, the formal solution can be expressed in terms of eigenvalues and eigenfunctions from equation (27.13) as

$$U(y) = \sum_i \frac{1}{\lambda_i}\langle 1, \psi_i\rangle\psi_i(y, z) = \sum_{i=1}^{\infty} \frac{(-1)^{i-1}16}{(2i-1)^3\pi^3}\cos\left[(2i-1)\frac{\pi y}{2}\right]. \tag{27.25}$$

This gives the average velocity from equation (27.14) as

$$\langle U\rangle = \sum_i \frac{\beta_i}{\lambda_i} = \sum_{i=1}^{\infty} \frac{32}{\pi^4}\frac{1}{(2i-1)^4} = \frac{32}{\pi^4}\frac{\pi^4}{96} = \frac{1}{3} \tag{27.26}$$

which matches the result from direct solution (equation (27.18)). Similarly, the friction factor can be obtained from equation (27.8) or (27.15) as

$$f\,\mathrm{Re} = \frac{8}{\langle U\rangle} = 24 \tag{27.27}$$

that matches the direct solution (equation (27.19)).

27.3 Specific case: elliptical ducts

Consider the elliptical duct with the transverse domain given by

$$\Omega' \equiv \frac{y'^2}{a^2} + \frac{z'^2}{b^2} - 1 < 0; \quad -a \le y' \le a,\ -b \le z' \le b; \quad \sigma = \frac{a}{b} \tag{27.28}$$

where a and b are the lengths of semimajor and semiminor axes, and σ is the aspect ratio of the ellipse ($\sigma = 1$ for a circle). The momentum equation (27.1) can be solved in many ways. Here, we utilize a single trial function that vanishes at the boundary $\partial\Omega'$ and expresses the velocity profile as follows:

$$u_x = \beta\left(1 - \frac{y'^2}{a^2} - \frac{z'^2}{b^2}\right) \tag{27.29}$$

The model equation (27.1) leads to

$$\mu\nabla'^2 u_x = \left(\frac{\Delta P}{L}\right) \Longrightarrow \left(\frac{\Delta P}{\mu L}\right) = -2\beta\left(\frac{1}{a^2} + \frac{1}{b^2}\right) \tag{27.30}$$

$\Longrightarrow$ the characteristic velocity u^* from equation (27.3) can be expressed as

$$u^* = R_\Omega^2\left(\frac{-\Delta P}{\mu L}\right) = 2\beta R_\Omega^2\left(\frac{1}{a^2} + \frac{1}{b^2}\right). \tag{27.31}$$

The average velocity can be obtained from equation (27.29) as follows:

$$\begin{aligned}
\langle u_x\rangle &= \frac{1}{A_{\Omega'}}\int_{\Omega'} u_x(y', z')\,dy'\,dz' \\
&= \frac{4\beta}{\pi ab}\int_0^a \left[\int_0^{b\sqrt{1-\frac{y'^2}{a^2}}} \left(1 - \frac{y'^2}{a^2} - \frac{z'^2}{b^2}\right) dz'\right] dy' \\
&= \frac{4\beta}{\pi ab}\int_0^a \frac{2b}{3}\left(1 - \frac{y'^2}{a^2}\right)^{3/2} dy' \\
&= \frac{8\beta}{3\pi}\int_0^{\pi/2} \cos^4(\theta)\,d\theta \quad (\text{taking } y' = a\sin\theta) \\
&= \frac{8\beta}{3\pi}\cdot\frac{3\pi}{16} = \frac{\beta}{2}
\end{aligned} \tag{27.32}$$

Thus, the friction factor can be expressed from equations (27.8), (27.31) and (27.32) as

$$f\,\mathrm{Re} = \frac{8}{\langle U\rangle} = \frac{8u^*}{\langle u_x\rangle} = 32R_\Omega^2\left(\frac{1}{a^2}+\frac{1}{b^2}\right) \tag{27.33}$$

where $R_\Omega = \frac{A_{\Omega'}}{P_{\Omega'}}$ is the hydraulic radius of the duct. The cross-section area of the duct is $A_{\Omega'} = \pi ab$ while the wetted perimeter $P_{\Omega'}$ is given by the integral:

$$P_{\Omega'} = \int_{\partial\Omega'} ds = \int_{\partial\Omega'} \sqrt{dx^2 + dy^2}$$

Since $\partial\Omega' \equiv \frac{y'^2}{a^2} + \frac{z'^2}{b^2} - 1 = 0$, taking $y' = a\sin\theta$ and $z' = b\cos\theta$, the above expression for the perimeter can be simplified as

$$\begin{aligned} P_{\Omega'} &= 4\int_0^{\pi/2} \sqrt{a^2\cos^2\theta + b^2\sin^2\theta\, d\theta} \\ &= 4a\int_0^{\pi/2} \sqrt{\cos^2\theta + \frac{1}{\sigma^2}\sin^2\theta\, d\theta} \\ &= 4a\int_0^{\pi/2} \sqrt{1-\left(1-\frac{1}{\sigma^2}\right)\sin^2\theta\, d\theta} = 4a \quad E(\sigma). \end{aligned}$$

The hydraulic radius can be expressed as

$$R_\Omega = \frac{\pi ab}{P_{\Omega'}} = \frac{\pi b}{4E(\sigma)}; \quad \text{where } E(\sigma) = \int_0^{\pi/2} \sqrt{1-\left(1-\frac{1}{\sigma^2}\right)\sin^2\theta d\theta}. \tag{27.34}$$

Thus, equation (27.33) can be further simplified to express the friction factor as

$$\begin{aligned} f\,\mathrm{Re} &= 32\frac{R_\Omega^2}{b^2}\left(1+\frac{1}{\sigma^2}\right) \\ &= 2\left(1+\frac{1}{\sigma^2}\right)\left[\frac{\pi}{E(\sigma)}\right]^2; \quad \text{where } E(\sigma) = \int_0^{\pi/2} \sqrt{1-\left(1-\frac{1}{\sigma^2}\right)\sin^2\theta\, d\theta} \end{aligned} \tag{27.35}$$

Figure 27.2 plots friction factor for elliptical duct against its aspect ratio and few of these values are listed in Table 27.1. Note from Figure 27.2 that the friction factor curve is symmetric around $\sigma = 1$ on log-linear plot. This is because aspect ratio of σ and σ^{-1} have the same cross-sectional shape of the duct, and hence the same friction factor. It can also be seen from equation (27.35) that in the limit of $\sigma \to \infty$, $E(\sigma) \to \int_0^{\pi/2} \cos\theta\, d\theta = 1$,

Table 27.1: Friction factor and E-function for elliptical ducts of aspect ratio σ.

σ	$E(\sigma)$	f **Re**	
$\sigma \ll 1$	$\frac{1}{\sigma^2}$	$2\pi^2 = 19.74$	
0.1	10.16	19.314	
0.5	2.422	16.823	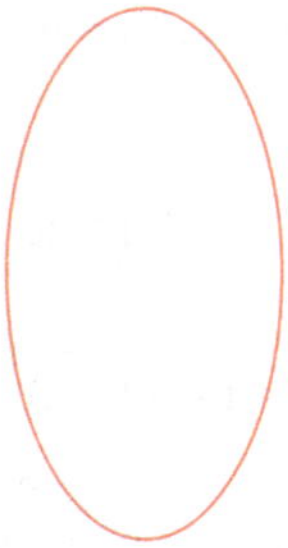
1.0	$\frac{\pi}{2} = 1.571$	16.0	
1.5	1.322	16.31	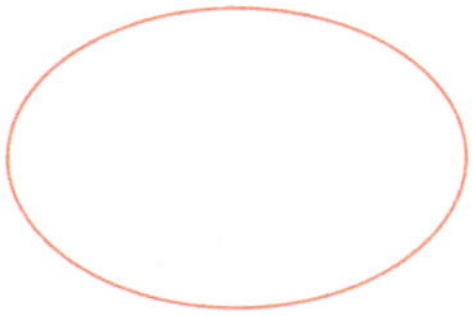
2.0	1.211	16.823	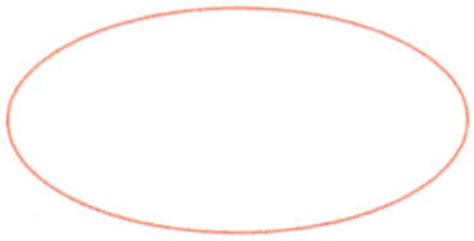
10.0	1.016	19.314	
∞	1	$2\pi^2 = 19.74$	

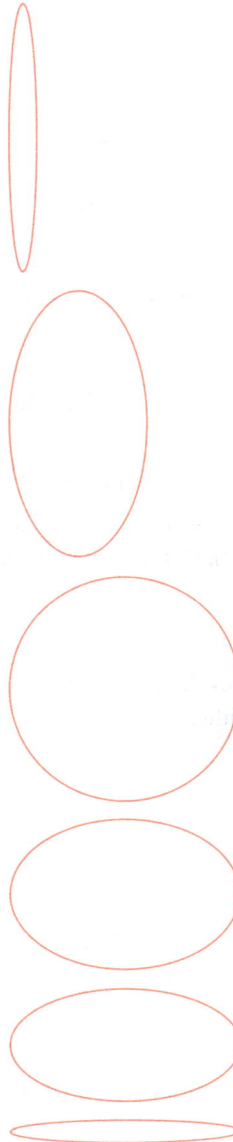

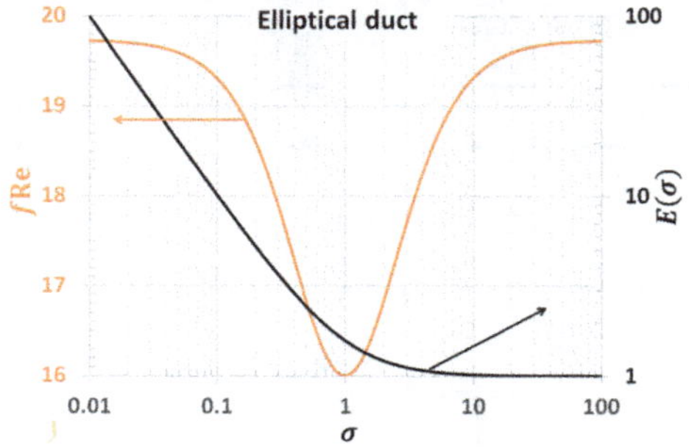

Figure 27.2: Friction factor and E-function for elliptical ducts against the aspect ratio σ.

and hence $\lim_{\sigma\to\infty} f\,\mathrm{Re} = 2\pi^2$. In addition, since $f\,\mathrm{Re}$ is symmetric around $\sigma = 1$ (on log-linear plot),

$$\lim_{\sigma\to 0} f\,\mathrm{Re} = \lim_{\sigma\to\infty} f\,\mathrm{Re} = 2\pi^2 \tag{27.36}$$

This also implies from equation (27.35) that in the limit of $\sigma \to 0$, $E(\sigma) \to \frac{2\pi^2}{f\,\mathrm{Re}}(1 + \frac{1}{\sigma^2}) \to \frac{1}{\sigma^2}$, which can also be seen from equation (27.35). We also note that for the case of a circular duct ($\sigma = 1$), $E(\sigma) = \frac{\pi}{2}$, which leads to $f\,\mathrm{Re} = 16$.

Problems

1. Consider the problem of laminar flow in a duct of equilateral triangular cross-section with side a and height $H = \sqrt{3}a/2$ as shown in Figure 27.3:

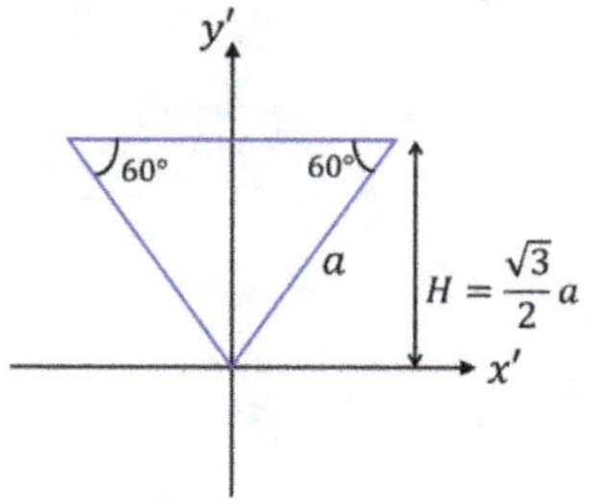

Figure 27.3: Schematic of an equilateral triangular duct.

(a) Verify that the expression

$$u_z(x', y') = c_0\left(\frac{y'}{H} - 1\right)\left(\frac{y'^2 - 3x'^2}{H^2}\right)$$

satisfies the no-slip condition, and hence represents a possible velocity profile. Determine the constant c_0 so that equation (27.1) is satisfied.

(b) Formulate the momentum balance in dimensionless form so that $R_\Omega = 1$.

(c) Use the results in (a) to show that $f\,\mathrm{Re} = \frac{40}{3}$.

2. (a) Determine the dimensionless velocity profile for a duct of circular cross-section (with $R_\Omega = 1$) and use it to slow $f\,\mathrm{Re} = 16$.

(b) Obtain the expression for $f\,\mathrm{Re}$ using FFT.

3. Consider the problem of steady laminar flow in a rectangular duct of height $2b$ and width $2a$. Formulate the problem and cast the model in dimensionless form. Use the Fourier transform method to show that the friction factor is given by

$$f\,\mathrm{Re} = \frac{24}{(1+\lambda)^2[1 - \frac{192\lambda}{\pi^5}\sum_{k=1}^{\infty}\frac{\tan h[\frac{(2k-1)\pi}{2\lambda}]}{(2k-1)^5}]}$$

where $\lambda(= a/b)$ is the aspect ratio. Use the above formula to evaluate and plot $f\,\mathrm{Re}$ as a function λ for $1 \le \lambda \le \infty$. Determine the numerical value of $f\,\mathrm{Re}$ for a square duct.

4. Solve the model of steady laminar flow in a square duct of side $2a$ to determine the velocity profile using FFT in both transverse directions. Obtain an expression for $f\,\mathrm{Re}$ as a double infinite sum having a numerical value of 14.23.

5. Consider the problem of pulsatile laminar flow in a tube of circular cross-section in which the applied pressure gradient varies sinusoidally around a mean value.

(a) Formulate the model equations that determine the velocity profile and cast them in dimensionless form.

(b) Obtain the solution to the velocity profile using FFT and simplify it for large times.

(c) Express the result in (b) as a mean flow pulsetile part and give a physical interpretation.

28 Multicomponent diffusion and reaction

Problems of multicomponent diffusion and reaction are of fundamental importance in chemical engineering. They arise in the design of catalysts and catalytic reactors, adsorption and separation processes as well as many other applications. We have already illustrated the application of linear analysis to these problems in Sections 4.5.6, 5.3 and Chapters 23 and 25. In this chapter, we show some further application to determine catalyst effectiveness factors and calculation of effluent concentrations in monolith reactors.

In the next section, we examine the problem of catalyst effectiveness factor. We also introduce the concept of internal mass transfer coefficient and its calculation for an arbitrary geometry. This is followed by a discussion of the multicomponent case and illustration of the calculations.

28.1 Generalized catalyst effectiveness factor problem

Consider a porous catalyst particle of arbitrary shape with activity profile given by $a(x',y',z')$ as shown in Figure 28.1.

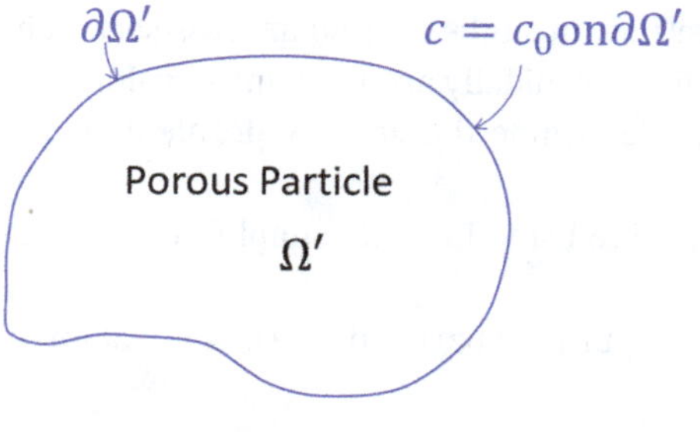

Figure 28.1: Schematic of a porous catalyst particle of arbitrary shape.

Assuming a single step reaction $A \to B$ with linear kinetics, the steady-state reactant concentration profile $C(x',y',z')$ satisfies the following diffusion–reaction problem:

$$D_e \nabla'^2 C = a(x',y',z') k_0 C \text{ in } \Omega'; \quad C = C_0 \text{ on } \partial\Omega' \tag{28.1}$$

where D_e is the effective diffusivity of the reactant in the porous particle; k_0 is the first-order rate constant (based on unit volume); and $a(x',y',z')$ is the normalized activity profile, i. e.,

$$\frac{1}{V_{\Omega'}} \int_{\Omega'} a(x',y',z')\, dV_{\Omega'} = 1 \tag{28.2}$$

https://doi.org/10.1515/9783111598055-034

Here, $V_{\Omega'}$ is the volume of the catalyst. Let $S_{\Omega'}$ is the external surface area of the catalyst particle, then the effective diffusion length R_Ω, Thiele modulus ϕ and other quantities can be defined as

$$R_\Omega = \frac{V_{\Omega'}}{S_{\Omega'}}; \quad (x,y,z) = \frac{(x',y',z')}{R_\Omega}; \quad c(x,y,z) = \frac{C(x',y',z')}{C_0};$$
$$\phi^2 = \frac{R_\Omega^2 k_0}{D_e}; \quad g(x,y,z) = a(R_\Omega x, R_\Omega y, R_\Omega z). \tag{28.3}$$

Thus, the model equation (28.1) can be expressed in dimensionless form as follows:

$$\nabla^2 c = g(x,y,z)\phi^2 c, \quad (x,y,z) \in \Omega; \quad c = 1 \text{ on } \partial\Omega \tag{28.4}$$

Here, the dimensionless activity profile $g(x,y,z)$ satisfies the normalized condition, i. e.,

$$\frac{1}{V_\Omega}\int_\Omega g(x,y,z)\,d\Omega = \frac{1}{V_\Omega}\int_\Omega g(x,y,z)\,dx\,dy\,dz = 1, \tag{28.5}$$

where V_Ω is the volume of particle in dimensionless units (i. e., of scaled domain). For the case of uniform catalyst activity, $g(x,y,z) = 1$. In the general case, we assume $g > 0$.

To solve equation (28.4), we consider the EVP:

$$L\psi = \frac{1}{g(x,y,z)}\nabla^2\psi = -\lambda\psi \text{ in } \Omega; \quad \psi = 0 \text{ on } \partial\Omega \tag{28.6}$$

which is self-adjoint with respect to the activity weighted inner product:

$$\langle\psi_i,\psi_j\rangle = \frac{1}{V_\Omega}\int_\Omega g(x,y,z)\psi_i(x,y,z)\psi_j(x,y,z)\,dx\,dy\,dz = \delta_{ij} \tag{28.7}$$

As shown in earlier chapters, the eigenvalues λ_j are real and positive. In addition, the Parseval's relation suggests that the normalized eigenfunctions and Fourier weights satisfy

$$\sum_i \beta_i = 1; \quad \beta_j = \text{Fourier weight} = \langle 1,\psi_j\rangle^2 \tag{28.8}$$

Equation (28.4) can be solved in terms of eigenvalues and eigenfunctions defined by equation (28.6). For this, we define a new variable $u = 1 - c$ that satisfies the following equation:

$$Lu = \frac{1}{g(x,y,z)}\nabla^2 u = \phi^2(u-1) \text{ in } \Omega; \quad u = 0 \text{ on } \partial\Omega \tag{28.9}$$

$\Longrightarrow$ after taking inner product with ψ_i

$$\langle Lu, \psi_i\rangle = \phi^2\langle u, \psi_i\rangle - \phi^2\langle 1, \psi_i\rangle = \langle u, L\psi_i\rangle = -\lambda_i\langle u, \psi_i\rangle$$

$\Longrightarrow$

$$\langle u, \psi_i\rangle = \frac{\phi^2\langle 1, \psi_i\rangle}{\phi^2 + \lambda_i}$$

$\Longrightarrow$

$$u = \sum_i \langle u, \psi_i\rangle\psi_i = \sum_i \frac{\phi^2\langle 1, \psi_i\rangle}{\phi^2 + \lambda_i}\psi_i$$

$\Longrightarrow$ The formal solution for the dimensionless concentration $c(x, y, z)$ is

$$c = 1 - u = 1 - \sum_i \frac{\phi^2\langle 1, \psi_i\rangle}{\phi^2 + \lambda_i}\psi_i(x, y, z) \tag{28.10}$$

28.1.1 Effectiveness factor

The effectiveness factor or the average reaction rate in the particle can be determined using the concentration profile. The effectiveness factor η is defined by

$$\eta = \frac{1}{V_\Omega}\int_\Omega g(x, y, z)c(x, y, z)\,dx\,dy\,dz$$

$$= \langle 1, c\rangle = c_m = \text{activity weighted average concentration}, \tag{28.11}$$

$\Longrightarrow$

$$\eta = \langle 1, c\rangle = 1 - \sum_i \frac{\phi^2\langle 1, \psi_i\rangle^2}{\phi^2 + \lambda_i} = 1 - \sum_i \frac{\phi^2\beta_i}{\phi^2 + \lambda_i} = \sum_i \frac{\lambda_i\beta_i}{\phi^2 + \lambda_i} \tag{28.12}$$

where β_i are Fourier weights defined by equation (28.8). The above expression for effectiveness factor can be expanded in power series of Thiele modulus ϕ^2 as

$$\eta = 1 + \sum_i \sum_{j=1}^{\infty}(-1)^j \frac{\beta_i}{\lambda_i^j}(\phi^2)^j, \tag{28.13}$$

where the coefficients in the power series are termed as Aris numbers (Balakotaiah [5]):

$$\eta = 1 + \sum_{j=1}^{\infty}(-1)^j Ar_j(\phi^2)^j; \quad Ar_j = \sum_i \frac{\beta_i}{\lambda_i^j} \tag{28.14}$$

The Aris numbers depend only on the geometric shape of the catalyst particle. These can also be obtained by using a perturbation method, by expanding the concentration in powers of ϕ^2 as

$$c = \sum_{j=0}^{\infty}(-1)^j c_j(x,y,z)(\phi^2)^j; \tag{28.15}$$

where

$$\begin{aligned} Ar_j &= \langle 1, c_j\rangle; \quad c_0(x,y,z) = 1; \\ \nabla^2 c_j &= -g(x,y,z)c_{j-1}(x,y,z) \text{ in } \Omega; \quad c_j = 0 \text{ on } \partial\Omega, j \geq 1. \end{aligned} \tag{28.16}$$

Equation (28.12) gives the effectiveness factor for a catalyst particle in terms of eigenvalues, Fourier weights and ϕ^2.

28.1.2 Sherwood number (for internal mass-transfer coefficient)

The internal mass-transfer coefficient k_{ci} for species exchange between the boundary and interior of the catalyst particle can be defined as follows:

$$k_{ci} = \frac{\frac{1}{S_{\Omega'}}\int_{\partial\Omega'} D_e \mathbf{n}.\nabla C \, dS_{\Omega'}}{C_0 - C_m} \tag{28.17}$$

where C_m is the activity weighted average concentration, defined by

$$C_m = \frac{1}{V_{\Omega'}}\int_{\Omega'} a(x',y',z')C(x',y',z')\, dV_{\Omega'}. \tag{28.18}$$

Thus, by using the divergence theorem:

$$\int_{\Omega'} \nabla.(\nabla C)\, dV_{\Omega'} = \int_{\partial\Omega'} \mathbf{n}.\nabla C \, dS_{\Omega'} = \frac{k_0 V_{\Omega'} C_m}{D_e}, \tag{28.19}$$

the internal Sherwood number Sh_i (or dimensionless mass transfer coefficient) can be expressed as

$$\text{Sh}_i = \frac{k_{ci}R_\Omega}{D_e} = \frac{C_m}{C_0 - C_m}\frac{k_0 R_\Omega^2}{D_e} = \frac{\eta\phi^2}{1-\eta} = \frac{\left(\sum_{i=1}^{\infty}\frac{\lambda_i\beta_i}{\phi^2+\lambda_i}\right)}{\left(\sum_{i=1}^{\infty}\frac{\beta_i}{\phi^2+\lambda_i}\right)} \tag{28.20}$$

or

$$\eta = \frac{1}{1 + \frac{\phi^2}{\text{Sh}_i}}. \tag{28.21}$$

28.1.3 Exact expressions for Sh_i for some common geometries

The exact expressions for the internal Sherwood number for a slab, an infinite cylinder and a sphere for uniform activity can be obtained as (see Chapters 23 and 25)

$$\mathrm{Sh}_i = \left(\frac{1}{\phi \tanh \phi} - \frac{1}{\phi^2}\right)^{-1} \quad \text{(slab)} \tag{28.22}$$

$$\mathrm{Sh}_i = \frac{\phi^2 I_1(2\phi)}{\phi I_0(2\phi) - I_1(2\phi)} \quad \text{(infinite cylinder)} \tag{28.23}$$

$$\mathrm{Sh}_i = \frac{3\phi^3 \coth(3\phi) - \phi^2}{3\phi^2 - 3\phi \coth(3\phi) + 1} \quad \text{(sphere).} \tag{28.24}$$

Several limiting cases of equations (28.12) and (28.20) for the general case are of interest. For $\phi^2 \to 0$, we have

$$\eta = 1 - Ar_1 \phi^2 + O(\phi^4), \tag{28.25}$$

and

$$\mathrm{Sh}_i \triangleq \mathrm{Sh}_{i\infty} = \frac{1}{Ar_1} = \left(\sum_{i=1}^{\infty} \frac{\beta_i}{\lambda_i}\right)^{-1}. \tag{28.26}$$

For $\phi^2 \gg 1$, the sum defining η may be replaced by an integral. In this limit, it can be shown that $\eta \to \frac{1}{\phi}$ while $\mathrm{Sh}_i \to \phi$. Using these limits, Sh_i for any ϕ may be approximated by

$$\mathrm{Sh}_i = \mathrm{Sh}_{i\infty} + \phi \tanh[\Lambda^* \phi], \tag{28.27}$$

where the constants $\mathrm{Sh}_{i\infty}$ and Λ^* depend only on the geometric shape of the catalyst particle. These constants can be related to $Ar_j (j = 1, 2, \dots)$ or λ_i and β_i. Values of these constants for the case of uniform activity and various common geometries can be found in the literature (Tu et al. [30]; Sarkar et al. [27]). A plot of Sh_i for infinite slab ($\mathrm{Sh}_{i\infty} = 3$ and $\Lambda^* = 0.2$), cylinder ($\mathrm{Sh}_{i\infty} = 2$ and $\Lambda^* = 0.33$) and sphere ($\mathrm{Sh}_{i\infty} = \frac{5}{3}$ and $\Lambda^* = 0.43$) is shown in Figure 28.2. The shape of the Sh_i versus ϕ curve for any arbitrary geometry is similar to the curves shown in Figure 28.2.

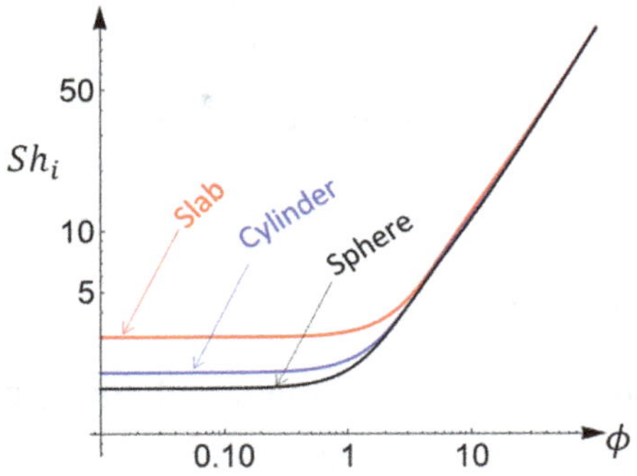

Figure 28.2: Internal Sherwood number versus Thiele modulus for slab, infinite cylinder and a sphere.

28.2 Multicomponent diffusion and reaction in the washcoat layer of a monolith reactor

So far, we have examined only the case of a single reaction in the catalyst particle or layer. We now consider the case of multiple reactions with the following assumptions: (i) steady state, (ii) diffusion in the washcoat/catalyst is governed by Knudsen mechanism, (iii) catalyst layer is thin so that gradients are important in only one direction.

Consider R (independent) reactions among S species:

$$\sum_{j=1}^{S} \nu_{ij} A_j = 0; \quad i = 1, 2, \ldots, R \tag{28.28}$$

Let r_i be the rate of reaction i, then the steady-state species balance for A_j gives

$$D_{ej} \frac{d^2 c_{wj}}{dy^2} + \sum_{i=1}^{R} \nu_{ij} r_i(\mathbf{c}_w) = 0 \tag{28.29}$$

$$c_{wj} = c_{sj} \ @ y = 0 \tag{28.30}$$

$$\frac{dc_{wj}}{dy} = 0 \ @ y = \delta_c. \tag{28.31}$$

In vector form, equations (28.29) and (28.30)–(28.31) may be written as

$$\mathbf{D}_e \frac{d^2 \mathbf{c}_w}{dy^2} + \boldsymbol{\nu}^T \mathbf{r}(\mathbf{c}_w) = \mathbf{0}; \quad \mathbf{c}_w = \mathbf{c}_s \ @ y = 0; \quad \frac{d\mathbf{c}_w}{dy} = \mathbf{0} \ @ y = \delta_c, \tag{28.32}$$

where $\mathbf{D}_e$ is the diagonal matrix of Knudsen diffusivities; $\boldsymbol{\nu}$ is the stoichiometric coefficient matrix; $\mathbf{r}$ is a $R \times 1$ vector of reaction rates, $\mathbf{c}_w$ is the $S \times 1$ vector of species concentrations and δ_c is the thickness of the catalyst layer [Remark: If $S > R$, the number of independent species balances can be reduced to R using the *stoichiometric invariants* but we do not make that simplification here].

We consider the case of linear kinetics, i. e.,

$$\mathbf{r}(\mathbf{c}_w) = \widehat{\mathbf{K}}\mathbf{c}_w, \quad \widehat{\mathbf{K}} = R \times S \text{ matrix of rate constants}$$

$\Longrightarrow$

$$-\boldsymbol{\nu}^T \mathbf{r}(\mathbf{c}_w) = -\boldsymbol{\nu}^T \widehat{\mathbf{K}}\mathbf{c}_w = \mathbf{K}_\nu \mathbf{c}_w,$$

where $\mathbf{K}_\nu = S \times S$ matrix of effective rate constants (for species consumption). Equation (28.32) may be written as

$$\mathbf{D}_e \frac{d^2\mathbf{c}_w}{dy^2} = \mathbf{K}_\nu \mathbf{c}_w; \quad 0 < y < \delta_c.$$

Defining

$$\xi = \frac{y}{\delta_c}, \quad \boldsymbol{\Phi}_w^2 = \delta_c^2 \mathbf{D}_e^{-1}\mathbf{K}_\nu$$

$\Longrightarrow$

$$\frac{d^2\mathbf{c}_w}{d\xi^2} = \boldsymbol{\Phi}_w^2 \mathbf{c}_w, \quad 0 < \xi < 1; \quad \frac{d\mathbf{c}_w}{d\xi} = \mathbf{0} \;@\; \xi = 1; \quad \mathbf{c}_w = \mathbf{c}_s \;@\; \xi = 0. \tag{28.33}$$

Here, $\boldsymbol{\Phi}_w$ is the Thiele matrix, which may be shown to have real and nonnegative eigenvalues for the case of monomolecular kinetics (Wei and Prater [31]).

Equation (28.33) may be solved using the standard matrix methods (as discussed in Chapter 5):

$$\mathbf{c}_w = \cosh[\boldsymbol{\Phi}_w(1-\xi)].(\cosh \boldsymbol{\Phi}_w)^{-1}\mathbf{c}_s. \tag{28.34}$$

The observed or effective rate vector is given by

$$\begin{aligned} \mathbf{r}_{\text{obs}} &= \int_0^1 \mathbf{K}_\nu \mathbf{c}_w(\xi)\, d\xi \\ &= \mathbf{K}_\nu (\tanh \boldsymbol{\Phi}_w)\boldsymbol{\Phi}_w^{-1}\mathbf{c}_s = \mathbf{K}_\nu^* \mathbf{c}_s \end{aligned} \tag{28.35}$$

where $\mathbf{K}_\nu^*$ is known as the diffusion-disguised rate constant matrix, and is given by

$$\mathbf{K}_\nu^* = \mathbf{K}_\nu \mathbf{H}; \quad \mathbf{H} = (\tanh \boldsymbol{\Phi}_w)\boldsymbol{\Phi}_w^{-1},$$

where $\mathbf{H}$ is the effectiveness factor matrix.

It is easily verified that

$$\mathbf{K}_\nu^* = \mathbf{K}_\nu \quad \text{for } \delta_c \to 0 \text{ or } \|\boldsymbol{\Phi}_w\| \to 0, \tag{28.36}$$

while we have

$$K_v^* = \frac{1}{\delta_c}\mathbf{K}_v(\mathbf{D}_e^{-1}\mathbf{K}_v)^{-1/2}, \quad \|\mathbf{\Phi}_w\| \gg 1. \tag{28.37}$$

Sherwood matrix for internal mass transfer

We define the internal mass transfer coefficient matrix $\mathbf{k}_{ci}$ by

$$\mathbf{k}_{ci}(\mathbf{c}_s - \overline{\mathbf{c}_w}) = -\mathbf{D}_e \frac{\partial \mathbf{c}_w}{\partial y}\bigg|_{y=0} = \mathbf{j}_{f-wc} \tag{28.38}$$

where the average concentration vector in the washcoat $\overline{\mathbf{c}_w}$ and the species flux vector at fluid-washcoat interface $\mathbf{j}_{f-wc}$ are given by

$$\overline{\mathbf{c}_w} = \mathbf{H}\mathbf{c}_s; \quad \mathbf{j}_{f-wc} = \delta_c \mathbf{K}_v \mathbf{H}\mathbf{c}_s = \delta_c \mathbf{K}_v^* \mathbf{c}_s, \tag{28.39}$$

$\Longrightarrow$

$$\mathbf{k}_{ci} = \delta_c \mathbf{K}_v \mathbf{H}(\mathbf{I} - \mathbf{H})^{-1} \tag{28.40}$$

Defining the Sherwood matrix of dimensionless mass trasnfer coefficients by

$$\mathbf{Sh}_i = \delta_c \mathbf{D}_e^{-1}\mathbf{k}_{ci}, \tag{28.41}$$

we obtain

$$\begin{aligned}\mathbf{Sh}_i &= \delta_c^2 \mathbf{D}_e^{-1}\mathbf{K}_v \mathbf{H}(\mathbf{I} - \mathbf{H})^{-1} \\ &= \left[\mathbf{\Phi}_w^{-1}(\tanh \mathbf{\Phi}_w)^{-1} - \mathbf{\Phi}_w^{-2}\right]^{-1},\end{aligned} \tag{28.42}$$

which is a generalization of the expression for the scalar case (slab or parallel plate geometry). [Remark: For arbitrary shaped catalyst particle, the expression of internal Sherwood number matrix can be approximated as

$$\mathbf{Sh}_i = \mathrm{Sh}_{i\infty}\,\mathbf{I} + \mathbf{\Phi}_w \tanh[\Lambda^* \mathbf{\Phi}_w], \tag{28.43}$$

which is a generalization of the scalar case (equation (28.27)). Here, the scalars $\mathrm{Sh}_{i\infty}$ and Λ^* depend only on the shape/geometry of the washcoat/catalyst.]

The $\mathbf{k}_{ci}$ or $\mathbf{Sh}_i$ matrix can be calculated using the spectral theorem or the Cayley–Hamilton theorem for calculating functions of a matrix. A sample calculation is shown in the next section.

28.3 Isothermal monolith reactor model for multiple reactions

Consider a single channel of a monolith reactor schematically shown in Figure 28.3.

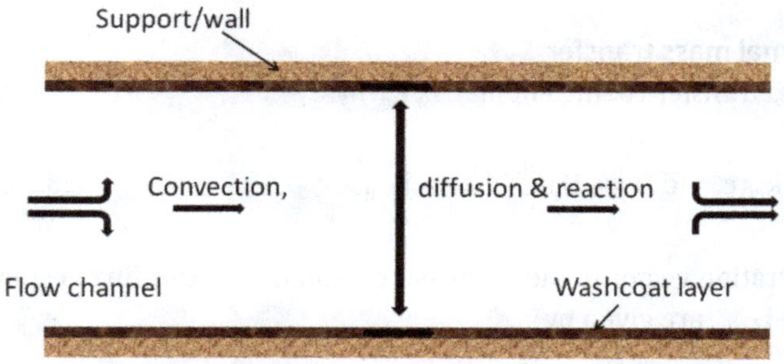

Figure 28.3: Schematic diagram of a monolith reactor with washcoat/catalytic layer.

Considering the isothermal case, the steady-state reactor model that couples convection in the channel to transverse diffusion and reaction in the catalyst layer can be expressed as

$$\bar{u}\frac{d\mathbf{c}_f}{dx} = -a_v\mathbf{k}_{ce}(\mathbf{c}_f - \mathbf{c}_s); \quad \mathbf{c}_f = \mathbf{c}_{f,\text{in}} \ @\ x = 0 \tag{28.44}$$

$$\mathbf{j}_{f-wc} = \mathbf{k}_{ce}(\mathbf{c}_f - \mathbf{c}_s) = \mathbf{k}_{ci}(\mathbf{c}_s - \overline{\mathbf{c}_w}) \tag{28.45}$$

$$= \mathbf{k}_{co}(\mathbf{c}_f - \overline{\mathbf{c}_w}) = \delta_c\mathbf{R}_v(\overline{\mathbf{c}_w}) \tag{28.46}$$

where

$$\mathbf{k}_{co}^{-1} = \mathbf{k}_{ce}^{-1} + \mathbf{k}_{ci}^{-1}, \tag{28.47}$$

is the overall mass transfer coefficient matrix, $\mathbf{k}_{ce}$ and $\mathbf{k}_{ci}$ are the external and internal mass transfer coefficient matrices, respectively. For linear kinetics,

$$\mathbf{R}_v(\overline{\mathbf{c}_w}) = \mathbf{K}_v\overline{\mathbf{c}_w} \tag{28.48}$$

and we can write equation (28.44) with $z = \frac{x}{L}$ as

$$\frac{d\mathbf{c}_f}{dz} = -\frac{\tau}{R_\Omega}\delta_c\mathbf{K}_v(\mathbf{k}_{co} + \delta_c\mathbf{K}_v)^{-1}\mathbf{k}_{co}\mathbf{c}_f = -\mathbf{Da}\mathbf{c}_f; \tag{28.49}$$

$$\mathbf{c}_f = \mathbf{c}_{f,\text{in}} \ @\ z = 0. \tag{28.50}$$

Here, L is the channel length and $\tau = \frac{L}{\bar{u}}$ is the contact (space) time. Thus, the species exit concentration vector is given by

$$\mathbf{c}_{fe} = \mathbf{c}_f(z = 1) = \exp[-\mathbf{Da}]\mathbf{c}_{f,\text{in}} \tag{28.51}$$

Here, **Da** is the Damköhler matrix and it can be computed from $\mathbf{K}_v$ and $\mathbf{k}_{co}$ as

$$\mathbf{Da} = \frac{\tau}{R_\Omega} \delta_c \mathbf{K}_v (\mathbf{k}_{co} + \delta_c \mathbf{K}_v)^{-1} \mathbf{k}_{co} \tag{28.52}$$

$$\mathbf{k}_{ce} = \frac{1}{4R_\Omega} D_m \mathbf{Sh}_e, \tag{28.53}$$

$$\mathbf{k}_{ci} = \delta_c \mathbf{K}_v [\Phi_w (\tanh \Phi_w)^{-1} - \mathbf{I}]^{-1}. \tag{28.54}$$

Note that in dilute mixture, $\mathbf{Sh}_e$ may be assumed to be a diagonal matrix of external Sherwood numbers.

Using these expressions, we can determine the impact of external mass transfer as well as that of pore diffusion on the yield (or selectivity) of intermediate products.

28.3.1 Example: reversible sequential reactions

Consider the reaction scheme shown in Figure 28.4 among four components A, B, C and D, with rate constants as shown.

$$A \underset{k/2}{\overset{k}{\rightleftharpoons}} B \underset{k/4}{\overset{k/2}{\rightleftharpoons}} C \underset{k/8}{\overset{k/4}{\rightleftharpoons}} D$$

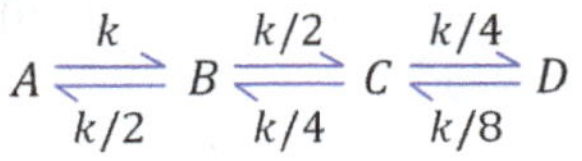

Figure 28.4: Reaction scheme (reversible consecutive reactions with 4 components).

For simplicity, assume that the diffusivities of all species (in gas phase and washcoat) are equal (but the diffusivity in the catalyst layer is smaller than that in the gas phase).
$\Longrightarrow$

$$\mathbf{K}_{ce} = k_{ce} \mathbf{I}, \quad k_{ce} = \mathrm{Sh}_{e\Omega} \frac{D_m}{R_\Omega}; \tag{28.55}$$

$$\mathbf{k}_{ci} = \frac{D_e}{\delta_c} \mathbf{Sh}_i; \quad \mathbf{K}_v = k\mathbf{A}, \quad \mathbf{A} = \begin{pmatrix} 1 & -\frac{1}{2} & 0 & 0 \\ -1 & 1 & -\frac{1}{4} & 0 \\ 0 & -\frac{1}{2} & \frac{1}{2} & -\frac{1}{8} \\ 0 & 0 & -\frac{1}{4} & \frac{1}{8} \end{pmatrix}, \tag{28.56}$$

$\Rightarrow$ from equation (28.52),

$$\mathbf{Da} = \frac{\delta_c k \tau}{R_\Omega} \mathbf{A} (\mathbf{k}_{co} + \delta_c k \mathbf{A})^{-1} \mathbf{k}_{co} \tag{28.57}$$

where the overall mass-transfer coefficient matrix is given by

$$\mathbf{k}_{co}^{-1} = \frac{1}{k_{ce}}\mathbf{I} + \frac{\delta_c}{D_e}\mathbf{Sh}_i^{-1}$$
$$= \mathrm{Sh}_{e\Omega}^{-1}\frac{R_\Omega}{D_m}\mathbf{I} + \frac{\delta_c}{D_e}\mathbf{Sh}_i^{-1}$$

$\Rightarrow$

$$\mathbf{k}_{co}\frac{\tau}{R_\Omega} = \frac{D_m\tau}{R_\Omega^2}[\mathrm{Sh}_{e\Omega}^{-1}\mathbf{I} + \mu\mathbf{Sh}_i^{-1}]^{-1}; \quad \mu = \frac{D_m}{D_e}\frac{\delta_c}{R_\Omega}, \tag{28.58}$$

where μ is the ratio of diffusion velocity in the fluid to that in the washcoat. This expresses the Damköhler matrix (equation (28.57)) as

$$\begin{aligned}
\mathbf{Da} &= \mathbf{A}\left(\frac{1}{k\delta_c}\mathbf{k}_{co} + \mathbf{A}\right)^{-1}\frac{D_m\tau}{R_\Omega^2}(\mathrm{Sh}_{e\Omega}^{-1}\mathbf{I} + \mu\mathbf{Sh}_i^{-1})^{-1} \\
&= \mathbf{A}\left(\frac{D_m}{k\delta_c R_\Omega}[\mathrm{Sh}_{e\Omega}^{-1}\mathbf{I} + \mu\mathbf{Sh}_i^{-1}]^{-1} + \mathbf{A}\right)^{-1}\frac{D_m\tau}{R_\Omega^2}(\mathrm{Sh}_{e\Omega}^{-1}\mathbf{I} + \mu\mathbf{Sh}_i^{-1})^{-1} \\
&= \frac{D_m\tau}{R_\Omega^2}\mathbf{A}\left\{\frac{D_m}{k\delta_c R_\Omega}\mathbf{I} + (\mathrm{Sh}_{e\Omega}^{-1}\mathbf{I} + \mu\mathbf{Sh}_i^{-1})\mathbf{A}\right\}^{-1}, \\
&= \left[\frac{R_\Omega}{k\tau\delta_c}\mathbf{A}^{-1} + \frac{R_\Omega^2}{D_m\tau}(\mathrm{Sh}_{e\Omega}^{-1}\mathbf{I} + \mu\mathbf{Sh}_i^{-1})\right]^{-1}.
\end{aligned} \tag{28.59}$$

Limiting cases

1. No external and internal resistances to mass transfer:
   In this case, the Damköhler matrix (equation (28.59)) reduces to

$$\mathbf{Da} = \frac{\delta_c}{R_\Omega}k\tau\mathbf{A} \tag{28.60}$$

2. No pore diffusion resistance:
   In this case, the overall mass transfer matrix consists of only the external mass-transfer resistance, i. e.,

$$\mathbf{k}_{co} = \mathbf{k}_{ce} = \frac{D_m}{R_\Omega}\mathrm{Sh}_{e\Omega}\mathbf{I}$$

   and the Damkohler matrix (equation (28.59)) reduces to

$$\mathbf{Da} = \left\{\frac{R_\Omega}{\delta_c k\tau}\mathbf{A}^{-1} + \frac{R_\Omega^2}{D_m\tau}\frac{1}{\mathrm{Sh}_{e\Omega}}\mathbf{I}\right\}^{-1}. \tag{28.61}$$

   Note that in the limit of fast kinetics, $k \to \infty$ (or $\phi^2 \gg 1$), the Damköhler matrix simplifies to diagonal form

$$\mathbf{Da} = \frac{D_m \tau}{R_\Omega^2} \mathrm{Sh}_{e\Omega} \mathbf{I}, \tag{28.62}$$

which is independent of kinetics (and hence temperature if average value of diffusivity is used).

3. No external mass transfer resistance:
   In this case, the overall mass transfer matrix consists of only the internal mass-transfer resistance, i. e.,

$$\mathbf{k}_{co} = \mathbf{k}_{ci} = \frac{D_e}{\delta_c} \mathbf{Sh}_i$$

and the Damköhler matrix simplifies to

$$\begin{aligned} \mathbf{Da} &= \left[\frac{R_\Omega}{k\tau\delta_c} \mathbf{A}^{-1} + \frac{R_\Omega^2 \mu}{D_m \tau} \mathbf{Sh}_i^{-1} \right]^{-1}. \\ &= \frac{\delta_c k\tau}{R_\Omega} (\mathbf{A}^{-1} + \phi^2 \mathbf{Sh}_i^{-1})^{-1}. \end{aligned} \tag{28.63}$$

If $\mathbf{Sh}_i = \mathrm{Sh}_{i\infty} \mathbf{I}$ (i. e., asymptotic approximation), the Damköhler matrix reduces (equation (28.63)) to

$$\mathbf{Da} = \frac{\delta_c k\tau}{R_\Omega} \mathbf{A} \left(\mathbf{I} + \frac{\phi^2}{\mathrm{Sh}_{i\infty}} \mathbf{A} \right)^{-1}.$$

For $\phi^2 \gg 1$ (i. e., at high temperature or fast kinetics), the asymptotic approximation of internal Sherwood number reduces the above expression for Damköhler matrix to a diagonal form

$$\mathbf{Da} = \frac{D_e \tau}{R_\Omega \delta_c} \mathrm{Sh}_{i\infty} \mathbf{I},$$

which is independent of kinetics, and hence temperature (except the temperature dependence may be come through the diffusivity term). This is also true in general even when external mass-transfer resistance is present (see limiting case 4 below).

4. Asymptotic internal Sherwood number (independent of kinetics):
   If internal Sherwood number can be expressed by only asymptotic value, i. e., $\mathbf{Sh}_i = \mathrm{Sh}_{i\infty} \mathbf{I}$, then the overall mass transfer coefficient matrix is diagonal, i. e.,

$$\begin{aligned} \mathbf{k}_{co} &= \left(\frac{R_\Omega}{D_m \mathrm{Sh}_{e\Omega}} + \frac{\delta_c}{D_e \mathrm{Sh}_{i\infty}} \right)^{-1} \mathbf{I} \\ &= \frac{D_m}{R_\Omega} \left(\frac{1}{\mathrm{Sh}_{e\Omega}} + \frac{\mu}{\mathrm{Sh}_{i\infty}} \right)^{-1} \mathbf{I} \end{aligned}$$

and the Damkohler matrix reduces from equation (28.59) to

$$\mathbf{Da} = \left[\frac{R_\Omega}{k\tau\delta_c}\mathbf{A}^{-1} + \frac{R_\Omega^2}{D_m\tau}\left(\frac{1}{\mathrm{Sh}_{e\Omega}} + \frac{\mu}{\mathrm{Sh}_{i\infty}}\right)\mathbf{I}\right]^{-1}. \tag{28.64}$$

If the first term is negligible or $k \to \infty$, the Damköhler matrix can be further simplified from equation (28.64) to

$$\mathbf{Da} = \frac{D_e\tau}{R_\Omega^2}\left(\frac{1}{\mathrm{Sh}_{e\Omega}} + \frac{\mu}{\mathrm{Sh}_{i\infty}}\right)^{-1}\mathbf{I},$$

which is independent of kinetics, and hence temperature as discussed earlier. However, for fast kinetics, the internal Sherwood number usually depends on kinetics and asymptotic representation may be oversimplification.

5. Pore diffusion control limit with no external mass-transfer resistance:
If there is no external mass-transfer resistance and kinetics is fast ($\phi^2 \gg 1$ or $k \to \infty$), the reactive transport process may be pore diffusion controlled. In this case, the internal Sherwood number may be simplified from equation (28.42) or equation (28.43) with $\mathbf{\Phi}_w^2 = \phi^2\mathbf{A}$ as

$$\mathbf{Sh}_i \equiv \mathbf{\Phi}_w = \phi\sqrt{\mathbf{A}} = \delta_c\sqrt{\frac{k}{D_e}\mathbf{A}},$$

which is the fast kinetics asymptote. Thus, in this limit, the Damkohler matrix (equation (28.59)) reduces to

$$\mathbf{Da} = \frac{D_e\tau}{\delta_c R_\Omega}\mathbf{Sh}_i = \frac{D_e\tau}{\delta_c R_\Omega}\phi\sqrt{\mathbf{A}} = \frac{\sqrt{kD_e}}{R_\Omega}\tau\sqrt{\mathbf{A}}. \tag{28.65}$$

Note that for the reaction network and rate constants shown in Figure 28.4, the matrix $\mathbf{A}$ is given in equation (28.56), which leads to

$$\sqrt{\mathbf{A}} = \begin{pmatrix} 0.9033 & -0.2951 & -0.0435 & -0.0174 \\ -0.5901 & 0.8598 & -0.1997 & -0.0413 \\ -0.1738 & -0.3994 & 0.5929 & -0.1749 \\ -0.1393 & -0.1653 & -0.3497 & 0.2336 \end{pmatrix}. \tag{28.66}$$

The positive square root of matrix $\mathbf{A}$ (when it has all positive eigenvalues) can be obtained either using spectral method or Caley–Hamilton theorem (see Chapter 5). Thus, while original reaction constant matrix is $\mathbf{K}_v = k\mathbf{A}$, the observed (pore diffusion disguised) reaction rate constant matrix becomes $\mathbf{K}_{\mathrm{obs}} = \frac{\sqrt{kD_e}}{R_\Omega}\sqrt{\mathbf{A}}$. Figure 28.5 shows the effect of pore diffusion on reaction network suggesting the total number of observed reaction to be 12 in contrast to 6 original reactions (i. e., 6 new reactions appear—three reversible reactions between species A and D, A and C and B and D).

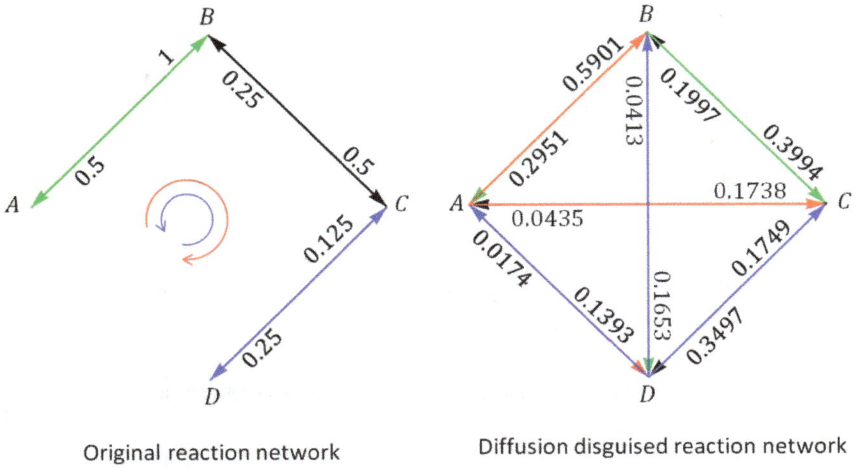

Figure 28.5: Original and pore diffusion disguised reaction networks.

In this case, the exit concentration vector can be expressed from equation (28.51)–(28.54) as

$$\mathbf{c}_{fe} = \exp\left[-\frac{\sqrt{kD_e}}{R_\Omega}\tau\sqrt{\mathbf{A}}\right]\mathbf{c}_{f,\text{in}}. \tag{28.67}$$

Taking the temperature dependence of the rate constant and other parameters values as

$$k = 10^{12}\exp\left(\frac{-12000}{T}\right)\,\text{s}^{-1}; \quad T \text{ in K}$$
$$R_\Omega = 100\,\mu\text{m}; \quad D_e = 10^{-7}\,\text{m}^2/\text{s}; \quad \tau = 1\,\text{ms},$$

and inlet concentration vector as

$$\mathbf{c}_{f,\text{in}} = c_{\text{in}}\begin{pmatrix} 0.97 & 0.01 & 0.01 & 0.01 \end{pmatrix}^T,$$

the exit concentration of each of the four components is calculated (using equation (28.51)) and plotted against temperature in Figure 28.6.

As expected, the concentration of A decreases and the concentration of D increases monotonically, while the concentration of the intermediate components can vary nonmonotonically with temperature and may exhibit maxima. In this specific case, the concentration of component C exhibits maxima near $T = 915.6$ K.

The above calculations may be repeated for the case of negligible pore diffusional resistance or external mass transfer to show that the presence of either external or internal mass transfer reduces the yield (or maximum attainable concentration) of intermediate products.

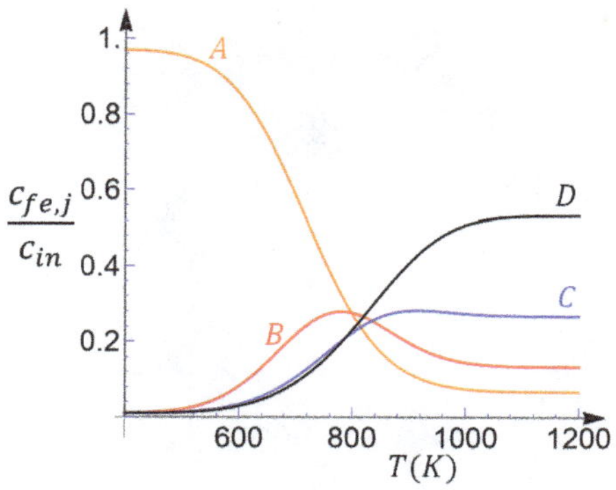

Figure 28.6: Exit concentrations of the four components in pore diffusion controlled limit.

Problems

1. Consider the three regular geometries of a sphere of diameter $2a$, cylinder of height and diameter $2a$ and a cube of side $2a$.
   (a) Show that the effective diffusion length (R_Ω) is the same for all three geometries
   (b) Formulate the diffusion–reaction problem with linear kinetics and uniform activity for the three geometries and obtain the solution.
   (c) Determine and plot the effectiveness factor and comment on the shape of the curves.
   (d) Determine and plot the internal Sherwood number for the three cases.
2. Consider a porous catalyst particle in the form of a hollow cylinder of length $2L$, inside radius μa ($0 \leq \mu < 1$) and outside radius a as shown in Figure 28.7.

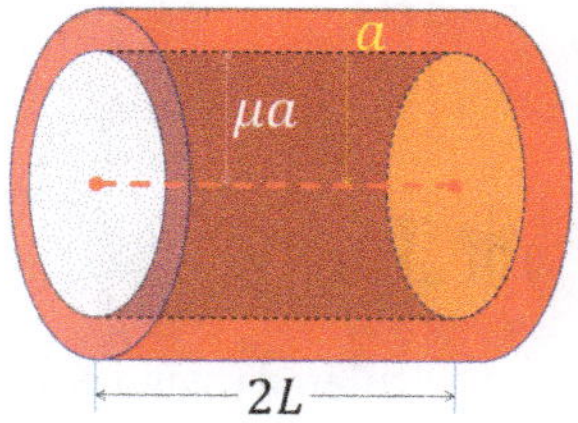

Figure 28.7: Schematic of a porous catalyst particle in form of hollow cylinder.

   (a) Formulate the diffusion–reaction problem for the case of Dirichlet BC at the interior and exterior surface.
   (b) Solve the model for the case of linear kinetics and determine an expression for the effectiveness factor η.
   (c) Determine the internal Sherwood number Sh_i and the various limits of η and Sh_i for $\mu \to 0$, $\mu \to 1$, $\frac{L}{a} \to \infty$ and $\frac{L}{a} \to 0$.

3. The steady-state concentration of a reactant inside a cylindrical pore with wall reaction is described by the partial differential equation

$$\frac{1}{r}\frac{\partial}{\partial r}\left(r\frac{\partial C}{\partial r}\right)+\frac{\partial^2 C}{\partial x^2}=0;\quad 0<r<R,\ 0<x<L,$$

with boundary conditions

$$\frac{\partial C}{\partial x}=0\ @\ x=0;\quad C=C_0\ @\ x=L;\quad C=\text{finite}\ @\ r=0;\quad -D\frac{\partial C}{\partial r}=kC\ @\ r=R.$$

Here, R is the radius of pore, L is the length, k is the reaction rate constant, D is the molecular diffusivity of the reactant and C_0 is the concentration of the reactant in the gas at the pore mouth.

(a) Cast the equations in dimensionless form and obtain a formal solution to the concentration profile

(b) The quantity of interest is the pore effectiveness factor defined by

$$\eta=\frac{D}{2\pi RLkC_0}\int_0^R\frac{\partial C}{\partial x}(L,r)2\pi r\,dr.$$

Determine an expression for the effectiveness factor

(c) Simplify the expression in (b) for the limiting case in which the pore radius is very small compared to the length.

4. Consider the nonlinear boundary value problem describing multicomponent diffusion and reaction in a catalyst layer:

$$\mathbf{D}\frac{d^2\mathbf{c}}{dx^2}=\mathbf{r}(\mathbf{c}),\quad 0<x<L$$

$$\mathbf{c}=\mathbf{c}_s\ @\ x=L;\quad \frac{d\mathbf{c}}{dx}=0\ @\ x=0.$$

Show that linearization of the BVP at the surface conditions leads to

$$\frac{d^2\mathbf{u}}{dx^2}=\Phi^2\mathbf{u}-\alpha,\quad \mathbf{u}'(0)=0,\quad \mathbf{u}(1)=0,$$

where

$$\mathbf{u}=\mathbf{c}_s-\mathbf{c},\quad \Phi^2=L^2\mathbf{D}^{-1}\mathbf{J},\quad \alpha=L^2\mathbf{D}^{-1}\mathbf{r}(\mathbf{c}_s)$$

and $\mathbf{J}$ is the Jacobian of the rate vector $\mathbf{r}(\mathbf{c})$ evaluated at $\mathbf{c}=\mathbf{c}_s$. Solve the linearized problem and show that the flux vector

$$\mathbf{j}_s=\mathbf{D}\left.\frac{d\mathbf{c}}{dx}\right|_{x=L}$$

may be approximated by

$$\mathbf{j}_s = L\mathbf{DHD}^{-1}\mathbf{r}(\mathbf{c}_s), \quad \mathbf{H} = \mathbf{\Phi}^{-1}\tanh\mathbf{\Phi}.$$

Discuss how you would use this result in a reactor model.

5. Consider the problem of diffusion and reaction in an isothermal spherical catalyst particle of radius a. When several first-order reactions occur, the concentration vector satisfies the equations

$$\mathbf{D}\frac{1}{r^2}\frac{d}{dr}\left(r^2\frac{d\mathbf{c}}{dr}\right) = \mathbf{Kc}, \quad 0 < r < a$$
$$\mathbf{c} = \mathbf{c}_0 \quad @r = a$$
$$\mathbf{c} \text{ finite at } r = 0$$

where $\mathbf{D}$ is a positive definite matrix of diffusivities, $\mathbf{K}$ is a nonnegative definite matrix of rate constants, and $\mathbf{c}_0$ is the vector of concentrations at pellet surface.

(a) Cast the equations in dimensionless form and obtain a formal solution. Determine the relationship between the concentration vectors at the pellet center and surface.

(b) If the quantity of interest is the observed (or diffusion-disguised) rate vector defined by

$$\mathbf{r}_{obs} = \frac{1}{(\frac{4}{3}\pi a^3)}\int_0^a 4\pi r^2\mathbf{Kc}(r)dr,$$

then show that

$$\mathbf{r}_{obs} = \mathbf{K}^*\mathbf{c}_0,$$

where the diffusion-disguised rate constant matrix $\mathbf{K}^*$ is given by

$$\mathbf{K}^* = \mathbf{KH},$$

with $\mathbf{H} = 3(\mathbf{\Phi}\coth\mathbf{\Phi} - \mathbf{I})\mathbf{\Phi}^{-2}$ and $\mathbf{\Phi}^2 = \mathbf{D}^{-1}\mathbf{K}a^2$ ($\mathbf{H}$ is the effectiveness matrix).

(c) Consider the special case in which the diffusivity matrix is given by

$$\mathbf{D} = d\,\mathbf{I};$$

and let

$$\mathbf{K} = k\mathbf{A}; \quad \mathbf{K}^* = k\mathbf{A}^*; \quad \mathbf{\Phi}^2 = \phi^2\mathbf{A}; \quad \phi^2 = \frac{ka^2}{d}$$

where $\mathbf{A}$ (or $\mathbf{A}^*$) is the relative rate constant (or diffusion-disguised relative rate constant) matrix. Use the result in (b) to determine an expression for $\mathbf{A}^*$. Discuss the two limiting cases of negligible pore diffusion resistance ($\phi \to 0$) and strong pore diffusion resistance ($\phi >> 1$). Calculate the diffusion-disguised relative rate constant matrix when $\phi = 1$ for the case of three consecutive irreversible reactions among 4 species when all rate constants are equal.

29 Packed-bed chromatography

A packed-bed is used to carry out reactions and separations. In Chapter 17, we examined the problem of heat storage and transfer in a packed-bed. Here, we examine a closely related problem of mass transfer with chromatography.

29.1 Model formulation

Consider a packed-bed of particles as shown in Figure 29.1 (top) with a schematic of the porous particle packing and fluid flow around it.

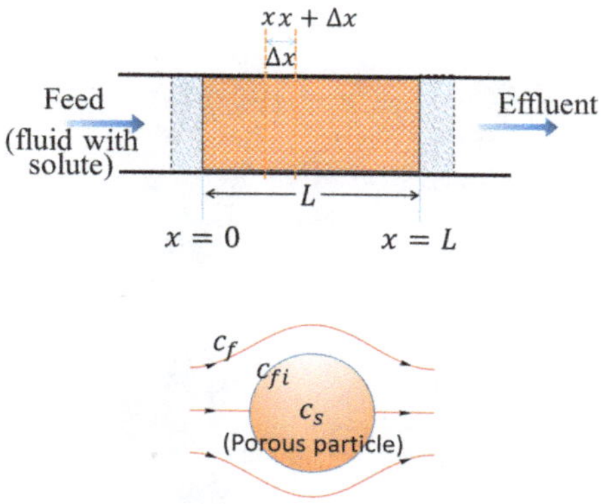

Figure 29.1: Schematic of a packed-bed of porous particles for adsorption of solute from a fluid.

In developing a model, the following assumptions are made: (i) plug flow of fluid, (ii) uniform and constant transport properties, (iii) small particles (i. e., no internal gradients), (iv) dilute solution and (v) negligible axial dispersion in fluid phase. With these assumptions, the species (solute) balance can be expressed in fluid phase for the small volume element (between x and $x + \Delta x$) as follows:

$$\frac{\partial}{\partial t}[A_c \Delta x \varepsilon c_f] = A_c \varepsilon u_0 c_f|_x - A_c \varepsilon u_0 c_f|_{x+\Delta x} - A_c \Delta x a_v k_c (c_f - c_{fi}) \tag{29.1}$$

$\Longrightarrow$

$$\varepsilon \frac{\partial c_f}{\partial t} = -\varepsilon u_0 \frac{\partial c_f}{\partial x} - k_c a_v (c_f - c_{fi}) \tag{29.2}$$

where ε is the bed porosity, A_c is the bed cross-section area, u_0 is fluid interstitial velocity, c_f is the solute concentration, k_c is the mass transfer coefficient, a_v is the fluid-solid interfacial area per unit bed volume, c_{fi} is the solute concentration in fluid phase at the

https://doi.org/10.1515/9783111598055-035

solid-fluid interface and c_s is the solute concentration in the solid phase. Similarly, the solid phase species balance can be expressed as

$$\frac{\partial}{\partial t}[A_c \Delta x(1-\varepsilon)c_s] = A_c \Delta x a_v k_c (c_f - c_{fi}) \tag{29.3}$$

$\Longrightarrow$

$$(1-\varepsilon)\frac{\partial c_s}{\partial t} = k_c a_v (c_f - c_{fi}). \tag{29.4}$$

The above model (equations (29.2) and (29.4)) is not closed until c_{fi} is represented in terms of c_s or c_f. For this, we can utilize the adsorption isotherm as described below.

29.1.1 Adsorption isotherm

The rate of adsorption of the solute from fluid to solid phase r_a can be expressed as

$$r_a = k_a c_{fi}(c_{s0} - c_s) \tag{29.5}$$

where c_{s0} is the saturation concentration in the solid phase (adsorption capacity of the solid) and k_a is the adsorption rate constant. Defining the fractional adsorbed concentration θ as

$$\theta = \frac{c_s}{c_{s0}} = \text{fraction of adsorbed sites}, \tag{29.6}$$

the rate of adsorption can be expressed as

$$r_a = k_a c_{fi} c_{s0}(1-\theta). \tag{29.7}$$

The rate of desorption (from solid to fluid phase) may be expressed as

$$r_d = k_d c_s = k_d c_{s0}\theta. \tag{29.8}$$

If we assume local equilibrium, or assume that adsorption/desorption rates are much faster as compared to convective mass transfer, we have

$$k_a c_{fi} c_{s0}(1-\theta) = k_d c_{s0}\theta$$

or

$$\frac{\theta}{1-\theta} = \frac{k_a}{k_d} c_{fi} = K_{\text{eq}} c_{fi}; \quad K_{\text{eq}} = \frac{k_a}{k_d} = \text{adsorption equilibrium constant} \tag{29.9}$$

$\Longrightarrow$

$$\theta = \frac{c_s}{c_{s0}} = \frac{K_{\text{eq}} c_{fi}}{1 + K_{\text{eq}} c_{fi}} \tag{29.10}$$

Equation (29.10) gives the so-called Langmuir isotherm, which is also plotted in Figure 29.2.

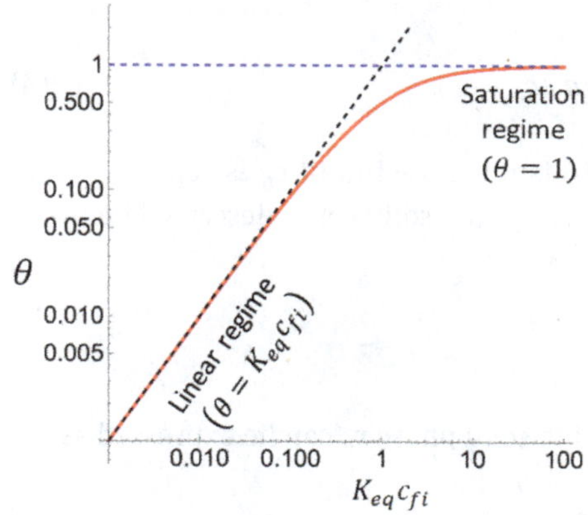

Figure 29.2: Langmuir isotherm demonstrating the linear regime ($K_{eq} c_{fi} \ll 1$) and saturation regime ($K_{eq} c_{fi} \gg 1$).

When $K_{\text{eq}} c_{fi} \ll 1$, we can write

$$c_s = (K_{\text{eq}} c_{s0}) c_{fi} = K c_{fi}; \quad K = K_{\text{eq}} c_{s0}, \tag{29.11}$$

which shows that the isotherm can be linearized in this regime (see Figure 29.2). Here, K is the dimensionless adsorption equilibrium constant. In this linear regime, the model (equations (29.2) and (29.4)) becomes closed and linear, and can be expressed as follows:

$$\varepsilon\left(\frac{\partial c_f}{\partial t} + u_0 \frac{\partial c_f}{\partial x}\right) = -k_c a_v (c_f - c_{fi}) \tag{29.12}$$

$$(1 - \varepsilon)\frac{\partial c_s}{\partial t} = k_c a_v (c_f - c_{fi}) \tag{29.13}$$

$$c_s = K c_{fi} \tag{29.14}$$

along with initial and inlet conditions:

$$c_f(x, t = 0) = c_{f0}(x); \quad c_s(x, t = 0) = c_{s0}(x) \tag{29.15}$$

$$c_f(x = 0, t) = c_{fin}(t) \tag{29.16}$$

For consistency, we should have initial conditions related as $c_{s0}(x) = K c_{f0}(x)$. For a column that is free from adsorbate (solute) initially, we can take $c_{f0}(x) = c_{s0}(x) = 0$. Below we consider only this specific case.

29.1.2 Nondimensional form

We can define the dimensionless spatial coordinate and time as follows:

$$z = \frac{x}{L}; \quad \tau = \frac{u_0 t}{L}; \quad L = \text{column length} \tag{29.17}$$

and the following dimensionless groups:

$$\alpha = \frac{K(1-\varepsilon)}{\varepsilon} = \text{capacitance ratio} \tag{29.18}$$

$$p = \frac{\varepsilon u_0}{k_c a_v L} = \frac{1/(k_c a_v)}{L/(\varepsilon u_0)} = \text{local (or transverse) Peclet number} \tag{29.19}$$

where $1/(k_c a_v) = t_m$ represents the characteristic time for external mass-transfer; $\varepsilon u_0 = \langle u \rangle$ is the superficial velocity and $L/\langle u \rangle = L/(\varepsilon u_0) = t_c$ represents the convection (or space) time; the local Peclet number p is the ratio of external mass-transfer time to space time. Thus, the model (equations (29.12)–(29.16)) can be expressed in nondimensional form as follows:

$$p\left(\frac{\partial c_f}{\partial \tau} + \frac{\partial c_f}{\partial z}\right) = -(c_f - c_{fi}), \tag{29.20}$$

$$\alpha p \frac{\partial c_{fi}}{\partial \tau} = (c_f - c_{fi}), \quad \tau > 0, \quad 0 < z < 1 \tag{29.21}$$

along with initial and inlet conditions:

$$c_f(z, \tau = 0) = 0; \quad c_{fi}(z, \tau = 0) = 0; \tag{29.22}$$

$$c_f(z = 0, \tau) = c_{in}(\tau) \tag{29.23}$$

Note that c_s is eliminated by using the local equilibrium relation equation (29.14). Similarly, c_{fi} can be eliminated from equation (29.20) and can be expressed in terms of c_f as

$$c_{fi} = c_f + p\left(\frac{\partial c_f}{\partial \tau} + \frac{\partial c_f}{\partial z}\right) \tag{29.24}$$

which can be substituted in equation (29.21) to obtain

$$\alpha p\left(\frac{\partial c_f}{\partial \tau} + p\frac{\partial^2 c_f}{\partial \tau^2} + p\frac{\partial^2 c_f}{\partial \tau \partial z}\right) = -p\left(\frac{\partial c_f}{\partial \tau} + \frac{\partial c_f}{\partial z}\right)$$

$\Longrightarrow$

$$\frac{\partial c_f}{\partial \tau} + \frac{1}{1+\alpha}\frac{\partial c_f}{\partial z} + \frac{\alpha p}{1+\alpha}\frac{\partial^2 c_f}{\partial \tau \partial z} + \frac{\alpha p}{1+\alpha}\frac{\partial^2 c_f}{\partial \tau^2} = 0 \tag{29.25}$$

$$c_f(z, \tau = 0) = 0; \quad c_f(z = 0, \tau) = c_{in}(\tau) \tag{29.26}$$

[Note also that $\frac{\partial c_f}{\partial \tau}(z, \tau = 0) = 0$ for the initial conditions given by equation (29.22)]. The above model (equations (29.25)–(29.26)) is a single second-order PDE (of hyperbolic type) for $c_f(z, \tau)$. The exit concentration, i. e., $c_f(z = 1, \tau)$, can be solved and plotted as a function of time for any given input (or initial condition). The response to a unit step input is referred to as the *breakthrough curve* while response to a Dirac delta function (or pulse) input is referred to as the *dispersion curve.*

29.1.3 Limiting case: $p \longrightarrow 0$

In the limiting case of $p \longrightarrow 0$, the external mass transfer resistance is neglected and $c_{fi} = c_f$. In this limit, the model equations (29.25)–(29.26) reduce to a single hyperbolic equation (plug flow) as

$$\frac{\partial c_f}{\partial \tau} + \frac{1}{1+\alpha}\frac{\partial c_f}{\partial z} = 0, \quad \tau > 0, \quad 0 < z < 1 \tag{29.27}$$

$$c_f(z, \tau = 0) = 0; \quad c_f(z = 0, \tau) = c_{in}(\tau) \tag{29.28}$$

The above model can be solved by Laplace transform (LT). Let

$$\widehat{c_f}(z, s) = \mathcal{L}[c_f(z, \tau)] = \text{Laplace transform of } c_f(z, \tau) \tag{29.29}$$

Taking LT, equations (29.27)–(29.28) give

$$\frac{d\widehat{c_f}}{dz} = -(1+\alpha)s\widehat{c_f}; \quad \widehat{c_f}(z = 0) = \widehat{c_{in}}(s)$$

$\Longrightarrow$

$$\widehat{c_f} = \exp[-(1+\alpha)sz]\widehat{c_{in}}(s), \tag{29.30}$$

which after taking inverse LT, we get

$$c_f(z, \tau) = c_{in}(\tau - (1+\alpha)z). \tag{29.31}$$

Thus, for unit-step input given by the Heaviside's unit-step function

$$c_{in}(\tau) = H(\tau) = \begin{cases} 1, & \tau > 0 \\ 0, & \tau < 0, \end{cases} \tag{29.32}$$

the concentration profile can be obtained from equation (29.31) as

$$c_f(z,\tau) = H(\tau - (1+\alpha)z) = \begin{cases} 1, & \tau > (1+\alpha)z \\ 0, & \tau < (1+\alpha)z \end{cases} \tag{29.33}$$

Thus, the step input or discontinuity moves with a dimensionless speed $\frac{dz}{d\tau} = \frac{1}{1+\alpha}$, or the adsorption front moves with a velocity

$$\frac{dx}{dt} = \frac{u_0}{1 + \frac{K(1-\varepsilon)}{\varepsilon}}. \tag{29.34}$$

It can also be seen from Figure 29.3, which shows the profiles of unit-step input and the concentration. However, for finite but small values of p, the front velocity remains the same but the front is not sharp due to dispersion or mass transfer effects.

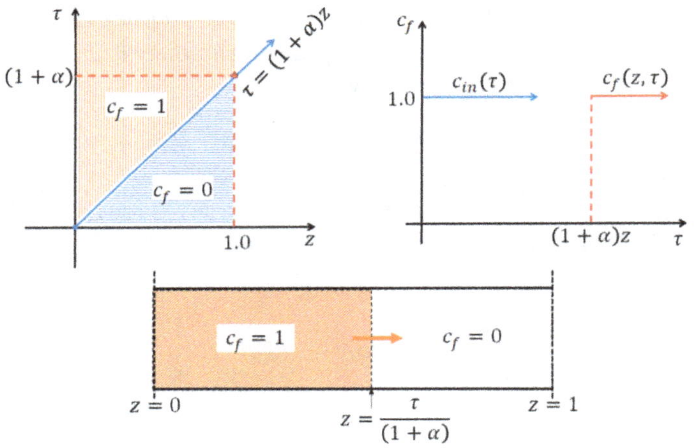

Figure 29.3: Solution profile with unit step input for packed-bed chromatography in the plug flow limit.

Equation (29.34) is of fundamental importance in chromatographic separations. It shows that a solute that interacts with the packing and having an adsorption equilibrium constant of K moves through the bed at a speed of $\frac{u_0}{1+\alpha}$, $\alpha = K(\frac{1-\varepsilon}{\varepsilon})$. Thus, if a pulse containing several solutes (having different K_i) is injected at the inlet to the column, the solutes move with different velocities, with the one having high K_i value moving slowly. Thus, the solutes are separated with the separation distance increasing along the column.

It also follows from equation (29.34) that when the adsorbate/particles/solid in the bed is initially free of the solute and a step input is introduced, the bed gets saturated with the solute and breakthrough occurs (as a step function) at $t = \frac{L}{u_0}(1+\alpha)$. As we show below, when mass transfer and dispersion are present, the breakthrough curves (and separation of solutes) are not sharp.

29.2 Similarity with heat transfer in packed-beds

As shown in Section 17.5.4, the analogous heat transfer problem is described by

$$p_h\left(\frac{\partial\theta_f}{\partial\tau}+\frac{\partial\theta_f}{\partial z}\right)=-(\theta_f-\theta_s); \tag{29.35}$$

$$\alpha_h p_h\frac{\partial\theta_s}{\partial\tau}=(\theta_f-\theta_s),\quad \tau>0,0<z,1 \tag{29.36}$$

$$\theta_f(z,0)=0;\quad \theta_s(z,0)=0 \tag{29.37}$$

$$\theta_f(0,\tau)=\theta_{in}(\tau) \tag{29.38}$$

Here, the local Peclet number p_h is the ratio of local solid to fluid heat transfer time t_h to the space time t_c. We note that the model (equations (29.35)–(29.38)) for heat transfer in packed-bed is identical to that for chromatography in packed-bed, where θ_f, θ_s, α_h, p_h and θ_{in} can be replaced by c_f, c_{fi}, α, p and c_{in}, respectively.

29.3 Impact of interphase mass transfer

Now consider the packed-bed chromatography model given by equations (29.25)–(29.26). For small values of p, i. e., $p\to 0$, the model leads to

$$\frac{\partial c_f}{\partial\tau}\approx-\frac{1}{1+\alpha}\frac{\partial c_f}{\partial z}+O(p). \tag{29.39}$$

Using the above approximation in equation (29.25), we get

$$(1+\alpha)\frac{\partial c_f}{\partial\tau}+\frac{\partial c_f}{\partial z}+\alpha p\left(\frac{\partial^2 c_f}{\partial\tau\partial z}-\frac{1}{1+\alpha}\frac{\partial^2 c_f}{\partial\tau\partial z}+O(p)\right)=0$$

$\Longrightarrow$

$$(1+\alpha)\frac{\partial c_f}{\partial\tau}+\frac{\partial c_f}{\partial z}+\frac{\alpha^2 p}{1+\alpha}\frac{\partial^2 c_f}{\partial\tau\partial z}=0+O(p^2), \tag{29.40}$$

which is the hyperbolic form of the model. The same approximation (equation (29.39)) can be used again in equation (29.40) that leads to

$$(1+\alpha)\frac{\partial c_f}{\partial\tau}+\frac{\partial c_f}{\partial z}+\frac{\alpha^2 p}{1+\alpha}\left(\frac{-1}{1+\alpha}\frac{\partial^2 c_f}{\partial z^2}+O(p)\right)=0+O(p^2)$$

$\Longrightarrow$

$$(1+\alpha)\frac{\partial c_f}{\partial\tau}+\frac{\partial c_f}{\partial z}-\frac{\alpha^2 p}{(1+\alpha)^2}\frac{\partial^2 c_f}{\partial z^2}=0+O(p^2)$$

$\Longrightarrow$

$$\frac{\partial c_f}{\partial \tau} + \frac{1}{(1+\alpha)}\frac{\partial c_f}{\partial z} - \frac{\alpha^2 p}{(1+\alpha)^3}\frac{\partial^2 c_f}{\partial z^2} = 0 + O(p^2) \tag{29.41}$$

which is the parabolic form of the model where the dimensionless dispersion coefficient is $\frac{\alpha^2 p}{(1+\alpha)^3}$.

The model equation can also be expressed in interfacial concentration mode (c_{fi}) for chromatography or solid temperature (θ_s) for heat transfer. For this, let us consider the two-mode model (equations (29.20)–(29.23)). We can use equation (29.21) to express c_f in terms of c_{fi} as

$$c_f = c_{fi} + \alpha p\frac{\partial c_{fi}}{\partial \tau} \tag{29.42}$$

and substitute it into equation (29.20), which leads to

$$p\left(\frac{\partial c_{fi}}{\partial \tau} + \frac{\partial c_{fi}}{\partial z} + \alpha p\frac{\partial^2 c_{fi}}{\partial \tau^2} + \alpha p\frac{\partial^2 c_{fi}}{\partial \tau \partial z}\right) = -\alpha p\frac{\partial c_{fi}}{\partial \tau}$$

$\Longrightarrow$

$$\frac{\partial c_{fi}}{\partial \tau} + \frac{1}{(1+\alpha)}\frac{\partial c_{fi}}{\partial z} + \frac{\alpha p}{(1+\alpha)}\frac{\partial^2 c_{fi}}{\partial \tau \partial z} + \frac{\alpha p}{(1+\alpha)}\frac{\partial^2 c_{fi}}{\partial \tau^2} = 0, \tag{29.43}$$

which is the same equation as that satisfied by c_f as given in equation (29.25). Therefore, using equation (29.11), $c_{fi} = \frac{c_s}{K}$ can be substituted in above equation (29.43), which leads to equation in solid phase concentration c_s as

$$\frac{\partial c_s}{\partial \tau} + \frac{1}{(1+\alpha)}\frac{\partial c_s}{\partial z} + \frac{\alpha p}{(1+\alpha)}\frac{\partial^2 c_s}{\partial \tau \partial z} + \frac{\alpha p}{(1+\alpha)}\frac{\partial^2 c_s}{\partial \tau^2} = 0 \tag{29.44}$$

Thus, all concentration modes c_f, c_{fi} and c_s satisfy the same equation (see equations (29.25), (29.43) and (29.44)).

Thus, in the limit of $p \to 0$, the front velocity is the same in solid and fluid phases. In addition, dispersion of the front in the two phases is also the same. In general, this is not true when axial conduction or intraparticle gradients are included.

[Remark: Though the differential equations satisfied by $c_f(z,\tau)$, $c_{fi}(z,\tau)$ and $c_s(z,\tau)$ are the same, the initial and inlet conditions are different for these variables.]

29.3.1 Pseudo-homogeneous model

We define the volume averaged concentration by

$$c_m = \varepsilon c_f + (1-\varepsilon)c_s = \varepsilon c_f + K(1-\varepsilon)c_{fi}. \tag{29.45}$$

Since, both c_f and c_{fi} satisfy the same linear differential equation, we multiply the c_f-equation by ε and c_{fi}-equation by $K(1-\varepsilon)$, and add, which leads to

$$(1+\alpha)\frac{\partial c_m}{\partial \tau} + \frac{\partial c_m}{\partial z} + \alpha p\frac{\partial^2 c_m}{\partial \tau \partial z} + \alpha p\frac{\partial^2 c_m}{\partial \tau^2} = 0 \tag{29.46}$$

For small values of p, the above hyperbolic model can be expressed in parabolic form as

$$\frac{\partial c_m}{\partial \tau} + \frac{1}{(1+\alpha)}\frac{\partial c_m}{\partial z} - \frac{\alpha^2 p}{(1+\alpha)^3}\frac{\partial^2 c_m}{\partial z^2} = 0$$

$\Longrightarrow$ in dimensional form as

$$\frac{\partial c_m}{\partial t} + \frac{u_0}{(1+\alpha)}\frac{\partial c_m}{\partial x} - \frac{\alpha^2}{(1+\alpha)^3}u_0^2\frac{\varepsilon}{k_c a_v}\frac{\partial^2 c_m}{\partial x^2} = 0$$

$\Longrightarrow$

$$\frac{\partial c_m}{\partial t} + u_{\text{eff}}\frac{\partial c_m}{\partial x} - D_{\text{eff}}\frac{\partial^2 c_m}{\partial x^2} = 0; \tag{29.47}$$

$$u_{\text{eff}} = \frac{u_0}{1+\alpha}; \quad D_{\text{eff}} = \frac{\varepsilon}{k_c a_v}\frac{\alpha^2 u_0^2}{(1+\alpha)^3} \tag{29.48}$$

where u_{eff} and D_{eff} are the effective velocity and effective dispersion coefficient, respectively. This leads to the effective axial Peclet number Pe_{eff} as

$$\text{Pe}_{\text{eff}} = \frac{u_{\text{eff}}L}{D_{\text{eff}}} = \frac{(1+\alpha)^2}{\alpha^2}\frac{Lk_c a_v}{u_0\varepsilon} = \frac{(1+\alpha)^2}{\alpha^2}\frac{1}{p} \tag{29.49}$$

29.4 Solution of the hyperbolic model by Laplace transform

Consider the two-phase hyperbolic model (equations (29.20)–(29.23)). Taking the Laplace transform, we can write equation (29.21) as

$$\alpha p s\widehat{c_{fi}} = (\widehat{c_f} - \widehat{c_{fi}}) \Longrightarrow \widehat{c_{fi}}(z,s) = \frac{\widehat{c_f}(z,s)}{1+\alpha p s} \tag{29.50}$$

Thus, equation (29.20) gives

$$\frac{d\widehat{c_f}}{dz} = -s\widehat{c_f} - \frac{as}{1+aps}\widehat{c_f}; \quad \widehat{c_f}(z=0,s) = \widehat{c_{in}}(s)$$

$\Longrightarrow$

$$\widehat{c_f}(z,s) = \exp\left[-sz - \frac{asz}{1+aps}\right]\widehat{c_{in}}(s). \tag{29.51}$$

We have already discussed the inversion of equation (29.51) in Section 17.5.4.

For a pulse input, the dispersion curve $E(\tau)$ is given by

$$E(\tau) = \begin{cases} \exp[-\frac{1}{p} - \frac{(\tau-1)}{ap}].[\sqrt{\frac{1}{ap^2(\tau-1)}}I_1(2\sqrt{\frac{(\tau-1)}{ap^2}}) + \delta(\tau-1)], & \tau > 1 \\ 0, & \tau < 1 \end{cases} \tag{29.52}$$

Long time asymptote

In the limit of $u \gg 1, I_1(u) \approx \frac{\exp[u]}{\sqrt{2\pi u}}$. Therefore, the dispersion curve in the limit of $\frac{\tau-1}{ap^2} \gg 1$ (or $\tau \gg 1 + ap^2$) can be obtained from equation (29.52) as follows:

$$\begin{aligned} E(\tau)|_{\tau\gg 1+ap^2} &= \sqrt{\frac{1}{4\pi p}}\frac{1}{[a(\tau-1)^3]^{1/4}}\exp\left[-\frac{1}{p} - \frac{(\tau-1)}{ap} + \frac{2}{p}\sqrt{\frac{(\tau-1)}{a}}\right] \\ &= \left(\frac{1}{16\pi^2p^2a(\tau-1)^3}\right)^{1/4}\exp\left[-\frac{(\sqrt{(\tau-1)}-\sqrt{a})^2}{ap}\right]. \end{aligned} \tag{29.53}$$

The above expression can be simplified to evaluate the E-value at the effective or mean residence time $\tau = 1 + a$ as

$$E(\tau)|_{\tau=1+a} = \frac{1}{a\sqrt{4\pi p}}. \tag{29.54}$$

Figure 29.4 shows the dispersion curves $E(\tau)$ for specific value of $a = 5$ and p-values from 0.01 to 0.1. It can be seen from this figure that for small p-values (i. e., $p \to 0$), the dispersion curve is symmetric around $\tau = 1 + a$ with maximum value given in equation (29.54). For large p-values, the dispersion curve is asymmetric and may have long tail.

For a unit-step input, $c_{in}(\tau) = 1$ (or $\widehat{c_{in}}(s) = \frac{1}{s}$), the breakthrough curve is given by

$$F(\tau) = \begin{cases} 0, & \tau < 1 \\ c_f^*(\tau,1), & \tau > 1 \end{cases} \tag{29.55}$$

$$c_f^*(\tau,1) = \exp\left[-\frac{1}{p}\right]\left[\begin{array}{l} \exp[-\frac{\tau-1}{ap}].I_0(2\sqrt{\frac{(\tau-1)}{ap^2}}) \\ \quad + \frac{1}{ap}\int_0^{\tau-1}\exp[-\frac{u}{ap}].I_0(2\sqrt{\frac{u}{ap^2}})\,du \end{array}\right]. \tag{29.56}$$

The breakthrough curve (i. e., exit concentration versus time plot) at various times is shown in Figure 29.5 for $a = 5$ and p-values in the range of 0.01 to 0.2. Here, breakthrough

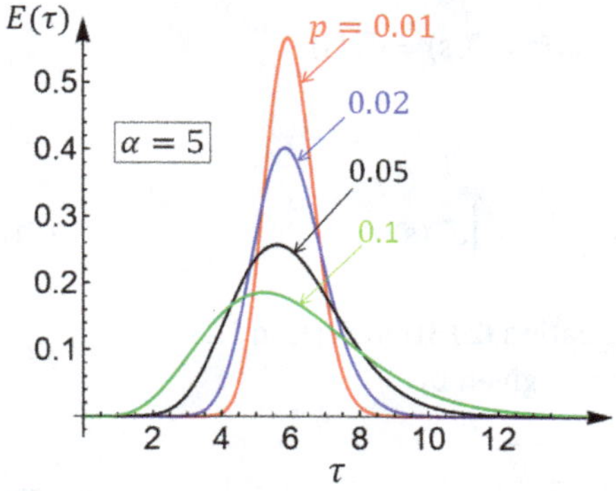

Figure 29.4: Dispersion curves $E(\tau)$ for packed-bed chromatography for $\alpha = 5$ and p-values varying from 0.01 to 0.1.

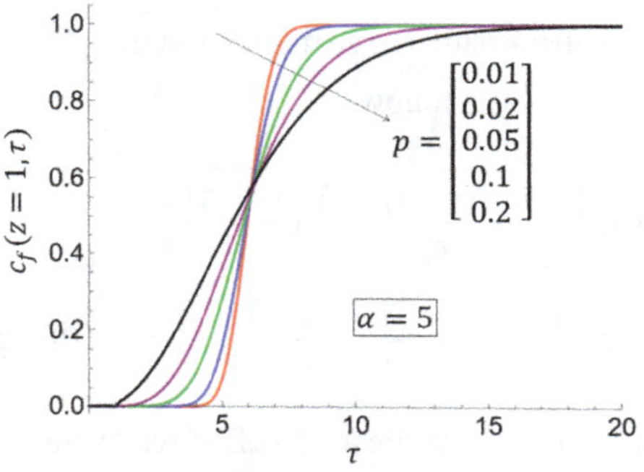

Figure 29.5: Breakthrough curve for packed-bed chromatography for various transverse Peclet number p.

occurs at $\tau = 1 + \alpha = 6$ (as expected), where all the curves (for small p) pass through the point $(6, 0.5)$. However, as p increases, the breakthrough curves are not symmetric.

29.5 Chromatography model with dispersion in fluid phase

When dispersion in the fluid phase is included, the model can be expressed as

$$p\left(\frac{\partial c_f}{\partial \tau} + \frac{\partial c_f}{\partial z}\right) = -(c_f - c_{fi}) + \frac{p}{\mathrm{Pe}_{mf}} \frac{\partial^2 c_f}{\partial z^2} \tag{29.57}$$

$$\alpha p \frac{\partial c_{fi}}{\partial \tau} = (c_f - c_{fi}), \quad \tau > 0, \quad 0 < z < 1 \tag{29.58}$$

with initial conditions:

$$c_f(z, 0) = c_{fi}(z, 0) = 0 \quad \text{(zero ICs)}, \tag{29.59}$$

inlet boundary condition:

$$\frac{1}{\mathrm{Pe}_{mf}}\frac{\partial c_f}{\partial z} = c_f - c_{in}(\tau) \ @ \ z = 0 \tag{29.60}$$

and exit boundary condition:

$$\frac{\partial c_f}{\partial z} = 0 \ @ \ z = 1. \tag{29.61}$$

Here, we have three dimensionless parameters: α, p and Pe_{mf}. The new parameter is the axial Peclet number $\mathrm{Pe}_{mf} = \frac{u_0 L}{D_{xe}}$, where D_{xe} is the effective axial dispersion coefficient.

29.5.1 Limiting cases

1. $\mathrm{Pe}_{mf} \to \infty$ (i. e., negligible dispersion in fluid phase). In this case, the model reduces to hyperbolic form and is discussed earlier.
2. $\mathrm{Pe}_{mf} \to 0$ (i. e., fluid phase is well mixed or no axial gradient in fluid or solid phase). In this case, the model reduces to a set of ODEs:

$$p\frac{dc_f}{d\tau} = -(c_f - c_{fi}) + p[c_{in}(\tau) - c_f] \tag{29.62}$$

$$\alpha p\frac{dc_{fi}}{d\tau} = (c_f - c_{fi}), \quad \tau > 0, \tag{29.63}$$

$$c_f = c_{fi} = 0 \ @ \ \tau = 0. \tag{29.64}$$

   This is also referred to as lumped model (or single stage model).
3. $p \to 0$ or $\alpha \to 0$. In this case, the model reduces to axial dispersion model, which is discussed in detail in previous chapters.

29.5.2 Lumped model for $p \to 0$

The lumped model (equations (29.62)–(29.64)) in the limit of $p \to 0$ (i. e., neglecting difference between c_f and c_{fi}) reduces to

$$p\frac{dc_f}{d\tau} + \alpha p\frac{dc_f}{d\tau} = p[c_{in}(\tau) - c_f]; \quad c_f(\tau = 0) = 0$$

$\Longrightarrow$

$$(1+\alpha)\frac{dc_f}{d\tau} = c_{in}(\tau) - c_f, \quad \tau > 0; \quad c_f = 0 \ @ \ \tau = 0 \tag{29.65}$$

LT ($\tau \to s$) gives

$$(1+\alpha)s\widehat{c_f} = \widehat{c_{in}}(s) - \widehat{c_f}$$
$$\Longrightarrow \widehat{c_f} = \frac{\widehat{c_{in}}(s)}{1+s(1+\alpha)} = \frac{\widehat{c_{in}}(s)}{(1+\alpha)[s+\frac{1}{(1+\alpha)}]} \tag{29.66}$$

For a unit step input: $\widehat{c_{in}}(s) = \frac{1}{s}$, we get

$$\widehat{c_f} = \frac{\widehat{c_{in}}(s)}{(1+\alpha)s[s+\frac{1}{(1+\alpha)}]} = \frac{1}{s} - \frac{1}{s+\frac{1}{(1+\alpha)}} \tag{29.67}$$

$\Longrightarrow$ the exit concentration or breakthrough curve is given by

$$c_f(\tau) = 1 - e^{-\frac{\tau}{1+\alpha}}. \tag{29.68}$$

29.5.3 Lumped model for $p > 0$

Consider the lumped model for $p > 0$ as given in equations (29.62)–(29.64). Taking LT gives

$$sp\widehat{c_f} = -(\widehat{c_f} - \widehat{c_{fi}}) + p(\widehat{c_{in}}(s) - \widehat{c_f}) \tag{29.69}$$
$$asp\widehat{c_{fi}} = \widehat{c_f} - \widehat{c_{fi}} \Longrightarrow \widehat{c_{fi}} = \frac{\widehat{c_f}}{1+aps} \tag{29.70}$$

Substituting equation (29.70) in equation (29.69) and simplifying leads to

$$\widehat{c_f} = \frac{(1+aps)\widehat{c_{in}}(s)}{[1+as+aps+s+aps^2]}. \tag{29.71}$$

Thus, the breakthrough curves can be obtained by considering a unit step input ($\widehat{c_{in}}(s) = \frac{1}{s}$) as

$$\widehat{F}(s) = \frac{(1+aps)}{s[1+as+aps+s+aps^2]}, \tag{29.72}$$

while the dispersion curve can be obtained by considering a unit impulse input ($\widehat{c_{in}}(s) = 1$) as

$$\widehat{E}(s) = \frac{(1+aps)}{[1+as+aps+s+aps^2]}. \tag{29.73}$$

Note that the LT of the dispersion curve (equation (29.73)) has two poles for $p \neq 0$ given by

$$s_1, s_2 = \frac{-(1+a+ap) \pm \sqrt{\triangle}}{2ap}; \tag{29.74}$$

$$\begin{aligned}\triangle &= (1+\alpha+\alpha p)^2 - 4\alpha p \\ &= (1+\alpha)^2 + \alpha^2 p^2 + 2\alpha(\alpha-1)p. \end{aligned} \tag{29.75}$$

It can be shown that $\triangle > 0$ for all $\alpha > 0$ and $p > 0$, and s_1, s_2 are always real and negative.

Write

$$\begin{aligned}\widehat{F}(s) &= \frac{1}{s}\widehat{E}(s) = \frac{1+\alpha ps}{s\alpha p(s-s_1)(s-s_2)} \\ &= \frac{s+\frac{1}{\alpha p}}{s(s-s_1)(s-s_2)}. \end{aligned} \tag{29.76}$$

$\Longrightarrow$

$$\text{Residue } e^{st}\widehat{F}(s)|_{s=0} = \frac{1}{\alpha p s_1 s_2} = 1 \tag{29.77}$$

$$\text{Residue } e^{s\tau}\widehat{F}(s)|_{s=s_1} = \frac{e^{s_1\tau}(s_1+\frac{1}{\alpha p})}{s_1(s_1-s_2)} = \beta_1 e^{s_1\tau} \tag{29.78}$$

$$\text{Residue } e^{s\tau}\widehat{F}(s)|_{s=s_2} = \frac{e^{s_2\tau}(s_2+\frac{1}{\alpha p})}{s_2(s_2-s_1)} = \beta_2 e^{s_2\tau} \tag{29.79}$$

$\Longrightarrow$

$$F(\tau) = 1 + \beta_1 e^{s_1\tau} + \beta_2 e^{s_2\tau},$$

where β_1 and β_2 are given by equations (29.78) and (29.79). An example of the breakthrough curve is plotted in Figure 29.6 corresponding to $p = 0.01$ and $\alpha = 5.0$.

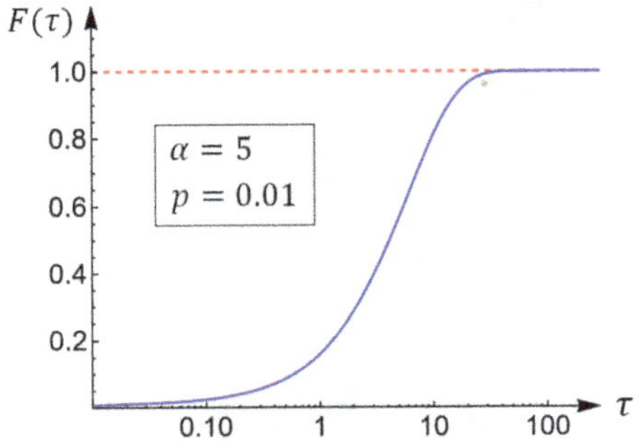

Figure 29.6: A plot of the breakthrough curve from lumped model with $\alpha = 5$ and $p = 0.01$.

29.5.4 Chromatography model with dispersion in fluid phase for unit impulse input

Consider the chromatography model with dispersion in fluid phase (equations (29.57)–(29.61)) with unit impulse input $c_{in}(\tau) = \delta(\tau)$. LT of equations (29.57)–(29.61) gives

$$p\left(s\widehat{c_f} + \frac{d\widehat{c_f}}{dz}\right) = -(\widehat{c_f} - \widehat{c_{fi}}) + \frac{p}{\mathrm{Pe}_{mf}}\frac{d^2\widehat{c_f}}{dz^2} \tag{29.80}$$

$$aps\widehat{c_{fi}} = (\widehat{c_f} - \widehat{c_{fi}}) \Longrightarrow \widehat{c_{fi}} = \frac{\widehat{c_f}}{1+aps} \tag{29.81}$$

$\Longrightarrow$

$$\frac{1}{\mathrm{Pe}_{mf}}\frac{d^2\widehat{c_f}}{dz^2} - \frac{d\widehat{c_f}}{dz} - s\widehat{c_f} - \frac{as}{1+aps}\widehat{c_f} = 0, \tag{29.82}$$

with boundary conditions:

$$\frac{1}{\mathrm{Pe}_{mf}}\frac{d\widehat{c_f}}{dz} = \widehat{c_f} - 1\,@\,z = 0; \qquad \frac{d\widehat{c_f}}{dz} = 0\,@\,z = 1. \tag{29.83}$$

Note that for $p = 0$ or $a = 0$, the above model reduces to axial dispersion model while in the limit of $\mathrm{Pe}_{mf} \to \infty$, it reduces to the hyperbolic model. Here, we consider the mixed case where a, p and Pe_{mf} are finite. In this case, the LT of the dispersion curve can be expressed by solving equations (29.82)–(29.83) as

$$\widehat{E}(s) = \widehat{c_f}(z=1,s) = \frac{4qe^{\frac{\mathrm{Pe}_{mf}}{2}}}{\left[(1+q)^2 e^{\frac{\mathrm{Pe}_{mf}q}{2}} - (1-q)^2 e^{\frac{-\mathrm{Pe}_{mf}q}{2}}\right]} \tag{29.84}$$

where

$$q = \sqrt{1 + \frac{4s}{\mathrm{Pe}_{mf}}\left(1 + \frac{a}{1+aps}\right)}. \tag{29.85}$$

The above equations (29.84)–(29.85) can be numerically inverted. Figure 29.7 shows a plot of dispersion curves from numerical LT inversion for $a = 5$, $p = 0.01$ and for various values of Pe_{mf} in the range of 1 to 1000.

The LT solution (equations (29.84)–(29.85)) can also be expanded as power series in s to obtain the temporal moments without Laplace inversion. Expanding $\widehat{E}(s)$ gives

$$\widehat{E}(s) = 1 - (1+a)s + \frac{s^2}{2}M_2 + O(s^3) \tag{29.86}$$

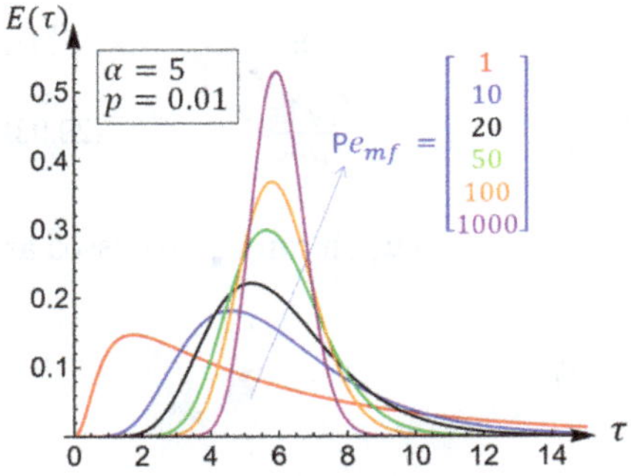

Figure 29.7: Dispersion curves for chromatography model with dispersion in fluid phase for $\alpha = 5$, $p = 0.01$ and various values of $\mathrm{Pe}_m f$ in the range of 1 to 1000.

where

$$M_2 = 1 + 2\alpha + \alpha^2 + 2p\alpha^2 + \frac{(1+\alpha)^2}{\mathrm{Pe}_{mf}^2}[2\,\mathrm{Pe}_{mf} + 2e^{-\mathrm{Pe}_{mf}} - 2] \tag{29.87}$$

Thus, the zeroth and first moments are

$$M_0 = 1; \quad M_1 = (1+\alpha) \tag{29.88}$$

and the dimensionless variance is

$$\sigma_\theta^2 = \frac{M_2 - M_1^2}{M_1^2} = \frac{2p\alpha^2}{(1+\alpha)^2} + \left[\frac{2}{\mathrm{Pe}_{mf}} - \frac{2}{\mathrm{Pe}_{mf}^2}(1 - e^{-\mathrm{Pe}_{mf}})\right], \tag{29.89}$$

where the first term (in equation (29.89)) is the dispersion contribution due to external mass transfer and second term (in square bracket) is the dispersion contribution due to axial mixing in the fluid phase.

29.5.5 Finite stage chromatography model

Consider again the hyperbolic model (equations (29.20)–(29.23)). When the upwinding first-order discretization scheme is used for the convection term, the discretized model can be expressed as

$$\frac{dc_{f,j}}{d\tau} + \frac{c_{f,j} - c_{f,j-1}}{\Delta z} = -\frac{(c_{f,j} - c_{fi,j})}{p}; \quad c_{f,0} = c_{in}(\tau); \tag{29.90}$$

$$\alpha\frac{dc_{fi,j}}{d\tau} = \frac{(c_{f,j} - c_{fi,j})}{p}; \quad c_{f,j}(\tau = 0) = 0 = c_{fi,j}(\tau = 0); \quad j = 1, 2, \ldots, N \tag{29.91}$$

Taking LT gives

$$ps\widehat{c_{f,j}} + (\widehat{c_{f,j}} - \widehat{c_{f,j-1}})Np = -(\widehat{c_{f,j}} - \widehat{c_{fi,j}}); \quad \widehat{c_{f,0}} = \widehat{c_{in}}(s); \tag{29.92}$$

$$aps\widehat{c_{fi,j}} = (\widehat{c_{f,j}} - \widehat{c_{fi,j}}) \Longrightarrow \widehat{c_{fi,j}} = \frac{\widehat{c_{f,j}}}{1+aps} \tag{29.93}$$

where $\Delta z = \frac{1}{N}$, and N is the number of stages. Thus, the LT solution can be expressed as

$$\widehat{c_{f,j}} = \frac{\widehat{c_{f,j-1}}}{1 + \frac{s}{N} + \frac{as}{(1+asp)N}}; \quad \widehat{c_{f,0}} = \widehat{c_{in}}(s) \tag{29.94}$$

$\Longrightarrow$

$$\widehat{c_{f,N}} = \frac{\widehat{c_{in}}(s)}{[1 + \frac{s}{N} + \frac{as}{(1+asp)N}]^N} \tag{29.95}$$

For large N, i. e., when $N \to \infty$, equation (29.95) reduces to

$$c_f(s) = e^{-(s+\frac{as}{1+asp})}.\widehat{c_{in}}(s) \tag{29.96}$$

The LT solution (equation (29.95)) can be expanded as power series in s to determine the moments as discussed earlier.

29.6 Impact of intraparticle gradients

The packed-bed chromatography model examined in the previous sections does not include gradients that could exist within the particles, and is valid only for beds of small particles. We now extend the model to include intraparticle gradients. Let the bed be composed of spherical particles of diameter $2a$ and let D_e be the effective intraparticle diffusivity of the solute. Using the same notation, the model may be expressed as

$$\frac{\partial c_f}{\partial \tau} + \frac{\partial c_f}{\partial z} = -\frac{1}{p}(c_f - c_{fi}) \tag{29.97}$$

$$\left(\frac{1-\varepsilon}{\varepsilon}\right)\frac{\partial \overline{c_s}}{\partial \tau} = \frac{1}{p}(c_f - c_{fi}) \tag{29.98}$$

$$\Lambda\frac{\partial c_s}{\partial \tau} = \frac{1}{\xi^2}\frac{\partial}{\partial \xi}\left(\xi^2\frac{\partial c_s}{\partial \xi}\right), \quad 0 < \xi < 1 \tag{29.99}$$

$$\frac{\partial c_s}{\partial \xi} = 0 \;@\; \xi = 0; \quad \left.\frac{\partial c_s}{\partial \xi}\right|_{\xi=1} = \mathrm{Sh}(c_f - c_{fi}) \tag{29.100}$$

$$c_s = c_{si} = Kc_{fi} \;@\; \xi = 1 \tag{29.101}$$

with initial and boundary conditions

$$c_f(z,0) = 0; \quad c_s(\xi, z, 0) = 0; \quad c_f(0,\tau) = c_{in}(\tau). \tag{29.102}$$

The new dimensionless groups appearing above are given by

$$\Lambda = \frac{a^2 u_0}{LD_e}; \quad \mathrm{Sh}^* = \frac{k_c a}{D_e}; \quad \mathrm{Sh} = \frac{\mathrm{Sh}^*}{5}.$$

Here, $\overline{c_s}$ is the average concentration of the adsorbed solute in the particle:

$$\overline{c_s} = \int_0^1 3\xi^2 c_s(\xi, z, \tau)\, d\xi. \tag{29.103}$$

It may be seen that for $(a^2/D_e) \to 0$, the gradient inside the particle becomes negligible and $\overline{c_s} = c_{si} = Kc_{fi}$ and the model reduces to the hyperbolic two-phase model. When the gradient inside the particle is small, by simplifying equations (29.99)–(29.101) and (29.103), it may be shown that

$$\overline{c_s} = \left(1 + \frac{\mathrm{Sh}}{K}\right) c_{si} - \mathrm{Sh}\, c_f. \tag{29.104}$$

[Remark: In the literature, this approximation is known as the linear driving force or parabolic profile approximation.] Using equation (29.104), the model (equation (29.98)) may be expressed as

$$\alpha\left(1 + \frac{\mathrm{Sh}}{K}\right)\frac{\partial c_{fi}}{\partial \tau} - \left(\frac{1-\varepsilon}{\varepsilon}\right)\mathrm{Sh}\,\frac{\partial c_f}{\partial \tau} = \frac{1}{p}(c_f - c_{fi}). \tag{29.105}$$

Equations (29.97) and (29.105) along with the initial and inlet conditions:

$$c_f(z,0) = 0; \quad c_{fi}(z,0) = 0; \quad c_f(0,\tau) = c_{in}(\tau), \tag{29.106}$$

define the model that includes intraparticle gradients.

The LT method may be used to obtain an analytic solution to equations (29.97), (29.105) and (29.106). We note that elimination of c_{fi} gives

$$(1+\alpha)\frac{\partial c_f}{\partial \tau} + \frac{\partial c_f}{\partial z} + \alpha p\left(1 + \frac{\mathrm{Sh}}{K}\right)\left(\frac{\partial^2 c_f}{\partial \tau^2} + \frac{\partial^2 c_f}{\partial z \partial \tau}\right) = 0. \tag{29.107}$$

Using leading order approximation, equation (29.107) may be written as

$$\frac{\partial c_f}{\partial \tau} + \frac{1}{(1+\alpha)}\frac{\partial c_f}{\partial z} + \frac{\alpha^2 p}{(1+\alpha)^3}\left(1 + \frac{\mathrm{Sh}}{K}\right)\frac{\partial^2 c_f}{\partial z^2} + O(p^2) = 0. \tag{29.108}$$

Thus, as can be expected, including of intraparticle gradients does not change the speed of the adsorption front but the effective axial Peclet number becomes

$$\frac{1}{\mathrm{Pe}_{\mathrm{eff}}} = \frac{\alpha^2}{(1+\alpha)^3} p\left(1 + \frac{\mathrm{Sh}}{K}\right) \tag{29.109}$$

or in dimensional form

$$D_{\mathrm{eff}} = \frac{\varepsilon u_0^2 a^2}{(1+\alpha)^3}\left[\frac{1}{k_c a_v} + \frac{a^2}{15 D_e}\right]. \tag{29.110}$$

We note that the first term in the parentheses of equation (29.110) is the external mass transfer time while the second term represents the intraparticle diffusion time.

Problems

1. Consider the problem of solute uptake by a spherical particle (of radius a), which is initially of solute free and surface exposed to a time varying concentration $c_{si}(t)$. Solve the intraparticle diffusion problem and show that the concentration within the particle is given by

$$c_s(r,t) = \frac{2D_e}{a}\sum_{n=1}^{\infty}\frac{(-1)^{n+1} n\pi}{r}\sin\left[\frac{n\pi r}{a}\right]\int_0^t e^{-\frac{D_e n^2\pi^2}{a^2}(t-t')} c_{si}(t')\,dt'$$

2. Obtain the solution, and hence the breakthrough curves for the chromatography model with intraparticle gradients defined by equations (29.97), (29.105) and (29.106) of Section 29.6.
3. Determine the dimensionless second central moment of the dispersion curve for the chromatographic model that accounts for external mass transfer, dispersion in the fluid phase and intraparticle gradients.
4. Consider the problem of chromatography in a coated circular tube in which the adsorbent layer thickness is very small (so that transverse gradients in the layer can be neglected).
   (a) Formulate a model similar to that in section 29.1 assuming flat velocity profile and neglecting axial diffusion.
   (b) Cast the model in dimensionless form, identify the dimensionless groups and compare the same with that of packed-bed.
   (c) Discuss the impact of velocity profile (flat versus parabolic) on the observed breakthrough curve in the linear regime.
5. Extend the chromatography model of problem 4 above to include axial dispersion in the fluid phase and also gradients in the adsorbent layer. Discuss the impact of these extensions on the observed breakthrough curves.

30 Stability of transport and reaction processes

In this chapter, we discuss two problems in which the stability of a base solution is determined by examining the eigenvalues of the linearized system of differential equations.

30.1 Lapwood convection in a porous rectangular box

30.1.1 Model formulation

Consider a fluid filled porous medium in a closed rectangular box of dimensions $L \times H$ as shown schematically in Figure 30.1.

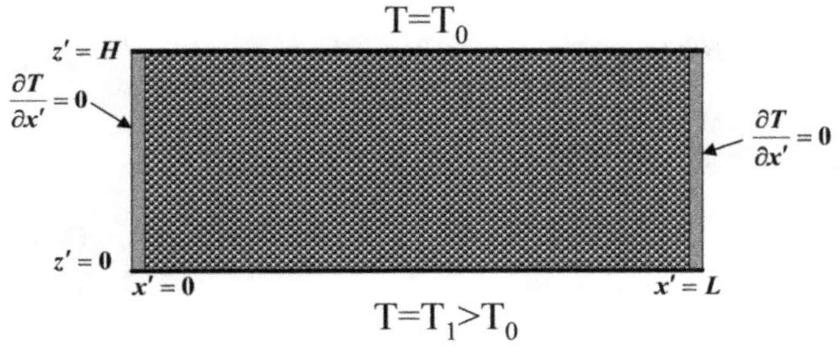

Figure 30.1: Schematic diagram illustrating Lapwood convection in a porous rectangular box.

The (continuity, momentum/Darcy's law and energy balance) equations describing the velocity, pressure and temperature of the fluid and porous medium (Figure 30.1) are given by

$$\nabla' \cdot \mathbf{u}' = 0 \quad \text{(continuity equation)} \tag{30.1}$$

$$\nabla' p' = -\rho g \mathbf{e}_z - \frac{\mu}{\kappa}\mathbf{u}' \quad \text{(Darcy's law)} \tag{30.2}$$

$$\sigma \frac{\partial T}{\partial t} = -\mathbf{u}' \cdot \nabla' T + \lambda_e \nabla'^2 T \quad \text{(energy balance)} \tag{30.3}$$

where $\mathbf{u}'$, p' and T are velocity vector, pressure and temperature, respectively; $\sigma = \frac{\rho_m c_{pm}}{\rho_f c_{pf}}$ is the dimensionless heat capacity ratio; κ is the permeability; λ_e is the effective thermal diffusivity; g is the acceleration due to gravity; $\mathbf{e}_z$ is the unit vector in the direction of gravity and ρ is the density, which may vary with temperature and can be expressed using the Boussinesq approximation as

$$\rho = \rho_0 \cdot [1 - \beta(T - T_0)]. \tag{30.4}$$

https://doi.org/10.1515/9783111598055-036

The boundary and initial conditions are given by

$$T = T_1 \,@\, z' = 0; \quad T = T_0 \,@\, z' = H \tag{30.5}$$

$$\frac{\partial T}{\partial x'} = 0 \,@\, x' = 0, L; \quad T = T^*(x', z') \,@\, t = 0 \tag{30.6}$$

$$\mathbf{n}.\mathbf{u}' = 0 \,@\, z' = 0, \quad H \text{ and } x' = 0, L \tag{30.7}$$

Here, **n** is the unit normal to the boundary.

Dimensionless form

We define the following dimensionless variables and parameters:

$$z = \frac{z'}{H}, \quad x = \frac{x'}{H}, \quad a = \frac{L}{H}, \quad \theta = \frac{T - T_0}{T_1 - T_0}, \quad \mathbf{u} = \frac{H}{\lambda_e}\mathbf{u}', \quad \nabla' = \frac{1}{H}\nabla, \tag{30.8}$$

$$\tau = \frac{\lambda_e t}{\sigma H^2}, \quad Ra_d = \frac{\kappa g \rho_0 H \beta (T_1 - T_0)}{\mu \lambda_e}, \quad p = \frac{Ra_d}{\beta(T_1 - T_0)}\left[\frac{p'}{\rho_0 g H} + z\right] \tag{30.9}$$

and express the model equations in the following dimensionless form:

$$\nabla \cdot \mathbf{u} = 0 \tag{30.10}$$

$$\nabla p = \mathrm{Ra}_d\, \theta \mathbf{e}_z - \mathbf{u} \tag{30.11}$$

$$\frac{\partial \theta}{\partial \tau} = -\mathbf{u} \cdot \nabla \theta + \nabla^2 \theta \tag{30.12}$$

with boundary conditions,

$$\theta = 1 \,@\, z = 0; \quad \theta = 0 \,@\, z = 1; \tag{30.13}$$

$$\frac{\partial \theta}{\partial x} = 0 \,@\, x = 0, a; \quad \theta = \theta^* \,@\, \tau = 0 \tag{30.14}$$

$$\mathbf{u} \cdot \mathbf{n} = 0 \,@\, z = 0,1 \quad \text{and} \quad x = 0, a, \tag{30.15}$$

where Ra_d is known as the Darcy–Rayleigh number. Note that Darcy's law does not permit the specification of tangential velocity at a boundary. We can only specify the normal component of the velocity to be zero. In component form, the model may be written as

$$\frac{\partial u_x}{\partial x} + \frac{\partial u_z}{\partial z} = 0 \tag{30.16}$$

$$\frac{\partial p}{\partial x} = -u_x \tag{30.17}$$

$$\frac{\partial p}{\partial z} = \mathrm{Ra}_d\, \theta - u_z \tag{30.18}$$

$$\frac{\partial \theta}{\partial \tau} = -u_x \frac{\partial \theta}{\partial x} - u_z \frac{\partial \theta}{\partial z} + \frac{\partial^2 \theta}{\partial x^2} + \frac{\partial^2 \theta}{\partial z^2}. \tag{30.19}$$

We can satisfy the continuity equation and remove the pressure and velocity variables from these equations by introducing the stream function $\psi(x, z)$ as follows:

$$u_x = -\frac{\partial \psi}{\partial z}, \quad u_z = \frac{\partial \psi}{\partial x} \tag{30.20}$$

$\Longrightarrow$

$$\frac{\partial p}{\partial x} = -u_x = \frac{\partial \psi}{\partial z} \Rightarrow \frac{\partial^2 p}{\partial x \partial z} = \frac{\partial^2 \psi}{\partial z^2} \tag{30.21}$$

$$\frac{\partial p}{\partial z} = \mathrm{Ra}_d\, \theta - \frac{\partial \psi}{\partial x} \Rightarrow \frac{\partial^2 p}{\partial z \partial x} = \mathrm{Ra}_d \frac{\partial \theta}{\partial x} - \frac{\partial^2 \psi}{\partial x^2} \tag{30.22}$$

$\Longrightarrow$

$$\frac{\partial^2 \psi}{\partial x^2} + \frac{\partial^2 \psi}{\partial z^2} = \mathrm{Ra}_d \frac{\partial \theta}{\partial x} \tag{30.23}$$

Therefore, the model equations may be expressed in terms of two scalar variables $\psi(x,y)$ and $\theta(x,y)$ as

$$\nabla^2 \psi = \mathrm{Ra}_d \frac{\partial \theta}{\partial x} \tag{30.24}$$

$$\frac{\partial \theta}{\partial \tau} = \frac{\partial \psi}{\partial z} . \frac{\partial \theta}{\partial x} - \frac{\partial \psi}{\partial x} . \frac{\partial \theta}{\partial z} + \nabla^2 \theta, \quad 0 < x < a, \quad 0 < z < 1 \tag{30.25}$$

with the boundary conditions

$$\left.\frac{\partial \psi}{\partial z}\right|_{x=0,a} = 0 \quad \text{and} \quad \left.\frac{\partial \psi}{\partial x}\right|_{z=0,1} = 0 \tag{30.26}$$

$$\left.\frac{\partial \theta}{\partial x}\right|_{x=0,a} = 0; \quad \theta|_{z=0} = 1; \quad \theta|_{z=1} = 0, \tag{30.27}$$

and appropriate initial conditions for the time-dependent case. The boundary conditions on ψ can also be taken as

$$\psi = 0 \ @\ x = 0,\ a \text{ and } z = 0, 1 \tag{30.28}$$

instead of equation (30.26). Both sets of boundary conditions give the same solution for the stability boundary, and in what follows we use the second set given by equation (30.28).

30.1.2 Conduction state and its stability

The steady-state model can be obtained by setting $\frac{\partial}{\partial \tau} = 0$ and may be written as

$$F\begin{pmatrix} \psi \\ \theta \end{pmatrix} = \begin{pmatrix} \nabla^2 \psi - \mathrm{Ra}_d \frac{\partial \theta}{\partial x} \\ \nabla^2 \theta + \frac{\partial \psi}{\partial z}\frac{\partial \theta}{\partial x} - \frac{\partial \psi}{\partial x}\frac{\partial \theta}{\partial z} \end{pmatrix} = \begin{pmatrix} 0 \\ 0 \end{pmatrix} \tag{30.29}$$

$$\psi = 0 \,@\, x = 0,\ a; \quad \textit{and} \quad z = 0, 1 \tag{30.30}$$

$$\left.\frac{\partial \theta}{\partial x}\right|_{x=0,a} = 0; \quad \theta|_{z=0} = 1; \quad \theta|_{z=1} = 0; \tag{30.31}$$

Note that the only nonlinear terms in the model equations are the quadratic convection terms in the energy balance of equation (30.29). The base state or conduction solution that exists for all values of Ra_d is given by

$$\psi_0(x,z) = 0 \quad \text{and} \quad \theta_0(x,z) = 1 - z \tag{30.32}$$

As the Rayleigh number Ra_d increases, the buoyancy force overcomes the viscous force and the fluid begins to move or convection sets in. Our aim is to determine the critical value of Ra_d at which the conduction state loses stability leading to convective states. We also note that if $\begin{pmatrix} \psi(x,z) \\ \theta(x,z) \end{pmatrix}$ is a solution of equations (30.29)–(30.31) then so is $\begin{pmatrix} -\psi(a-x,z) \\ \theta(a-x,z) \end{pmatrix}$. Thus, the convective solutions appear in pairs having reflectional symmetry in the domain. Let $\mathbf{v}$ and $\mathbf{u}_0$ given by

$$\mathbf{v} = \begin{pmatrix} v_1(x,z) \\ v_2(x,z) \end{pmatrix}, \quad \mathbf{u}_0 = \begin{pmatrix} \psi_0(x,z) \\ \theta_0(x,z) \end{pmatrix} = \begin{pmatrix} 0 \\ 1-z \end{pmatrix}, \tag{30.33}$$

denote the perturbation to the base state and the base state, respectively. To determine the stability of the base state $\mathbf{u}_0$, we linearize the model equations:

$$\begin{aligned}
&DF(\psi_0, \theta_0, \mathrm{Ra}_d) \cdot \mathbf{v} \\
&= \lim_{s\to 0} \frac{\partial}{\partial s} F(\mathbf{u}_0 + s\mathbf{v}) = \lim_{s\to 0} \frac{\partial}{\partial s} \begin{pmatrix} F_1(\psi_0 + s v_1, \theta_0 + s v_2) \\ F_2(\psi_0 + s v_1, \theta_0 + s v_2) \end{pmatrix} \\
&= \lim_{s\to 0} \frac{\partial}{\partial s} \begin{pmatrix} \nabla^2\psi_0 + s\nabla^2 v_1 - \mathrm{Ra}_d \frac{\partial \theta_0}{\partial x} - s\,\mathrm{Ra}_d \frac{\partial v_2}{\partial x} \\ \nabla^2\theta + s\nabla^2 v_2 + (\frac{\partial \psi_0}{\partial z} + s\frac{\partial v_1}{\partial z})(\frac{\partial \theta_0}{\partial x} + s\frac{\partial v_2}{\partial x}) - (\frac{\partial \psi_0}{\partial x} + s\frac{\partial v_1}{\partial x})(\frac{\partial \theta_0}{\partial z} + s\frac{\partial v_2}{\partial z}) \end{pmatrix} \\
&= \begin{pmatrix} \nabla^2 v_1 - \mathrm{Ra}_d \frac{\partial v_2}{\partial x} \\ \nabla^2 v_2 + \frac{\partial \theta_0}{\partial x}\frac{\partial v_1}{\partial z} + \frac{\partial \psi_0}{\partial z}\frac{\partial v_2}{\partial x} - \frac{\partial \theta_0}{\partial z}\frac{\partial v_1}{\partial x} - \frac{\partial \psi_0}{\partial x}\frac{\partial v_2}{\partial z} \end{pmatrix} = \begin{pmatrix} \nabla^2 v_1 - \mathrm{Ra}_d \frac{\partial v_2}{\partial x} \\ \nabla^2 v_2 + \frac{\partial v_1}{\partial x} \end{pmatrix}
\end{aligned}$$

Thus, the linearization around the base solution is given by

$$L\mathbf{v} = \begin{pmatrix} \nabla^2 v_1 - \mathrm{Ra}_d \frac{\partial v_2}{\partial x} \\ \nabla^2 v_2 + \frac{\partial v_1}{\partial x} \end{pmatrix} = DF(\psi_0, \theta_0, \mathrm{Ra}_d) \cdot \mathbf{v} \tag{30.34}$$

The boundary conditions on v_1 and v_2 are obtained in a similar way from equations (30.30) and (30.31) by setting

$$v_1 = \psi - \psi_0 \quad \text{and} \quad v_2 = \theta - \theta_0, \tag{30.35}$$

and are given by

$$v_1(0,z) = v_1(a,z) = v_1(x,0) = v_1(x,1) = 0 \tag{30.36}$$

$$\frac{\partial v_2}{\partial x}(0,z) = \frac{\partial v_2}{\partial x}(a,z) = v_2(x,0) = v_2(x,1) = 0 \tag{30.37}$$

Thus, new steady-state solutions (or bifurcation from trivial solution) can occur only if the (linearized homogeneous) equation

$$L\mathbf{v} = \mathbf{0}$$

has a nontrivial solution. Now, we look at the linear homogeneous equation $L\mathbf{v} = \mathbf{0}$ in component form:

$$\frac{\partial^2 v_1}{\partial x^2} + \frac{\partial^2 v_1}{\partial z^2} - \mathrm{Ra}_d \frac{\partial v_2}{\partial x} = 0 \tag{30.38}$$

$$\frac{\partial^2 v_2}{\partial x^2} + \frac{\partial^2 v_2}{\partial z^2} + \frac{\partial v_1}{\partial x} = 0. \tag{30.39}$$

The spatial operator in z-direction is

$$-\frac{d^2\phi}{dz^2}, \quad \phi(0) = \phi(1) = 0,$$

which has eigenfunctions

$$\phi_n = \sin n\pi z \tag{30.40}$$

corresponding to eigenvalues $n^2\pi^2$, $(n = 1, 2, 3, \ldots)$. Thus, we write

$$v_1(x,z) = w_1(x)\phi_n(z) \tag{30.41}$$

$$v_2(x,z) = w_2(x)\phi_n(z) \tag{30.42}$$

and substitute (30.40)–(30.42) into (30.38)–(30.39) to obtain the following eigenvalue problem for the x-direction eigenfunctions:

$$w_1'' - n^2\pi^2 w_1 - \mathrm{Ra}_d\, w_2' = 0, \quad w_1(0) = w_1(a) = 0 \tag{30.43}$$

$$w_2'' - n^2\pi^2 w_2 + w_1' = 0, \quad w_2'(0) = w_2'(a) = 0 \tag{30.44}$$

These equations may be combined to give a single fourth-order boundary value problem:

$$\frac{d^4 w_1}{dx^4} + (\mathrm{Ra}_d - 2n^2\pi^2)\frac{d^2 w_1}{dx^2} + n^4\pi^4 w_1 = 0 \tag{30.45}$$

$$w_1(0) = w_1(a) = \frac{d^2 w_1}{dx^2}(0) = \frac{d^2 w_1}{dx^2}(a) = 0 \tag{30.46}$$

This is a linear equation with constant coefficients and can be solved easily. By inspection, we see that

$$w_1(x) = \sin\left(\frac{m\pi x}{a}\right), \quad m = 1, 2, \ldots \tag{30.47}$$

satisfies equations (30.45)–(30.46) iff

$$\frac{m^4\pi^4}{a^4} + (\mathrm{Ra}_d - 2n^2\pi^2)\left(\frac{-m^2\pi^2}{a^2}\right) + n^4\pi^4 = 0$$
$$\Rightarrow \mathrm{Ra}_d = \frac{\pi^2(m^2 + n^2a^2)^2}{m^2a^2}. \tag{30.48}$$

30.1.3 Neutral curve and critical Ra_d

We are interested in determining the smallest value of the Darcy–Rayleigh number for which there is a nontrivial solution. Since Ra_d is monotonically increasing with n but nonmonotonic with m, we take $n = 1$ (In physical terms, this implies that it is the first vertical mode that is always destabilized in this specific problem):

$$\Rightarrow \mathrm{Ra}_d = \frac{\pi^2(m^2 + a^2)^2}{m^2a^2} \tag{30.49}$$

Equation (30.49) is plotted for different values of m (= 1, 2 and 3) in Figure 30.2.

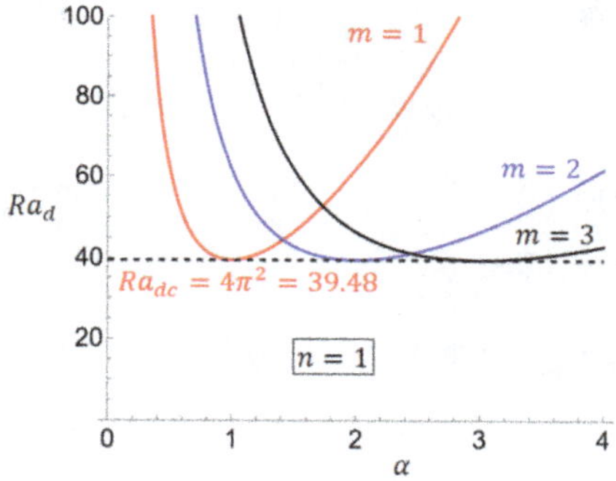

Figure 30.2: Neutral stability curves (bifurcation set) for the Lapwood convection problem: First vertical mode and different horizontal modes.

Also, note that

$$\frac{d\,\mathrm{Ra}_d}{da^2} = 0 \Rightarrow a^2 = m^2 \quad \text{and} \quad \left.\frac{d^2\,\mathrm{Ra}_d}{d(a^2)^2}\right|_{a^2=m^2} = -\frac{2\pi^2}{m^4} < 0 \tag{30.50}$$

$\Longrightarrow$

$$\mathrm{Ra}_d \geq \mathrm{Ra}_{dc} = \frac{\pi^2(a^2 + a^2)^2}{a^4} = 4\pi^2. \tag{30.51}$$

Thus, the smallest value of Ra_d is $\mathrm{Ra}_{dc} = 4\pi^2$ and is attained when $a = m = 1, 2, 3, \ldots$. We also note that when $a = \sqrt{m(m+1)}$, the value of Ra_d given by equation (30.49) is the same for m and $m+1$. This is called a bicritical point where two horizontal modes are destabilized at the same time.

In practice, only the lower envelope of the above curves (which is part of the bifurcation set) is of interest as it defines the boundary between "conduction only" solutions and "convective" solutions. We can also determine the shape of the bifurcating solution by noting that the nontrivial solution may be written in a parametric form:

$$\mathrm{Ra}_d = \mathrm{Ra}_{dc} + O(|\varepsilon|) \quad \text{and} \tag{30.52}$$

$$\begin{pmatrix} \psi(x,z) \\ \theta(x,z) \end{pmatrix} = \begin{pmatrix} 0 \\ 1-z \end{pmatrix} + \varepsilon y_0 + O(|\varepsilon|^2) \quad \text{as } \epsilon \to 0 \tag{30.53}$$

where y_0 is the eigenfunction corresponding to zero eigenvalue of the linear operator L. In this case, we have

$$y_0 = \begin{pmatrix} v_1(x,z) \\ v_2(x,z) \end{pmatrix} = \begin{pmatrix} \sin\pi z \cdot \sin\pi x \\ \frac{2\pi}{\mathrm{Ra}_d} \cdot \sin\pi z \cdot \cos\pi x \end{pmatrix} \tag{30.54}$$

Thus, we can determine the streamlines and isotherms using equations (30.52)–(30.53) and (30.54). The eigenfunctions from equation (30.54) are plotted in Figure 30.3 for $a = 2$.

Note that there are two circulation cells (the symmetric pair has cells rotating in opposite direction). For $a = n$, the solution has n circulation cells.

Remark. The above solution can be modified easily for an infinite layer ($a \to \infty$), In this case, $m\pi/a$ becomes a continuous variable (often called the wave number and denoted by k) and equation (30.49) modifies to

$$\Rightarrow \mathrm{Ra}_d = \frac{(n^2\pi^2 + k^2)^2}{k^2} \tag{30.55}$$

Thus,

$$\frac{d\,\mathrm{Ra}_d}{dk^2} = 0 \Rightarrow k^2 = n^2\pi^2 \Rightarrow \mathrm{Ra}_d \geq \mathrm{Ra}_{dc} = \frac{4n^4\pi^4}{n^2\pi^2} = 4n^2\pi^2 \tag{30.56}$$

Once again, the smallest Ra_d occurs for $n = 1$ and is given by

$$\mathrm{Ra}_{dc} = 4\pi^2 = 39.48\ldots \quad \text{(critical Rayleigh number)}$$

while the critical wave number is given by $k_c = \pi$. The corresponding eigenfunction is given by

$$v_1(x,z) = \sin\pi z \cdot \sin k_c x = \sin\pi z \cdot \sin\pi x \tag{30.57}$$

$$v_2(x,z) = \sin\pi z \cdot \cos\pi x/(2\pi) \tag{30.58}$$

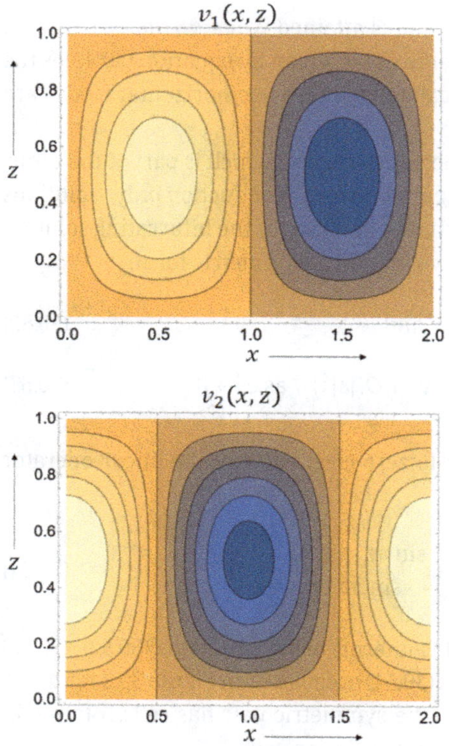

Figure 30.3: Contour plots of the eigenfunctions (streamlines and isotherms) for the Lapwood problem.

The corresponding flow pattern is similar to that shown in Figure 30.3 except now the cells are square-shaped.

The temperature θ for the convective solutions is of the form (from equations (30.52)–(30.53) and (30.54)),

$$\theta(x,z) = 1 - z \pm \frac{2\pi}{\text{Ra}_{dc}}\varepsilon \cdot \sin \pi z \cdot \cos \pi x \tag{30.59}$$

where ε is the amplitude of the convective branch. The isotherms of these convective branches

$$\theta_1 = 1 - z + \frac{2\pi}{\text{Ra}_{dc}}\varepsilon \cdot \sin \pi z \cdot \cos \pi x$$
$$\theta_2 = 1 - z - \frac{2\pi}{\text{Ra}_{dc}}\varepsilon \cdot \sin \pi z \cdot \cos \pi x$$

are shown in Figure 30.4 for $\varepsilon = 1$.

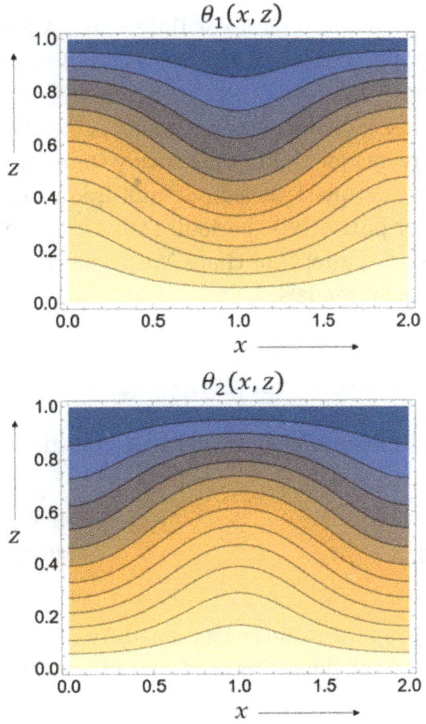

Figure 30.4: Isotherms of the bifurcating convective branches for the Lapwood convection problem.

30.2 Chemical reactor stability and dynamics

Mathematical models of chemical reactors are obtained by writing down the species, energy and momentum balances, and combining them with the constitutive relations for the various rate processes. These equations are nonlinear (in most cases) as the reaction rates are usually nonlinear functions of concentration and/or temperature. These models may be expressed in the form

$$\begin{aligned} &\mathbf{C}\frac{\partial \mathbf{u}}{\partial t} = \mathbf{F}(\mathbf{x}, \mathbf{u}, \nabla\mathbf{u}, \nabla^2\mathbf{u}, \mathbf{p}) \quad \text{in } \Omega \\ &\text{BCs:} \quad \boldsymbol{\beta}(\mathbf{x}, \mathbf{u}, \nabla\mathbf{u}) = \mathbf{0} \quad \text{on } \partial\Omega, \quad t > 0 \\ &\text{I.C.:} \quad \boldsymbol{\Gamma}(\mathbf{x}, \mathbf{u}, \nabla\mathbf{u}) = \mathbf{0} \quad \text{in } \Omega\ @\ t = 0. \end{aligned} \tag{30.60}$$

When spatial dependence of the state variables is ignored, we get the so-called "lumped resistance" models or simply called lumped models. In this case, equations (30.60) are of the form:

$$\mathbf{C}\frac{d\mathbf{u}}{dt} = \mathbf{F}(\mathbf{u}, \mathbf{p}), \quad t > 0; \quad \mathbf{u} = \mathbf{u}_0\ @\ t = 0 \tag{30.61}$$

and is a set of (non)linear Ordinary Differential Equations (ODE) or Differential Algebraic Equations (DAE) system. Here, **C** is the capacitance matrix, **u** is the vector of state variables and **p** is a vector of system parameters.

The models described by equation (30.60) are known to exhibit complex steady-state and transient/dynamic behavior. These may include (i) multiple steady states, (ii) periodic states in time, (iii) periodic states in space, (iv) quasi-periodic or multifrequency states in time and/or space, (v) traveling waves or pulses, (vi) aperiodic states in space and time, (vii) complex and irregular spatiotemporal patterns or chaos. We discuss here the first two of these using the simplest of the reactor models.

30.2.1 Model of a cooled continuous-flow stirred tank reactor (CSTR)

Consider an ideal CSTR (shown schematically in Figure 30.5) in which N_R reactions among N_s species, represented by

$$\sum_{j=1}^{N_S} \nu_{ij} A_j = 0; \quad i = 1, 2, \dots, N_R \tag{30.62}$$

are taking place. Assuming (a) constant physical properties and constant density, (b) volume of reactor and volumetric flow rate being constant, the species and energy balances are given by

$$\frac{dc_j}{dt} = \frac{c_{j,\text{in}}(t) - c_j}{\tau_c} + \sum_{i=1}^{N_R} \nu_{ij} r_i(\mathbf{c}, T), \quad j = 1, 2, 3, \dots N_s \tag{30.63}$$

$$\text{Le}_R \frac{dT}{dt} = \frac{T_{\text{in}}(t) - T}{\tau_c} + \sum_{i=1}^{N_R} \frac{(-\Delta H_{R,i})}{\rho_f C_{pf}} r_i(\mathbf{c}, T) - \frac{UA_h}{V_R \rho_f C_{pf}} (T - T_c(t)), \tag{30.64}$$

where

$$\text{Le}_R = 1 + \frac{(MC_p)_{\text{wall}}}{V_R \rho_f C_{pf}} = \text{reactor Lewis number}$$

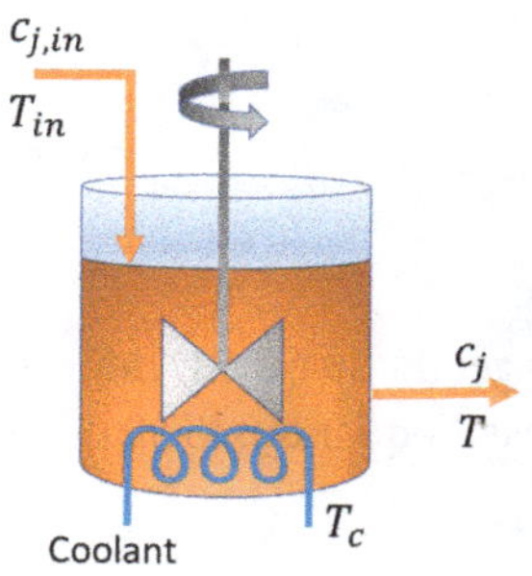

Figure 30.5: Schematic of an ideal CSTR (continuous-flow stirrered tank reactor).

and U is the overall heat transfer coefficient for heat exchange between reactor contents and coolant; A_h is heat transfer area; V_R is the volume of reactor; $\tau_c = \frac{V_R}{q_0}$ is the space (residence or convection) time, q_0 is volumetric flow rate, T_{in} is the feed temperature, T_c is the coolant temperature, $c_{j,in}$ is the concentration of species j in the feed, $r_i(c, T)$ is the rate of reaction i and $\Delta H_{R,i}$ is the heat of reaction i.

Equations (30.63) and (30.64) represent $(N_s + 1)$ nonlinear ODEs that describe the variation of reactor composition and temperature with time. These equations have to be integrated numerically with appropriate initial conditions:

$$c_j = c_{j0} \quad \text{and} \quad T = T_0 \ @\ t = 0 \tag{30.65}$$

Denoting

$$\mathbf{C} = \begin{pmatrix} 1 & 0 & \cdots & 0 & 0 \\ 0 & 1 & \cdots & 0 & 0 \\ \vdots & \vdots & \vdots & \vdots & \vdots \\ 0 & 0 & \cdots & 1 & 0 \\ 0 & 0 & \cdots & 0 & \mathrm{Le}_R \end{pmatrix}, \quad \mathbf{u} = \begin{pmatrix} c_1 \\ c_2 \\ \vdots \\ c_{N_s} \\ T \end{pmatrix} \tag{30.66}$$

and considering only the special case in which the inputs $c_{j,\mathrm{in}}(t)$, $T_{\mathrm{in}}(t)$ and $T_c(t)$ are independent of time, we can write equations (30.63), (30.64) and (30.65) in the autonomous form given by equation (30.61). If inputs vary with time, equation (30.61) can be modified to

$$\mathbf{C}\frac{d\mathbf{u}}{dt} = \mathbf{F}(t, \mathbf{u}, \mathbf{p}), \quad t > 0; \quad \mathbf{u} = \mathbf{u}_0 \ @\ t = 0 \tag{30.67}$$

The above form (equation (30.67)) of the lumped model is known as the forced or nonautonomous system. Here, we consider only the autonomous case.

30.2.2 Dimensionless form of the model for a single reaction

Assuming a single step exothermic reaction of the form $A \to B$ with linear kinetics, we can express the rate of reaction (for disappearance) of the species A as

$$r = k(T)c_A = k_0 \exp\left(-\frac{E_a}{RT}\right)c_A \tag{30.68}$$

where k_0 is a preexponential factor, E_a is activation energy and c_A is the concentration of species A. Since $0 \le c_A \le c_{A,\mathrm{in}}$ while absolute temperature T can be a large number, it is convenient to use dimensionless quantities that have the same order of magnitude. Thus, we define

$$\begin{gathered} \tau = k(T_{\mathrm{in}})t; \quad \chi = 1 - \frac{c_A}{c_{A,\mathrm{in}}}; \quad y = \frac{T - T_{\mathrm{in}}}{T_{\mathrm{in}}}; \quad \mathrm{Da} = k(T_{\mathrm{in}})\tau_c; \\ \Delta T_{\mathrm{ad}} = \frac{(-\Delta H_R)c_{A,\mathrm{in}}}{\rho_f C_{pf}}; \quad \gamma = \frac{E_a}{RT_{\mathrm{in}}}; \quad \beta = \frac{\Delta T_{\mathrm{ad}}}{T_{\mathrm{in}}}; \quad \tau_h = \frac{\rho_f C_{pf} V_R}{UA_h}; \end{gathered} \tag{30.69}$$

$$\alpha = \frac{1}{k(T_{\text{in}})\tau_h}; \quad \chi_0 = 1 - \frac{c_{A0}}{c_{A,\text{in}}}; \quad y_c = \frac{T_c - T_{\text{in}}}{T_{\text{in}}}; \quad y_0 = \frac{T_0 - T_{\text{in}}}{T_{\text{in}}}$$

where τ is the dimensionless time (scaled with reaction time at the inlet temperature); χ is the conversion; y is the dimensionless temperature of fluid; Da is Damköhler number at inlet temperature; γ is dimensionless activation energy; ΔT_{ad} is the adiabatic temperature rise; β is dimensionless adiabatic temperature rise and τ_h is the heat exchange time with the coolant (or cooling time); α is the ratio of characteristic reaction time at the inlet temperature to the cooling time; y_0 is the initial fluid temperature and χ_0 is the conversion corresponding to initial concentration. With these dimensionless quantities, the model equations (30.63) and (30.64) in dimensionless form reduce to two nonlinear ODEs:

$$\frac{d\chi}{d\tau} = -\frac{\chi}{\text{Da}} + (1-\chi)\exp\left(\frac{\gamma y}{1+y}\right); \tag{30.70}$$

$$\text{Le}_R \frac{dy}{d\tau} = -\frac{y}{\text{Da}} + \beta(1-\chi)\exp\left(\frac{\gamma y}{1+y}\right) - \alpha(y - y_c); \tag{30.71}$$

$$\chi = \chi_0 \quad \text{and} \quad y = y_0 \ @\ t = 0 \tag{30.72}$$

This more general model has three additional parameters Le_R, α and y_c compared to the simpler adiabatic case ($\alpha = 0$) with $\text{Le}_R = 1$ (or negligible reactor wall thermal capacitance).

30.2.3 Stability analysis

In what follows, we consider the special case of $y_c = 0$ (i. e., coolant and feed temperature are equal). For this case, the steady-state model reduces to

$$y_s = \frac{\beta\chi_s}{1 + \alpha\,\text{Da}} \tag{30.73}$$

$$g(\chi_s, \text{Da}) \equiv \chi_s - \text{Da}(1-\chi_s)\exp\left(\frac{\gamma\beta\chi_s}{1 + \beta\chi_s + \alpha\,\text{Da}}\right) = 0, \tag{30.74}$$

where χ_s and y_s are steady-state conversion and dimensionless temperature, respectively.

Note that for adiabatic case (i. e., $\alpha = 0$), the Damköhler number Da can explicitly be expressed in terms of χ_s as $\text{Da} = \frac{\chi_s}{1-\chi_s}\exp(-\frac{\gamma\beta\chi_s}{1+\beta\chi_s})$. However, in case of cooled CSTR, Da cannot be expressed explicitly in terms of χ_s or y_s. Thus, the determination of different types of conversion versus Da curves is more difficult. We summarize here the results without detailed derivations. It turns out that equation (30.74) has five different types of χ_s versus Da solution (bifurcation) diagrams, depending on the specific values selected for the parameters γ, β and α. The phase diagram for any $\gamma > 4$ is shown below schematically in Figure 30.6, along with the five different types of χ versus Da diagrams in Figure 30.7.

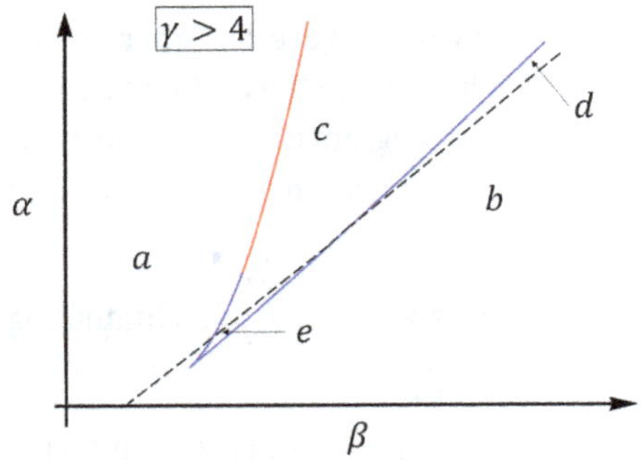

Figure 30.6: Schematic phase diagram for cooled CSTR. Letters in each region correspond to different qualitative behaviors.

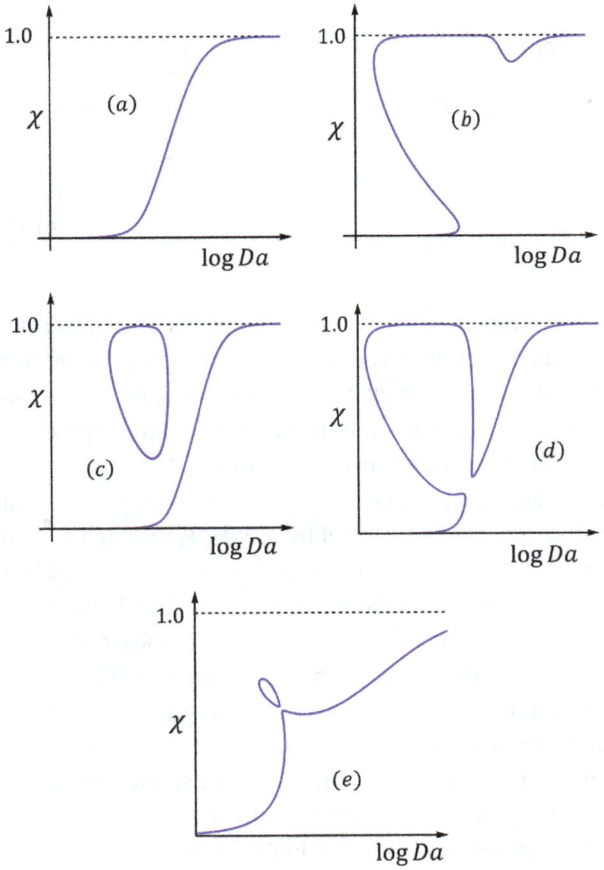

Figure 30.7: Different types of χ versus Da diagrams in each of the five regions denoted in Figure 30.6.

The solid and dashed lines in the phase diagram shown in Figure 30.6 are referred to as the isola and hysteresis locus, respectively. These loci divide the (α, β) plane into five regions in each of which a different type of χ versus Da diagram is obtained. Regions c and e can exhibit isola (or an isolated solution branch) as shown in Figure 30.7(c) and Figure 30.7(e).

Remark. The hysteresis locus is obtained by setting $g = \frac{\partial g}{\partial \chi_s} = \frac{\partial^2 g}{\partial {\chi_s}^2} = 0$ and eliminating χ_s and Da. The isola locus is obtained by setting $g = \frac{\partial g}{\partial \chi_s} = \frac{\partial g}{\partial \mathrm{Da}} = 0$.

To determine the stability of the steady state, we write equations (30.70)–(30.72) for the case of $y_c = 0$, as

$$\frac{d\chi}{d\tau} = F_1(\chi, y) = -\frac{\chi}{\mathrm{Da}} + (1-\chi)\exp\left(\frac{\gamma y}{1+y}\right); \tag{30.75}$$

$$\frac{dy}{d\tau} = F_2(\chi, y) = \frac{1}{\mathrm{Le}_R}\left[\frac{-y}{\mathrm{Da}} + \beta(1-\chi)\exp\left(\frac{\gamma y}{1+y}\right) - \alpha y\right] \tag{30.76}$$

We linearize these equations around the steady state and determine the eigenvalues of the linearized matrix:

$$\mathbf{A} = \left.\begin{pmatrix} \frac{\partial F_1}{\partial \chi} & \frac{\partial F_1}{\partial y} \\ \frac{\partial F_2}{\partial \chi} & \frac{\partial F_2}{\partial y} \end{pmatrix}\right|_{(\chi_s, y_s)} \tag{30.77}$$

In this case, the eigenvalues can be complex for $\mathrm{Le}_R \geq 1$ and the trace of the matrix can change sign leading to periodic solutions in time. Usually this occurs when the reactor is cooled strongly. In such cases, even when a single steady state exists, it could be locally unstable leading to sustained oscillations of the exit conversion and temperature (though the inlet concentration and temperature remain constant).

For example, we consider the following parameters: $\beta = 1$, $\gamma = 30$, $y_c = 0$ and $\alpha = 35$, where the steady-state diagram can be obtained by solving $F_1 = 0$ and $F_2 = 0$ and is shown in Figure 30.8 (these parameters lie in region b in Figure 30.6 but just below the dashed isola locus). The top plot shows the conversion while the bottom plot shows the dimensionless temperature at steady state. At each point of the steady-state curve in Figure 30.8, the eigenvalues of the matrix $\mathbf{A}$ defined by equation (30.77) can be obtained. When the real part of any of the two eigenvalues is positive, the solution becomes unstable (such states are shown in Figure 30.8 by the dashed lines). The system has a stable solution only when real parts of both eigenvalues are negative (shown in Figure 30.8 by solid lines). The singular points, where the solution transitions from stable to unstable regime, are shown by diamond marker points in Figure 30.8.

Remark. The singular points consist of limit (or turning) points (where a real eigenvalue changes its sign) and Hopf bifurcation points (where a complex pair of eigenvalues cross

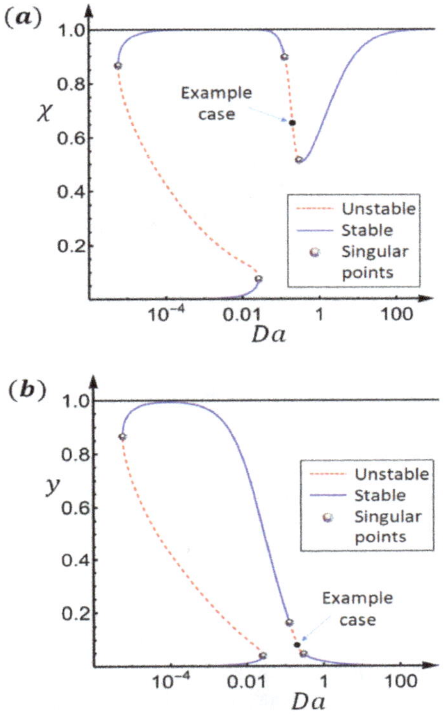

Figure 30.8: Steady-state diagram of (a) conversion (χ) versus Damköhler number (Da) and (b) dimensionless temperature (y) versus Damköhler number (Da) for the cooled CSTR corresponding to $\beta = 1, \gamma = 30$, $y_c = 0$ and $\alpha = 35$. Dashed curve corresponds to the unstable region while the solid curve corresponds to the stable region.

the imaginary axis). At a limit point, new stead-state solutions appear or disappear while at a Hopf bifurcation point, periodic states appear or disappear.

To be specific, if we choose a point denoted by the black circle in Figure 30.8 as an example case, which corresponds to the cooled region with Da = 0.2 and the steady-state conversion and temperature as

$$\chi_s = 0.6705; \quad y_s = 0.0838.$$

Taking $\mathrm{Le}_R = 1.5$ and computing the eigenvalues of the linearized matrix, we find that they are complex with a positive real part, i. e.,

$$\lambda_{1,2} = 7.62033 \pm 7.82384i.$$

Thus, the steady state is unstable. Integrating the full nonlinear equations numerically, we find that the conversion and temperature oscillate with time, as shown in Figure 30.9. In this figure, the top plot shows the transient oscillation of conversion and tempera-

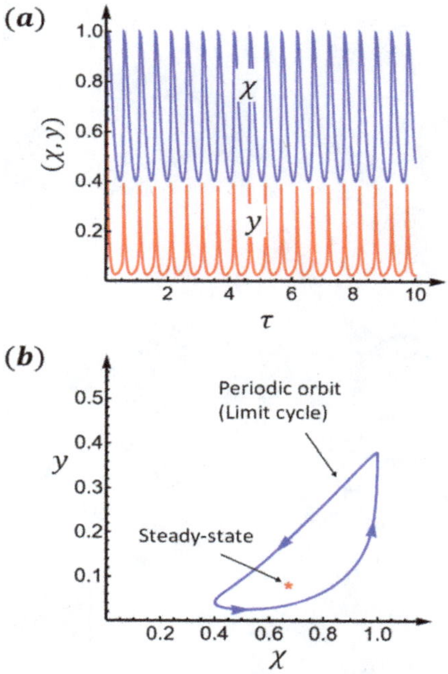

Figure 30.9: Transient solution corresponding to the parameters $\beta = 1$, $\gamma = 30$, $y_c = 0$, $\alpha = 35$, $\mathrm{Da} = 0.2$ and $\mathrm{Le}_R = 1.5$, demonstrating (a) conversion and temperature variation with time (b) periodic orbit in phase plane.

ture, while the bottom plot shows the periodic orbit (i. e., limit cycle) in the $\chi - y$ phase plane.

A similar analysis can be performed in other regions and for the more general case when the feed and coolant temperatures are different.

Problems

1. *Bifurcation set for discrete model of thermohaline convection*: Consider the following discrete model of thermohaline convection (where the fluid density is assumed to vary both with temperature and salt concentration):

$$\frac{dx}{dt} = \mathrm{Pr}(y - x + u) = f_1$$

$$\frac{dy}{dt} = -xz + \frac{4}{27\pi^4}\,\mathrm{Ra}_T\,x - y = f_2$$

$$\frac{dz}{dt} = xy - \frac{8}{3}z = f_3$$

$$\frac{du}{dt} = -xv - \mathrm{Le}\,\frac{4}{27\pi^4}\,\mathrm{Ra}_c\,x - \mathrm{Le}\,u = f_4$$

$$\frac{dv}{dt} = xu - \frac{8}{3}\,\mathrm{Le}\,v = f_5$$

Here x, y, z, u, v represent the amplitudes for velocity, temperature and concentration eigenfunctions. The parameters Pr, Le, Ra_T and Ra_c are the Prandtl, Lewis, thermal Rayleigh and concentration Rayleigh numbers, respectively. Assuming the vector of variables $\psi = (x, y, z, u, v)^T$, write the above equations as

$$\frac{d\psi}{dt} = \mathbf{f}(\psi) = (f_1, f_2, f_3, f_4, f_5)^T \tag{30.78}$$

(a) Show that equation (30.78) has a trivial steady-state solution, i. e., $\psi_s = (x, y, z, u, v)^T = \mathbf{0}$ is one of the steady-state solutions.

(b) Determine the Jacobian $\mathbf{J} = \{\frac{\partial f_i}{\partial \psi_j}\}$ of the function $\mathbf{f}$ and show that it is given by

$$\mathbf{J} = \left\{\frac{\partial f_i}{\partial \psi_j}\right\} = \begin{pmatrix} -\mathrm{Pr} & \mathrm{Pr} & 0 & \mathrm{Pr} & 0 \\ \frac{4}{27\pi^4}\,\mathrm{Ra}_T - z & -1 & -x & 0 & 0 \\ y & x & \frac{-8}{3} & 0 & 0 \\ \mathrm{Le}\,\frac{4}{27\pi^4}\,\mathrm{Ra}_c - v & 0 & 0 & -\mathrm{Le} & -x \\ u & 0 & 0 & x & -8\frac{\mathrm{Le}}{3} \end{pmatrix} \tag{30.79}$$

(c) *Neutral curve near trivial solution:* Determine the neutral curve near the trivial solution ψ_s (by setting the determinant of the Jacobian $\mathbf{J}$ at ψ_s to zero) and show that it is given by

$$\begin{aligned} \mathbf{J}|_{\psi_s} &= -\frac{64\,\mathrm{Le}^2\,\mathrm{Pr}}{243\pi^4}(27\pi^4 + 4\,\mathrm{Ra}_c - 4\,\mathrm{Ra}_T) = 0 \\ \Rightarrow \mathrm{Ra}_T &= \mathrm{Ra}_c + \frac{27\pi^4}{4} \end{aligned} \tag{30.80}$$

(d) Give a physical interpretation of the neutral curve.

2. Consider the Lapwood problem in a porous rectangular box open at the top and insulated on the sides (Figure 30.1).

(a) Formulate the mathematical model and cast it into dimensionless form. State any assumptions clearly.

(b) Determine the conduction state. Show that this state loses its stability when there is a nontrivial solution to the following boundary value problem:

$$\frac{d^2\phi}{dz^2} - k^2\phi + \psi = 0$$

$$\frac{d^2\psi}{dz^2} - k^2\psi + \mathrm{Ra}_d\,k^2\phi = 0$$

$$\phi(0) = \phi(1) = 0; \quad \psi(0) = \frac{d\psi}{dz}(1) = 0; \quad k^2 = \frac{m^2\pi^2}{a^2}$$

where a is the aspect ratio (width to height of the box) and Ra_d is the Darcy–Rayleigh number.

(c) Determine the marginal/neutral stability boundary and compare the critical Ra_d to that of the box closed at the top.

3. Consider the homogeneous boundary value problem (BVP)

$$\frac{d^2w}{dx^2} = -\lambda w, \quad 0 < x < 1; \quad w'(0) = 0 = w(1)$$

(a) What is the smallest value of λ for which the BVP is compatible?

(b) If $\lambda = \lambda_1$ is the value determined in (a), show that for $-\infty < \lambda < \lambda_1$, the only solution to the BVP is the trivial one.

(c) Now, consider the nonlinear boundary value problem

$$\frac{d^2w}{dx^2} = -f(w), \quad 0 < x < 1; \quad w'(0) = 0 = w(1)$$

and reason that it has only one solution if the maximum value of $f'(w) < \lambda_1$. [Hint: Consider the case that it has two solutions and take their difference and use the mean value theorem of calculus.]

(d) Use the result in (c) to determine the maximum value of ϕ^2 for which the following nonlinear BVP has only one solution:

$$\frac{d^2w}{dx^2} = -\phi^2(B - w)e^w, \quad 0 < x < 1; \quad w'(0) = 0 = w(1)$$

Here, B is a positive constant and $0 < w < B$.

4. Consider the Glass–Mackey equation

$$\frac{dz}{dt} = \beta - \frac{z(t-\tau)^3}{1 + z(t-\tau)^3} z(t); \quad t > 0, \ z(t) = z_0 \text{ for } -\tau \le t \le 0$$

where β and τ are positive constants.

(a) Determine the steady state(s) and the equation that determines the stability of the steady state.

(b) Determine the locus in the (β, τ) plane for which the system becomes unstable.

(c) Integrate the equation for a set of parameter values in the unstable region and plot the solution.

5. (Rayleigh–Bernard Convection): Consider a viscous Newtonian fluid confined to a rectangular box (as shown in Figure 30.1 but without a porous medium) and subject to an unstable density stratification.

(a) Assuming incompressible flow, formulate the governing equations under the Boussinesq approximation and cast them in dimensionless form.

(b) Determine the conduction state and the linearized equations that determine the stability of the conduction state to small perturbations.

(c) Verify that the resulting eigenvalue problem is the same as that given in Problem 5 of Chapter 20.

Bibliography

[1] Abate J, Valkó PP. Multi-precision Laplace transform inversion. Internat. J. Numer. Methods Engrg. 2004;60(5):979–93.

[2] Abramowitz M, Stegun IA. Handbook of mathematical functions with formulas, graphs, and mathematical tables. US Government printing office; 1964.

[3] Amundson NR. Mathematical Methods in Chemical Engineering: Matrices and their application. Prentice-Hall; 1966.

[4] Aris R, Balakotaiah V. Asymptotic effectiveness of a catalyst particle in the form of a hollow cylinder. AIChE J. 2013;59(11):4020–4.

[5] Balakotaiah V. On the relationship between Aris and Sherwood numbers and friction and effectiveness factors. Chem. Eng. Sci. 2008;63(24):5802–12.

[6] Balakotaiah V, Gupta N. Controlling regimes for surface reactions in catalyst pores. Chem. Eng. Sci. 2000;55(17):3505–14.

[7] Bender CM, Orszag SA. Advanced mathematical methods for scientists and engineers. 1978.

[8] Bronson R. Schaum's Outline of Matrix Operations. McGraw-Hill; 2011.

[9] Bronson R, Costa GB. Matrix methods: Applied linear algebra. Academic Press; 2008.

[10] Carslaw HS, Jaeger JC. Conduction of heat in solids. Oxford, at the Clarendon Press; 1947.

[11] Churchill RV. Fourier Series and Boundary Value Problems. McGraw-Hill; 1969.

[12] Churchill RV. Operational Mathematics. McGraw Hill; 1972.

[13] Coddington EA, Levinson N. Theory of Ordinary Differential Equations. McGraw Hill; 1965.

[14] Cole RH. Theory of ordinary differential equations. Appleton-Century-Crofts; 1968.

[15] Courant R, Hilbert D. Methods of mathematical physics. New York: Interscience Publication, 1953.

[16] Crank J. The mathematics of diffusion. Oxford university press; 1979.

[17] Defreitas CL, Kane SJ. The Numerical Inversion of the Laplace Transform in a Multi-Precision Environment. Appl. Math. 2022;13(5):401–18.

[18] Gantmacher FR. The theory of matrices. New York, 1964.

[19] Gundlapally SR, Balakotaiah V. Heat and mass transfer correlations and bifurcation analysis of catalytic monoliths with developing flows. Chem. Eng. Sci. 2011; 66(9):1879–92.

[20] Halmos PR. Introduction to Hilbert space and the theory of spectral multiplicity. Courier Dover Publications; 2017.

[21] Kamke E. Differential Gleichungen. J. W. Edwards; 1943.

[22] Lipschutz S, Lipson M. Schaum's outline of linear algebra. McGraw Hill; 2017.

[23] Morse PM, Feshbach H. Methods of Theoretical Physics. New York, 1953.

[24] Naylor AW, Sell GR. Linear operator theory in engineering and science. Springer Science & Business Media; 1982.

[25] Ramkrishna D, Amundson NR. Linear operator methods in chemical engineering with applications to transport and chemical reaction systems. Prentice Hall; 1985.

[26] Ratnakar RR, Balakotaiah V. Coarse-graining of diffusion–reaction models with catalyst archipelagos. Chem. Eng. Sci. 2014;110:44–54.

[27] Sarkar B, Ratnakar RR, Balakotaiah V. Multi-scale coarse-grained continuum models for bifurcation and transient analysis of coupled homogeneous-catalytic reactions in monoliths. Chem. Eng. J. 2021;407:126500.

[28] Sneddon IN. Fourier transforms. Courier Corporation; 1995.

[29] Spiegel MR. Theory and Problems of Complex Variables, Schaum's Outline Series in Mathematics; 1964.

[30] Tu M, Ratnakar R, Balakotaiah V. Reduced order models with local property dependent transfer coefficients for real time simulations of monolith reactors. Chem. Eng. J. 2020;383:123074.

[31] Wei J, Prater CD. The structure and analysis of complex reaction systems. In: Advances in Catalysis (Vol. 13, pp. 203–392). Academic Press; 1962.

https://doi.org/10.1515/9783111598055-037

Index

https://doi.org/10.1515/9783111598055-038

www.ingramcontent.com/pod-product-compliance
Lightning Source LLC
Chambersburg PA
CBHW082101220326
41598CB00066BA/4525
* 9 7 8 3 1 1 1 5 9 7 3 8 6 *